16th Process Plant Safety Symposium 2014

(PPSS)

Topical Conference at the 2014 AIChE Spring Meeting and 10th Global Congress on Process Safety

New Orleans, Louisiana, USA
30 March - 3 April 2014

ISBN: 978-1-63439-158-0

Printed from e-media with permission by:

Curran Associates, Inc.
57 Morehouse Lane
Red Hook, NY 12571

Some format issues inherent in the e-media version may also appear in this print version.

Printed by Curran Associates, Inc. (2014)

For permission requests, please contact AIChE
at the address below.

AIChE
3 Park Avenue
New York, NY 10016-5991

Phone: (203) 702-7660
Fax: (203) 775-5177

www.aiche.org

Additional copies of this publication are available from:

Curran Associates, Inc.
57 Morehouse Lane
Red Hook, NY 12571 USA
Phone: 845-758-0400
Fax: 845-758-2634
Email: curran@proceedings.com
Web: www.proceedings.com

TABLE OF CONTENTS

354012 Possible vs. Practical: Engineers Must Lead the Development of Practical Technologies

Abstract:

Chemical Engineering is need today more than ever. There is a strong societal desire for sources of energy and feedstocks that can sustainably support a high standard of living for all. Thermodynamic, material and economic challenges stand in the way. Most of the general public does not have the basic understanding of the energy flows, chemical transformations, and the scale of the systems that support our current way of life. These concepts are fundamental to chemical engineering. Chemical engineers should be on the vanguard of separating hype from truly practical options yet occationally we let society down. This talk will present several examples where the fundamentals of chemical engineering can be used to access the practicality of proposed solutions to energy and water issues, as well as explain why the venture capital model struggles in the materials market.

Improving Recognition of Change

Donald K. Lorenzo
ABS Consulting
10301 Technology Drive
Knoxville, TN 37932
dlorenzo@absconsulting.com

Della Wong
Canadian Natural Resources Limited
1800, 324 - 8 Avenue, SW
Calgary, Alberta T2P 2Z2
della.wong@cnrl.com

Mark Suyama
Valero Energy Corporation
One Valero Way
San Antonio, TX 78249
mark.suyama@valero.com

Prepared for Presentation at
American Institute of Chemical Engineers
2014 Spring Meeting
10th Global Congress on Process Safety
New Orleans, LA
March 30 – April 2, 2014

UNPUBLISHED

Improving Recognition of Change

Donald K. Lorenzo
ABS Consulting
10301 Technology Drive
Knoxville, TN 37932
dlorenzo@absconsulting.com

Della Wong
Canadian Natural Resources Limited
1800, 324 - 8 Avenue, SW
Calgary, Alberta T2P 2Z2
della.wong@cnrl.com

Mark Suyama
Valero Energy Corporation
One Valero Way
San Antonio, TX 78249
mark.suyama@valero.com

Keywords: Management of Change, Hazard Identification

Abstract

Even if a management of change (MOC) system is administered perfectly, it will still fail if changes are not recognized and entered into the system. Thus, an essential element of a successful MOC system is teaching front-line workers to recognize changes in the course of their everyday duties. Such changes include undocumented improvements (changes) in procedures, improvised solutions (changes) that work around equipment limitations, and external circumstances (changes) that affect day-to-day operations. This paper discusses ways to improve the recognition of process hazards as a practical way to improve the identification and analysis of changes. Equipping the workers with basic process safety information and examples drawn from previous incidents dramatically improves their ability to recognize changes in the workplace and helps ensure that those risks are properly managed.

1. Introduction

When a Pareto analysis of a company's management of change (MOC) system failures is performed, "failure to recognize change" is typically one of the top reasons for the failures. However, when one interviews the individuals involved, their explanation is often that they did not believe there were any hazards associated with the change. So, despite organizations' clearly written policies that define changes subject to their MOC system, the practical definition applied by front-line workers is "no hazard, no change." This lack of recognition sets up the chicken-and-egg conundrum: the primary purpose of an MOC system is to manage the hazards associated with process changes, yet front-line workers often fail to submit a formal request to review a change precisely because they do not believe there are any hazards associated with differences they observe or create.

The authors attempted to improve their MOC systems by inverting the MOC paradigm – focusing on hazard recognition to improve the MOC system rather than focusing on the MOC system to improve hazard recognition. This approach resulted in improvements not only in MOC analyses, but also in process hazard analyses, job safety analyses, mechanical integrity, and planned inspections of work areas.

2. Technical definitions of change

2.1 Regulatory definitions

Since May 26, 1992, the Occupational Safety and Health Administration (OSHA) has required covered employers to implement a process safety management program. Specifically, 29 CFR 1910.119(l)(1)[1] requires employers to establish and implement written procedures to manage changes (except for "replacements in kind") to "process chemicals, technology, equipment, and procedures; and, changes to facilities that affect a covered process." The Environmental Protection Agency (EPA) followed in 1996 with its risk management program rule. In 40 CFR 68.75(a)[2], it similarly requires the owners or operators of covered stationary sources to establish and implement written procedures to manage changes. In 2011, the Pipeline and Hazardous Materials Safety Administration began requiring pipeline operators to manage changes that affect control rooms. However, the definition of change in 49 CFR 192.631(f) and 195.446(f)[3] includes "participation in planning prior to implementation of significant pipeline hydraulic or configuration changes," so the practical scope of required MOCs extends far beyond control rooms.

2.2 Corporate definitions

When companies implement an MOC program, whether required by regulations or not, their policies typically echo the relevant regulatory language. That language works reasonably well for the technical staff whose work is organized around projects. If a new bypass line is needed, it is obviously a change. If a new chemical will be added, it is obviously a change. If a procedure must be rewritten, it is obviously a change.

It is equally important that corporate policies define what is **no**t a change. Here, again, most corporate policies simply echo the regulatory language that a "replacement-in-kind" is not a change. A replacement-in-kind means a replacement that satisfies the design specification. This exception also works well for the technical staff. If one is replacing a piece of carbon steel pipe with one just like it, it is obviously not a change. If one is changing the feed rate of a chemical within established safe operating limits, it is obviously not a change. If one is simply making a typographic correction to a procedure, it is obviously not a change.

2.3 Gaps in the definitions

The problem of unrecognized changes primarily arises in the day-to-day activities of the front-line workers. Ideally, their work is proceduralized and their activities follow those procedures, so there is no change. Realistically, workers use their skills to get the job done in a safe and efficient manner under conditions that differ slightly every day. Their challenge is to recognize when "difference" rises to the level of "change" that warrants MOC review. Defining replacement-in-kind too narrowly can essentially stop work, while defining it too broadly can allow changes that exceed the organization's risk tolerance to proceed with no formal hazard review. Several times each day, workers are expected to decide whether a difference is a change, including differences involving equipment, process, personnel, and environmental conditions:

2.3.1 Differences in equipment conditions

An inspector asks permission to temporarily remove the external insulation from a heat exchanger, which is a code-rated pressure vessel. The operator knows that the surface temperature of the shell will not pose a burn hazard if someone touches its surface while the insulation is removed. If the request is not subjected to an MOC review, will anyone realize that the design basis of the relief valve was the external fire case and that the designer took credit for the insulation in sizing the relief valve? Thus, the heat exchanger may have inadequate pressure relief should a fire occur while the insulation is removed. But if this is a change, then should the removal of any insulation for any length of time be subject to MOC review?

Some refractory-lined hot metal piping in a pipe rack is normally inaccessible. In preparation for maintenance work, a field safety supervisor requests that additional insulation be wrapped around the piping to reduce the risk of a worker being burned if he accidentally touches the pipe. If the request is not subjected to MOC review, will anyone realize that the added insulation will increase the metal temperature above its design rating? Thus, the piping will be weaker and may burst as a result of the added insulation. But if this is a change, should the addition of any insulation on any piping be subject to MOC review?

2.3.2 Differences in process conditions

The instrument air system is temporarily switched to a nitrogen backup. If the use of nitrogen is not subjected to an MOC review, will anyone realize that a worker entering the transmitter room could be asphyxiated? But if this is a change, should any use of inert gas be subject to MOC review?

A chemical plant switched to a different supplier of a commodity chemical (50% caustic). If the new material is not subjected to an MOC review, will anyone realize that the trace chlorine in the city water used by the new supplier could crack the stainless steel piping in the caustic system? But if this is a change, does that mean every change in commodity suppliers should be subjected to MOC review?

2.3.3 Differences in personnel conditions

The supervisor is called away to attend to a family medical emergency, but no one is immediately available to fill the position. If the supervisor's absence is not subjected to an MOC review, will anyone realize that the remaining operator will be overloaded if there is a process upset? But if this is a change, does that mean that an MOC review is required every time the supervisor takes a break?

The supervisor is called away to attend to a family medical emergency, but no one is immediately available to fill the position. The decision is made to suspend the unit startup until a replacement can be called in. However, if the delay is not subjected to an MOC review, will anyone realize that continued circulation at this point in the startup could create a water emulsion that will cause a steam explosion when the startup resumes? But if this is a change, does any interruption in startup or process activities require an MOC review?

2.3.4 Differences in environmental conditions

Due to high summer temperatures, the cooling water system is overtaxed. The operator is scheduled to make a batch of the most exothermic product. If the request is not subjected to an MOC review, will anyone realize that the exotherm will exceed the heat removal capacity of the kettle, given the higher cooling water temperature? But if this is a change, should the production of every batch be reviewed to consider diurnal and seasonal temperature variations?

A tank must be steamed out in preparation for maintenance. If the request is not subjected to an MOC review, will anyone realize that the open vent will not be capable of relieving the vacuum? But if this is a change, does every tank steamout activity require an MOC review?

All of these examples contributed to real incidents. Given 20/20 hindsight, it is obvious that all were changes with adverse effects. The challenge is to help workers recognize changes with potentially adverse effects before making the change. Unfortunately, attempts to eliminate the gray areas by making the definitions of change more complex or restrictive have been generally unsuccessful. For example, a company might say any equipment that is not the same make and model must undergo MOC review. Really? So every time a vendor changes the model number, an MOC review must be performed? While the purist could correctly argue that such changes have resulted in process safety incidents, the pragmatist will correctly argue that given finite resources, strict adherence to that definition will overwhelm the MOC system, rendering it nonfunctional. Once the MOC system is overwhelmed and paralyzed, workers may revert to using their good judgment regarding hazards of changing conditions and only use the MOC system when they believe a hazard is not adequately addressed.

3. Improving recognition of change

The historic approach to improving the recognition of change has been to clarify the technical definitions and train workers in those definitions. That approach was successful in the early years of MOC, and there is an ongoing need for such training. However, the authors believed a new approach was needed to help front-line workers recognize changes more effectively.

The essence of this new approach was to include the entire front-line staff in the hazard recognition training already being given to MOC reviewers. The content was revised to emphasize observable conditions relevant to field personnel as opposed to technical discussions of corrosion or reactive chemistry. The content was streamlined to fit within an 8-hour training session, and a combination of lectures, videos, and hands-on exercises was used to maintain participant interest.

3.1 Recognizing hazards from first principles

A fundamental principle of successful adult education is to explain why a requested action is important to the individual and to those he cares about. Thus, we needed to go beyond a simple recitation of rules, such as "Never enter a confined space without respiratory protection," because sometimes the hazard (asphyxiation) extends beyond the technical boundary of a "confined" space. We wanted to provide the participants with a framework for recognizing potential hazards without overburdening them with all the technical details that would be necessary for an MOC reviewer to evaluate the risk.

We achieved this by organizing hazards into major categories of energy (such as fire or mechanical energy), materials (such as toxics or asphyxiants), and environment (such as weather or nearby work). We then briefly discussed the technical principles associated with each category of hazards and illustrated them with specific examples of conditions the workers might observe in the field.

3.2 Recognizing hazards from previous incidents

Throughout human history, stories have been told to help us remember important events, and none are more powerful than our own family stories. We used this same principle to help front-line workers remember important principles of hazard recognition from their own organizations. In some cases these were painful losses; in other cases they were near misses. Regardless, the incidents highlighted actual industry losses that others suffered, or could have suffered under slightly different circumstances, to emphasize the importance of recognizing process safety hazards.

3.3 Recognizing hazards from missing barriers

James Reason's "Swiss cheese" model[4] has been widely used to explain how barriers can break down and allow hazards to propagate to loss events. We used a simplified version of this model and grouped all engineered and administrative controls into three barrier types: (1) people (human training and behavior), (2) plant (physical protections), and (3) process (management systems). We then used this model to encourage critical thinking by the participants. First we asked, "If a barrier exists, what hazard is it protecting you from?" Then the follow-up question was, "So if you observe or create a hole in that barrier, isn't that a change?"

4. Example applications of the hazard recognition approach

As previously discussed, we used real-world examples to help participants recognize differences in their everyday activities, equipment, and environment that really warrant a more formal MOC review. The following are two examples.

4.1 Recognizing hazards in the context of a previous incident

Stored energy was one category of hazards we discussed from first principles. We showed participants the photograph in Figure 1 and asked, "What happened"? The workers needed to remove a leaky tube bundle for repair, but the hydraulic bundle puller was not strong enough for the task. Workers decided to heat the shell with 450-psig steam to expand it and free the stuck bundle. A steam hose was connected to one of the drains on the bottom of the shell, and a bleeder valve on the top of the shell was opened so any uncondensed steam would be vented, because there was no intention to build any pressure. No pressure, no hazard; no hazard, no change.

Figure 1. Tube bundle ejected 60 ft. from shell

Unfortunately, the tube bundle remained stuck, and the workers decided they needed more heat on the shell. Since no hotter steam was readily available, the workers decided to use ceramic heaters normally used for post-weld heat treating. The workers failed to recognize that condensate had accumulated in the shell and would be boiled by the added heat. The additional steam quickly overwhelmed the vent and increased pressure in the shell, ultimately ejecting the bundle 60 feet. Fortunately, the workers had left during shift change and no one was hurt.

The impact of this lesson on the participants was immediately evident. In the discussion that followed, participants shared their own stories of unexpected stored energy in systems that had been locked, tagged, and were presumably de-energized. Participants also discussed the potentially disastrous process safety implications of creative uses of ceramic heaters to temporarily add process heat or to temporarily compensate for missing insulation.

4.2 Recognizing hazards in the context of a field observation

The photographs in Figures 2 through 5 (courtesy of CNRL, 2013) were taken before class in the participants' normal work areas. After discussing the hazard category of explosion, we asked participants what change was evident in Figure 2 and did it imply any process hazard. The scaffolding was easily recognized as a change, and knowledgeable participants shared that it had been erected to support tarpaulins and block winter wind. Further discussion revealed that the change was needed to shelter the pilot lights that had been routinely extinguished by high wind. So, the change was necessitated by poorly designed pilot lights, and the tarpaulins were an improvised workaround to reduce the chance of a firebox explosion. However, the tarpaulins also enclosed the area under the burners and could provide enough confinement to cause an explosion if there were a gas leak. Also, there was a previous incident in which similarly arranged tarpaulins had been blown against, and ignited by, a propane-fired space heater. The burning tarpaulins had overheated the propane tank to the point of venting through its relief valve; fortunately, the fire was controlled before a boiling liquid expanding vapor explosion (BLEVE) of the tank occurred.

Figure 2. Scaffolding was erected to support tarpaulins

Figure 3. The relief valve pilot line insulation was removed

We showed the participants Figure 3 after discussing the hazards of high pressure. In the photograph, the relief valve inlet and outlet are insulated, but the pilot line insulation had been removed. Why the change? Knowledgeable participants explained that even when insulated, the pilot line was cold enough to act as a heat pipe in which moisture would condense and freeze in the winter. Without pressure in the pilot line, the relief valve would spuriously open; therefore, operators had removed the insulation so they could use a steam hose as needed to thaw the pilot line. The change was necessitated by a poorly designed relief system. The process safety implication was that a worker might be tempted to close the block valve underneath the relief valve during an actual emergency, mistakenly believing that the pilot line was frozen again.

Figure 4. The chain was removed

As a test, we showed the participants Figure 4 after discussing mechanical hazards. Participants quickly saw the missing chain and recognized the personal safety hazard it could cause if it fell off and hit someone. So what was the change? Leaving the chain unrepaired meant that the valve operator was inaccessible without scaffolding. So the process safety implication was that in an emergency, the valve could not be used to isolate equipment and minimize the size of a release.

Figure 5 was shown to the participants after a discussion of corrosion hazards. The change is difficult to see in the photograph because it is clear plastic film wrapped around the flange in the foreground. This change was instigated by the leak detection technicians who did it to improve the sensitivity of the gas detector. Since the leak test must be performed regularly, it is faster and easier to leave the duct tape or plastic wrap on the flange. Most participants were surprised by the process safety implications of a flange leak underneath the plastic wrap. Sulfur compounds in the gas react with humidity in the air, producing acidic conditions that attack the carbon steel flange bolts. If the corrosion continued undiscovered, the entire flange could separate and cause a hazardous release of toxic gases and hydrocarbons.

Figure 5. The flange was wrapped with plastic film

After the classroom discussions of hazards and examples, the participants were divided into multi-disciplinary teams and asked to perform field inspections in process units other than their normal work area.

Each team was accompanied by a worker from the inspected unit and followed all the appropriate unit access and safe work procedures. When the class reconvened about 2 hours later, each team shared descriptions and/or photographs of changes they observed as a result of their enhanced mindfulness of process hazards.

As further reinforcement, we gave the participants photographs of field hazards in the form of 4 x 6 in. laminated flash cards on a key ring. A photograph was on one side of each card, and a description of the process hazard was on the back. Participants took these back to their work areas as an ongoing reminder of things to look for in their daily rounds and field inspections that might be indicative of a process change.

5. Results

The initial results of this new approach to recognizing field changes that warrant MOC review was overwhelmingly positive. As workers have been trained at each site, there has been a significant increase in the recognition of process hazards, both within the MOC system and within other process safety elements.

5.1 Increased reporting of field changes

The significant increase in the number of MOC review requests was one of the most identifiable metrics of thc impact on the MOC system. The actual number of MOCs only increased about 10 to 15%, but the number of requests increased by more than 33%. We also interpreted that difference as a success, because many of the proposed changes were immediately rejected when the process safety implications were discussed with knowledgeable personnel.

5.2 Increased reporting of process safety hazards

Another immediate benefit of the approach was the dramatic increase in process safety hazards reported in housekeeping audits. Prior to the training, over 95% of observed hazards were occupational safety hazards (unsecured hoses, loose grating, sharp edges, etc.). After the training, up to 40% of the reported hazards had process safety implications (damaged insulation, worn hoses, missing labels, temporary equipment not fit-for-purpose, etc.).

5.3 Improved hazard analyses

A less quantifiable benefit of the approach was an improvement in the quality of MOCs as well as other hazard analyses such as job safety analyses (JSAs) and process hazard analyses. The difference was most evident in JSAs that had traditionally focused on protection against personal safety hazards such as cuts and falls. After the training, we observed JSAs were also recommending things such as hose whip checks to protect against pressure hazards and field readings to confirm control room indications.

6. Conclusion

The challenge of managing change is comparable to the challenge of managing an ocean voyage. The effects of discrete changes in course are relatively easy to recognize compared to the subtle effects of daily differences in winds, waves, currents, and tides. Each small change becomes the new normal, so the next change seems only a small difference from that. Unfortunately, it is often only after the ship hits an obstruction that the captain realizes how severely the ship has drifted off course.

The approach described herein is intended to help the front-line workers (the "sailors") recognize hazards and changes. The authors implemented this approach at six different sites. The sites varied widely in geographic location, regulatory environment, labor relations, size, age, and process activity (e.g., mining, upgrading, refining). In all cases, the approach was well received and produced similar success in closing gaps in the MOC systems. Those wishing to try a similar approach would benefit from *A Practical Approach to Hazard Identification for Operations and Maintenance Workers*[5] and the related computer-based training modules offered by the Center for Chemical Process Safety (CCPS). The *Process Safety Beacon* published monthly by CCPS is also a valuable resource. Equipping front-line workers with basic process safety information and examples drawn from previous incidents definitely improves their ability to recognize changes in the workplace and helps ensure that those risks are properly managed.

7. References

[1] U.S. OSHA, 29 CFR 1910.119, Process Safety Management of Highly Hazardous Chemicals

[2] U.S. EPA, 40 CFR 68, Risk Management Program for Chemical Accident Release Prevention

[3] U.S. DOT, 49 CFR 192.631 and 195.446, Pipeline Safety: Control Room Management/Human Factors

[4] Reason, J., *Human Error*, Cambridge: Cambridge University Press, 1990

[5] CCPS, *A Practical Approach to Hazard Identification for Operations and Maintenance Workers,* Center for Chemical Process Safety, New York, NY, 2010

The HA-MOC Interaction – Getting the Best out of Both Processes

Rainer Hoff, Ph.D., P.Eng.
Gateway Consulting Group, Inc,
8610 Transit Road, Suite 100, East Amherst, NY, 14051
rhoff@gatewaygroup.com

Paul Gathright
INVISTA
4123 East 37th St. N., Wichita, KS, 67220
paul.gathright@invista.com

Keywords: Management of Change, MOC, Hazard Analysis, HA, Process Hazard Analysis, PHA.

Abstract

The safety community, bolstered by the PSM regulations, has emphasized the importance of proper management of change, "MOC", to ensure safe, reliable and economic operation of chemical plants and refineries. Various authors [1-9] have stressed the importance of a systematic process of conducting MOCs. Hoff [10-11] has taken this further by proposing the use of a standard lifecycle model for MOC, and has demonstrated how the lifecycle model can be used as a basis for an optimal MOC practice. A typical lifecycle for a permanent MOC consists of 8 "states"—Initiation > Scoping > Change Design > Impact Analysis > Approvals > Implementation > PSSR > Close-out—although other state sequences are possible.

An important aspect of MOC is the consideration of the "Impact on safety and health" [4]. These impacts are normally addressed by performing a hazards analysis (HA) on the proposed change, using an appropriate methodology [4].

Hazard analyses take on many forms in operating plants. An initial Process Hazards Analysis, for example, is a regulatory requirement (OSHA 1910.119 [4]) for PSM covered facilities. The OSHA PSM standard also requires revalidation of PHAs per paragraph (e)(6). A properly executed MOC process will keep the PHA evergreen by performing a HA on each MOC. The timing of the HA within the MOC lifecycle impacts the effectiveness of managing hazards and the efficiency of the MOC process. In assisting facilities in implementing MOC systems, Hoff has observed that there is an astonishing diversity of when HAs are performed: some sites perform HA as early as the Initiation state, others as late as the PSSR state, and most sites perform the HAs somewhere in-between.

This paper discusses the importance of HA timing in the MOC context, and extracts best practice recommendations, to the extent possible. In so doing, the paper discusses subsidiary concerns including the merits of risk prescreening MOCs, streamlining the selection of HA methodology in the MOC context and the management of follow-up items. Examples are drawn from experience at actual plants.

1. Introduction

Managing change effectively has always been important at chemical plants and refineries. Lack of a proper MOC program is the prime cause in a number of well-known accidents, including the Flixborough explosion [1], and a contributing factor in other well-known accidents [2, 3]. Since the promulgation of OSHA 1910.119 [4]—the Process Safety Management regulations—proper Management of Change is a regulatory requirement, for Covered Processes (as defined in the regulation).

The Process Safety Management regulations cover 14 elements, two of which are PHA and MOC. As a result, when the regulations were first promulgated in 1992, substantial effort was needed to bring most plants into compliance, and that included:

- ensuring that Process Safety Information "PSI" was up-to-date
- performing an initial Process Hazards Analysis, "PHA", and revalidating it periodically
- implementing a Management of Change, "MOC", procedure

When an MOC is conducted, it is necessary to conduct a hazards analysis. The HA methodology can be typical of those used in the PHA. Hence there is an important link between PHAs and MOCs.

1.1 MOC Overview

A high-level model of MOC, referred to as a "lifecycle" model, is shown in Figure 1. The diagram depicts a permanent, non-emergency MOC process. A lifecycle is composed of "states": Initiation, Scoping, Change Design, etc. While the MOC is active, it could be considered to be "in" a state: e.g. the MOC is in the Approvals state.

Further details of lifecycle models can be found in the references [10].

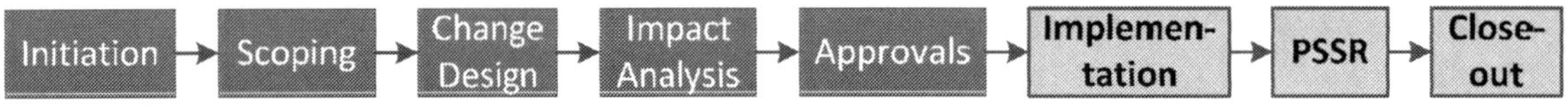

Figure 1. Typical permanent, non-emergency MOC lifecycle.

The "Impact Analysis" state is a logical time to conduct the HA. However, based on field experience at several dozen different companies, not every facility's MOC process triggers the HA during the Impact Analysis state. In fact, different companies have implemented MOC procedures that trigger the HA during:

- Initiation
- Scoping/Change Design
- Impact Analysis
- Implementation

The purpose of this paper is to assess the various HA timing alternatives and determine which is optimal. Ideally, one case would be best in all circumstances, rather than just best in limited circumstances.

2. HA Timing

2.1 HA During Impact Analysis

The MOC process flow, used in many organizations, is shown in Figure 1. The logic of this process is as follows:

1) Initiation:
   a) The initiator completes an MOC form (regardless of whether this is paper or electronic MOC)
   b) MOC initiation requires identifying:
      i) the location of the change (e.g. area, unit)
      ii) timing information (permanent vs. temporary; emergency vs. routine)
      iii) a description of the change
      iv) the justification for the change
      v) other administrative or housekeeping information
   c) The MOC initiator conducts a "risk prescreening". This is usually accomplished by answering a small number (often about 10) questions, which categorize the change as, for example, a "low", "medium" or "high" risk change.
2) Scoping:
   a) The purpose of scoping the MOC is to create a list of action items for the MOC
   b) An unscoped MOC simply has the header information provided during Initiation
   c) A scoped MOC typically has 20 – 50 action items that need to be accomplished during the lifecycle of the MOC
   d) The scoping process is usually assisted by checklists. Many of the checklists are very detailed and embody considerable knowledge and expertise that is potentially helpful to a person designing a change.
   e) Some action items will not be triggered by a checklist (because it is impractical to contain all knowledge in checklists), so those action items will still need to be created by the person conducting the MOC scoping.
3) Change Design:
   a) The change is designed, either by redlining existing drawings or creating new sketches or drawings.
   b) Documents, other than drawings, may also be drafted or modified.
   c) The change design activities must document the proposed change so that the necessary information is available to allow a PHA to proceed
4) Impact Analysis:
   a) If the MOC pertains to a covered process, then process safety impacts must be analyzed using an acceptable HA methodology
   b) Depending on the nature of the change, other impacts may also be analyzed:
      i) Environmental
      ii) Financial
      iii) Industrial hygiene
      iv) Personnel safety
5) Approvals:
   a) The design change and the resultant impact analyses are reviewed by subject matter experts

   b) The design change and resultant impact analyses are approved by the designated approvers, in compliance with the written MOC procedure
6) Implementation:
   a) In some cases, additional detailed design work may occur. This augments the conceptual or lesser detail work conducted during Change Design
   b) The change is implemented in the facility
   c) As-built information is captured
   d) Documentation needed to safely startup and operate is made available
   e) Notification and training of affected persons is conducted
7) PSSR:
   a) The Pre-startup Safety Review is conducted
   b) If there are any pre-startup follow-up items, they must be closed out during the PSSR state
   c) Once all the pre-startup items are complete, then an authorization to startup is provided, and the facility/unit/equipment may be started again
8) Close-out:
   a) Post-startup follow-up items are completed
   b) Redlines are incorporated into their respective design basis documentation

2.2 HA During Initiation

Some companies advocate for conducting the HA during MOC Initiation. The process flow during Initiation therefore becomes:

1) Initiation:
   a) The initiator completes an MOC form (regardless of whether this is paper or electronic MOC)
   b) MOC initiation requires identifying:
      i) the location of the change (e.g. area, unit)
      ii) timing information (permanent vs. temporary; emergency vs. routine)
      iii) a description of the change
      iv) the justification for the change
      v) other administrative or housekeeping information
   c) Conduct HA

Note that it wouldn't be reasonable to conduct the risk prescreening, identified in 2.1(1)c), if a HA is performed at this time.

2.2.1 Stated "Pros"

The stated advantages[1] of this approach are:

- **High risk avoidance:** A HA can reveal what might be a proposal for a high-risk change, and the company can decide to avoid this change
- **Low benefit avoidance:** A HA can reveal what might be a proposal for a change that doesn't provide a significant benefit, and the company can decide not to proceed with this change

[1] The "stated advantages" are those put forward by the proponents of a particular approach. The authors do not necessarily agree with the statements, but they are presented herein to inform the discussion.

- **Gap identification:** There may be gaps in the design of existing facilities that the HA may find.
- **Enhanced innovation:** Early HAs promote brainstorming and encourage innovative approaches, new ideas and alternative solutions to address the given problem
- **Scheduling convenience:** There may be a regularly-scheduled interdisciplinary meeting (e.g. daily, weekly). Since the "team" is already assembled, this would be a good time to initiate an MOC and do the HA.

2.2.2 "Cons"

A critical review of this approach by the authors has identified the following shortcomings:

- **Insufficient detail:** Not enough detail is known about the change to conduct a proper hazard review. The danger is that key risk items may be overlooked.
- **HA change management needed:** If the HA is done during Initiation then, as the details of the change evolve, say during the Change Design activities, the HA itself would require rigorous change management. As the design changes/evolves, the basis for the HA changes/evolves, and therefore the HA would need to change/evolve.
- **Missing expertise:** HAs are intended to be interdisciplinary. During Initiation, it may not be known which disciplines are needed for the HA, which may lead to key risks being overlooked.
- **Inefficient HA:** The fixed-team approach (as in the "Scheduling convenience" point, above) is not as effective as an HA team based on a review of the capabilities needed for a given HA because the fixed team may have unneeded capabilities. This often leads to wasted time or non-value-added actions.
- **Inefficient brainstorming:** There is nothing stopping any number of interested parties from meeting and brainstorming at any time during an MOC. Adding the formality of a hazard review to a brainstorming session is unneeded and wasteful.
- **Overdesign:** Premature HAs tend to lead to overdesign. Depending on the nature of the change, the HA would anticipate the existence of various interlocks and other safety features. The design team would then view these items as constraints and tend not to search for a simpler or better design, or exploit opportunities for inherently safer design. Design teams tend not to overrule or challenge hazards analysis recommendations.
- **"Anchoring" problems:** If the HA is done during Initiation then, as the details of the change evolve, the HA itself requires updates, as indicated above. Tversky and Kahneman [12] identified the psychological phenomenon of "anchoring" whereby prior data unduly influences future decision-making. Wolf and Wasileski [13] label this effect "complacency" whereby existing data and conditions are not rigorously challenged because they "always worked in the past". A Hazards Analysis that is performed too early in the process, will always suffer from information deficiencies when compared to a later HA—the problem is that, once an HA has been done, it may establish an invalid benchmark (an "anchor") that others are unwilling to, or don't feel a need to, challenge.

2.3 HA During Scoping/Change Design

Some companies advocate for conducting the HA during Scoping. The argument is as follows:

- Scoping usually involves the exercise of various checklists.
- The checklists usually involve design considerations, such as:
  - the pressure safety valve checklist
  - the control system checklist
  - the electrical checklist
  - etc.
- the checklists usually involve safety considerations as well, such as:
  - the process safety checklist
  - the new chemical checklist
  - the environmental checklist
  - etc.
- If one is considering all of these issues during MOC scoping, why not just extend the safety checklist activity a bit, and perform the HA at that point?
- After all, aren't "checklists" and "what-if + checklist" considered acceptable HA methodologies according to the PSM regulations?

The process flow during Scoping therefore becomes:

2) Scoping:
   a) The purpose of scoping the MOC is to create a list of action items for the MOC
   b) An unscoped MOC simply has the header information provided during Initiation
   c) A scoped MOC typically has 20 – 50 action items that need to be accomplished during the lifecycle of the MOC
   d) The scoping process is usually assisted by checklists. Many of the checklists are very detailed and embody considerable knowledge and expertise that is potentially helpful to a person designing a change.
   e) Some action items will not be triggered by a checklist (because it is impractical to contain all knowledge in checklists), so those action items will still need to be created by the person conducting the MOC scoping.
   f) Conduct HA

2.3.1 Stated "Pros"

The stated advantages[2] of this approach are:

- **High risk avoidance:** Same as for Initiation
- **Low benefit avoidance:** Same as for Initiation
- **Gap identification:** Similar to Initiation, except that during Scoping there is a better chance of finding these gaps, because the scoping/checklist process leads the analyst through consideration of the various relevant disciplines

[2] The "stated advantages" are those put forward by the proponents of a particular approach. The authors do not necessarily agree with the statements, but they are presented herein to inform the discussion.

❖ **Scheduling convenience:** There may be a regularly-scheduled interdisciplinary meeting (e.g. daily, weekly). Since the "team" is already assembled, this would be a good time to scope the MOC, as a group, and then do the HA.

2.3.2 "Cons"

A critical review of this approach by the authors has identified the following shortcomings:

❖ **Insufficient detail:** Similar to Initiation, but lower magnitude
❖ **HA change management needed:** Similar to Initiation, but lower magnitude.
❖ **Inefficient HA:** Similar to Initiation, but lower magnitude.
❖ **Anchoring problems:** Similar to Initiation, but lower magnitude.

2.4 HA During Implementation

Some companies advocate for conducting the HA during Implementation. The argument is as follows:

- Some would argue that, an HA only becomes a requirement prior to the facility/unit/equipment being started up. So there's nothing prohibiting the HA during Implementation
- Oftentimes the Process Safety Information, "PSI", is out-of-date or has errors. There's no point in doing a HA if the baseline documentation is invalid. Once the change is implemented and the as-built information is captured, this is the ideal time to perform the HA.
- Sometimes it's not even clear that an MOC is needed until a work order is being implemented, at which point someone notices that a change is taking place.

The process flow during Implementation therefore becomes:

6) Implementation:
   a) In some cases, additional detailed design work may occur. This augments the conceptual or lesser detail work conducted during Change Design
   b) The change is implemented in the facility
   c) As-built information is captured
   d) Conduct HA
   e) Documentation needed to safely startup and operate is made available
   f) Notification and training of affected persons is conducted

2.4.1 Stated "Pros"

The stated advantages[3] of this approach are:

❖ **Compensating for poor PSI:** In the event that the PSI is out-of-date, performing the HA on as-built data provides a more accurate basis for conducting the HA.
❖ **Compensating for missing MOC:** In the event that one discovers during Implementation that an MOC or HA is actually needed, then there's still time to get it done.

[3] The "stated advantages" are those put forward by the proponents of a particular approach. The authors do not necessarily agree with the statements, but they are presented herein to inform the discussion.

- **Scheduling convenience:** There may be a regularly-scheduled interdisciplinary meeting (e.g. daily, weekly). Since the "team" is already assembled, this would be a good time to scope the MOC, as a group, and then do the HA.

2.4.2 "Cons"

A critical review of this approach by the authors has identified the following shortcomings:

- **Overlooking construction hazards:** Hazards associated with a change can pertain to:
  - Hazards existing while the facility/unit/equipment is in operation
  - Hazards existing while the facility/unit/equipment is under construction or modification.

  It is reasonable to expect that an HA assesses both the operational and construction hazards, and provides guidance for both process safety and consequential personnel safety. If the HA is performed after the change is implemented, then there is no opportunity for the HA to identify hazards present during construction, and no opportunity to suggest suitable mitigations
- **PSI deterioration:** If there is reliance on "catching everything as as-builts", then the quality of Process Safety Information will deteriorate, since less emphasis is placed on good design documentation.
- **MOC/PHA degradation:** If there is a reliance on identifying something as a change during Implementation, then the benefits of systematic MOC and HA processes are not achieved.
- **Acceptance of greater risk:** Once implementation has occurred, and the sunk costs have been expended, there is economic pressure to accept the implementation "as is", even if it introduces more risk into the process than necessary, since any rework would bring associated costs.

3. Relative Merits of Different HA Timing

So far the discussion has centered on the pros and cons of conducting HAs during a state other than the standard Impact Analysis state. This chapter attempts to identify the "optimal" approach by looking at four criteria:

- Loss Prevention
- Direct costs (time and materials)
- Indirect costs
- MOC efficiency

3.1 Loss Prevention Concerns

Losses are avoided when the applicable risks are identified and mitigated. Risk identification and mitigation are more effective when more information is available. Table 1 ranks the different HA timing approaches, according to their impact on loss prevention. Four stars are better than one star.

Table 1. Relative ranking of loss prevention concerns, depending on HA timing.

HA Timing	Score	Explanation
Impact Analysis	★★★★	The best information is available during Impact Analysis because the design data has already been developed during the Change Design state
Scoping	★★★	Less information available than Impact Analysis
Initiation	★★	Less information available than Scoping
Implementation	★	Given that the costs have already been expended, there is economic pressure to accept the change "as is". This constrains the HA exercise into a "make it work" activity, rather than a search for the optimum solution.

3.2 Direct Costs

The direct costs associated with the HA itself and its immediate consequences are:

- Cost of rework, as design errors or deficiencies are discovered,
- Direct costs of implementing sub-optimal designs, when better designs could reasonably have been created, and,
- Costs of managing the changes to the HA as circumstances in the facility change or evolve.

The cost of rework is generally far greater than the cost of updating the HA, so rework becomes the controlling factor. Table 2 ranks the different HA timing approaches, according to their impact on direct costs.

Table 2. Relative ranking of direct costs, depending on HA timing.

HA Timing	Score	Explanation
Impact Analysis	★★★★	The best information is available during Impact Analysis because the design data has already been developed. The HA is more accurate and would lead to the least rework.
Scoping	★★★	As described in §2.2.2, HAs performed prior to design tend to prescribe design elements, that the design team tends not to challenge. This could lead to sub-optimal design.
Initiation	★★	The same logic as the "Scoping" explanation, except that the problem may even be more severe due to even earlier HAs
Implementation	★	If the HA team does a thorough job, it may identify design deficiencies that have safety implications. Since the change has already been implemented at this point, the facility will experience the maximum cost of rework.

3.3 Indirect Costs

The indirect costs associated with the HA timing are largely due to:

- The loss of production due to delays, and,
- The loss of business due to delays.

Lost production may potentially be made up by running additional shifts, following the delay. However, lost business (i.e. lost, discontented customers) are much harder to make up.

The fundamental issue is therefore the delays caused by different HA timing scenarios, as summarized in Table 3.

Table 3. Relative ranking of indirect costs, depending on HA timing.

HA Timing	**Score**	**Explanation**
Impact Analysis	★★★★	The best information is available during Impact Analysis because the design data has already been developed. The HA is more accurate and would lead to the fewest delays.
Scoping	★★★	As described in §2.2.2, HAs performed prior to Change Design completion may require updates. This would necessitate another cycle of assembling the HA team and conducting the work, which may cause delays.
Initiation	★★	The same logic as the "Scoping" explanation, except that the problem may even be more severe due to even earlier HAs
Implementation	★	HAs conducted during Implementation are the most likely to cause the greatest amount of rework. Time is required to do the actual rework. Also, delivery times of supplies become an issue during rework cycles.

3.4 MOC Efficiency

The most efficient work processes are generally those with the fewest steps. Any time that steps need to be repeated, the efficiency is reduced. In this context, steps may need to be repeated because:

- New design information becomes available, which impacts the HA. If the HA is already complete, when the new information becomes available, then some or all of the HA may need to be repeated
- Implementation work may need to be redone due to shortcomings identified by a HA.

The fundamental issue is therefore the probability of redoing work as a function of different HA timing scenarios, as summarized in Table 4.

Table 4. Relative ranking of MOC efficiency, depending on HA timing.

HA Timing	Score	Explanation
Impact Analysis	★★★★	The best information is available during Impact Analysis because the design data has already been developed. The HA is more accurate and would lead to the least rework.
Scoping	★★★	As described in §2.2.2, HAs performed prior to Change Design completion may require updates. This would necessitate another cycle of HA activity.
Initiation	★★	The same logic as the "Scoping" explanation, except that the problem may even be more severe due to even earlier HAs
Implementation	★	HAs conducted during Implementation are the most likely to cause the greatest amount of rework. Rework is inherently inefficient

4. Summary and Conclusion

Tables 1 through 4 present a relative ranking of the merits of performing HAs at different times during the MOC lifecycle. In every case, the relative ranking is the same:

HA Timing	Rank
Impact Analysis	Always first
Scoping	Always second
Initiation	Always third
Implementation	Always the worst

The fact that each dimension—loss prevention, direct costs, indirect costs, MOC efficiency—produces the same ranking suggests that performing the HA during the Impact Analysis state is optimal, and not dependent on context or selection of parameters.

The key points, revealed by this study are:

1) The optimal time for a HA is during the Impact Analysis state
2) HAs performed too early create a need for managing the HA itself
3) HAs performed too early may anchor or bias the thinking of subsequent design teams, leading to sub-optimal designs
4) The more fully developed the PSI, the higher the probability of properly evaluating risk

Finally, it is important to underscore that critical thinking and hazard awareness are important at each stage in the MOC lifecycle.

6. References

[1] Parker, R. J., et al., 1975, "Formal Investigation into Accident on 1st June 1974 at the Nypro Factory in Flixborough," London.

[2] Baker III, J. A., Bowman, F. L., Erwin, G., Gorton, S., Hendershot, D. C., Leveson, N., Priest, S., Rosenthal, I., Tebo, P. V., Wiegmann, D. A., and Wilson, L. D., 2007, "The Report of the BP U.S. Refineries Independent Safety Review Panel," Report, Washington.

[3] Travers, I., 2010, "Learning the lessons from Buncefield," AIChE National Spring Meeting, AIChE, San Antonio, TX, p. 10pp.

[4] OSHA, 1992, "Process safety management of highly hazardous chemicals," 29CFR1910.119, OSHA, Washington.

[5] Board, U. S. C. S. a. H. I., 1998, "Investigation Report, Explosives Manufacturing Incident, Sierra Chemical Company," Chemical Safety Board, Washington, DC.

[6] OSHA, 2007, "Directive CPL 03-00-004 Petroleum Refinery Process Safety Management National Emphasis Program," OSHA, ed.Washington, p. 104.

[7] OSHA, 2010, "Directives -10-05 (CPL 02) - PSM Covered Chemical Facilities National Emphasis Program," OSHA, Washington.

[8] Herber, J. W., "Why Smaller Companies Struggle with Process Safety," Proc. 8th Global Congress on Process Safety, AIChE.

[9] Goddard, S., "A Pre-Plan for Process Safety Management," Proc. 8th Global Congress on Process Safety, AIChE.

[10] Hoff, R., 2007, "Lifecycles," MOC Best Practices, 1(2).

[11] Hoff, R., 2011, "Quantitative Models for Assessing MOC Performance," AIChE National Spring Meeting, 7th Global Congress on Process Safety, AIChE, Chicago, p. 15pp.

[12] Tversky, A., and Kahneman, D., "Judgement under Uncertainty: Heuristics and Biases", *Science*, 185 (4157) p. 1124-1131.

[13] Wolf, E.M., and Wasileski, R.F., 2014, "The Illusion of Attention: Are There Gorillas in Your Plant?," AIChE National Spring Meeting, 10th Global Congress on Process Safety, AIChE, New Orleans, p. 12pp.

Using Metrics to Make Improvements in Management of Change Systems

Mason D. Martin
Phillips 66
Spur 119 North
Borger, Texas 79007
Mason.D.Martin@P66.com

Prepared for Presentation at
American Institute of Chemical Engineers
2014 Spring Meeting
10th Global Congress on Process Safety
New Orleans, LA
March 30 – April 2, 2014

UNPUBLISHED

Using Metrics to Make Improvements in Management of Change Systems

Mason D. Martin
Phillips 66
Spur 119 North
Borger, Texas 79007
Mason.D.Martin@P66.com

Keywords: Management of Change, metrics, MOC, improvement, implementation, audits

Abstract

Process Safety awareness continues to increase the importance of safely, responsibly, and reliably producing critical components to our global economy. Today, nothing is more valuable than those things that can be applied at a site to improve the site's process safety performance. Management of Change (MOC) is a vital system, connecting to all other process safety elements. MOC systems keep all other programs (procedures, training, process safety information, emergency response, etc.) current and pave the way for operational excellence in all elements. MOC is unique to each site, and each site is unique in where it needs to improve. This paper shows how a simple process can be implemented at any site to improve the site's MOC system and performance. This simple process to map potential errors, audit the MOC system for these errors, and use the error rates to target efforts for improvements will have significant impacts on the site's execution of its MOC system as well as other elemental programs.

1. Introduction

1.1 Process Safety is not Academic

In many ways, much of what one sees presented in papers and at conferences is cutting edge technology. Academia has done a fantastic job of pushing the limits of computational complexity to the point that there is a perceivable lack of applied knowledge in the field. There is a need for things that the field can implement for improvement at the point need (like information at the point of use). There is a need for something that can be implemented anywhere, despite recognition that no two sites are exactly alike. There is a need for methods to be proven and reproducible. There is a need for improved Management of Change (MOC).

Process Safety is important to us all. Awareness of this fact continues to grow. Executive Order 13650 signed by President Barak Obama on August 1, 2013 is another sign that society wants industry to have command of its business. Industry is taking measures to cooperate and network at an unprecedented level in order to share learnings and leverage resources to ensure the ties that bind continue to strengthen our businesses rather than weigh them down. Industry understands the inherently safer model of reliable operation and continues to pursue improvement in this area. The market financially rewards consistent operators and penalizes those who incur unplanned outages. All of these factors tie Process Safety to both our individual lives as well as the global economy.

Unlike most Process Safety elements, MOC touches all others. The unrelenting march of time will cause even the best of programs to become outdated without change; therefore, change is inevitable. The modern workplace embraces change and recognizes that the need for change comes from a variety of sources. Economic conditions, societal expectations, regulatory requirements, competition, and other pressures force both individuals and organizations to change frequently. Proper and effective management of these changes, regardless of the source, becomes a key to sustaining safe and reliable operation.

While all PSM MOC programs must address the same regulatory requirements, MOC programs must be uniquely designed to handle different business models and organizational structures. However, all effective processes adhere at some level to the tenets of the MOC Lifecycle[1]. As processes go, then process improvement techniques still apply. Few process improvement techniques are simpler than the Plan-Do-Check-Adjust (PDCA or Deming) cycle.[2] The good news is that the Do step is happening all of the time. This paper focuses on how to establish a Check step and how to Adjust and Plan for the next cycle of Do.

Fig. 1 – Deming Cycle Diagram[3]

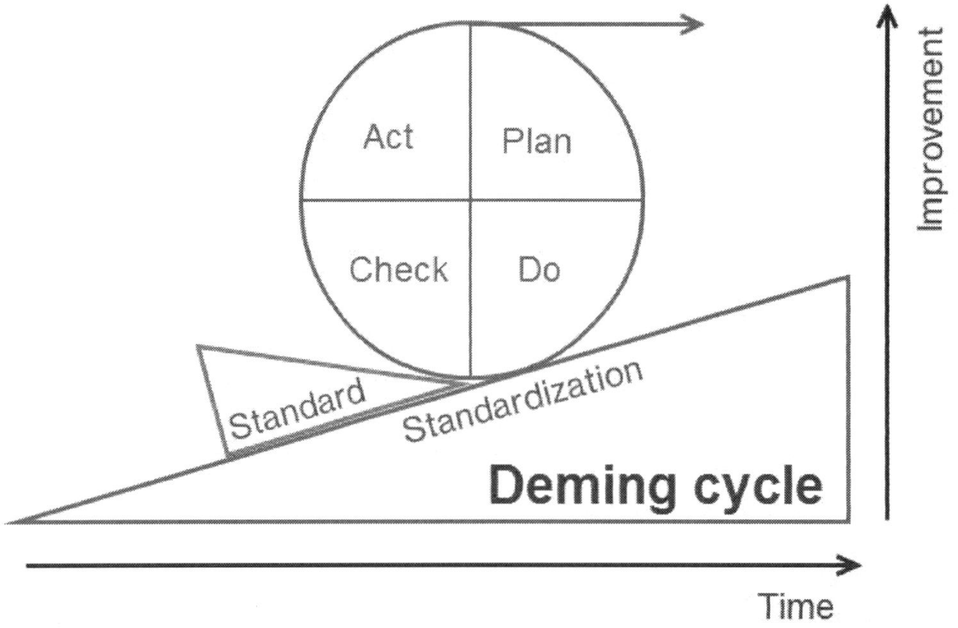

Sometimes, people are caught up in believing that their problems are so unique that no one can help them. Sometimes, people think their problems are so big that they don't know where to begin. Sometimes, people need to be reminded they are not alone and there is no better place to begin than where they are.

1.2 Seeing is Believing

Site A Profile	
MOC System	Electronic
Power Users	Limited
Organization	Stable

The process outlined in this paper was developed and implemented over a period of time at Site A (ca. 2003-2005). Unfortunately, early data showing the organization's initial performance and improvement trends were not retained. Data presented here begins in 2008 with Site A at an elevated level of performance (above 90% correct execution), but performance was not always as high. The organization detected issues with MOC execution and recognized the need for a formal feedback loop to users.

The mature, stable culture at the site contributed to the speed of improvement. The electronic MOC system had been in place for approximately five years and required minimal changes for maintenance and improvement. Power users consisted of a relatively small group of people per production area (3-5 people per work group). These people were easily held accountable, and their experience gave them credibility in changing the performance of those around them.

Fig. 2 – Site A – Relative Timeline of Events

Year	Event
2003	**100% auditing of MOCs begins.**
2004	**Major procedure revision (March).**
2005	**Major procedure revision (December).**
2006	**High levels of performance achieved plant-wide.**
2007	**High levels of performance sustained.**
2008	**High levels of performance sustained (see Fig. 3).**
2009	**High levels of performance sustained (see Fig. 4).**

High levels of performance were seen within various areas in a matter of months. Consistently high performance in various areas took longer. Consistently high levels of performance across the entire site took a number of years. The important point is that this site proved that elevated performance is not only achievable but sustainable over a significant period of time.

Fig. 3 – Site A – 2008 Performance

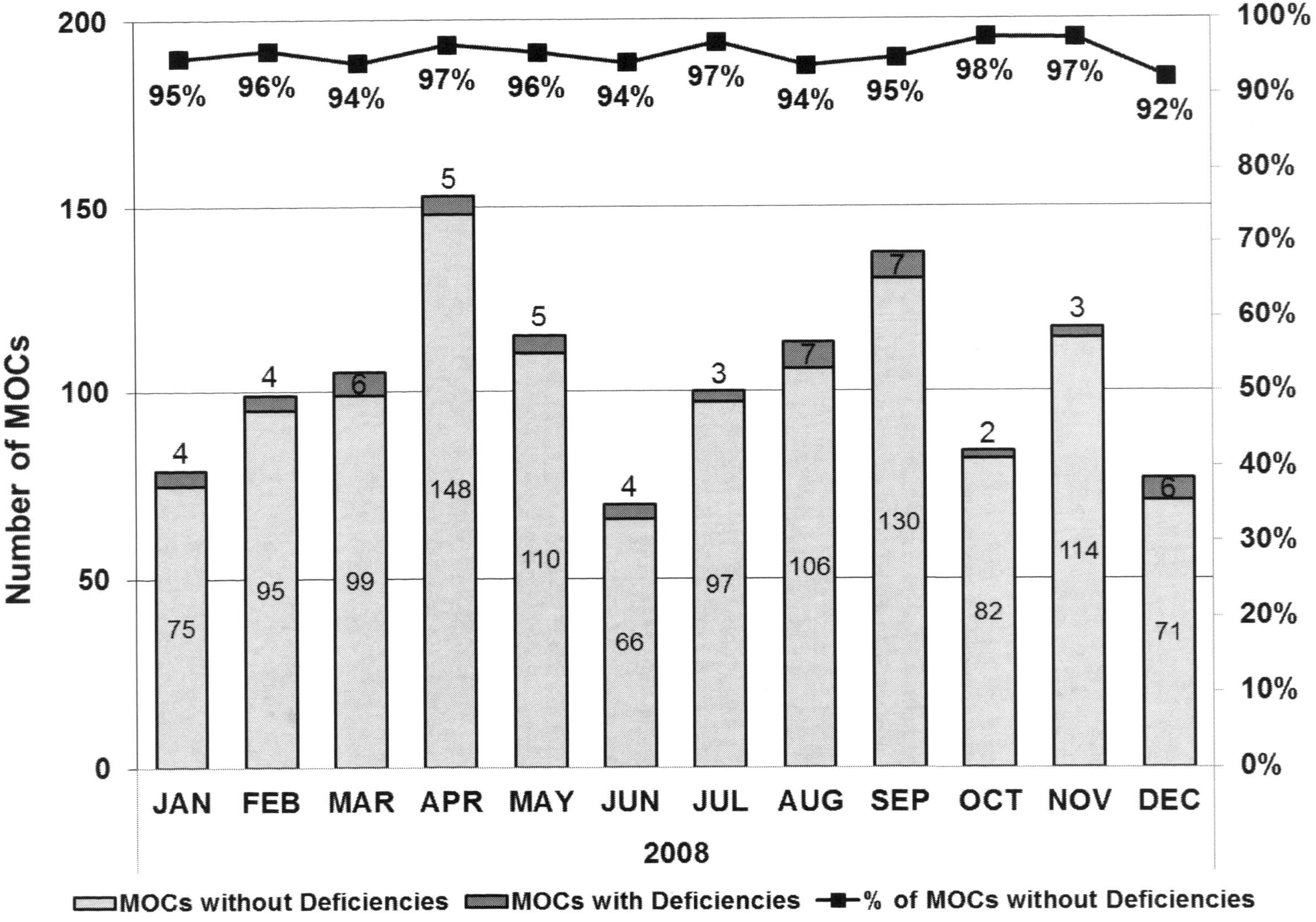

Fig. 4 – Site A – 2009 Performance

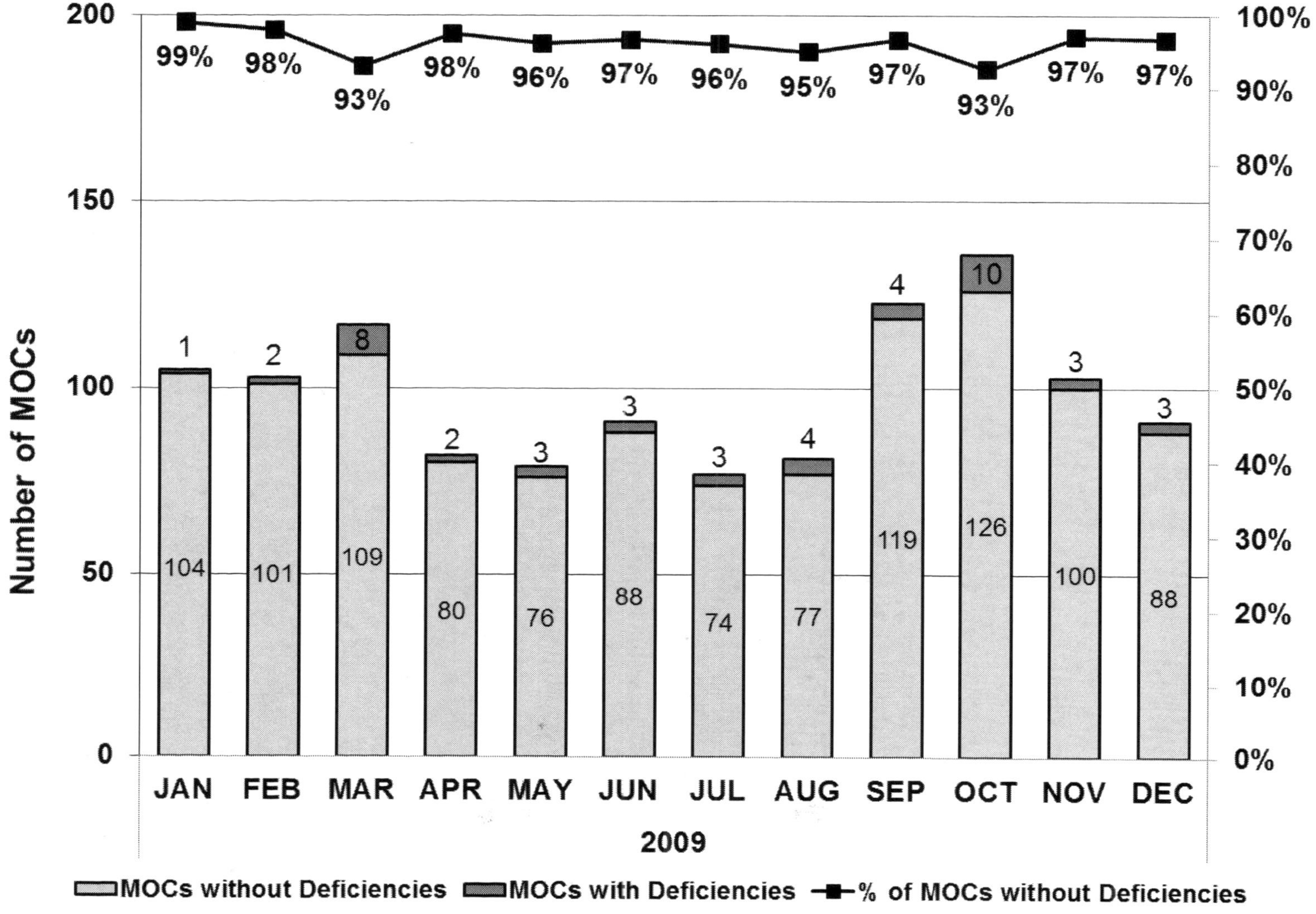

2. Process

What is required for a site to achieve similar results? The simple answer is to apply a similar methodology. Outlined within this section is the six-step process first used at Site A (ca. 2003-2010). The same process was implemented at Site B (2009-present) and Site C (2011-present). The Data and Results sections of this paper focus on implementations at the two latter sites.

2.1 Step 1: Build a Strong Foundation for Change

2.1.1 Support & Alignment

Get a member of senior leadership to support the work. This doesn't have to be formal, but it needs to be effective, or part of your job will become persuading the site leadership that this is important. In my case, I probably took it for granted. Fortunately, site leadership has always responded positively to the initiatives for improvement. Others might not be so lucky on their first attempt.

2.1.2 Communication

It is also good to think about how to communicate with the rest of the organization regarding this initiative. The communication plan should remain flexible to accommodate feedback and anything that gets overlooked.

2.1.3 Map the MOC Process

This may already be done. You may have a map embedded in your procedure or in an electronic workflow, but map the process for yourself to ensure that the map is accurate. It doesn't need to be perfect, but it needs to be understandable, accurate, and subject to revision. It needs to show all options at each step. A good process map can be used later to facilitate and perhaps improve training.

Fig. 5 – Example MOC Process Map (swimlane format)

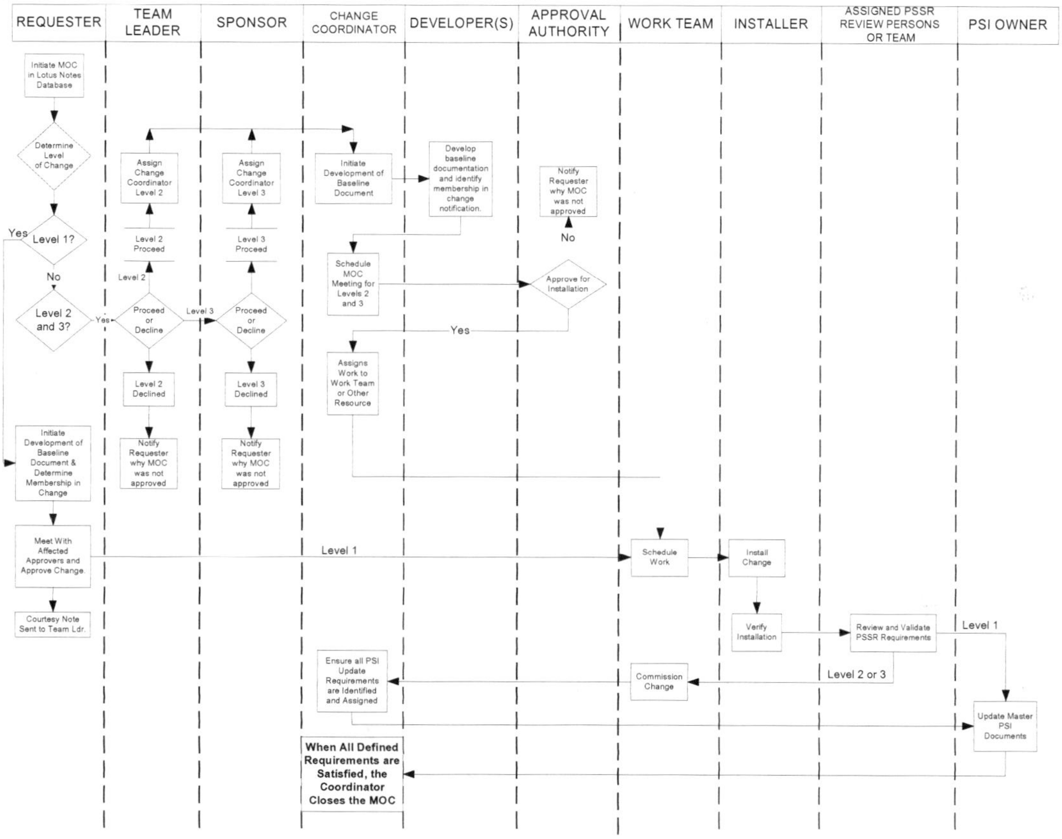

2.2 *Step 2: Identify and Rank the Hazards*

This step is the creation of an audit process (or what will become the Check step). Make a list of things that can go wrong in the process at each phase of the MOC process, paying particular attention to things that cause problems. Keep this list refreshed for any new issues that arise. Sometimes, we recognize later that certain things are connected back to a previous mistake. Be flexible enough to make improvements in the tracking system.

Develop a priority for errors. It is advised that this be risk-based, but alternatives can work too. Also keep this list flexible. What you thought was important may change through the feedback process.

Fig. 6 – Example List of Hazards with Risk Rank

MOC Section	MOC Step	Risk Rank - Category
Section 1 - MOC Definition	1.1 Needs Documentation	3 - MOC Desc/Doc
	1.2 Needs More Detail	2 - MOC Desc/Doc
	1.3 MOC & Field do not match	4 - PSSR
	1.4 Should be Level II	2 - Review Requirements
	1.5 Should be Level III	4 - Hazard Review/PHA
	1.6 Should be Temp	3 - Review Requirements
Section 2	2.1 Membership Incomplete	2 - Notification Requirements
Section 3	3.3 Approval Pending	1 - Review Requirements
Section 5 - Impact Review	5.1 AMOCC Personnel missing	3 - Review Requirements
	5.2-3 Hazd or Env Table inaccurate	3 - Hazard Review/PHA
	5.4 PSI Checklist incomplete/Missing	1 - Req Checklist
	5.5-6 Missing PSI or Other Actions	3 - Req Checklist
	5.7 Needs PHA review	4 - Hazard Review/PHA
	5.8 Impact Review signoff incorrect	2 - Impact Review Signoff
Section 7 - Work Complete	7.1-4 PSSR Check not completed	4 - PSSR
	7.5-8 PSSR Ck not req'd w/o reason	2 - PSSR
	7.9 Section Incomplete	1 - Misc
Section 9 - PSSR	9.1 PSSR Docs not attached	3 - PSSR
	9.2 PSSR Incorrect	4 - PSSR
	9.3 Pre S/A Actions incomplete	4 - Pre Startup Actions
Section 10 - Closeout	10.1 Temp Closed w/o Permanent	3 - Review Requirements
	10.2 MOC Closed, actions incomplete	2 - Review Requirements

Fig. 7 – Example Risk Rank Table

Risk Rank	Discrepancy Category
4 - SEVERE	Hazard Review/PHA not conducted
	Pre Startup Actions incomplete
	PSSR incorrect
	PSSR not complete
3 - HIGH	Impact Review Requirements not complete
	Missing Documentation
	Missing Post Startup Assignments
	Missing Pre Startup Assignments
	PSSR not attached
	Temporary MOC Completion Requirements
	Required Checklist inaccurate/incomplete
2 - MEDIUM	Completion Requirements not complete
	Impact Review Signoff missing
	Missing Detail in Description
	Notification Requirements not complete
	Missing PSSR Checklist was not Required but missing explanation
1 - LOW	Missing Business Manager Approval
	PSI Checklist incomplete
	Miscellaneous
0	No Discrepancies

2.3 Step 3: Audit MOCs and Correct the Errors

This is the Check step. Note each error contained in each MOC. MOCs can have multiple errors. Ensure the errors are corrected.

2.4 Step 4: Provide Feedback

This is the Act step. Develop a process for providing feedback to those who commit errors. It can evolve over time. Start with individual coaching. Be sure to emphasize the why behind the process (refer back to what can go wrong). Ease leadership into the conversation. Be prepared to get feedback and adjust procedures and training appropriately.

2.5 Step 5: Track Progress and Make Changes Based on Data

This is the Plan step. Track progress, issue metrics, and report to senior leadership. (Your champion on the site leadership team becomes important in bringing along their colleagues and driving the right attitude through the organization.)

2.6 *Step 6: Maintain Alignment & Communication*

This is the cyclic nature of the PDCA cycle. Dust yourself off and get ready for the next round. Remember that the Do step is constantly happening. The only separations between cycles are the changes you and your team make to the process. Keep the wheel turning and keep the organization on track. If you don't have alignment at this point, you'll know about it.

3. Data

3.1 *Site B*

This process was implemented at Site B beginning January 1, 2009. Data presented in the graphs below (Figures 8-13) cover the full implementation over a five-year period. Similar to Site A, the organization detected issues with MOC execution and recognized the need for a formal feedback loop to users.

Site B Profile	
MOC System	Paper
Power Users	Unlimited
Organization	Stabilizing

3.1.1 Paper or Plastic

Site B used a paper MOC system, so all audit data accumulated for 2009 (Fig. 8) was after the fact, causing a delay in the feedback to the power users. A power user may have created 5 more MOCs with the same error before learning of the error from the first one completed.

3.1.2 Everyone a King

Site B also let anyone be a power user. Consequently, it can be difficult for individuals responsible for MOCs only once or twice each year to get enough repetition to create and maintain good MOC habits.

3.1.3 We're not in Kansas Anymore

The site had experienced a high degree of turnover, so there was broad diversity in the work history of the personnel. Some were used to the process the way it was while others were accustomed to different practices. Many of these differences worked themselves out with time but created points of contention and reasons for misunderstandings.

3.1.4 Into Every Life, a Little Rain Must Fall

Improvements in MOC execution first seen in the fourth quarter of 2009 (Fig. 8) provided a good platform for improvement in early 2010 (Fig. 9). However, a large, planned outage for maintenance (turnaround) loomed in the fourth quarter of 2010. As attention became more focused on getting the work done, the attention to detail in the MOC process diminished. This also highlighted an opportunity to conduct refresher training before turnarounds and improved MOC tracking and support during and through the completion phase of turnarounds.

Fig. 8 – Site B – 2009 Performance

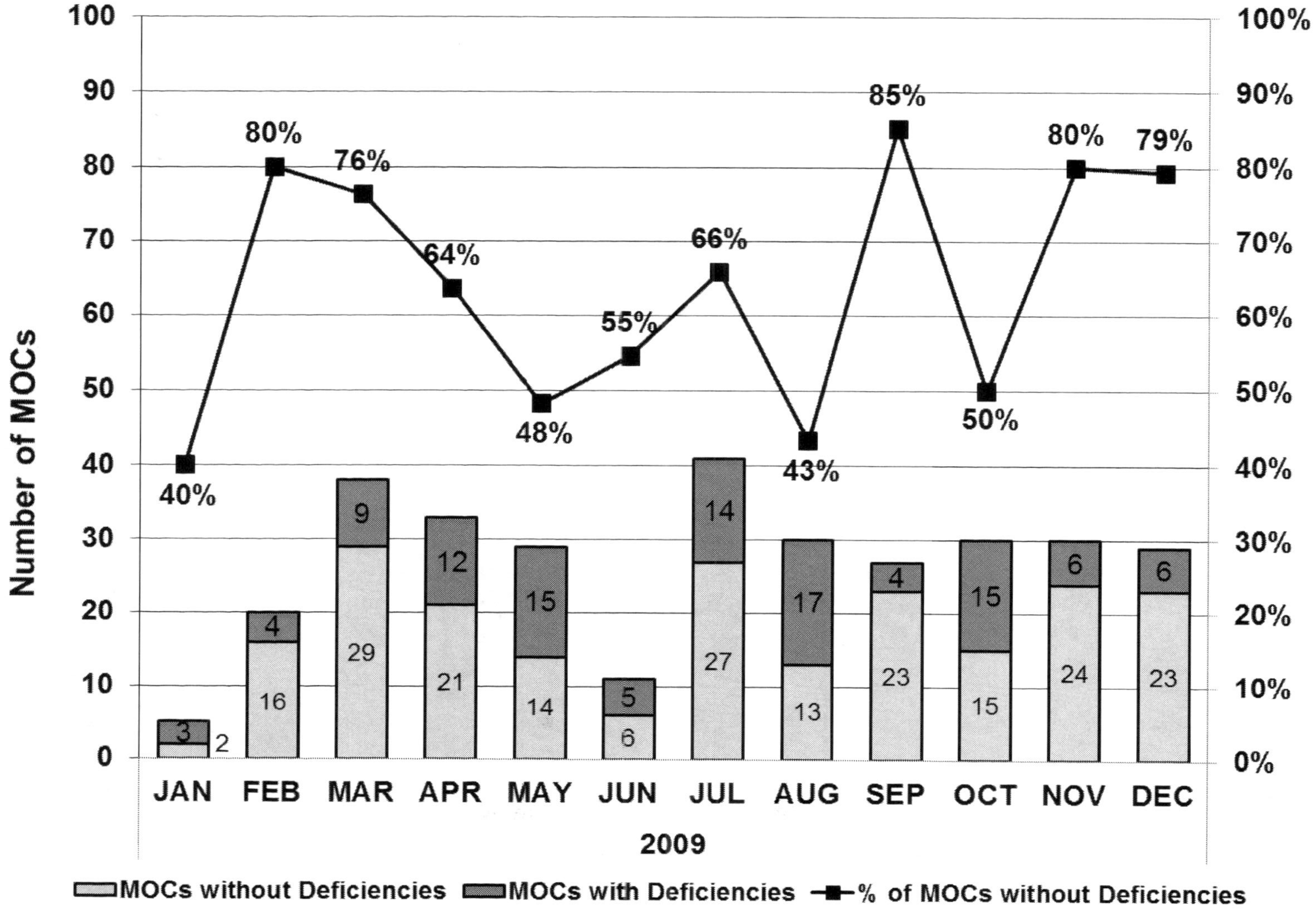

Fig. 9 – Site B – 2010 Performance

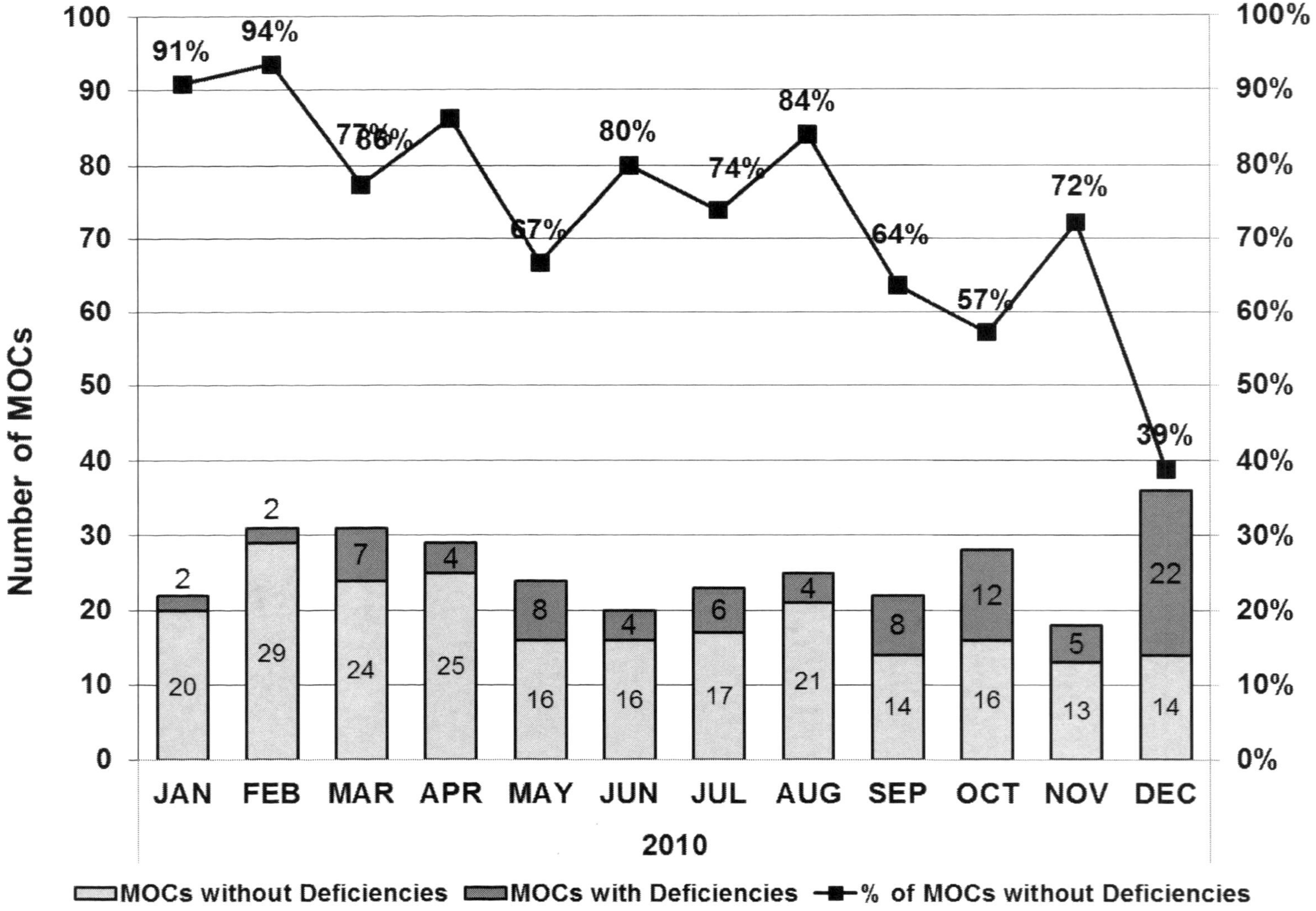

3.1.5 2011: A New Year, a New MOC System

Site B implemented an electronic MOC system in the first and second quarters of 2011 (Fig. 10). On March 30, 2011, the first electronic MOC was created and MOCs ceased to be originated in the paper process. However, paper MOCs were allowed to complete naturally until March 1, 2013. For almost two years, personnel at Site B gradually migrated from the paper system to the electronic system.

Fig. 10 – Site B – MOC Activity During Transition from Paper to Electronic System

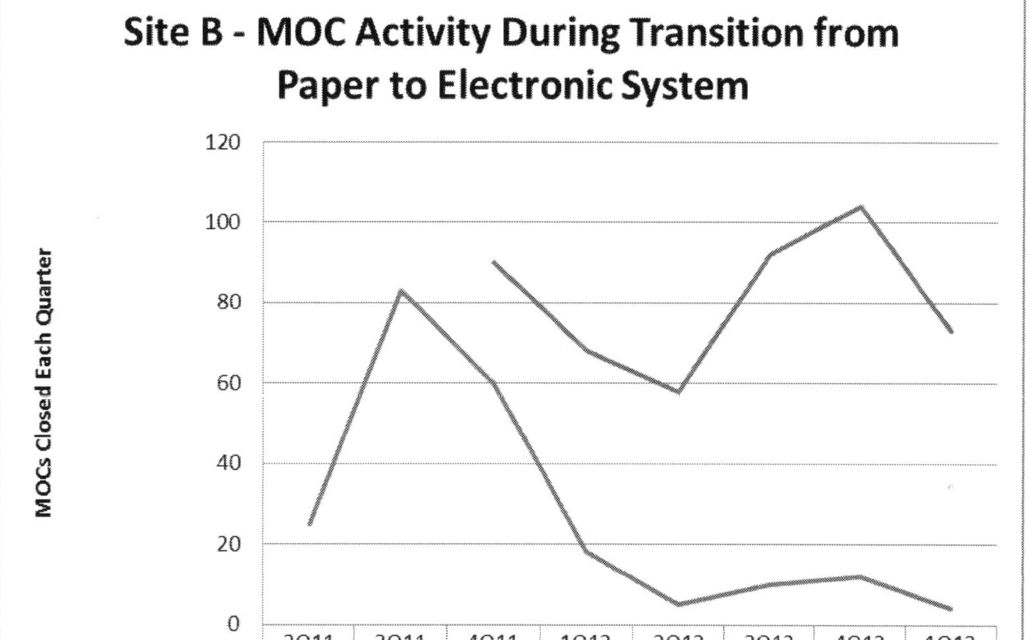

	2Q11	3Q11	4Q11	1Q12	2Q12	3Q12	4Q12	1Q13
Paper MOCs	25	83	60	18	5	10	12	4
Electronic MOCs			90	68	58	92	104	73

3.1.6 To Infinity and Beyond

As Site B transitioned fully to the electronic system, the organization was able to focus more on execution within the electronic system. Training improvements and continued use of the system reinforced good habits developed by power users in the electronic system. Site performance has steadily increased since 2011. Performance above 80% has been sustained since July 2012. With a single, minor exception, performance above 90% has been sustained since August 2013.

Fig. 11 – Site B – 2011 Performance

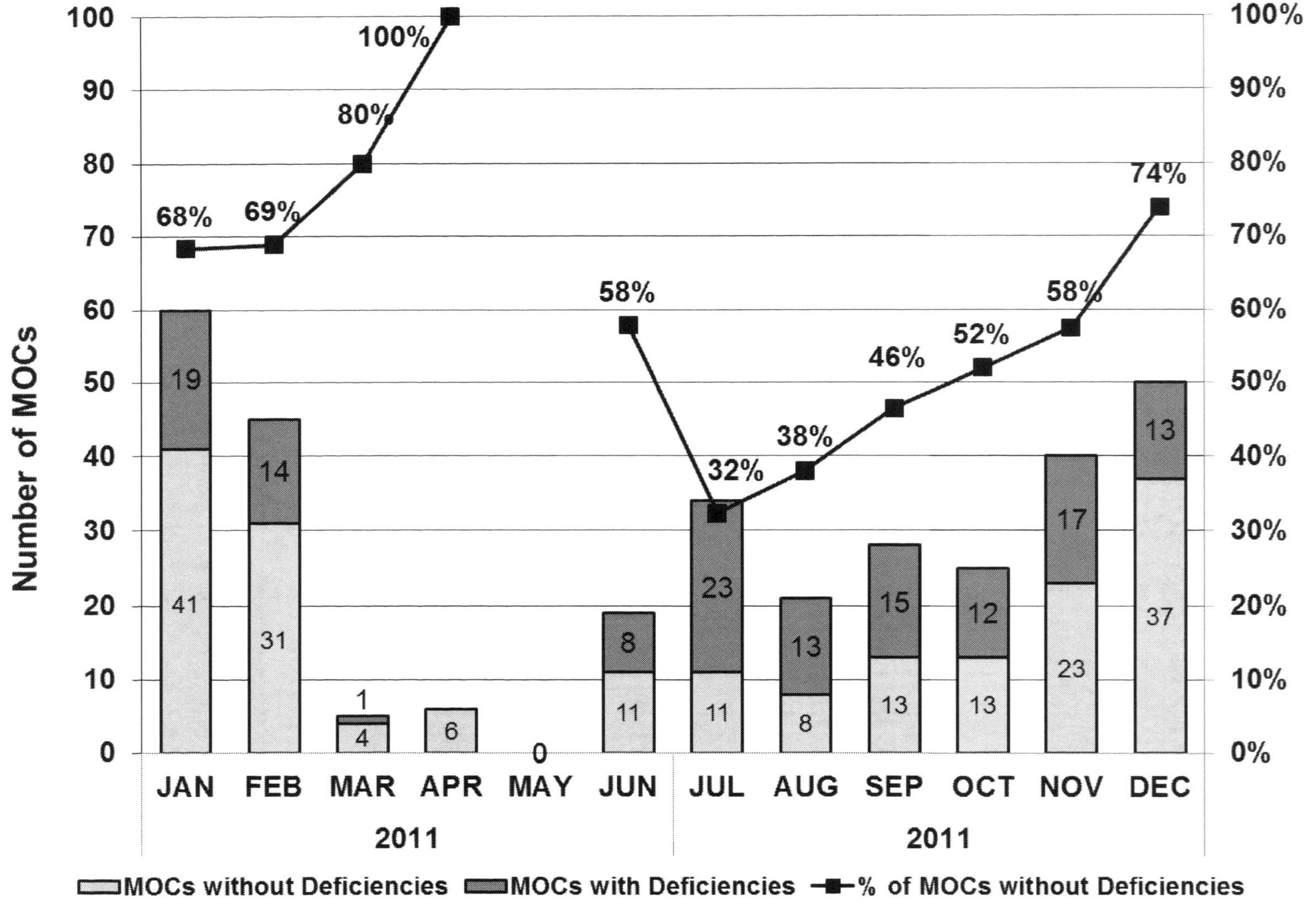

Fig. 12 – Site B – 2012 Performance

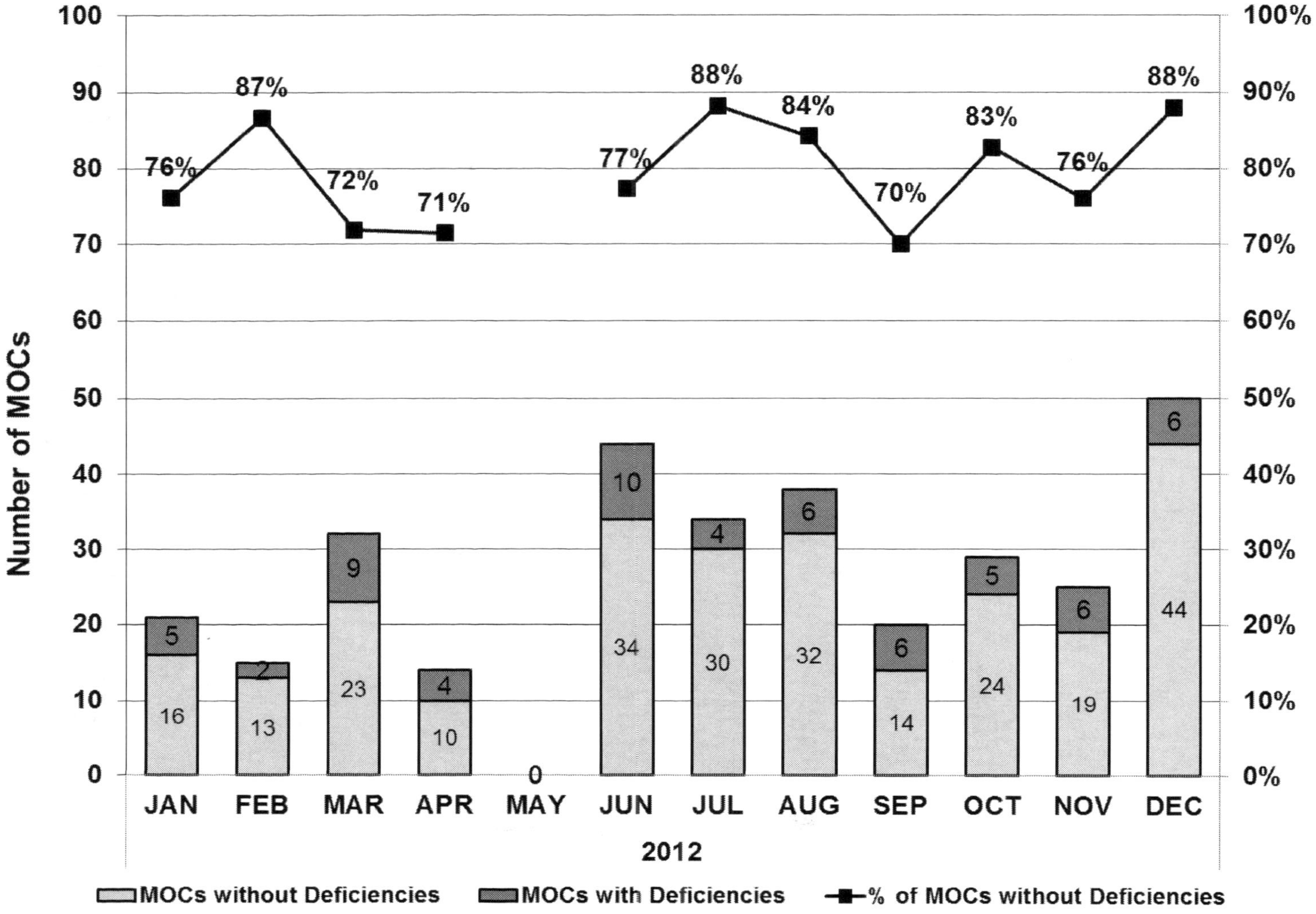

Fig. 13 – Site B – 2013 Performance

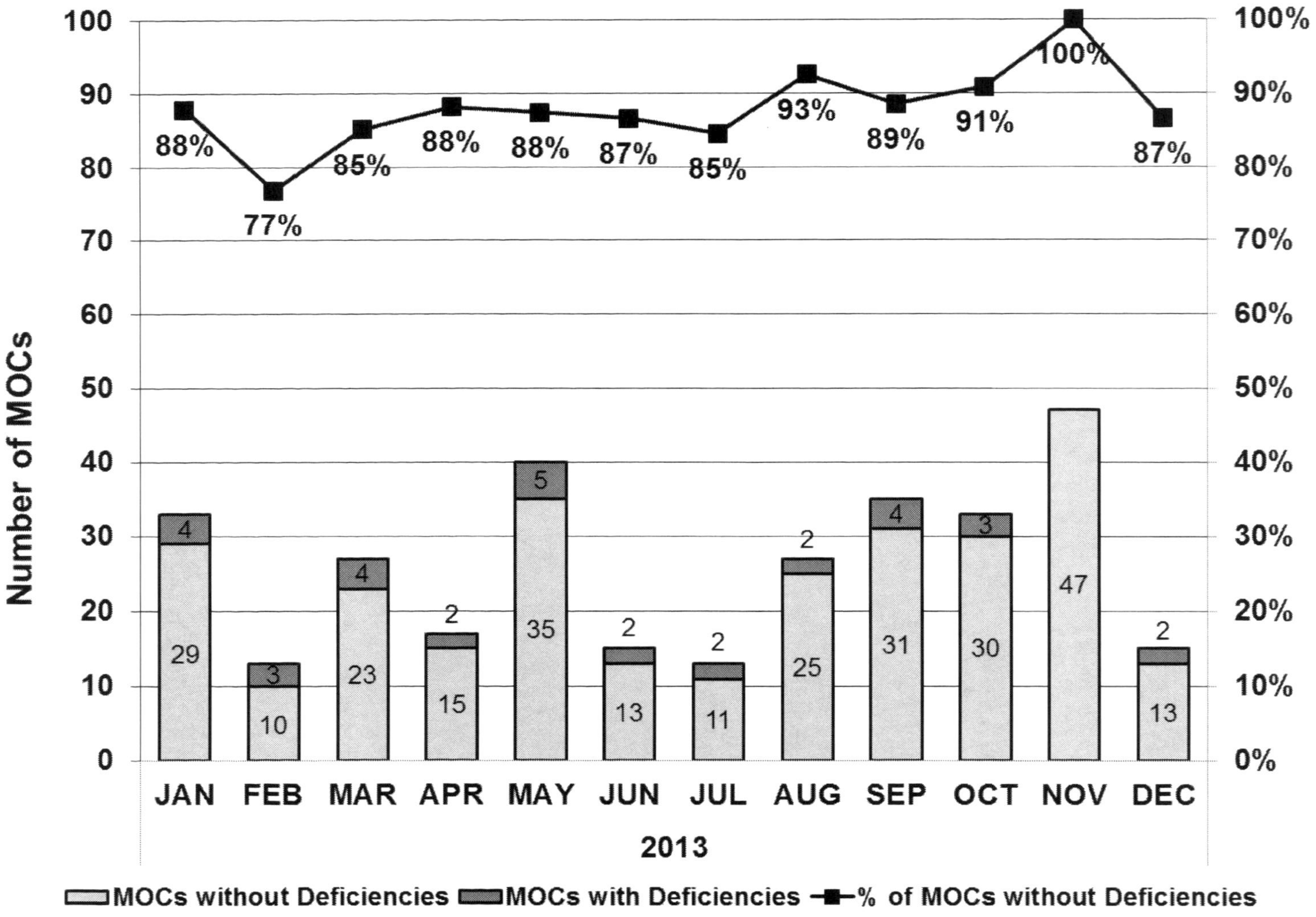

3.2 Site C

Site C Profile	
MOC System	Electronic
Power Users	Unlimited
Organization	Stabilizing

This process was implemented at Site C beginning June 1, 2011. Data presented in the graphs below cover the full implementation (Figures 14-18). Similar to Sites A and B, the organization detected issues with MOC execution and recognized the need for a formal feedback loop to users.

Implementation at Site C was conducted softly and with little fanfare. After approximately nine months of data had been accumulated, the MOC improvement team shared the data with Site Management along with a path forward for improvement. Not only was data developed on the frequency of errors per month (Fig. 14), but data was also available for the frequency of various errors that occurred (Fig. 15). These errors were risk ranked, allowing the team to see which errors presented more risk than others and target initiatives to reduce risk to the site. Site Management embraced the plan.

Fig. 15 – Site C – Discrepancy Evaluation

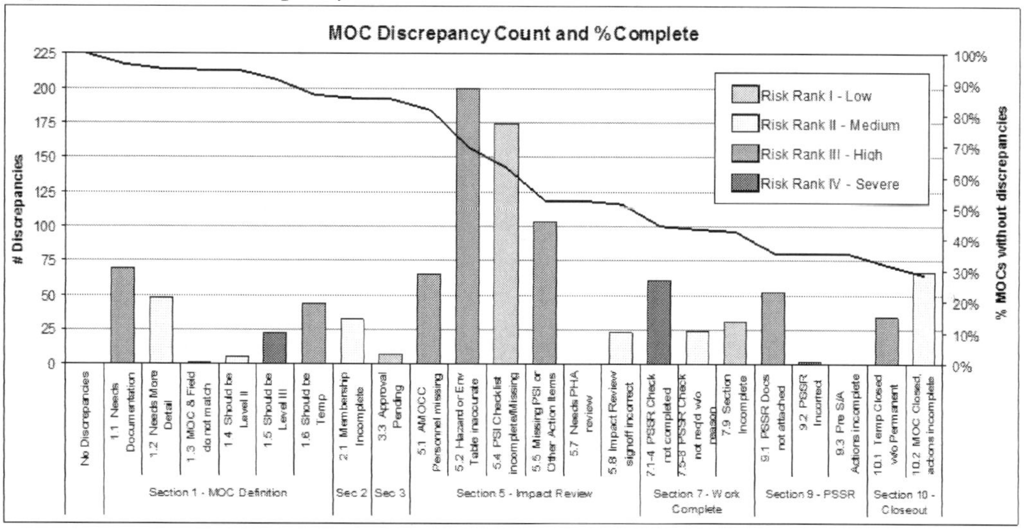

3.2.1 Hold on Tight

As part of the improvement plan, communication of errors to power users was formalized and included mid-management. This allowed for alignment within the organization and mid-management to reinforce the messages from Site Management during daily MOC execution. The MOC training was revamped, specifically with an increased focus on MOC execution, why MOC execution is important, and the hazards associated with particular mistakes. Multiple documents associated with the MOC process were streamlined and consolidated into two documents—a formal procedure for MOCs and a step-by-step guideline to navigate the electronic system.

Fig. 14 – Site C – Initial Performance (3Q11 – 1Q12)

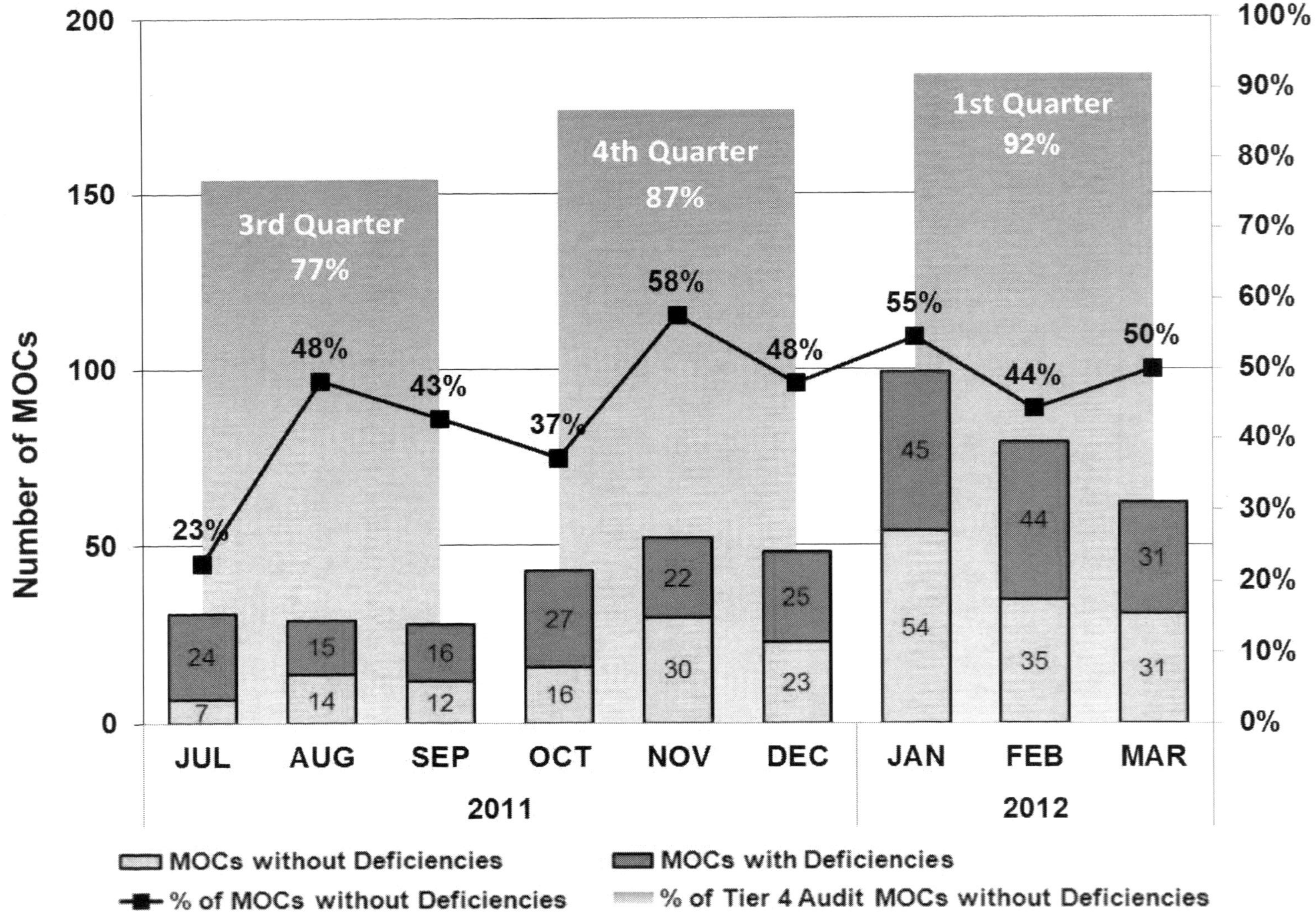

3.2.2 Storms on the Horizon

A turnaround was planned for 4Q12. Following lessons learned from Site B in 2010, the MOC process was further streamlined. Error-prone steps in the procedure were eliminated or redesigned where possible. Refresher training was provided prior to the turnaround. Enhanced MOC tracking was implemented on a daily basis, tying all turnaround MOCs directly to turnaround activity lists. Communication emails were sent daily to the organization on the status of turnaround MOCs related to scheduled unit startups. Field walk-down specialists were sent with PSSR teams to help identify, document, and address potential hazards prior to the startup of each unit. These efforts did not result in perfect MOC execution throughout the turnaround, but they did provide more consistent execution in 4Q12 (Fig. 16) and eliminated the sharp drop in execution performance experienced by Site B in 4Q10.

Fig. 16 – Site C – 2012 Performance

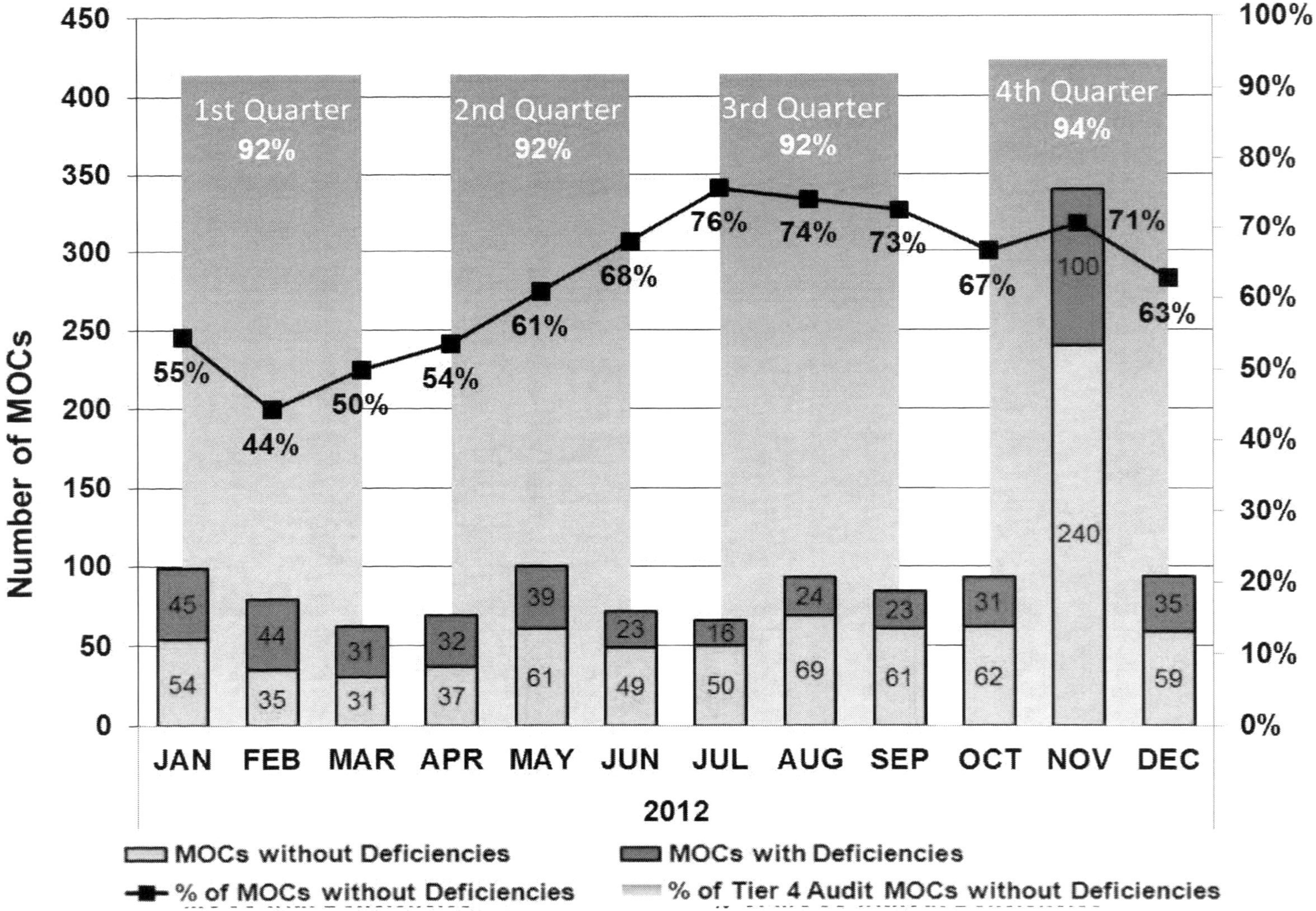

3.2.3 What's Next?

Following the turnaround, improvement of performance was limited. The metrics were reviewed and streamlined based on risk. Higher risk errors were kept on 100% review, and lower risk errors were left to the API 754 sampling process. The MOC software was stagnant and had become obsolete. A pre-planned migration to new software was already scheduled for later in the year. Moving to new software offered the opportunity to redesign the process from the beginning without the same built-in pitfalls. The MOC improvement team and Site Management used the discrepancy data to take aim at the remaining issues (Fig. 17). These issues became the new focus of the team responsible for the system transition and the training for site personnel.

Fig. 17 – Site C – December 2012 Discrepancy Evaluation

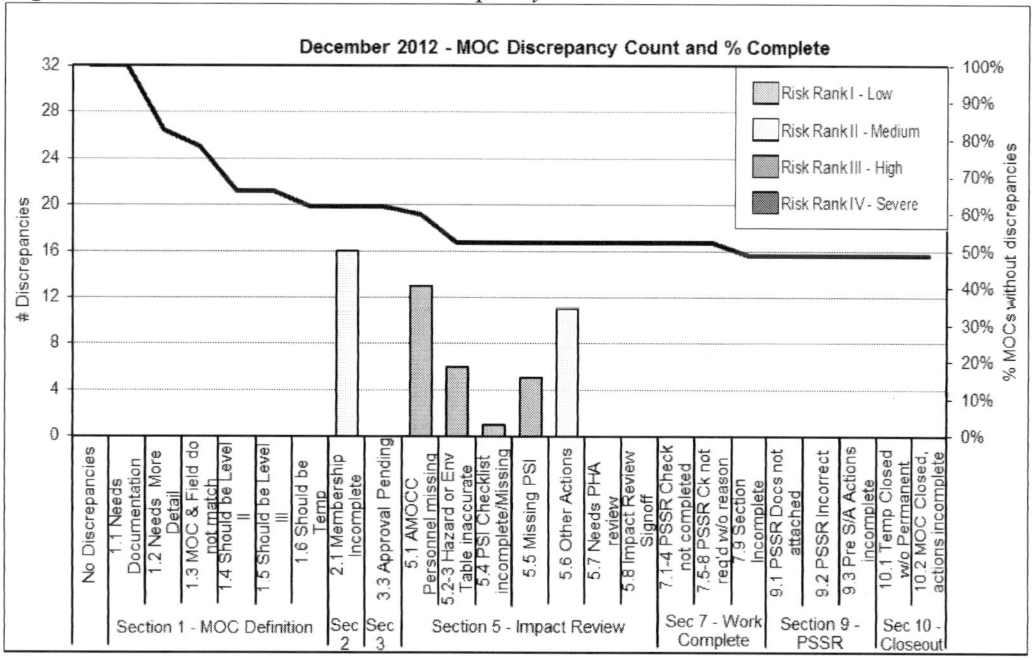

3.2.4 New MOC System Goes Live

The new MOC system was installed and the old system is being allowed to run until its natural end in July 2014. The metrics were revised to accommodate data from both systems and focus is maintained on the errors. Below are the numbers for 2013 performance (Fig. 18). Performance above 80% has been sustained since January 2013. With a single, minor exception, performance at or above 90% has been sustained since April 2013.

Fig. 18 – Site C – 2013 Performance

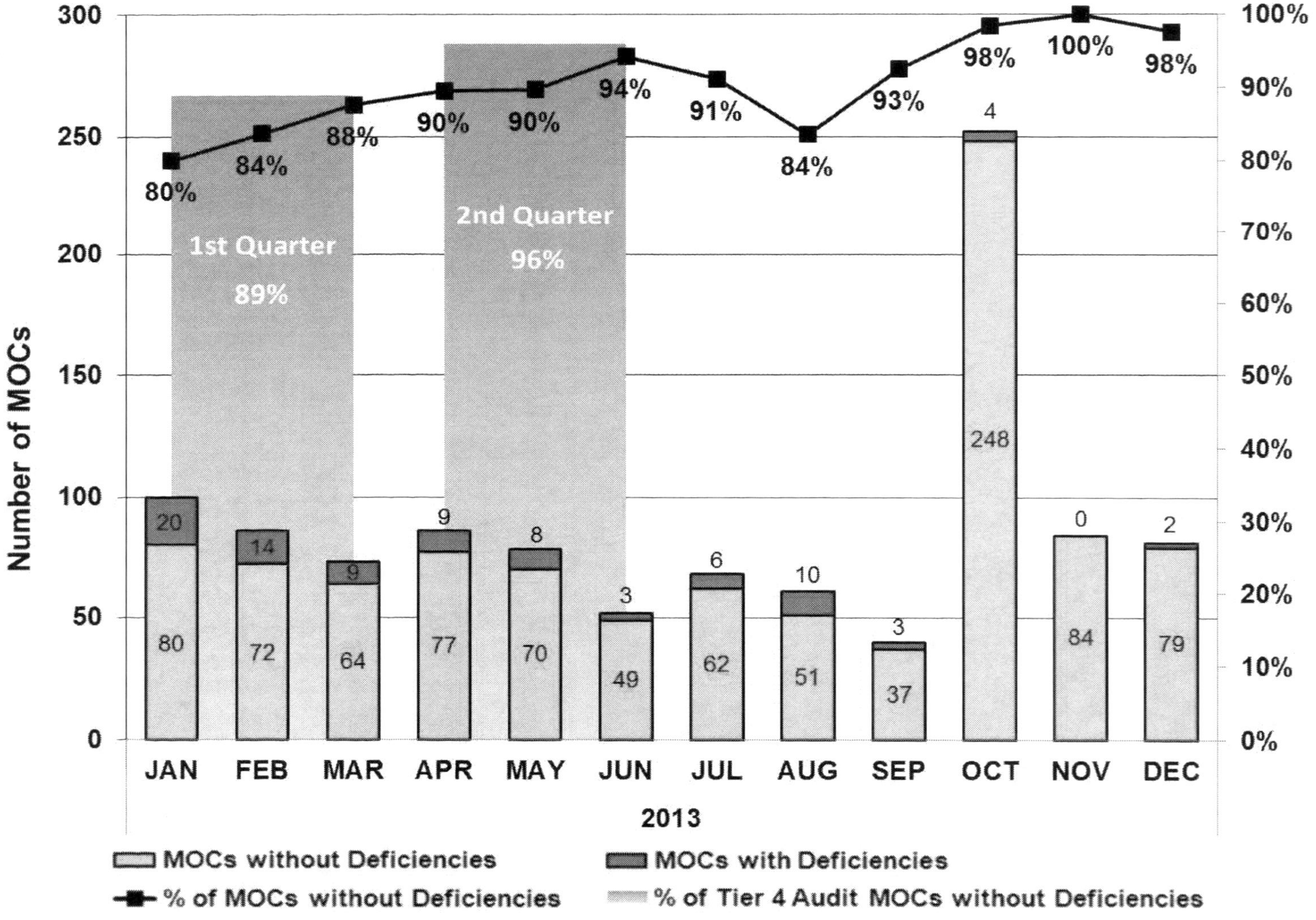

4. Results

Results vary from site to site initially, but the goal and the end results are the same. It is also necessary to keep the metrics review process in place.

4.1 Site B

It took time for Site B to achieve a consistently high level of performance with regard to MOC execution. A lessons learned was the importance of why avoiding errors within MOCs is vital and how that relates to safe, reliable operation of the plant. It is not enough to say procedures must be followed for their own sake. Procedures (as well as other program documents) are written in the blood of our predecessors. It is their wish for us to not repeat the past that leads to change in our programs.

Site B also provided a learning opportunity in the form of MOC management prior to, during, and immediately following turnaround activities. It is easy for organizations to be distracted with other priorities, but we should never been in such a hurry that we do not take the time to do our work safely. It is also easy within support groups to watch others do the work. There is much work a support group can do in order to prepare others for what is coming and facilitate and track work processes to completion.

4.2 Site C

Site C was afforded the benefit of lessons learned from Site B. Training focused on why good MOCs are important and what good looks like. Site C also took advantage of opportunities to improve alignment and communication and communication with the organization, especially around turnarounds. Site C also benefited from an opportunity to escape an obsolete software program and rebuild their process with components that make good MOCs. This leveraged the progress made in improved training on what makes an MOC good and why good MOC execution is important.

5. Conclusions

If your site does not have a comprehensive audit process for its MOCs, then you probably don't have to think hard to imagine the things that can go wrong with MOC execution. Your organization is likely reminded of it from time to time due to outdated procedures or inaccurate process safety information. Perhaps it feels as if there is so much wrong that they can never be fixed. One thing is certain—they can never remain fixed unless someone teaches the organization to quit breaking and abusing them.

This isn't rocket science. You are not alone. No hill is too steep. This is a challenge of yards measured one inch at a time. Simple steps can provide repeatable process. The results here can be achieved anywhere, but as Henry Ford once said, "Whether you think you can, or you think you can't--you're right. The difference is you. Are you going to make the difference for your organization?

6. References

[1] Hoff, Ranier, "MOC Best Practices", Gateway Consulting Group, Inc., East Amherst, NY, May 2007, p. 1.

[2] PDCA, Wikipedia.com, January 2014.

[3] Standardization Method with Deming Cycle, comindwork.com, January 2014.

OVERCOMING CHALLENGES IN USING LAYERS OF PROTECTION ANALYSIS (LOPA) TO DETERMINE SAFETY INTEGRITY LEVELS (SILS)

Paul Baybutt
Primatech Inc.
Columbus, Ohio, USA
paulb@primatech.com

Prepared for Presentation at
American Institute of Chemical Engineers
2014 Spring Meeting
10th Global Congress on Process Safety
New Orleans, LA
March 30 – April 2, 2014

UNPUBLISHED

OVERCOMING CHALLENGES IN USING LAYERS OF PROTECTION ANALYSIS (LOPA) TO DETERMINE SAFETY INTEGRITY LEVELS (SILS)

Paul Baybutt
Primatech Inc.
Columbus, Ohio, USA
paulb@primatech.com

Key words: Layers of protection analysis; risk tolerance criteria; risk allocation: safety instrumented system; safety instrumented function; safety integrity level; IEC 61511, ISA 84.

Abstract

The international standard on safety instrumented systems (SISs), IEC 61511 / ISA 84, requires the determination of safety integrity levels (SILs) for safety instrumented functions (SIFs) in SISs. Arguably, this is the most important requirement in the standard. Various methods are used for SIL determination including risk matrices and risk graphs, although layers of protection analysis (LOPA) is used increasingly. In employing these methods, a determination must be made as to whether existing risk levels for a process are tolerable by comparing risk estimates with risk tolerance criteria. If a risk gap is found, the SIL required to close the gap is specified. This determination requires setting risk tolerance criteria and making risk estimates.

LOPA is used to calculate the risk of individual hazard scenarios. Consequently, in order to simplify the determination of SILs using LOPA, overall risk tolerance criteria often are allocated to individual hazard scenarios so that risk estimates for scenarios can be compared with the allocated criteria. Sometimes hazardous events are used and the risks of scenarios that involve the same hazardous event are aggregated. Unfortunately, allocation of facility risk tolerance criteria to scenarios or events is fraught with difficulties and use of LOPA with allocated risk tolerance criteria can produce inaccurate results within a facility and inconsistent results across facilities.

A procedure is provided in this paper for SIL determination using LOPA in which facility risk tolerance criteria are used meaningfully and justifiably. The procedure entails the calculation of overall facility risk and an examination of contributions to the risks to individuals and groups of people from sources of risk such as different types of hazards, different modes of process operation, and different units and processes. The procedure is illustrated with an example.

1. Introduction

The industry standard, ANSI/ISA-S84.00.01-2004, (IEC 61511 Mod) Functional Safety: Safety Instrumented Systems for the Process Industry Sector [1 - 3], called IEC 61511 / ISA 84 herein, requires that safety integrity levels (SILs) be determined for safety instrumented functions (SIFs) that make up safety instrumented systems (SISs). Risk / safety matrix, risk graph, layers of protection analysis (LOPA), and other more quantitative methods are identified in the standard as acceptable approaches for determining SILs, depending on the circumstances [3]. Risk matrices and risk graphs are the simplest of the methods and some companies favor them for that reason. Unfortunately, these simple methods pose difficulties [4]. Fortunately, the difficulties can be overcome by using LOPA. However, LOPA is susceptible to errors in using risk tolerance criteria and considerable care must be taken when using it for SIL determination.

This paper identifies issues in using LOPA for SIL determination, provides a procedure that addresses the issues, and illustrates it with an example.

2. SIL Determination and Risk Tolerance Criteria

Process incidents originate with an initiating event that produces various intermediate events leading to an impact on one or more receptors such as people, the environment or property. The complete sequence of events is referred to as a hazard scenario. At some point in the scenario, a hazardous condition may activate a SIF to protect against a hazardous event that would adversely impact receptors. Thus, hazardous conditions and hazardous events are intermediate events in a hazard scenario. The same hazardous event may be produced by different hazard scenarios, e.g. ones with different initiating events, and they may activate the same SIF. Therefore, multiple hazard scenarios may be protected by the same SIF and contribute to its challenge frequency. Indeed, the IEC 61511 / ISA 84 standard states, "The integrity level specified for the (safety instrumented) function should take into account the cumulative total of all the risks involved if the function fails on demand" [3].

Risk tolerance criteria are used to determine if the existing risks of a process are tolerable by comparing estimates of the risks with the criteria to determine if there is a risk gap. If a gap exists, recommendations are developed for risk reduction measures to close the gap. Risk estimates are made for hazard scenarios and used to determine risk measures for comparison with criteria.

This section addresses issues that affect the overall approach to SIL determination using LOPA.

2.1 IEC 61511 / ISA 84 Requirements

The IEC 61511 / ISA 84 standard describes the determination of SILs for SIFs with reference to hazardous events. Indeed, SIFs protect against specific hazardous events. The standard calls for risk tolerance criteria to be established for them. However, the term "hazardous event" is not defined in the standard and an invariant definition is not possible [5]. Attempts to assign risk tolerance criteria to hazardous events likely will result in inconsistent and misleading SIL determinations. Some practitioners use risk tolerance criteria for hazard scenarios but the approach suffers from the same problem as using hazardous events because there is no invariant definition of what is meant by a hazard scenario.

In order to address this issue, SIL determinations are best performed using a risk model that employs facility risk tolerance criteria [6, 7] or, at least, that allocates overall facility risk tolerance criteria appropriately [8].

2.2 *Facility Risk*

Facilities may contain multiple processes and units. Risks can be evaluated for entities such as individual processes, units, hazardous events or hazard scenarios. However, individuals who work in a facility, or members of the public living in the vicinity of a facility, will be concerned about the total risk to which they are exposed from all hazards within the facility, not just the risk from one hazard scenario, hazardous event, unit or process. Companies will be concerned about the risk to all employees and members of the public from all hazards within the facility, as well as the risk to individuals to ensure no one person is exposed to undue risk. The determination of these risks requires that the risks of hazard scenarios be aggregated over all processes, process units, process modes, and hazard types for the facility. In other words, cumulative risk estimates are needed for comparison with cumulative risk criteria, i.e. overall facility risk tolerance criteria. Indeed, these are the type of criteria used by regulators, such as the U.K. Health and Safety Executive (HSE) [9]. Only total facility risk for specific receptors at a specific facility from all hazards at the facility has real meaning.

2.3 *Allocation of Risk Tolerance Criteria*

Some practitioners mistakenly use an overall facility risk tolerance criterion as the risk tolerance criterion for hazard scenarios or hazardous events which results in much higher levels of risk than should be tolerated. Overall facility risk tolerance criteria should not be applied to individual processes, units, hazardous events or hazard scenarios without suitable allocation. Clearly, less risk is tolerable from individual processes, units, hazardous events, and hazard scenarios than from all processes, units, hazardous events, and hazard scenarios at a facility. Unfortunately, risk tolerance criteria documented in the literature often do not clarify the entity to which they apply.

Some practitioners use risk tolerance criteria for hazard scenarios or hazardous events in the belief that it is easier to calculate their risk rather than the overall risk of a facility. However, risk tolerance criteria for individual hazard scenarios or hazardous events have no meaning by themselves [8]. They must be derived by allocating or apportioning overall facility risk to the scenarios or events by dividing overall facility risk tolerance criteria by the estimated numbers of scenarios or events. However, this practice is problematic [8]. The numbers of scenarios and events must be estimated but these numbers are subject to uncertainty resulting in questionable criteria. Moreover, it is difficult to define and use scenarios and events equivalently from the perspective of risk allocation. Furthermore, both individual and group risk should be addressed for people and separate allocations are needed but often they are not employed.

A better approach is to aggregate the risks of individual scenarios for comparison with facility risk tolerance criteria. Of course, reliance solely on meeting overall facility risk tolerance criteria may result in the inequitable distribution of risk across a facility. While the overall facility risk criterion may be met, there may be receptors that bear the brunt of the risk from processes, areas, units, process modes, etc. owing to the disproportionate allocation of risk across the facility. Consequently, overall risk determination should be followed by a check that risk has been allocated equitably within and across facilities, particularly to all receptors as ultimately that is what matters.

Process hazard analysis (PHA) studies are performed on a process but there are no standards on how a process is defined as part of a facility. One company may break their facility into more processes than another company. If facility risk tolerance criteria are applied to each process, the company that has defined more, smaller processes will accept a higher overall level of facility risk than the company that has defined fewer, larger processes. Of course neither approach is correct. The risk from all processes should be summed and compared with the facility risk tolerance criteria.

3. Procedure for Using LOPA to Determine SILs

A risk model should be constructed that manages risk across all processes, units, and all their operating modes for a facility in order to use LOPA for SIL determination. The risk model should assign the following attributes to hazard scenarios: process, process unit, operating mode, consequence type, consequence severity, and consequence receptor. Other attributes may be assigned, such as hazard type and receptor location. Appropriate summations of risk can then be made and the estimates compared with compatible risk tolerance criteria. If the criteria are not met, those scenarios that contribute most risk can be identified and adjustments made, for example, the addition of SIFs or an increase in the SILs of existing SIFs.

A step-by-step approach for performing the analysis is now described.

Step 1. Define receptors at risk

Receptors are the entities affected by a hazard scenario, i.e. people, the environment, property, etc. People include those who are onsite at a facility (e.g. workers and contractors) and those who are offsite (i.e. members of the public). Property includes equipment and buildings onsite as well as property offsite that is owned by others such as members of the public. The receptors to be addressed in the analysis must be defined. They are determined by the scope and objectives of the SIL determination study. At a minimum, studies usually address risks to people onsite and offsite, and the environment.

Step 2. Determine type of risk to use

Various types of risk measures can be used [10, 11]. For people, both risk to individuals and risk to groups of people are important risk measures and should be addressed in risk analysis. However, some practitioners address only individual risk which is unfortunate as the HSE has found that individual risk tolerance criteria “rarely bite”, and that societal, or group, criteria often play a far greater role in deciding whether a risk is acceptable [9]. Both types of criteria should be used to ensure facility risk is managed properly for people.

Individual risk is the frequency at which an individual may experience a given level of harm as a result of exposure to one or more hazards, and group, or societal, risk is the relationship between frequency and the number of people in a given population experiencing a specified level of harm from exposure to one or more hazards [11]. Individual risk does not describe completely situations where one or more events could cause multiple casualties, hence the need for group risk. Individual and group risk tolerance criteria specify limits to these risks for individuals and groups of people. Individual risk criteria protect any single individual from bearing too large a share of the risk. Group risk criteria protect populations of people from bearing undue risk. Both individual and group risk measures are calculated from the same set of frequency and consequence data for the set of hazard scenarios used in a risk analysis [10]. They are related through the individual risk to which receptors are exposed.

The type of risk estimates employed must match the risk tolerance criteria used for a valid comparison to be made and sound decisions to be taken. A common error is to use individual risk criteria but calculate group risk measures which can lead to more stringent risk reduction measures than needed.

Step 3. Determine form of risk to use

Various forms of risk measures can be used [10, 11]. Arguably, the preferred form of individual risk is the actual risk to each individual determined by calculating individual risk at every geographical location where people are present and accounting for time spent at the location. Such a measure allows risk estimates to reflect variations in exposure durations, protections available, individual susceptibility to harm, etc. Unfortunately, many calculations are needed to address the risk to each exposed individual and the resources required quickly become prohibitive and the analysis of the results becomes complex. However, the concept of a representative receptor can be used to simplify the analysis [12]. A number of representative receptors may be needed to represent different categories of people with different exposures, protections, susceptibilities, etc.

Usually, PHA is used to identify hazard scenarios for analysis with LOPA [13]. The scenario consequences identified in PHA are intended to be actual impacts on receptors. Consequently, actual risks are calculated by LOPA when PHA scenarios are analyzed.

Step 4. Specify consequence severity levels to be addressed

Hazard scenarios of varying severity levels can occur. For example, for people, impacts ranging from fatalities to first-aid cases may be possible. Arguably, SIL determination should be based not only on the risks of scenarios with the highest severity level but also lower-severity-level scenarios where the consequences are still significant. Thus, for people, not only should fatalities be of concern but also injuries.

Step 5. Specify risk tolerance criteria for receptors

The tolerability of risk for various types of receptors is different, for example, facility personnel and members of the public [9]. Therefore, separate risk tolerance criteria should be established for each type of receptor of concern. Values of risk tolerance criteria can be based on regulatory and industry precedents, and public decisions on the tolerability of major hazards. Usually, decisions on individual risk tolerance criteria for facility personnel are made with reference to the risk levels from accidents that actually have been tolerated in the workplace and decisions on individual risk tolerance criteria for the public are made with reference to the risk levels from non-work-related accidents that actually have been tolerated. Regulatory precedents are a good starting point, for example, criteria from the HSE [9].

Risk tolerance criteria for members of the public should be set lower than for facility personnel as members of the public do not voluntarily accept the risks from a facility. The same is true for off-site and on-site property, if property damage is to be addressed in the analysis.

Group risk is defined for the population exposed to hazards from facilities. It can be calculated for the public and facility personnel separately, or in combination. Group risk for facility personnel has been addressed infrequently, although logically it should be included in risk-informed decision making. A problem that can occur when the two types of risk are used in combination is that one or the other may exceed tolerable criteria for its type owing to the effects of averaging. Regulatory group risk tolerance criteria usually are established only for the public. In using such criteria as reference points for setting company criteria, it is important to know which group risk is being calculated. Usually, a distinction is made in PHA between impacts on the public and facility personnel. Therefore, group risk can be calculated from the hazard scenarios for either the public or facility personnel, or both combined together. The risk measure calculated must be paired with the correct risk tolerance criterion. If PHAs do not make this distinction, group risk will be calculated for facility personnel and the public combined. This may have the effect of requiring more stringent risk reduction measures than necessary.

Risk tolerance criteria should reflect the views of stakeholders including company owners, shareholders, directors and personnel; the public; regulators; and insurers. Of course, particular attention should be paid to the views of those people exposed to the risk such as facility personnel and members of the public.

Step 6. Determine offsets to risk tolerance criteria

In process safety, the concern is with risk from hazardous processes in facilities where catastrophic accidents can result from the realization of flammable, explosive, reactive and/or toxic hazards (so-called *major hazards*). However, facilities may also pose risks to people from such other hazards as working at height, confined space entry, asphyxiants, corrosives, hot gases, hot liquids, cryogenics, electricity, pinch points, vehicle accidents, etc. Overall facility risk tolerance criteria usually are intended to address risks from all hazards at a facility. Consequently, overall facility risk tolerance criteria should be offset to account for casualties from any sources excluded from a risk analysis. These other sources of casualties can be significant contributors to the overall facility risk and therefore the offsets can be significant.

Offsets for casualties from other sources will vary according to the severity of the consequences. For example, it is likely there will be more injuries than fatalities in a facility from other sources. Similarly, for group risk, it is likely that the offset will be higher for lower numbers of casualties, such as single fatalities, than for higher numbers of casualties because multiple fatality accidents likely are less probable from these other sources.

Commonly, hazard scenarios taken from PHA studies are used with LOPA to determine SILs for SIFs. However, no PHA study can be guaranteed to identify hazard scenarios completely and, unfortunately, practitioners are unaware of how many scenarios may have been missed and what contribution they may make to the facility risk. Lack of completeness in identifying hazard scenarios could result in significant underestimates of risk and the tolerance of higher risk than desired in SIL determination studies. This is an inherent limitation of current PHA approaches. Furthermore, actual PHA studies may be of poor quality and identify fewer scenarios than may be possible or may be deficient in other ways that impact SIL determination. The issues of PHA completeness and quality are not addressed in IEC 61511 / ISA 84, although problems in either area may undermine efforts to comply with the standard. Practitioners should consider offsetting overall facility risk tolerance criteria by a conservative amount to address this issue.

Step 7. Specify risk tolerance criteria for various consequence severities

Usually, PHA studies identify scenarios that result in impacts on receptors of varying severity. Therefore, it is appropriate to assign risk tolerance criteria to the various consequence severities of importance. Regulatory criteria address fatalities. Ideally, risk tolerance criteria for injuries should be set in a similar way to those for fatalities [9] by analyzing data on injuries to workers from industrial accidents and to members of the public from non-work-related accidents. However, that has not yet been done.

An alternative approach is to use an equivalence concept. Risk tolerance criteria for injuries to people can be developed by employing the tolerable frequency of a fatality as a reference point. For example, if ten incapacitating injuries are equated to one fatality, and the tolerable frequency of a fatality is 1×10^{-3}, the tolerable frequency of a single incapacitating injury would be set at 1×10^{-2} (risk = frequency x severity of consequence, so equivalent tolerable frequency = $1 \times 10^{-3} / (1/10) = 1 \times 10^{-2}$). A lesser injury category could be defined, for example, as hospitalization for a certain period, and the tolerable frequency reduced by another order of magnitude to 1×10^{-1}. Although such criteria are convenient and easy to assign, the equivalences are debatable because value judgments must be made. It is also possible to set proportionately more restrictive criteria for more serious consequence severities to account for aversion to high severity consequences.

Experience shows that process safety accidents that produce fatalities often produce accompanying injuries which are often much more numerous than fatalities for catastrophic accidents. Thus, injuries can be a significant component of the harm caused by catastrophic accidents and the exclusive use of fatality risk tolerance criteria for people in such situations is not completely satisfactory. PHA studies should identify the scenario consequences that include all important impacts. Of course, risk tolerance criteria need to be adjusted in such cases. An equivalence concept similar to that described above can be used.

Distinctions between the degree of damage are of lesser concern for other receptors such as equipment and property. For example, equipment damage is converted into monetary terms and aggregated for all levels of damage produced by an incident to provide an overall measure of the severity of equipment damage. Typically, separate criteria are established for different consequence severity levels for all types of receptors of concern.

Step 8. Decide on risk allocations and scaling

Both individual and group risk tolerance criteria can be allocated within a facility [8, 11]. Risk tolerance can be distributed across processes, units, operating modes, and receptors. However, it is also possible to allocate group risk across multiple facilities as group risk is a function not only of facility hazards but also the population exposed to risks under the control of a company, i.e. it depends on the risk exposure of all receptors from all facilities. Individual risk cannot be allocated across multiple facilities. Some companies set group risk tolerance criteria for the entire company and allocate risk across all their facilities. The allocation of risk to a facility can be adjusted according to a measure of its number of operations and size compared to other facilities operated by a company in a process called scaling [11]. Practitioners should decide on allocations and scaling that are to be used.

Step 9. Identify hazard scenarios

The hazard scenarios that contribute to the risks of a facility must be identified. Usually, they are obtained from PHA studies [13].

Most facilities comprise multiple processes within which hazard scenarios can occur. For receptors located within a process, while the risk to them is usually dominated by hazard scenarios that originate within the process, a contribution to the risk may be made by hazard scenarios that originate in other processes. For example, toxic releases from one process may result in employee impacts in an adjacent process. Usually, PHA identifies impacts on receptors wherever they are located. Consequently, risk analysis for those scenarios will provide the risk contribution from the process to the overall facility risk. However, the risk to receptors calculated from the scenarios for the process is not a complete measure of risk either for the receptors within the process or those in other processes that are impacted. In order to correctly determine risk for receptors within the process, contributions from hazard scenarios in other processes must be included. Similarly, the correct determination of risk for receptors within other processes must include contributions from hazard scenarios in the process together with those from any other processes that may contribute to the risk. Consequently, an aggregation of risk for scenarios across all processes at a facility is needed using a risk model that includes hazard scenarios for all the processes and their units at the facility.

Furthermore, hazard scenarios may occur during all modes of process operation and the same SIF may be challenged during different modes of operation. Consequently, risk contributions from all modes of process operation should be included in the risk model. Thus, hazard scenarios are required from PHA studies that address all processes and all their modes of operation for a facility.

PHA assumes the worst-case consequence scenario, i.e. all safeguards fail and usually just this scenario is recorded [14]. The assumption is made that this scenario is the worst-case risk scenario. However, there will be other scenarios involving combinations of safeguard failures *and successes* that contribute to the total risk for a process. Generally, these scenarios are not included in a LOPA study. Thus, the LOPA risk as normally calculated is not the total risk from the process, although it is likely the major contributor. This suggests that an attempt should be made to include some of these additional scenarios in the risk evaluation. The most obvious candidates are those in which mitigation safeguards work successfully. Since they are the last line of defence before the scenario impact is experienced, they may have a significant impact on the total risk.

Process incidents may have more than one outcome. For example, a leak of a volatile flammable material from a process may result in a jet fire, flash fire, or vapor cloud explosion, or it may simply disperse. These various incident outcomes may be further categorized as incident outcome cases which are differentiated by the weather conditions and wind direction at the time of the release. Hazard scenarios taken from PHA may not include these variations. However, the risks of these scenario variants may be substantially different and could have a significant impact on the overall evaluation of risk. They can be addressed in a LOPA study by considering them as individual scenarios and modifying their frequencies by factors that account for the fraction of releases that result in a particular incident outcome or incident outcome case [15].

The successful operation of a safety function may produce a safe state for the process with respect to the hazardous event that it protects against. However, its operation may result in a different hazardous event. For example, a SIF that closes a block valve to stop flow in a process may deadhead a pump resulting in seal failure and a release of process materials. Safety functions may also activate spuriously, i.e. when no hazardous event has occurred. Such activation may produce a different type of hazardous event. For example, the spurious opening of vent valves designed to avoid an over-pressure failure in a process may result in a release of process materials. Such scenarios contribute to the overall risk of the process and should be addressed when performing SIL determination studies.

Some hazard scenarios, such as those initiated by the loss of utilities for a process, e.g. electric power, may trigger the simultaneous operation of SIFs in various parts of a process. If these SIFs fail to protect the process, a global hazardous event occurs, possibly with greater consequences than any of the hazardous events addressed by the individual SIFs. Consequently, all SIFs that may be involved in a global hazardous event must be addressed as a set when determining their required SILs. Global hazard scenarios should be addressed in PHA and they should not be overlooked when performing SIL determination studies.

Step 10. Calculate scenario risks

The risks of hazard scenarios are calculated using LOPA so they can be combined into suitable risk measures for comparison with appropriate risk tolerance criteria. Procedures for performing LOPA studies are available in the literature [16, 17].

All scenarios that are protected by a SIF must be evaluated and their risk tolerability determined regardless of their risk and consequence level. It is conceivable that the required SIL of a SIF to protect against an operator fatality may be less than that to protect against injuries because there are typically more hazard scenarios for the lesser consequence. Of course, the lower risk tolerance requirement for injuries may outweigh the larger number of scenarios.

Scenarios that are not protected by a SIF may be protected by other means but they still make a contribution to the risks of a process and should be included in the risk model.

Step 11. Calculate individual and group risks

The risk estimates from LOPA for individual hazard scenarios must be combined into suitable risk measures for comparison with the risk tolerance criteria that have been set.

For individual risks to people, the risks of all scenarios that could impact an individual, regardless of the number of people impacted by the scenario, contribute to the risk. For example, if a hazard scenario can result in the single fatality of Operator A and another hazard scenario can result in 5 fatalities, one of which is Operator A, both scenarios would contribute to the individual risk to Operator A. It is a mistake to consider only the risk from scenarios that cause a single fatality when calculating individual risk.

Group risk is often expressed as f-N or F-N curves [10, 11], although F-N curves are more commonly used. f-N curves display the frequencies of all events that result in N casualties. F-N curves display the cumulative frequencies of N or more casualties from a single event summed for all events that lead to N or more casualties. When f-N or F-N curves are used for group risk tolerance criteria they are called limit lines [18]. Group risk tolerance criteria can be presented as an F-N limit line or as a tabulation of the numbers of casualties and their corresponding maximum tolerable frequencies. Data usually are displayed for categories representing intervals of casualty numbers, for example numbers of people impacted, e.g. 1, 2 - 9, 10 - 99, etc. because risk analysis methods usually organize consequence severities into bins for computational convenience and uncertainties in the analysis obviate the need for more precision. They can be displayed in the form of a histogram. Commonly, fatalities are used as impacts although other impacts such as injuries can be used.

A common misconception is that group risk applies only to scenarios that involve more than one casualty. Group risk is actually calculated beginning with the consideration of single casualties. For example, the risk of a fatality of an individual person at a facility may be calculated and that is a measure of individual risk. However, the risk of a single fatality for any person at the facility may also be calculated. It is the sum of the risk of all possible single fatalities.

Some practitioners calculate group risk but compare risk estimates with risk tolerance criteria for individuals. Obviously, this practice is incorrect and can lead to more stringent risk reduction measures than needed.

For other types of receptors, such as the environment and property, distinctions usually are not made between impacts on individual and multiple receptors. Thus, if a process contains multiple pumps, no distinction would be made between the tolerable risk of ten pumps being destroyed in individual scenarios versus ten pumps in a single scenario, so long as the risk was the same; that is the total loss was equally likely for both cases. Naturally, in the case of ten pumps being destroyed at the same time, the actual loss may be greater than the replacement cost of the pumps owing to the impacts of such a loss on the process. Such effects can be captured in the estimation of the overall damage costs and the tolerability of such losses is expressed using criteria that define tolerable frequencies as a function of the monetary loss.

Similarly, the concept of individual risk for people is usually not extended to other receptors. For example, normally, the risk to individual pieces of equipment or property is not of concern; rather, it is total risk of harm that matters.

Risk measures for hazard scenarios that impact receptors such as the environment and property are usually expressed by converting impacts such as property losses and environmental remediation into monetary terms and specifying risk as the frequency of the monetary loss. Difficulties may be encountered if the impacts cannot be converted into monetary terms, for example, in the case of permanent environmental impacts and the loss of irreplaceable property. There are no commonly-used risk tolerance criteria for these receptors.

Step 12. Make risk comparisons

The calculated facility risks must be compared with facility risk tolerance criteria and the allocation of risk across the facility should be reviewed to ensure it is distributed equitably. These comparisons are needed for those receptors and types of risk and consequence levels that have been specified for the study.

Step 13. Formulate risk reduction measures

In cases where risk estimates exceed specified risk tolerance criteria, recommendations are made to reduce the risk by modifying existing safety functions, including SIFs, implementing new ones, or taking other actions to reduce risk. Any one safety function may impact the risk of multiple hazard scenarios in a process and across its operating modes. Also, hazard scenarios from multiple processes contribute to the facility risks. Thus, a risk model that incorporates linking of safety functions is invaluable for the analysis.

Step 14. Update PHA and LOPA Studies

Implementation of recommendations for risk reduction may eliminate or modify the risk of existing hazard scenarios or make possible new scenarios. Some recommendations may interact with each other. Consequently, LOPA and other SIL determination studies should be updated to reflect any changes made to the process to ensure risk tolerance criteria continue to be met. This can be accomplished using the risk model that incorporates all hazard scenarios for the process.

4. Example of Using LOPA to Determine SILs

The example uses a toluene storage and charging process with a high level shutdown system that is separate from the basic process control system. It is intended to be illustrative and not complete nor necessarily accurate in every way. The purpose of the example is to illustrate the application of the procedure for SIL determination described in the previous section.

The example considers impacts of hazard scenarios on facility personnel and members of the public (step 1). For simplicity, no distinctions are made between facility personnel or between members of the public, that is, a single representative receptor is considered for each one. Both risk to individuals and to groups of people should be considered (step 2). This example will present results only for groups of people for economy of space. Estimates of actual risk will be used as the scenarios employed in the analysis are taken from a PHA study (step 3). Two fatality consequence severity levels were used for group risk (step 4). Values of risk tolerance criteria were set arbitrarily for purposes of the example (step 5). For simplicity, no offsets were taken (step 6), no adjustments were made for any accompanying injuries (step 7), and no allocation or scaling of the limit lines across facilities is used (step 8). They would simply adjust the values of the risk tolerance criteria. Allocation of risk between modes of operation is addressed (step 8).

Hazard scenarios for the process are taken from a PHA study (step 9). Risk contributions from other processes in the facility are not addressed in this example for the sake of simplicity. However, they contribute to risk in the same way as scenarios from within the process so there would be no fundamental difficulty in including them. The process has two modes of operation: tank filling and transfer. It is assumed that tank filling occurs 10% of the time and transfer the rest of the time. Hazard scenarios may result in fires and explosions. Examples of initiating events that may occur during filling are:

- Level transmitter, LT TK-104, fails to detect high level
- Level indicating controller, LIC TK-104, fails to stop pump, P100
- Pump, P-100, fails on

Examples of initiating events that may occur during transfer are:

- Pump, P-101, mechanical seal failure
- Unloading line failure due to corrosion
- Transfer line rupture due to impact of dropped object

Some initiating events may occur during either mode of operation, for example:

- Tank, TK-104, corrosion failure

Various SIFs are present in the process, including:

- High level shutoff for tank, TK-104
- Deluge system to protect against a fire

Other safeguards that are present include:

- Operator action to stop pump, P-100
- Operator action to close tank, TK-104, isolation valve
- Plant fire brigade
- Plant evacuation
- Public evacuation
- Hazardous material response team
- Periodic line inspection
- Periodic tank inspection
- Policy on crane use

Some of these safeguards are credited as independent protection layers (IPLs) [16, 17] in the analysis.

Several enablers were addressed in the analysis including conditional modifiers. Values of initiating event frequencies, enabler multipliers, and IPL probabilities of failure on demand (PFDs) are intended to be illustrative only for purposes of the example. The logic of the values used is as follows. The enabler, lack of preventive maintenance (PM) on level transmitter, LT TK-104, was assigned a multiplier of 5 to increase the frequency of the initiating event, level transmitter, LT TK-104, fails to detect level. The enabler, lack of PM on pump, P-100, was assigned a multiplier of 2 to increase the frequency of the initiating event, Pump, P-100, fails on. Generally, for fire scenarios, the conditional modifier, probability of ignition, was assigned a value of 0.5 based on the assumption that ignition was not guaranteed and the conditional modifier, probability of personnel in affected area, was assigned a value of 0.5 for tank filling and 0.1 for transfer based on average operator occupancies, and the conditional modifier, probability of harm from exposure, was conservatively assigned the value of 1. The conditional modifier, probability of an explosion, was assigned a value of 0.1 because it was viewed as less likely than a fire. Also, the probability of plant personnel or the public being in the affected area for an explosion and the probability of harm from exposure were conservatively assumed to be 1. For scenarios involving an object dropped on the transfer line, it was assumed that either a fire or an explosion would result. Fire was considered more likely and assigned a probability of 0.9 so that explosion was assigned a probability of 0.1. Different analysts may make different assumptions.

The risk of each scenario was calculated (step 10) using the software tool, LOPAWorks®. Figure 1 shows an example of a completed LOPA worksheet for a hazard scenario. The scenario risks were used to calculate group risk for employees and for members of the public (step 11). Figure 2 shows these risks aggregated for the process and including both modes of operation. The figure also shows a comparison with risk tolerance criteria (step 12). It can be seen that the risk to employees is tolerable but that to the public is not. Options for needed risk reduction can be explored (step 13) For example, the highest consequence level for the public requires risk reduction by a factor of 7.2. It can be seen in Figure 2 that scenario 3 contributes 69.4% to the total risk. The scenario is protected by two IPL's, one of which is a SIF. Options for risk reduction include strengthening either one or adding another IPL. The effect of these options on the risk analysis can be explored. For example, if the SIF SIL is changed from 1 to 2, the risk becomes tolerable for the highest consequence level (see Figure 3). The breakdown of contributions to the total process risk also can be calculated (see Figure 4) to obtain insights into the principal sources of risk and to help ensure appropriate allocation of risk. Once decisions have been made on suitable measures, the PHA and LOPA studies can be updated (step 14).

5. Conclusions

The application of IEC 61511 / ISA 84 poses numerous challenges within the context of risk analysis [19]. Not least of these challenges is the determination of SILs using techniques such as LOPA. Various issues affect the use of LOPA for this purpose including the nature of hazardous events and hazard scenarios, and the significance, allocation, and use of overall facility risk tolerance criteria. Not only do risk tolerance criteria help to determine the extent of harm that is viewed as tolerable, but also they influence the allocation of resources and the technologies used in facilities.

A procedure was described that addresses these issues. It involves the use of a risk model for a facility that can be used to calculate risk measures for comparison with overall facility risk tolerance criteria which are the only form with real meaning. The procedure was illustrated using an example.

The software tool, LOPAWorks®, was used to perform the calculations described in this paper and computer screen captures were made using it. LOPAWorks is a registered trademark of Primatech Inc.

6. References

[1] ANSI/ISA—84.00.01—2004 Part 1 (IEC 61511-1 Mod), Functional Safety: Safety Instrumented Systems for the Process Industry Sector — Part 1: Framework, Definitions, System, Hardware and Software Requirements.

[2] ANSI/ISA—84.00.01—2004 Part 2 (IEC 61511-2 Mod), Functional Safety: Safety Instrumented Systems for the Process Industry Sector — Part 2: Guidelines for the Application of ANSI/ISA—84.00.01—2004 Part 1 (IEC 61511-1 Mod).

[3] ANSI/ISA—84.00.01—2004 Part 3 (IEC 61511-3 Mod), Functional Safety: Safety Instrumented Systems for the Process Industry Sector — Part 3: Guidance for the Determination of the Required Safety Integrity Levels — Informative.

[4] Baybutt, P., The Use of Risk Matrices and Risk Graphs for SIL Determination, Process Safety Progress, in press.

[5] Baybutt, P., Using risk tolerance criteria to determine safety integrity levels for safety instrumented functions, Journal of Loss Prevention in the Process Industries, 25 (6), pps 1000 - 1009, 2012.

[6] Baybutt, P., Letter to the editor, Process Safety Progress, Vol. 31, No. 4, pps 328–329, December 2012.

[7] Baybutt, P., Risk tolerance criteria and the IEC 61511/ISA 84 standard on safety instrumented systems, Process Safety Progress, Vol. 32, Issue 3, pps. 307–310, September 2013.

[8] Baybutt, P., On the Allocation of Risk Tolerance Criteria, Process Safety Progress, in press.

[9] Reducing Risks, Protecting People, United Kingdom Health and Safety Executive Books, 2001.

[10] Guidelines for Chemical Process Quantitative Risk Analysis, *Center for Chemical Process Safety / American Institute of Chemical Engineers*, New York, NY, 2nd Edition, 2000.

[11] Guidelines for Developing Quantitative Safety Risk Criteria, *Center for Chemical Process Safety / American Institute of Chemical Engineers*, New York, NY, 2009.

[12] Baybutt, P., Risk tolerance criteria for layers of protection analysis, Process Safety Progress, Vol. 31, No. 2, pps.118–121, June, 2012.

[13] Baybutt, P., Conducting process hazard analysis to facilitate layers of protection analysis, Process Safety Progress, Vol. 31, No. 3, pps 282–286, September, 2012.

[14] Baybutt, P., Analytical Methods in Process Safety Management and System Safety Engineering – Process Hazards Analysis, in Handbook of Loss Prevention Engineering, Haight, J. M., (ed), Wiley-VCH, 2013.

[15] Baybutt, P., Addressing Enablers in Layers of Protection Analysis, Process Safety Progress, in press.

[16] Layer of Protection Analysis, Center for Chemical Process Safety / American Institute of Chemical Engineers, 2001.

[17] Baybutt, P., Analytical Methods in Process Safety Management and System Safety Engineering – Layers of Protection Analysis, in Handbook of Loss Prevention Engineering, Haight, J. M., (ed), Wiley-VCH, 2013.

[18] Cox, D. C., and Baybutt, P., Limit Lines for Risk, *Nuclear Technology*, Vol. 57, pps. 320 - 330, June 1982.

[19] Baybutt, P., The interface of functional safety with process safety and risk analysis, Process Safety Progress, Vol. 32, Issue 4, pages 346–350, December 2013.

[20] Baybutt, P., The ALARP Principle in Process Safety, Process Safety Progress, Vol. 33, Issue 1, Pages: 36–40, March 2014.

[21] HID's Approach to ALARP Decisions, SPC/Permissioning/39, Annex, http://www.hse.gov.uk/foi/internalops/hid_circs/permissioning/spc_perm_39.htm

Figure 1. Example of a Completed LOPA Worksheet for a Hazard Scenario

Number	1		
Description	Tank level transmitter fails and overfill tank, TK-104, with fire and employee impacts.		
Process	Toluene Storage and Charging		
Process Mode	Tank filling		
Consequence	Description	Type	Level
	Overfill tank, TK-104	EMP	2
Hazard Type	Fire		
Events	Item	Type	Value
	Initiating Event		**Frequency**
	Level transmitter, LT TK-104, fails to detect high level	EQP	1×10^{-1}
	Enablers (regular, at-risk factors, and conditional modifiers)		**Value**
	Time in tank filling mode	ARF	1×10^{-1}
	Lack of PM on level transmitter LT TK-104	RFG	5
	Probability of ignition	CM	5×10^{-1}
	Probability of personnel in affected area	CM	5×10^{-1}
	Probability of harm from exposure	CM	1
	Independent Protection Layers		**PFD**
	High level shutoff for TK-104	SIF	1×10^{-1}
	Operator action to stop pump, P-100	HUM	1×10^{-1}
	Safeguards (non-IPL)		
	Plant fire brigade	HUM	
Summary	Item		Value
	Frequency of Mitigated Consequence		1.3×10^{-4}
LOPA Recommendations	Recommendation	By	Due Date
	<no recommendations for the scenario>		
Notes	<no Notes for the Scenario>		

Figure 2. Total Risk for Process

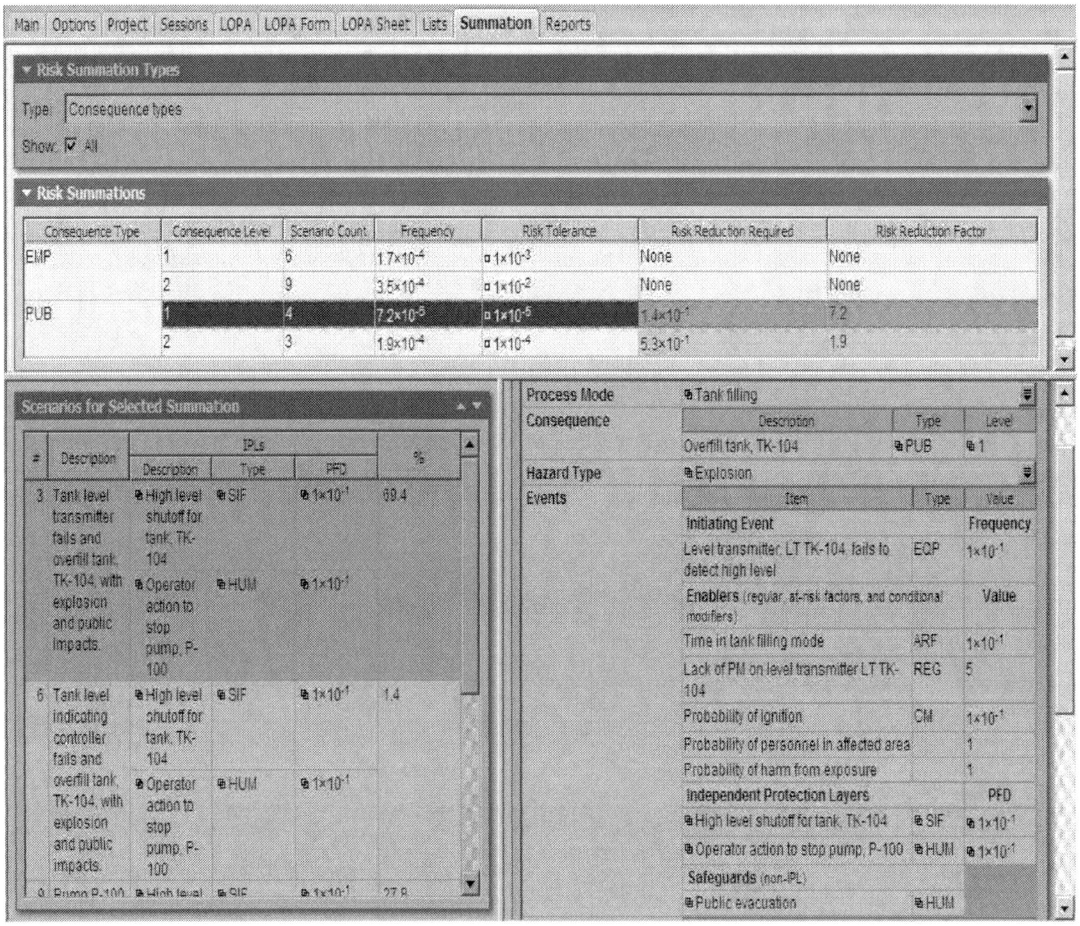

Figure 3. Adjusted Total Risk for Process

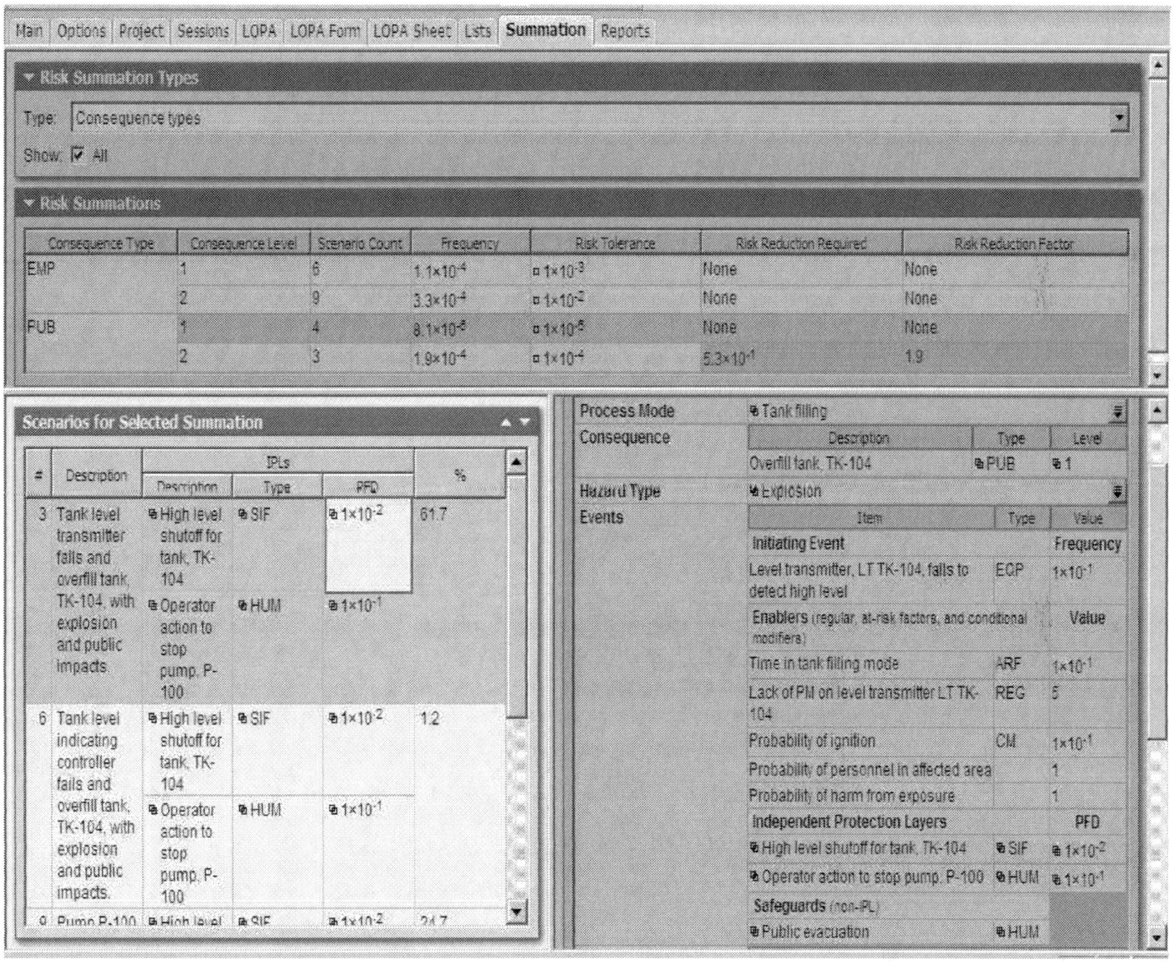

Figure 4. Risk Breakdown for Process

▼ Risk Summations

Process M...	Consequence Type	Consequence Level	Scenario Count	Frequency	Risk Tolerance	Risk Reduction Required	Risk Reduction Factor
Tank filling	EMP	1	4	7.2×10^{-5}	□ 1×10^{-3}	None	None
		2	4	1.8×10^{-5}	□ 1×10^{-2}	None	None
	PUB	1	4	7.2×10^{-5}	□ 1×10^{-5}	1.4×10^{-1}	7.2
Transfer	EMP	1	2	9.9×10^{-5}	□ 1×10^{-3}	None	None
		2	5	3.3×10^{-4}	□ 1×10^{-2}	None	None
	PUB	2	3	1.9×10^{-4}	□ 1×10^{-4}	5.3×10^{-1}	1.9

▼ Risk Summations

Hazard T...	Consequence Type	Consequence Level	Scenario Count	Frequency	Risk Tolerance	Risk Reduction Required	Risk Reduction Factor
Fire	EMP	2	8	2.6×10^{-4}	□ 1×10^{-2}	None	None
Explosion	EMP	1	6	1.7×10^{-4}	□ 1×10^{-3}	None	None
		2	1	9×10^{-5}	□ 1×10^{-2}	None	None
	PUB	1	4	7.2×10^{-5}	□ 1×10^{-5}	1.4×10^{-1}	7.2
		2	3	1.9×10^{-4}	□ 1×10^{-4}	5.3×10^{-1}	1.9

▼ Risk Summations

Process...	Hazard...	Consequence Type	Consequence Level	Scenario Co...	Frequency	Risk Tolerance	Risk Reduction Required	Risk Reduction Factor
Tank filling	Fire	EMP	2	4	1.8×10^{-5}	□ 1×10^{-2}	None	None
	Explosion	EMP	1	4	7.2×10^{-5}	□ 1×10^{-3}	None	None
		PUB	1	4	7.2×10^{-5}	□ 1×10^{-5}	1.4×10^{-1}	7.2
Transfer	Fire	EMP	2	4	2.4×10^{-4}	□ 1×10^{-2}	None	None
	Explosion	EMP	1	2	9.9×10^{-5}	□ 1×10^{-3}	None	None
			2	1	9×10^{-5}	□ 1×10^{-2}	None	None
		PUB	2	3	1.9×10^{-4}	□ 1×10^{-4}	5.3×10^{-1}	1.9

Lessons Learned from Application of LOPA throughout the Process Lifecycle

William G. Bridges, President
Process Improvement Institute, Inc. (PII)
1321 Waterside Lane
Knoxville, TN 37922
Phone: (865) 675-3458
Fax: (865) 622-6800
e-mail: wbridges@piii.com

Prepared for Presentation at
10th Global Congress on Process Safety
New Orleans, LA
March 31-April 2, 2014

UNPUBLISHED

10TH GLOBAL CONGRESS ON PROCESS SAFETY

Lessons Learned from Application of LOPA throughout the Process Lifecycle

William G. Bridges, President
Process Improvement Institute, Inc. (PII)
1321 Waterside Lane
Knoxville, TN 37922
Phone: (865) 675-3458
Fax: (865) 622-6800
e-mail: wbridges@piii.com

Keywords: layer of protection analysis, LOPA, independent protection layer, IPL, initiating event, PFD, IEF

Abstract

LOPA has been implemented throughout major capital projects, on existing facility PHAs, and in PHA revalidations and management of change risk reviews. This paper discusses lessons learned for implementing LOPA in each phase of a process lifecycle and outlines some of the ways to optimize the use of LOPA. The paper describes how implementation of standards for IPLs and initiating event maintenance is necessary in each company. The paper also covers consolidation of SIL evaluation into the related PHA and LOPA at each life cycle phase. Special emphasis is given to optimizing the application of LOPA and SIL evaluation through the various phases of a major capital project.

1. Brief History of LOPA

The initial development of layer of protection analysis (LOPA) was done internally within individual companies. However, once a method had been developed and refined, several companies published papers describing the driving forces behind their efforts to develop the method, their experience with LOPA, and examples of its use (Bridges, 1997[1]; Dowell, 1997[2]; Ewbank and York, 1997[3]). In particular, the papers and discussion among the attendees at the October 1997 CCPS (Center for Chemical Process Safety, part of AIChE), International Conference and Workshop on Risk Analysis in Process Safety, brought agreement that a book describing the LOPA method should be developed.

In parallel with these efforts, discussions took place on the requirements for the design of safety instrumented systems (SIS) to provide the required levels of availability. United

States and international standards (ISA S84.01 [1996], IEC [1998, 2000])[4,5,6] described the architecture and design features of SISs. Informative sections suggested methods to determine the required safety integrity level (SIL), but LOPA was not mentioned until the draft of International Electrotechnical Commission (IEC) 61511, Part 3, which appeared in late 1999. These issues were summarized in the CCPS workshop on the application of ISA S84, held in 2000.

The first LOPA book was developed by a CCPS committee from 1997 through 2000 and was published in 2001[7] (William Bridges and Art Dowell were the two co-originators and principal authors of the book). LOPA became widely used following the publication and most companies around the world have used LOPA, with some companies having used LOPA a lot. During roughly 15-years of widespread use of LOPA, and especially during the last 10-years, use of LOPA has greatly accelerated. It is likely that several million LOPAs have been performed. During this same period, many abuses of LOPA have been noted and several innovations have occurred.

In 2007, CCPS commissioned a new guideline book (1) to expand the list of independent protection layers (IPLs) and initiating events (IEs) and (2) to try to remedy some of the major issues noted in the use of LOPA. The new book is discussed in another paper at this conference; this book is due into publication this year (2014; William Bridges was the primary contractor/author of this book until April 2012). Another companion book on the related topics of conditional modifiers and enabling events and conditions was published in 2013 (Mr. Bridges was committee member and contributed to this book as well).

2. Intent of LOPA

LOPA is one of many methods for assessing a given scenario to determine if the risk is acceptable. It uses rigid rules to simplify and standardize the definitions of independent protection layers (IPLs) and initiating events (IEs). If these rules are followed, then the simplified risk assessment math of LOPA is valid and the risk assessment should give an order-of-magnitude approximation of the risk of a given cause-consequence pair (scenario). The rules also cover the minimum criteria for maintaining features and task executions that relate to IEs and IPLs.

LOPA is only one option for judging risk. The most common and still the best method for judging the risk of most scenarios is the process hazard analysis (PHA) team; their judgment is qualitative, but the "fuzzy" math of the individual team members usually coalesces into excellent judgment of risk for nearly all accident scenarios.

NOTE: As stated in all books and papers on the topic, LOPA does not find accident scenarios. Typically only the qualitative hazard evaluation methods (such as HAZOP, What-if, and FMEA) can find new accident scenarios.

3. Relationship to SIL determination

LOPA started with and continues to have a unique relationship with SIS, and particularly to SIF identification and SIL assignment (sometimes called SIL determination). Some of the originators of LOPA needed LOPA to defend against an arbitrary assignment of safety instrumented functions (SIFs) for systems that were already "adequately" safeguarded by other means. This became apparent in the mid-1990s with the early development of SIS standards within chemical companies and by (at that time) the Instrument Society of America (ISA). Some of these early standards would have imposed a minimum SIL for a given consequence, without much regard for the number and value of other IPLs that already existed or were viable alternatives to the SIFs. Much of these arbitrary requirements for SIS have disappeared, but some remain.

For the most part today, LOPA is seen as one tool (in many parts of the world, the preferred tool) for determining if a SIF is necessary and if it is the correct choice of risk reduction; and LOPA is the preferred method for determining what SIL is necessary, if an SIF is chosen as the risk reduction method. With that said, PHA teams are also allowed by IEC 61508, 61511, and related TR from ANSI/ISA to make these same determinations. Per ISA TR 84.00.02, 2002 (and 2004), Section 3.8[8]:

> *A qualitative method may be used as a first pass to determine the required SIL of all SIFs. Those which are assigned a SIL 3 or 4 by this method should then be considered in greater detail using a quantitative method to gain a more rigorous understanding of their required safety integrity.*

However, some organizations use LOPA to answer the question: "What SIL is needed to lower the risk to the risk target?" without first asking, "Are we at tolerable risk already?" or "Are there better alternatives for lowering the risk?" This leads to a huge over-specification of SIFs (and the wasting of resources to design, implement, and maintain these SIFs) and to many spurious shutdowns of units (which also waste money and increase the risk of accidents that can occur during re-start of the process).

Note that the system (loop) boundary for an instrumented safety system is defined differently by SIS standards versus LOPA (see Figure 1). As illustrated below, the system boundary for calculating the SIL for a given SIF includes only the instrumented components of the system. This omits the systemic failures possible from the process itself and more importantly omits the specific human errors of leaving the system in bypass or dependent errors of mis-calibrating multiple sensors in high SIL SIFs. LOPA however, requires that the system boundary for any IPL include all aspects of the IPL. This difference in system boundary definitions can make the difference between an SIF being a SIL 1 or a SIL 3 (installed predicted performance versus instrument only reliability).

Figure 1: Boundary for SIF ***(courtesy of Process Improvement Institute, Inc.)***

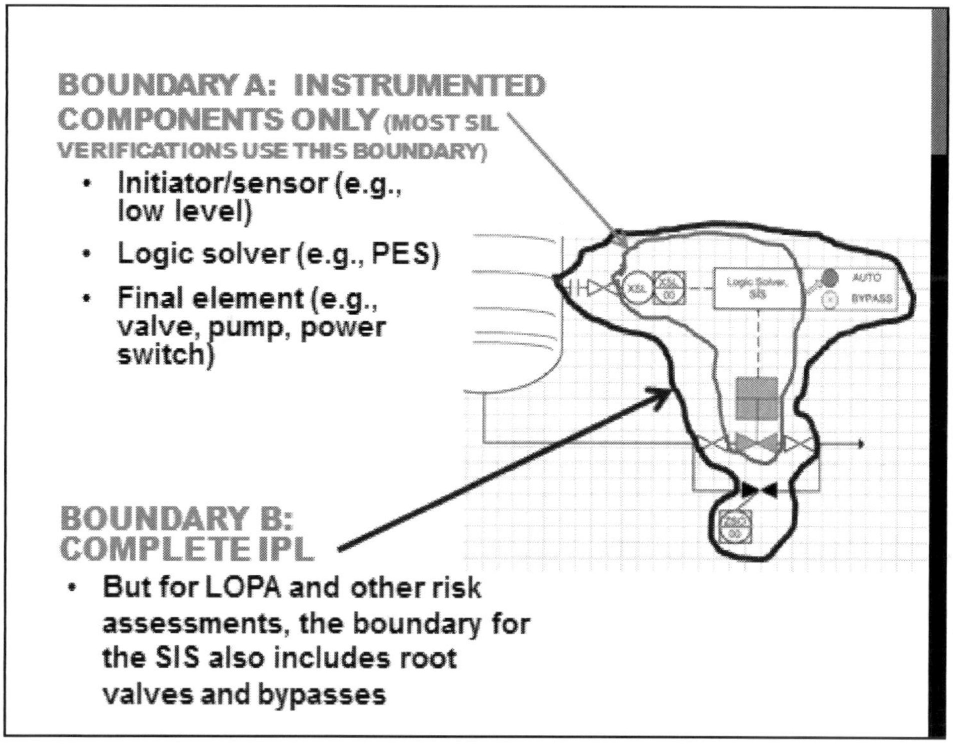

4. When to Use LOPA (in general)

The originators thought LOPA would be used a lot less frequently than it is currently. As shown in Figure 2, it was anticipated that LOPA would be used on 1-5% of the scenarios uncovered in PHAs. It was also anticipated that LOPA would eventually be used "after" a PHA team meeting, since that is how the originators were using it. Various examples of overuse are discussed below *(Bridges 2007[10], discussed these issues in detail):*

- **Using LOPA within PHAs - a bad idea as it detracts from brainstorming.** Many of us on the original LOPA book authorship considered LOPA a single analyst job, after a PHA/HAZOP, for just a few scenarios (maybe after 100 HAZOP nodes, you would do 1-10 LOPA). Instead, the trend appears to be that companies (or perhaps their consultants) make LOPA part of the PHA (in-situ). If the PHA/HAZOP team is properly disciplined on what qualifies as a safeguard (a qualitative definition of an IPL from LOPA), then performing LOPA in situ is usually overkill. In most situations, a qualitative team (HAZOP team) can make just as good or better judgment than provided by LOPA. LOPA is just another way to make a decision, has many pitfalls, and doesn't work for many types of scenarios. Other issues with use of LOPA within a PHA setting is that it distracts the team from brainstorming and it adds to team burnout because it takes time away from what is critical for the PHA team to do: *Identify scenarios for ALL modes of operation.*
- **Use for every Medium and High Risk Scenario --** Like in the point above, increasing the number of scenarios that must go through LOPA, reduces the resources

available to find (in a PHA/HAZOP/What-if) the undiscovered scenarios and to manage existing layers of protection.

- **When to use LOPA** The following is the guide we use to decide when a LOPA is required (Category 6 is equivalent to consequences greater than $100,000,000 and/or with potential multiple fatalities):

FIGURE 2: When to use LOPA ***(courtesy of Process Improvement Institute, Inc., 2004)***[11]

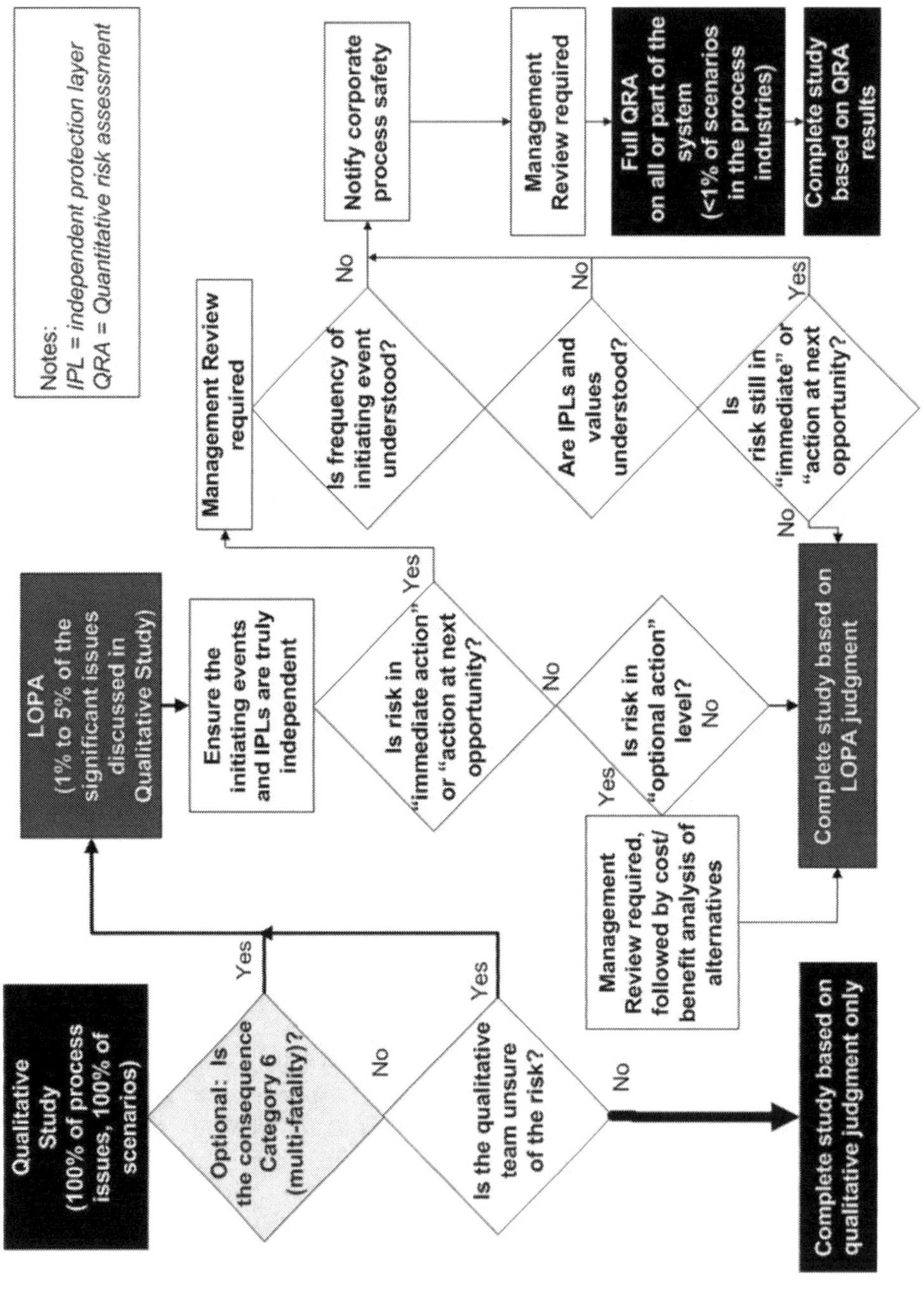

4.1 Hazard Evaluation at Different Plant Lifetime Stages

Process hazards can be identified from the onset of a project through its end (or assimilation into another project), and process hazards of changes are evaluated in mini-PHAs during the operating phase of the lifecycle. Hazard evaluations are also required for decommissioning. Industry has found that hazard evaluation must be used in all phases of a unit life cycle to ensure safe operations.

As usual, LOPA can be used in conjunction with any of these hazard evaluations to help estimate the risk of selected scenarios.

The following plant lifetime stages will be used in this section:

- Process development (perhaps several hazard evaluations during lab and pilot phases)
- Capital Projects
  - Conceptual and Preliminary design (1-2 hazard evaluation phases)
  - Detailed design (1-3 hazard evaluation phases)
  - Construction and start-up (completion of the initial and post-startup PHA)
- Operating lifetime (revalidations and MOC risk reviews)
- Extended shutdowns
- Decommissioning

4.2 Process Development Phase (Technology Phase) of Lifecycle

During the process development phases, various hazard reviews are necessary to help understand the risks and more and more is known of the chemistry and operations. HAZOP, What-if, and checklists are very useful during process development.

LOPA may be required for selected scenarios during normal, startup, switch-overs, modifications, etc., especially for large pilot plants.

4.3 Major Capital Project Phase of Lifecycle

There are various sizes and structures of projects, depending on the scope of the endeavor, the urgency, the nature of the business, the company culture, the company sophistication, and many other factors. The two most important project types (factors) for purposes of this paper are (1) project size (expressed usually in expenditure expected or size and number of pieces of equipment to be installed) and (2) type of financial control for the project.

4.3.1 Typical Project Sizes:

Table 1 below provides a basic definition of projects sizes and the typical number of risk reviews conducted during the project:

Table 1: Project Size/Scopes, Relative Project Length, and Risk Review Effort

Project Size	Example Project Scope	Example Project Length/Duration (concept→ commissioning)	Number of Risk Reviews
Major	Major projects handled **external** to an affiliate/plant, such as expansions and new facilities	12-36 months	4-7
Moderate	Works **engineered by an affiliate/ plant** (installing a new design of knockout pot for a feed to a unit)	6-9 months	2-3
Small	**Minor** affiliate/plant works (installing piping to bypass a control valve)	1-2 months	1

4.3.2 Scope of Risk Reviews and PSM Development For Each Project Phase

As mentioned earlier, major projects can have 4 to 7 or more phases and these can be spread over 12 months to 36 months or more depending on the project size. However, decisions for controlling risk made during the project phases echo through the next 20 to 50 years of operation, because design features, automated control features, and human interactions must be managed continuously to control the inherent hazards of chemical processes.

Figure 4 presents an example of a major capital project's phases for a large new chemical process unit or plant, with six "in-project" phases and one "post-project" phase. Though not universal, this approach appears to be a widely accepted view of major project phases. For smaller projects, condense this approach to 5, 4, or a minimum of two phases. **Table 2** on the following pages provides much more detail on the risk review for each project phase.

Note the six hazard evaluations (risk reviews) that are required. The six build on the prior risk review; at the end, the site has a complete PHA/HAZOP for use as the initial PHA for the new process. Further note that LOPA would only be used on the middle 3-4 hazard reviews, and only for scenarios that the PHA/HAZOP team has trouble in making tolerable risk decisions.

Figure 4: Example Project Phases and Related Scope of Risk Reviews (RR), which in this instance means the same as hazard evaluation[12]

	Phase 1 Conceptual Design	**Phase 2** Feasibility & Detailed Specification	**Phase 3** Preliminary Design	**Phase 4** Detailed Design	**Phase 5** Construction	**Phase 6** Pre-Commissioning	**Phase 7** Post-Commissioning
	Project Responsibility						Plant/Unit Responsibility
Project Concerns and Deliverables	Fit to business strategic plan Fit to existing operations Review of available technology Inherent safety options Site planning Raw material resourcing options	Detailed feasibility study (availability of technical staff; marketing plan) Detailed technology review and specification development Preliminary plot plan and tie-in plan +/- 40% cost estimate Preliminary schedule & milestones	Preliminary construction & operation plans Initial process flow diagrams (PFDs) Initial material and energy balances Raw material planning Utility planning Candidate vendors for major components Fire protection plan	Process design and generation of Rev 0 P&IDs; continuing to Rev 5 (nominal) in this phase. Revise material & energy balance Technical specs for all components Basics design of process controller Fabrication started and major components ordered	Execute fabrication and installation plan Develop detail plans for tie-ins to OSBL Continue to develop specs for OEM manuals and operating procedures Set up CMMS/ database for ITPM Draft PSM management systems/procedures	Complete fabrication/ installation Complete development and closure of PSSR and other punchlists Commission/ validate equipment (dry, wet, with HHC) Populate CMMS and other PSM/ reliability databases Complete operating and MI procedures (by SMEs); validate	Assist commissioning team Ensure training by vendors/OEMs are completed in the field and proficiency of plant staff is validated Complete performance measure of initial operation (to ensure contract commitments are met) Manage changes Closeout project
Risk Reviews	**Conceptual RR** Strategic plans Inherent safety Plot plan review for Facility Siting; consequence modeling for major releases Begin Human Factor consideration		**Preliminary Design RR** What-if analysis of each major unit operation HAZOP/FMEA of selected scenarios LOPA of selected scenarios & review options for inherent safety	**Detailed Design RR** HAZOP/FMEA of most nodes (focusing on continuous mode of operation) LOPA of 1-5% of scenarios Final SIL (if needed) determination	**Final Detailed Design RR** HAZOP/FMEA of changes since previous RR, including rec. resolutions; place special attention to changes in field Begin human factor and facility siting (HF&FS) checklists	**Commissioning RR ("Initial PHA for new unit")** HAZOP/FMEA of changes HAZOP and/or What-if of non-routine operating modes (startup, emergency shutdown, etc.) Complete HF&FS checklists	**Post-Startup RR (3 to 6 months after startup)** Close any recommendations that were rated as post-startup issues Review each MOC for its impact on the "Initial Unit PHA" Perform critique of risk review efforts during project.

Table 2: Details of Process Safety Development Phases of a Major Capital Project (with LOPA highlighted)[12]

Project Phase #	Phase Name	Goals of RISK REVIEW	RISK REVIEW Methodology	RISK REVIEW Team Membership (in addition to leader & scribe)
1&2	Conceptual	Choose inherently safer option, ensure overall feasibility, estimate impact on neighbors	• Consequence modeling (to help on next project phase) • What-If (no guidewords) • Selected checklist for judging inherent safety	• Senior operator for unit or from similar unit • Senior process engineer for unit or from similar unit • Process/design engineer from project • Process Safety specialist (if not already listed above)
3	Preliminary Design	Identify and resolve most expensive design alternatives, including layout of plant, facility siting concerns, environmental protection issues, and major tie-ins	• What-If (no guidewords) • HAZOP/FMEA of selected scenarios • **LOPA of selected scenarios (note that sufficient information for LOPA will likely be missing for 20-30% of accident scenarios at this phase)**	• Senior operator for unit or from similar unit • Senior process engineer for unit or from similar unit • Process/design engineer from project • Process Safety specialist (if not already listed above) • Possibly I&E Engineer • **LOPA should be performed by a single analyst**
4	Detailed Design	Begin detailed identification of potential accident scenarios, primarily focused on normal (usually continuous) mode of operation. Begin risk assessment for scenarios with large residual risk	• HAZOP/FMEA of equipment nodes, focusing on normal (usually continuous) mode of operation • What-If of lower consequence & lower complexity systems • **LOPA of 1-5% of the scenarios; determine SIL, as necessary**	• Senior operator for unit or from similar unit • Senior process engineer for unit or from similar unit • Process/design engineer from project • Process Safety specialist (if not already listed above) • Possibly I&E Engineer • Possible vendor of unique equipment • **LOPA should be performed by a**

Project Phase #	Phase Name	Goals of RISK REVIEW	RISK REVIEW Methodology	RISK REVIEW Team Membership (in addition to leader & scribe)
				single analyst
5	Final Design	Update results of previous RR for new details, identify potential accident scenarios for nodes not previously reviewed, primarily focused on normal (usually continuous) mode of operation. Resolve most previous recommendations Complete risk assessment for scenarios with large residual risk	• Complete HAZOP, FMEA, or What-If for nodes started in previous RR • Perform HAZOP, FMEA, What-If for nodes not covered in previous RR (due to previously missing information) • Begin Human Factors and Facility Siting checklist • Perform general Utility Failure checklist • **LOPA of 1-5% of the scenarios; determine SIL, as necessary**	• Senior operator for unit or from similar unit • Senior process engineer for unit or from similar unit • Process/design engineer from project • Process Safety specialist (if not already listed above) • Possibly I&E Engineer • Possible vendor of unique equipment • **LOPA should be performed by a single analyst**
6	Commis-sioning	Conduct full hazard/risk review of operating procedures to control risk of errors during startup, shutdown, emergency shutdown, and other non-routine modes of operation Close out previous RISK REVIEW issues (from earlier phases of project) and complete the human factors & facility siting checklist This RISK REVIEW creates the "Initial PHA" of the process	• HAZOP (2 guideword or 8 guideword) or What-If (no guideword) of operating procedures (choose method based on hazard and complexity of each task) • Complete HAZOP, FMEA, or What-If for nodes started in previous risk reviews • Perform HAZOP, FMEA, What-If for nodes not covered in previous risk reviews (due to previously missing information) • **LOPA of 1-5% of the scenarios; determine SIL, as necessary** • Complete Human Factors and Facility Siting checklist	• Senior operator for unit or from similar unit • New/junior operator for unit • Senior process engineer for unit or from similar unit • Process/design engineer from project • Process Safety specialist (if not already listed above) • Possibly I&E Engineer • **LOPA should be performed by a single analyst**
7	Post-Startup	Conducted 3-6 months after startup similar to the future Revalidations, but with the goal of compensating for	• Audit of MOCs (and P&IDs and SOPs) since "Initial PHA" (since commissioning RR) to ensure nothing has been missed by MOC	• Senior operator for unit or from similar unit • New/junior operator for unit • Senior process engineer for unit or

Project Phase #	Phase Name	Goals of RISK REVIEW	RISK REVIEW Methodology	RISK REVIEW Team Membership (in addition to leader & scribe)
		weaknesses in MOC process at the initial startup of the new unit/process	• HAZOP or What-If of missed or poorly reviewed changes • Update PHA for the entire set of changes (looking at whole picture for effect of all changes) • Close any pending recommendations (if possible)	from similar unit • Possibly project/design engineer (for QA of project)

4.3.4 Risk Control – Initial Phases of a Major Capital Project

As described in **Figure 4** and **Table 2**, the first two project phases are critical for establishing the inherent safety of the process, and therefore an opportunity for company leadership to show their true colors.

There typically is no call for LOPA during the conceptual stage or very preliminary design phase of a major capital project.

4.3.5 Risk Control – Detailed Design Phases of a Major Capital Project

The risk reviews are a major risk control feature of the design phases of a project as well. These risk reviews can be one to three progressive efforts over one to three project phases, depending on the size of the "major capital project," with the Risk Review report building toward the "initial" official hazard review report for the process unit (discussed in the next section).

LOPA occur frequently during this Detailed Design project phase

The Risk Reviews during the detailed design phases can typically include:

- Using HAZOP, FMEA, and/or What-if (brainstorming methods) in progressively more detail
- Initiating and then progressively improving (from phase to phase) the risk review record (HAZOP tables, What-If tables, checklist tables)
- As mentioned in **Figure 1** and **Table 2**, the risk reviews during detailed engineering will evaluate the risk of any design modifications and/or newly identified hazardous scenarios, which have been added since the previous reviews.
- Maximize inherently safer design in the selected process
- Addressing damage mechanisms and ensuring materials and equipment location and engineered safeguards minimize the likelihood or effects of external impacts
- Performing a final review of equipment, ventilation, containment, and environmental safeguards, including instrumentation, interlocks, fail-safe decisions, detailed layouts, and fire protection provisions
- Begin the Human Factors risk review (checklist-based)
- Continue the Facility Siting risk review initiated earlier (checklist-based and modeling-based risk reviews)

- **Apply Layers of Protection Analysis (LOPA) to complex risk scenarios and use this to define Safety Instrumented Systems' needs. These phases of the project are where LOPA gets used the most. With that said, only 1-5% of potential accident scenarios will require LOPA in the risk decision making process. Exceptions will be for novel technology:**

Example: ***We performed a PHA/HAZOP over a six (6) week period for a new way to make an explosive. There had only been limited pilot testing of some of the technology before scale up. The PHA/HAZOP normally would have taken two (2) weeks, but the PHA/HAZOP was doubling as a design review. In addition, since the controls and safeguarding for this process were all novel, about 7% of the accident scenarios went to LOPA. This is more than the 1-5% of the scenarios that normally go to LOPA. About 30 LOPA in all were performed.***

As before, the Risk Reviews during the detailed engineering phase require intensive participation by operations' senior staff, including operators, supervisors, and process engineers.

And as before, the LOPA is usually performed by a single analyst, outside of the qualitative risk review (PHA/HAZOP).

4.3.6 Risk Control – Pre-Commissioning / Commissioning (Initial Startup) Phase of a Major Capital Project

The pre-commissioning Risk Review builds upon the previous Risk Reviews in the project. As the equipment design is completed the fabrication and construction begins. During this same period, initial training of the new or transferred staff occurs, using the procedures mentioned in the previous section. The pre-commissioning Risk Review can begin just prior (4-6 weeks prior) to start-up of a new facility, or a little earlier if possible. The key consideration for this project phase is to complete the risk review of non-routine modes of operations. The project Risk Reviews to this point will not have covered these modes of operation very well. (Note that in perfect world, the risk review of the non-routine modes of operations, which uses the operating procedures as a basis, would be completed before training begins. However, in most cases, the training begins as the procedures are being completed and as the risk review is done.) The risk review of non-routine operating modes can be performed using a full 8 guideword HAZOP, a streamline 2 Guideword approach (which is what was used before HAZOP was invented in the 1960s), or a No Guideword What-if.[13] All of these approaches are described elsewhere[13] and will be explained in some detail in the 3rd Edition of the Guidelines for Hazard Evaluation Procedures, CCPS[14]. This procedural analysis is to ensure that hazards due to human error in association with the process design have been identified and analyzed.

During this final risk review before start up, the project team must also ensure that all the PSM requirements for initial PHAs have been met. PHAs must address the hazards of the process; therefore hazards during all modes of operation must be analyzed. The resulting report will be the "initial PHA" of the process unit, which is required to meet PSM standards.

This pre-commissioning risk review should not be confused with the pre-startup safety review (PSSR; also referred to as Operational Readiness Review [ORR]), which is also necessary but the purpose of the PSSR is to validate that the process design and specifications have been met.

This final Risk Review session before startup consists of:

- Reviewing and evaluating changes made during construction, ensuring that no new hazards have been added since the last hazard review. High priority is given to detecting details which may have been overlooked, and to concentrating on the adequacies of plans to cope with operating emergencies that might arise
- Maximizing inherently safer design in the selected process, such as planning for rework of initial product
- Completing reviews for Facility Siting and access issues
- Completing the review for Human Factors issues
- Reviewing (HAZOP/What-If) of start-up, shutdown, emergency shutdown, and on-line maintenance procedures. Some believe this risk review to be another "validation" review of procedures, to ensure they are correct. But that is not the purpose of this risk review of procedures. Our aim in the HAZOP/What-If of the procedures is to ensure we have adequate safeguards (hardware, interlocks, SIL, and/or independent administrative safeguards) to offset the errors of skipping steps and performing steps wrong – such human errors WILL occur, it is just a matter of when.[13]
- **Performing any LOPA, if necessary for the new scenarios that are uncovered during the hazard review of non-routine modes of operation.**

Although LOPA is valuable during this project phase, the lack of risk review of non-routine mode of operations, such as by HAZOP or What-if of procedural steps, is the most frequently observed weakness in the project risk review cycle

After this risk review, the project team can proceed to close recommendations, decide which (if any) of the recommendations can be deferred until after initial startup, close the PSSR (not part of the risk review, but part of PSM in general), and finalize the initial PHA report for the new process unit. Typically, the plant MOC system begins to take over control of new risks after the pre-commissioning Risk Review meeting is closed.

There are of course many deliverables from the project team, including the finished equipment, ready to commission and then smoothly commissioned, operating and maintenance procedures, populated databases for mechanical integrity (MI), Process Safety Information - files of all necessary design bases for relief valves, completed drawings, complete equipment files of all types, etc.

4.4 Operating Phase of Lifecycle

4.4.1 Management of Change

Change is an inevitable and necessary feature for all organizations. When changes occur in an operation that contains hazards of any sort, it is necessary that the change process be managed to understand and control those hazards. Most organizations have some "Management of Change" (MOC) policies and procedures in place to address the wide range of issues related to changes. Regardless of the type of change, the risk of change must be analyzed. All of the hazard evaluation methods are applicable to MOC risk reviews and the criteria for selecting the appropriate technique is the same as for other risk reviews. However, one nuance is that the choice of technique may depend also on how the unit hazard evaluation will be revalidated (updated). For instance, if the unit hazard evaluation (called a Process Hazard Analysis in the US) was accomplished using primarily HAZOP, then long-term it might be best if the risk review of the "change" also be documented in HAZOP format; this will make rolling-up the data into the next revalidation that much easier for the company.

Some of the MOC risk reviews may affect an existing LOPA or the risk review team (best understood as a mini-PHA team) may recommend a LOPA for a new scenario resulting from the change. For the case of an MOC effecting and existing LOPA, the prior LOPA should be updated, with a notation on the LOPA documentation that makes it easy for the change to be incorporated in the next Revalidation cycle. Again, LOPA will likely only be required for 1-5% of the scenarios.

4.4.2 Cyclic Reviews (Revalidations)

Even as process changes never end during the life of a facility, there will always be the necessity to continue hazard evaluations. Periodic updating or revalidation of the hazard study is the method used to maintain adequate safeguards. The timing of these cyclic reviews depends on factors such as regulations, the rate of process changes, and the nature of those changes. A significant change outside the fence line can also trigger the need for a hazard review. Examples of such changes are:

- Population changes such as new residential housing nearby
- Land/water/air traffic pattern changes
- Necessary community emergency and security adjustments
- New buildings such as schools or commercial establishments
- Demolished buildings.

Although MOCs may require a few LOPA, Revalidations almost never require new LOPA. But, if one or more of the MOCs affect existing LOPAs or the MOC risk review team required a new LOPA, then related changes or notation are likely necessary for the Revalidation report.

4.4.3 Restarts

During the normal operating phase, units restarted following:

- a normal shutdown
- an emergency (abnormal) shutdown
- a "turnaround" or maintenance phase.

LOPA may be required for the new scenarios that are uncovered during the hazard review of these unique, non-routine modes of operation.

Each has safety issues associated which could be unique to the shutdown event. Start-up from a normal or scheduled shutdown (without maintenance action) should already have been addressed as a procedure-based hazard evaluation (HAZOP, What-if, or 2 Guide Word technique) of non-routine operations. Less likely to have had an established safety review is the start-up from an emergency or abnormal shutdown, since not all possible emergency scenarios would have been predicted. These are frequently considered to be operated like a normal start-up once the system has reached a safe state. A careful restart review (usually 2 Guide Word or What-if) is advised, since by the (hopefully) unique nature of the emergency, experience with the specific type of restart is limited.

4.5 Extended Shutdowns

Although LOPA is likely not needed for preparation for extended shutdown, a restart back to the "normal" mode of operation may need a new PHA/HAZOP and therefore some LOPA may be required.

Mothballing a plant or a unit within a plant site goes beyond the steps taken to shut down a process to an established safe state. Tanks, lines, and valves must all be drained and any residual reactive materials neutralized. Most of these operations will likely be performed only once during the lifetime of the operation, and hence will have no history to help guide the safe implementation of the mothballing effort. A potentially riskier activity is a restart from an extended shutdown. In that event, the condition and intended operation of all equipment and instruments must be checked. In particular, the safety systems must be re-verified and validated. The restart of a mothballed unit within and perhaps attached to an active production site should include a review of the hazard evaluation for the connected active units.

4.6 Decommissioning Phase of Lifecycle

An active plant or unit that is slated for decommissioning would go through the stages of a normal shutdown, then cleanout in preparation for mothballing, followed by disassembly. In the refining industry as well as in other process industries, a hazard review is conducted before decommissioning. Health, safety, and environmental issues would be related to uncontrolled pressure releases, workers potentially exposed to noxious or toxic vapors, and spills during line separation. A plant or unit that has been previously mothballed would have all the issues associated with the decommissioning of an active plant or unit, with the added potential hazard associated with corrosion products. Over a period of time, residues can change such that these new compounds may be unknown and have unknown health and environmental effects or thermal decomposition sensitivities.

LOPA would only rarely be required for decommissioning of an entire process; but if a part of a process is decommissioned; this may affect existing LOPA scenarios for the portion of the process that will remain in operation.

6. Conclusions

LOPA is one risk assessment tool that has proven useful in helping organizations judge if selected accident scenarios have sufficient safeguards. Throughout the process lifecycle, from the technology phase through the end of ongoing operations, LOPA has and continues to be used. If the general rules outlined in this paper are followed and if LOPA is not arbitrarily applied, then it will continue to benefit industry as a valuable tool for risk assessment.

7. ABBREVIATIONS and GLOSSARY

HAZOP – Hazard and Operability; as in HAZOP Analysis or HAZOP Study
IE – Initiating Event
IEC – International Electrotechnical Commission
IEF – Initiating Event Frequency
IPL – Independent Protection Layer
ISA – International Society of Automation
LOPA – Layer of Protection Analysis
MOC – Management of Change
PFD – Probability of Failure on Demand
PHA – Process Hazard Analysis
P&ID – Piping & Instrumentation Diagram

PSM – Process Safety Management
SIF – Safety Instrumented Function
SIL – Safety Integrity Level
SIS – Safety Instrumented System

8. References

1. Bridges, William G., and Tom R. Williams (1997), "Risk Acceptance Criteria and Risk Judgment Tools Applied Worldwide within a Chemical Company," *International Conference and Workshop on Risk Analysis in Process Safety*, October 21–24, 1997, Atlanta, GA, pp. 13–28. New York: American Institute of Chemical Engineers.
2. Dowell, A. M., III (1997), "Layer of Protection Analysis: A New PHA Tool, After HAZOP, Before Fault Tree," *International Conference and Workshop on Risk Analysis* in Process Safety, October 21–24, 1997, Atlanta, GA, pp. 13–28. New York: American Institute of Chemical Engineers.
3. Ewbank, Rodger M., and Gary S. York (1997), "Rhône-Poulenc Inc. Process Hazard Analysis and Risk Assessment Methodology," *International Conference and Workshop on Risk Analysis in Process Safety,* October 21–24, 1997, Atlanta, GA, pp. 61–74, New York: American Institute of Chemical Engineers.
4. IEC 61508, *Functional Safety of Electrical Electronic/Programmable Electronic Safety-Related Systems*, The International Electrotechnical Commission, 2010.
5. IEC 61511, *Functional Safety: Safety Instrumented Systems for the Process Industry Sector - Part 1: Framework, Definitions, System, Hardware and Software Requirements*, International Electrotechnical Commission, 2003.
6. ANSI/ISA 84.00.01-2004 (IEC61511-1 Mod), *Functional Safety: Safety Instrumented Systems for the Process Industry Sector - Part 1: Framework, Definitions, System, Hardware and Software Requirements,* 2004.
7. *Layer of Protection Analysis: Simplified Process Risk Assessment,* CCPS/AIChE, 2001.
8. ISA TR84.00.02, *Safety Instrumented Functions (SIF) – Safety Integrity Level (SIL) Evaluation Techniques*, International Society of Automation, 2002.
9. *Guidelines for Initiating Events and Independent Protection Layers*, 2014 (pending), CCPS/AIChE.
10. Bridges, W. and Clark, T., "Key Issues with Implementing LOPA (Layer of Protection Analysis) – Perspective from One of the Originators of LOPA,", *5th Global Congress on Process Safety,* April 2009, AIChE.
11. *Layer of Protection Analysis*, Training Course Notebook, Process Improvement Institute, Inc., 2004-2014.

12. Bridges, W. and Tew, R., "Controlling Risk During Major Capital Projects," *24th CCPS International Conference and Workshop on Process Safety (CCPS/AICHE)*, New Orleans, LA, April 2008.

13. Bridges, W. and Clark, T., "How to Efficiently Perform the Hazard Evaluation (PHA) Required for Non-Routine Modes of Operation (Startup, Shutdown, Online Maintenance)," *7th Global Congress on Process Safety*, Chicago, AIChE, March 2011.

14. *Guidelines for Hazard Evaluation Procedures*, 3rd Edition, 2008, CCPS/AIChE.

Leading LOPA Forward: A CCPS Evergreen LOPA Database

J. Wayne Chastain P.E.
Eastman Chemical Company
P.O. Box 511, B18
Kingsport, TN 37663
chastain@eastman.com

John Murphy P.E.
Center for Chemical Process Safety
2304 Kenya Lane
Punta Gorda, FL 33983
hamjfm@embarqmail.com

Prepared for Presentation at
American Institute of Chemical Engineers
2014 Spring Meeting
10th Global Congress on Process Safety
New Orleans, LA
March 30 – April 2, 2014

UNPUBLISHED

Leading LOPA Forward: A CCPS Evergreen LOPA Database

J. Wayne Chastain P.E.
Eastman Chemical Company
P.O. Box 511, B18
Kingsport, TN 37663
chastain@eastman.com

John Murphy P.E.
Center for Chemical Process Safety
2304 Kenya Lane
Punta Gorda, FL 33983
hamjfm@embarqmail.com

Keywords: Layer of Protection Analysis, LOPA

Abstract

With the publication of *Guidelines for Enabling Conditions and Conditional Modifiers in Layer of Protection Analysis* (2013), *Guidelines for Determining the Probability of Ignition of a Released Flammable Mass* (2013), and *Guidelines for Initiating Events and Independent Protection Layers in Layers of Protection Analysis* (2014), CCPS has made significant strides in documenting much of the progress that has occurred in the Layer of Protection Analysis community since the publication of *Layer of Protection Analysis Simplified Process Risk Assessment* in 2001. During the development of these companion texts, however, CCPS reached the conclusion that a more timely response to changes in the technology was required to serve the LOPA community.

To address this need, CCPS is announcing an exciting new project; an Evergreen LOPA Database is being developed to keep practitioners apprised of new developments, new data, and other changes in the field. The Evergreen LOPA Database will enable new data and analysis to be incorporated into the most up to date source available on initiating events, independent protection layers, enabling conditions and conditional modifiers. The Evergreen LOPA database will also allow the community to more easily propose new initiating events and IPLs and have the values and management practices associated with these proposed items peer reviewed and presented to the broader community. This paper describes the current plans for the database project including the content that will be available, the proposed management practices, and the ability for discussion and continuing development by the community.

1. Introduction

The roots of Layer of Protection Analysis were established in the 1980s with the publication of the Responsible Care Process Safety Code of Management Practices by the then Chemical Manufacturers Association which mentioned that "sufficient layers of protection" should be present in an effective process safety management system [1]. Since that time the method has seen significant development with the next major milestone reached in the publication of the first edition of *Guidelines for Safe Automation of Chemical Processes* [2]. The method was standardized across the industry with the publication of *Layer of Protection Analysis Simplified Process Risk Assessment.*

After the publication of that text, significant development of the technique occurred in individual companies with sharing of that knowledge through the multiple sessions held at a variety of safety conferences and the publication of papers in various journals supporting the chemical industry. Recently, CCPS has sought to provide additional guidance to users of the technique by establishing two subcommittees focused on producing texts clarifying and extending different aspects of the technique. First, a subcommittee was established to develop a text focusing on Independent Protection Layers. This subcommittee determined that an appropriate extension of their mandate was to provide guidance to users on initiating events as well. The product of this subcommittee will be published this year as *Guidelines for Initiating Events and Independent Protection Layers for Layer of Protection Analysis.*

The development of a text concentrating on initiating events and Independent Protection Layers inevitably lead to the conclusion that additional guidance was needed by the community focusing on two other important parameters of Layer of Protection Analysis; enabling conditions and conditional modifiers. A CCPS subcommittee was established to produce a text for these additional important factors in LOPA. This subcommittee produced the text *Guidelines for Enabling Conditions and Conditional Modifiers for Layer Protection Analysis* published in 2013 [3].

An important aspect of the new LOPA texts produced by CCPS is a focus on helping users avoid misuse of the technique. Appropriate validation of the values used for all of the factors in LOPA is stressed in both of the new texts. Appropriate use of LOPA can increase the safety of chemical facilities, but inappropriate use of the technique can result in poor risk decisions and the potential for elimination of tried and true risk management solutions. Both the books try to help LOPA practitioners in making justifiable decisions in their LOPAs to support safe design practices.

The CCPS technical steering committee was presented with another proposal related to LOPA during the period that these two books were being written. The purpose of this proposal was not the publication of an additional book on LOPA, but instead a different means of delivering information on the technique by providing an online resource which could be used by practitioners to obtain the most up to date information available on a given factor. Due to many of the subject matter experts on the technique being engaged in the production of the two aforementioned CCPS Guidelines books, the subcommittee was not established for this online project until early in 2014. The charge for this subcommittee is to develop an online database for datasheets much like those provided in *Guidelines for Initiating Events and Independent Protection Layers for Layer of Protection Analysis.* In addition similar datasheets may be provided for conditional modifiers and enabling conditions. These entries are to be kept evergreen by a perpetual CCPS subcommittee charged with maintaining the technical content of the database. In addition, the online database may provide access to LOPA tools potentially including blank datasheets and LOPA templates.

2. Existing Online Resources

The growth of the Internet has been tremendous since its inception, often cited as the connection of the first two nodes that would become ARPANET on October 29th, 1969 [10]. Since the millennium, the internet has become ubiquitous. For example, the following chart shows the percentage of internet users in the United States between the years of 2000 and 2012.

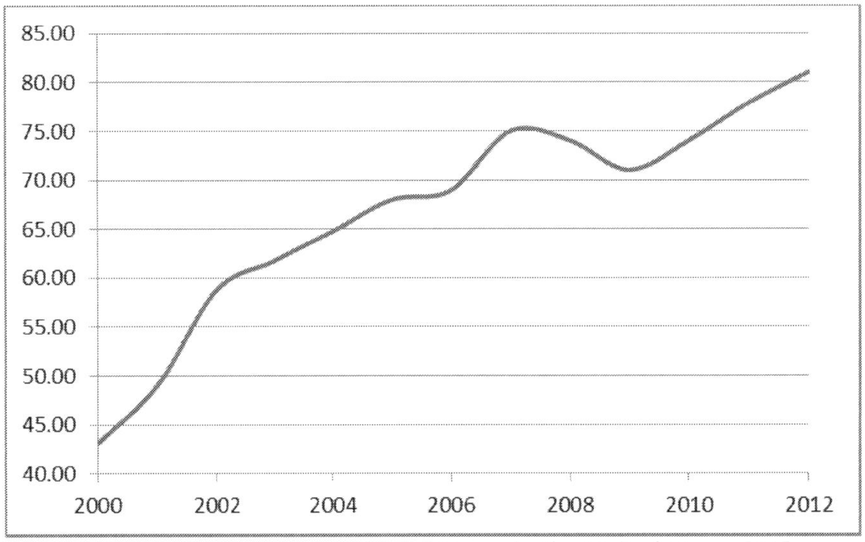

Figure 1: Percentage of Internet Users in the United States [11]

The current values of internet usage in the United States are comparable to the vehicle ownership per capita which as of 2010 was 769 vehicles in operation per 1000 people [12].

Access to information is greater today that it has ever been before. Wikipedia is one of the best known sites on the internet and fifth most frequently visited. It is currently the largest online encyclopedia and the largest encyclopedia ever created with over 4.4 million articles in the English language version alone. It also continues to grow at a very fast rate, adding 280,000 articles in 2013 [13]. Wikipedia has been successful in its astounding growth primarily because of the way it is structured as a site developed by crowd sourcing. Any user of Wikipedia can become an editor and the community itself acts as a means of keeping the content of the encyclopedia up to date [14].

2.1 Online Engineering Resources

Providing online resources for engineers is well established in the marketplace. Engineering Village, currently maintained by Elsevier, was an important early provider of online information used by engineers. The Engineering Index (Ei) was founded in 1884 at Washington University in St. Louis, MO. This archive of an index and abstracts of "articles of permanent value" was maintained over time and produced in an electronic form as the "Current Information Tape for Engineers (CITE)" in 1967. In 1969, CITE was made into a machine-readable program. This program was first available online via the World Wide Web in 1995 as EngineeringVillage.com and continues to be a valuable source of engineering information online [4].

Engineers, like all demographics, are adopting the use of social media. More than 50% of engineers today use social networks, however, the proportion is significantly skewed toward younger engineers with more than 80% of engineers under the age of 30 using social networks [5]. Much of this use is truly social through sites like Facebook. However, a significant number of engineers use social networks such as LinkedIn to make and manage contacts with other professionals. Social networks are used by engineers for assisting in professional networking and job searching, but in a community like LinkedIn many groups have formed for discussion in certain areas of subject matter expertise. A subset of the groups on LinkedIn with focus on various areas of process safety are shown below:

Table 1. LinkedIn Groups Focused on Process Safety by Members [6]

LinkedIn Group	Members
EHSQ Elite (No. 1 IN SAFETY) Environmental Health Safety Sustainability Security Quality Elite	49551
Oil and Gas HSE Practitioners	32248
Process Safety Management	10955
HAZOP	6882
Process Safety	5233
CCPS – The Global Community Committed to Process Safety	4515
Safety Automation Forum	3627
CFSE	1843
Combustible Dust Policy Institute Group	1340
Mary Kay O'Conner Process Safety Center	1257
AIChE Safety & Health Division	829
Process Safety Forum	780
Process Safety Hazops LOPA & SIL Analysis & Study	770
PHA	513
Pressure Relief	389
Instrument Reliability Network – Public Forum	290
DIERS	44
LOPA	33

Groups like these are available for members to ask questions of the community to receive guidance within the subject matter area, provide an opportunity for benchmarking, and provide networking opportunities within the niche.

Other significant engineering resources are also available online. Some of these websites provide additional communities for addressing the questions of members in an open forum as well as providing a variety of tools for engineers to use. One such site focused on the discipline of chemical engineering is cheresources.com. As of the writing of this article, the site which bills itself as "Your Chemical Engineering Community" had 85,815 members and boasted 77,802 community messages [7].

Professional organizations are also moving their content online. IHS is one portal which manages online standards access for over 370 standards developing organizations [8]. Access to standards through IHS is a paid service. Organizations like the National Fire Protection Association are providing their standards online through their own website. NFPA provides free access to all of their standards in a read only format, supporting the cost of development of additional standards through the sale of hard copy media of the same content. NFPA, which has provided online access to their codes and standards for over 10 years, indicated that providing the information through this media and at no charge is the best way to accomplishing their mission. Other standards organizations such as ASTM International and ASHRAE have followed the lead of NFPA and offer their content online and offer some portion of their content at no charge [9].

Knovel is one major provider of technical content online. This site acts as an online library providing access to over 2500 science and engineering encyclopedias, handbooks, texts, and databases [15]. Access to all of the texts produced by CCPS can be obtained from Knovel, making it an important source of guidance for chemical process safety.

3. Benefits of an Online LOPA Database

With all of the sources already mentioned, why would CCPS consider setting up a subcommittee to provide an online LOPA database? What are the benefits that could be derived from an online database for LOPA as opposed to CCPS following its standard business model and producing new and revised texts on the topic?

3.1 Accessibility

Based on the previous discussion, the public at large and engineers as a subgroup are already moving online for access to data. This trend is even more apparent for younger engineers. As our society and profession move online, access to information online as opposed to in a hard copy format will become more important. Providing information on LOPA in an online format allows engineers to access the information from any location where they have access to the internet. As internet access becomes almost ubiquitous, accessing the information online may very well become easier than accessing the information in a hard copy.

3.2 Frequency of updating content

The argument could be made that engineers will have access to the latest and greatest content on LOPA from CCPS with a Knovel subscription that grants them access to the CCPS library and this will be true in 2014. However, 12 years passed between the publication of the first CCPS LOPA text and *Guidelines for Enabling Conditions and Conditional Modifiers for Layer of Protection Analysis*. The CCPS process for producing new Guidelines books or revising existing texts is slow by its very nature. Moving the content online and putting it in the form of a database as opposed to a textbook will more easily allow the revision of existing content.

The chemical industry is faced with a changing landscape. New standards are produced by a various standard setting organizations at a faster pace than CCPS produces new texts. While any engineer working in process safety should remain apprised of changes to the standards which impact their area of expertise, LOPA methodologies should stay up to date with a broad range of standards and regulations since initiating events and IPLs are derived from many diverse areas. A CCPS subcommittee with expertise from a broad range of specialties can help to keep an evergreen online database up to date.

To meet the update requirement, changes to the online database may occur anywhere from several times per year to every few years depending on the rate of change in the industry. Regardless of the frequency of updates to the existing information, the capability for a quick update, which matches the pace of development of standards and regulations, is possible with the online database. Contrasting this update cycle with the approximate 10 year update cycle associated with the production of CCPS Guidelines books, the benefit becomes clear.

3.2 Frequency of new additions

Not only should the existing content in the online database be kept up to date, but new content should be added. Not all initiating events and Independent Protection Layers are covered in *Guidelines for Initiating Events and Independent Protection Layers in Layer of Protection Analysis*. Likewise, not all enabling conditions or conditional modifiers are covered in the companion text focused on those factors. Maintaining this information in an online database will allow new information to be added with much greater frequency than releases of revised Guidelines texts.

In addition, users of the online LOPA database will be able to see which factors are currently available for use and submit their own requests and information to the CCPS subcommittee for potential inclusion in the database. In general, when CCPS texts are being revised, the group which is available to submit information for inclusion in the text is the subcommittee and the limited group of peer reviewers of the text. While CCPS subcommittees do typically include some of the best subject matter experts on the topic being developed, a greater number of ideas will be generated from the larger body of global users of an online database. The ability to receive feedback from the users on a timely and frequent basis and respond with changes to the database will only increase the utility of the online database as compared to the less frequently updated Guidelines books. It is this process that will truly help to make the database an evergreen project.

4. What the Evergreen Online LOPA Database is Not Going to Be

The subcommittee for this project has only recently been formed so many of the details remain to be established; however, several issues that define the parameters of the project have been decided.

4.1 Use of Wikipedia Model

Wikipedia was mentioned above as an online encyclopedia that has seen tremendous growth, primarily due to the model that it uses in which any user can be an editor and add or change information to the site. While this model lowers the activation energy for change and growth of the product, there have been criticisms of the accuracy of the source due to this operational model. A 2005 report in *Nature* indicated that Wikipedia "comes close" to the Encyclopedia Britannica in terms of scientific accuracy. Although this report was disputed, *Nature* stood by the results [15]. However, Wikipedia itself indicates that it "is not considered a credible or authoritative source" and that "citation of Wikipedia in research papers may be considered unacceptable" [16]. Part of this is due to Wikipedia, like all encyclopedias, being a tertiary source of information, but it is also due to the means by which the content of Wikipedia is developed.

CCPS Guidelines books all contain a Limit of Liability / Disclaimer of Warranty which indicates that the publisher and author have used their best efforts in preparing the book, but that there is no warranty that the contents of the book are accurate or complete. While there can most certainly be errors in CCPS Guidelines texts, the reader can be assured that the subcommittee charged with the production of the text has spent many hours reading, rereading, discussing, and editing all of the content.

While adopting a Wikipedia style model for the Evergreen Online LOPA Database might result in faster growth and editing of the content in the database, it would remove some of the assurance that a reader has about the quality of the data present in the database. It is the intent of the subcommittee that the content of the online database will receive the same level of attention and review that would occur for the content in a Guidelines book. For this reason, the Evergreen Online LOPA Database will not follow a Wikipedia model for development.

4.2 Presence of a Forum

The subcommittee discussed the inclusion of a forum as part of the Evergreen Online LOPA Database. Such a forum would allow discussion of LOPA and related topics among the users of the database and the members of the subcommittee. After discussion of the inclusion of this type of tool, the subcommittee decided against it. Several factors entered into this decision. There are several other locations where such forums already exist, as documented earlier in this paper. Thus, the need for a new forum space for this type of discussion among members was not apparent. In addition, with a forum hosted on a CCPS site, the discussion and answers that individuals provide could be misinterpreted as having the support of the CCPS subcommittee of subject matter experts. Even if only members of the CCPS subcommittee could respond to posts on such a forum, even their answers would not have been subject to the screening and evaluation process which will be in place before any addition or modification is made to the online database.

Even without a forum, one of the strengths of the Evergreen Online LOPA Database will be the capability of community providing input. There will be a "yet to be determined" mechanism for individuals to submit both proposals for new content, differences of opinion on existing content, and questions that do not seem to be addressed by the content on the site. These submissions may result in changes to the material provided on the site or new material being placed on the site. In addition, the subcommittee has discussed the inclusion of a Frequently Asked Questions area to address any issues which seem to arise from the end users. Answers on this FAQ will be subject to the same review process as other information provided on the website.

5. Format of the Database

As previously indicated the work by the subcommittee has only just started. However, the likely format of the database entries has already been established by the work done by the subcommittee for *Guidelines for Initiating Events and Independent Protection Layers in Layer of Protection Analysis*. This subcommittee developed a template that is used for all of the initiating events and a second very similar template used for all Independent Protection Layers. An example of the template used for a spring loaded pressure relief valve is given as Figure 2.

As can be seen from the data table provided, a significant amount of information is provided around each IPL in addition to a suggested probability of failure on demand (PFD). The various notes, special considerations, generic validation methods, and the provided basis for the both the PFD and the validation method will provide significant guidance to the user. This method of providing a data sheet associated with each item in the database also allows the user to focus only on the specific item in the database that they are interested in and still obtain all of the relevant information associated with that initiating event or IPL. Although the information in *Guidelines for Enabling Conditions and Conditional Modifiers for Layer of Protection Analysis* did not present the information in this format, it may be reformated to fit this template for presentation in the online LOPA database.

The theme of the data presented in the Evergreen Online LOPA Database will be assisting LOPA practitioners in making the best analysis and providing a product that supports consistent improvement of risk management across the industry. The database, much like the books before it, will attempt to protect end users from misuse of the technique, which can lead to poor risk decisions.

6. Issues to be Investigated

The subcommittee has established a direction for the project, however, there are still signficant issues to be investigated and decided. These are issues that the subcommittee will be facing in the coming year as the details for the database are established. These are also areas where input from the broader LOPA community can be very valuable in producing a product that meets the needs of the largest number of LOPA practictioners. Questions which need to be answered include:

1) What data should be included in the database? Is there a signficant desire by the LOPA community to have information on enabling conditions and conditional modifiers in the database?
2) What is the process for vetting entries into the database? Will the subcommittee act as the approval body for changes or will peer review of new content be required?

3) Should the database include only datasheets or should broader information be included such as best practices for conducting LOPA?

IPL Description
Spring-operated pressure relief valve
Generic PFD suggested for use in LOPA
0.01 for failure to open enough at set pressure (100% of rating) If there is an isolation valve (block valve) upstream or downstream of the relief device, then the suggested PFD is 0.1, unless there is a management system in place to ensure that valves are returned to service in their proper positions after maintenance and that they remain in the appropriate state during operation. *NOTE: If fire cladding was assumed to be in place on the protected vessel in the calculation of the size of a relief valve, then the combined PFD of the insulation plus relief valve is 0.01.*
Special considerations for use of generic PFD for this IPL
• The PRV is sized for the scenario being considered. • The inlet and outlet piping are sized correctly and are mechanically adequate for relief flow. • The relief valve is in clean service, and the metallurgy is corrosion-resistant to the particular service. • The service under evaluation does not have the potential for freezing of the process fluid before or during relief; if freezing is possible, then adequate heat tracing of the relief valve and piping is installed and maintained.
Generic validation method
• The ITPM frequency is set in accordance with the manufacturer's recommendations and/or code requirements and may be adjusted based on the results of previous inspections. • The relief valve is periodically removed and bench-tested by a certified individual. • The inlet and discharge piping are inspected to ensure that there is no blockage or corrosion that could impede proper functioning. • An internal inspection is performed to detect the onset of failure (such as corrosion, damaged internal components, or fouling/plugging). • The relief valve is returned to like-new condition prior to its return to service.
Basis for PFD and generic validation method
Consensus of the *Guidelines* subcommittee, based in general on *Guidelines for Pressure Relief and Effluent Handling Systems* (CCPS 1998b), Chapter 2, and recent published data (Bukowski and Goble 2009).

Figure 2: Example Data Table

4) How will the database maintenance be paid for? Will CCPS members have free access? Will everyone pay to access the database on a yearly basis?
5) How is the database management going to be managed long term? Is it possible to have the structure of the database designed so that an external contractor is not needed and the management can be handled by CCPS staff?

These are all questions that the subcommittee will be dealing with as the details for the project are put in place. Any input on these or other issues related to the Evergreen Online LOPA Database process should be sent to the one of the authors of this paper.

8. Conclusion

The case has been made that online data sources are becoming or have become extremely important. In the engineering field, the transition to online sources of data is also occurring, although younger engineers are in the vanguard of this change. CCPS has produced two new texts that extend the information available to LOPA practitioners significantly. A new committee has been formed to take this information, determine the best means of providing it in an online reference, provide appropriate mechanisms for keeping the information evergreen, and maintain the high level of quality associated with products from CCPS. This Evergreen Online LOPA database is an exciting new project with the potential to provide this critical information to engineers in a form that meets their needs in the modern marketplace and improves the safety of our industry.

9. References

[1] CCPS. *Layer of Protection Analysis Simplified Process Risk Assessment.* Center for Chemical Process Safety. New York, NY. 2001

[2] CCPS. *Guidelines for Safe Automation of Chemical Processes.* Center for Chemical Process Safety. New York, NY. 1993

[3] CCPS. *Guidelines for Enabling Conditions and Conditional Modifiers for Layer Protection Analysis.* Center for Chemical Process Safety. New York, NY. 2013

[4] "History | Elsevier – Engineering Village," http://www.elsevier.com/online-tools/engineering-village/history (accessed Feb. 11, 2014)

[5] "Expert Webinar – Engineers Professional Use of Social Networks Today and Where It's Heading," http://why.knovel.com/resources/expert-webinars/engineers-professional-use-of-social-networks-today-and-where-its-heading.html (accessed Feb. 11, 2014)

[6] LinkedIn, http://www.linkedin.com/ (accessed Feb. 11, 2014)

[7] ChEResources.com, http://www.cheresources.com/ (accessed Feb. 11, 2014)

[8] IHS Standards Expert, http://www.ihs.com/products/design/industry-standards/standards-expert.aspx (accessed Feb. 11, 2014)

[9] NFPA, https://www.nfpa.org/codes-and-standards/free-access (accessed Feb. 11, 2014)

[10] Internet, http://en.wikipedia.org/wiki/Internet (accessed Feb. 11, 2014)

[11] "ICT Facts and Figures 2013," http://www.itu.int/en/ITU-D/Statistics/Pages/stat/default.aspx (accessed Feb. 11, 2014)

[12] Motor Vehicle, http://en.wikipedia.org/wiki/Motor_vehicle (accessed Feb. 11, 2014)

[13] Wikipedia: Size of Wikipedia, http://en.wikipedia.org/wiki/Wikipedia:Size_of_Wikipedia (accessed Feb. 11, 2014)

[14] Wikipedia: Who writes Wikipedia, http://en.wikipedia.org/wiki/Wikipedia:Who_writes_Wikipedia (accessed Feb. 11, 2014)

[15] "Special Report Internet encyclopedias go head to head" http://www.nature.com/nature/journal/v438/n7070/full/438900a.html (accessed Feb. 14, 2104)

[16] Wikipedia: Academic Use http://en.wikipedia.org/wiki/Wikipedia:Academic_use (accessed Feb. 14, 2014)

Are You Being Honest With Yourself Regarding IPL Integrity?

Mr. Andrew C. Madewell, P.E.
Cognascents Consulting Group, Inc.
11777-A Katy Freeway, Ste 438
Houston, TX 77079
andrew.madewell@cognascents.com

Prepared for Presentation at
American Institute of Chemical Engineers
2014 Spring Meeting
10th Global Congress on Process Safety
New Orleans, Louisiana
March 30 – April 3, 2014

UNPUBLISHED

Are You Being Honest With Yourself Regarding IPL Integrity?

Mr. Andrew C. Madewell, P.E.
Cognascents Consulting Group, Inc.
11777-A Katy Freeway, Ste 438
Houston, TX 77079
andrew.madewell@cognascents.com

Keywords: Independent Protection Layer (IPL), Process Safety Management (PSM), Layer of Protection Analysis (LOPA), Process and Instrumentation Drawing (P&ID), Process Hazard Analysis (PHA), Process Safety Lifecycle, Safety Integrity Level (SIL).

Abstract

Independent Protection Layers (IPLs) are critical pieces of armor designed to protect against process upsets that may harm people, the environment, and/or commercial interests. IPLs play a key role in any Process Safety Management (PSM) program. They are often used to close the risk gap between elimination/mitigation measures and associated hazard scenario causes and consequences; hence, regulations exist that require companies to demonstrate IPL integrity and adequacy.

When conducting a Process Hazard Analysis (PHA) using the Layer of Protection Analysis (LOPA) methodology, IPLs are used to close the risk gap between elimination/mitigation measures for a given hazard scenario and its cause, consequence, and conditional modifiers. In order for an IPL to be "available", it must meet certain criteria defined by industry standards and company-specific guidance documents.

Demonstrating IPL effectiveness, or adequacy, involves multiple pieces of information that are not always linked together and kept "evergreen". In addition, the various data repositories and tasks required to maintain the integrity of an IPL are "owned" by several functional roles. Required data for IPL adequacy include the following: IPL design information, IPL integrity level analyses, maintenance and function testing data, and process safety time and IPL response time analyses.

Companies do not approach IPL integrity the same way. For example, companies perform IPL function testing and maintenance using different criteria; some employ a standard function testing methodology of confirmation that the IPL acts within the designated time window per regulatory requirements, while other companies actually measure the specific time it takes an IPL to respond to get to its process safe condition. Either of these approaches provides compliance as to whether the IPL will act adequately given a process upset; however, vulnerabilities may exist depending on the approach employed.

The author posits that use of an evergreen IPL lifecycle reduces potential vulnerabilities in the design and function of an IPL. In this paper, the author presents the advantages and disadvantages of using an evergreen lifecycle approach to establish and maintain the integrity and "availability" of IPLs. The author also provides recommendations to enhance the robustness of maintaining IPL adequacy throughout the lifecycle of the protection layer. Specifically, the author provides examples of IPL integrity successes through the use of an evergreen lifecycle methodology, enhanced operational insight, and potential pitfalls when not employing a holistic IPL lifecycle approach in maintaining IPL integrity and adequacy.

The target audience for this paper includes project managers, project or process engineers, EH&S managers, PSM coordinators, and operators; however, anyone involved with small or large capital projects may also benefit from this paper's content.

1. Background and Purpose

Demonstrating IPL integrity plays a critical role in mitigating or eliminating undesired events across the oil, gas, and chemical industries. These safety devices are strategically designed and installed to reduce the risk involved during process operation. The design and operational effectiveness of a particular safety device involves multiple pieces of information/data. Required information includes design basis and process description, Safety Integrity Level (SIL) calculations, IPL testing and maintenance philosophy, and process safety time analysis.

Protective instruments are typically specified during early facility design (i.e. Define stage); however, the author has found that ongoing review and analysis of each installed IPL at a given facility throughout the facility lifetime does not always occur, leaving potential safety protection shortfalls. Something as simple as changing an alarm or trip set-point may affect the integrity of an installed IPL, and the author can reference multiple instances where minor changes like this have proven an IPL to be inadequate at providing proper protection.

One way to ensure IPL integrity throughout the lifetime of a facility is the implementation of an evergreen lifecycle approach to capture the "real-time" process risk. This paper outlines such a methodology, documenting the benefits and potential hardships. Specifically, the author outlines how all pieces of information surrounding the integrity of an IPL relate to one another, and how a small change in one process parameter may have profound changes to IPL integrity, and ultimately the risk profile. Without the implementation of an evergreen lifecycle approach, vulnerabilities may exist leaving a system inadequately protected against a process excursion.

2. IPL Criteria

When conducting a PHA using the LOPA methodology, IPLs are used to close the risk gap between elimination/mitigation measures for a given hazard scenario and its cause, consequence, and conditional modifiers. In order for an IPL to be "available", it must meet certain criteria defined by industry standards and company-specific guidance documents. The required criteria based on Recognized and Generally Accepted Good Engineering Practices (RAGAGEP) is given below.

2.1 Effectiveness

In order to take credit for an IPL in a LOPA study, the system, device, or action must be deemed effective at either mitigating or eliminating the undesired event from occurring. The following is a list of questions that should be asked when discussing potential IPLs during a LOPA session. Note that the list of questions below is not a complete exhaustive list.

- Will the potential IPL detect the unwanted condition?
- Will the potential IPL act quick enough to prevent the undesired consequence?
- Does the potential IPL have adequate capacity to protect against the undesired scenario (i.e. relief device capacity)?

2.2 Independence

For a safeguard to be considered an IPL, the LOPA team must ensure that the IPL is independent of the hazard scenario initiating event, as well as independent from any additional IPLs credited for the same event. For example, if the initiating event is an inadvertent valve opening due to human error, credit should not be taken (as an IPL) for operator response to an alarm, regardless of the time required to reach an unsafe condition. Furthermore, if credit is taken for a high pressure interlock that is acting as part of the safety shutdown system, credit should not be given to any additional IPLs installed on the same safety shutdown system.

2.3 Auditability

An IPL must be auditable in order to demonstrate that it meets the necessary risk mitigation criteria. This audit process must confirm the effectiveness and independence of the IPL(s) in question as outlined above. The auditability ties directly into the evergreen IPL lifecycle outlined below in that if one incorporates an evergreen lifecycle approach to their IPLs, then the auditability aspect becomes part of the evergreen lifecycle.

3. Data Required for IPL Integrity

When designing a protective system or device, the following information/data is required. This information is considered the bare minimum required when implementing protective devices to mitigate and/or eliminate undesired events.

3.1 Design Basis

A protective device, or IPL, is designed and installed for a specific function in the process, and design information ensures that the device specified will meet those requirements. Design information includes the following:

- Process conditions, including equipment and process limitations
- Drawings and diagrams
- Installation criteria
- Potential hazard scenarios the device is designed to protect against

3.2 Safety Integrity Level (SIL) Analysis

The quantitative analysis documenting a IPLs capability to protect against a hazard scenario is the SIL calculation analysis. Major inputs to these calculations include process conditions (including equipment and process limiting criteria), as well as applicable hazard scenarios for which the IPL is designed and installed to protect against (as identified in all applicable LOPA analyses). These inputs are used to generate a specific Probability of Failure on Demand (PFD) for the IPL in question. This PFD value will encompass all the demands placed on the IPL (i.e.

the PFD value will reflect whether the IPL is installed to protect against one hazard scenario or four hazard scenarios).

3.3 IPL Operational Surveillance

The output PFD from the SIL calculations is typically based on a low-usage rate for the IPL; therefore operational surveillance surrounding IPL trips needs to be conducted to track the instances where the IPL is being called upon to act. This data should be closely monitored as the integrity of the IPL may change if the IPL trip rate increases over time and exceeds the low-usage rate criteria limit.

3.4 IPL Testing and Maintenance Philosophy

The testing and maintenance philosophy and records are typically driven by regulatory and/or corporate specific compliance standards, as well as the SIL of the IPL. For example, an IPL with a PFD of $1x10^{-1}$ will have a different testing and maintenance schedule than an IPL with a calculated PFD of $1x10^{-3}$. The testing a maintenance programs around protective layers become stricter, and the frequencies of testing increase with increasing PFD values.

3.5 Process Safety Time (PST) and Response Time Analyses

The PST calculations document the time it takes for a process to deviate from normal conditions to an upset (i.e. the time required for the process to deviate from the maximum operating pressure to exceeding design pressure in the event of a blocked outlet). Process safety time criteria may deviate slightly across the industry, but the goal of the exercise is to compare the process safety time against the response time of the IPL to ensure that the protective device will act fast enough to avoid an unsafe condition and return the process to a safe state. The response time is the time it takes an IPL to act from the initial IPL initiation to the point of a safe condition.

4. Evergreen Lifecycle for IPL Integrity

All of the data previously identified around an IPL is typically overseen by multiple disciplines. The design information and SIL calculations are usually owned by the instrumentation discipline, while the testing and maintenance activities are employed by operations and/or maintenance disciplines. The process surveillance and process safety times are typically owned by the process engineering discipline. This multi-discipline ownership creates a potential area of concern when ensuring the adequacy of an IPL as a lack of communication may cause an IPL to become inadequate in regards to its integrity rating due to changing process operating conditions.

A solution to this potential shortfall is an evergreen lifecycle approach to IPL integrity. The lifecycle starts with the design and installation of the IPL and continues through the decommissioning process. The use of an evergreen lifecycle reduces potential vulnerabilities in the design and function of an IPL. When established correctly, an evergreen lifecycle effectively ensures that when one set of data or information is updated or modified, all other sets of data are reviewed and/or updated during the process.

One main challenge which the author has witnessed across the industry is ensuring adequate cross-communication between disciplines. All too often separate disciplines (i.e. instrumentation and process engineering) operate in a semi-vacuum and do not effectively communicate process design or operation changes. This creates the potential for an IPL and its associated design to not be maintained and updated properly through the design change process, leading to the question of whether or not an installed IPL is really protecting against an unsafe condition.

With an evergreen lifecycle approach to IPL integrity, this communication gap is reduced as separate functional teams are required to communicate with one another. For example, if operating conditions are modified to increase process throughput, then instrumentation and operations personnel are responsible for consulting with each other to evaluate whether the operating condition change will affect the design or SIL values for the IPL installed. The collaboration would then help determine whether or not the IPL will still provide adequate protection against undesired events, or if modifications are required.

A second challenge that has been witnessed surrounds process safety times. Multiple instances have occurred in which process safety times have been examined when reviewing an IPL and it has been determined that the process safety time calculations were outdated and not based on current operational conditions. If the process safety times are not accurate, then the adequacy of an IPL may be in question as the process safety time may be lower than the device response time. Using an evergreen lifecycle, a change in process conditions would require a review of the process safety times. The reviewed and/or updated process safety times would then be compared with the IPL response time to ensure adequate IPL integrity.

A third challenge is the lack of operational surveillance around IPL trips. A SIL calculation is based on a low usage rate for the IPL in question. If process surveillance is not being conducted to determine when and how many trips an IPL has over a specified time duration, a SIL calculation and resulting PFD value may be inaccurate and no longer valid. If trip rates exceed the low-usage rate criteria in the SIL calculations, then the resulting PFD may in fact be lower (i.e. $1x10^{-1}$ instead or $1x10^{-2}$) leaving the system in a lower state of protection. Employing an evergreen lifecycle approach would facilitate the comparison of such pieces of information and provide feedback for the ongoing integrity of an IPL.

An evergreen approach to IPL analysis ensures that all aspects of the IPL lifecycle are being analyzed and maintained in "real-time", with the added benefit of allowing continual feedback. If a change in operating conditions occurs (i.e. alarm or trip set-point change), then the process safety times are revalidated based on the change to reflect accuracy. This will then be used to revalidate the SIL calculations to achieve the new PFD for the IPL. The updated SIL rating will drive any necessary changes to the IPL testing and maintenance frequencies, along with changes to the risk profile resulting from the updated PFD values in the LOPA analyses. Figure 1 below shows this inter-relationship, specifically outlining how each piece of data has the capability to affect any and all other information surrounding an IPL.

Figure 1: Evergreen IPL Integrity Lifecycle

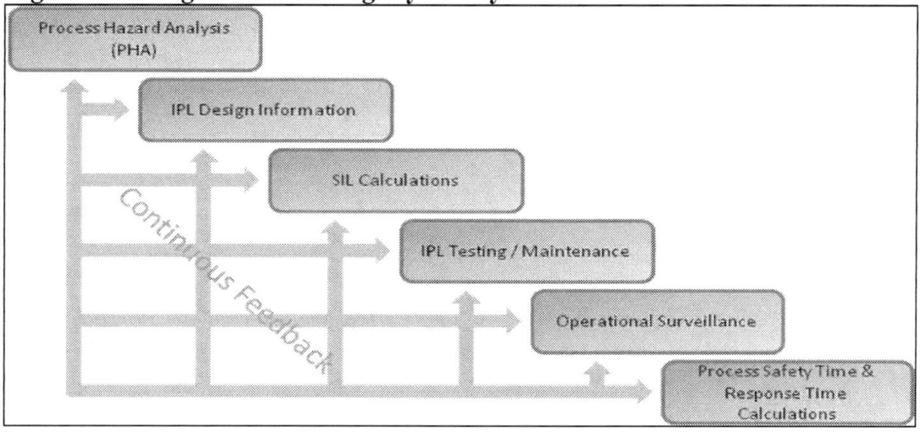

The evergreen lifecycle approach will also facilitate greater operational insight to the process, which may lead to future Inherently Safer Design (ISD) opportunities. For example, if an IPL is determined to be tripping at an elevated rate (through operational surveillance), one may determine that reducing operating conditions may be an inherently safer option than ensuring that the IPL SIL calculations are valid for the elevated usage rate (may require IPL design changes). This type of analysis is at the heart of process safety and falls in alignment with process safety principles.

A potential hardship of employing an evergreen lifecycle approach to IPL integrity may be the additional time required to complete multiple comparisons and analyses when ensuring integrity compliance. However, this additional time (man hours and cost) can be weighed against the potential impacts of not using such a methodology. The author argues that the safety and financial impacts of not using such a philosophy is far greater than the cost of following an evergreen approach.

5. Conclusion

The author recommends implementing an evergreen lifecycle approach to IPLs, which will help aid cross-discipline communication. This is a main challenge routinely observed when looking into IPL integrity at different facilities. Although all necessary IPL information may be available, it is common to observe pieces of data that are based on slightly different design and operational parameters due to a lack of effective communication among necessary shareholders.

More specifically, the author suggests that an evergreen lifecycle approach to IPL integrity will help ensure that IPLs (1) are designed adequately, (2) will effectively provide protection against process upsets throughout the life of the process, and (3) are accurately reflected during LOPA analyses to determine the "real-time" risk of the process being evaluated. This approach will also ensure compliance with Process Safety Management (PSM) regulations and audits. Failure to implement a lifecycle approach or similar methodology to IPLs may leave one susceptible to inadequate process safety protection and/or potential process incidents, something industry seeks to avoid.

6. References

[1] Center for Chemical Process Safety, *Layer of Protection Analysis - Simplified Process Risk Assessment*, Copyright © 2001, American Institute of Chemical Engineers, 3 Park Avenue, New York, NY 10016-5991.

[2] Center for Chemical Process Safety, *Guidelines for Process Safety Documentation*, Copyright © 1995, American Institute of Chemical Engineers, 345 East 47th Street, New York, NY 10017.

[3] Center for Chemical Process Safety, *Guidelines for Safe and Reliable Instrumented Protective Systems*, Copyright © 2007, Published by John Wiley & Sons, Inc., Hoboken, 111 River Street, New Jersey 07030.

IPL/CMS- Integrity Management of Non-SIS Independent Protection Layers after the LOPA

Ron Nichols
aeSolutions
10375 Richmond Avenue, Suite 800
Houston, Texas 77042
Ron.Nichols@aesolns.com

Prepared for Presentation at
American Institute of Chemical Engineers
2014 Spring Meeting
10th Global Congress on Process Safety
New Orleans, LA
March 30 – April 2, 2014

UNPUBLISHED

IPL/CMS- Integrity Management of Non-SIS Independent Protection Layers after the LOPA

Ron Nichols
aeSolutions
10375 Richmond Avenue, Suite 800
Houston, Texas 77042
Ron.Nichols@aesolns.com

Keywords: Non SIF IPL, Identification, Selection, Implementation, Life-cycle management, MOC, Security.

Abstract

This paper discusses the identification, selection, implementation and management of Non-SIF IPLs through the process lifecycle.

1. Layer of Protection Analysis

Layers of Protection Analysis (LOPA) in conjunction with the Process Hazard Analysis (PHA) is now a key tool used by the chemical, oil and gas industries to assist companies in identifying, implementing and managing the critical safeguards needed to achieve their risk tolerance targets. The LOPA is used to identify the number of Independent Protection Layers (IPLs) and their integrity needed to reduce the likelihood to an acceptably low frequency that an initiating cause will progress to an undesired consequence.

2. Lifecycle Management of IPLs

Since the acceptance of ISA 84.00.01/IEC-61511, the life cycle management of safety instrumented systems is now being implemented throughout industry. The required safety integrity level (dependability) for the safety instrumented functions (SIFs) are obtained by closing the LOPA gaps between the existing mitigated event likelihood (MEL) and the company's target mitigated event likelihood (TMEL). Often a SIF is combined with non-SIF IPLs to achieve the risk reduction gap closure, reducing the SIL requirement assigned to that SIF.

To maintain acceptable risk targets, all IPLs, not just SIFs, must be managed through the lifecycle of the process. This is because many LOPA gaps are closed by only non-SIF IPLs and the SIL assignment for many SIFs depend on the use of non-SIF IPLs used in that LOPA.

3. From IPL Identification through Lifecycle Management

IPL identification begins in the PHA where the worksheet captures the deviation initiating causes, hazard consequences (without credit for active safeguards/IPLs), consequence severity and existing IPLs (engineered and administrative safeguards). Based on the company's LOPA selection criteria, consequences assigned either high severity or high risk are selected for LOPA to determine if their MEL is adequate to meet the company's risk tolerance criteria. In the typical LOPA, the initiating cause has a frequency assigned and existing IPLs are identified and credited with a probability to fail on demand (PFD) for calculating the current mitigated event likelihood (MEL). LOPA quantifies if there is a gap between the MEL and the Target Mitigated Event Likelihood (TMEL) established to meet the company's risk tolerance requirements. If the LOPA finds a gap where the envisioned consequence is more frequent than the company's risk targets, the team will propose additional IPLs to close the risk gap or design changes to reduce the credible consequence severity. After identifying how many IPLs are needed to close a LOPA gap, existing IPLs must be validated as meeting the IPL criteria. If additional IPLs are needed to close a gap, they must be identified, validated to meet IPL criteria and implemented. After selected for implementation, all IPLs needed to meet the company's TMEL must be placed under a program for lifecycle management.

4. Requirements to Credit a Safeguard as an IPL

To classify a safeguard as an IPL it must meet the following criteria:

- Effectiveness/Specificity - this barrier by itself can prevent the cause from progressing to the undesired consequence.
- Independence - the IPL is independent from the initiating cause and other IPLs used in the LOPA to meet the targeted minimum event likelihood (TMEL) required by the company's risk tolerance requirements. Common cause failures between this IPL and the initiating cause or other IPLs will not defeat the protection.
- Dependability - the IPL can be relied on to function with an expected probability to successfully prevent the undesired consequence when placed in demand.
- Auditability - the IPL can be routinely audited/tested at an adequate frequency through the process lifecycle to maintain its' dependability.
- Access Security - there are controls in place that prevent unauthorized changes to the IPLs.

5. IPL Types

IPLs are often grouped into two categories: Preventative and Consequence Mitigation.

Preventative

Preventative IPLs will prevent the cause from progressing to the undesired consequence. Examples of preventative IPLs include alarms that initiate operator intervention, standard operating procedures (independent of the cause) that can prevent the consequence from occurring, basic process control interlock functions and safety instrumented systems.

Consequence mitigating

Consequence mitigating IPLs often referred to as consequence mitigation systems (CMS) reduce the severity of the undesired consequence to a level that meets the companies risk tolerance. Examples of CMS include pressure relief devices, dikes and instrument activated deluge systems.

6. Efficient Documentation for the IPL Lifecycle

The efficiency of the IPL implementation work process improves when the documentation and procedures for information capture, validation, selection and integrity management are integrated through the different phases. The IPL implementation process can involve different teams and stakeholders operating over an extended period of time. Developing an IPL implementation documentation protocol for consistent capturing of the key information and standardizing the work practice between different individuals or teams early in the process improves the efficiency of the IPL implementation and subsequent lifecycle management.

7. IPL Documentation during Initial PHA/LOPA study

For the initial PHA and LOPA studies, it is helpful for the company to provide a protocol to capture key information in the PHA Worksheet that can be efficiently referenced. Consistent capture of key details of the initiating causes and existing IPLs in the PHA worksheets assists later teams through the steps of IPL identification, validation, selection, implementation and lifecycle management.

Examples of initiating cause details to capture in the PHA worksheet can include:

- BPCS Control Loop: Tag numbers of instrument inputs and outputs, P&IDs where the input and output instruments are located.
- Human Operations: Frequency the step is routinely performed, reference to document number and Step # for that operation, reference of procedure for independent checks of the procedure, guidance on probability of failure.

Examples of IPL information to capture in the PHA Worksheets can include:

- BPSC or SIF: Instrument tag numbers inputs/outputs, logic solver ID (if multiple), P&IDs the instruments are located on.
- Alarms: Instrument tag number, setpoints and reference to documented operator response requirement.
- SOP: Document number and step for the action that prevents the progression to the undesired consequence.

7.1 IPL Register

While the numbers of instruments, procedures and other safeguards used on a site can be enormous, the fraction classified as IPLs is relatively few. The development of a central IPL register can be a useful tool to aid the implementation, management and update of the site's IPLs. This register can provide a cross reference to the PHA/LOPA source(s) that uses the IPL and reference to the documents that support the IPL's validation.

7.2 Assigning PFD to IPL

To provide consistency across different PHA/LOPA teams, procedures should provide the teams with a common set of IPL dependability values (PFDs) with a reference to any literature source and guidance on their selection. The following are some common sources of published sources for dependability values used in industry:

- Oreda
- Exida
- Nureg
- CCPS.

7.3 Team Development of Probability of Failure on Demand (PFD) Values

Where no published PFDs are available, it is useful for the site to provide guidance for an IPL study team to develop and document PFDs based on the site's operating experience or maintenance records.

8. IPL Validation

8.1 Consequence Severity Verification

During the IPL validation step, performing a more rigorous consequence assessment may allow a severity reduction to the hazard reducing the number or even need for IPLs. The use of these tools may not be available or practical for use during the initial PHA/LOPA study resulting in the team assigning an overly conservative consequence severity. Often the use of these tools is driven by the challenge to provide adequate IPLs to meet the TMEL for the identified consequence. Some examples of more rigorous consequences modeling tools include:

- PHAST
- HYSIS/ Aspen
- Computational Fluid Dynamics.

9 Confirming Existing and Proposed IPLs Meet Design Criteria

During the validation stage, the IPLs are reviewed to confirm compliance with the following IPL requirements:

9.1 Independence

Independence between the initiating cause and IPL(s) or between different IPLs used to mitigate the event likelihood must be confirmed. If the site procedure allows two BPCS credits within a single LOPA then:

- The BPCS inputs and outputs must not be shared between the initiating cause and an IPL or between the two different IPLs.
- The IPLs or initiating cause must not share common sensors, elements, or I/O cards.

9.2 Effectiveness/Specificity

The IPL should provide documentation supporting how this IPL can by itself prevent the initiating cause from reaching the undesired consequence, such as:

- PSV documented with design basis and calculations to support sizing meets required relief rate for the consequence identified in the LOPA.
- Response times for the operator response to alarms are studied and documented to show there is adequate time to diagnose the event and take the required actions to prevent the undesired consequence
- BPCS interlock response time and outputs are adequate to prevent the undesired consequence
- Dikes are adequately sized to hold the contents of the loss of containment.

9.2.1 Dependability

Adherence to appropriate recognized and generally accepted good engineering practice (RAGAGEP) is useful to support the PFD credit given during the initial LOPA. The following are some common RAGAGEP providing guidance on the design of IPLs that can support their dependability and auditability:

- PSV: Boilers ASME I, Unfired Equipment ASME VIII, API 520, API 521, API 5210/5210A,
- PVSV: NFPA 30, API 2000,
- BMS: NFPA 85, NFPA 86, NFPA 87, API 556
- SOPs: 29 CFR 1910(f)(1)(i)(D), HSE Human Factors Toolkit
- Alarms ISA 18.2, NUREG/CR-1278,
- Deluge: NFPA 15,
- Explosion Prevention NFPA 69,
- Explosion Venting: NFPA 68
- etc.

9.3 Eliminating or Modifying IPLs

If during the validation process identified IPLs are determined to be invalid, then recommendations must be made to modify or replace IPLs as needed to close the gap. The company should define a protocol to replace or modify IPLs that includes:

- Written Management of Change Procedures and Practices
- Who is involved in accepting and approving the change, i.e. stakeholder/team composition?
- What is the documentation procedure and how will the original PHA/LOPA be updated?

10 Final Selection of IPLs

From experience, the final selection of IPLs to be implemented and managed through the process lifecycle involves team meetings that include the original PHA/LOPA team and the facility controls/SIS engineer. The team reviews the IPL options and makes final IPL selections based on the following criteria:

- Operability
- Process Safety time
- Maximum allowable Response Time
- Hardware Fault tolerance for SIFs
- Financial analysis

This team meeting also provides the opportunity to close the recommendations associated with consequences reduced by performing a more rigorous consequence assessment or by implementing inherently safe design change.

The output of this meeting is the IPL list for implementation and management that provides their basis for criticality or design.

10.1 Excess IPLs

Occasionally more IPLs are identified by the team than are needed to close the TMEL gap. This can occur if multiple valid IPLs are identified during the PHA/LOPA or if a SIF integrity level design is driven by a higher gap scenario reducing the number of supporting IPLs needed to close the risk reduction gap.

How will your procedures handle excess IPLs?

- Will all IPLs be kept or only those needed to close the gap?
- What will be done with excess IPLs removed from the IPL list; deleted or converted to non IPL Safeguards?

10.2 IPL Priority

For consistency between the different PHA/LOPA teams and through IPL management, the company may wish to provide a philosophy of priority for selecting IPLs. For example:

- Prevention versus Mitigation,
- Engineered response verse operator response etc.

11 IPL Integrity Management (Auditability) through the Process Lifecycle

11.1 Auditability/Integrity Management Procedures

After IPL selection, the company must place the IPLs under a testing program (auditing) to support their dependability. This testing program must include:

- Defining test frequency and maintenance priority
- Documenting test results in maintenance system
- Defining corrective action requirements for test results showing nonconformance, e.g.:
- Firewater pumps don't pass test
- PSV fail testing, nozzles repeatedly found plugged
- Alarms nonfunctioning or defeated
- SOP refresher testing fails
- Defining maximum time allowed to repair and steps for authorization to operate with deficient IPL, or risk gap, until repair is made.

11.2 RAGAGEP that Supports IPL Testing

Adherence to a RAGAGEP for testing IPLs can provide a good basis to support the safeguards dependability. The following are examples of a few RAGAGEP that provide guidance on inspection and testing of IPLs:

- PSV: ASME 576
- BMS- NFPA 85, NFPA 86, NFPA 87, API 556
- SIF-ISA 84.00.01/IEC-65111.

Where there is no guidance from RAGAGEP, the site should develop a procedure/protocol to define appropriate testing that each IPL needs to support its' dependability.

To meet the requirements of 29 CFR1910.119for engineered and administrative safeguards, the site must have written management systems to identify the IPL auditing tests, schedule the required tests, maintain test records and initiate the required correction of nonconformance.

12 Tools for IPL Validation

12.1 Checklist to Support the IPLs Validation Process

A valuable tool to provide the IPL validation team is a set of checklists to guide and document the requirements for the different IPL types. These checklists confirm and document the independence, effectiveness, dependability, auditability and security for the different IPLs being implemented. The following are examples of several IPL checklists to demonstrate the concept.

Table 1. Alarm SOP

Description	Guidance/Comments
Independence	
Verify the consequence severity was based on no active safeguards. The contribution of this IPL was not considered when assigning the severity.	
Confirm the alarm that initiates operator response is independent of the LOPA's initiating cause	If the cause is a BPCS failure there can be no shared instrument loops, including the I/O card. Define where this is documented.
Confirm the alarm and required operator action(s) are fully independent from all other IPLs credited in the same LOPA scenario(s).	Instrument inputs to other IPLs must be independent of this alarm. The operator actions must be independent of other IPL outputs to prevent the consequence. If two uses of the BPCS are allowed for either initiating cause and IPL or as 2 separate IPLs, they must not share the same I/O card.
Effectiveness/Specificity	
Verify the response time for alarm setpoint provides the operator adequate time to diagnose and take action to prevent the consequence.	Provide location of documented study supporting adequate response time.
Verify the operator actions for alarm response are documented	Reference the document detailing required response to the alarm.
Confirm the service which this Alarm's sensor(s) is in is not prone to problems from plugging, polymerization, fouling or external environmental conditions.	
Performance or Dependability	
Verify documented basis for Probability of Failure on Demand (PFD) used in LOPA calculation.	Reference source/basis for PFD.
If no published basis, provide documentation on how the team developed the PFD used in the LOPA.	Provide procedures on PHA/LOPA team PFD development and documentation for retrieval and reference.
Define required repair time for the alarm as needed to maintain dependability. If a backup alarm is identified as needed it is also placed under IPL management.	
Check if design is per a RAGAGEP	ISA 18.2

Description	Guidance/Comments
Auditability	
Verify the following: • The management system records the alarm setpoint. • The alarm's logic solver and sensors are identified in the sites management systems as safety critical. • That requirement for periodic testing of the alarm is specified • There is documentation linking this IPL to the consequence in the PHA/LOPA it protect against	Reference SOPs and work orders for managing alarm and response procedure testing.
The data sheet for this Alarm's sensor(s) is complete and available.	Document where are datasheets located to reference? How are they updated and who is the custodian?
Verify the operator actions to be taken in response to the part of a the periodic retraining program	Reference to the procedure listing the frequency for required operator retraining on this procedure.
The SOP(s) which describes the required Operator response to this Alarm is periodically reviewed by the department.	1910.119(f)(1)(i)(D)
Verify the alarm's sensors are calibrated and inspected in accordance with site procedures	Reference the site: Procedures for testing. Procedure to archive test results. Procedures to require correcting nonconformance.
Testing meeting appropriate RAGAGEP.	CFR 1910.119 (f)(3)
Security	
Verify the alarm setpoints are secure.	Alarms are not resettable by the operator without authorization and management of change.
All Software changes which impact this Alarm are reviewed and authorized under "Management of Change" protocol.	

Table 2. SOP

Description	Guidance/Comments
Independence	
Verify the consequence severity was based on no active safeguards. The contribution of this SOP or any other active IPL did not reduce the severity assigned to the consequence.	
This SOP, the related Operator Action and the final element(s) that the operator may use are independent of the LOPA scenario initiating cause(s).	Example: The primary operator response can't be to put a control valve in manual and close it to isolate the flow if the initiating event was a BPCS of a loop operating that valve.
This SOP, the related instruments in use during the execution of the steps or final elements related to the Operator Action are fully independent and separate from all other IPLs credited in the same LOPA scenario(s).	Example: The operator can't actuate a valve closed that is being closed by a BPCS interlock.
The potential for Common Cause Failure between this SOP and other IPLs credited in the same LOPA scenario(s) has been ruled out.	
Performance or Dependability	
The documentation source(s) supporting the Probability of Failure on Demand (PFD) selected for this IPL is referenced.	

Description	Guidance/Comments
Specificity	
This SOP is designed to fully prevent the LOPA scenario consequence(s).	Document basis for action of this SOP preventing the consequence The procedure is of sufficient detail to support consistent execution by any operator.
There is ample time for the operator to execute the procedure as intended and described.	An independent operator reviewing in the field on rounds has enough response time for a slow developing event Independent person performing checklist verification prior to start up verifies actions of prior operator
Auditability	
The SOP(s) which describes the required Operator response to this Alarm is periodically reviewed by the department.	
The SOP is inventoried on the IPL register. The affected personnel receive required training and periodic retraining The SOP receive periodic review and authorization to verify it is current and effective.	CFR 1910.119 (f)(3)
Security	
Changes to the procedural steps in this SOP are reviewed and authorized under "Management of Change" protocol.	

Table 3. BPCS IPL

Description	Guidance/Comments
Independence	
Verify the consequence severity was based on no active safeguards. The contribution of this BPCS Interlock or any other active IPL did not reduce the severity assigned to the consequence.	
The failure of this BPCS Loop's logic solver, sensors and control elements are independent of the LOPA scenario initiating cause(s).	
This BPCS Loop's logic solver, sensors and control elements are fully independent and separate from all other IPLs credited in the same LOPA scenario(s).	If two BPCS loops are allowed for the LOPA (e.g. initiating cause and IPL or as 2 separate IPLs) they must not share the same I/O card.
Performance or Dependability	
Verify documented basis for Probability of Failure on Demand (PFD) used in LOPA calculation.	Reference source/basis for PFD.
Design is per a RAGAGEP.	
Specificity	
Document how BPCS Interlock is designed to fully prevent the LOPA scenario consequence(s).	
When de-energized, this BPCS Loop's control elements will transition to their "safe" states for the credited LOPA scenario(s).	

Description	Guidance/Comments
The service which this BPCS Loop's sensors and control elements perform do not have records of instrumentation problems from plugging, polymerization, fouling, or external environmental conditions.	
When this BPCS Interlock detects the hazardous process excursion, it can fully transition its control elements to their safe states within the Maximum Allowable Response Time.	
Define required repair time for the BPCS as needed to maintain dependability.	
Auditability	
This BPCS Loop's logic solver, sensors and control elements are properly documented as IPL's per the site's procedures.	
The data sheet for this BPCS Loop's sensors and control elements are complete and available.	
This BPCS Loop's sensor(s) and control element(s) calibration and proof tested procedures are complete and available.	
This BPCS Loop's calibration is in accordance with site procedures and records are well maintained, current, available and includes both "as found" and "as left" conditions.	
This BPCS Loop's proof test records are well maintained, current, available and includes both "as found" and "as left" conditions.	
Security	
All Software changes which impact this BPCS Interlock are reviewed and authorized under "Management of Change" protocol.	
This BPCS Loop's logic block, sensors and control elements are only bypassed with heightened administrative controls and the bypassed state is clearly indicated on the HMI and the bypass is authorized per site procedures.	

Table 4. Rupture Disk IPL

Description	Guidance/Comments
Independence	
Verify the consequence severity was based on no active safeguards. The contribution of this rupture disk or any other active IPL did not reduce the severity assigned to the consequence.	
The failure of this Rupture Disk is independent of the LOPA scenario initiating cause(s).	Condition that can plug the nozzle to the rupture disk (e.g. polymerization, freezing, coking).
This Rupture Disk is fully independent and separate from all other IPLs credited in the same LOPA scenario(s).	
Performance or Dependability	
The Probability of Failure on Demand (PFD) selected for this IPL is based on a published source	Document source for PFD
If this Rupture Disk is coupled with a Pressure Relief Valve, the interstitial space is monitored to detect leakage.	If the interstitial space is field monitored there is a procedure for routine monitoring and documentation of the interstitial space observations. If a pressure alarm monitors the interstitial space then this alarm receive periodic testing. Procedure initiates corrective action if the rupture disk is determined to be leaking.
If this Rupture Disk is in potentially freezing service, then measures have been implemented to counter these effects (e.g. heat tracing).	If nozzle and rupture disk are heat traced, then procedures or instruments are in place to detect its' failure
Procedure is in place to inspected/replace rupture disk if it is suspected to have been challenged.	
Design is per a RAGAGEP	ASME VIII, API 520, API 521
Specificity	
This Rupture Disk is only bypassed with heightened administrative controls. Valves to the rupture disk are under a management system	
This Rupture Disk is designed to fully prevent the LOPA scenario consequence(s).	Relief device design basis for the rupture disk includes the LOPA consequences it is protecting against.
The outlet of this Rupture Disk has been designed for safe discharge (e.g., sized for adequate flow, process sewer, above congested process equipment, to a flare header).	
Auditability	
This Rupture Disk is periodically removed from service, tested and inspected for signs of corrosion.	
If this Rupture Disk is installed in polymerizing or fouling service, or there are extreme external conditions, it is removed for inspection during a shutdown.	
The data sheet for this Rupture Disk is complete and available.	
This Rupture Disk proof test procedure is complete, available	

Description	Guidance/Comments
and includes both "as found" and "as left" conditions. It also includes inspecting and verifying the information on the Rupture Disk's tongue.	
Testing meets appropriate RAGAGEP.	
This Rupture Disk's inspection records are well maintained, current and available.	
Security	
If there are Maintenance valves in line with the rupture disk, they are on a periodic inspection to verify they are in the open position	
Change to the rupture disk requires MOC.	
If the process changes made can affect relief requirements, then the MOC updates relief device design basis, calculations and initiated changes as needed	

Table 5. PSV IPL

Description	Guidance/Comments
Independence	
Verify the consequence severity was based on no active safeguards. The contribution of this PSV or any other active IPL did not reduce the severity assigned to the consequence.	
The failure of this Pressure Relief Valve is independent of the LOPA scenario initiating cause(s).	
Performance or Dependability	
The Probability of Failure on Demand (PFD) selected for this IPL is based on a published value.	PFD is based on published source and nature of service (e.g. clean or fouling, risk of corrosion etc.).
All manual valves which can block this Pressure Relief Valve are managed (locked or car-sealed) open.	
If this Pressure Relief Valve is coupled with a Rupture Disk, a pressure gauge is installed in the interstitial space to detect leakage.	If the interstitial space is field monitored there is a procedure for routine monitoring and documentation of the interstitial space observations. If a pressure alarm monitors the interstitial space then this alarm receive periodic testing. Procedure initiates corrective action if the rupture disk is determined to be leaking.
If this Pressure Relief Valve is installed in polymerizing service or fouling service, or there are extreme external conditions, it is periodically inspected during a shutdown to verify that nozzles are open.	
Design is per a RAGAGEP.	
Specificity	
This Pressure Relief Valve is only bypassed with heightened administrative controls and the bypass is authorized per site procedures.	
This Pressure Relief Valve is designed to fully prevent the LOPA scenario consequence(s).	Design basis documents that the PSV is adequately sized for the LOPA consequences it is to protecting against.

The outlet of this Pressure Relief Valve has been designed for safe discharge.	
Auditability	
This Pressure Relief Valve is properly documented as an IPL per the site's procedures.	
The data sheet for this Pressure Relief Valve is complete and available.	
There is instrumentation which will indicate when this Pressure Relief Valve has been challenged. This Pressure Relief Valve is to be inspected if it is suspected to have been challenged.	
This Pressure Relief Valve's proof test procedure is complete and available.	
This Pressure Relief Valve is periodically removed from service, tested and inspected for signs of corrosion.	
Testing meeting appropriate RAGAGEP.	API 576
This Pressure Relief Valve's inspection records are well maintained, current, available and includes both "as found" and "as left" conditions.	
Security	
Change to the PSV requires an MOC	
If a process changes made can affect relief requirements, then the MOC updates relief device design basis, calculations and initiates required design changes.	
If there are maintenance valves in line with the rupture disk they are on a periodic inspection to verify they are in the open position	

12.2 Drill Down Audit of a Statistical Sample of IPLs

The "drill down" audit is a useful tool to validate the IPL management system effectiveness. The drill down audit involves periodic selection of a random sample of the different IPLs at the site. The audit trail selects a starting location to obtain the sample the IPLs, i.e. the originating PHA/LOPA to the IPLs documented or the site IPL register. The following outlines examples of drilldown audits for different IPLs.

Auditing Alarm IPLs

In a drill down audit, the following features are verified for Alarm IPLs:

- The setpoints are properly documented and set on the BPCS.
- These alarms are under management to prevent unauthorized changes and no history of unauthorized changes have been made.
- There is a PM in the maintenance/MI system to periodically test the alarm.
- The documented instructions for operator alarm response are reviewed and confirmed as current.
- These documented procedures have been periodically reviewed per site procedure.
- Operators are interviewed to verify their knowledge of the actions to be taken.

- The training program is reviewed to verify the training program includes both initial training and periodic retraining of the affected operators.
- Operator training records are reviewed to show the training of all affected operators is up to date.
- Maintenance records are reviewed confirming the periodic alarm tests have been performed and any corrective actions implemented for any nonconformance.
- Any changes to the alarm IPLs or associated initiating cause/consequences were correctly managed under MOC.

<u>SOP IPLs</u>

The sample of SOPs identified as IPLs are obtained for drill down audit to verify:

- These SOPs are periodically reviewed per site requirements.
- The training program includes these SOPs for both initial training and periodic retraining of the affected operators.
- Operator training records are reviewed to show that the training of all affected operators is up to date for initial and refresher training.
- Operators are interviewed to verify their knowledge of the actions to be taken.
- If gaps are found in the SOP or operator training, corrective actions are implemented to address nonconformance,
- Any changes to the SOP IPLs or the associated initiating cause/consequences were correctly managed under MOC.

<u>BPCS interlock</u>

The sample of BPCS interlocks identified as IPLs are obtained for drill down audit to verify the following:

- Inputs and outputs are properly documented and controlled to prevent unauthorized changes.
- The BPCS interlocks have a PM work order in the in the maintenance/MI system to perform a functional test.
- This test includes the testing of the input instrument to verify it activates the interlock at the correct setpoint.
- The test include verifying the outputs of the interlock correctly functions to prevent the event, i.e. the valve closes and is confirmed to isolate the flow, pump is shut down and flow is confirmed to be stopped.
- Records are reviewed confirming the tests have been performed and any corrective actions implemented for any nonconformance.
- Any changes to the IPLs or the associated initiating cause/consequences were correctly managed under MOC

PSV/Rupture disks

The sample of PSVs identified as IPLs are obtained for drill down audit to verify the following:

- The PSVs/Rupture disks have a have a PM work order in the in the maintenance/MI system to perform a functional test.
- There are records showing these tests have been performed per the required test interval.
- Records are reviewed confirming the tests have been performed and any corrective actions implemented for any nonconformance.
- Any changes to the IPLs or the associated initiating cause/consequences were correctly managed under MOC.

13 Access Security/MOC

The site must maintain the security of the IPL throughout its lifecycle. Systems must be in place that prevent unauthorized changes to the IPLs. For example:

- Alarms or interlocks on a BPCS are controlled to prevent unauthorized changes or disabling (e.g. password, key locks and physical locks to access equipment etc.) are in place to prevent changes.
- Systems are in place to manage change to SOP or alarm response procedures. SOPs are under MOC control and periodic auditing is performed to verify that procedures are understood and being followed. Training includes adequate understanding of role for this procedure in preventing the unwanted consequence and is trained as a mandatory action.
- Systems are in place to authorize the temporary bypass of IPLs and follow-up on their return to service.

When the site makes a change that affects a process, the site must assess if the changes affect the severity of consequence identified from the PHA/LOPA or any IPLs used to mitigate these consequences. The site should give guidance on what triggers a review of LOPA/PHA consequence from a MOC.

For MOCs the site should consider:

- How will the site flag that BPCS LOOP components, procedures, alarms, dikes, PSVs being affected are associated with an IPL?
- How the MOC/Maintenance systems will include these flags to assist the MOC process?
- How the MOC process initiates review of how the proposed changes affects the PHA/LOPA initiating causes or consequence severity that lead to IPL implementation?

14 Summary

- Non-SIF IPLs must be managed through the lifecycle of the process the same as SIF IPLs.
- Efficient flow through the process for IPL identification, validation, selection, implementation and integrity management is aided by procedures to define information capture, document work flow for IPL Identification, validation, selection and implementation.
- The IPL must be validated as part of the initial implementation and audited through its lifecycle to show compliance with the IPL requirements of independence, efficiency/specificity, dependability, auditing/testing and security.
- Tools are needed to identify and review the effect of MOC on the process IPLs and their originating LOPA.

Auditing IPLs – Using Safety Critical Functions Manuals

Alex J. Sellers
Safety Consultant, Bluefield Process Safety, LLC
3101 Cherokee Street
Saint Louis, Missouri 63118, USA
Alex.Sellers.BluefieldSafety@gmail.com

Michael S. Schmidt, P.E., CFSE
Principal, Bluefield Process Safety, LLC
3101 Cherokee Street
Saint Louis, Missouri 63118, USA
Mike.Schmidt.BluefieldSafety@gmail.com

Prepared for Presentation at
American Institute of Chemical Engineers
2014 Spring Meeting
10th Global Congress on Process Safety
New Orleans, LA
March 30 – April 2, 2014

UNPUBLISHED

Auditing IPLs – Using Safety Critical Functions Manuals

Alex J. Sellers
Safety Consultant, Bluefield Process Safety LLC.
3101 Cherokee Street
Saint Louis, Missouri 63118, USA

Michael S. Schmidt, P.E., CFSE
Principal, Bluefield Process Safety LLC

bluefieldsafety@gmail.com

Keywords: LOPA; safety critical; administrative control; IPL

Abstract

The three criteria for an independent layer of protection (IPL) are effectiveness, independence, and auditability. The effectiveness and independence of an IPL typically are evaluated as part of a Layer of Protection Analysis (LOPA), and normally are confirmed by the time the LOPA is finalized. Auditability, on the other hand, is typically not addressed during the LOPA, and frequently is not addressed until someone is charged with the task of actually auditing the IPL, assuming that even occurs. For a Safety Instrumented Function (SIF) with a Safety Integrity Level to be installed in a Safety Instrumented System (SIS), the requirements for auditing are listed in the standards such as IEC 61511 and ANSI/ISA S84 and the relevant information is contained in a Safety Requirements Specification (SRS).

However, an SRS for an SIS does not address IPLs that are not SIFs: administrative functions, non-instrumented functions, and non-rated control functions, even though they are also critical to assuring sufficient risk reduction for specific hazards. The auditability, and hence the auditing, of IPLs not installed in an SIS too often becomes an afterthought.

A Safety Critical Functions Manual (SCFM) can pick up where the SRS leaves off. It defines requirements for testing, maintenance, and training for IPLs not included in the SIS. The very act of creating an SCFM forces the issue of auditability to be addressed, and defines the items that need to be audited. While an SCFM could manifest itself in different ways, it generally can be expected to include datasheets which address the operation, testing, maintenance, and training requirements for each safety critical function. Although it would be analogous to an SRS, an SCFM typically would be simpler.

The paper includes a discussion of an approach to SCFMs that has been successful at a number of different organizations for documenting and maintaining those safety critical IPLs that are not already covered by one of the SIS standards.

1. Introduction

The Center for Chemical Process Safety (CCPS) book on Layer of Protection Analysis (LOPA) describes an Independent Layer of Protection (IPL) as "a device, system, or action that is capable of preventing a scenario from proceeding to its undesired consequence independent of the initiating event or the action of any other layer of protection associated with the scenario." [1] The book goes on to say that "The effectiveness and independence of an IPL must be auditable". Between these two statements, three criteria emerge as requirements for using functions (including instrumented and non-instrumented devices and procedural and administrative controls) as IPLs: effectiveness, independence, and auditability.

Of these three criteria, two (effectiveness and independence) normally are evaluated and confirmed before the LOPA is finalized. Auditability, meaning a determination of whether or not the function is able to be audited, can be determined during the LOPA as well, but the implied extension of the auditability requirement—actually auditing the functions to ensure continued effectiveness and independence—must be carried out throughout the life of the process or until the function is no longer required as an IPL. At the very least, this auditing is necessary to assure that the IPL has a Probability of Failure on Demand (PFD) no higher than what was assumed during the LOPA. The CCPS LOPA guidebook provides a more thorough description of what is expected:

> "An organization must establish a system to periodically assess (audit) the elements (components and human interventions) identified as IPLs to ensure that the IPLs remain in service at the anticipated PFD. In some cases this will require functional testing of the devices (SIFs-interlocks, relief systems, etc.) or the human interventions. In other cases it could include inspections, such as for passive protections like dikes, drainage systems, fire walls, etc. For some IPLs, replacement or preventive maintenance may be required at a specified frequency.
>
> "In all cases, the organization must ensure that the testing, inspection, preventive maintenance, procedure drills, etc., are accomplished at the appropriate frequency and with the appropriate amount of rigor. These assurance steps are necessary to achieve the PFD assigned for the IPL. The results of these assurance steps (proof tests) must be recorded, including any corrective actions taken. These records must be available to the LOPA analyst(s)." [1]

This excerpt makes it very clear that some organized system should be in place to ensure that the necessary audits, maintenance, etc. occur and to facilitate those processes.

For a Safety Instrumented Function (SIF) with a Safety Integrity Level (SIL) that is or will be installed in a Safety Instrumented System (SIS), the requirements for auditing are listed in the standards such as IEC 61511 and ANSI/ISA S84 and the relevant information is contained in a Safety Requirements Specification (SRS). However, an

SRS for an SIS does not address IPLs that are not SIFs—administrative functions, non-instrumented functions, and control functions without a SIL rating—even though they are also critical to assuring sufficient risk reduction for specific hazards. The auditability, and hence the auditing, of non-SIS IPLs too often receives little or no thought during the five year period between Process Hazard Analysis (PHA) revalidations.

To aid in the process of auditing non-SIS IPLs, this paper proposes a Safety Critical Functions Manual (SCFM). There are a number of approaches that can be and have been used to decide which functions are safety critical. These include engineering judgment and LOPA, which consider hazards on a case-by-case basis, as well as organizational or regulatory approaches. For the purpose of this paper, the SCFM will be discussed in terms of validating assumptions made in a LOPA. However, the format of the SCFM can be used to address any function that is deemed "safety critical" by the organization using it.

The decision to include a safeguard in an SCFM is not trivial. Functions that are included are subject to a higher degree of testing, maintenance, training, and documentation than other functions, and so impose a burden on the organization. Likewise, with the establishment of an SCFM, functions that are not included are not as likely to be treated as safety critical.

An important premise of an SCFM is that all safety critical functions and only safety critical functions comprise the manual. When a function is safety critical, an operator or supervisor, engineer or technician should be able to find it in the SCFM. On the other hand, when a function is not safety critical, it should not be included in the SCFM. Key to this, then, is deciding which functions are safety critical.

While using a process involving a thorough PHA followed by a LOPA review for scenarios of high consequence can be an effective means of identifying safety critical functions, it should be understood that each individual organization has different policies and procedures for classifying scenarios and safeguards as "safety critical". Regardless of the method chosen for identifying safety critical functions, the SCFM should be a valuable tool for compiling the information necessary to ensure the functions are operated and maintained properly.

2. Safety Critical Functions Manual Structure

An SCFM can be divided into two distinct parts: a section that outlines the scope of the manual, the purpose of the manual, and general instructions for the manual's continued use and upkeep, and a section consisting of datasheets that contain the required information for each individual function.

2.1 Sections

2.1.1 Report

The first section of the SCFM is the SCFM report. The SCFM report should tell the user why the SCFM exists and what it is intended to do. This includes an explanation of the methodology used to decide which safeguards to include in the SCFM. This information can be captured in a purpose statement such as:

> "The purpose of this manual is to aid the technical, operating, and maintenance staff responsible for the operation of (insert unit or plant name) in recognizing those functions that have been identified as critical to the safe operation of the unit/plant. These components have all been identified in a (Layer of Protection Analysis (LOPA) or other identification method) as integral to reducing the risk of (insert unit or plant name) to a level that can be tolerated. Additionally, it is the purpose of this manual to describe the operation of these safety critical functions and the measures necessary to maintain the integrity of these safety critical functions."

The SCFM report should also state who is responsible for the upkeep of the manual and provide instructions for when and how it should be updated.

2.1.2 Datasheets

The second section of the SCFM contains the datasheets for individual functions. The intention of the safety critical function datasheets is to collect all relevant information pertaining to a safety critical function in one place. This is a key advantage, as any time employees have a question or concern about a particular safety critical function, they know exactly where to go to find answers.

Ideally, a datasheet would capture this information on a single page. It is helpful to separate the datasheets by function type (SIFs, instrumented functions without SIL ratings, non-instrumented functions, and procedures and administrative controls) and to include a table of contents with a numbering system to organize the datasheets and make the manual accessible and easy to use.

2.2 Practical Implementation

The manual can be prepared as a physical document or as an electronic database or both. Organizations have used Microsoft Word templates to record both the report section and the section containing datasheets, although a database designed for this purpose of compiling and generating datasheets can provide additional advantages for data usage. Depending on the facility, it may make sense to prepare a site-wide manual or to prepare separate manuals for each unit.

The SCFM should be a living document that is updated and reviewed as necessary. The frequency at which updating is required depends on the system that an organization uses to classify safety critical devices. Typical occasions to update include the exercise of a Management of Change procedure and during the five-year PHA revalidation required for compliance with the Process Safety Management (PSM) standard of the Occupational Safety and Health Administration (OSHA). [2]

3. Information Required for Individual Functions

To facilitate the audit process, it is essential to compile relevant information to provide the auditor with a general background of the role the function plays in preventing the hazardous outcome, as well as to provide specific requirements related to upkeep, operation, and testing to ensure the device remains effective. Some of this information will be similar for all types of functions used as IPLs, while other types of information will be relevant only for specific types of functions. General information that is necessary for all types of functions is discussed here. Information specific to different types of functions is discussed later in the paper, using four specific function types: SIFs, instrumented functions without SIL ratings, non-instrumented functions, and procedural and administrative functions.

3.1 Information Required for All Functions

3.1.1 Hazard Description

The hazard description should take a concise, unambiguous form and follow a template that uses the same basic structure for all scenarios. An example of such a template is:

> "Deviation" in "Equipment Name", "Equip. No.", resulting in "event", leading to "safety impact", "community impact", and "environmental impact".

Putting a commonly found scenario into this template gives the following result:

> "High pressure in Example Reactor One, R-001, resulting in catastrophic vessel rupture, leading to an employee fatality, a community hospitalization, and a reportable environmental release".

Each impact description should be based on the organization's Risk Tolerance Criteria (RTC). Applying this type of template and the RTC consistently to describe hazards increases the utility of the SCFM across the organization.

Documenting the hazard description for each function in this fashion will be helpful for end users and auditors alike. The auditor clearly can understand the hazard against which the function is meant to protect. This gives the auditor insight into why the function must be audited, as well as an indication of which other devices might be protecting against that hazard, in order to verify independence. For end users, this additional background information also helps transform the mindset of those that interact with safety critical functions from "inspect and test this device or train on this procedure because it is policy" to "inspect and test this device or train on this procedure because it helps prevent an explosion that could result in a fatality". The difference in these statements is not trivial; it is the difference between compliance on the one hand and engagement and responsibility on the other. This in turn promotes and increases the level of care shown during the operation and maintenance of safety critical functions and training on those safety critical functions.

3.1.2 How the Function Prevents or Limits the Consequence(s)

Not only should it be clear what hazard a function protects against, it should also be clear how the function prevents or limits the consequences. This description should include all the actions that must be accomplished or conditions that must be met in order to keep the consequence from occurring.

It also should be noted when a function only addresses some of the consequences associated with a hazard. For example, a rupture disk will prevent a vessel from rupturing and throwing shrapnel, but may not prevent an environmental release from occurring. It is important to include this sort of information to reduce the potential for confusion in the future about the function's intent and capabilities.

3.1.3 References

Two types of references are important when it comes to upkeep and validation of safety functions. The first type documents the process used to identify the function as safety critical. This documentation certainly depends on the method used by the organization for identifying safety critical functions. It also includes references to a PHA or LOPA scenario, and in the case of SIFs, an SRS. This background information will allow personnel that did not participate in the entire the cycle of determining that a function is safety critical to retrace those steps, giving them a more complete understanding of why the safety critical classification was used.

The second type of reference is to the method or methods used to schedule and direct the work of operating and maintaining the function, i.e. work orders. This facilitates the audit process. The method used to schedule and issue work orders often involves Enterprise Resource Planning (ERP) software such as SAP. This reference is often vital when auditing the functions in the SCFM and determining whether or not the requirements set forth in the SCFM actually are being met. The corresponding instructions in the ERP software or other scheduling method should refer back to the SCFM as well so that the root of the task is traceable and the task is denoted as safety critical to alert those that will carry out the task of its importance.

3.2 Information Required for Specific Function Types

3.2.1 Safety Instrumented Functions

The requirements associated with SIFs are outlined clearly in standards such as IEC 61508/61511 and ANSI/ISA S84 and have been discussed at great length in numerous papers. [3, 4] Therefore, it is not the purpose of this paper to revisit the information necessary for an SRS. However, the SCFM is a convenient place to record and implement the requirements that are contained within the SRS.

An example of a function datasheet template that could be used to assist plant staff in the proper operation, maintenance, and testing for a SIF is shown below in Figure 1. Text that is highlighted in green is meant to change based on specific function information. Text highlighted in yellow denotes places where different options are available and should be removed prior to final implementation.

Figure 1

XXX-SIF-01

SIF No. **SIF Name** P&ID Dwg. No.: **12345**

Hazard: "Deviation" in "Equipment Name", "Equip. No.", resulting in "event", leading to "safety impact", "community impact", and "environmental impact".

Operation: On "condition" in "Equipment Name", "Equip. No.", "action" "specific final control element".
This protects against "deviation" *or* "event" *or* "impact" by "operating principle".
Always enabled. *or* Enabled during "Step 1", "Unload" ...
MooN voting by sensor(s). X sec *or* no delay. Single block valve on each inlet gives MooN architecture on final elements.
On sensor fault, SIS treats fault as non-voting input, degrades architecture to MooN on the remaining good sensors, and notifies operator of the fault. *or* a vote to trip, degrades architecture to MooN on the remaining good sensors, and notifies operator of the fault *or* a vote to trip, trips the SIF, and notifies operator that the trip is based on a fault. On final control element fault, SIS notifies operator of fault.
Bypasses of this function are not allowed *or* by sensor *or* by input to a voting bloc in the SIS *or* of the entire SIF. They may be cleared manually, or will clear automatically after X hours.

Causes	Set Point Units	Effects	Safe Action

Response: The condition that trips this SIF is normal, and no response is required. The SIF will reset automatically after this normal condition clears. *or* The condition that trips this SIF is unsafe, and an operator should investigate the cause of the trip. The SIF will *or* may require a manual reset, but only after the trip condition clears and the operator is sure that it is safe to proceed. *or* The control system will automatically reset when the trip condition clears and the output delay has been met. A trip of this SIF requires an incident investigation, and the trip should be noted in the history of this SIF.
A fault should be repaired as soon as possible. Calculations assume a mean time to repair of 72 hours. The fault should be noted in the history of this SIF.

Maintenance: The design of this SIF is based on proof test of the sensors once every year, a proof test of the logic solver once every year, and a proof test of the final control elements once every year.
The sensor proof test should show that the sensor detects the trip condition accurately, and that the set point is correct.
The logic solver proof test should show that delays are correct, voting logic is correct, that a trip condition cannot be defeated by other systems, that the response to fault is correct, and that the response to bypass is correct.
The final control element proof test should show that a trip signal results in the correct action, that the final control element performs as designed, and that the absence of a trip signal results in the correct action.
Every proof test should be recorded, showing "as found", "as left", date of proof test, and by whom the test was performed.

References: HazOp Title, Section, Date
LOPA Title, LOPA Worksheet No., Date
Other references (SRS)

In addition to the three elements common for all functions (hazard description, how the function protects against the hazard, and references), the SIF datasheet also contains other information required specifically for SIFs. In order to describe the operation of the SIF, information including voting architecture for sensors and the architecture for Final Control Elements (FCEs), the delay time incorporated into the function, how the logic solver is to treat faults in the field devices, and how bypasses of the SIF are to be handled is recorded in the section on operation. The datasheet then uses a table to describe causes and effects of the function, which is meant to include sensor and FCE tag numbers, the setpoint of sensors, and the safe action of the FCEs.

The section on safe action should describe what the FCE should do when the function is tripped. A block valve, for example, would have a safe action of either "open" or "close". Additional rows should be added to the table to accommodate all devices used in the function.

The section on response is meant to describe what workers in the unit or area are expected to do in the event the SIF is tripped. In the case of SIFs that act as permissives, the SIF "trips" under normal operating conditions, so no response is required when it trips and the function should reset automatically. In the case of SIFs that serve as interlocks, instructions on when and how the process is to be returned to service should be outlined in the datasheet. This section should also state whether demands on the function should be recorded or whether an incident report describing the event should be completed.

The section on maintenance includes general instructions on the intervals that sensors, logic solver, and FCEs should be proof tested. These intervals will have been determined during the SIF design phase and should be provided in the SRS. Because the function datasheet is intended to be a single page document, it is unreasonable to expect the datasheet to include specific proof testing instructions. However, the specific proof testing instructions should exist, and be incorporated by reference, meaning that the document containing the proof testing instructions should be listed in the reference section.

3.2.2 Instrumented Functions

Of the other three types of safety critical functions, instrumented functions that are not SIL-rated bear the greatest resemblance to SIFs. To create a datasheet for these functions, one need only start with the SIF datasheet and remove sections that are unnecessary. In many cases, the information about how to handle faults in sensors and bypasses are not necessary for non-rated instrumented functions. On the other hand, the table that records field device tag numbers, setpoints, and safe actions will be identical to the table found in the SIF datasheet.

The section on response also should contain much of the same information listed for SIFs. An important difference between SIFs and other safety critical instrumented functions is that other safety critical instrumented functions may include alarms that require operator action to be effective. The datasheet for alarms necessarily must include instructions on the actions an operator should take when the alarm sounds; otherwise, the alarm is just noise. These instructions should be specific enough that any operator in the unit or area will understand what is expected of them, but general enough that an operator's choices to respond are not restricted unnecessarily. If the scenario is high temperature in a reactor, for example, there may be several different responses, depending on the circumstance: turn off steam, stop charging raw material that promotes an exothermic reaction, turn on cooling water, turn on deluge water, etc.). When the specific course of action clearly is defined, then it should be listed. When the best course of action is not entirely clear or is dependent on conditions, then instructions should be appropriately general—as general as necessary, but no more general than necessary.

The datasheet section on maintenance should also be very similar to that in the SIF datasheet. Below is an example of a datasheet for instrumented functions that are not SIL-rated.

Figure 2

XXX-CF-01

Control Function Name

Sequence No. P&ID Dwg. No.: **12345**

Hazard: **"Deviation" in "Equipment Name", "Equip. No.", resulting in "event", leading to "safety impact", "community impact", and "environmental impact".**

Operation: **On "condition" in "Equipment Name", "Equip. No.", "action" "specific final control element".**
This protects against "deviation" *or* **"event"** *or* **"impact" by "operating principle".**
Always enabled, *or* **Enabled during "Step 1", "Unloading" ...**
MooN voting by sensor(s). X sec *or* **no delay. Single block valve on each inlet gives MooN architecture on final elements.**

Causes	Set Point	Units	Effects	Safe Action

Response: **This control loop** *or* **control sequence** *or* **interlock** *or* **alarm trip condition is normal, and no response is required.** *or* **The condition that trips this interlock** *or* **alarm is unsafe, and an operator should investigate the cause of the trip. The control system will** *or* **may require a manual reset, but only after the trip condition clears and the operator is sure that it is safe to proceed.** *or* **The control system will automatically reset when the trip condition clears and the output delay has been met. A trip of this interlock** *or* **alarm requires an incident investigation.**

Maintenance: **Any detected fault should be repaired as soon as possible.**
Because this control loop *or* **control sequence** *or* **interlock** *or* **alarm is safety critical, its performance should be audited annually. This includes an audit of the sensor performance, an audit of the logic in the control system, and an audit of the final control elements.**
The sensor audit should show that the sensor detects the trip condition accurately, and that the set point is correct.
The logic audit should show that the code and the description of the code match.
The final control element proof test should show that a trip signal results in the correct action, that the final control element performs as designed, and that the absence of a trip signal results in the correct action.
Every audit should be recorded, showing "as found", "as left", date of audit, and by whom the audit was performed.

References: **PHA Title, Section, Date**
LOPA Title, LOPA Worksheet No., Date
Other references

Other Notes: **N/A**

A question that frequently arises concerning instrumented functions that are not SIL-rated is about proof test frequency. Inspection and testing should occur with sufficient frequency and rigor to justify continued use as an IPL. In the absence of manufacturer's recommendations, annual testing is a commonly suggested frequency for instruments and other electrical, electronic, and mechanical devices, although a thorough statistical analysis may justify extending the test frequency to as little as once every five years. [5] Testing and inspection frequency depends on a number of factors, including accuracy requirements, inherent stability of the specific instrument or device, and environmental factors affecting that stability. [6]

3.2.3 Non-instrumented Functions

"Non-instrumented functions" is a very broad classification that includes many different types of functions. This group includes anything that is not an instrumented, administrative, or procedural protection. This includes, but is not limited to, relief valves, rupture disks, conservation vents, fire suppression, dikes, check valves, flame arrestors, mechanical overspeed trips, metered nitrogen sweeps, and excess flow valves. These devices have some requirements similar to those of instrumented functions, although there are certainly differences. The SIS standards are not applicable to non-instrumented functions, but there are other standards that apply to specific functions. For instance, NFPA 25 applies to aqueous fire suppression systems. [7] These standards should be applied for maintenance and testing requirements. Where such standards are not available, an inspection and testing period of one year is often a reasonable baseline to be adjusted for specific device types, based on plant service and maintenance history, or the on the formal application of technique such as Failure Modes and Effects Analysis.

A datasheet similar to those for SIFs and instrumented functions that are not SIL-rated can be modified for use with most non-instrumented functions. An example of a non-instrumented function datasheet is shown in Figure 3. This datasheet, like the others, begins by explaining the hazard and how the function protects against it. It may also contain a causes-and-effects table, which is a useful place to record set points and safe actions. Some non-instrumented protections, such as dikes, will require no information in this regard. Other non-instrumented protections, such as conservation vents, will require much of the same information as instrumented functions.

A section on response will be necessary and should be used to record the actions that staff will need to take following an event that causes a non-instrumented function to perform. As is the case with instrumented functions, there are some non-instrumented functions that perform their actions routinely and therefore require no staff response when they act. A conservation vent is an example of this type of function. Others may require specific instructions on the staff response required following action by the non-instrumented functions, such as the bursting of a rupture disk, which requires instruction on action to take to prevent further releases or other hazards from developing, as well as actions to take until the burst rupture disk can be replaced.

The section on maintenance is structured the same as the section on maintenance for other IPLs, but may need to be adjusted to reflect each individual function type. Some non-instrumented functions, such as dikes, will only require inspection. Other non-instrumented functions, such as metered nitrogen sweeps, will require full functional testing. There are many industry standards and regulations that use a one year inspection or testing interval [8, 9, 10, 11], so that is a good place to start in the absence of manufacturer or industry standards. However, it is certainly possible for longer or shorter intervals to be justified on the basis of inspection history. [12]

Figure 3

Non-Instrumented Function Name

Tag No. — P&ID Dwg. No.: **12345**

Hazard: **"Deviation" in "Equipment Name", "Equip. No.", resulting in "event", leading to "safety impact", "community impact", and "environmental impact".**

Operation: **On "condition" in "Equipment Name", "Equip. No.", "action" "specific final control element".**
This protects against "deviation" *or* **"event"** *or* **"impact" by "operating principle".**
Always enabled, *or* **Enabled during "Step 1", ...**
Revise or delete the following table to the extent necessary for it to be meaningful, given the type of non-instrumented function.

Causes	Set Point	Units	Effects	Safe Action

Response: **This "non-instrumented function" performs its function routinely, not in response to a hazardous condition, so no response from an operator is required when it acts.** *or* **The condition that puts this "non-instrumented function" into use is unsafe, and an operator should investigate the cause of "the unsafe condition". The "non-instrumented function" should only be restored to its ready state after "the unsafe condition" clears and the operator is sure that it is safe to proceed. A demand on this "non-instrumented function" requires an incident investigation.**

Maintenance: **Any detected fault should be repaired as soon as possible.**
Because this non-instrumented function is safety critical, its performance should be audited annually. This includes an audit of the sensor performance, an audit of the logic in the control system, and an audit of the final control elements. *Revise or delete the previous sentence and the following three sentences to the extent necessary for them to be meaningful, given the type of non-instrumented function.*
The sensor audit should show that the sensor detects the trip condition accurately, and that the set point is correct.
The logic audit should show that the code and the description of the code match.
The final control element proof test should show that a trip signal results in the correct action, that the final control element performs as designed, and that the absence of a trip signal results in the correct action.
Every audit should be recorded, showing "as found", "as left", date of audit, and by whom the audit was performed.

References: **PHA Title, Section, Date**
LOPA Title, LOPA Worksheet No., Date
Other references

Other Notes: **N/A**

3.2.4 Procedures and Administrative Controls

Procedures and administrative controls that are used as IPLs can also vary across a wide spectrum, from lock out/tag out procedures to operator environmental rounds, from valve car-seal programs to standard operating procedures. All of these practices require initial training and periodic refresher training, and must be audited to ensure they remain effective and to validate their assumed PFD.

The first step in this process is to ensure that any procedure or administrative control for which credit has been taken actually is documented in some fashion. Taking credit for unwritten rules and practices automatically violates the auditability requirement for IPLs.

It is important to note the formal name and identifying number of any document that describes a procedure or administrative control that is credited as an IPL. It is also important to delineate the specific steps of that practice that apply to preventing the critical hazardous event from occurring. Not only is it important to identify the document on a safety critical functions datasheet, it is important to ensure that the safety critical steps of the practice are highlighted in some way in the documents to denote that they are used as IPLs and therefore safety critical.

The section in the datasheet on operation shown in Figure 4 includes a line to describe when the procedure is performed. This section should also include a section on response. For many procedures and administrative controls, this section will not be necessary, since the practice will be performed routinely, not in response to a hazardous condition. Even practices that are performed routinely, however, will benefit from a description of the practice's impact on safety. For instance, operator environmental rounds are used to identify hazardous situations, so it would be beneficial to describe specifically what an operator is looking for and what actions that operator should take when hazardous conditions are discovered.

The section on maintenance and testing becomes a place to record training and audit requirements for the procedure or administrative control. Depending on the nature of the procedure or administrative control, different types of training may be optimum. This may be self-study, classroom training, informal reviews, on-the-job training, or other methods. Regardless of how formal or informal the training, the type of training to employ should be chosen for each procedure and administrative control. Another important aspect of training is ensuring that trainees walk away with sufficient understanding of the training material. Depending on the type of training given and the complexity of the material, understanding can be demonstrated in a number of ways: declaration by trainee that material was read and understood, discussion with instructor, written exams, observation of successful demonstration by qualified personnel, or other types. The method chosen to demonstrate understanding for each procedure and administrative control should be recorded on each datasheet.

The datasheet should also state the frequency at which refresher training is required. OSHA requires three year refresher training on operating procedures in process areas that are covered by the PSM standard. Because safety critical procedures demand a higher standard of care than other procedures, a more frequent schedule is recommended. Lock out/Tag out procedures, which are often identified as IPLs, have an annual training requirement per the OSHA Lockout/Tagout Standard [13], so annual training for IPLs is a reasonable starting point.

Finally, audits of procedures should validate that they are conducted as described, in order to validate the PFD assumed in the LOPA. This may be in association with or independent of training. Audits should also confirm that all affected personnel are being trained as scheduled.

Figure 4

XXX-PRO-01

Procedure Name	P&ID Dwg. No.: **12345**
Hazard:	**"Deviation" in "Equipment Name", "Equip. No.", resulting in "event", leading to "safety impact", "community impact", and "environmental impact".**
Formal Procedure:	*Insert the name and number of the formal procedure.*
Operation:	*Summarize the key steps of the procedure, being sure to list the actions, equipment name, equipment number, and person or position responsible for those steps.* **This protects against "deviation"** *or* **"event"** *or* **"impact" by "operating principle".** **This procedure is performed** *state basis for performing this procedure, e.g. every shift, every Saturday, every batch, etc.*
Response:	**This procedure is performed routinely, not in response to a hazardous condition.** *If there are hazardous conditions that the procedure can be expected to reveal, describe what should be done in response to those findings.*
Training:	**Because this procedure is safety critical, it should be reviewed annually for accuracy and revised accordingly. The procedure should get a new revision date with each annual review, even if no changes are made. At any time, the most recent revision date should be less than one year previous.** **Personnel required to perform this procedure should receive training on the procedure prior to performing it for the first time and then annually, as well as anytime changes are made to the procedure. Training should be in the form of classroom training** *or* **informal safety meeting review** *or* **on the job training** *or other*, **and demonstration of understanding should be in the form of written tests** *or* **verbal checks** *or* **observation by qualified personnel. Training records should be kept in accordance with Afton Chemical's training policies.**
References:	**PHA Title, Section, Date** **LOPA Title, LOPA Worksheet No., Date** **Other references**
Other Notes:	**N/A**

4. Other criteria for inclusion in a Safety Critical Functions Manual

When considering whether a function is safety critical, the authors have found it valuable to consider the following criteria:

- "Safety critical" is limited to scenarios involving major hazards or catastrophic events: those that result in fires, explosions, or toxic releases and lead to catastrophic impacts—fatalities in the workplace, serious injuries in the community, or environmental disasters.
- "Safety critical" applies to safeguards that are relied upon to reduce the risk of a major hazard to a tolerable level.
- "Safety critical" also applies to components, the failure of which can trigger a catastrophic event.

However, there are other criteria that organizations may choose to use when deciding to include functions in their SCFM, including organizational policy, regulatory requirements, and causes.

4.1 Organizational policy

There are many corporate or plant procedures that require the implementation of particular IPLs or other risk reduction measure as a matter of organizational policy. Even though the implementation of that IPL or other risk reduction measure may not rise to the level of safety critical in a specific instance, to violate these policies can put the organization at legal risk, either contractual or regulatory. Because of this, IPLs or other risk reduction measures that are required as a matter of policy, rather than as a matter of detailed analysis, are often included in SCFM.

4.2 Regulatory requirements

There is generally a correlation between measures that are installed to reduce risk and measures that are installed to comply with regulatory requirements. The correlation is not perfect, however, and there are occasions when a rigorous analysis will not identify risk reductions measures required to comply with regulatory requirements as safety critical. Nonetheless, complying with regulatory requirements is compelling in and of itself. For this reason, many organizations include in their SCFM IPLs and other risk reduction measures required for regulatory compliance that would not otherwise qualify as safety critical.

4.3 Causes

In addition to safeguards that are relied upon to reduce the risk of a major hazard to a tolerable level, the term "safety critical" also applies to components, the failure of which can trigger a catastrophic event. Because of that, a few organizations also include the causes identified during the LOPA of a safety critical LOPA scenario in their SCFM.

Most do not. The failure rate associated with the cause of a safety critical LOPA scenario already reflects how the failed component is operated and maintained. The argument, then, is that nothing extraordinary is required to assure that the cause fails at the normal rate, so inclusion of causes in an SCFM is not necessary, and needlessly clutters the SCFM.

5. Conclusion

IPLs need to be effective and independent, and that effectiveness and independence must be auditable. An organized system ensures that safety critical IPLs are not only auditable, but audited, and that necessary operation, maintenance, and training all occur as required. An SCFM provides this organized system, and in so doing, provides a tool to validate assumptions made in a LOPA. It is useful to format the SCFM in two distinct parts: a report section that defines general requirements and a section consisting of datasheets for each safety critical function included in the SCFM.

The report of general requirements should outline the scope of the manual, the purpose of the manual, and general instructions for the manual's continued use and upkeep. The datasheets would be part of a database that captures the specific requirements of each individual safety critical function in the SCFM. The SCFM can be prepared as a physical document (report and datasheets) or as a set of electronic documents (text document and database) or both.

All safety critical function datasheets should include a hazard description, a description of how the function prevents or limits the consequences, and references that link the datasheet to other documents. The hazard description should be concise and unambiguous form and follow a template that uses the same basic structure for all scenarios. Each safety critical function datasheet should link the function to a reference PHA or safety critical LOPA scenario at the front end of the audit trail, and to the tracking method used to schedule and issue work orders related to the function at the back end of the audit trail.

In addition to the three elements common for all functions (hazard description, how the function protects against the hazard, and references), safety critical function datasheets should also address how personnel are to respond to a demand on the function, and how the function should be maintained. The datasheets for SIFs and instrumented functions that are not SIL-rated will be very similar. Datasheets for non-instrumented functions and for procedures and administrative controls will differ from these and each other.

The use of an SCFM accomplishes much in fulfilling the promise of LOPA. It gives a structure for allocating finite safety resources. It collects information pertaining to safety critical functions in one place to facilitate employee or auditors with a question or concern about safety critical functions finding answers. Finally, it provides a framework for ensuring that functions are operated and maintained properly.

6. References

[1] Dowell, A., et al.. *Layer of Protection Analysis: Simplified Process Risk Assessment*. New York: Center for Chemical Process Safety of the AIChE. 2001.

[2] Occupational Safety and Health Administration. *29 CFR 1910.119, Process Safety Management of Highly Hazardous Chemicals*. 1992.

[3] Crosland, Andy. "The importance of a clear Safety Requirements Specification as part of the overall Safety Lifecycle", *2011 Safety Control Systems Conference*. IDC Technology, held in Manchester, United Kingdom. May 24-26, 2011. http://www.yumpu.com/en/document/view/5564355/the-importance-of-a-clear-safety-requirements-idc-technologies (accessed 22-Feb-2014).

[4] Hedberg, Johan. *Safety requirements specification Guideline*. Version 1.0 SP Swedish National Testing and Research Institute. 27-Oct-2005. http://www.sp.se/sv/index/services/functionalsafety/Documents/Safety%20requirements%20specification%20guideline.pdf (accessed 22-Feb-2014).

[5] Underwriters Laboratory. *Equipment Calibration Intervals*. 2012. http://www.ul.com/global/documents/offerings/services/programs/dap/tools/00-OP-C0045.pdf. (accessed 15-Aug-2013).

[6] National Institute of Standards and Technology. *Recommended Calibration Interval*. January 2010. http://www.nist.gov/calibrations/recommendedcalibrationinterval.cfm . . (accessed 15-Aug-2013).

[7] National Fire Protection Association (2011). *NFPA 25: Standard for the Inspection, Testing, and Maintenance of Water-Based Fire Protection Systems*. Quincy, Massachusetts.

[8] Intersafe electrical testers. *Testing Intervals*. Intersafe, Ltd. 2014. http://www.fixedtesting.com/testing-intervals (accessed 22-Feb-2014).

[9] National Fire Protection Association (2013). *NFPA 72: National Fire Alarm and Signaling Code*. Quincy, Massachusetts.

[10] National Fire Protection Association (2012). *NFPA 101: Life Safety Code*. Quincy, Massachusetts.

[11] Occupational Safety and Health Administration. *29 CFR 1910.157(e)(3), Portable Fire Extinguishers*. 1980.

[12] Brookhaven National Laboratory. "Pressure Vessel and Relief Valve Inspection Frequency Table", *Standards-Based Management System*, November 6, 2008. https://sbms.bnl.gov/sbmsearch/subjarea/171/171_exh2.cfm?ExhibitID=7147 (accessed 22-Feb-2014).

[13] Occupational Safety and Health Administration. *29 CFR 1910.147(c)(6)(i), Control of Hazardous Energy (Lockout/Tagout)*. 1989.

349805 How to Better Manage the Combined Effects of Barrier (IPL) Impairment and Ongoing Work Hazard Risks

Abstract:

In recent years there has been a growing recognition that the way we identify and manage cumulative risk in our offshore installations needs to be modified to address the frequency of change of the sources of risk. As one of those main sources of risk, monitoring and maintaining the condition of our independent layers of protection (ILP's) has become a major challenge given the number of protection and mitigation systems on a typical offshore installation. This is especially true for aging assets where maintaining integrity of these systems has become a never ending battle in more mature basin areas. Adding to the complexity of this challenge is the necessity to account not only for the condition of these systems but also for the ongoing day to day work on the plant. We need to understand how this impacts on the effectiveness of protection layers, the new hazards which it introduces and be in a position to judge whether the resultant cumulative risk posed by a work plan is acceptable or not. This paper will explain this challenge in detail and talk about ways in which technology can help manage such risks. It will discuss the operations reality of day to day work on a plant, the not so uncommon conditions of protection layers and the necessity to defeat or remove many of these systems in order to carry out maintenance and repair work.

Rigorous quantitative risk assessments techniques do not lend themselves to the dynamically changing conditions of day to day operations. The assessment of these risks and the judgement of what activities are safe to execute therefore relies on the experience and competence of the people at the front line. There are now emerging, new technology platforms where much improved tools can be used to help support these "in field" decisions. The paper will explore how these tools can be used on a day to day basis and how as an industry we can start to use more meaningful real time data to provide a better picture of the sources and magnitude of risk. The status of protection layers can be shown graphically alongside ongoing or planned work to show their potential impact on each other. As a result, decisions on what is safe to execute today or to plan for tomorrow can be made with much better insight on risk

For nearly 12 years our software systems have been used for managing the risks associated with work offshore (in the UKCS North Sea our systems are used on 90% of platforms). They have been used to help operations people manage task hazard risk as well as to account for risks associated with non performing equipment and impaired independent protection layers.

349102 Comparative Consequence Analysis Between LNG Import and Export Terminals through the Use of PHAST and GIS

Abstract:

According to the US Energy Information Administration (EIA), the natural gas consumption in the US was 1,726 billion cubic feet in January 2013, and the demand has been increasing steadily around the world as natural gas becomes the fuel of choice for electric power providers. In order to meet this demand, the US economic interest was focused on the Liquefied Natural Gas (LNG) import terminals installation. However, since 2009, changes in domestic shale production and other commercial changes, the focus in the US has shifted from the construction and operation of LNG import terminals to the creation of new export terminals in pursuit of exporting LNG. Based on this shifting focus, risk assessments for LNG terminals have also shifted towards potential hazards caused by flammable and cryogenic characteristics of LNG such as vapor cloud, flash fires, pool fires and Boiling Liquid Expanding Vapor Explosion. Potential hazards associated with LNG export terminals during liquefaction and storing processes require more in-depth study.

The implementation of Computational Fluid Dynamics (CFD) modeling plays an important role in identifying the contrasting consequences of incidents associated with potential scenarios at LNG export terminals. It is intended to determine the exclusion zones for these scenarios through the utilization of CFD in order to estimate the areas where people, property, or the environment would be more severely affected. Additionally, this project seeks to generate high accuracy consequence predictions by including the implementation of a Geographic Information System (GIS) which will allow the use of geo-spatial factors. This computational approach will help to develop a more robust analysis, which will predict the consequences of an incident. Not only will GIS facilitate the integration of topographical information into the model, but it will also provide a significant input for the emergency mitigation aspect. The location of emergency response personnel would be incorporated, and the time of response would be assessed. Consequently, a wider number of variables would facilitate the performance of the consequence analysis as realistically as achievable. The main goal of this project is the combination of these two computational tools (CFD and GIS) to fill a gap between the consequence estimation of LNG catastrophic events and the final representation of the results.

350020 Challenges of Explosion Risk Management in Artic Environments

Abstract:

Fully enclosed platforms utilized in artic climates can present challenges for managing explosion risk. Generally, modules on modern platforms are built with an "open" wall design. These open wall designs provide: (1) increased ventilation, which helps to prevent the accumulation of flammable gases within the module; (2) open pathways for flammable gases to dissipate; and (3) open pathways for the explosion to vent, thus reducing the overpressures. Due to the harsh conditions in artic environments, the outer walls of the platform are solid and enclosed, which prevents flammable gas from escaping during unintentional releases. Unlike open modules, nearly all mass flow leak rates (including the high frequency of occurrence, low mass flow rate leaks) can fill an entire process area with near stoichiometric gas concentrations at some time during the release. When ignited, these clouds can produce devastating explosion overpressures. Enclosed modules also increase the duration of time for which these clouds are present, which can increase the probability of ignition as mechanical ventilation is the only means of diluting and removing the gas vapor. This study will use FLACS CFD explosion and dispersion modeling to analyze the application of various mitigation techniques to help manage the explosion risk in these enclosed platforms. More specifically, the effectiveness of soft barriers/partitions (to reduce cloud sizes) as well as partially opening the outer enclosing walls (to reduce cloud sizes and the time that gas vapor may remain in the flammable range) will be evaluated. Exceedance curves will be produced for various geometry changes to analyze the 10^{-4} and 10^{-5} per year frequency of occurrence overpressures.

Experimental study on the relationship between the charge amount of polypropylene granules and electrostatic discharges while silo loading

Kwangseok Choi
National Institute of Occupation Safety and Health, Japan
1-4-6 Umezono, Kiyose, Tokyo, Japan
choiks@s.jniosh.go.jp

Tomofumi Mogami
Kasuga Denki INC
2-4 Shinkawasaki, Saiwai, Kawasaki, Kanagawa, Japan

Teruo Suzuki
Kasuga Denki INC
2-4 Shinkawasaki, Saiwai, Kawasaki, Kanagawa, Japan

Mizuki Yamaguma
National Institute of Occupation Safety and Health, Japan
1-4-6 Umezono, Kiyose, Tokyo, Japan

Prepared for Presentation at
American Institute of Chemical Engineers
2014 Spring Meeting
10th Global Congress on Process Safety
New Orleans, LA
March 30 – April 2, 2014

UNPUBLISHED

Experimental study on the relationship between the charge amount of polypropylene granules and electrostatic discharges while silo loading

Kwangseok Choi
National Institute of Occupation Safety and Health, Japan
1-4-6 Umezono, Kiyose, Tokyo, Japan
choiks@s.jniosh.go.jp

Tomofumi Mogami
Kasuga Denki INC

Teruo Suzuki
Kasuga Denki INC

Mizuki Yamaguma
National Institute of Occupation Safety and Health, Japan

Keywords: Electrostatic charges, Electrostatic discharges, Granules, Polypropylene, Silo

Abstract

This paper is a report on the relationship between the charge amount (charge to mass ratio) of polypropylene (PP) granules and the electrostatic discharges that occur while loading a metal silo. The feedback control system was used in order to control the charge to mass ratio of PP granules. The electrostatic discharges inside the silo were also observed using a conventional image-intensifier system. The results obtained from the experiments show that (1) two kinds of electrostatic discharges were clearly observed inside a metal silo while loading PP granules, i.e., brush discharges and incendiary bulk surface discharges; (2) the number of brush discharges and incendiary bulk surface discharges increased with the increase in the charge to mass ratio of PP granules, almost reaching saturation; and (3) brush discharges and incendiary bulk surface discharges began to occur at the -1.16 μC/kg and -2.33 μC/kg points in the charge to mass ratio, respectively, in this study.

1. Introduction

A variety of combustible powders are currently being used in the food and pharmaceutical industries and in the materials used in electronic devices. Dust explosions and/or fires attributed to electrostatic discharges often occur in various processes of industries. There is no risk of ignition by electrostatic discharges with granules that are greater than 500 μm [1]. However, mixtures of granules and fine powders that are more concentrated than the low explosion limit are potentially hazardous in industrial situations. In fact, our previous studies observed visually the electrostatic discharges generated inside a metal silo while loading polypropylene granules. Some results obtained were noteworthy and, thus, published [2]. However, the question still remains as to what amount of charged polypropylene granules is needed to generate electrostatic discharges. Therefore, the main focus of this present paper is to investigate clearly the relationship between the charge amount of polypropylene granules and the electrostatic discharges that occur while loading a metal silo. A primary objective of this study is to communicate new information about safety management that will be important to individuals working in the field of accident prevention and mitigation.

2. Experimental

2.1 Pneumatic Powder Transport Facility and Sample Granules

An actual-sized pneumatic powder transport facility was used in this study (Fig. 1). It consists of an earthed cylindrical silo (stainless steel; diameter, 1.5 m; body length, 3 m; capacity, 4.8 m^3), a pipeline (stainless steel; diameter, 0.1 m; total length, 23 m), an air blower (11 m^3/min) driven by an inverter motor, and an air conditioning unit controlling the temperature (30 ± 2 °C) and humidity (30 ± 2% RH) of the blowing air. The metal silo had a rotary valve (13 rpm) driven by an inverter motor to discharge sample granules from the bottom of the silo. The discharged sample granules were driven back into the silo from the top through a pipeline with a length of 20 m.

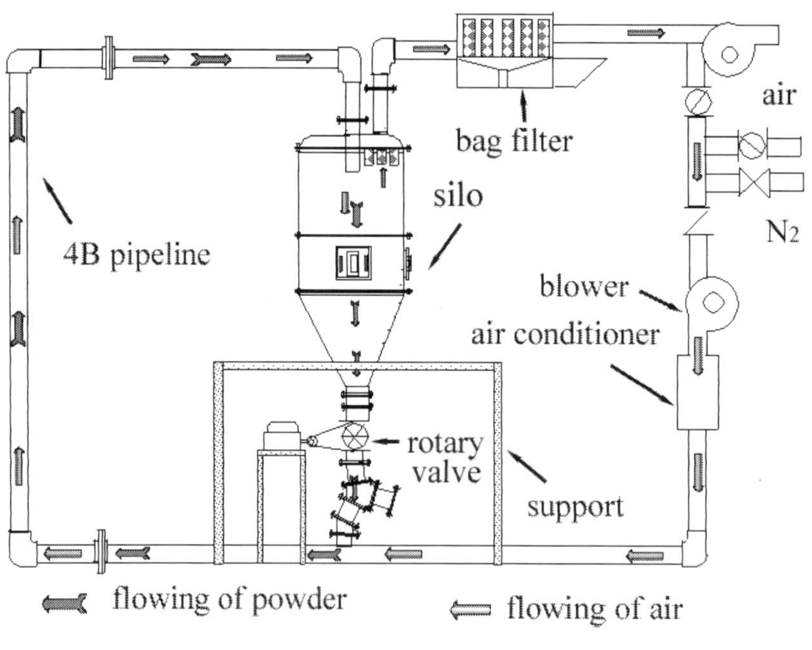

Fig. 1 Pneumatic powder transportation facility used in this experiment.

As a sample, 300 kg of polypropylene (PP, Fig. 2) granules was employed in this experiment, since PP has been widely used in chemical industries. The PP granules used in this study were the same used in our previous studies. The PP granules had a typical particle size of 2 to 3 mm. The apparent volume resistivity, R [Ω•m], of the PP granules was on the order of 10^{15} Ω•m.

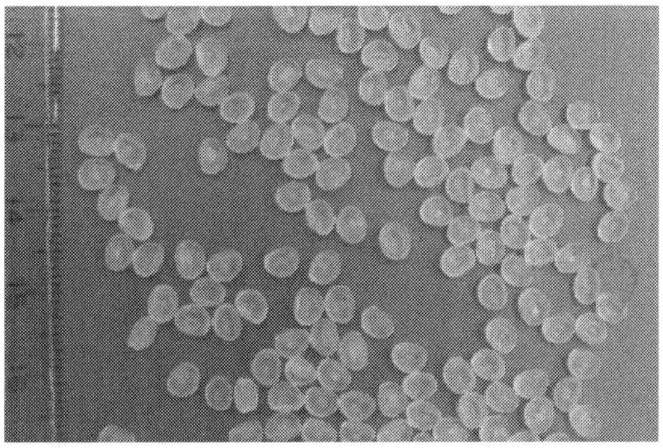

Fig. 2 Polypropylene (PP) granules used in the experiment.

2.2 *Observation of Electrostatic Discharges*

For observing electrostatic discharges inside the silo, an image-intensifier unit (Hamamatsu Photonics, Ltd.; C9016-02; max. gain, 1,000,000) was set on the windowpanes of the silo roof. The unit consists of an image-intensifier head, a remote controller, a CCD camera or a single-lens reflex camera, a wide-angle lens (18–35 mm), a power supply, an AC adapter, a PC, and a digital image recorder. The image intensifier was operated and controlled using the remote controller or a PC via the USB interface connector provided on the rear panel of the image-intensifier head. The image-intensifier unit was used with block-out curtains placed around the silo to prevent sunlight from entering.

2.3 *Controlling the charge amount of PP Granules*

The feedback control system was used to control the charge amount (charge-to-mass ratio, q/m) of PP granules. This feedback control system has been developed from our previous study and published [3]. The system is composed mainly of an ionizer (NKF10, Kasuga Denki Inc., flange-type, Fig. 3), an electrostatic air-blow field meter (KSF-0201, Kasuga Denki Inc., Fig. 4), a high-voltage power source, compressed air with zero relative humidity, and computer control equipment. The ionizer and the electrostatic air-blow field meter were attached to the end of the loading pipe inside the silo. The feedback control system imitates the process whereby the electrostatic air-blow field meter detects and monitors the electrostatic field E_e [kV/m] that occurs as a result of the charged granule particles. In this way, the feedback of the monitored data to the ionizer causes the ionizer to generate ionized air until the charged powders have reached the desired value.

The relationship between the E_e and the q/m is shown in the following equation (1) [3]:

$$q/m = \frac{2\varepsilon_0 v \pi r E_e}{u}, \qquad \text{[Eq. 1]}$$

where u [kg/s] is the flow rate of granules, v [m/s] is the velocity of granules, r [m] is the radius of the pipe for the electric field, and ε_0 [F/m] is the permittivity of free space (8.854×10^{-12} F/m).

It should be noted that u is 0.42 kg/s, v is 7 m/s, and r is 0.05 m in this study.

Fig. 3 Flange-type ionizer used in the experiment.

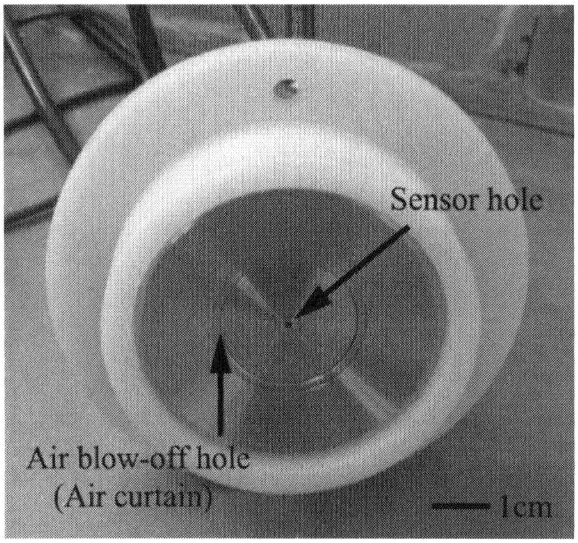

Fig. 4 Electrostatic air-blow field meter used in the experiment [4].

3. Results and Discussion

Two kinds of electrostatic discharges inside a metal silo while loading PP granules, i.e., brush discharges and incendiary bulk surface discharges known as cone discharges, were clearly observed as shown in Fig. 5. This fact was agreed upon in our previous studies, and their electrostatic hazards were discussed [2]. Specifically, it is well known that the energy released in incendiary bulk surface discharges depends on the silo diameter and the particle size of the products forming the powder heap. For silos with diameters in the range of 0.5 m to 3.0 m and powders with a median range of 0.1 mm to 3.0 mm, the energy released in bulk surface discharges can be estimated using the following equation (2) [1]:

$$W = 5.22\, D^{3.36}\, d^{1.46}, \qquad \text{[Eq. 2]}$$

where W is the upper limit of the energy of the bulk surface discharge (mJ), D is the diameter of the silo (m), and d is the median of the particle size distribution (mm).

The D of the silo and the d of the PP were 1.5 m and 2 to 3 mm, respectively, in this study. Following from the above equations, W was 56 mJ to 101 mJ. These energy values can easily ignite combustible powders, including fine particles.

The question still remains as to what amount of charged PP granules is needed to generate these electrostatic discharges. In this study, the PP granules were negatively charged using the feedback control system, since PP granules were negatively charged in the pneumatic powder transport facility. The charging control range for PP granules was from 0 to -12 μC/kg in the q/m.

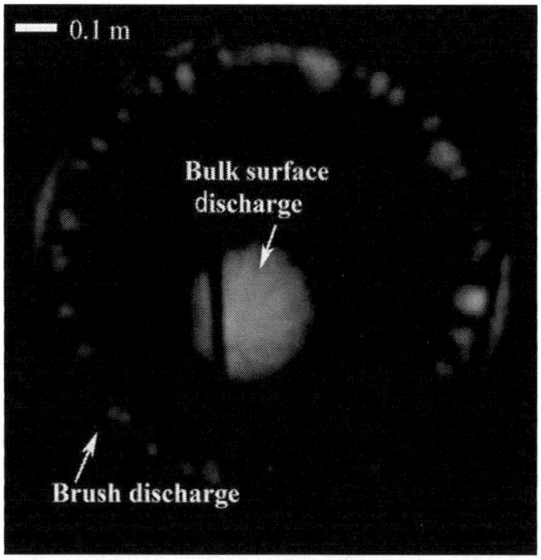

Fig. 5 Electrostatic discharges that occurred while loading PP granules.

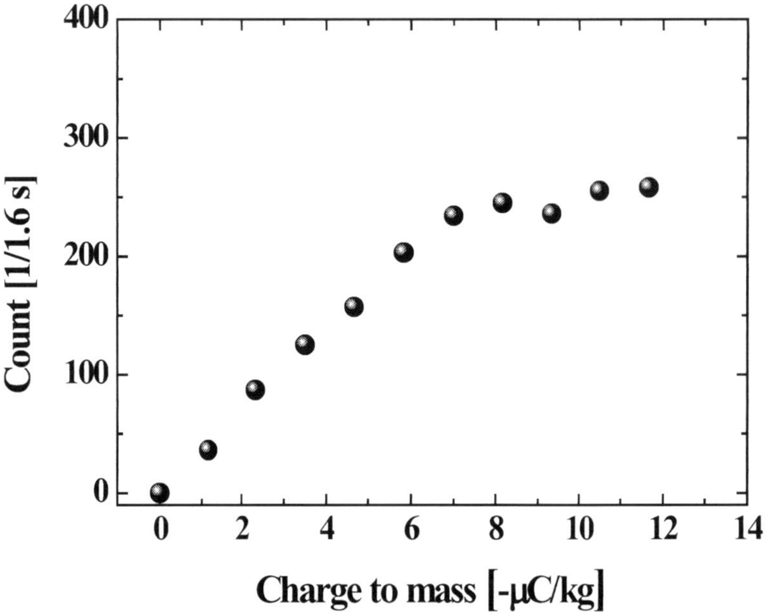

Fig. 6 Relationship between the amount of charged PP granules and the brush discharges that occurred while loading.

Figure 6 shows the relationship between the *q/m* of PP granules and the brush discharges that occurred while loading. Each dot represents the total number of brush discharges out of fifty random photographs. The exposure time was 2 s. Figure 7 shows the relationship between the *q/m* of PP granules and the incendiary bulk surface discharges that occurred while loading. Each dot represents the total number of bulk surface discharges that occurred during the five minutes. The results obtained from the experiments led to the following conclusions: (1) Neither brush discharges nor bulk surface discharges occurred when the *q/m* of PP granules was 0 μC/kg, as expected; (2) The generation of brush discharges was about ten times larger than that of bulk surface discharges; (3) The *q/m* significantly affected the generation of electrostatic discharges, with higher *q/m* resulting in the occurrence of more brush discharges and bulk surface discharges, almost reaching saturation; (4) Brush discharges and incendiary bulk surface discharges began to occur at the -1.16 μC/kg and -2.33 μC/kg points, respectively, in this study. This means that the threshold that must be exceeded to begin brush discharges is between greater than 0 and -1.16 μC/kg, and that of bulk surface discharges is between greater than -1.16 μC/kg and -2.33 μC/kg. It is well known that the threshold that can cause electrostatic charging problems with powders, especially electrostatic agglomeration and adhesion in process manufacturing, is 0.10 μC/kg (absolute value) [5]. However, the threshold related to electrostatic discharges has not been determined. Therefore, the findings mentioned above are noteworthy for communicating new information about electrostatic risk management that will be important for individuals working in the field of dust explosion (and/or fire) prevention and mitigation. Especially, it is noteworthy that bulk surface discharges, known as incendiary electrostatic discharges, did not occur when the q/m was -1.16 μC/kg in this study. If the real industry process is similar to the experimental condition in this paper, we suggest that the q/m of PP granules while loading must be kept below -1 μC/kg to prevent dust explosions and/or fires attributed to incendiary bulk surface discharge. Many different testing conditions, such as the kinds and sizes of granules and the size of the silo, may affect the electrostatic charging and discharging of granules; this needs to be examined in the future.

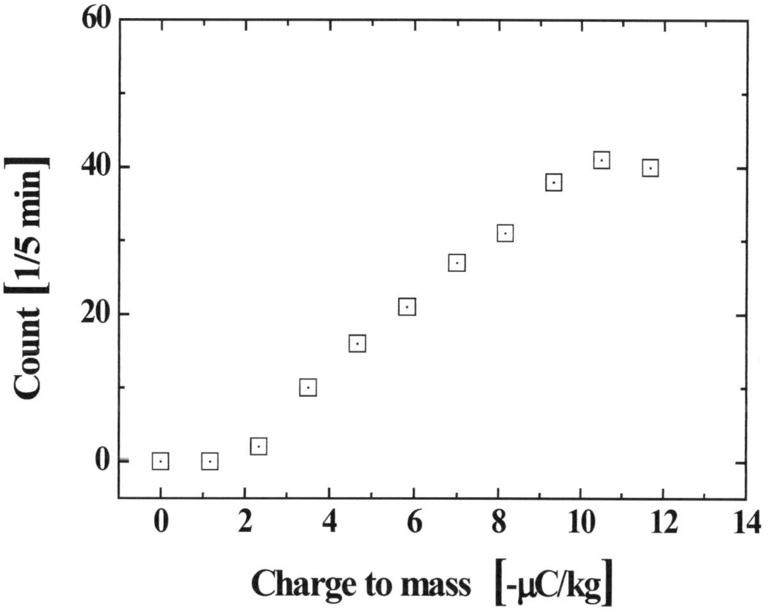

Fig. 7 Relationship between the amount of charged PP granules and the bulk surface discharges that occurred while loading.

4. Conclusion

To achieve a better understanding of electrostatic hazards, the relationship between the charge-to-mass ratio and electrostatic discharges generated inside a metal silo while loading polypropylene granules was investigated experimentally. The key results are summarized below:

(1) Two kinds of electrostatic discharges inside a metal silo while loading PP granules, i.e., brush discharges and incendiary bulk surface discharges, were clearly observed.
(2) Both brush discharges and incendiary bulk surface discharges increased with the increase in the charge-to-mass ratio of polypropylene granules, almost reaching saturation.

(3) Brush discharges and bulk surface discharges began to occur at the -1.16 µC/kg and -2.33 µC/kg points, respectively, in this study.

5. References

[1] IEC50404,"Electrostatic-Code of practice for the avoidance of hazards due to static electricity," 2003.

[2] K. Choi, T. Mogami, T. Suzuki, S. Kim, and M. Yamaguma,"Charge reduction on polypropylene granules and suppression of incendiary electrostatic discharges by using a novel AC electrostatic ionizer, " J. of Loss Prevention in the Process Industries, Vol. 26, NO.1, pp.255–260, 2013.

[3] T. Mogami, T. Suzuki, K. Choi, and T. Ikehata," Control of electrostatic charge for powder by using feedback control-type ionizer system, "J. of Loss Prevention in the Process Industries, Vol.23, NO.2, pp.237–241, 2010.

[4] K. Choi, S. Kim, J. Chung, "Experimental Study on Electrostatic Field of Polymer Powders in Freeboard Region of Fluidized Bed Reactor, " Advanced Science Letter, Vol. 19, pp.113-117, 2013.

[5] T.B. Jones and J.L. King: Powder Handling and Electrostatics, Lewis Publishers, p.34, 1986.

350342 Management of Change at Shell

Abstract:

Management of Change or lack thereof has been cited in many industrial accident investigation reports as a contributing factor to the cause of those accidents. Shell considers Management of Change to be a critical business control process and require that all businesses within Shell implement a Management of Change process. This paper will discuss how Shell Exploration and Production Company, SEPCo, implemented their Management of Change process and the steps associated with that process. This paper will also discuss what types of activities are covered by the Management of Change process. Lastly, this paper will look at the implementation challenges, lessons learned, key success factors, and Key Performance Indicators associated with Management of Change.

Inherently Safer Design: Lessons Learned About the Principle of Simplification

Andrew R. Carpenter
Russell A. Ogle
Brenton L. Cox
Sean J. Dee
Exponent
4580 Weaver Parkway
Warrenville, IL 60555
acarpenter@exponent.com, rogle@exponent.com, bcox@exponent.com, sdee@exponent.com

Prepared for Presentation at
American Institute of Chemical Engineers
2014 Spring Meeting
10th Global Congress on Process Safety
New Orleans, LA
March 30 – April 2, 2014

UNPUBLISHED

Inherently Safer Design: Lessons Learned About the Principle of Simplification

Andrew R. Carpenter
Exponent
4580 Weaver Parkway
Warrenville, IL 60555
acarpenter@exponent.com

Russell A. Ogle
Brenton L. Cox
Sean J. Dee
Exponent
4580 Weaver Parkway
Warrenville, IL 60555

Keywords: Inherently Safer Design; Simplification; hazardous release, fire, explosion

Abstract

The American Institute of Chemical Engineers (AIChE), the Chemical Safety Board (CSB), and the Occupational Safety and Health Administration (OSHA) have all emphasized the importance of advancing inherently safer design concepts into chemical process plants. Incident investigations offer an important opportunity to identify, evaluate, and correct potential shortcomings in the design, construction, operation, and maintenance of a chemical process unit that has experienced a release.

This paper focuses on the inherently safer design principle of simplification. The case studies share the common theme of storing or handling liquids. Each case study illustrates how a design flaw led to unintended flow, reverse flow, or overfilling of a vessel which ultimately led to a fire, explosion, or hazardous release. Based on the incident investigation experiences of the authors, the paper illustrates how application of the simplification principle could have prevented these incidents.

1. Introduction

Process risk management has four generic strategies for reducing risk in chemical processes: Inherently Safer, Passive, Active and Procedural. Inherently safer design (ISD) eliminates potential process hazards through the use of safer materials or operating conditions. [1] ISD can be considered during initial design conceptualization and typically results in solutions with low operating costs, high reliability and low levels of complexity when compared with the three alternative risk reduction strategies. [2] While it is not always feasible to eliminate hazards, many professional organizations and regulatory groups have advocated that ISD should be considered before accepting process hazards and relying on safety systems for hazard management. [3-4] Therefore, ISD is commonly placed at the top of the risk reduction design solution hierarchy. [5-6]

This study focused on the examination of the principle of simplification within ISD through case studies. Simplification, along with Minimization, Substitution, and Moderation, are the four main approaches to designing inherently safer processes. In the context of ISD, simplification refers to the elimination of unnecessary complexity thereby reducing the probability of error or misinterpretation. Herein, four case studies are presented involving the storage and handling of liquids: Chlorine Release at a Commercial Water Park, Fire and Explosion at a Specialty Chemical Manufacturer, Chemical Release at an Ethanol Production Facility, and Mixing of Incompatible Wastes at a Waste Treatment Facility. All four cases highlight the role of the principle of simplification in creating inherently safer processes.

2. Methods of Risk Reduction and Design

The prevention of the progression of a hazardous event into undesired consequences can be controlled through various layers of protection. [7] These layers of protection can be grouped into three categories. At the outer edges of the layers of protection are response layers, which correspond to actions taken after a hazard event has taken place. Response layers intend to reduce the severity of the consequence of an event, but their effectiveness is often unknown. Underneath the response layers, are control layers which are intended to keep the hazards of a process within safe allowable operating limits. Control layers include basic controls, alarm systems, safety instrumented systems, automated shutdown systems, and physical protection. The impact of these systems can influence both likelihood and severity of an event several orders of magnitude. The design core lies at the core of the layers of protection. This design layer determines the inherent hazards within the process and is the most important layer because it has the capability to completely eliminate hazards through ISD. All the additional layers of protection around the design core served to mitigate risks and hazards that cannot be eliminated through ISD. Therefore, when performing a layer of protection analysis, inherently safety design is the preferred method for reducing risk when compared with adding independent protection layers.

3. Principles of Inherently Safer Design

Inherent safety is at the top level of the hierarchy of process risk management strategies, above passive, active and procedural approaches. ISD is the process of identifying and implementing solutions that result in the permanent and inseparable elimination of a process hazard. [8] Reduction of process hazards are also included in ISD, and are referred to as second order approaches to inherent safety.

ISD includes the following four design strategies:

- Minimization – The use of smaller quantities of hazardous materials.

- Substitution – The replacement of a hazardous material with a less hazardous material.
- Moderation – The concept of limiting the impact of an event by using less hazardous conditions, less hazardous forms of a material, or modifying the facility
- Simplification – The elimination of unnecessary complexity in the design of a facility, resulting in the reduction in the number of errors or an tolerance of process operation fluctuations.

While the concepts of minimization, substitution, and moderation appear simple to apply to a process, simplification can be much more intimidating. Simplification involves the deliberate questioning of the necessity of every aspect of a process and a broad knowledge of all potential alternatives. Simplification also requires individuals to discount the operational history or a process or unit and consider that safer approaches were either disregarded in the initial design, or not considered all together. The primary reason complexity exists in plant designs stems from the need to control hazards. [9] Unnecessary complexity can also result from the need to control hazards, the desire for technical elegance, failure to carry out a hazard analysis until too late in the design process, following standards or specification that are not applicable or necessary, or creating flexibility and redundancy. The strategy of simplification implies a reduction in the complexity of safeguards while simultaneously maintaining or increasing the level of risk reduction.

Understanding the appropriate definition of risk is paramount to the utilization of the concept of simplification in ISD. Risk is the product of an event likelihood and consequence severity. It is important to remember that process designs can be made inherently safer though strategies that affect both likelihood and consequence. Process equipment can often be designed to withstand overpressures and fires at much lower costs than complex safety instrumentation systems. This eliminates the likelihood of a loss of containment event. However, there are circumstances where a loss of containment event cannot be eliminated through process design. In those situations, hazard consequences can sometimes be mitigated by reducing potential consequences through siting. For example, if a vessel cannot be designed to withstand the overpressure generated from an explosion, consequences can be mitigated by ensuring the vessel is not located next to critical pieces of equipment, storage of additional hazardous materials, or occupied buildings. Thus, by being conscious of siting, the risk can be mitigated through reduction of an event consequence.

4. Case Studies: Storage and Handling of Liquids

The desire to reduce risk though complex control schemes can create new hazards that can be mitigated through inherently safer design. The handling and storage of hazardous chemicals at facilities is an example of a routine task that can become unnecessarily complex. These case studies illustrate three common accident scenarios that can arise: unintended transfers caused by siphoning, receiving vessel overflows, and unexpected accumulation in low points of the flow system. In each case, the principles of inherently safer design could have served to reduce the likelihood or severity of the associated incidents.

4.1 Chlorine Release at a Commercial Water Park

This case study involved a chlorine gas release at a public swimming pool. Over one hundred patrons were exposed to the chlorine gas release. The chlorine gas was generated by the inadvertent mixing of sodium hypochlorite solution and muriatic acid (31% aqueous solution of hydrochloric acid). The swimming pool involved in this incident was a lazy river type pool that had a total volume of water greater than 1,000,000 liters. The water treatment system for the pool consisted of filtration, disinfection with bleach, and pH control with hydrochloric acid. The water treatment system operated continuously and had a design capacity of 2,400 liters per minute.

A maintenance contractor was performing some non-routine maintenance on the water treatment system. This activity required that he shut the system down. What the contractor did not know was that while the system was idle, both the concentrated bleach solution and the hydrochloric acid began to siphon from the feed tanks and drain into the mixing tee of the circulation pipe. The mixing tee was a short vertical pipe spool at the lowest point in the mechanical room. The orientation of the injection ports for the chemicals resulted in the lighter solution, muriatic acid, forming a stable layer above the denser solution of sodium hypochlorite. When the main circulation pump was restarted, the superchlorinated water was discharged into the pool area releasing chlorine gas.

Two design improvements would have resulted in an inherently safer design. The first improvement would be to lower the elevation of the chemical feed tanks to reduce the potential for siphoning. The second improvement would be to reverse the installation of the injection ports so that, in the event of siphoning, the heavier solution would enter the mixing tee above the less dense solution, thus encouraging the mixing of the two chemicals.

4.2 *Fire and Explosion at a Specialty Chemical Manufacturer*

A specialty chemical manufacturer produced a very high purity chemical for the electronics and optics industries. The process involved several stages of batch distillation and batch vacuum distillation operating at near total reflux conditions to achieve the very high purity required. A typical batch distillation could run for several days producing only milliliters of high purity distillate product per hour. The distillate was water reactive, and when allowed to oxidize produced a highly unstable, shock sensitive solid.

A control system failure occurred in the evening while the facility was not staffed. The control system alerted the operators via email that the system had shut down, but the control system did not place the system into a safe state. The control system failure left the heating system in operation and opened the product withdrawal valve on the distillation system. The product withdrawal valve normally operates by opening and closing such that a very slow flow rate of product flows into a receiver. The distillation system typically processes a batch of 30 to 40 liters, while the receiving vessel was approximately 5 liters in volume. With the heating system on and the product withdrawal valve open the contents of the distillation system overflowed the receiving vessel. When the operators arrived at the site the following morning, they observed that the receiving vessel had overflowed and a significant volume of distillate had spilled onto the production floor. While attempting to clean the spill, the product ignited resulting in a fire that destroyed the entire facility and caused the rupture of several chlorine tanks resulting in an environmental release. Fortunately, the operators who were cleaning the spill had just exited the area prior to the fire and were not injured.

While the extent of damage precluded the ability to determine the specific nature of the control system failure, a simple design improvement would have resulted in an inherently safer design that would have prevented the release and the fire. The receiving vessel was significantly smaller in volume that the volume of material being processed. By increasing the size of the receiving vessel to contain the entire batch contents, the overflow condition resulting in the fire and chlorine release would not have occurred.

4.3 Chemical Release at an Ethanol Production Facility

A corn based ethanol producer sold a variety of animal feed stocks from the residual products left over from the fermentation process. One of these products, wet-cake, is a combination of distillers dry grain (ddg) and corn syrup. The wet cake is produced by adding hot corn syrup at 180°F into the ddg as it exits a grain dryer using a configuration similar to that shown in Figure 1. Hot corn syrup was injected into the ddg conveyor system near the top of an inclined conveyor transition. This location allowed hot corn syrup to pool in the bottom of the conveyor. This build up would progress and eventually plug the vertical transition, preventing ddg from entering the inclined conveyor. To clear the blockage, the facility operator cut a porthole near the bottom of the inclined transition. When an operator attempted to clear the blockage by opening the porthole and rodding through the plug, a stream of hot corn syrup was released which resulted in significant burn injuries to the operator.

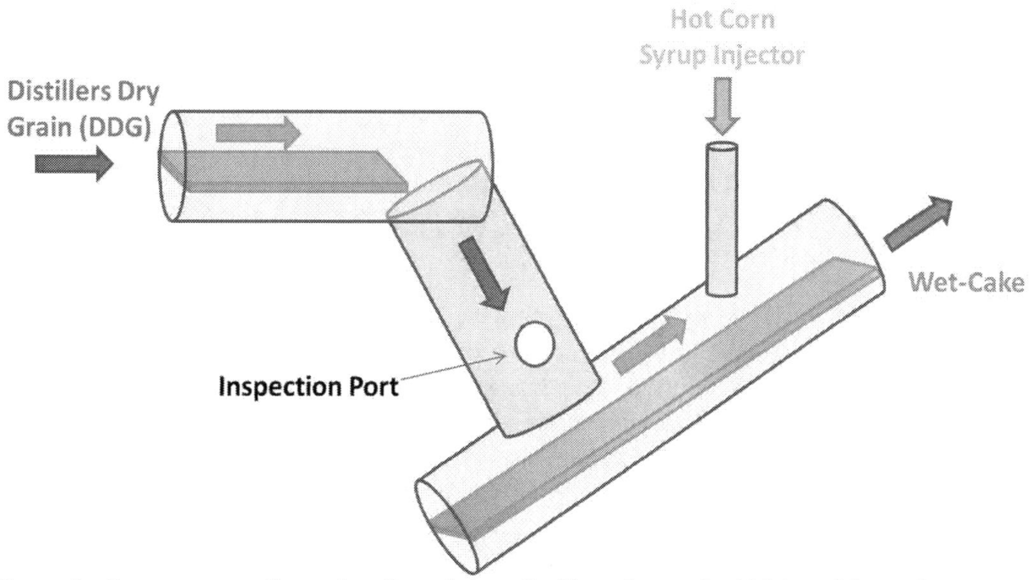

Figure 1: Conveyor configuration for mixing distillers dry grain (ddg) and heated corn syrup to produce wet-cake.

Two design improvements would have resulted in an inherently safer design. The first would be to place the conveyor inspection hatch at the top of the conveyor system such that an operator would not have to stand below the opening while attempting to clear a blockage. The second was to move the corn syrup injection point to the front of the horizontal conveyor that feeds into the inclined conveyor. This location would allow for the ddg and the corn syrup to mix and produce a wet cake and eliminate the pooling which was causing the blockage to occur.

5. Conclusion and Lessons Learned

The storage and handling of liquids requires the use of relatively simple technologies. But even simple technologies can experience failures. If the liquids are hazardous, the consequences of these potential failures must be controlled. These case studies illustrate three common accident scenarios that can arise: unintended transfers caused by siphoning, receiving vessel overflows, and unexpected accumulation in low points of the flow system.

In the case studies presented above, relatively minor changes could have mitigated, or potentially eliminated, the likelihood or severity of the hazards associated with each incident. In the first case study, both system layout and mixing point location alterations could have moderated or prevented the release of superchlorinated water. In the second case, a larger receiving vessel could have prevented the release of the hazardous chemical, and, in turn, the subsequent fire and explosion. In the third case, relocating the an inspection port would have kept the operator out of harm's way during a hazardous procedure, and relocating the injection point for the hot corn syrup may have eliminated the need for the procedure altogether.

Note that, while these alternatives may be obvious in hindsight, the alternatives, or even the hazards, may not have been apparent to the engineers at the time of the design. However, when designing or reviewing a system, let these incidents serve as a reminder to ask, "Are there modifications to the system that could reduce or eliminate any associated hazards?"

6. References

[1] Lees, F. P. *Loss Prevention in the Process Industries: Hazard Identification Assessment and Control.* 2nd Edition, Volume 1. 1996.

[2] CCPS. *Guidelines for Engineering Design for Process Safety.* Center for Chemical Process Safety. 2012.

[3] "CSB Releases New Safety Video on Inherently Safer Design and Technology: 'Inherently Safer: The Future of Risk Reducation' Examines how Industry Can Eliminate or Reduce Hazards," Jul 11, 2012, http://www.csb.gov/csb-releases-new-safety-video-on-inherently-safer-design-and-technology-inherently-safer-the-future-of-risk-reduction-examines-how-industry-can-eliminate-or-reduce-hazards/ (accessed Jan. 20, 2014).

[4] Hendershot, D. C. "Inherently Safer Design: The Fundamentals," CEP, January 2012.

[5] Crowl, D. A. *Understanding Explosions.* Center for Chemical Process Safety/AIChE. 2003.

[6] Johnson, R. W.; Rudy, Steven W.; Unwin, Stephen D. *Essential Practices for Managing Chemical Reactivity Hazards.* 2003.

[7] CCPS. *Guidelines for Safe and Reliable Instrumented Protective Systems.* Center for Chemical Process Safety. 2007.

[8] CCPS. *Inherently Safer Chemical Processes: A Life Cycle Approach.* Center for Chemical Process Safety. 2009.

[9] Kletz, T. A. *Process Plants: A Handbook for Inherently Safer Design.* 1998.

On What Do You Base the Safety of Your Process?

John C. Wincek
Croda, Inc.
Mill Hall, PA
john.wincek@croda.com

Prepared for Presentation at
American Institute of Chemical Engineers
2014 Spring Meeting
10th Global Congress on Process Safety
New Orleans, LA
March 30 – April 2, 2014

UNPUBLISHED

On What Do You Base the Safety of Your Process?

John C. Wincek
Croda, Inc.
Mill Hall, PA
john.wincek@croda.com

Keywords: Basis of Safety; Process Safety Information

Abstract

Process Plants often require certain safeguards to satisfy management that the risk of operating the plant is sufficiently low. The required safeguards are based upon the credible consequences of a process incident, and the events that cause them. Safeguards can be preventative (preventing the incident from occurring), protective (e.g. safety relief valves) or mitigative (controlling and minimizing damage after the incident has occurred). In each category, the safeguards might be split into those that are "have to have" (i.e. the process will not be operated without this safeguard or a suitable replacement) and "nice to have" (i.e. those without which the process could run for defined periods of time). When the number of safeguards becomes large, it is difficult for the process operator to know them all, understand why they exist and what they do. It can be difficult to prioritize the safeguards when developing Asset Integrity plans. Management of Change (MOC) items can be difficult to evaluate without knowing if a "have to have" vs. a "nice to have" safeguard is impacted. This article will present a method of organizing and prioritizing safeguards. The key to this effort is determining which safeguards are required to meet a minimum level of safety. By identifying the subset of "have to have" safeguards, a concise Basis of Safety can be documented for the process. This document can then be used for operator training, maintenance and asset integrity prioritization, MOC review, and identification of safety-critical equipment such as sensors, regulators, and valves.

1. Introduction

Many companies subject their hazardous processes to one or more hazard studies before operation [1]. This might be a single Hazard and Operability (HazOp) study, another type of study, or several studies throughout design and construction. Existing facilities have often undergone one or more hazard studies as well. These studies are typically updated at some scheduled frequency to ensure any changes to the process risks have been reviewed and accounted for.

Hazard Study reports are typically large documents with summaries of findings, detailed worksheets and recommendations. There is also a significant amount of Process Safety Information (PSI) [2] for the process, including Piping and Instrumentation Drawings (P&IDs),

chemical safety data, equipment data and process technology data. Because there is so much information available it can be difficult for anyone, especially a front-line operator, to determine what information is most important for safely operating the process.

The Basis of Safety (BoS) document provides a summary of the most important information about the process. In a single, concise document, it contains a description of the process hazards, and the causes and consequences. This is followed by a list of the safety objectives that must be met to operate the process, and the minimum safeguards that must be in place to meet these objectives. Finally, it lists the Safe Operating Envelope for the process.

2. Process Knowledge

Anyone who operates, manages or interacts with a process should be familiar with the hazards and potential consequences it is capable of causing. This will help them understand the importance of operating the process safely. However to operate safely, one must know the potential causes of these incidents and the means to prevent them. The minimum safeguards that must be in place to meet the safety objectives must be known. For complex processes, this can be an overwhelming amount of information. When you add to this the safe operating limits of the process and other PSI such as chemical hazards, it easily can be too much information to expect employees to remember.

3. Equipment Safeguards

Safeguards are used to achieve the safety objectives for the process. For example, the safeguards used to meet the safety objective of "inerting with nitrogen" might include a vessel inerting procedure, a system for maintaining positive nitrogen pressure on the vessel, and nitrogen makeup when material is discharged from the tank. There might also be an oxygen meter in the vessel headspace, and a low pressure alarm on the tank. Safeguards can be classified into three categories: *Preventive*, *Protective* and *Mitigative* [3].

Preventive safeguards are those that preclude an incident from occurring through early detection and action. One example is a high level switch on a storage tank that prevents tank overflow by alerting an operator or automatically stopping the filling operation. Another example is a high temperature shutoff switch on a reactor, stopping reactant feeds to prevent a runaway reaction.

Protective safeguards usually prevent a serious incident, but often allow an incident with less severe consequences to occur. An example is a relief valve on a pressurized storage tank. In a fire impingement scenario, the relief valve likely prevents high pressure-induced tank failure by relieving excessive pressure. If the valve discharges to atmosphere, the valve opening allows a (hopefully less severe) release of material to the atmosphere at a planned, safer location.

Mitigative safeguards act to minimize the severity of an incident that has already occurred. Examples a tank dike, fire suppression systems, and water sprays on a chemical leak.

4. Minimum Required Safeguards

Some companies have established Quantitative Safety Risk Criteria as described by the Center for Chemical Process Safety [4]. These often specify a maximum allowable risk to employees in terms of fatal injuries. Other companies use varying methods of measuring risk. In many cases, however, companies do not simply stop when they have reached this level. Many additional safeguards are added based on industry practices, regulatory requirements and cost-effectiveness. While safeguards are important and must be maintained, one could separate them into "have-to-have" and "desirable-to-have" safeguards.

The "have-to-have" safeguards are those that are required to reach a minimum level of safety, such as those needed to meet the company's Quantitative Safety Risk Criteria. As will be shown later, this is typically a small subset of the existing safeguards that, without the benefit of the "desirable-to-have" safeguards, reduce the risk below whatever maximum allowable risk criteria are used.

One way to differentiate between "have-to-have" and "desirable-to-have" safeguards is through the use of Layer of Protection Analysis (LOPA). As LOPA practitioners know, not all safeguards can be counted as an Independent Protection Layer (IPL). The typical general requirements used to qualify IPLs are [5]:

1. It (solely) must be effective in preventing the consequences when it functions.
2. It must be independent of the initiating event and the components of any other IPL claimed for the scenario being evaluated.
3. It must be auditable (i.e. capable of being validated by some means)

When a scenario is evaluated using LOPA, many of the existing safeguards cannot be counted as contributing to the safety of the system because they do not meet the above rules. If LOPA is being used to demonstrate that the maximum risk criteria are met, and only safeguards that qualify as IPLs can be considered to reduce risk, then the safeguards claimed as IPLs must be sufficient to reduce the risk below the maximum risk criteria. These would be classified as the "have-to-have" safeguards. System safeguards not claimed as IPLs would be classified as "desirable-to-have."

As an example of this in use, consider the risk of overfilling the hydrocarbon storage tank in Figure 1. To prevent this from occurring, the following are in place:

1. An operator monitors the tank level using Level Transmitter A, and closes Valve B when the tank reaches 90% full.
2. The Basic Process Control System (BPCS) alarms at 92% full, as measured by Level Transmitter A.
3. The BPCS has an interlock that closes Valve C when Level Transmitter A reaches 93% full.
4. The BPCS closes Valve E if Level Switch D activates (at 94% full)
5. A Safety Instrumented System (SIS), independent of the BPCS, closes Valve F if Level Switch F is activated at the 95% full level.
6. The tank is contained in a diked area.

Figure 1. Example Classification of Safeguards for a Hydrocarbon Storage Tank

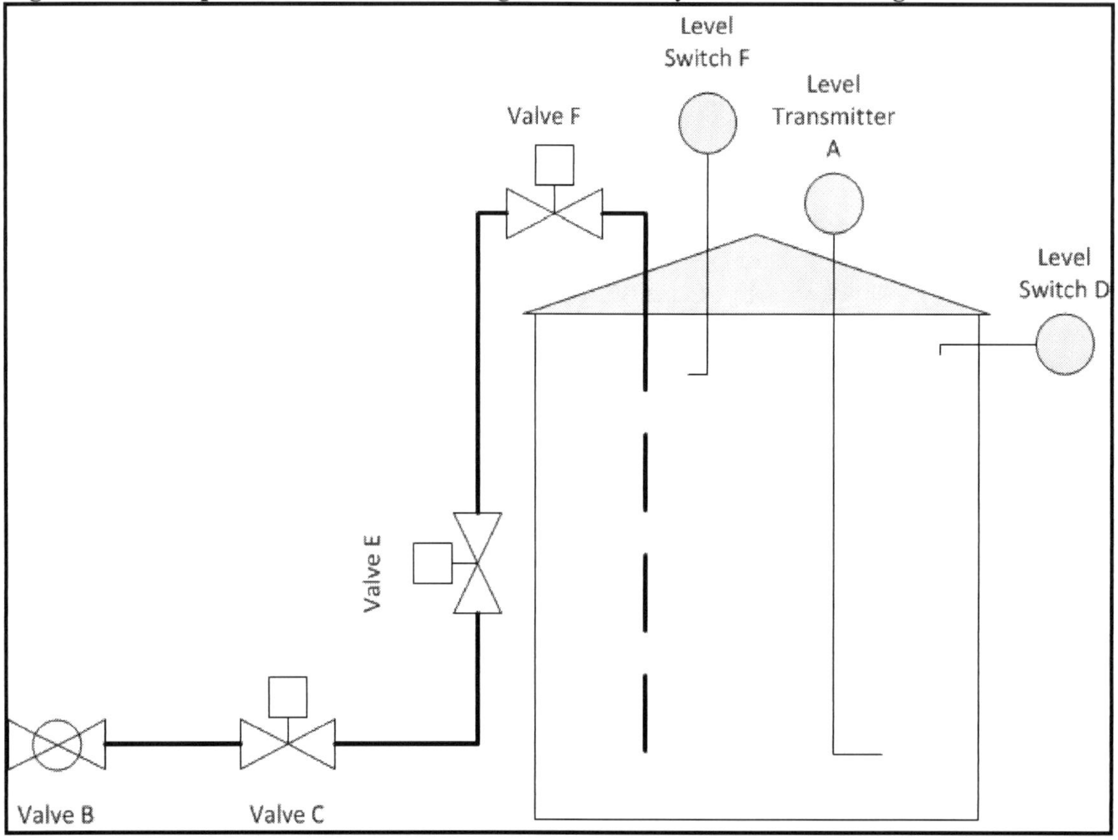

Table 1 shows which of the above could be considered IPLs in a scenario where the operator becomes distracted and does not close Valve B when required.

Table 1. Qualification of Independent Protection Layers

Potential Safety Mechanism	IPL?
# 1 above could not be considered a safeguard, as it is not independent of the initiating event (i.e. it has already failed)	No
#2 relies on the operator (who initiated the event by failing to close Valve B) to take action. This is not independent of the initiating event; it could not be considered an IPL.	No
#3 and 4 rely on the BPCS to function; they are not independent of each other. Therefore, only one of them would count as an IPL.	Yes (but only one of them)
#5 - The SIS is independent of the BPCS, its sensors and valves – it would qualify as an IPL.	Yes
#6 - The dike is independent of all other events and safety devices; it would qualify as an IPL. (Not all LOPA practitioners would count the dike as an IPL. They may also evaluate a second scenario where material is spilled to the dike instead of spreading over a wide area).	Yes

The Layer of Protection Analysis, therefore, considers only that information shown in Table 2.

Table 2. LOPA Form Entries

LOPA Entry	Description
Initiating Event	The initiating event (employee fails to close Valve B).
IPL	Level Switch D activates at 94% full, closing Valve E. (alternatively, #3 could be used as an IPL)
IPL	The SIS closes Valve F upon activation of Level Switch F.

If the LOPA outcome shows that the maximum risk criteria have been met, no further safeguards to meet this requirement are needed. Numbers 1, 4 and 5 would be considered "have-to-have." numbers 2, 3 and 6 (if not counted in LOPA) would be considered "desirable-to-have."

5. Basis of Safety

The information labeled number 1 through 6 above pertains to one very simple scenario – the overfilling of a tank. Imagine that information multiplied by the number of different scenarios in a plant or unit, and you can easily see how one could be overwhelmed. On the operator level, we haven't even considered the task procedures he/she must know. At the maintenance level, there is no clear way to prioritize maintenance tasks – it can't all be critical to safety, can it? The Basis of Safety is designed to minimize, if not solve, these problems.

The Basis of Safety for a process should contain the following information (Shown in Figures 2 and 3):

- **Process Description** - a brief description of the process
- **Chemical Hazards** - a brief guide to the hazards of the main chemicals used in the process
- **Hazards Table** – Showing the Hazards, Causes and Worst-case Consequences
- **Safety Objectives** – the things which, if achieved, will allow the reasonably safe operation of the process.
- **Safeguards** (Have-to-haves only)
  - Preventive
  - Protective
  - Mitigative
- **Safe Operating Envelope** – the limits (e.g. temperature, pressure) within which the process is designed.
- **Appendix (optional)** - contains the "desirable-to-have" safeguards

Figure 2. Sample Portion of a Basis of Safety Document

Process Description

Epichlorohydrin is stored in a 10,000-gallon storage tank located in the flammable tank farm on the south side of the facility. The tank is rated for 60 psig, and constructed of 304 stainless steel. The storage tank sits on a concrete, diked pad, segregated from other storage tanks by dike walls. The storage tank is equipped with an automatic foam fire suppression system. Epichlorohydrin is unloaded from a tank truck by pump, with vapor balancing, into the storage tank. The tank pressure is maintained at approximately 12 psig with nitrogen. Epichlorohydrin is transferred to multiple synthesis vessels using a transfer pump and single flow meter, limiting the system to supplying a single vessel at a time.

Hazardous Substances

NFPA Rating	Health	Fire	Reactivity
Epichlorohydrin	3	3	2

Epichlorohydrin is a corrosive liquid. It can cause corrosive injury to the eyes, skin, and respiratory tract, and can also cause sensitization – an allergic-like reaction that occurs after repeated exposures to the substance. Systemic effects from long-time exposure can include liver and kidney damage. Epichlorohydrin is listed as a potential human carcinogen by OSHA and is also listed by the International Agency for Research on Cancer.

At high temperatures, epichlorohydrin (Epi) can react violently and sometimes explosively with other chemicals, such as acids, alcohols, alkalis, amines and salts. Epi will form a two-phase mixture with water. Given time, an exothermic reaction will start at the interface between the epi and water layers, which can generate enough heat to create a pressure build-up in the vessel.

Worst Case Scenarios

Hazards	Causes	Consequences
1. Storage tank failure due to nitrogen overpressure.	A. Failure of nitrogen control system.	Toxic release, vapor cloud explosion or flash fire, on/off site fatalities.
2. Release through storage tank relief valve.	B. Failure of nitrogen control system.	Toxic release, vapor cloud explosion or flash fire, on/off site fatalities.
3. Release through tank wagon pressure relief valve.	C. Failure of nitrogen control system on epi storage tank.	Toxic release, vapor cloud explosion or flash fire, on/off site fatalities.
4. Piping leak.	D. Confinement (blocking in) of epi in piping segment.	Toxic exposure to on-site employees; potential pool fire/flash fire.
	E. Gasket failure	Toxic exposure to on-site employees; potential pool fire/flash fire.
	F. Piping Corrosion	Toxic release, pool fire, on/off site fatalities.

Figure 3. Safety Objectives, Safeguards and Safe Operating Envelope for Basis of Safety

Safety Objectives

1. Inerting with nitrogen.
2. Avoid ignition sources.
3. Avoid loss of containment.

Control measures for Basis of Safety

These control measures (or suitable alternatives) must be in place for the process to operate.

Prevention (including Process Control)

- Temperature switch (TSH-773-018) on unloading line shuts down unloading pump on high temperature. 11O LAST LINE OF DEFENSE
- Temperature sensor TI-773-017 shuts down transfer pump on high temperature. 12P LAST LINE OF DEFENSE
- Low storage tank pressure alarm at -5 psig. (PI-773-002). 6I, 15S LAST LINE OF DEFENSE for 15S
- Power monitor (JSL-773-021) shuts down unloading pump on high or low power draw. 11O
- Recirculation line on transfer pump discharge dissipates heat buildup if pump is blocked in. 12P
- Level transmitter LT-773-094 alarms at 184". 7J

Protection

- Pressure relief valve PSV-077310/11 on epi storage tank, set at 60 psig. 1A, 1B, 5G, 5H, 6I SYSTEM RELIEF
- Relief regulator (PRV-077382) opens at 15 psig to relieve excess nitrogen pressure. 1A, 1B, 1C

Safe Operating Envelope

- Oxygen concentration in Epichlorohydrin storage tank less than 7.9 vol. % (60% of minimum oxygen concentration required for ignition).
- Storage tank pressure less than 60 psig.
- Epi storage tank level < 194" (90% of tank volume)

6. Uses for the Basis of Safety

The Basis of Safety (BoS) document contains the final results of all of the hazard studies, risk analyses, LOPAs, etc. that have been performed for the process. It contains only the most critical information for process safety. If, after all of the studies are complete, you were asked "Where are the results?" you could answer by handing them the Basis of Safety. If someone asks "How do you keep your process from causing a fire/explosion/toxic release?" you could hand them a Basis of Safety.

Site Directors, Unit Managers and others who need to understand the risks of the process and how they are controlled can review the Basis of Safety for each process on site. From the BoS, they can tell how bad the consequences might be, and can identify key risk control measures. They also can use it to quiz operators on process knowledge related to process safety.

Shift Managers and Operators must be aware of the potential consequences of an incident. However, they must also have a good understanding of the Safety Objectives and the safety systems used to achieve them. The BoS can help them to understand if a process upset could be the beginning of a catastrophic incident, and it can help determine if a process can continue to run if a sensor is malfunctioning.

Maintenance personnel, using the BoS, now have a smaller list of the highest-priority equipment to work from. Fewer jobs are competing for attention because fewer of the repairs will be to "have-to-have" equipment.

The likelihood of a major oversight when reviewing MOCs can be decreased if the BoS is reviewed. The reviewer can ensure that the safeguards in the Basis of Safety are not compromised by the change. One might even require escalation of an MOC for approval at higher levels of the organization if the BoS is being changed.

The BoS is an excellent training tool for many levels of the organization. Anyone who operates or is responsible for a process should be trained and tested on the BoS, and the BoS documents should be made available to the employees for quick reference

7. Last Line of Defense

The Last Line of Defense (LLoD) is a concept that arose from the Basis of Safety and LOPA. It provides a useful metric of process safety performance. Consider the LOPA form entries from Table 2:

A. The initiating event (employee fails to close Valve B).
B. Level Switch D activates at 94% full, closing Valve E. (alternatively, #3 could be used as an IPL)
C. The SIS closes Valve F upon activation of Level Switch F.

The Initiating Event and IPLs are listed in order of occurrence / activation. The employee should close Valve B when the tank is full. If that doesn't happen, the BPCS should close Valve E when the Level Switch D activates. If not, the SIS should stop the incident by closing Valve F. These three actions are sometimes called Lines of Defense. In the scenario described above there are three Lines of Defense, the last of which is the SIS. All three Lines of Defense must fail for the incident to occur.

If the SIS fails to function, the incident (overfilling the tank) will occur because there are no safeguards to stop the incident after it fails. It is, therefore, the Last Line of Defense. If the Last Lines of Defense for the various scenarios in your plant are identified, they can be used as a Key Performance Indicator. If an incident sequence is stopped by your Last Line of Defense, you came very close to actually having a potentially catastrophic incident. But for the last safeguard (LLoD), the incident would have occurred. Any preceding safeguards that should have stopped the sequence have failed. The entire event, including the initiating event, failure of the safeguards that precede the LLoD, and the activation of the LLoD, should be thoroughly investigated.

8. Process Safety-Critical Equipment

If the BoS lists the most critical safety systems in the plant, then the associated equipment (sensors, valves, etc.) should be included in a list of Process Safety-Critical Equipment. Calibration, maintenance, inspection, breakdowns, etc. on this equipment must be completed with the highest priority and utmost care. If deficiencies or other unexpected findings are discovered, they must be thoroughly inspected. Operations must know that if a piece of Process Safety-Critical Equipment fails, it must be addressed immediately. Safety systems not listed in LOPA would be considered less critical.

Plant personnel should be aware, however, that the BoS will not exhaustively or explicitly list some Process Safety-Critical Equipment. For example if the initiating event in a LOPA is a regulator failure, that regulator must be counted as Process Safety-Critical. Also, if the regulator failure will result in overpressure and rupture of a tank, then the tank should also be considered Process Safety-Critical. If we fail to identify these types of critical equipment, much of the equipment we depend on for containment of hazardous materials will be overlooked, as will equipment whose failure would initiate a sequence of events leading to a catastrophic loss.

9. Other Uses for the Basis of Safety Documentation

It is important that the information in the BoS remains fresh in employees' minds. If we can immerse them with constant reminders during their shift, and help them to recall information, they will be much more successful at running the plant safely. Some ways to achieve this include:

- Identify Process Safety-Critical Equipment – Give the components (valves, sensors, etc.) a unique appearance (color, shape, etc.) on operator display screens.
- Tag Alarms - If a LLoD activates, include in the alarm notification that it is a LLoD that activated.
- Record LLoD activation – If a LLoD activates, log it to a report that is then periodically viewed by unit management or engineering. This can help ensure that these events are thoroughly investigated.
- Physically Identify Equipment – Use a unique tag, paint color or other means to clearly mark Process Safety-Critical Equipment in the field. If an operator finds a device hanging by its wires, or a mechanic is called to trouble shoot an actuated valve, they should immediately know the criticality as Process Safety-Critical Equipment through identifying it in the field.
- Identify Process Safety-Critical Equipment on P&IDs. This emphasizes it to engineers who are marking up changes, and to Hazard Study Team members reviewing the drawings during a hazard study.

10. Conclusion

If information critical to avoiding catastrophic incidents is not retained after training, the likelihood of a major process incident increases. The Basis of Safety document can provide the most critical information to operators in a concise format that can be used both for training and as a reference. The list of potential incidents, the causes, and the most critical safeguards preventing these incidents can be presented in a logical order, enabling better retention due to both the organization of the document and the relatively small amount of information included. The document serves as a summary of all prior hazard studies / risk assessments.

The document can be leveraged to serve many purposes beyond operations. Senior managers, new engineers and others can use portions of the document to understand the major process risks at a facility. Each safeguard listed might be treated as "have-to-have," without which an alternative protective measure is required to continue operating. The document can also serve as a framework for identifying Process Safety-Critical Equipment, Last Lines of Defense, and to identify critical alarms, interlocks and devices in the field, on operator screens and drawings.

11. References

[1] Kaszniak, M. (2010), Oversights and omissions in process hazard analyses: Lessons learned from CSB investigations. Process Safety Progress, 29: 264–269. doi: 10.1002/prs.10373.

[2] Process Safety Management of Highly Hazardous Chemicals; US Occupational Safety and Health Administration; 29 CFR 1910.119(d).

[3] CCPS; Guidelines for Safety and Reliable Instrumented Protective Systems; 2009; Wiley-Interscience; Hoboken, NJ; Appendix B.

[4] CCPS; Guidelines for Developing Quantitative Safety Risk Criteria; 2009; John Wiley and Sons, Inc.; Hoboken, NJ.

[5] CCPS; Layer of Protection Analysis – Simplified Process Risk Assessment; American Institute of Chemical Engineers; 2001.

350026 The Need for a Unified Process Safety Map

Abstract:

Knowing Where You're Going -- The Need for a Unified Process Safety Map

Mr. Teddy Bucher Mr. John T. Perez, P.E. Cognascents Consulting Group, Inc. 11777-A Katy Fwy Ste 438 Houston, TX 77079

teddy.bucher@cognascents.com

Keywords: Process Hazard Analysis (PHA), Process Safety Management (PSM), Process Safety Information (PSI), effective communication, Independent Protection Layers (IPL), Hazard and Operability Study (HAZOP), hazard identification, Layer of Protection Analysis (LOPA), risk assessment, risk management.

Abstract

Since the passage of the OSHA Process Safety Mandate, the various methods of Process Hazard Analysis (PHA) have become commonplace. Process Safety Management (PSM) systems and the corresponding Process Safety Information (PSI) that feed into them have proliferated and permeated numerous aspects of production companies. As this has occurred, constituent groups within those companies have further developed the specific subsets of PSI relevant to them, yielding data sets that are increasingly useful and that allow them to operate more efficiently.

Yet with every PHA revalidation cycle, approaches in both preparation and execution frequently tend to concern themselves with the immediate - the presence of the direct inputs needed for the hazard analysis and the production of the direct outputs of the analysis. Viewing these separate components of PSM as distinct elements has a tendency to result in little consideration being given to the overall quality of the inputs, with the assumption that issues with the design of the input will be captured on the next revalidation cycle for that input. This approach of separate revalidation cycles for each element can be effective for the capture of design changes arising from major modifications to the process itself. However, it can also introduce the possibility for multiple changes to be made to a system in parallel, potentially rendering the input inadequate. It also overlooks the possibility that changes to one element of a PSM program could introduce new demands on another element previously thought to be adequate.

The authors posit that implementation of a unified process safety map that reflects the true interconnected nature of the various elements of a PSM program would minimize this potential for systemic error and allow for a more efficient PSM system. In this paper, the authors discuss the structure of an effective PSM program which utilizes a process safety map. This map would highlight how common PSM elements serve as both inputs to and outputs from the same processes through comprehensive communication of changes throughout the whole PSM program. In addition, the authors also provide examples of breakdowns in communication of PSI to illustrate the importance of a PSM program which continually improves as a whole, as a result of modifications to only one part of the program.

The target audience for this paper includes project managers, project engineers, EH&S managers, PSM coordinators, and operators; however, anyone involved with small or large capital projects may also benefit from this paper's content.

350100 How Is Your Process Safety Vision? - Try Pursuing "Perfect Process Safety"

Abstract:

Many companies have learned that only pursuing regulatory compliance is not enough to manage risks. Economic cycles, organizational changes, and aging assets and people all put great stress on companies and degrade their ability to improve and sustain safety performance. In occupational safety and HSE, some companies have adopted one form or another of "driving bad frequencies to zero" in order to keep their organizations focused on driving safety performance improvement. In process safety, why can't we do the equivalent - let's go after "Perfect Process Safety." This paper will describe a practical framework for pursuing perfect process safety:

- A culture based on proper ownership of process safety empowered by visionary leadership
- Risk-informed sensitivity that guides everything
- Effective, fit-for-purpose management systems
- PS practices embraced and followed with good operational discipline at ALL levels
- Learning from ALL sources – internal, external and outside industry group
- Well-formed/visible performance pyramid; metrics at every level that drive intended behaviors
- Goals and actual performance that improves

Effective process safety leadership throughout organizational levels and learning via management of a company's wisdom life cycle are two vital components of this pursuit strategy. This paper will define ways of measuring process safety leadership based on 15 key attributes of effective leaders and how to identify learning-disabled organizations.

348737 Using Metrics to Improve Emergency Management

Abstract:

10Th Global Conference on Process Safety

Using Metrics to Improve Emergency Management

Rixio E. Medina

AcuTech Consulting Group

September 25, 2013

As part of the process industry's efforts to manage process risk, facilities' emergency planning and response plans are essential. These plans have traditionally being developed by experienced fire chiefs or other emergency response personnel using professional and trade organizations' guidelines, local requirements and equipment manufacturers' recommendations. The OSHA Process Safety Management standard and EPA Risk Management Program regulation established minimum requirements for emergency planning and response for process safety. Unfortunately, the Chemical Safety and Hazard Investigation Board (CSB) continue to identify emergency planning and response gaps in some of the accidents they investigate. Industry's ultimate goal is to have emergency planning and response systems that address all credible accident scenarios they can potentially encounter and that the system is capable of delivering the intended results every time it is activated to minimize the consequences of events. To that end, quality assurance have been attempted by conducting period reviews of systems, equipment and resources using audits, checklists, emergency drills and exercises to identify the level of readiness.

The Center for Chemical Process Safety (CCPS) Risk-Based Process safety (RBPS) Guidelines (2007) provide additional insight on how to achieve an effective emergency management performance for PSM.

The author will examine the RBPS approach for measurement and metrics of emergency management and will offer advice to improve emergency management practices via metrics.

Flame Propagation in Dust/Air Mixtures Under Reduced Pressure Conditions

Hannes Kern
Montanuniversität Leoben
Franz-Josef Strasse 18, 8700 Leoben
hannes.kern@unileoben.ac.at

Gerald J. Wieser, Harald Raupenstrauch
Montanuniversität Leoben
Franz-Josef Strasse 18, 8700 Leoben
gerald.wieser@stud.unileoben.ac.at, harald.raupenstrauch@unileoben.ac.at

Prepared for Presentation at
American Institute of Chemical Engineers
2014 Spring Meeting
10th Global Congress on Process Safety
New Orleans, LA
March 30 – April 2, 2014

UNPUBLISHED

Flame Propagation in Dust/Air Mixtures Under Reduced Pressure Conditions

Hannes Kern
Montanuniversität Leoben
Franz-Josef Strasse 18, 8700 Leoben
hannes.kern@unileoben.ac.at

Gerald J. Wieser, Harald Raupenstrauch
Montanuniversität Leoben
Franz-Josef Strasse 18, 8700 Leoben
gerald.wieser@stud.unileoben.ac.at, harald.raupenstrauch@unileoben.ac.at

Keywords: dust explosion, flame propagation, non-atmospheric conditions

Abstract

Industrial processes are often operated at non-atmospheric conditions. Nevertheless, the development of dust explosions and flame propagation under reduced pressure conditions has not received much attention. One explanation could be that standard laboratory equipment like the 20 l Siwek chamber does not allow investigations at very low pressures.

An experimental device was developed for the investigation of ignition and flame propagation under reduced pressure (0.25 bar to atmospheric pressure). Experiments have been carried out with lycopodium at dust concentrations of 100 g/m³, 200 g/m³ and 300 g/m³. Flame propagation was investigated using a video analysis system and the actual flame speed was measured by optical sensors.

Flame shape and velocity strongly differs from results at atmospheric conditions. Effects like buoyancy of hot gases during ignition and flame propagation are less strong than at atmospheric conditions. For the investigated dust concentrations, flame velocity reached speeds that are nearly an order of a magnitude higher than at ambient conditions. Results gained from these experiments are contributing to the setup of mathematical models for the description of gas explosions and ignition.

1. Introduction [1]

Dust explosion research still faces fundamental problems in the explanation and evaluation of basic parameters of dust explosions, such as minimum ignition energy and temperature, as well as those that define flame propagation, such as flame velocities. Much experimental and theoretical work has already been done but the sheer complexity of the problem still leaves many questions unanswered. Dust explosion testing, for example, still relies on methods that are only able to determine explosion effects under very specific circumstances. The deduction of physical or chemical mechanisms from such experiments is usually not possible. A more knowledge-based approach towards the fundamentals of ignition and combustion is needed for a better description of dust explosions. This paper should contribute to the understanding of dust explosion processes under conditions that deviate from standard ambient conditions. The results gained from these investigations should also lead to a better understanding of dust explosions in general.

2. Experimental Work

The experimental investigation of dust/air flames is rather difficult because dust/air mixtures are inherently turbulent. Any kind of mixing process that creates undefined turbulence levels strongly influences the results of the experiments. Different experimental approaches to determine the flame velocity of dust/air flames are known. For the present experiments, Lycopodium dust was used as a reference substance due to the well-investigated combustion characteristics and its physical parameters. Lyopodium shows a very narrow particle size distribution with a median value of around 30 µm. The shape of the lycopodium spores is nearly spherical with a fine structure that provides a big reactive surface.

2.1 Basic experimental Setup [2]

In the development of dust explosions, turbulence is a key factor in addition to the specific characteristics of the dust itself. Therefore, an experimental setup had to be found that allowed the creation of a homogeneous dust cloud on the one hand and low turbulence conditions on the other hand. Apparatuses described by Krause et al. [3] are able to provide a good particle distribution by using a constant gas flow through a sinter plate under constant but comparatively high turbulence conditions. Other apparatuses use dust feed systems at the top of the testing device, which requires accepting an inferior particle distribution and the disadvantage of not being able to investigate dusts that tend to strongly agglomerate [4, 5, 6]. Pre-tests showed that the dispersion of lycopodium in air could be realized quite easily [7]. Therefore, a top feed arrangement for testing seemed adequate, especially in terms of low turbulence levels. Various dust concentrations are achievable with proper dust feeding systems. A top feed system also meets the requirements for investigations under reduced pressure conditions because no gas stream is required for dust dispersion.

Experiments in closed vessels often do not allow the visual observation of flame propagation and explosion development. For investigations of flame development, optical systems allowing the determination of flame propagation as well as flame geometry are obligatory. Experiments with pressure levels below atmospheric conditions required a completely sealed combustion section.

For further experiments a main explosion tube (length 2000 mm) made of Plexiglas® was chosen. A tube with an inner diameter of 140 mm and a wall thickness of 5 mm fulfilled the requirements for experiments at atmospheric as well as below atmospheric pressure conditions.

Basically the apparatus consists of a bottom section connected to a vacuum pump and housing a spark ignition system, the main combustion tube and a top section with the dust feeding device. Measurement and detection is conducted using an optical system with photodiodes and a video camera for the measurement of the flame shape (Figure 1).

The head element houses the dust input system and is equipped with a flame arrestor and an explosion flap. The dust input is carried out by a screw conveyor that can be regulated by the operating voltage. The input system allows the generation of dust concentrations from 50 g/m³ to 500 g/m³ (Lycopodium). The uniform distribution of the dust over the whole cross section of the tube was monitored by a light extinction of two lasers in a height of 100 mm and 900 mm above the igniton point. The particle distribution was also simulated by computational fluid dynamics and showed that particles are distributed evenly over the whole cross section 500 – 700 mm below the feed point.

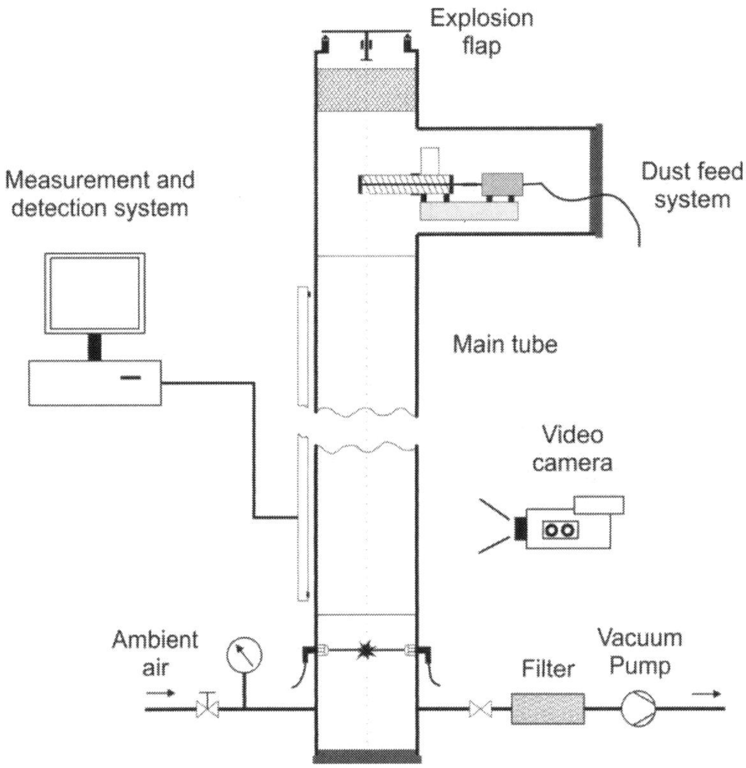

Figure 1: Testing assembly for the determination of flame velocity below atmospheric pressure conditions

The vacuum pump used for the creation of the proper pressure conditions allowed a minimum absolute pressure of around 90 mbar. The total volume of the assembly is around 40 litres. Due to Bartknecht's [3] findings that an ignition below 20 mbar is not likely and the fact that the spark igniter used provided considerably low ignition energies (~10 mJ), the pump characteristics proved to be satisfactory.

Since the experiments dealt with dust explosions that could cause injury to persons and significant damage to equipment, experimental safety was a key issue during the experiments. Main considerations were:

- control of flame spread
- pressure design
- relief systems for uncontrolled pressure build up
- fire protection

After a critical safety analysis, several modifications were made to the apparatus: flame arresters to keep the dust flame from spreading outside the combustion section and a pressure relief system to prevent the apparatus from uncontrolled overpressures.

2.2 Ignition System

For the ignition of the dust/air mixture inside the combustion tube, a spark ignition system was used. The ignition source provided an electrical spark with an ignition energy of around 10 mJ, which was sufficient for the ignition of the dust/air mixture down to around 250-300 mbar. The circuit diagram of the ignition generator is demonstrated in Figure 2.

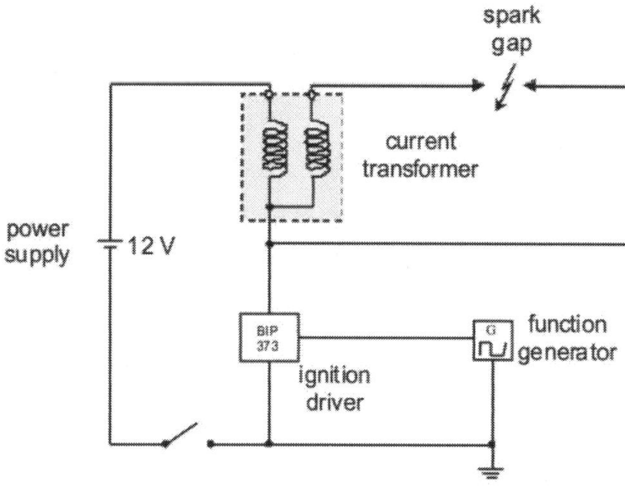

Figure 2: Circuit diagram of the spark generator used for the ignition of the investigated dust/air mixtures.

Two pointed steel-electrodes with a diameter of 2 mm were used as spark igniters. The electrodes were placed such as to form a gap of 6 mm right in the centre of the combustion tube. This electrode assembly is normally also used for dust explosion testing [8].

2.3 *Results*

The main goal of the present work was to investigate flame propagation under non-atmospheric conditions. Investigations on phenomenological variations of the combustion process itself as well as the parameter of flame velocity were of special interest. The ignition process was also of certain interest, although in-depth research on its mechanisms was not possible with the chosen experimental setup.

2.2.1 Flame shape

At ambient conditions the flame forms a parabolic shape due to wall effects [9]. During the experiments under reduced pressure conditions, a change in the appearance of the flame of the flame was noticed. The closed flame shape that is formed at ambient pressure becomes distorted with the reduction of pressure. At the given concentration, flames at 450 mbar and 250 mbar show nearly the same flame speed of around 7 m/s but appear visually as completely different.

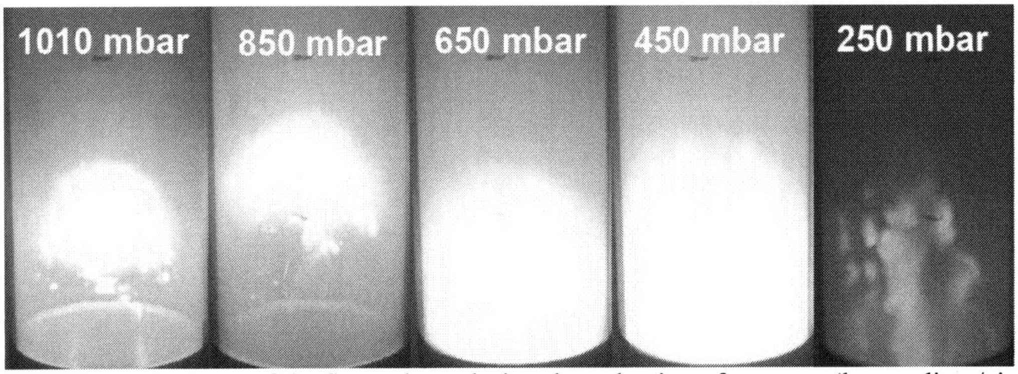

Figure 3: Development of the flame shape during the reduction of pressure (lycopodium/air at 100 g/m³)

At 250 mbar the flame does not form a uniform flame surface anymore and is dominated by single spot flames. In addition, the light intensity of the flames in the low pressure area is much lower than at ambient pressure, which may be caused by lower combustion temperatures.

2.2.2 Flame speed

Investigations on flame speed were conducted in the apparatus described in figure 1. For igniting the dust/air mixture the spark generator described (IE ~10 mJ) was used. Flame speeds measured by Han et al. [5] lie at 0.5 m/s and below, but are not directly comparable with the results achieved during this experiments. The systems used works with a closed bottom end, which leads to a stronger acceleration of the flame due to a stronger increase of pressure behind the flame front.

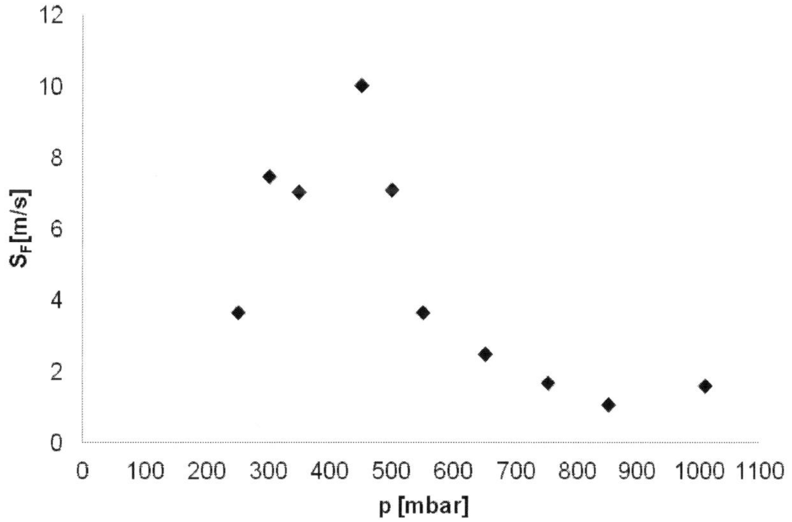

Figure 4: Pressure vs. flame velocity at 100 g/m³ lycopodium

For the experiments, the apparatus was evacuated down to the designated pressure level and ignited after around 30 s of dust feeding to assure homogeneous dust dispersion. Experiments were carried out at dust concentrations of 100 g/m³, 200 g/m³ and 300 g/m³. Figure 4 shows the results of the first set of experiments carried out at 100 g/m³.

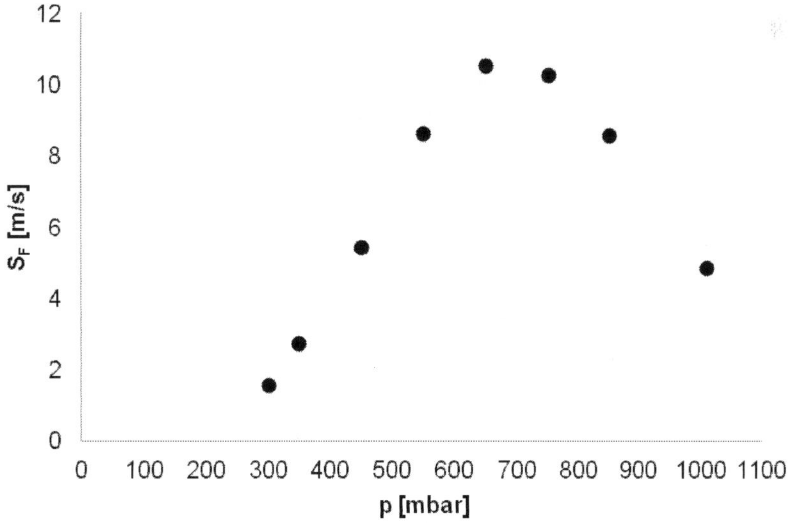

Figure 5: Pressure vs. flame velocity at 200 g/m³ lycopodium

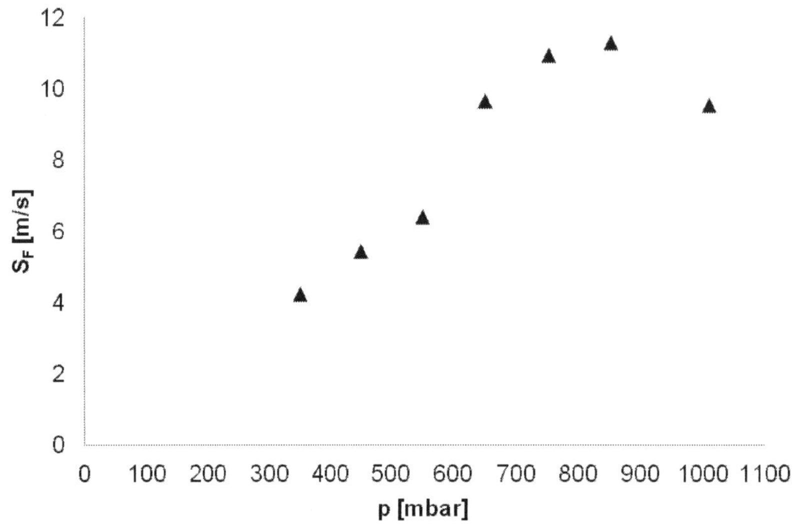

Figure 6: Pressure vs. flame velocity at 300 g/m³ lycopodium

A relatively stable flame speed was found between 1010 mbar and 750 mbar. Below 750 mbar the flame speed increases significantly reaching a maximum at 450 mbar. Further reduction of the pressure slows down flame propagation. With the spark generator used, ignition below a pressure of 250 mbar was impossible. As described above, the formation of a parabolic flame shape was not observed at very low pressures. The considerably high flame speed and the not very bright flame also led to a stronger variation of the values measured at reduced pressures.

Figure 5 shows flame speeds measured at a concentration of 200 g/m³ lycopodium in air. Corresponding to the previous set of experiments, the flame speed rises during the reduction of pressure. A section of constant flame speed was not found. Maximum flame speed was observed at around 650 mbar. Ignition became impossible at a pressure of 300 mbar. As can be seen in Figure 6, the maximum flame speed shifted further towards ambient pressure when the dust concentration was raised to 300 g/m³. The limit of ignitability shifted further and was reached at 350 mbar.

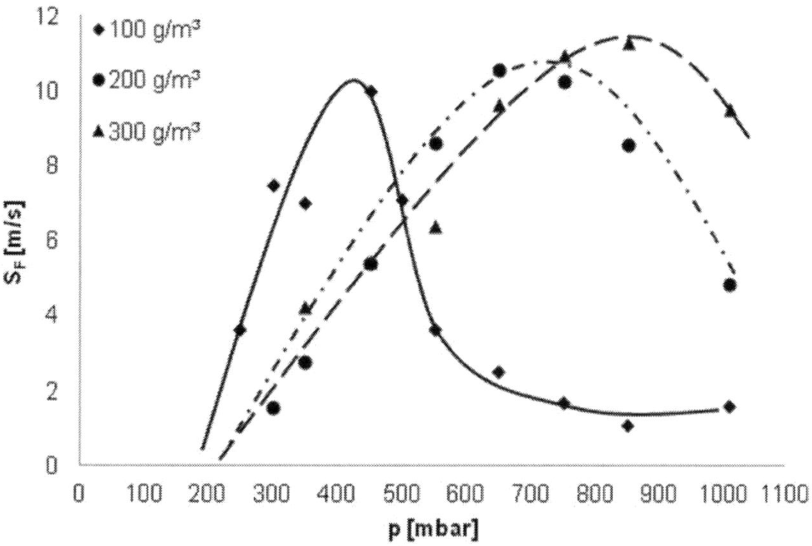

Figure 6: Pressure vs. flame velocity at 100 g/m³, 200 g/m³ and 300 g/m³ lycopodium

A comparison of all three datasets reveals that maximum flame velocities can be achieved at different pressures depending on the fuel concentration (Figure 6). The slope of the increasing as well as the decreasing flank of each curve also varies with the dust concentration. The curve at 100 g/m³ is more narrow than at the other two concentrations. The curves spread with increasing dust concentrations. Interestingly, the falling flanks of all three curves seem to join at a value of around 200 mbar when extrapolated. 1 m³ of air at 200 mbar contains the amount of oxygen that is necessary for the total combustion of 20-25 g of lycopodium. The lower explosion level of lycopodium lies right in this area. Therefore, the result seems consequent.

3 Conclusion

Flame velocity is not a very common parameter for the evaluation of combustible dust/air mixtures, but the experimental settings chosen in this work allowed investigations of flame spread under reduced pressure conditions. In contrast to present testing equipment, the equipment used for this study allowed a closer look at the combustion process itself. It was possible to investigate the influencing parameters on and the conditions at the ignition process in a more detailed way than possible with the 20 l Siwek chamber or the MIKE 3 apparatus.

The experiments below atmospheric conditions revealed effects that had not been predictable in the first place. Flame speeds generally increased during pressure reduction up to a certain peak value and decreased after passing that maximum speed. Flame speeds measured at low pressure conditions reached an order of magnitude higher than at ambient conditions for the investigated dust concentrations of 100 g/m³, 200 g/m³ and 300 g/m³. The velocity maximum shifted towards atmospheric pressure with increasing dust concentrations. Also the flame shape and appearance differs strongly from atmospheric conditions. Within the measurement and detection system there is still room for improvement but the results presented in this paper revealed some new perspectives in the research of flame propagation. Especially the speed measurement system has to be improved to be able to provide better results especially in the lower pressure regions.

The results may be helpful for the design of systems working below atmospheric pressure, but further research on the mechanisms and influencing parameters of this phenomenon have to be done. From the knowledge thus gained, theories of combustion processes under atmospheric conditions can be improved.

4 References

[1] Kern, H., Explosible Dust/Air Mixtures - Investigations on flame propagation under non atmospheric conditions, PhD Thesis, Montanuniversität Leoben, Leoben, 2013.

[2] Wieser, G., Untersuchung der Flammenfortpflanzung in Staub/Luft-Gemischen unter reduzierten Druckbedingungen, Master Thesis, Montanuniversität Leoben, Leoben, 2013.

[3] Steen, H., Handbuch des Explosionsschutzes, WILEY-VCH, Weinheim, 2000.

[4] Proust, C., A few fundamental aspects about ignition and flame propagation in dust clouds, Journal of Loss Prevention in the Process Industries 19 (2006), 104–120.

[5] Han, O. S., Yashima, M., Matsuda, T., Matsui, H., Miyake, A., Ogawa, T., Behavior of flames propagating trough lycopodium dust clouds in a vertical duct, Journal of Loss Prevention in the Process Industries 2000 (13), 449–457.

[6] Palmer, K. N., Tonkin, P. S., Coal dust explosions in a large-scale vertical tube apparatus, Combustion and Flame 1971, 17, 159–170.

[7] Kern, H., Held, K., Raupenstrauch, H., Investigations on the influence of the oxygen concentration on the flame propagation in lycopodium/air mixtures, ACHEMA Congress 2012, Frankfurt 2012

[8] Bartknecht, W., Zwahlen, G., Explosionsschutz, Springer, Berlin [etc.], op. 1993.

[9] Eckhoff, R. K., Dust explosions in the process industries, Butterworth-Heinemann, Oxford [u.a.], 1997.

Using Incident Risk Analysis to Learn from Near Misses

Sharon Tinker
ExxonMobil Research and Engineering Company
4500 Bayway Drive
Baytown, Texas 77520
sharon.k.tinker@exxonmobil.com

Kelly Keim
ExxonMobil Research and Engineering Company
4500 Bayway Drive
Baytown, Texas 77520
kelly.k.keim@exxonmobil.com

Prepared for Presentation at
American Institute of Chemical Engineers
2014 Spring Meeting
10th Global Congress on Process Safety
New Orleans, LA
March 30 – April 2, 2014

UNPUBLISHED

Using Incident Risk Analysis to Learn from Near Misses

Sharon Tinker
ExxonMobil Research and Engineering Company
4500 Bayway Drive
Baytown, Texas 77520
sharon.k.tinker@exxonmobil.com

Other Author(s)
Affiliation(s)

Keywords: incident risk analysis, incident investigation

Abstract

Ineffective or missing layers of protection or protective barriers can result in incidents with consequences that range from insignificant to catastrophic. We usually refer to incidents with minor or insignificant consequences as a near miss. Investigation findings from these near misses can be as valuable to organizational learning and incident prevention as the findings from serious accident investigations. The challenge is to determine which near misses will provide valuable lessons and justify a full investigation. This paper will present a process to identify which near misses to investigate by analyzing the risk associated with each near miss.

The risk analysis considers the actual consequences of the incident, as well as the potential consequences of the incident. The more severe the actual and potential consequences, the higher the risk associated with the incident. Effective protective barriers that prevented the incident from escalating to the potential consequences are also identified and considered. The absence of effective barriers indicates higher risk, while numerous effective barriers reduce the risk. Near misses that are analyzed as high risk receive a full investigation, similar to catastrophic accidents, to capture the invaluable lessons learned.

1. Introduction

ExxonMobil Chemical[1] wanted to apply the safety pyramid concept to process safety, similar to evaluating personnel safety performance, in the early 2000s. Large fires and explosions were visible incidents and should be the top layers of the process safety pyramid. But how could the company capture other process safety incidents and near misses to populate the lower levels of the pyramid?

---

[1] References to "ExxonMobil", "ExxonMobil Chemical", "ExxonMobil Research and Engineering", "we" and "our" and any references to countries and other geographic areas are used for convenience and may refer to one or more of Exxon Mobil Corporation and its affiliates.

There was also a need to create a leading metric to monitor process safety performance. At the time, there were no process safety performance measures used widely in the industry. The low number of catastrophic process safety incidents did not provide a meaningful indication or trend of process safety performance.

ExxonMobil Chemical Company developed the Process Safety Measurement System (PSMS) in 2001 as a risk-based method to analyze the risk of each process safety incident – whether it had actual consequences or was a near miss. The risk analysis considered actual consequences, potential consequences, and barriers that prevented the potential consequences for the incident. The analysis was named the Incident Risk Analysis Tool or IRAT.

The incident risk, as indicated by the IRAT score, has been used for the following in the last several years:

- Determine investigation requirements
- Identify risks not found by the Process Hazard Analysis (PHA) process
- Analyze incidents for common causes to identify areas for improvement

IRAT was implemented at ExxonMobil's chemical plants first, and then use was expanded to the ExxonMobil refineries in 2005. The implementation of industry lagging and leading metrics for process safety by API RP-754 has provided some standardization of process safety metrics, but has not replaced the use of IRAT within ExxonMobil to evaluate incident risk. IRAT is still used to focus on the lower levels of the process safety pyramid.

This paper will briefly explain the application of the IRAT, demonstrate how the creation of the leading metric has been applied to increase focus on process safety risk in ExxonMobil, and compare the difference in incident evaluation between RP-754 Process Safety Events (PSEs) and our incident risk analysis.

2. Incident Risk Assessment Using IRAT

Chuck Fellman, Scott Ostrowski, and Don Williams, from ExxonMobil Chemical, have previously presented a paper titled "A Process Safety Measurement System" at the Mary Kay O'Connor Process Safety Center's International Symposium in 2005. It demonstrated the four-step process of IRAT scoring. This is the same information that I will present first to explain the application of the Incident Risk Assessment Tool.

2.1 Determine the Actual Consequences

Figure 1: IRAT Actual Consequence Level

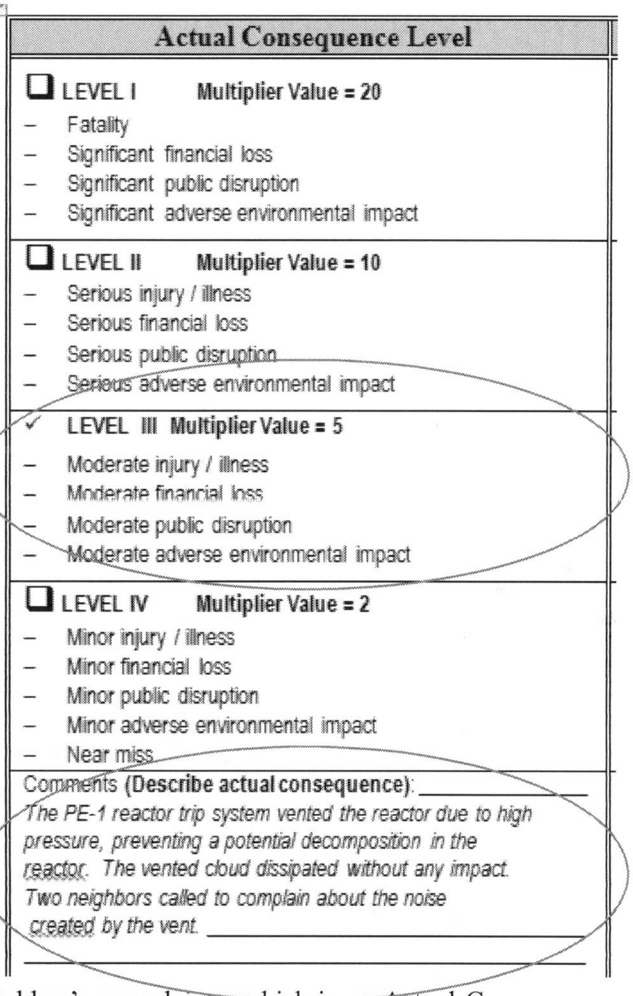
Actual Consequence Level

❑ LEVEL I **Multiplier Value = 20**
- Fatality
- Significant financial loss
- Significant public disruption
- Significant adverse environmental impact

❑ LEVEL II **Multiplier Value = 10**
- Serious injury / illness
- Serious financial loss
- Serious public disruption
- Serious adverse environmental impact

✓ LEVEL III **Multiplier Value = 5**
- Moderate injury / illness
- Moderate financial loss
- Moderate public disruption
- Moderate adverse environmental impact

❑ LEVEL IV **Multiplier Value = 2**
- Minor injury / illness
- Minor financial loss
- Minor public disruption
- Minor adverse environmental impact
- Near miss

Comments **(Describe actual consequence)**:

The PE-1 reactor trip system vented the reactor due to high pressure, preventing a potential decomposition in the reactor. The vented cloud dissipated without any impact. Two neighbors called to complain about the noise created by the vent.

The first step is to determine the actual consequence for the incident or near miss. The actual consequence is evaluated against the levels in the corporate risk matrix and considers the impact on safety and health, the environment, the public, and direct cost financial costs. Each level is assigned a multiplier for the risk analysis calculation.

An IRAT Worksheet provides guidance for determining the IRAT score. An accurate description is written for the incident so that anyone not present will be able to understand what happened. This description is documented at the bottom of the Actual Consequence Column in the IRAT Worksheet in Figure 1. The IRAT Worksheet provides guidance to determine the level of actual consequence. If there are multiple consequences, the highest level is chosen for the analysis.

The example incident shown in Figure 1 is the activation of a safety instrumented system which prevents the incident from escalating. The highest level actual consequence is the public impact from neighbor's complaints, which is an Actual Consequence Level III. The multiplier for Actual Consequence Level III is 5.

2.2 Determine Potential Consequence Level

Figure 2: IRAT Potential Consequence Level

Potential Consequence Level
✓ LEVEL I **Multiplier Value = 20** - Fatality - Significant financial impact - Significant public disruption - Significant adverse environmental impact
❑ LEVEL II **Multiplier Value = 10** - Serious injury / illness - Serious financial loss - Serious public disruption - Serious adverse environmental impact
❑ LEVEL III **Multiplier Value = 5** - Moderate injury / illness - Moderate financial loss - Moderate public disruption - Moderate adverse environmental impact
❑ LEVEL IV **Multiplier Value = 2** - Minor injury / illness - Minor financial loss - Minor public disruption - Minor adverse environmental impact - Near miss
Comments: **(Describe assumed potential scenario & its consequence)** *The PE-1 reactor experienced a high pressure excursion and the trip system did not vent the reactor. The reactor pressure continued to increase and initiated an ethylene decomposition reaction. The decomp reaction caused the reactor pressure to exceed design pressure resulting in a reactor rupture. The blast wall around the reactor failed when the hydrocarbons released from the reactor ignited in a VCE and caused fatal burn injuries to three maintenance technicians working on the PE-2 recycle gas compressor.*

The second step is to determine the potential consequence level by developing potential consequence scenarios. These scenarios include the failure of barriers that did work or were available to prevent the incident from escalating to the potential consequence. The potential consequence scenario must be credible and explain the progression of the incident from the point at which the incident actually ended to the potential consequence. If there are multiple potential consequence scenarios with different consequences, each should be considered and the potential consequence scenario with the highest resulting IRAT score should be selected to determine the investigation level and for reporting.

The potential consequence scenario is documented at the bottom of the Potential Consequence Level column in the IRAT Worksheet as shown in Figure 2. The IRAT Worksheet provides the same guidance from the corporate risk matrix for Potential Consequence Level.

The potential consequence for the example incident is shown in Figure 2. It includes failure of the safety instrumented system and failure of a blast wall resulting in a fatality of a worker in the area. The highest potential consequence level is the fatality, which is a Potential Consequence Level I. The multiplier for Potential Consequence Level I is 20.

2.3 Determine Barriers

The third step is to determine barriers to the potential consequence. The type of barriers that can be counted are human intervention, safety instrumented systems (SIS), safety relief valves, automatic response of the process control computer, and post-release mitigation. These are shown on the IRAT Worksheet in Figure 3. The barriers should be identified in the potential consequence scenario description. The multiplier value for the IRAT calculation is determined by the number of barriers.

Figure 3: IRAT Barriers to Potential Consequence

Barriers to Potential Consequence

Note: To be counted, the barriers must have or could have helped prevent, alert, or mitigate the actual incident from becoming the potential incident scenario. They should be independent from the initiating event and from each other.

❑ **Human intervention** (supported by appropriate alarms, procedures, training, response PPE, etc.) (1 barrier)
- Alarms to alert (including gas detectors)
- Emergency Block Valves
- Manual Overspeed Trip
- Emergency shut-down systems (Manual activation)
- Manually activated depressuring systems
- Operator response to process computer alarm / data

✓ **Safety Instrumented Systems** (SIS) (1 barrier)
- Safety Critical shut-down / protective systems (**Automatic** response action required)

❑ **Safety relief valves and rupture discs** (1 barrier)

❑ **Process control computer**, automatic response action.

❑ **Post-release mitigation** (1 barrier)
- Fireproofing, properly designed & maintained
- Firewater, firefighting equipment, & fire team
- HAZMAT / spill control team (Trained & equipped)
- Safe Havens
- Closed disposal, spill containment system, secondary containment system (dike, process sewer, flare, etc.)
- Emergency medical response

✓ **Personal Protective Systems** (1 barrier)
- Personal Protective Equipment (PPE)
- Equipment guards & interlocks

Other Independent Barriers (list):
❑ ____________________

Total barriers	Multiplier Value
0	20
1	10
2	5
3 +	2

Barriers must meet the following guidelines to be counted in the IRAT calculation:

- Independent of the initiating event of the incident
- Independent of other barriers
- Reliable, with an availability of greater than 90%

In the example, the safety instrumented function that vented the reactor and stopped the incident from progressing is counted as a barrier. The blast wall that surrounds the reactor and would have directed the fire and blast wave away from people in the potential consequence scenario is also counted as a Personal Protective Systems barrier, even though it was not needed in the actual incident it was available and would have prevented the potential consequences. The process control computer could not respond fast enough to prevent the reaction from progressing to decomposition, so would not be counted as a barrier. This results in two barriers and a Multiplier Value of 5.

2.4 Calculate the Score to Determine Risk Level

The final step is multiply the multiplier values to calculate the IRAT score and determine the incident risk level. The risk level categories were developed based on historical incidents. The example incident has multipliers of 5, 20, and 5 resulting in an IRAT score of 500, as shown in Figure 4. This would be a Medium Risk incident.

Figure 4: IRAT Calculation and Risk Level

Actual Multiplier Value __5__ X Potential Multiplier Value__20__ X Barriers Multiplier Value __5__ = __500__ IRAT Score

Lower Risk = 399 and below Medium Risk = 400-799 Higher Risk = 800 and above

2.5 Use of the Incident Risk Data

Incidents with severe actual consequences are highly visible and are investigated to satisfy the demands of stakeholders at the site and outside of the site. The IRAT score provides guidance on which near misses are significant and should be fully investigated to identify learning opportunities. This is a risk-based approach to assigning limited resources to near miss investigation. As noted in the book *Incidents That Define Process Safety* by Atherton and Gil:

> "Sadly, there are many cases where the second accident occurs before the investigation of the first, or preceding near miss, has taken place. Not every event construed as an accident, or near miss, can be investigated to the extent that a major fatal accident is, although the process for doing so should be the same. This is why assessing the true potential of a near miss or minor accident is vital to ensuring that safety critical issues are identified and effort can be placed to resolve these in a timely fashion."

The immediate use of the IRAT score is to determine the level of investigation. Higher Risk and Medium Risk incidents require an investigation team and use of the company's root cause analysis tool to identify causal factors and related management system gaps. The example incident is a Tier 3 PSE and has two barriers, but the potential consequences increase the risk to a Medium Risk and indicate an investigation with full root cause analysis is justified.

Another benefit of IRAT is the increased awareness of potential consequences and barriers in the first-line supervisors and technicians. In particular, the discussion of effective barriers has helped the operating teams understand the purpose of multiple barriers and the importance of their effectiveness in preventing the Level I Potential Consequences from becoming Level I Actual Consequences. This has created a risk screening mind-set of identifying potential consequences and what is preventing those consequences.

The Medium Risk and Higher Risk incidents are also evaluated to determine if a risk has been found that was not identified by the site's PHA work processes. The incident data used to determine the IRAT score and investigation findings are entered in the corporate incident reporting database. This allows site-wide and company-wide analysis of incidents for common causes by risk level.

3. Comparing Incident Risk Analysis with Process Safety Events

ExxonMobil has completed an annual analysis of process safety events in Refining and Chemicals since about 2007. One aspect of the analysis is the health of process safety incident reporting and whether the lower risk, low visibility, incidents are being captured.

Figure 5: 2012 Process Safety Pyramid by Incident Risk

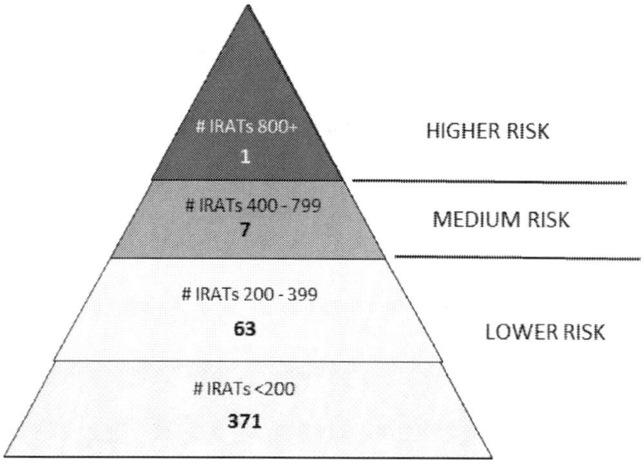

The shape of the process safety pyramid for 2012 based on process safety IRAT scoring is shown in Figure 5. The pyramid shows very healthy reporting with seven Medium Risk incidents for every Higher Risk incident and 434 Lower Risk incidents for every Higher Risk incident. This is similar to the shape of the pyramid in 2011. Both 2011 and 2012 had a much broader base than 2010 and earlier, indicating improvement in miss reporting from earlier years.

Stable and healthy incident reporting is required to use the incident risk scores as a leading indicator of process safety health. Previously, upward trends in cumulative IRAT score were an indication of increased reporting of near misses and lower risk incidents. With stable incident reporting, an upward trend in cumulative IRAT scores could indicate deterioration in management system effectiveness and an increase in site risk. This would indicate further analysis of investigation findings and management system gaps is required.

Figure 6: 2012 Process Safety Pyramid by PSE Tier

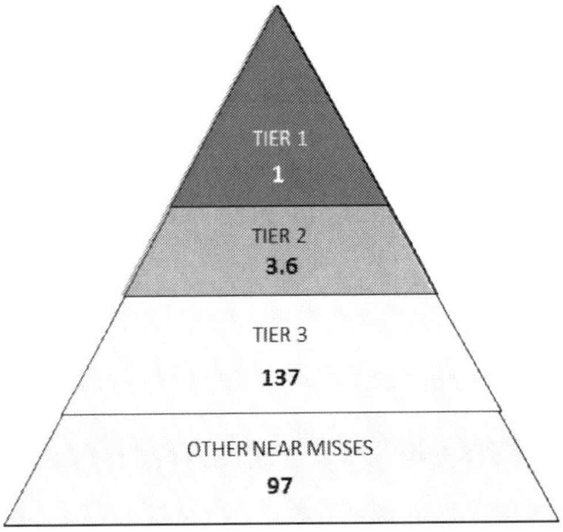

The data analysis team also looked at the same incidents by Process Safety Event tier. The process safety pyramid in Figure 6 shows the ratio of incidents by PSE Tier. There were 3.6 Tier 2 PSEs for every Tier 1 PSE. There were 234 near misses for every Tier 1 PSE and 65 near misses for every Tier 2 PSE. Over 40% of the near misses did not meet the definition of a Tier 3 Process Safety Event.

A comparison between the tiered process safety events and incident risk analysis shows that about half of the higher risk incidents were also Tier 1 Process Safety Events, but 29% of the higher risk incidents were Tier 3 Process Safety Events or other near misses, as shown in Figure 7. Figure 8 shows the contribution of near misses to risk is even higher for the medium risk incidents, with 81% of them being from near misses.

Figure 7: 2012 Higher Risk Incidents

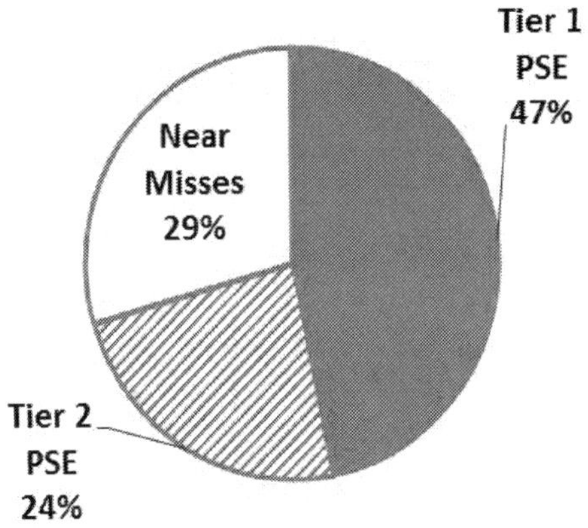

Figure 8: 2012 Medium Risk Incidents

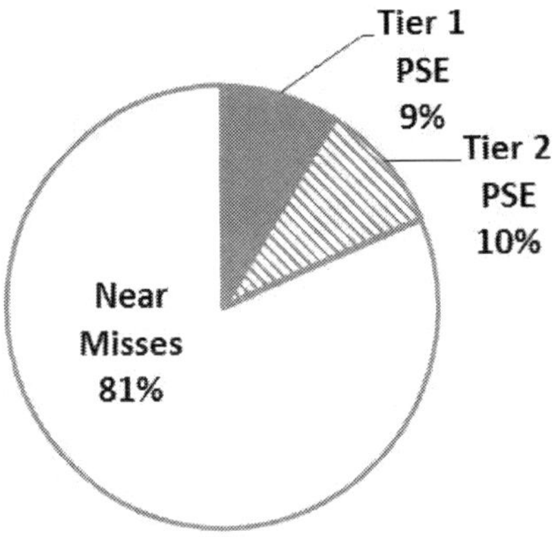

This demonstrates the value of the IRAT tool for identifying risk by using the combination of potential consequences and barriers in addition to the actual consequences that are captured by the APR RP-754 lagging indicators. Investigating only the Tier 1 and Tier 2 PSEs would be missing the opportunities to learn and improve from the higher risk and medium risk near misses and potentially allow repeat incidents that could have severe consequences.

Figure 9: Center Cut of Incidents for Common Cause Analysis in PSE Pyramid

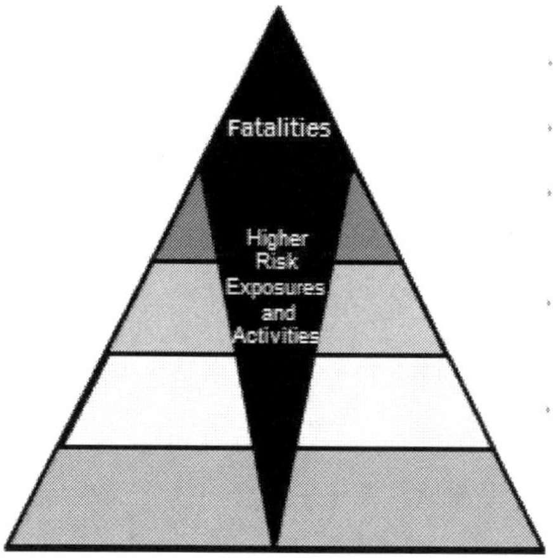

Using the incident risk to determine which incidents are included in the common cause incident analysis focuses the analysis resources on the subset of incidents that provide the most learning opportunity, as shown by the black diamond in the center of the process safety pyramid. This is extending the risk-based approach to improving process safety.

4. Conclusion

ExxonMobil has developed an incident risk assessment process that has been fully deployed at the chemicals and refining manufacturing sites for several years. The immediate, short-term benefits of the tool were to provide a structured method to identify higher risk and medium risk near miss incidents that justified investigation to determine root causes and corrective actions to prevent repeat incidents that could result in more serious consequences. Use of the IRAT also increased the awareness of potential consequences and barriers and the understanding of risk in the people most exposed to the risk – the technicians and first-line supervisors.

A quote from Trevor Kletz, a well-known authority on chemical process safety, on reporting and investigating near misses in the Chemical Safety Board Video *Anatomy of a Disaster* captures the importance of identifying the near misses that should be investigated:

> "Well, it is widely recognized now that you should investigate near misses, or really near accidents that are called near misses, as thoroughly as you investigate accidents. For it is only a matter of luck whether a leak of flammable liquid or gas catches fire or not."

The longer term benefit of IRAT is the development of a leading process safety metric to expand the process safety pyramid. The increased and stable reporting of all incidents, including near misses, and use of IRAT to evaluate the risk has produced a useful leading indicator to evaluate process safety health. We are beginning to evaluate trends in cumulative IRAT scores and average IRAT score by incident across the company and at individual sites to identify focus areas for improvement.

The comparison between incident risk and tiered Process Safety Events demonstrates that risk can be found before it results in severe consequences. This provides additional risk identification outside of the PHA processes. By focusing on identified risk before the risk progresses to an incident with serious consequences, the company is using IRAT to be proactive at improving management system performance and reducing risk, This is always better than being reactive after a serious incident.

5. References

[1] Fellman, C. H., Ostrowski, S. W., and Williams, D.D., "A Process Safety Measurement System," *Mary Kay O'Connor Process Safety Center 8th Annual Symposium Proceedings*, 2005

[2] Atherton, John and Gil, Fredric. *Incidents That Define Process Safety.* Center for Chemical Process Safety, New York, NY, 2008. page 210.

[3] U. S. Chemical Safety Board. *Anatomy of a Disaster.* 2008. accessed at www.csb.gov.

350205 Use of Kpi's for Process Safety

Abstract:

INTRODUCTION

In recent years, the concept of Key Performance Indicators (KPI) and their application to Process Safety has taken prominence following incidents like the ones at BP Texas City Refinery in the USA and Buncefield Oil terminal in the UK.

KPIs could be leading or lagging indicators and their proper usage could help prevent incidents involving hazardous materials and manage Process Risk.

TEXT

The intent of this paper is to briefly walk though the concepts of KPIs, define and explain Leading and Lagging indicators, explanation of the Four Tiers for Process Safety Events (PSE) as identified in API RP 754, usage of Tier 1 and 2 PSE KPIs to asses Company performance, and Tier 3 and 4 PSE KPI's, to monitor weaknesses in Independent Protection Layers (IPL) with the help of Pro-active and Reactive KPIs and application of these KPIs during the lifecycle management of Safety systems.

An example from the Process industry will be used to explain the concepts and how the KPIs could be identified and optimally used to reduce and manage the Process Safety Events in a Process plant. In particular, for Safety Instrumented Systems (SIS) designed based on IEC61511 / ISA84.01, the paper will indicate how such KPIs help measure the 'real' performance at any given time, rather than theoretical performance based on reliability calculations during the design phase, at both the Plant and Corporate level.

CONCLUSION

Use of Key Performance Indicators (KPI) to reduce all levels of Process Safety Events (PSE) is gaining a lot of importance globally. It is understood that this will lead to reduction in Process related accidents and reduce harm to Personnel and the Environment. This paper hopes to re-iterate this message.

349873 Pssr: The Easiest and Sometimes Forgotten PSM Element

Abstract:

PSSR: The easiest and sometimes forgotten PSM Element

Pre-Startup Safety Reviews (PSSR) is to Process Safety what Personal Protective Equipment is to Personal Safety. Often, the PSSR is the last line of defense to ensure that a change or other major plant activity has not introduced an unintended or unmitigated risk. The principles behind the PSSR process are straightforward. The actual PSSR effort is basic and logical. And everyone says they do them and that they are easy. Is there data to support these statements? In this presentation, not only will PSSR best practices be reviewed but common PSSR weaknesses and improvement opportunities will be exposed along with suggested remedies and solutions.

Kinetic identification and risk assessment based on non-linear fitting of calorimetric data

Charles Guinand
ISIC-GSCP
EPFL
Lausanne, Switzerland

HES-SO, Ecole d'ingénieurs et d'architectes de Fribourg
Institut ChemTech
Fribourg, Switzerland,
charles.guinand@epfl.ch

Michal Dabros
HES-SO, Ecole d'ingénieurs et d'architectes de Fribourg
Institut ChemTech
Fribourg, Switzerland,
michal.dabros@hefr.ch

Bertrand Roduit
AKTS AG
Sierre, Switzerland
b.roduit@akts.com

Thierry Meyer
ISIC-GSCP
EPFL
Lausanne, Switzerland
thierry.meyer@epfl.ch

Francis Stoessel
Swissi Process Safety GmbH
Basel, Switzerland
francis.stoessel@swissips.com

Prepared for Presentation at
American Institute of Chemical Engineers
2014 Spring Meeting
10th Global Congress on Process Safety
New Orleans, LA
March 30 – April 2, 2014

UNPUBLISHED

Kinetic identification and risk assessment based on non-linear fitting of calorimetric data

C. Guinand[1, 2], M. Dabros[1], B. Roduit[3], T. Meyer[2], F. Stoessel[4]

[1]HES-SO Haute école spécialisée de Suisse occidentale, Ecole d'ingénieurs et d'architectes de Fribourg, Institut ChemTech, 1705 Fribourg, Switzerland

[2]Ecole Polytechnique Fédérale de Lausanne, 1015 Lausanne, Switzerland

[3]AKTS Inc., www.akts.com, TECHNOArk 1, 3960 Sierre, Switzerland

[4]Swissi Process Safety GmbH, Schwarzwaldallee, 4058 Basel, Switzerland

Abstract

Chemical processes safety requires a thorough knowledge of the kinetic and thermal parameters regarding the chemical reactions involved. The main goal of this work was to develop a method in order to extract the kinetic parameters using different calorimetric technics, experimental design and optimization tools. In the first stage, the esterification of acetic anhydride by methanol was used as a case study for the kinetic identification. In the second stage, a risk assessment was performed using the identified reaction model. Thus, the global kinetic parameters of this reaction, as well as its activation energy and reaction orders were characterized.

Keywords: Calorimetry; Reaction kinetic; Design of Experiment; Thermal Safety; Risk assessment, Optimization

1. Introduction

In most cases, chemical reaction incidents are caused by loss of control, resulting in runaway reactions caused by process deviations in different units. These incidents could be foreseen and avoided, or at least their frequency decreased, with an appropriate process development and risk assessment tool [1].

Reaction characterization is aimed at providing a better understanding of reaction pathways and thermal behavior [2]. In order to achieve this goal efficiently, the system dynamics can be reproduced by a mathematical model whose parameters are estimated based on measured data. Such models are based on ordinary differential equations describing, on one hand the reaction kinetic law (reaction kinetic model) and on the other hand, the measurement model.

After an initial estimation, the measurement model does not represent the overall reality, but only the one in the measurement conditions. For this reason, several experiments at different scales have to be done to maximize the disturbances and explore the overall reaction kinetics [3].

Nevertheless, even with this method, the investigation can be very long due to the increasingly complex models, involving a large number of parameters and needing a huge amount of measurements. Therefore, a new approach to solve these problems has to be developed.

This approach can be summarized in two points: the first one is to plan the measurements in an optimal way to cover the experimental space [4-6]. The second is to develop a numerical method, able to extract the reaction kinetic parameters, for a given model structure, from the measurements available [7].

2. Theoretical part

2.1. *Reaction kinetic model [1, 8-11]*

Assuming a reaction j taking place in a homogeneous solution, the reaction rate r is given by the kinetic law:

$$r_j = k_j \cdot \prod_{i=1}^{I} C_i^{m_i} \qquad \text{[Eq. 1]}$$

The rate constant k is described by the Arrhenius law:

$$k_j = k_{j,0} \cdot \exp\left(-\frac{E_{a,j}}{RT_r} \right) \qquad \text{[Eq. 2]}$$

The concentration C of a component i is calculated in function of the number of mole N in the reaction mass m_r:

$$C_i = \frac{N_i}{m_r} \qquad \text{[Eq. 3]}$$

The transformation rate R of a component i is:

$$R_i = \frac{dC_i}{dt} = \sum_{j=1}^{J} \nu_{ij} r_j \qquad \text{[Eq. 4]}$$

where the sum is made over all J compounds. The thermal behavior is directly linked to the reaction kinetics; indeed any reaction implies a production or consumption of energy which can be detected as heat q by calorimetric techniques and is described as:

$$q_{rx,j} = m_r \left(r_j \cdot \left(-\Delta_r H_j \right) \cdot \sum_{i=1}^{I} \nu_{ij} M_i \right) \qquad \text{[Eq. 5]}$$

The reaction heat flow depends on many factors as for example the reaction rate r [Eq. 1], its enthalpy $\Delta_r H$ and the reaction mass m_r. The molar mass M_i is used to transform the reaction enthalpy in [$J \cdot mol^{-1}$] into the specific heat of reaction given in the units of [$J \cdot g^{-1}$].

2.2. *Optimization Non-linear fitting [12]*

The system considered here is a least squares optimization problem. Its objective consists of adjusting the parameters θ of a model in order to fit in the best way a data set, e.g using a least squares approach defined as:

$$F(\theta)=\frac{1}{2}\sum_{p=1}^{P}\left(f_p(\theta)\right)^2 \quad \text{[Eq. 6]}$$

Where F is the objective function and f_i: $\mathbb{R}^n \longrightarrow \mathbb{R}$, $p=1,\ldots,P$ the residuals. p are the studied functions (e.g. the measurement simulation) with $p \geq n$ and n the number of parameters in the model.

It is assumed that it exists a vector of parameters $\hat{\theta}$ which allows the model to describe the response y as:

$$y_i = Model\left(\hat{\theta},t_i\right)+\varepsilon_i \quad \text{[Eq. 7]}$$

where ε_i is the measurement error compared to the model.

For any choice of θ, the residuals can be calculated as:

$$f_i(\theta)=y_i-Model(\theta,t_i) \quad \text{[Eq. 8]}$$

For least squares fits, the optimal vector $\hat{\theta}$ is considered as the minimizer of the sum of residual squared of the model:

$$\hat{\theta}=\arg\min_{\theta}\{F(\theta)\} \quad \text{[Eq. 9]}$$

In general, such problems are very hard to solve due to non-linearity; however, some methods are presented by K. Madsen et al.[7]. The one retained for this work is the secant version of the Levenberg-Marquardt (L-M) algorithm [7, 13, 14]. This method is used to optimize the model parameters (in this case, the kinetic parameters) without the need for analytical derivative solutions, in order to quickly find the optimal minimum using gradient methods. Indeed, they are numerically estimated.

3. Experimental part

3.1. Studied reaction

The chemical reaction chosen to apply the method proposed above is the exothermic esterification reaction of acetic anhydride and methanol [15]. This reaction is widely used as a test for safety-oriented studies and as such, is adequate for the modeling and simulation of dynamic processes during runaway scenarios.

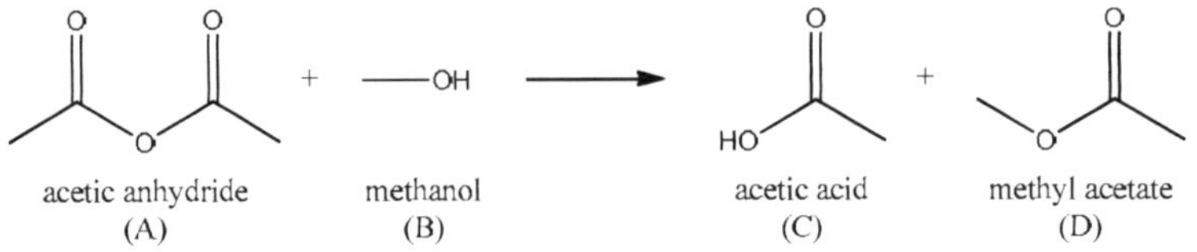

A two-step reaction scheme was proposed by Bohm et al [15] considering the autocatalytic effect of the acetic acid on the kinetic of this esterification as:

$$A + B \rightarrow C + D$$
$$A + B + C \rightarrow 2C + D$$

This reaction scheme will be used as hypothesis to define the different kinetic parameters for the reaction model.

3.2. *Determination of the work space*

Two non-isothermal measurements at different heating rates between 25 and 300°C have been carried out at 0.1 and 4°C/min respectively in Setaram C80® and Mettler Toledo® DSC. Considering the esterification reaction, the temperature range should be below the boiling point, but nevertheless sufficiently high to have an appropriate experiment duration and with a high range of experimental temperatures. On the basis of these measurements, the working temperature range was determined in order to keep the kilo scale measurements safe and clean. Moreover no significant secondary decomposition reaction was detected in the temperature range of the measurement.

Other factors are relevant at the kilogram scale, for example the number of moles fed into the reactor ($n_{i,fd}$) and the time to feed (t_{fd}). Consequently, in order to determine all the different ranges, a rough optimization of the reaction model parameters was performed on the preliminary measurements. Then, using the physical properties (boiling points, heat capacities) and the kinetic parameters, it was possible to predict the process behavior at some different temperatures (Table 1).

Table 1. Variable ranges for the Design of Experiment

	Temperature ($T_{r,set}$)	Time to feed ($t_{fd,B}$)	Number of added moles ($n_{feed,B}$)
Units	[°C]	[min]	[mol]
Min	40	60	0.7
Max	60.3	120	1.5

3.3. *Design of Experiment (DoE)*

An experimental design, based on the Wootton, Sergent and Phan-Tan-Luu's (WSP) algorithm and a Space Filling Design method [16] was used to choose six experimental points, as shown in Table 2. Each of the three factors shown in Table 1 appears at a minimum of three levels – a criterion which was needed to extract the value of the activation energy.

3.4. *RC measurements*

The RC experiments are performed under isothermal fed-batch conditions in a Mettler Toledo® RC1e where the methanol (B) is fed at different rates. The acetic anhydride (A) is initially present in the reactor, up to 1.5 moles, and the acetic acid (C) and methyl acetate (D) are produced during the process. The results are shown in Table 2.

Table 2. DoE study: Space Filling Design results resulting from WSP algorithm[16] performed on 10000 points in the candidate set randomly generated with minDist = 0.8500 and Cov = 0.0020. Results: Feed rate, time of experiment Obtained heat (Q_r), average heat capacity ($c_{p,average}$) and reaction enthalpy ($\Delta_r H$) for the measurements at RC scale.

	DoE plan			**Results**			
Exp.	$T_{r,set}$	$n_{feed,B}$	t_{fd}	m_{fd}	$c_{p,average}$	Q_r	$\Delta_r H$
	[°C]	[mol]	[s]	[g·min^{-1}]	[J·K^{-1}·g^{-1}]	[KJ]	[kJ·mol^{-1}]
1	50.15	1.09	5425	0.39	2.08	-61.30	-56.17
2	60.30	0.69	3843	0.35	2.05	-36.26	-52.65
3	41.35	0.69	7247	0.18	2.09	-43.59	-63.58
4	40.00	1.49	7034	0.41	2.05	-76.38	-51.13
5	60.04	1.48	3776	0.76	2.10	-84.10	-56.65
6	60.03	0.69	7200	0.19	2.10	-40.78	-58.69
Average					**2.08**		**-56.5 ± 4.7**

The determined mean reaction enthalpy corresponds to -56.5 ± 4.7 [kJ·mol^{-1}], a lower value than the one found in literature, 67.3 ± 3.0 [kJ·mol^{-1}] [17] as well as 65.99 [kJ·mol^{-1}] based on online tabulated standard enthalpies of formation [18].

This deviation could be explained by a too high working temperature, causing heat losses by evaporation considering a boiling point of 56°C and 64°C for methyl acetate and methanol respectively. Nevertheless, these data are a good input to create the data basis for the optimization and to create a reaction model.

4. Results and discussion

4.1. Model

Using the non-linear optimization based on the Levenberg-Marquardt algorithm on the obtained calorimetric data (DSC, RC) and the reaction scheme proposed by Bohm[15], it was possible to determine the kinetic parameter values (Table 3).

Table 3. Obtained kinetic parameters for the esterification of acetic anhydride with methanol considering an autocatalytic effect of the acetic acid (product).

Reaction 1				
	k_0	E_a	$\Delta_r H$	Orders
Units	[g$^{mA+mB-1}$·s^{-1}·mol$^{-(mA+mB-1)}$]	[J·mol^{-1}]	[J·g^{-1}]	[-]
Value	4.462e8	71678	-662.3	m_A=0.96 m_B=0.71
Reaction 2				
Units	[g$^{(mA+mB+mC-1)}$·s^{-1}·mol$^{-(mA+mB+mC-1)}$]	[J·mol^{-1}]	[J·g^{-1}]	[-]
Value	3.293e7	51815	-260.1	m_A=1.16 m_B=0.6 m_C=0.52

4.2. *Validation*

In order to validate the model, the simulation was confronted to two different measurements at the milligram (DSC) and kilogram (RC1) scales, as shown in Figure 1.

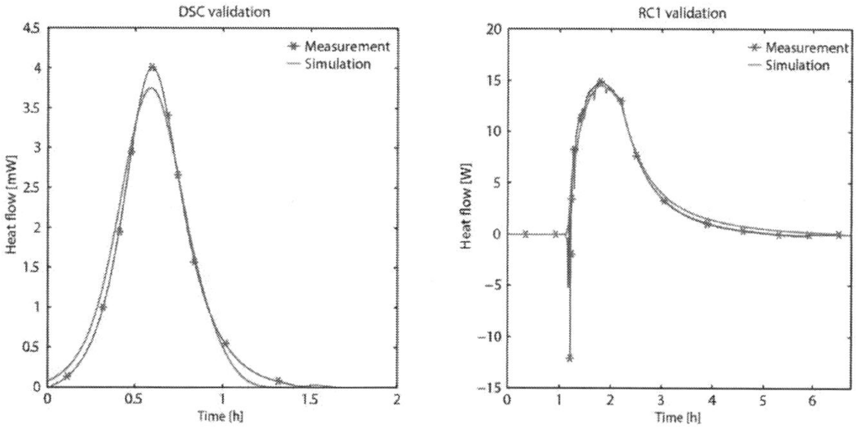

Figure 1 Validation: comparison of the measurements (DSC at 1.5K/min: left; RC1 at 62°C: right) versus their respective simulations using the optimized reaction kinetic model.

It can be observed that the model fits the RC measurement somewhat better than the DSC data. This can be explained by a large number of measurements at this scale and focused on a narrow temperature range. In other words, the DSC leads to an extrapolation (large temperature range) where RC1 measurement are closer to interpolation. Hence, the optimization is more disturbed by these latter which leads to an optimum displaced toward the kilogram scale. Other facts as micro-mixing or difficulties in the baseline definition may have caused the slight misfit observed in the DSC validation.

Another point to keep in mind is the fact that a model is valid only in the range where it was defined. In this case, the goal is to perform a scale-up of this reaction; consequently, a good fit at the kilogram scale points out the right direction.

4.3. *Risk assessment*

In order to get a safe process, several points regarding the thermal safety should be addressed. The best way to assess the risk, is the cooling failure scenario [1].

The main variable to consider in a process risk assessment is the maximum temperature that can be reached in the reaction vessel as a result of a cooling failure (Maximum Temperature of the Synthesis Reaction, MTSR). Would this temperature be sufficiently high to initiate secondary reactions? Would the consequences be worse? What would the level of temperature be after such events? Is it the worst moment during the process for cooling failure to happen?

4.4. *Batch process*

Based on the reaction kinetic model, the thermal behavior of the autocatalytic esterification can be evaluated. Thus, the worst case can be assessed depending on the kind of reactor and cooling system used. A simulation with a standard reactor of 2.5m^3 was used to demonstrate the thermal risk assessment procedure.

4.4.1. Cooling failure scenario

The scenario in case of loss of control from a starting temperature of 25°C would lead to a MTSR of 236°C (ΔT_{ad} = 211°C). Such a temperature would lead to loss of containment with spillage of flammable compounds and possibly a secondary explosion. Moreover, the Time to Maximum Rate in case of adiabatic conditions is very short (TMR_{ad} = 1.36h). The two limit temperatures defining the thermal stability of the reaction mixture were determined using simulation for 24h and 8h are respectively -8°C (T_{D24}) and 4°C (T_{D8}). Thus, the lack of control of the reactor temperature results in a thermal explosion.

Another point to consider is the boiling point of the reaction which is about 60°C at the end of the process. This temperature is reached during the runaway, and the energy available at this temperature is sufficient to result in a pressure increase, leading eventually to a burst of the reactor. Consequently the Maximum Technical Temperature (MTT) should be based on the boiling point of the components involved.

All these points lead to a final conclusion: in order to work under safe conditions, the reaction cannot be performed as a batch without emergency measures. It requires a reliable and powerful cooling system or should preferably be performed in the fed batch mode allowing a better control of the reaction rate.

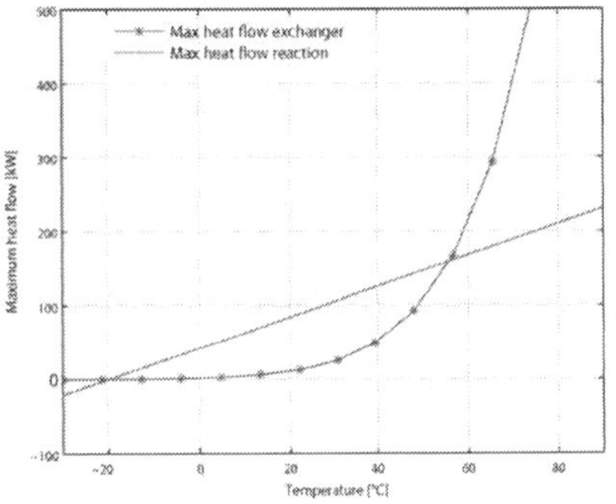

Figure 2 Semenov diagram: comparison of the maximum heat release of the autocatalytic esterification with the heat removal of the cooling system, considering standard reactor of 2.5m^3 (Area = 4.6 m^2; T_c = -20 °C; U = 500 W·K^{-1}·m^{-1}) with an equimolar mixture (15kmol of A and B).

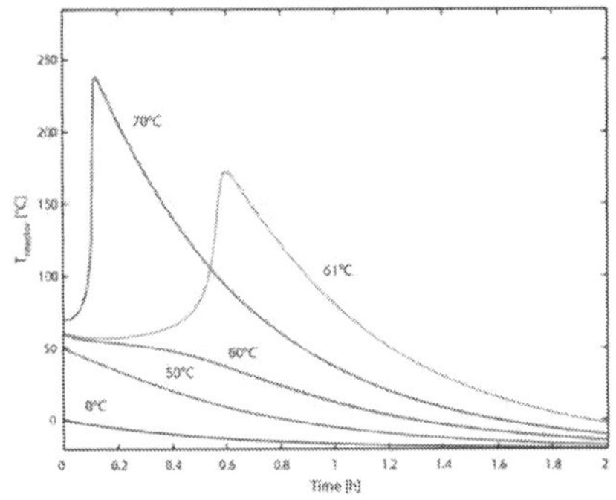

Figure 3 Influence of the initial temperature of the process for a constant temperature of the coolant (T_c = -20 °C)

4.4.2. Semenov diagram and assessment of cooling capacity

This criterion allows the discrimination of stable operation conditions from a runaway situation. The kinetic model was used to evaluate the maximum heat flow leading to the Semenov diagram (Figure 2) which can be used to define a working area for this reaction.

Nevertheless, it should be kept in mind that a too high temperature can produce small deviations resulting in excess heat production and leading to a runaway. In case of an autocatalytic system, a small deviation in temperature will produce more product, thereby accelerating the main reaction and leading to a quicker runaway.

Moreover, the kinetic model can be used to determine the influence of the initial temperature under isoperibolic control as shown in Figure 3.

The results of these simulations show that a too high initial temperature can lead to a runaway. In such conditions for the autocatalytic esterification, the Semenov Diagram predicts a runaway region around 60°C which is confirmed here.

Moreover, this temperature is higher than the boiling point of the final reaction mixture. This latter can be used as a barrier only if the process is designed to work under such conditions (reflux). Consequently, the risk of a pressure increase leading to an explosion is not negligible in the present case.

4.4.3. How to control this reaction

Different types of temperature control are available for batch processes such as adiabatic, isoperibolic or polytropic. In this project, only the polytropic control will be discussed, whereby the reaction temperature is first brought to a setpoint which is maintained by varying the jacket temperature.

As shown above (Figure 3), the initial temperature is a primordial factor. Hence, if this latter is poorly designed, it can lead to a loss of control and runaway reaction. Nevertheless, in order to explore the process behavior, knowledge of the dynamic behavior of an industrial reactor is necessary. Consequently, to define this last point, the different time constants for the heating and cooling were determined at different scales [19].

The time constants show how the cooling system responds to a deviation in temperature. Indeed, two factors need to be taken in account: the heat exchange variables (U, A, c_p and mass) and the controller (usually PID). In practice, even if the controller is designed to deliver a fast response, the heat carrier response will be limited due to the thermal inertia of the system at the industrial scale, therefore, the adjustment is not instantaneous.

Consequently, the reaction temperature must be chosen appropriately to obtain an economically reasonable process time, meaning as short as possible and sufficiently low to control the heat release rate. In this case, a low temperature would be preferred in order to keep the process safe. Unfortunately, maintaining a low temperature is not economically viable. Moreover, in order to keep isothermal conditions, the jacket temperature has to respond really quickly. Consequently, depending on the plant available, this kind of conditions may be difficult to achieve in practice. If this is the case, a possible way to manage this kind of reaction is to carry out a step control at the beginning: start at a low temperature and then let the reaction progresses adiabatically until the temperature reaches a switch value (T_s), at which time the cooling system is started (Figure 4).

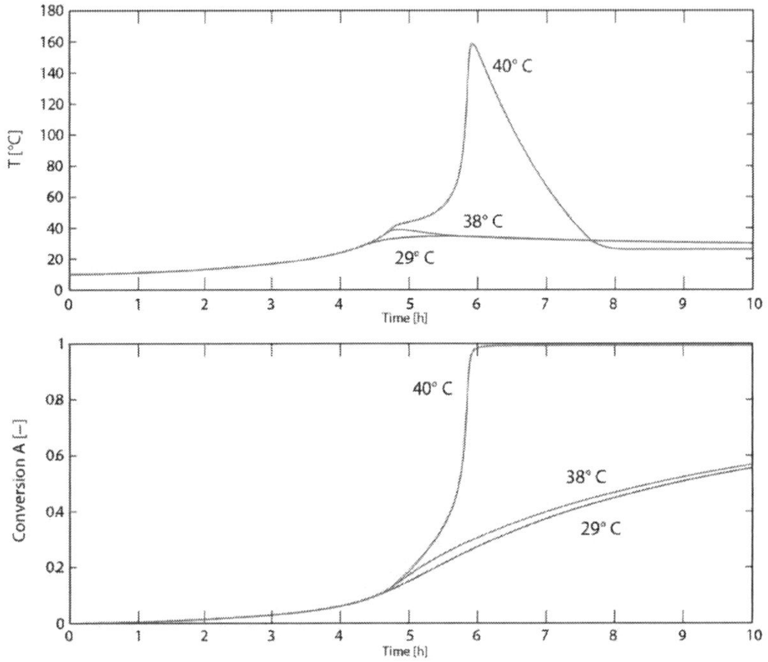

Figure 4 Reactor temperature as a function of time and respective conversion for different switching temperatures of the cooling system from a starting temperature at 10°C.

As it can be observed, even with this method, the conversion remains poor for batch processes under control. Another critical point remaining is the maximum Temperature reached in case of Cooling Failure (T_{cf}). Indeed, as mentioned previously, the boiling temperature of the final reaction mixture is low (around 60°C). Consequently, in order to avoid a rise of pressure, the T_{cf} should remain under the boiling point. Unfortunately, this condition can never be achieved in case of batch processes; only the reactor temperature can be adjusted to keep the temperature under control.

This assessment remains global and does not take in account supplementary equipment such as burst disk or condenser. Nevertheless, it was shown that for this kind of reaction, batch processes are not the most suitable. Moreover, the worst moment for a cooling failure is, in the case of this process, present from the start, as the accumulation is maximal, making this mode not safe at all for such autocatalytic reaction.

Considering all these points, in order to find a viable a safe solution, semi-batch process will be explored from a thermal safety point of view.

4.5. *Semi-Batch process*

4.5.1. *How to control this reaction?*

In semi-batch operation, many strategies can be used to improve the process safety: the temperature control strategy, the feed control strategy, and also the choice of reactant(s) to be initially added and, respectively, to be fed [1]. The most stable reactant is generally chosen to be initially added. Consequently, due to its higher boiling point, the acetic anhydride was chosen. Thus, the methanol is used as the fed reactant in order to limit the accumulation and evaporation. Moreover, in order to maintain isothermal conditions, the use of polytropic control is preferred.

The feed may be controlled in different ways such as:

- Constant feed rate: should be sufficiently low to avoid or at least to minimize reactant accumulation.
- Working under constraints based on physical considerations (volume at the end of the operation, flow rate,...) and safety considerations (the system must remain safe under normal operating condition,...) [20].

Nevertheless, the choice of the reactor temperature and feed strategy remains of primary importance during the thermal safety assessment. The next section will discuss a semi-batch process working under constraints based on safety considerations.

4.5.2. *How to determine an optimal feed profile working under constraints*

As mentioned previously in the batch process section, to maintain isothermal conditions, the heat exchanger system should be able to remove the heat release rate of the reaction at any moment, especially at the maximum. In order to control the reaction, the temperature that would be reached in case of a cooling failure T_{cf} can be used as constraint: if the calculated T_{cf} is greater than the maximum safe temperature (T_{max}), the feed is stopped until the content reaches a stable point anew [1]. The T_{max} is usually defined by the process itself or on the data basis of the reaction mixture (boiling point, decomposition temperature…).

The constraint is applied as follows: the simulation is performed under normal conditions and then every 60 seconds, the simulation is performed under cooling failure conditions (adiabatic). If the T_{cf} is higher than the set point T_{max}, the feed is stopped. The result is the evaluation of the best feed profile satisfying the constraint as shown in Figure 5.

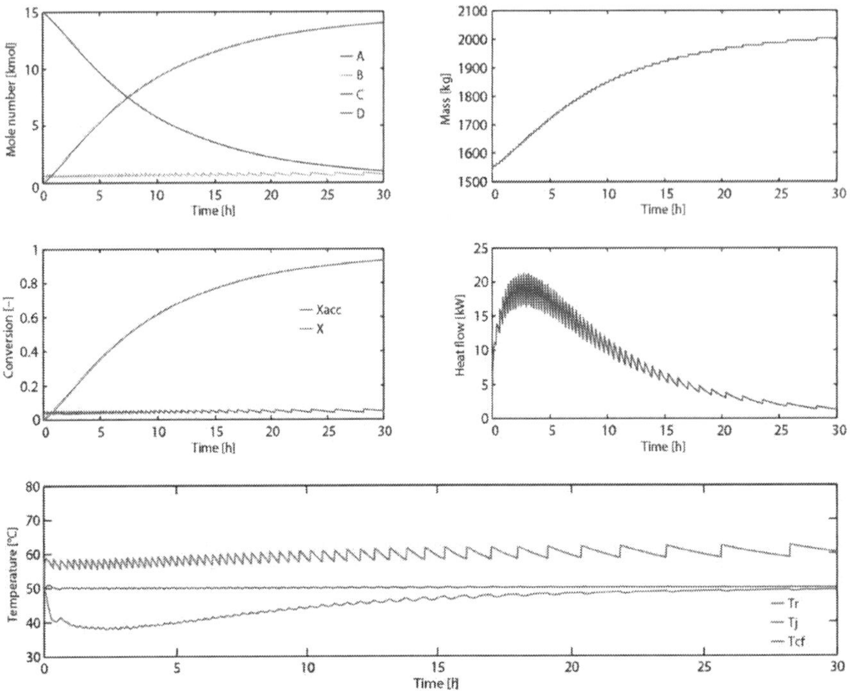

Figure 5 Concentration, mass, conversion (of A, and accumulation), heat flow profiles obtained for a semi-batch process working under T_{cf} constraint

With this strategy, the feed profile is optimized in order to ensure that T_{max} may not be surpassed even in case of failure and considering the autocatalytic reaction.

5. Conclusion

A first insight into a way of investigating reaction kinetics using non-linear model fitting has been presented. It was possible with limited experimental data to determine the reaction's kinetics and to predict its behavior under different operation conditions.

Industrial reactors dynamics were simulated, in order to optimize the thermal safety of a semi-batch reactor. Furthermore, the application of constraints has led to an optimized feed profile keeping the process thermally safe.

In order to validate the optimal feed profile, the next step in this study involves performing an experiment at industrial scale. To achieve this purpose, it is necessary to apply the scale-down approach presented by Zufferey [19]; a method used to simulate industrial reactor at laboratory scale.

In the future, an interesting study might involve the extension of this investigation procedure to other kind of reaction, leading to a generalized method to get reaction model and kinetics from calorimetric measurements.

6. Symbols

A	Area	(m^2)
C	Concentration	(mol·kg^{-1})
c_p	Specific heat capacity	(J·kg^{-1}·K^{-1})
Cov	Covariance	(-)
DoE	Design of Experiment	
E_a	Activation energy	(J·mol^{-1})
F	Objective function	(-)
k	rate constant	($\propto$)
k_0	Frequency factor	($\propto$)
M	Molar mass	(kg·mol^{-1})
m	Mass	(kg)
$MinDist$	Minimal distance between the points	(-)
$MTSR$	Max. Temperature of Synthesis Reaction	(K)
MTT	Max. Temperature for Technical reasons	(K)
N	Number of moles	(mol)
n	Number of parameters in the model	(-)
p	Studied function	(-)
PID	Proportional-Integral-Derivative	(-)
Q	Thermal energy	(J)
q	Heat flow	(W)
r	Reaction rate	(mol·kg^{-1}·s^{-1})
R	Transformation rate	(mol·kg^{-1}·s^{-1})
TMR_{ad}	Time to maximum rate adiabatic	(s)
T	Temperature	(K)
t	Time	(s)
U	Overall heat transfer coef.	(W·m^{-2}·K^{-1})
WSP	Wootton, Sergent, Phan-Tan-Luu's algorithm	
X	Conversion	(-)
$\Delta_r H$	Enthalpy of reaction	(J·mol^{-1})
ΔT_{ad}	Adiabatic temperature rise	(K)

Greek

ε	Error
θ	Parameters
ν	Stoichiometric coefficient

Subscript

acc	Accumulation
cf	Cooling failure
f	Final
fd	Feed
i	Component
j	Number of the reaction
r	Reactor
rx	Reaction
s	Switch
set	Set point

Superscript

m	Partial reactant order

7. References

[1] Stoessel, Francis. *Thermal Safety of Chemical Processes: Risk Assessment and Process Design*. Wiley. Weinheim. 2008.

[2] Blackmond, Donna G. "Reaction Progress Kinetic Analysis: A Powerful Methodology for Mechanistic Studies of Complex Catalytic Reactions," *Angewandte Chemie International Edition,* 44, 28, 2005

[3] Stegmaier, Johannes, Skanda, Dominik, et al. "Robust Optimal Design of Experiments for Model Discrimination Using an Interactive Software Tool," *PLoS ONE,* 8, 2, 2013

[4] Zaglauer, Susanne "Bayesian design of experiments for nonlinear dynamic system identification," Proceedings of the 5th International ICST Conference on Simulation Tools and Techniques, Desenzano del Garda, Italy, Institute for Computer Sciences, Social-Informatics and Telecommunications Engineering, 2012.

[5] Lazic, Z.R. *Design of Experiments in Chemical Engineering*. Wiley. 2004.

[6] Tinsson, W. *Plans d'expérience: constructions et analyses statistiques*. Springer. 2010.

[7] Madsen, K., Nielsen, N. B., et al. *Methods for non-linear least squares problems (2nd ed.)*, IMM, 2004.

[8] Ingham, J., Dunn, I.J., et al. *Chemical Engineering Dynamics: An Introduction to Modelling and Computer Simulation*. Wiley. 2008.

[9] Levenspiel, O. *Chemical Reaction Engineering*. 1999.

[10] Fogler, H.S. *Elements of chemical reaction engineering*. Prentice Hall PTR. 2006.

[11] Capellos, C. and Bielski, B.H.J. *Kinetic systems: mathematical description of chemical kinetics in solution*. R. E. Krieger Pub. Co. 1980.

[12] AKTS-*Thermokinetics Software, Reaction Calorimetry Version*, http://www.akts.com/reaction-calorimetry.html

[13] Nocedal, J.A. and Wright, S.J. *Numerical optimization*. Springer-Verlag New York. 1999.

[14] Wojciechowski, B. and Rice, N. *Experimental Methods in Kinetic Studies*. Elsevier Science. 2003.

[15] Bohm, Stefan, Hessel, Günther, et al. "Auto-catalytic effect of acetic acid on the kinetics of the methanol/acetic anhydride esterification," *Institute of Safety Research*, 2005

[16] Santiago, J., Claeys-Bruno, M., et al. "Construction of space-filling designs using WSP algorithm for high dimensional spaces," *Chemometrics and Intelligent Laboratory Systems,* 113, 2012

[17] Duh, Yih-Shing, Hsu, Chang-Chia, et al. "Applications of reaction calorimetry in reaction kinetics and thermal hazard evaluation," *Thermochimica acta,* 285, 1, 1996

[18] "National Institute of Standards and Technology (NIST)", http://webbook.nist.gov/chemistry (accessed February 2014).

[19] Zufferey, Benoît. *Scale-down approach: chemical process optimisation using reaction calorimetry for the experimental simulation of industrial reactors dynamics*. Swiss Federal Institute of Technology, Lausanne. 2006.

[20] Ubrich, O. *Improving safety and productivity of isothermal semi-batch reactors by modulating the feed rate*. Swiss Federal Institute of Technology, Lausanne. 2000.

349619 Integration of Learned Knowledge into the Technical Information System

Abstract:

How to integrate acquired knowledge of experienced employees into the technical information systems to drive continuous process safety improvement through training and risk assessment. Creation of mentoring programs and networking will minimize organization knowledge loss when key personnel are lost due to retirement, organization changes, or asset divestitures. This will create interaction locally, throughout the organization, and industry. It will enhance employee training and development, understanding of hazard and risk identification and mitigation systems, improve process technical understanding, and reduce the potential for a catastrophic event.

349712 Control Systems Integrity Review - Selective Application of Controls Systems HAZOP (CHAZOP) Study

Abstract:

A Control Systems Hazard and Operability Study (CHAZOP) provides a means of systematically reviewing the design and operation of a control and safety system to identify potential hazards and/or operability problems. It focuses on how a control or safety system may deviate from the design intent, and draws on the expertise of the team members and their past experience with the design and/or operations of similar facilities.

The CHAZOP technique stemmed from the evolution of control and protection system designs, from single-element analog devices a couple of decades ago, to the increased application of electronics, logic solvers with redundant processors and voting capabilities, and feedback loops to identify various sensor and end-device failures, including covert. Although the control and protection systems were evolving, many of the HAZOP Study applications and Teams did not always involve participants that could properly weave these new control and protection system features into the HAZOP Study scenarios. To address this gap, CHAZOP was developed to help ensure that the safety evaluation objectives of the HAZOP were achieved.

For high-quality HAZOP Studies that are supported by individuals with the correct "skill sets" and thoroughly address instrumentation issues (to the extent necessary for the conservative scenarios evaluated for the HAZOP Study), it can be argued that a complete CHAZOP isn't necessary. CHAZOP can also be a significant effort that is difficult to justify to management, if a HAZOP Study has been performed. However, there can still be value in reviewing some of the key characteristics associated with the Integrated Control and Safety Systems (ICSS) that are applied at modern process facilities. With the HAZOP Study Team (and any necessary supplemental instrumentation specialists), there can often be value in a general review of network architecture, workstations, switching networks, key communications/power cable routing, data transfer strategies, human-machine interface issues, back-up strategies, and power sources/reliability. This Control Systems Integrity Review can be implemented in a focused manner to review single failures and potential vulnerabilities.

The objective of this paper is to review appropriate applications of CHAZOP and to define a streamlined approach for Control Systems Integrity Reviews that can benefit many capital and instrumentation retrofit projects in process facilities. Tips will be provided to maximize the effectiveness of the Team's efforts when performing the Control Systems Integrity Review.

349110 Process Safety Culture Best Practices

Abstract:

This session shares industry best practices for establishing a process safety culture.

It will discuss how to align senior leaders to a common definition through metrics and ho to sustain a safety culture.

equion

Energía para la vida

Are We in Control of Our Safety Critical Equipment in Drilling Operations?

Claudio Castaneda– April/2014

equion
Energía para la vida

From hazard to major accident

10TH GLOBAL CONGRESS ON PROCESS SAFETY

Well barriers for Major Hazards risk

HARD → SOFT

PLANT	PROCESS	PEOPLE
Safety Critical Equipment – SCE	GPS	Engineering authorities - TA
Protective systems	LEP's & DWOP	Clear R & Rs
Integrity of design, drilling, completion & well intervention	IM Strategies	Competency assessment & training
Assets integrity & reliability	Risk Management	Contractor's key positions
Equipment dossiers	MoC	
	QA-QC	
	Contractor's Management	
	Incident investigation	
	OPS-ERP practices	

PERFORMANCE

Land rig map of SCEs

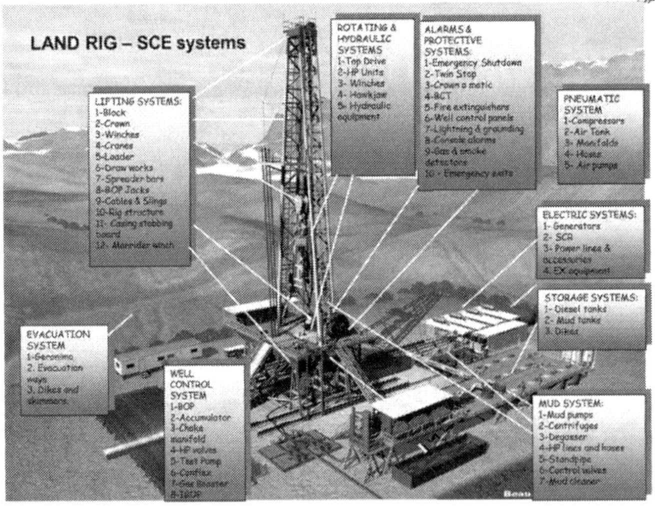

HOW ARE WE DOING?

10TH GLOBAL CONGRESS ON PROCESS SAFETY

LOWLIGHTS

- Asset integrity normally delegated to rig contractor.
- Backlog on PM tasks
- SCE concept and application barely applied.
- Criticallity of equipment based on operations rather than barriers for major incident risks
- Equipment modifications or alteration not following MoC
- Incident investigation- RCFA??

HIGHLIGHTS

- Drilling Contractors enforcing asset integrity – R.A.M. -
- Raising Awareness on key people competence and certification.
- OEM recommendations mainly followed
- Incorporating lessons learnt from industry
- Renewal / upgrade of rig fleets

SCE into PS model

10TH GLOBAL CONGRESS ON PROCESS SAFETY

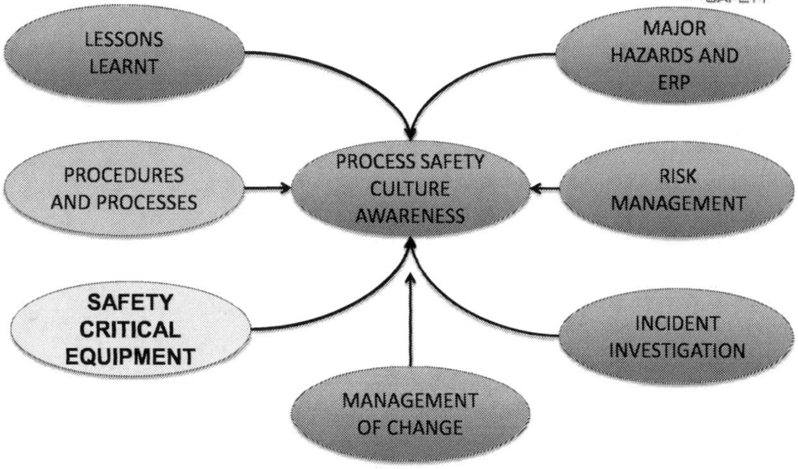

What´s next for SCEs !!

1. Define performance standards for rigs/platforms' SCEs addressing functionality, availability, reliability and failure modes, survival if designed for and interaction with other systems to operate.

2. Identify critical maintenance and test activities for SCEs and duly incorporate them into Contractor´s asset management systems.

3. Set key performance indicators – KPI – for maintenance and reliabilty related activities that lead performance and help to focus on compliance of PM tasks.

4. Define test frequencies and acceptance parameters for protective systems associated with SCEs.

5. Enforce SCE failure reporting, root cause analysis and sharing lessons

QUESTIONS ???

10TH GLOBAL CONGRESS ON PROCESS SAFETY

Development of low-charring nano-composites to aid in fundamental understanding of nano-composite flame-retardancy

Logan Hatanaka
Mary Kay O'Connor Process Safety Center
Artie McFerrin Department of Chemical Engineering, Texas A&M University
College Station, TX 77843
Lchatanaka@tamu.com

Sonny Sachdeva
Mary Kay O'Connor Process Safety Center, Texas A&M University,

Agustin Diaz
Department of Chemistry, Texas A&M University

Qinsheng Wang
Department of Chemical Engineering, Oklahoma State University

M. Sam Mannan
Mary Kay O'Connor Process Safety Center, Texas A&M University

Prepared for Presentation at
American Institute of Chemical Engineers
2014 Spring Meeting
10th Global Congress on Process Safety
New Orleans, LA
March 30 – April 2, 2014

UNPUBLISHED

Development of low-charring nano-composites to aid in fundamental understanding of nano-composite flame-retardancy

Logan Hatanaka[ab], Sonny Sachdeva[ab], Agustin Diaz[c], Qingsheng Wang[de], M. Sam Mannan[abf] and Zhengdong Cheng[abf]

[a] *Artie McFerrin Department of Chemical Engineering, Texas A&M University, College Station, TX 77843-3122, USA.*
[b] *Mary Kay O'Connor Process Safety Center, Artie McFerrin Department of Chemical Engineering, Texas A&M University, College Station, TX 77843-3122, USA.*
[c] *Department of Chemistry, Texas A&M University, College Station, TX 77843-3125, USA*
[d] *Department of Fire Protection & Safety, Oklahoma State University, Stillwater, OK 74078-4082, USA*
[e] *Department of Chemical Engineering, Oklahoma State University, Stillwater, OK 74078-5021, USA*
[f] *Material Science and Engineering, Texas A&M University, College Station, TX 77843-3003, USA*

Keywords: Fire, Retardant, Nano-composites, Polymer, Loss Prevention

Abstract

For decades, nano-sheets comprised of various different materials have been utilized to construct nano-composites with enhanced properties including high thermal stability, flammability reduction, and modified mechanical properties. Although research in this field has been actively pursued in the past, there is little understanding as to why these materials exhibit enhanced behavior compared to their pure polymeric counterparts.

Two major mechanisms for enhanced flame-retardancy have been proposed: (1) a physical barrier effect and (2) a catalytic charring effect. As the material burns, a physical barrier of carbonaceous char and stacked nanoparticles forms at the material surface, acting as a physical barrier for radiant, conductive, and convective heat transfer as well as a physical barrier for mass transfer of the fuel to the burning front. Secondly, the presence of nanoparticles can act as a catalyst for charring, increasing the char yield of the material and effectively altering the physical barrier that forms. By utilizing a nano-composite system with low charring it is possible to observe the physical effects caused by the presence of nano-sheets in the polymer matrix. As charring agents are added to the composite, the enhancement of flame-retardancy based on catalyzed charring can be observed.

Furthermore, the research into the field of thermally stable and flame-retardant nano-composites has only focused on a relatively narrow view of polymers and nanoparticles, so any advancement into new materials is useful to further understand how these systems work. The current research project incorporates α-zirconium phosphate, a material capable of forming atom-thick, nanoscale-width nano-sheets, and poly (methyl methacrylate), a widely used polymer in commercial application. Aside from enhanced thermal stability, these materials have altered glass-transition temperatures, excellent optical transparency with UV scattering, and a relatively low char yield, making them an excellent candidate to study the physical barrier effects and catalytic charring effects of polymer nano-composites.

1. Introduction to Flame-Retardant Nano-composites

Fires can affect anyone, whether at home or working in an industrial plant. Although it is important to understand the hazards of fires in order to prevent them, once a fire occurs it is necessary to protect property and life from further damage. For this purpose, flame retardant materials are critical in slowing the spread and severity of damage due to a fire. However, not everything about common flame-retardants is as good as they seem. A recent study by Duke University studied the types of materials being used as flame-retardant additives in commercial couches. With a sample size of 101 couches, the study found that 52% of couches contained tris (1,3-dichloroisopropyl) phosphate (TDCPP, commonly called "Tris")[1], a mutagen in animals[2,3] and a neurotoxin in mice cell[4].

The need for non-toxic and effective flame-retardant materials has promoted the research of technologies that aim to replace the antiquated materials of the past. Specifically, the last two decades have seen major advancements into polymeric nano-composites as flame-retardant materials. These materials are composed of a nano-particle dispersed within a polymer matrix and have the potential to replace flame retardant technologies of the past with materials that are more effective and safer to use.

2. Flame Retardant Mechanisms

Polymeric nano-composites have shown great potential as flame-retardant materials with high thermal stability, but a detailed understanding of the mechanisms is not currently available. However, it has been shown that two main mechanisms contribute to the flame-retardant effects found in these types of polymeric nano-composites.

2.1 The Catalytic Charring Effect

The first of the two main flame-retardant effects present when nano-composites burn is due to the formation of char. Upon heating, a linear (unbranched) polymer will go from a glassy to rubbery state near the polymer's glass transition temperature. If the polymer has crystalline domains, a melting transition will then occur at the polymer's melting temperature, or if the polymer is entirely amorphous, the rubbery state will slowly become more fluidic until the polymer is a liquid. Past this temperature, the polymer will begin to degrade and break apart in a process known as pyrolysis[5].

In non-charring polymers, pyrolysis creates lighter compounds from the larger original polymer chains that then vaporize and diffuse away from the material surface where they burn in the gas phase. This creates a region of pyrolysis at the surface of the material near the flame followed by a region of heated polymer melt and solid polymer away from the material surface[5]. However, if the temperature of the fire is high enough, more and more polymer will decompose over time, reducing the amount of solid and liquid polymer until essentially none is left.

In charring polymers, pyrolysis also forms lighter compounds from the original polymer chains. However, instead of diffusing away from the surface and then burning, these lighter components have the ability to cross-link the existing polymer chains. In this case, the flame impinges on the cross linked polymer, forming a char layer that covers the unburned cross linked material with an unreacted polymer melt and solid below it[5]. In order to burn, vaporized material must travel through this char layer to ignite in the gas phase, slowing the rate of burning and heat release in the polymer material. In addition, charring of the polymer leads to incomplete combustion of the polymer, reducing the overall heat released, and increasing flame-retardancy performance[5]. Some examples of non-charring polymers include poly (methyl methacrylate) (PMMA), polypropylene (PP), and ethylene-vinyl acetate (EVA), while an example of a charring polymer includes most cross-linked epoxies.

Up to this point, charring has only been considered for a pure polymer, but it has an important effect on flame-retardancy of polymeric nano-composites as well. The presence of nano-filler in the polymer can act as a charring agent similar to how Lewis acids and Friedel-Crafts alkylation reagents can increase the char yield of a normally non-charring polymer, increasing the overall flame-retardancy and heat resistance of the material[6-9]. For example, with a non-charring polymer filled with nano-filler, the addition of a carefully chosen nanoparticle could act as a catalytic charring agent to increase the over char-yield of a nano-composite, increasing flame-retardancy.

2.2) Physical Barrier Effect

The second effect that plays a role in the flame-retardancy of a nano-composite is the physical barrier effect. As a nano-composite burns, nanoparticles begin to agglomerate at the surface of the material near the flame, forming a physical barrier towards gas transport from the material to the flame. This layer also acts as an insulating layer between the flame and the unburned polymer below the physical barrier and a radiant heat barrier to prevent re-radiation of heat to the flame. The physical barrier formed consists of intumescent, stacked nano-sheets with small floccules of carbonaceous char trapped between plates[10-20]. These small floccules together make up the physical barrier[21].

2.3) Synergistic Effects

The physical barrier effect and the catalytic charring effect are coupled. The catalytic charring of the polymer effects the composition of the physical barrier and similarly, the physical barrier affects the conditions in which char formation occurs. Often in nano-composites, these two effects work together synergistically to produce flame-retardant properties but the interaction between these two effects is not well understood.

Cinausero *et al.* (2011) showed how these two effects act synergistically by first adding an alumina polyphosphate (AP) charring agent to PMMA and polystyrene (PS), then observing how the addition of silica and alumina nano-filler affected the thermal stability and flame-retardancy of PMMA[22]. The results showed that the limiting oxygen index (LOI) was increased for PMMA with AP char agents added when compared to only PMMA. To then observe the synergistic effects, nano-silica, organo-modified nano-silica, nano-alumina, and organo-modified nano-alumina were added to the polymer. The result was an increased LOI for all nano-fillers except organo-modified nano-alumina. Furthermore, the addition of AP to PMMA decreased the heat release rate (HRR), and further addition of the previously mentioned nano-filler reduced the HRR even more, showing synergistic effects between the charring agents and the nano-fillers. In addition to testing the synergistic effects of PMMA with AP and varying nano-filler, the same was done for PS. Similarly to PMMA, the addition of AP increased the catalyzed charring effect and increased the LOI. However, in the case of PS, only the addition of organo-modified silica increased the LOI further, while the other three nano-fillers (nano-alumina, organo-modified nano-alumina, and nano-silica) reduced the LOI and decreased flame-retardant performance. The cone calorimetry tests showed that PS and AP had a slightly reduced HRR, but the addition of any of the four nano-fillers further decreased the HRR, increasing the flame-retardant performance of the material. In this study, it was determined that a reaction occurs between the nanoparticle and charring agent, forming a new crystalline material which acts as a good physical barrier with catalytic charring abilities[22].

Laachachi (2007) has a similar study that utilizes titanium oxide and aluminum oxide as the physical barrier forming nanoparticle with a charring agent to provide charring effects to show synergistic effects[23]. Alone, aluminum oxide only provides a physical barrier effect, but when combined with charring agents, synergistic flame-retardant effects are also observed. This reinforces the idea that two separate mechanisms are at play, a physical barrier and a catalytic charring effect within the material. However, in order to fully understand how these effects work, each effect must be observed independently and synergistically[23].

3. Current Research

In order to further advance the flame-retardant properties of polymeric nano-composites, it is necessary to gain a deeper understanding of the flame-retardant mechanisms that make these materials effective. Observing the physical barrier effect separately from the catalytic charring effects and then together in varying degrees allows a complete understanding of the interactions between flame-retardant mechanisms in nano-composites. This knowledge would allow nano-composites to be optimized with a fine-tuned mix of physical barrier and catalytic charring effects. To start this research project, a physical barrier dominated nano-composite must be produced.

3.1 Synthesis of Physical Barrier Dominated Nano-composites

Initial experiments have successfully prepared nano-composites containing alpha-zirconium phosphate (ZrP) as the nano-sheet filler and PMMA as the polymer matrix. ZrP is useful in nano-composite research because it can readily form individual sheets of atomic-scale thickness and a controllable diameter [24-27].

ZrP can undergo intercalation and exfoliation under the correct chemical conditions. If an amine such as propylamine or tetrabutylammonium is present and at a controlled ratio of ZrP:amine, the ZrP becomes exfoliated and individual nano-plates of ZrP are formed[24-27]. Once individual ZrP nano-sheets are created, they can then be dispersed into a polymer to form a nano-composite. By mixing a dispersion of exfoliated ZrP with a dissolved polymer solution, a dispersion of ZrP and polymer in solvent can be established. The solvent is then dried to produce a

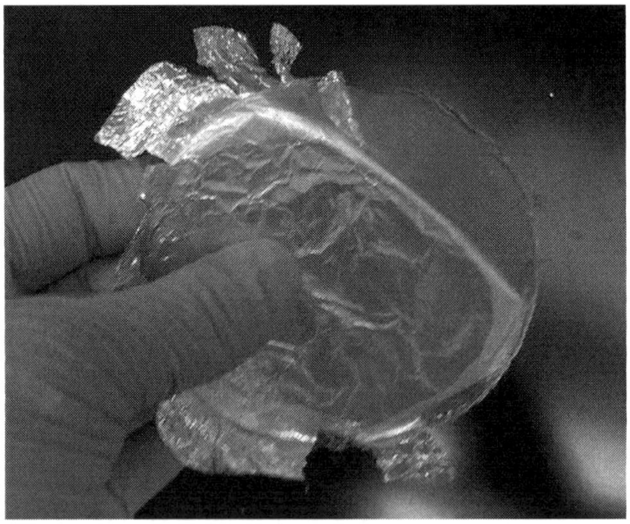

Figure 1: 5 wt% ZrP in PMMA nano-composite being removed from foil substrate.

ZrP/polymer nano-composite film.

This general method was used to produce ZrP/PMMA nano-composites. Figure 1 shows a 5 wt% ZrP in PMMA nanocomposite produced using this method. In the figure, a newly cured nanocomposite is peeled from a foil substrate, revealing an optically transparent ZrP/PMMA nanocomposite. Initial materials tests on a 1 wt% ZrP in PMMA composite have shown that intercalated and exfoliated nanoparticles are present, as seen in the X-Ray Diffraction (XRD) data in Figure 2. The agglomerated plate-to-plate distance of ZrP normally produces a peak at 11.6° which is absent in Figure 2, indicating that there is not a significant amount of agglomerated nanoparticles. In future work, scanning electron microscopy (SEM) and transmission electron microscopy (TEM) imaging will be used to confirm this. Instead, a peak at lower a 2Θ value indicates intercalated nanoparticles are likely present.

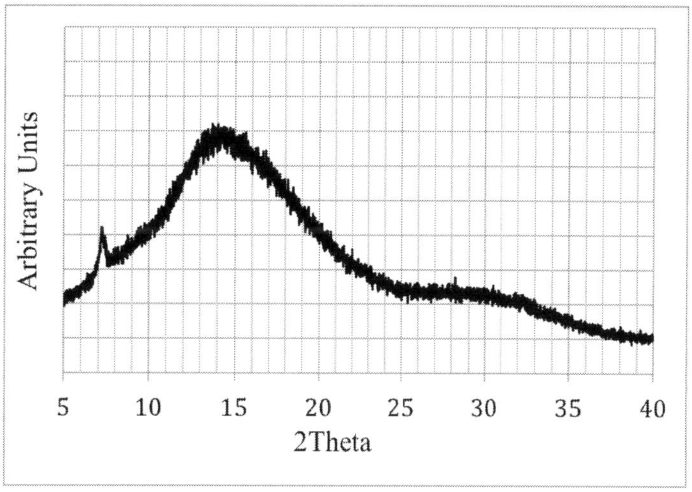

Figure 2: XRD results for a 1 wt% ZrP/PMMA nano-composite.

Initial thermogravimetric analysis (TGA) of ZrP/PMMA composites show that they most likely have a physical barrier dominated mechanism of heat resistance, as shown in Figure 3. Every composite shows the same 3 mass loss peaks, showing that the mechanism for degradation has not changed significantly. Additional cone calorimetry tests will be conducted to confirm if catalytic effects are also contributing to heat resistance during decomposition, but this is unlikely since the shape of each TGA curve is similar to PMMA.

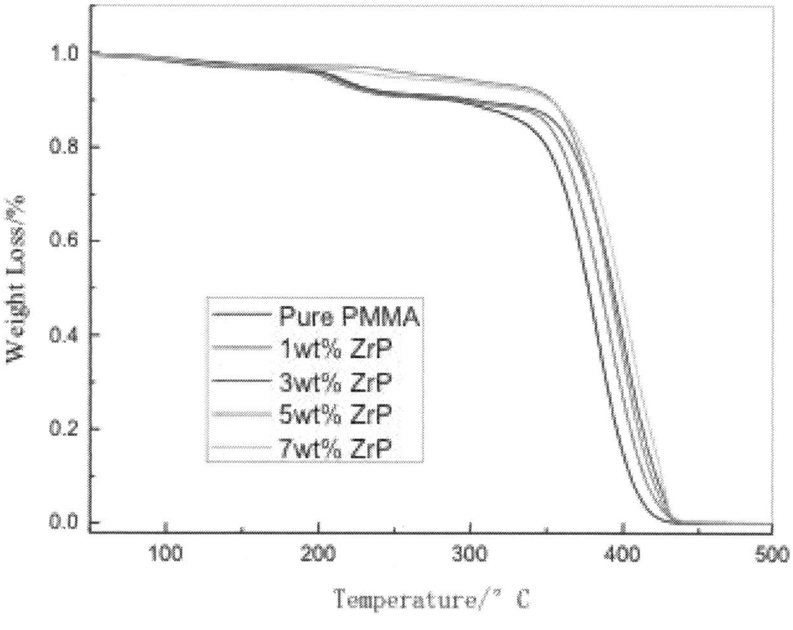

Figure 3: Initial TGA of various ZrP nano-composites.

3.2 Synthesis of Catalytic Char Dominated Nano-Composites

To view the opposite side of the spectrum, a material must be produced that is dominated by the catalytic charring effect. This nano-composite will be produced similarly to the first nano-composite, using the same polymer as the first composite for consistency. However, this charring nano-composite will utilize an entirely different nanoparticle that induces charring but is spherical rather than layered. This change in geometry is meant to hinder floccule formation while inducing polymer charring. Future research will utilize phosphotungstic acid, a noted charring nanoparticle, in PMMA in the same weight fraction as the nano-sheet in the physical barrier-dominated material[28].

3.3 Synthesis of Synergistic Nano-Composites

At this point, a physical barrier-dominated material and a catalytic char-dominated material has been established, but no composite exists to exhibit synergistic effects. Conveniently, the production of homogenous catalysts consisting of PTA and silica (a physical barrier forming nano-particle) can be through a sol-gel synthesis process[29]. Silica/PTA composites have been produced with up to 80 wt% PTA on silica, allowing a wide range of PTA to silica weight ratios to be examined. This could potentially uncover a wide range of synergistic effects with varying degrees of the physical barrier effect to the catalytic charring effect. It allows the chemical composition of the nano-filler to be finely tuned so that when it is introduced into the polymer, it gives precise control over the physical barrier and catalytic char behavior.

4. Conclusions:

Flame retardant nano-composites have shown great promise in providing non-toxic and effecive loss prevention materials. However, in order to increase the effectiveness of flame retardant nano-composites, the exact nature of the flame retardant mechanisms within these materials is necessary. Since the two mechanisms for nano-composite flame retardancy behave differently together than they do independently, it is important to observe the physical barrier effect and the catalytic charring effect independently and synergistically. Initial research has produced a ZrP/PMMA nanocomposite system which appears to be physical barrier dominated with little charring. This is the first step in developing nanocomposites which exhibit a complete range of nanocomposite flame-retardancy mechanisms. The next step is to produce nano-composites with catalytic char dominated behavior along with nano-composites with varying ratios of the physical barrier effect to the catalytic charring effect to observe the full range of flame retardant mechanisms.

Acknowledgements:

This material is based upon work supported by the National Science Foundation under Grant No. CBET1336162.

5. References

1. Stapleton, H.M., S. Sharma, G. Getzinger, P.L. Ferguson, M. Gabriel, T.F. Webster, and A. Blum, *Novel and High Volume Use Flame Retardants in US Couches Reflective of the 2005 PentaBDE Phase Out.* Environ Sci & Technol, 2012. **46**(24): p. 13432-13439.
2. Blum, A. and B.N. Ames, *Flame-retardant additives as possible cancer hazards.* Science, 1977. **195**(4273): p. 17-23.
3. Gold, M.D., A. Blum, and B.N. Ames, *Another flame-retardant, tris- (1,3-dichloro-2-propyl)-phosphate, and its expected metabolites are mutagens.* Science, 1978. **200**(4343): p. 785-787.

4. Dishaw, L.V., C.M. Powers, I.T. Ryde, S.C. Roberts, F.J. Seidler, T.A. Slotkin, and H.M. Stapleton, *Is the PentaBDE replacement, tris (1,3-dichloropropyl) phosphate (TDCPP), a developmental neuro- toxicant? Studies in PC12 cells.* Toxicol. Appl. Pharmacol. , 2011. **256**(3): p. 281-289.
5. Pochiraju, K.V., G.P. Tandon, G.A. Schoeppner, and A. Morgan, *Additives, Nanocomposites, and Barrier Coatings*, in *Long-Term Durability of Polymeric Matrix Composites*. 2012, Springer: US. p. 39-75.
6. Cai, Y., Y. Hu, L. Song, S. Xuan, Y. Zhang, Z. Chen, and W. Fan, *Catalyzing carbonization function of ferric chloride based on acrylonitrile-butadiene-styrene copolymer/organophilic montmorillonite nanocomposites.* Polymer Degradation and Stability, 2007. **92**(3): p. 490-496.
7. Jang, J., J. Kim, and J.-Y. Bae, *Effects of Lewis acid-type transition metal chloride additives on the thermal degradation of ABS.* Polymer Degradation and Stability 1999. **88**(2): p. 324-332.
8. Jang, J., J. Kim, and J.-Y. Bae, *Synergistic effect of ferric chloride and silicon mixtures on the thermal stabilization enhancement of ABS.* Polymer Degradation and Stability, 2005. **90**(3): p. 508-514.
9. Wang, Z., D.D. Jiang, C.A. Wilkie, and J.W. Gilman, *Further studies on fire retardant polystyrene by Friedel-Crafts chemistry.* Polymer Degradation and Stability 1999. **66**(3): p. 373-378.
10. Bras, M.L., G. Camino, S. Bourigot, and R. Delobel, *Fire Retardancy of Polymers: The use of intumesence.* 1998: Cambridge: Royal Society of Chemistry Information Services.
11. Bras, M.L., C.A. Wilkie, and S. Bourbigot, *Fire Retardancy of Polymers: New Applications of Mineral Fillers.* . 2005: Cambridge: Royal Society of Chemistry.
12. Fina, A., D. Tabuani, A. Frache, and G. Camino, *Polypropylene-polyhedral oligomeric silsesquioxanes (POSS) nanocomposites.* Polymer, 2005. **46**(19): p. 7855-7866.
13. Gilman, J.W., *Flammability and thermal stability studies of polymer layered-silicate (clay) nanocomposites.* Applied Clay Science 1999. **15**(1-2): p. 31-49.
14. Schartel, B., P. Potschke, U. Knoll, and M. Abdel-Goad, *Fire behaviour of polyamide 6/multiwall carbon nanotube nanocomposites.* European Polymer Journal 2005. **41**(5): p. 1061-1070.
15. Tidjani, A., *Polypropylene-graft-maleic anhydride-nanocomposites: II - fire behaviour of nanocomposites produced under nitrogen and in air.* Polymer Degradation and Stability 2005. **87**(1): p. 43-49.
16. Torre, L., E. Frulloni, J.M. Kenny, C. Manferti, and G. Camino, *Processing and characterization of epoxy–anhydride-based intercalated nanocomposites.* Journal of Applied Polymer Science, 2003. **90**(9): p. 2532-2539.
17. Zanetti, M., G. Camino, and R. Mulhaupt, *Combustion behaviour of EVA/fluorohectorite nanocomposites.* Polymer Degradation and Stability 2001. **74**(3): p. 413-417.
18. Zanetti, M., G. Camino, P. Reichert, and R. Mülhaupt, *Thermal Behaviour of Poly(propylene) Layered Silicate Nanocomposites.* Macromolecular Rapid Communications 2001. **22**(3): p. 176-180.
19. Zanetti, M., T. Kashiwagi, L. Falqui, and G. Camino, *Cone Calorimeter Combustion and Gasification Studies of Polymer Layered Silicate Nanocomposites.* Chemistry of Materials, 2002. **14**(2): p. 881-887.
20. Zhao, C., H. Qin, F. Gong, M. Feng, S. Zhang, and M. Yang, *Mechanical, thermal and flammability properties of polyethylene/clay nanocomposites.* Polymer Degradation and Stability, 2005. **87**(1): p. 183-189.
21. Gilman, J.W. *Flame Retardant Polymer Nanocomposites.* in *Pollution Prevention through Nanotechnology Conference.* 2007. Arlington, VA.
22. Cinausero, N., N. Azema, J.M. Lopez-Cuesta, M. Cochez, and M. Ferriol, *Synergistic effect between hydrophobic oxide nanoparticles and ammonium polyphosphate on fire properties of poly(methyl methacrylate) and polystyrene.* . Polymer Degradation and Stability, 2011. **96**(8): p. 1445-1454.
23. Laachachi, A., M. Cochez, E. Leroy, M. Ferriol, and J.M. Lopez-Cuesta, *Fire retardant systems in poly(methyl methacrylate): Interactions between metal oxide nanoparticles and phosphinates.* Polymer Degradation and Stability 2007. **92**(1): p. 61-69.
24. Alberti, G., M. Casciola, and U. Costantino, *Inorganic ion exchange pellicles obtained by delamination of alpha-zirconium phosphate crystals.* Journal of Colloid and Interface Science, 1985. **107**: p. 256-263.
25. Maclachlan, D.J. and K.R. Morgan, *Phosphorus-31 solid-state NMR studies of the structure of amine-intercalated .alpha.-zirconium phosphate. 2. Titration of .alpha.-zirconium phosphate with n-propylamine and n-butylamine.* J. Phys. Chem. , 1992. **96**: p. 3458-3464.
26. Sun, L., J.Y. O'Reilly, D. Kong, J.Y. Su, W.J. Boo, H.J. Sue, and A. Clearfield, *The effect of guest molecular architecture and host crystallinity upon the mechanism of the intercalation reaction.* Journal of Colloid and Interface Science, 2009. **333**(2): p. 503–509.
27. Clearfield, A. and Z. Wang, *Organically pillared microporous zirconium phosphonates.* Front cover Journal of the Chemical Society, Dalton Transactions,, 2002. **15**: p. 2937-2947.
28. Bourbigot, S. and S. Duquesne, *Fire retardant polymers: recent developments and opportunities.* Journal of Materials Chemistry, 2007. **17**: p. 2283-2300.
29. Kukovecza, Á., Z. Balogia, Z. Kónyaa, M. Tobab, P. Lentzc, S.-I. Niwab, F. Mizukamib, Á. Molnárd, J.B. Nagyc, and I. Kiricsia, *Synthesis, characterisation and catalytic applications of sol–gel derived silica–phosphotungstic acid composites.* Applied Catalysis A: General, 2002. **228**(1-2): p. 83-94.

349935 Best Practices in IPL Integrity Management - Case Study in Kuwait Oil Company

Abstract:

16th Process Plant Safety Symposium (PPSS) – Validation and Maintenance of IPLs

This paper presents Management of Independent Protection Layers at Kuwait Oil Company (KOC), Kuwait. Several methods and practices were initially followed by various project teams involved in SIL Assessment / Verification / Validation over several years since the SIL compliance was introduced. However, little interaction existed across projects and Operation & Maintenance personnel. As a result, valuable experience and observations from the actual plant operation were never fed back and were not integrated into the LOPA methodology. The conditional modifiers remained un-changed for almost a decade.

In order to revalidate the IPLs, a task was carried out to systematically compile, analyze and prepare in easily understandable form, failure modes and effects over the past seven (7) years from five major process plants. Mechanical, Electrical and Instrument failures were separately listed with their cause-consequence relations. The consequences are rationalized against the Company's LOPA criteria, in order to be able to sort them in the order of maximum to minimum consequence. For the high consequence events, the initiating events that have caused them were compared with the cause-consequence pairs listed in the Company standard. It was observed that for many of the high consequence events, the initiating events that were assumed were not the real ones as per the 7-year history. A detailed report that followed presented this analysis to the stake-holders for further consideration, to be reflected in a revised Company standard and also to make this method of revalidation of IPLs a Company practice.

An independent "Alarm Management" exercise issued a Company Alarm Management Philosophy document (APD). After implementing the Alarm Management per the new APD, several of the existing Process alarms were either removed, or modified to give them lower priority. The credits allowed by the LOPA standard now need revision to be in line with the APD. This issue was highlighted and soon, the LOPA standard will be revised to reflect the same.

The Flare & Blow-down systems in several of the company's process plants were provided with SIL approved SDVs under one major project. However, the credit for this SIF was not being reflected in most LOPA workshops. This issue was highlighted and was subsequently corrected.

The access security for a few SIFs was agreed to be given to the senior process operators, thus increasing the risk profile, as against the original assumption that only the Operations Manager will be approving changes. Hence these SIFs no longer deserve credits in LOPA and needed suitable modification in the LOPA standard.

The 'type' of equipment and thus their technology and reliability has changed significantly. For example, the Tank Vapour compression was changed to electric motor driven centrifugal / screw type from previous gas-engine driven reciprocating type, thus decreasing vibrations in the tanks and piping systems. Similar changes in other process units either decreased or increased the initiating events but these changes were not reflected in the frequencies and cause-consequence relationships of the initiating events.

All the above issues significantly impact the accuracy of the LOPA exercises and thus directly impact the Risk reduction in the process plants under consideration. Sustained efforts over several years resulted in noticeable change in perceptions of Operation and Maintenance personnel and the 7-year failure data has significantly contributed to rationalizing the Company's LOPA criteria to make it more meaningful for both existing and future process facilities.

349800 Multiple Perspectives on the Role of Safety Leadership in Major Hazard Organisations

Abstract:

There is widespread agreement between industry, regulators, and academics that leadership is a key component of a safe organisation. This view is widely supported by findings from almost all major incident inquiries and investigations, which have demonstrated that managerial failure is at least as important as technical failure and operator error. Managers play an important role because their decisions have a fundamental impact on the resources invested in safety and more broadly on how work is organised and managed. Despite the increased focus on managers', and particularly senior managers', roles and responsibilities both in legislation and in major incident enquiries, comparatively little research has been conducted on the types of behaviours and leadership styles that are effective in managing safety.

This paper will bring together different sources of evidence on the role of leadership particularly in major hazard organisations. First, based on a review of academic literature and an analysis of major incidents, the requirements for effective safety leadership will be presented. Specifically, a review of published evidence over the last ten years was carried out in order to capture current knowledge in the area of leadership. This was complemented by an analysis of accident reports and published papers for sixteen major incidents in order to identify any recurrent themes in the causation of these incidents, particularly those that represented leadership failures or have leadership implications for the management of safety. Second, an empirical study was conducted on the perceptions of senior managers regarding the roles and responsibilities of Board members in terms of safety leadership across three different major hazard organisations. The paper will discuss the implications of the findings from these different sources for the roles and responsibilities of managers, and what constitutes effective safety leadership in major hazard organisations.

Using Explicit Finite Element Analysis to Simulate the Dynamic Response and Predict the Structural Damage Associated with a Real-Life Process Equipment Failure Due to an Internal Detonation

Phillip E. Prueter
The Equity Engineering Group Inc., Shaker Heights, Ohio
peprueter@equityeng.com

Prepared for Presentation at
American Institute of Chemical Engineers
2014 Spring Meeting
10th Global Congress on Process Safety
New Orleans, LA
March 30 – April 2, 2014

UNPUBLISHED

Using Explicit Finite Element Analysis to Simulate the Dynamic Response and Predict the Structural Damage Associated with a Real-Life Process Equipment Failure Due to an Internal Detonation

Phillip E. Prueter
The Equity Engineering Group Inc., Shaker Heights, Ohio
peprueter@equityeng.com

Keywords: Explicit Finite Element Analysis, Blast Loading, Accidental Explosions, Internal Detonations, Structural Response, Heat Exchanger

Abstract

Simulating the realistic blast loading associated with an internal detonation occurring within a pressure vessel or heat exchanger is extremely challenging. Unlike evaluation of external blast loading on structures due to far-field explosions, where typical overpressure-time histories can be reasonably defined based on empirical data, investigating confined detonations presents additional complications such as accounting for internal blast wave reflection that can yield multiple blast waves and considering potential pressure build up due to the accumulation and temperature rise of gaseous explosion by-products. The subsequent impulsive peak reflected overpressure from confined detonations acting on a structure can be extremely high due to the often times, close proximity of the blast source to the vessel wall or pressure boundary. This establishes the possibility of significant structural damage for process equipment subjected to an internal detonation, even for relatively modest amounts of concentrated explosive products.

Performing dynamic computational simulations of internal detonations is valuable in assessing the structural response and possible failure modes of critical process equipment such as process piping, pressure vessels, or heat exchangers. Furthermore, assessing the potential damage from an internal detonation within such a structure and predicting ensuing damage to surrounding equipment if the containment vessel catastrophically fails can provide valuable information about protecting structures from explosions and implementing designs that promote blast damage mitigation and process safety. Additionally, determining a damage threshold for a given amount of concentrated chemical explosives can guide process engineers in mitigating the potential for accidental detonations that pose a significant risk to the structural integrity of pressure containing equipment.

This paper discusses the underlying theory of blast analysis and examines the practical application of non-linear, finite element based, explicit computational techniques for simulating the load acting on a structure due to internal and external blasts. The investigation of a recent, real-life industry failure of a heat exchanger due to a suspected internal detonation is discussed. This particular failure resulted in gross rupture of the pressure boundary and significant damage to nearby equipment. Explicit, three-dimensional blast analysis is performed on the heat exchanger in question, and an internal detonation is simulated to reasonably replicate the considerable damage actually observed in the field. This analysis permits the determination of an approximate amount of concentrated product that caused the accidental explosion; that is, the plausible equivalent amount of explosives is back-calculated based on the predicted damage to the finite element model of the equipment in question. Computational iterations of varying charge amounts are performed and the predicted amount of permanent damage is documented so sensitivity to the hypothesized charge amount can be quantified from a structural response standpoint. Furthermore, explicit blast analysis of nearby equipment is performed to predict the magnitude of plastic deformation actually observed due to external blast loading.

1. Introduction

In general, an explosion in air creates a blast wave as a result of the atmosphere surrounding the explosion being pushed back, and the resulting pressure of the compressed air at the blast wavefront decays as it moves away from the blast source. As discussed in references [1-6], a typical blast wave, observed at a location removed from the center of a chemical explosion, reaches a peak value of overpressure and begins to decay exponentially (as shown in Figure 1), eventually decreasing below atmospheric pressure. For conventional chemical explosions, the overpressure profile is divided into the positive pressure portion (positive phase) and negative pressure portion (negative phase). Furthermore, the amount of time it takes the peak overpressure to decay below atmospheric pressure is known as the positive phase duration and is generally, closely related to the overall damage capabilities of a given blast wave. Similarly, the amount of time it takes for the negative pressure to return to atmospheric pressure is the negative phase duration [3]. The area under the curve during the positive phase duration is the positive blast wave impulse. In general, peak overpressure observed at a location removed from the blast source decreases as the distance from the detonation center increases.

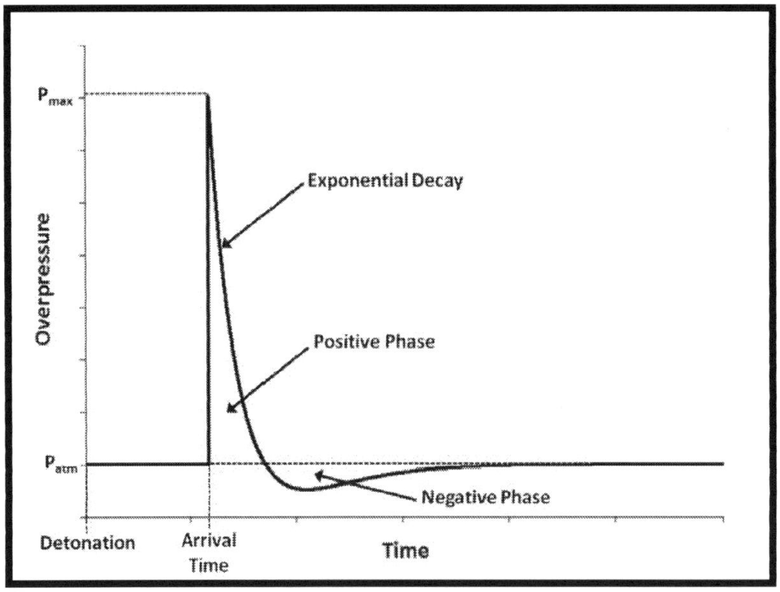

Figure 1: Typical overpressure amplitude for a chemical explosion [7].

This paper presents the underlying methodology for employing explicit finite element analysis to investigate a recent, real-life process industry failure of a vertical heat exchanger that experienced an unexpected internal detonation. The primary computational tool used to generate realistic blast wave profiles in this study is based on the Conventional Weapons Effects Blast Loading Model or CONWEP [8].

In this study, computational results for both the heat exchanger (subjected to internal blast loading) and surrounding equipment (subjected to external blast loading) are in good agreement with the measured plastic deformations and failure modes that were actually observed in the field. Commentary on the observed failure modes, supplemented by explicit computational finite element results, is provided. Additionally, an advanced finite element failure criterion is employed where portions of the computational model are removed from the simulation once a certain plastic strain threshold is reached. This approach facilitates simulation of the gross heat exchanger pressure boundary failure actually observed in this case. The explicit, three dimensional finite element based analyses discussed in this study reasonably predict the structural response and damage characteristics corresponding to a recent, real-life industry failure.

2. Blast Wave Reflection Effects

A complication associated with computational blast modeling is accounting for overpressure wave reflection effects. In general, the peak blast pressure that a structure is subjected to is greater than the peak incident overpressure and is known as the reflected pressure. This pressure is a function of the peak incident overpressure and the angle of incidence; that is, the angle between the blast wave loading direction and the surface normal of a given point on the loading surface.

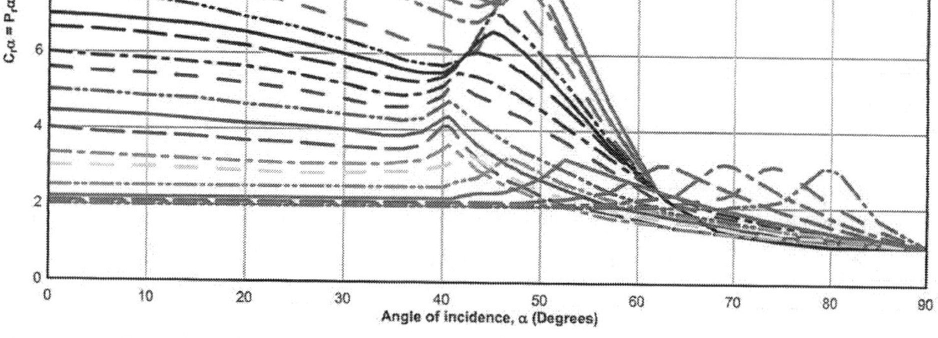

Figure 2: Reflection Coefficient vs. Angle of Incidence for Varying Peak Overpressures [3].

Figure 2 shows how the reflection coefficient changes with angle of incidence and peak incident overpressure. It can be observed that for extremely high overpressures, the actual reflected pressure that a structure experiences can be as high as 10 or 12 times the incident blast pressure alone. Neglecting incident wave reflection in a computational analysis can produce non-conservative results, particularly for high peak overpressures. On the contrary, assuming the maximum peak reflected pressure acts on an entire structure is potentially overly conservative. To this end, structural response due to impulsive loading is often times, sensitive to capturing the correct incident wave reflection. Accurately simulating realistic overpressure amplitudes acting on a structure for a given explosive scenario is challenging because the peak overpressure, positive phase duration, amount of explosives, and distance from the explosion source all affect the overpressure amplitude. Furthermore, accounting for reflection effects on a loading surface adds further complication to the blast loading model.

For internal detonations, accurately predicting blast wave reflection is extremely difficult because of the effects of confinement. As discussed in [9], an internal detonation within a containment vessel can be characterized by two distinct types of loading. First, immediately following a detonation, the structure is subjected to a transient impulsive pressure load. If this initial pressure wave does not cause catastrophic failure or rupture of the pressure boundary, a secondary, long-term quasi-static pressure build-up occurs in the vessel due to the evolution and heating of gaseous detonation by-products. Additionally, as discussed in [10], resonance effects between the vibration of the vessel and the reflected internal shock wave is a potential damage mechanism. In the analyses discussed herein, only the initial impulsive pressure wave is considered as a damage mechanism because it is believed that this type of loading caused gross failure of the pressure boundary for the actual equipment failure discussed. Furthermore, the interaction between any internal components and a propagating blast wave are not explicitly considered due to the complexity of the heat exchanger geometry and the computational cost of analyzing fluid-structure interaction in detail, which would require computational fluid dynamics (CFD) techniques.

3. CONWEP Blast Loading Model

The CONWEP blast loading model is based on a collection of conventional weapons effects calculations of TM 5-855-1 established by the United States Military [8]. The main advantage of using CONWEP is that realistic overpressure amplitudes and other blast wave parameters are calculated based on a defined charge amount, and the appropriate impulsive loading is automatically employed in the explicit finite element analysis. Additionally, overpressure amplitude decay and reflection formulations are incorporated into the model. The CONWEP blast loading model does not require modeling the fluid medium to account for overpressure reflection (eliminating the need for coupled structural-acoustic analysis and offering a significant improvement in computational efficiency, especially for complex simulations). For a more in-depth overview of the underlying theory behind the CONWEP blast loading model, refer to Reference [11]. An example of a computational study that employs CONWEP through explicit finite element analysis to investigate the response of storage tanks subject to external blast loading is provided in [12].

In order to develop the blast parameters required to generate realistic overpressure curves for conventional explosives, the equivalent TNT method is most commonly used. TNT equivalency of an explosive is simply the ratio of the mass of TNT to the mass of explosive such that both yield equal pressure and impulse [13]. As discussed in [3], the ratio of energy output of a chemical explosive material relative to a given amount of TNT is calculated following

$$W_{effective} = \left[\frac{H^d_{exp}}{H^d_{TNT}} \right] W_{exp}$$

where $W_{effective}$ is the effective charge weight, W_{exp} is the weight of explosive in question, H^d_{exp} is the heat of detonation of a given explosive, and H^d_{TNT} is the heat of detonation of TNT. In order to compare shock wave parameters to available published data, a scaled distance from a given explosion source to a structure may be used. As discussed in [14], the most common form of blast scaling is the Hopkinson-Cranz or cube-root scaling method following

$$Z = R/W^{(1/3)}$$

where Z is the scaled distance, R is distance from the explosion source to the loading point, and W is the weight of TNT. This formulation asserts that self-similar blast waves are produced at identical scaled distances when two different sized explosive charges (of similar geometry and chemical composition) are detonated in the same atmosphere. Using available empirical data [3], this scaled distance from the explosion source to a loading surface can be used to determine specific blast parameters such as peak reflected overpressure, arrival time, and positive phase impulse. The CONWEP blast loading model discussed herein is based on a TNT explosion (assumed surface blast in this study).

4. Recent Industry Failure Case Study

The results of an explicit finite element analysis investigating a recent industry failure of a vertical heat exchanger are summarized below. It is believed that this specific failure was the result of an accidental internal detonation within the heat exchanger. The overall damage to the exchanger involved gross failure/rupture of the shell-side and tube-side pressure boundaries and the fragmentation of numerous tubes and other internal components. Furthermore, after the detonation event, an extensive debris field was present in the vicinity of the exchanger and noticeable distortion to the bottom tubesheet and other internal components was observed. In addition, there was extensive damage and plastic deformation of nearby equipment, particularly the support skirt of an adjacent tower.

A schematic of the vertical exchanger is shown in Figure 3. The approximate diameter, of the exchanger is 4 feet with an overall length of approximately 15 feet. The exchanger tubesheets are approximately 4 inches thick, and there are roughly 1,000 tubes. The shell-side of the exchanger contains a flanged and flued expansion joint. The exchanger is supported vertically near the center where chair supports attach to surrounding structural steel.

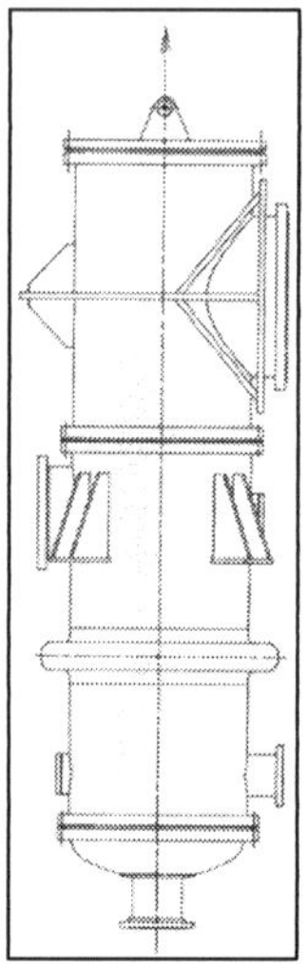

Figure 3: Schematic of the Vertical Heat Exchanger.

The goal of the analysis presented herein is to gain a better understanding of the detonation that occurred by looking at the structural response and failure modes of the heat exchanger and the surrounding equipment. In order to establish a reasonable approximation of the equivalent charge amount that led to the damage observed in the field, computational iterations are first performed for the adjacent tower and support skirt. A 3D finite element model of the tower and support skirt is constructed out of shell elements and subjected to external blast loading corresponding to varying charge amounts at the appropriate distance away from the tower (actual distance between the vertical exchanger and the tower). The overpressure amplitudes applied to the outside of the tower are calculated using CONWEP and account for reflection effects.

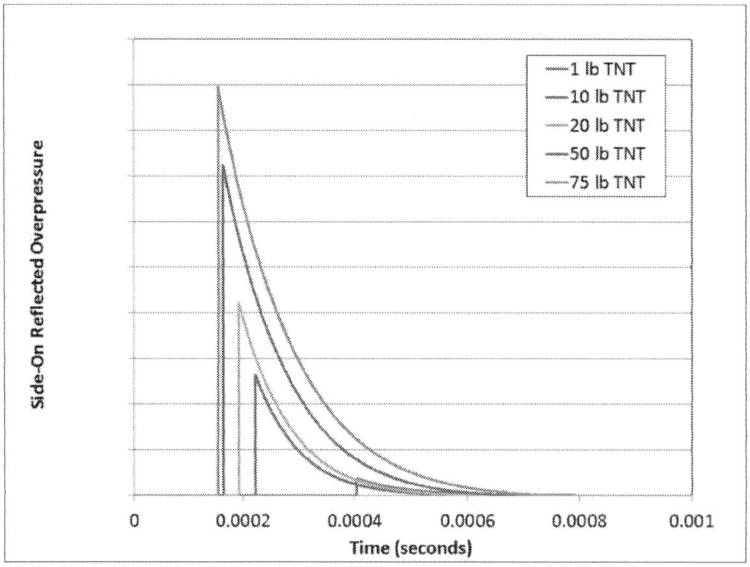

Figure 4: Peak Reflected Overpressures at the Tower Support Skirt for Varying Equivalent Charge Amounts.

The material model employed in the analysis is elastic-plastic and reflects quasi-static monotonic behavior (i.e. strain rate dependent material properties are not explicitly considered due to lack of available dynamic test data for the material in questions). Figure 4 shows the peak reflected overpressure amplitudes for varying equivalent amounts of TNT that would be observed at the base of the tower (note that specific overpressure values are intentionally omitted on the vertical axis of this figure). Based on these iterations, an equivalent charge amount is determined that approximates the amount of plastic deformation in the adjacent tower skirt that was measured in the field. Figure 5 shows contours of plastic deformations and equivalent plastic strains (based on this approximate charge amount) predicted in the tower skirt from explicit, non-linear FEA (again, specific contour values are intentionally omitted). This equivalent charge amount is then used to simulate an internal detonation from within the vertical heat exchanger shown in Figure 3, again using 3D explicit FEA techniques.

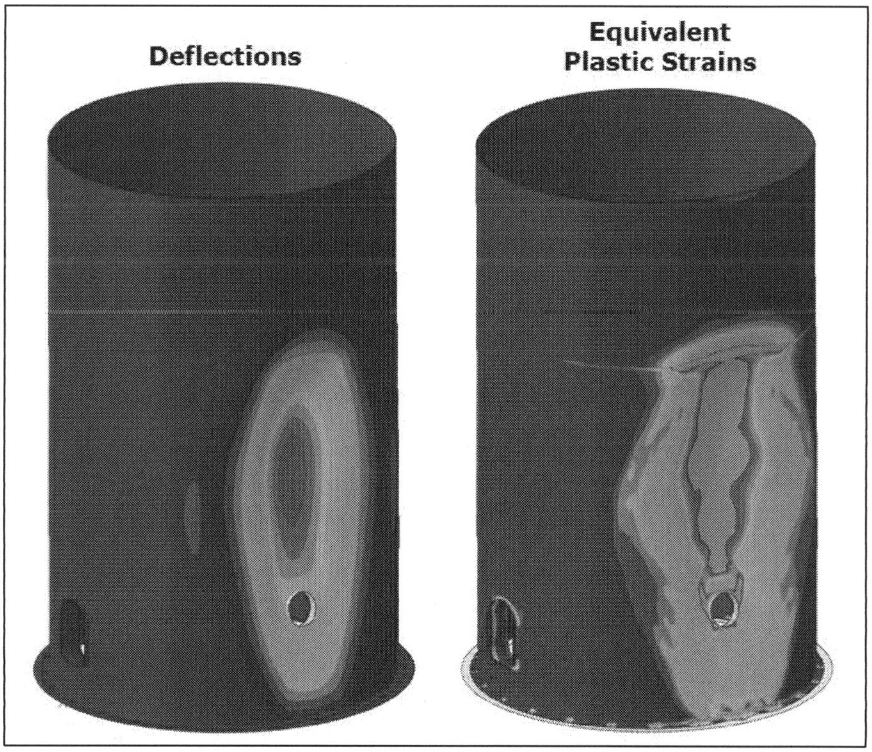

Figure 5: Permanent Deflections and Equivalent Plastic Strains in Adjacent Tower Skirt.

The analyses discussed herein employ explicit FEA formulations, which in general, are well suited for analyzing highly transient and non-linear behavior and permit evaluation of models with highly discontinuous events with large deformations. Explicit FEA simulations are computationally expensive due to the small time increments generally needed to achieve convergence. Explicit dynamic simulations on complex, 3D models would generally not be feasible or practical without the use of parallel computing optimization. For example, the simulations presented herein were run across 16 cpus or more simultaneously in order to achieve practical solution times.

A 3D finite element model of the bottom portion of the vertical heat exchanger is constructed using solid elements. The 3D model is shown in Figure 6 and includes the flanged and flued expansion joint. It is noted that the model is truncated at the shell-to-channel interface to facilitate computational efficiency and the holes in the bottom tubesheet are included. The top of the finite element model is constrained using a kinematic coupling definition in a local cylindrical coordinate system to permit radial dilation and maintain planar behavior at the end face (this reasonably mimics the restraint of the un-modeled shell). The detonation source is placed at a location within the exchanger that most reasonably matches the maximum deflection measured in the bottom tubesheet and observed in the shell; that is, based on the permanent damage observed in the field, a postulated location of the detonation center is assumed in the FEA simulation.

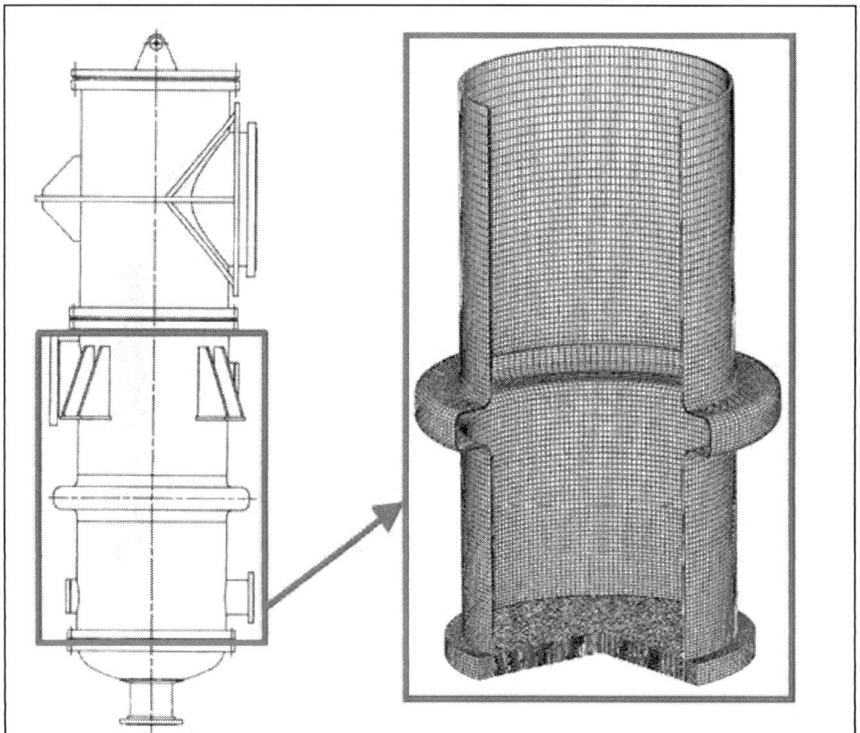

Figure 6: Portion of the Heat Exchanger Modeled and Subjected to Internal Blast Loading (Cutaway View of FEA Model Shown).

Explicit FEA is performed on the vertical exchanger where again, the CONWEP charge property is employed to generate the overpressure amplitudes applied to the inside surface of the exchanger. It is noted that in this case, a shear element failure criterion is implemented that is driven by plastic yielding where a given individual element loses stiffness and is removed from the model once it reaches a user-defined strain threshold. For the initial simulation presented herein, this threshold is defined as 100 percent equivalent plastic strain. This strain threshold is a somewhat arbitrary, but is used to bound critical behavior of the shell as discussed below.

In general, quasi-static material properties for carbon steel dictate that failure would occur between 25 and 30 percent strain. As discussed in Reference [15], increasing the strain rate in carbon steels generally results in a higher yield strength until eventually, the ultimate stress is approached without yielding. Essentially, higher rates of loading (impulsive blast loading results in very high loading rates) can cause steels to exhibit increased strain hardening and behave in a more brittle fashion (i.e. less ductility). In this case, because detailed dynamic material properties were not readily available for the material in question, some judgment was used in developing the explicit element failure criterion (with the intent being to conservatively bound failure). Figure 7 below shows the progression of damage in the pressure boundary of the vertical heat exchanger (contours of deflection are shown). It can be seen that a large portion of the shell fails (equivalent plastic strain threshold of 100 percent is exceeded) and is automatically removed from the model.

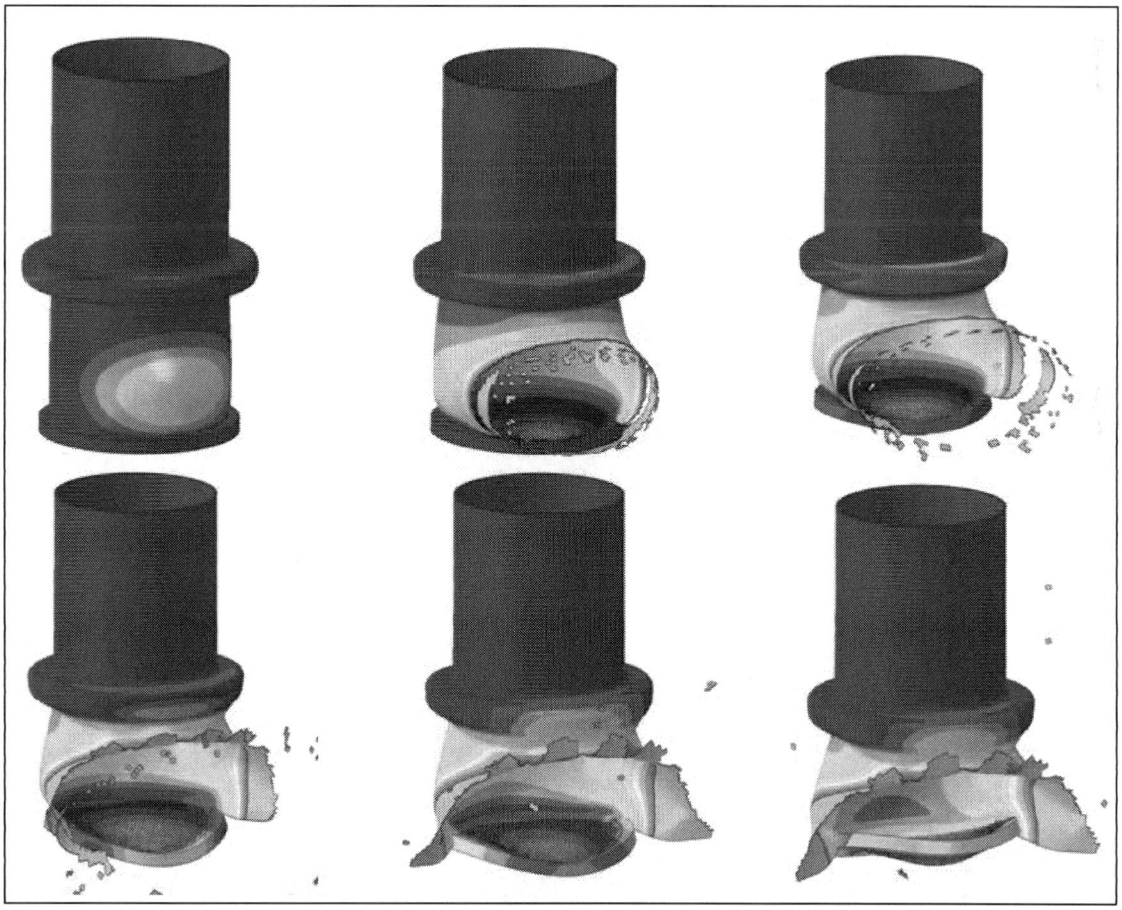

Figure 7: Contours of Displacement Showing Damage Progression of the Shell Due to Internal Blast Loading (Element Failure Criterion Employed at 100 Percent Strain).

Individual elements are also fragmented and thrown from exchanger in this simulation because all surrounding elements reached their user-defined strain threshold and were removed from the model; thus, there is no restraint surrounding these elements to hold them in place and the overpressure applied propels them outward from the blast center. It is noted that exchanger internals such as individual tubes, baffles plates, etc. are not explicitly modeled, and complex internal blast wave reflection from these components is neglected (as stated, investigating internal reflection effects due to complex geometry in detail would require the use of CFD analysis). In general, this simulation reasonably approximates the actual gross damage to the pressure boundary observed in the field; that is, the portion of the shell that was either fragmented or opened up following the presumed detonation event.

Sensitivity to the plastic strain threshold value used in the dynamic element failure criterion definition is investigated to examine the overall effect on structural response. Intuitively, lower user defined plastic strain thresholds predict more severe gross damage to the pressure boundary. Figure 8 shows contours of deflection for the exchanger shortly after the simulated internal detonation. In this case, the shear element failure criterion uses a strain threshold that is set to 30 percent equivalent plastic strain.

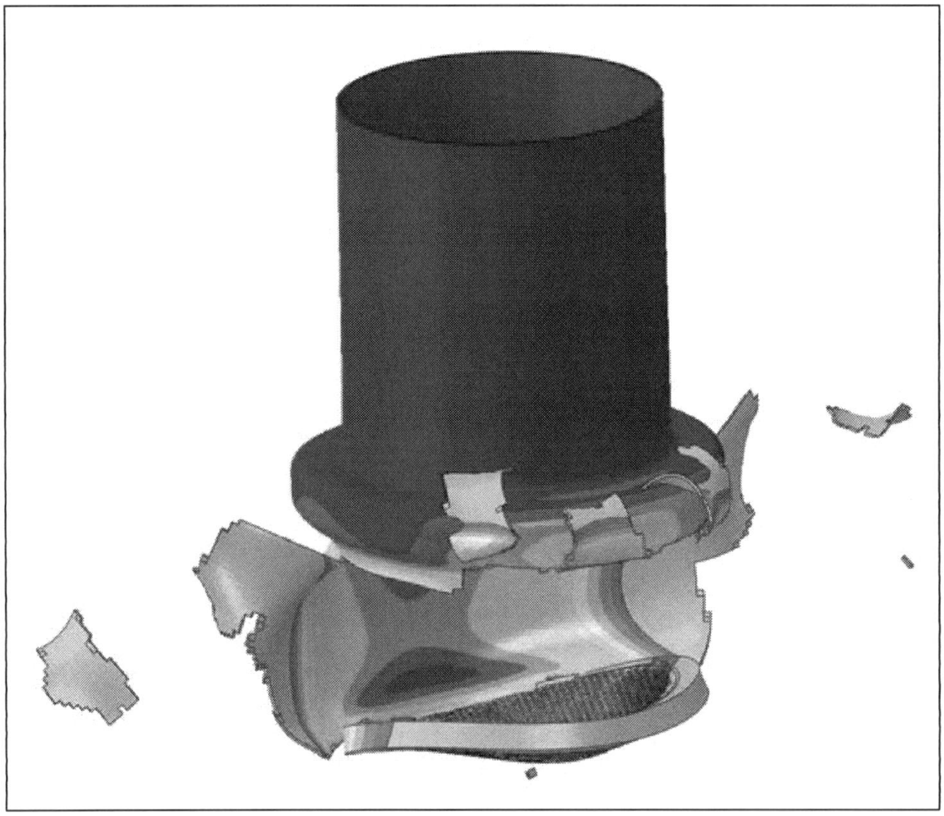

Figure 8: Contours of Displacement Showing Damage of the Shell Due to Internal Blast Loading (Element Failure Criterion Employed at 30 Percent Strain).

When compared to the damage progression depicted in Figure 7 (corresponding to a 100 percent equivalent plastic strain threshold), it can be observed that a greater portion of the shell is removed from the model. Furthermore, for this lower strain threshold, the overall damage extends further in the axial (element removal extends in the upward direction into the expansion joint) and circumferential directions, and plastic deformations are generally more severe in the shell that was not removed from the model.

In the explicit dynamic simulations discussed above, the stiffness of the tube bundle is neglected entirely. For this reason, the simulations over predict the amount of plastic deformation in the bottom tubesheet. Figure 9 shows the plastic deformation observed in the bottom tubesheet shortly after the simulated detonation. It can be observed that local to the assumed blast source within the exchanger, axial deflection in the tubesheet is predicted relative to the majority of the tubesheet area. This predicted deformation is consistent with the actual damage observed in the tubesheet in the field, although it is slightly higher than measured deformation. Additionally, there is some rigid body motion associated with the tubesheet after the simulated detonation because of the neglected stiffness of the tubesheet bundle from the model. After the removal of a large portion of the shell due to the assumed element failure criterion, there is little stiffness to hold the tubesheet in place. To this end, additional simulations are performed to investigate the effects of the tube bundle stiffness.

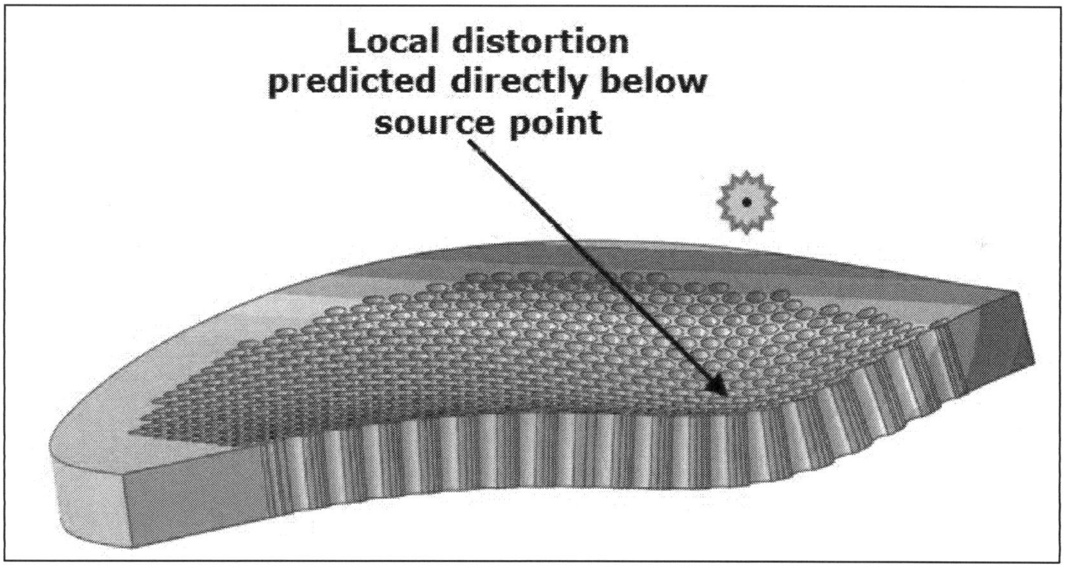

Figure 9: Exaggerated Tubesheet Deflections Local to the Assumed Blast Source.

In this case, explicitly modeling the tubes would be extremely computationally expensive and impractical due to required simulation times. In order to investigate the effect of the bundle on overall tubesheet deflections, equivalent beam elements are defined and coupled to each quadrant of the inside tubesheet face. This coupling definition permits the tubesheet to warp or bow while still distributing the stiffness of the bundle over each quadrant of the tubesheet face. Detailed discussion of this method of approximating the bundle stiffness is beyond the scope of this paper, but further explanation can be found in reference [16]. In this simulation, local tubesheet plastic deformations are under predicted (there is little local deformation of the tubesheet predicted directly below the assumed blast center). This occurs because after the detonation, the equivalent bundle stiffness is maintained and does not account for any damage/fragmentation (and ensuing removal) of individual tubes.

These results indicate that the amount of plastic deformation in the tubesheet actually observed in the field is in between the predicted amount of deformation in the model with the bundle completely neglected and the model with the equivalent bundle stiffness incorporated. The most rigorous approximation of tubesheet deformation would entail explicit modeling of the tubes and incorporation of progressive damage (and element deletion) such that the tubes in the immediate vicinity of the detonation source could be removed from the model as the overpressure wave propagates through the exchanger and once assumed strain limits are achieved. This simulation would be extremely computationally expensive and somewhat impractical. As stated, the overall predicted failure modes of the shell and tubesheet are in reasonable agreement with the actual damage observed in the field.

5. Conclusions

This study employs explicit finite element analysis to simulate a recent, real-life industry failure of a vertical heat exchanger subjected to an internal detonation. Realistic overpressure amplitudes are applied to 3D finite element models in accordance with the CONWEP blast loading model. Incident wave reflection is accounted for using this formulation without modeling the fluid medium and coupling it to the structural elements. The ensuing dynamic structural response and plastic deformations of both the vertical heat exchanger subjected to internal blast loading and the adjacent tower and support skirt subjected to external blast loading are quantified. This analysis permits the determination of an approximate amount of concentrated chemical product that caused the accidental explosion. Essentially, the approximate equivalent amount of explosives is back-calculated based on the predicted damage to the finite element model of the adjacent tower skirt. Computational iterations of varying charge amounts are performed and the predicted amount of plastic deformation is documented in each case.

Based on these iterations, the equivalent charge amount that most closely approximates the plastic damage of the nearby tower skirt is used to simulate internal blast loading on a 3D model of the vertical heat exchanger. An advanced element failure criterion is employed to simulate the damage of the pressure boundary observed in the field. According to this enforced formulation, elements lose stiffness and are removed from the model once a user-defined plastic strain threshold is reached. This dynamic simulation reasonably predicts the gross rupture and deflection of the pressure boundary observed and measured in the field. The explicit, three dimensional finite element based analyses discussed in this study permit the evaluation of the structural response and damage characteristics corresponding to this specific, recent industry failure. Explicit simulations, such as the ones discussed herein, can not only provide process safety engineers with a better understanding of the risks associated with accidental internal detonations, but can provide valuable insight into the mechanical design of fixed equipment that mitigates the risk of catastrophic rupture or failure due to internal or external blast loading.

6. References

1. Smith, P., and Hetherington, J., "Blast and Ballistic Loading of Structures, 1st ed," *Butterworth-Heinemann*, Oxford, 1994.
2. Kinney, G., and Graham, K., "Explosive Shocks in Air, 2nd ed," *Springer-Verlag*, Berlin, 1985.
3. U.S. Department of Defense, "Unified Facilities Criteria (UFC) 3-340-02: Structures to Resist the Effects of Accidental Explosions," *Department of Defense, United States of America*, 2008
4. Dusenberry, D., "Handbook for Blast-Resistant Design of Buildings, 1st ed," *John Wiley and Sons, Inc.*, Hoboken, New Jersey, 2010.
5. Bulson, P., "Explosive Loading of Engineering Structures, 1st ed," *Taylor and Francis Group*, London, 1997.
6. ASCE, "Design of Blast-Resistant Buildings in Petrochemical Facilities, 1st ed," Task Committee on Blast-Resistant Design, *American Society of Civil Engineers.*, Reston, Virginia, 1997.
7. Abaqus 6.10-1, DSS Simulia, "Analysis Users' Manual," 2010.
8. Hyde, D., "Users' Guide for Microcomputer Programs CONWEP and FUNPRO – Applications of TM 5-885-1," *U.S. Army Engineer Waterways Experimental Station*, Vicksburg, VA, 1988.
9. Duffey, T.A., Rodriguez, E.A., and Romero, C., "Detonation-Induced Dynamic Pressure Loading in Containment Vessels," Los Alamos National Laboratory: LA-UR-03-0741, 2003.
10. Dong, Q., Li, Q.M., and Zheng, J.Y., "Interactive Mechanisms Between the Internal Blast Loading and the Dynamic Elastic Response of Spherical Containment Vessels," *International Journal of Impact Engineering*, Volume 37, 4, 2010.
11. Prueter, P.E., Dewees, D.J., and Brown, R.G. "Examining Computational Techniques to Simulate the Effects of External Chemical Explosions Using Explicit Finite Element Analysis," *In ASME Pressure Vessels and Piping Conference.* PVP2012-78566, Toronto, Canada 2012.
12. Prueter, P.E. and Dewees, D.J., "Using Explicit Finite Element Analysis to Simulate the Effects of External Chemical Explosions on Single and Double-Walled Storage Tanks," *In Proceedings of the AIChE Spring Meeting and 9th Global Congress on Process Safety*, San Antonio, Texas, 2013.
13. Huntington-Thresher, W., and Cullis, I., "TNT Blast Scaling for Small Charges," *19th International Symposium of Ballistics*, 2001.
14. Baker, W., "Design Considerations for Toxic Chemical and Explosives Facilities – Chapter 1, Blast Pressure Effects: An Overview," *American Chemical Society*, 1987.
15. HSE, "Offshore Technology Report – OTO 1999 018: The Behaviour of Carbon Steels at High Strain Rates and Strain Limits," Health and Safety Executive, 1999.
16. Prueter, P.E., Macejko, B.R., Brown, R.G., and Dewees, D.J., "Assessing Flanged Joint Leakage and Tube-to-Tubesheet Joint Failure Due to Fluid Temperature Differentials in a Floating Head Heat Exchanger using Finite Element Analysis," *In ASME Pressure Vessels and Piping Conference,* PVP2013-97576, Paris, France 2013.

Analysis of the Potential Energy Sources of Risk of Hand Tools in Presence of Hazardous Area in Segment of Oil and Gas

Leandro Erthal
Petrobras S.A.
Av. Almirante Barroso, 81 suite 23 - Centro - Rio de Janeiro - Brazil
leandroerthal@petrobras.com.br
Universidade Federal do Rio de Janeiro (UFRJ) - Chemical Engineering Department (EQ)
Av. Um s/n - Cidade Universitária UFRJ - Rio de Janeiro - Brazil

Caetano Moraes
Universidade Federal do Rio de Janeiro (UFRJ) - Chemical Engineering Department (EQ)
Av. Um s/n - Cidade Universitária UFRJ - Rio de Janeiro - Brazil
caetano@eq.ufrj.br

Denize D. Carvalho
Universidade Federal do Rio de Janeiro (UFRJ) - Biochemical Engineering Department (EQ)
Av. Um s/n - Cidade Universitária UFRJ - Rio de Janeiro - Brazil
denize@eq.ufrj.br

Prepared for Presentation at
American Institute of Chemical Engineers
2014 Spring Meeting
10th Global Congress on Process Safety
New Orleans, LA
March 30 – April 2, 2014
UNPUBLISHED

Analysis of the Potential Energy Sources of Risk of Hand Tools in Presence of Hazardous Area in Segment of Oil and Gas

Leandro Erthal
Petrobras S.A.
Av. Almirante Barroso, 81 suite 23 - Centro - Rio de Janeiro - Brazil
leandroerthal@petrobras.com.br
Universidade Federal do Rio de Janeiro (UFRJ) - Chemical Engineering Department (EQ)
Av. Um s/n - Cidade Universitária UFRJ - Rio de Janeiro - Brazil

Caetano Moraes
Universidade Federal do Rio de Janeiro (UFRJ) - Chemical Engineering Department (EQ)
Av. Um s/n - Cidade Universitária UFRJ - Rio de Janeiro - Brazil
caetano@eq.ufrj.br

Denize D. Carvalho
Universidade Federal do Rio de Janeiro (UFRJ) - Biochemical Engineering Department (EQ)
Av. Um s/n - Cidade Universitária UFRJ - Rio de Janeiro - Brazil
denize@eq.ufrj.br

Keywords: Hazardous Area, Oil and Gas, Hand Tool

Abstract

One of the methods adopted to protect against explosion in the oil industry refers to the use of hand tools for non-ferrous materials, usually bronze, also known as non-sparking hand tools. Studies conducted by the American Petroleum Institute (API/USA) and European standards disagree as to the risks involved in the energy of sparks produced by mechanical tools. The API advises that operations with mechanical equipment at high speeds and high contact pressures are extremely difficult to produce sparks capable of causing ignition of gases and vapors in the oil industry, and concludes that the risks involved are low for hydrocarbon vapors. The IEC (International Electrotechnical Commission), however, has begun to formulate a set of standards on this topic. These standards will be adopted and may affect the operations in the oil and gas industry.

Concerning the incident in the British Petroleum Refinery, Texas City, USA, the "Report of the BP US Refineries independent Safety Review Panel" or "Baker Panel Report" [1] - a report produced by an independent commission of inquiry - recommends that the accident prevention process require permanent vigilance. Moreover, it states that the lack of accident for a long period is not an appropriate indicator of high level process safety and can contribute to a dangerous and growing sense of complacency.

Using data from the literature and laboratory tests, this paper aims to clarify and demonstrate the actual risks to the energy produced by ferrous and nonferrous hand tools.

1. Introduction

In the late 20s, seeking to discover the causes of fires in the mining industry, was published the article entitled "Ignition of Gases and Vapors - Review Ignition of Flammable Gases and Vapors by Friction and Impact" by F. Powell [2]. He reports the work done in investigating more than 82 accidents, but only some of the references involving hand tools and the author avoided making any conclusions about it. A study presented to the Group Session on Fire Protection at the annual meeting of the American Petroleum Institute, in 1941, described a series of test conducted over 15 years. According to the test, it would be very unlikely that the ignition of vapors of oil could be caused by sparks produced by the contact between steel and steel abrasive wheel, or a given mechanized equipment. The nature of the sparks was discussed and it was demonstrated that any material harder than iron could produce sparks when struck against it. Thus, it was concluded that the insistence on the use of special hand tools called "non-sparking tools" creates a false sense of security at the expense of other more important measures of security protection against fire. Comprehensive rules covering the use of such tools were recommended, many of which are against the industry interests. The use of non-sparking hand tools was not universal throughout the industry until 1941, but then, after the presentation of the article of Group Session on Fire Protection, many companies started a program to phase out the use of special tools. However, some companies associated with API believed that additional research was needed to obtain a definitive opinion, and an independent institution should do it. In 1948, the necessary funds had already been obtained and contracted in 1950 to conduct a series of tests at Underwriters Laboratories Inc, in Chicago, sponsored by the American Petroleum Committee on Accident Prevention and Fire Protection. Little progress, however, was done [3].

The Institute of Petroleum (UK) began to consider the importance of sparking hand tools as an ignition source in the early 60's through the Committee on Industrial Fires and Explosion of the Fire Research Board. These studies [4] included a comprehensive review of published information, but they did not contain any new experimental evidence. An introductory note prepared by Engineering Committee of the Petroleum Institute indicated that additional experimental studies were not necessary; accepting the conclusion that non-sparking tool materials do not significantly reduce the risk of ignition of oil vapors by sparks due to friction force, compared to coming from ferrous tools.

The American Petroleum Institute (API) issued in July 2004, the fourth edition of the report API RP 2214 - Sparks Ignition Properties of Hand Tools [5], concludes that nothing essentially new has been learned since the publication of the first document in 1956. This document complements recent publications such as the NFPA's Fire Protection Handbook, with the same conclusion of 40 years ago. Sparking produced by violent impact between some materials, including metals commonly termed as non-sparking, can actually cause the ignition of gases and vapors if generated enough energy. However, such conditions are rare to occur in operations using hand tools. The records on the fire at companies, which have never used or ceased to use non-sparking tools, confirm the position of the Safety Committee of API's Board of Directors in 1956.

According to PROUST [6], it has been well known for more than a century that the combustible dust suspended in air are responsible for part of the blasts, but available knowledge and practices about the explosions of flammable gases and vapors still contain a significant part of empiricism [7].

Over the last decade, many investigation committees have been created with a map road including several patterns with a set of standards to be developed for non-electrical equipment and components.

The aim of this work is to investigate what the real risk is and what parameters should be taken into account when manual hand tools are used in hazardous locations within the oil and gas industry.

2. Definitions

2.1 Hazardous locations

The manufacturing process, according to Garside [8], often involves the use, processing or storage of hazardous materials (flammable). Evidently, industrial plants associated with these operations need to be arranged and controlled so that these materials do not ignite or cause explosion, generating damage to people or the plant itself. Thus, the presence of gases, vapors, combustible dusts and fibers associated with the air, in optimal proportions, may occur in certain places and is referred as "explosive". From a chemical standpoint, oxidation, combustion and explosion are reactions of different speeds, being initiated by ignition or detonation.

In the "definitions" of API RP 500 [9], it is mentioned that any area where gases and vapors are present or may be present in the atmosphere in sufficient quantities to produce a flammable mixture concentrations is a "hazardous (classified) location."

In industrial plants, a technique of risk analysis called "classification of areas" is usually used.

An area classification or hazardous location is a study that analyzes the local industrial plant during the normal operation process that can contain flammables within its flammable range, with some frequency and therefore, requires that special care to be adopted to prevent that electrical or mechanical equipment be a source of ignition in such explosive atmospheres [10,11].

2.2 Risk of hazardous locations

There are numerous potential sources of ignition of an explosive atmosphere, according to Garside [8], such as the uncontrolled use of open flames, hot surfaces of the process itself, sparking, static electricity, including those generated by clothing and electrical discharges. Many of these sources can be controlled or minimized by a good arrangement process, management and control. However, there is an additional issue that are the electrical and mechanical equipment necessary to operate an industrial plant, such as pumps, fans, lights, instrumentation and other equipment are themselves potential sources of ignition.

Directive 1999/92/EC of European Committee [11] states that, "The installation, equipment, protection systems and associated connecting devices must only be put into service if the documents about the protections explosion indicate that it may be safely used in presence of explosive atmospheres". Another important point that determines Directive 1999/92/EC [11] is that the same conditions above must be:

> [...] It also applies to work equipment and associated connecting devices which are not considered equipment or protective systems in the design of Directive 94/9/EC if their incorporation into an installation can in itself lead to a risk of explosion. The necessary measures should be taken to avoid any incompatibility between connecting devices. "

Directive 1999/92/EC [11] provides as a protection against explosions:

> "All necessary measures should be taken to ensure that the workplace, work equipment and connecting devices made available to workers have been designed, manufactured, assembled and installed, and are maintained and operated as to minimize risk of explosion".

2.3 Certification

A certificate of conformity is to issue a statement based on a decision made after the critical analyses of the service to specified requirements has been demonstrated

2.4 Studies on the mechanical ignition risk

For the development of standards methods of explosion protection for non electrical equipment, according to PROUST [6], there is a significant gap in knowledge of their mechanisms of explosion. However, the key parameters were not identified in a precise manner.
There is some information provided to workers about the risks of fire and explosion of tools called "sparking", made from "materials harder", made of steel with index Vickers hardness greater than 230 HV, according to ASTM E92 (Standard Test Method for Vickers Hardness of Metallic Materials) [12].
For SCHUTH [13]:

> "When we refer to surveys conducted in Europe and USA, it was found that the source of the spark that causes more explosions relates to the use of common tools (steel) in places where there is risk of explosion ... According to experts, a spark from a tool steel reaches temperatures of 1550 °C to 1850 °C, being the product of a conversion of energy, i.e. the kinetic energy resulting from force applied to the tool or consequence of free fall, becomes the impact on thermal energy in both materials in collision ... It is clear that the risk of explosion exists and should be considered rigorously, and it is therefore avoided the use of steel tools where such risk may exist.

A study carried out by MADAKSON [14] shows that the friction of a material also depends on the test system. Replicate samples of the same material were distributed in different laboratories to measure friction under certain conditions. Using different measuring systems, each laboratory reported a value different from the friction. Between January 2001 and May 2005, the European Community developed a project called "MECHEX" (Mechanical ignition hazards in potentially explosive gas and dust atmospheres), with the French laboratory INERIS, to review of the risk of mechanical ignition (Ignition Hazard Assessments) and be able to apply the European standards (EN). This project sought to develop knowledge of the risks associated with these devices through two research projects: characteristics of sparks and hot spots produced under impact, friction or grinding. Simulated ignitions with the same characteristics of those found in industry to measure the sensitivity of an explosive environment ignition atmospheres and characterization with the goal of standardizing the ignitability sources of explosives were used. The results of these two projects were combined in order to categorize the equipment to use in explosive atmospheres based on the type of mechanical ignition sources expected to appear in a specific project, the method of operation and the environment capable of suffering ignition. Thus, the process of degradation of the mechanical energy into heat for the friction and impact was studied.

3. Overview of Associations and Companies

3.1 Certification of hand tools

The operating procedure called "Approval Standard for Sparks Resistant Tools" [15] was published in December 2007 by the certification body FACTORY MUTUAL APPROVAL (FM). The scope says that operating certification procedure is applied to any tool or device whose purpose is to function as a spark resistant. These devices can be manual or electric, such as hammer, screwdriver, pliers, sander or any other. It mentions that this procedure assesses the potential of these tools and devices, or material, in which it can be manufactured to generate a visible or cause ignition of flammable materials under certain conditions in laboratory test spark. It restricts the condition that it does not assess the practicality or suitability of the purpose of using the device, but it is limited to determining the potential to cause ignition of materials in normal operation.

According to WANG, in NFPA Memorandum [16], FM conducted tests described in "Report nº IN4A0.AF Job identification" in accordance with the requirements of the "Safety Tools Approval Test Procedure - 7910 class". It is used as a guide to assess tools designed for use in environments where the risk of ignition of flammable materials, dust or vapors by mechanical sparks generated by use of hand tools, iron and steel that may slip by or affecting a surface exists. This report has shown that:

- ✓ Sparks produced by steel plates caused explosion of the pad saturated with carbon disulfide in eight of eight tests, ethyl ether pads, five of eight tests and naphtha in four of eight tests;
- ✓ No sparks were observed and no ignition of flammable liquids mentioned above was certified security tools.

This document states that its requirements reflect tests and practices used to analyze the metallic characteristics to allow their use in the manufacture of spark resistant in order to obtain a certification tool.

The above-mentioned requirement to meet compliance of this operating procedure is that the tool should produce no visible spark when struck on a variety of surfaces. According to FM, this demonstrates a lack of ability to produce sparks, and thus reduces the risk of causing ignition of the material or atmosphere around them when the tool is used. In order to meet this requirement, no visible spark should be identified in the video recording of high speed. The spark must be generated by the impact of the alloy tested on a range of surfaces like those on which are expected in service. To meet the requirement of the test for flammable chemicals, particles generated by applying a sample of the material on the grinding wheel must not cause ignition of flammable vapors. The particles are released on a wad of cotton soaked with a flammable chemical. The cotton swab should not be ignited.

3.2 International Association for Hydrogen Safety (HySafe)

According to HySafe, the use of non-sparking tools is often recommended; however, the energy to ignite a flammable mixture of hydrogen in air is so small that these tools are still capable of causing ignition. Consequently, all tools need to be used judiciously to prevent landslides, shocks or falls because all these conditions can cause sparks. According to HAWKSWORTH [17] one, regardless of their material tool, dropping the height of a man is capable of causing ignition of an explosive atmosphere of hydrogen gas.

3.3 International Eletrotechnical Commission - IEC

3.3.1 Assessment of single impact sparks as Potential Ignition Sources

According to the criteria of the standard PT ISO/IEC 80079-36 [19] "single impacts between metal parts need not to be considered as potential ignition sources in the ignition hazard assessment if the following conditions are met.
Either:
a) the impact velocity is less than 1 m/s and the maximum potential impact energy is less than 500 J, and

- aluminium, titanium and magnesium in combination with ferritic steel is not used, or
- aluminium in combination with stainless steel (≥ 16,5 % Cr) can be only used if the steel cannot corrode and no iron oxide and/or rusty particles can be deposited on the surface (appropriate reference to the properties of the stainless steel shall be given in the technical documentation and instructions for use), or
- hard steel in combination with hard steel is not used, or
- hard steel is not used where it can impact on granite, or
- aluminium in combination with aluminium is only used if no iron oxide and/or rusty particles can be deposited on the surface.

NOTE 1 Hard steel is understood as being either all kinds of hardened steel (surface hardened or heat treated in another way to improve surface hardness) or other steel types with Vickers Hardness greater than 230 HV (acc. to ISO 6507 with test load ≥ 98 N).

Or

b) where a combination of non-sparking metals is used and the impact velocity is less than or equal to 15 m/s and the maximum potential energy is less than 60 J for gas/vapour-atmospheres or less than 125 J for dust-atmospheres.

3.3.2 Assessment of single impact sparks as effective ignition sources

Ignition sources generated by impact need not be considered as effective ignition sources if the impact velocity is less than 15 m/s and the maximum possible potential energy is less than the values given in Table 1

On the other hand, if the energies exceed the values in the Table 1, this does not necessarily mean that the ignition source will become effective. In this case, the ignition hazard assessment needs to assess all aspects and may show that the likelihood of the impact is low enough to be acceptable.

The impact energies need to be assessed if they are larger than those shown in the table. In this case, consideration shall be taken of when they occur and whether they are able to ignite the explosive.

Table 1. Single impact energy limits

Explosion group	Single impact energy limits	
	Non-sparking metals:	Other materials
IIC	60 J	5 J (Hydrogen) 3 J (Hydrocarbons incl. acetylene)
IIB	125 J	10 J
IIA	125 J	20 J
NOTE These criteria do not apply for atmospheres with flammable gases such as carbon disulfide, carbon monoxide and ethylene oxide		

3.4 Health and Safety Executive (HSE-UK)

In the publication HS (G) 140 [19], refers to the need not to rely too much on the use of non-sparking tools in areas of flammable, being necessary to remove all flammable materials before the service is performed, if possible, and if otherwise, the need to keep all the tools free of fouling.

3.5 National Fire Protection Association – NFPA

NFPA 30 Flammable and Combustible Liquids Code: 2003 [20] as well as the NFPA 1 Fire Code: 2008 [21], there was no recommendation on whether to use non-sparking hand tools.
The NFPA Technical Committee on Fundamentals cites in its opinion on the technical review conducted in consultation with the NFPA 30 Flammable and Combustible Liquids Code:2011[22] that the work of the American Petroleum Institute (API 2214) [23] refutes the need for using non-sparking hand tools, except in special circumstances. After the technical argument against the representative of a company non-sparking tools on the proposal no. 30-58, the Committee decided to reject the suggestion of inclusion of the proposed item: "6.5.6 Heating by friction or mechanical sparks, with the following text: iron and steel hand tools may produce sparks that can be a source of ignition of surrounding flammable substances. Where this risk exists, non-sparking tools should be used. "Thus, the decision was to include a new text A.6.5.1 in which to control ignition sources precautions should be taken to prevent ignition of flammable by sources such as heating by friction or sparks".
This same text was copied and expressed in NFPA 1 Fire Code: 2012 [24], and complemented the informational item related to this subject, A.66.6.5.1 item. It cites that for frictional heating or sparking, it is recognized that there is the need to control sources of ignition, including sparks mechanic hand tools; we have enough for the ignition of flammable vapors energy.

3.6 U.S. Occupational Safety & Health Administration - OSHA

In their training program "Office of Training and Education":1996, entitled " Construction Safety and Health Outreach Program" [25], in the chapter " Recognition of Risks ", it is mentioned that around flammable substances, sparks produced by the hand tools of iron and steel can be a dangerous ignition source. Where such hazards exist, non-sparking tools made by brass, plastic, aluminum or wood supply security.

Recognizing that steel tools can be a potential source of ignition in explosive atmospheres, OSHA has prepared a guidance through the booklet 3080 - Hand Tools and Power , Review 2002" [26], which describes that the steel hand tools and iron can produce sparks and that these can be a source of ignition of surrounding flammable materials. Where this hazard exists, non-sparking hand tools need to be used.
The OSHA inspection reports #309178523, #122021017, #313996118, #302920400 and #309178523 as reported security flaws lack of use of non-sparking hand tools. The OSHA report #313523375, for example, cites a violation non-compliance OSHA 29 CFR 1910.106 (h) (7) (i) (a) - Flammable Liquids and Combustive [27] that "precautions were not taken to prevent the ignition of flammable vapors. The employer did not use non-sparking hand tools in hazardous location classified as Class I Division I".
OSHA Regulation 29 CFR 1910.146 - Permit-Required Confined Spaces, appendix D - Confined Space Pre -Entry Check List [28], non-sparking hand tools are on the checklist of equipment, in clause 9, to enter into confined spaces.
OSHA 29 CFR 1910.146:1996 "Standard Interpretations, Interpretative guidance and Tackle approved for hazardous locations relative to the Permit -Required Confined Space", states that "... electrical equipment as well as other sources of ignition (non-sparking hand tools) needs to be considered and adopted to go into hazardous areas".
OSHA Office of Training and Education has a PowerPoint presentation about Hand and Power Tool Safety [29] that reports that environments around the flammable sparks produced by iron in contact with steel hand tools can be a dangerous ignition source. Moreover, it cites that non-sparking hand tools, made of brass, plastic, aluminum or wood, should be provided for safety where this risk exists.

3.6 Petroleum Companies

Table 2. Company versus recommendation to use non-sparking hand tools

Company	*Where*	*Recomendation*
ExxonMobil	MSDS-USA “gas” in land spill	Use
Chevron	MSDS-USA “gas”, section 6	Use
Valero	MSDS-USA “gas”, section 6, Accidental Release Measures	Use
CITGO Petroleum Corp.	MSDS conform OSHA 29 CFR 1910.1200. Section 6 - Accidental Release Measures [30]	Use
Marathon/Petrocom	MSDS-USA “gas”	Not informed
Shell	MSDS-USA “gas”	Not informed
ConocoPhillips	MSDS-USA “gas” - Section 7: Handling and Storage: Precautions for safe handling	Use
British Petroleum (BP)	MSDS-USA “gas” - Section 7: Handling and Storage: Handling:	Use
Total	MSDS-USA “gas” - Section 7: Handling and Storage: Precautions for safe handling	Not informed
Statoil	MSDS “gas” - Section 7: Handling and Storage: Technical measures	Use
Petrobras	MSDS “gas” - Section 7: Handling and Storage	Use

4. Method and type of search

4.1 Test method

Regarding the purposes of this project, the test method to be adopted is an adaptation of the ASTM E680-79 standard, revalidated in 2011[31]. This standard has original purpose as being a standardized test method to evaluate the sensitivity of hazardous materials in the solid phase to the impact of falling weights. Impact tests by falling weight specified in this standard will be performed using the method of Bruceton up and down.

Laboratories: At the first moment, the test will be conducted in a mechanical laboratory to verify "vibration" and energy. Next, the test will be in a laboratory to atmosphere explosive.

The research of the potential risk of power tools will be in relation to:

Hammer head (new and used) by stainless steel, steel carbon and others material called non-sparking.

Objective: To evaluate the ability of new tools and steel rust capable of causing ignition of gases from petroleum environments.

Phases:

1) Chemical analysis of the hammer's material;

2) Conducting impact tests for samples according to test procedure.

Materials:

1) Hammer head steel with Rockwell hardness according to ASTM E680-79 [31] (surface between 55 HRC and 59 HRC), industrial line without oxidation;

2) Model wooden handle;

3) Mass hammer considering only the mass of the steel head: 1.0 kg, 2.0 kg, 3.0 kg, 4.0 kg and 5.0 kg;

4) Cash test with pressure relief through plastic film covering;

5) Gases and reference concentrations for tests . The criteria set forth by the IEC 60079-1 [32] will be adopted: $(4{,}6 \pm 0{,}3)\%$ de propane and $(8 \pm 0{,}5)$ % de ethylene;

6) Three samples of hammer;

7) Mechanic impact;

8) High-speed video record camera;

9) Equipment for measuring temperature;

10) Humidity between 50% up 90%;

11) Temperature between 30 °C up 40 °C;

12) Atmospheric pressure between 86 kPa up 106 kPa (860 mbar up 1060 mbar);

13) Setting of test laboratory accredited by Inmetro (Brazil's official accreditation) which scope is to evaluate electrical equipment for explosive atmospheres.

Impact test :

1) The hammer shall be tested by running 300 strong , lilting and consecutive blows , with a difference of 10 seconds between impacts, simulating the use by a man of average height with a mass about 80 kg. The impact energy shall be up to 8.0 Kgf.m;

2) Address the impacts against a new target hardened steel with Rockwell hardness in the range of (55-59) HRC . The roughness of the target to be in the range of 6.3 to 8.0 μ μ .

3) Run the impacts against the surface of different materials target, like steel stainless, steel carbon and aluminum;

4) Evaluate if ignition up to 300 impacts occurs

4.2 Criteria for assessing the likelihood of ignition of hand tools

Four techniques will be used to evaluate the capacity of sparks:
a) Measurement of vibration;
b) Thermography;
c) Video recording in high speed;
d) Testing in explosive atmospheres.

4.2.1 Target for each technique

a) Measurement of vibration – by each type of hammer and height, a series of twelve tests (12) will be done to obtain maximum real value of energy caused by impact (taking into account the energy lost by friction resistances of the system) and adjust the scale of the equipment to test the values of energy. These values are measured by accelerometers installed on the hammer;
b) Thermography – to verify the maximum temperatures generated by impacts and to compare them with the ignition temperatures of flammable substances in the test;
c) Video recording in high speed – to check the capacity of impacts of a hammer to generate light energy on plates of different materials (stainless steel, carbon steel and aluminum);
d) Testing in explosive atmospheres – through a type of pass/no pass test to verify if the luminous and thermal energy released is sufficient to ignite an explosive atmosphere may be present in an environment of the oil industry.

4.2.2 Equipment test

The equipment is three-meter high, has a target that can change the position of impact, has an electrical system (drive keys, electrical protection and a frequency converter that drives an electric motor) and an instrumentation system (sensors associated with the frequency converter). When the system is turned on, the magnetic sensor at the end of the path identifies the presence of the hammer and fires the frequency converter. The motor starts slowly lifting the hammer. As the desired time is reached, another sensor to the converter indicates this position. The motor is turned off causing the free fall of the hammer. Upon reaching the end of the course, the first sensor detects the hammer and resumes operation. This condition can be programmed to repeat automatically the test for a fixed number of times and can be triggered manually or perform the test continuously.

Table 3. Photos of the equipment test

The path to the target

The hammer and the target (with regulators of angles)

The hammer at falling (freedom)

5. Discussion and critical analysis

According to PROUST [6], explosions of flammable gases and vapors have mechanisms that are not well known, and practices still seem to contain a significant portion of empiricism.
There are many different opinions about this matter but no confidence form to prove that a steel hand tool can be a source of ignition in an oil industry. The unique report confidence is the European program MECHEX that was dedicated to study the process of ignition during friction and impact with gas, improved by Ineris Laboratory, to attend the European Norm. This report concluded that "for "impacts", the relevant parameter for ignition is not the kinetic energy of the projectile, but its velocity and the nature of the materials. The experimental data are not enough to validate the model and more experiences need to be lead."
Regarding the document API 2214:1956 [3], it can be observed that experiments carried out by consolidated companies and research conducted in the 50's demonstrate that these called non-sparking tools do not improve the security of the premises. Initially, this document advised not to use this type of tool. In the 2004 version [5], it stated that the non-sparking tools had less energy and at certain points showed a dubious position about the topic, without being conclusive. This version cites the position of the NFPA to support its point of view, but NFPA is motivated by a demand for manufacturers of non-sparking tools, without scientific support, mainly based on the U.S. law. It can be seen that this legislation is guided by the historical results of accidents in mining segment in which the chemicals involved are different, extrapolated to all activities. It is not possible to say whether the manufacturers of this type of tool influenced the U.S. legislation, but by acting with NFPA, presumably, they act in the interests of their industry.
Other investigations also contradict API's conclusion on non-sparking tools. A paper published in the Fire Protection Journal, "The Danger of Incendive Sparks" [16], stated that after API's conclusion on "non-sparking" tools, the Institute of Petroleum sought the advice of the Fire Research Board Committee on Industrial Fires and Explosion. This committee came to certain conclusions, one of which was that "Until...evidence is obtained to show that the impact of steel hand tools on steel is safe, it is advisable to use non-sparking tools in situations where flammable atmospheres are unavoidable".
Another point worth mentioning is that during the period of studies and evaluations to issuing the first paper of this issue, the oil industry began operations with hydrogen, and it is not expected that this chemical element was being considered at that time. In tests conducted by BOEMRE, the impact between rusted steel and steel with no rust were capable of causing ignition of a flammable mixture; however, OLSON [33] showed that the threshold pressure for this fact was 100 psig, which corresponds to projectile speed 20 m/s, incompatible with the operating conditions of hand tools in the oil and gas industry. This fact was confirmed in the tests and procedures performed by Foyn and Moe, in PROST [6], where no ignition was recorded for impact between rusted steel and steel.
Companies in the oil and gas segment diverge on the topic for several reasons:
1) It must meet local legislation. In the United States: OSHA 29 CFR 1910.1200 (g) (2) (viii) standard [30];
2) The API reference shows little conviction in its recommendation without theoretical support and recent laboratory tests, repeating observations only practices a very different period. Professionals responsible for this activity tend to be conservative when they are not supported by laboratory tests or certifications of third party;
3) The oil industry' 50s had few processes that used hydrogen, unlike the current processes;

4) Europe has a long tradition in mining, which influences European companies and their employees;
5) World companies such as Shell and Total, as well as other companies do not recommend non-sparking tools in their MSDS. Others have recommended its use. Considering Petrobras, there is an orientation not to recommend non-sparking tools, but it is regarded in its large-scale application.

6. Conclusion

The tests with mechanical systems have a great variability of results due to the complexity of the reactions involved. The initial idea was to capture the thermal energy released by the decomposition of the system energy: total energy - (energy transferred in the form of vibration + energy transferred by deforming + sound energy + light energy) = thermal energy, but as the amount of thermal energy is low (in the order of milijoules) only mechanical variations unfeasible this analysis by decomposition of factors.
Measurements of thermograph are not sufficient for obtaining reliable data accuracy. For tests, the measured values were much lower than the ignition temperature of the gases.
It is not possible to identify visually the light energy being released in impacts. During the tests in the laboratory of explosive atmospheres, a high-speed camera will assist to identify possible release of light energy.
At this moment, there is no evidence that a steel hand tool can cause ignition of an explosive atmosphere in an oil industry. More tests are being carried out at this moment.

7. References

[1] THE REPORT OF THE BP U.S. REFINERIES INDEPENDENT SAFETY REVIEW PANEL - *2006 - Process Safety Culture Survey.*

[2] POWELL, F., "Ignition of Gases And Vapor – Review Of Ignition Of Flamamable Gases And Vapors By Friction And Impact." Industrial and Enginnering Chemistry. (1969). vol. 61 (12) pp. 29-37.

[3] AMERICAN PETROLEUM INSTITUTE. API 2214 - Spark Ignition Properties of Hand Tools. USA: API, 1956.

[4] RIDDLESTONE, H.G., BARTELS, A., "The Relative Hazards In The Use Of Ferrous And Non-Sparking Tools In The Petroleum Industry. American Petroleum Institute. 51 (495), 106-10.1965.

[5] AMERICAN PETROLEUM INSTITUTE, API RP 2214 - Spark Ignition Properties of Hand Tools. 4. ed. USA: API, 2004.

[6] PROUST, Ch., HAWKSWORTH, S., ROGERS, R., BEYER, M., LAKIC, D., RAVEAU, D., HERVÉ, P., PINA, V., PETITFRÈRE, C., LEFEBVRE, X., "Development Of A Method For Predicting The Ignition Of Explosive Atmospheres By Mechanical Friction And Impacts (MECHEX)". Journal of Loss Prevention in the Process Industries. (2006). Vol 19 p. 104-120. Elsevier Ltd.

[7] INTERNATIONAL CHAMBER OF SHIPPING (ICS) & THE OIL COMPANIES INTERNATIONAL MARINE FORUM (OCIMF) & INTERNATIONAL ASSOCIATION OF PORTS AND HARBORS (IAPH). International Safety Guide For Oil Tankers And Terminals (ISGOTT). London: ICS/OCIMF/IAPH. WITHERBY & CO. LTD. 5. ed., 2006.

[8] GARSIDE, R., "Electrical apparatus and hazardous areas". Aylesbury, Buckinghamshire, UK: Hexagon Technology Limited, 2.ed.rev., 1995a. 395 p.

[9] AMERICAN PETROLEUM INSTITUTE, API RP 500 - Recommended Practice for Classification of Locations for Electrical Installations at Petroleum Facilities Classified as Class I, Division 1 and Division 2. 2. ed. USA: API. 1997 and reaffirmed 2002. 131 p.

[10] ASSOCIAÇÃO BRASILEIRA DE NORMAS TÉCNICAS. ABNT NBR IEC 60050-426 - Vocabulário eletrotécnico internacional parte 426: Equipamentos para atmosferas explosivas. Rio de Janeiro. 2011. 53 p.

[11] PARLAMENTO EUROPEU. DIRECTIVA ATEX 1999/92/EC: EN Official Journal of the European Communities. 28.1.2000, L 23/57. Site: http://eur-lex.europa.eu/LexUriServ/LexUriServ/LexUriServ.do?uri=OJ:L:2000:023:0057:0064:EN:PDF. (accessed Jan. 15, 2012 at 13h40).

[12] AMERICAN SOCIETY FOR TESTING AND MATERIALS. ASTM E92-82 (2003) - Standard Test Method for Vickers Hardness of Metallic Materials. USA: 1982 reaffirmed at 2003. 9 p.

[13] SCHUTH, K. H., Palasio, C.M. Jr., "Ferramentas Anti Faiscantes". Net, Rio de Janeiro, Site: http://www.ecivilnet.com/artigos/ferramentas_antifaiscantes.htm. (accessed Jan. 02, 2012 at 13h53).

[14] MADAKSON, P.B., "The Frictional Behavior Of Materials". Wear. (1983). pp. 191-206.

[15] FACTORY MUTUAL APPROVAL. Procedure Class Number 7410 - Approval Standard for Sparks Resistant Tools. 2007. 18 p.

[16] NATIONAL FIRE PROTECTION ASSOCIATION: Memorandum: NFPA 30 ROC TC Letter Ballot (A2011). USA, Massachusetts, Quincy, 2010.

[17] HAWKSWORTH S., ROGERS R., PROUST, C., BEYER, M, SCHENK, S., GUMMER, J., RAVEAU, D., "Mechanical Ignition Hazards in Potentially Explosive Atmospheres - EC Project MECHEX". In: Int. ESMG Symposion on Process Safety and Industrial Explosion Protection (ExplosionRisk), Nuremberg, 2004.

[18] INTERNATIONAL ELECTROTECHNICAL COMMISSION: PT ISO/IEC 80079-36: Explosive atmospheres — Non-electrical equipment for explosive atmospheres – Basic methods and requirements, draft

[19] HEALTH AND SAFETY EXECUTIVE . Health and Safety Guidance: HS(G)140 - Safe use and handling of flamable liquids. United Kington. HSE, 1996. Site: http://www.hse.gov.uk/pubns/books/ hsg140.htm. (accessed Jan. 13, 2012 at 16h).

[20] NATIONAL FIRE PROTECTION ASSOCIATION. NFPA 30 - Flammable and Combustible Liquids Code, USA, Massachusetts, Quincy, Ed. 2003.

[21] NATIONAL FIRE PROTECTION ASSOCIATION. NFPA 1 – NFPA Fire Protection Handbook, USA, Massachusetts, Quincy, 15. ed., (2008).

[22] NATIONAL FIRE PROTECTION ASSOCIATION. NFPA 30 - Flammable and Combustible Liquids Code, USA, Massachusetts, Quincy, Ed. 2011.

[23] AMERICAN PETROLEUM INSTITUTE. API RP 2214 - Spark Ignition Properties of Hand Tools. 3ed. USA: API, 1989

[24] NATIONAL FIRE PROTECTION ASSOCIATION. NFPA 1 – Fire Code, USA, Massachusetts, Quincy, Ed. 2012.

[25] U.S. OCCUPATIONAL SAFETY & HEALTH ADMINISTRATION (U.S.OSHA). OSHA Office of Training and Education: Construction Safety And Health Outreach Program. (1996). Site: http://www.osha.gov/pls/imis/establishment. inspection_detail?id=304994304, (accessed Jan. 11, 2012 at 10h22).

[26] U.S. OCCUPATIONAL SAFETY & HEALTH ADMINISTRATION (U.S.OSHA). OSHA Office of Training and Education: OSHA book 3080 - Hand And Power Tools. (2002). Site: http://www.osha.gov/Publications/osha3080.html, (accessed Jan. 11, 2012 at 10h30).

[27] U.S. OCCUPATIONAL SAFETY & HEALTH ADMINISTRATION (U.S.OSHA). OSHA 29 CFR 1910.106: Flammable and combustible liquids. Site: http://www.osha.gov/pls/oshaweb/owadisp.show_document?p_id=9752&p_table=STANDARDS; (accessed Jan. 11, 2012 at 10h22).

[28] U.S. OCCUPATIONAL SAFETY & HEALTH ADMINISTRATION (U.S.OSHA). OSHA 29 CFR 1910.146: Permit-Required Confined Spaces, Appendix D - Confined Space Pré-Entry - Check List (1996). Site: http://www.osha.gov/pls/oshaweb/owadisp.show_document?p_table=INTERPRETATIONS&p_id=22234. (accessed Jan. 11, 2012 at 14h40).

[29] U.S. OCCUPATIONAL SAFETY & HEALTH ADMINISTRATION (U.S.OSHA). OSHA Office of Training And Education: Apresentation: *Hand and Power Tool Safety-Spark Resistant Tools*. (2008). Site: www.osha.gov/dte/.../module_5.ppt, (accessed Jan. 11, 2012 at 14h50)

[30] U.S. OCCUPATIONAL SAFETY & HEALTH ADMINISTRATION (U.S.OSHA). OSHA 29 CFR 1910.120: Hazardous waste operations and emergency Response. Site: http://www.osha.gov/pls/oshaweb/owadisp.show_document?p_table=STANDARDS&p_id=9765. (accessed Jan. 11, 2012 at 9h27)

[31] AMERICAN SOCIETY FOR TESTING AND MATERIALS. ASTM E680 - 79 - Standard Test Method for Drop Weight Impact Sensitivity of Solid-Phase Hazardous Materials. USA: 2007. (Reapproved 2011) 7p. Site: http://www.astm.org/Standards/E680.htm. (accessed Fev. 18, 2012 at 10h30).

[32] INTERNATIONAL ELECTROTECHNICAL COMMISSION: IEC 60079-1: Explosive atmospheres: Part 1: Equipment protection by flameproof enclosures "d". Geneve, 2007. 73 p.

[33] OLSON, D.L., MAROEF, I., and PARK, Y.D., "Feasibility Of Thermite Sparking With Impact Of Rusted Steel Onto Aluminum Coated Steel", Colorado School of Mines (2002). p66.

Increase Hazard Discovery and Minimize Errors in your Process Hazard Analyses

A Graph Theoretical Approach

R. Maqbool Qadir, PE
Enpro Solutions, Inc.
6500 Dublin Blvd., Suite 215, Dublin, CA 94568
Qadir@enprosolutions.com

Prepared for Presentation at
American Institute of Chemical Engineers
2014 Spring Meeting
10th Global Congress on Process Safety
New Orleans, LA
March 30 – April 2, 2014

UNPUBLISHED

Increase Hazard Discovery and Minimize Errors in your Process Hazard Analyses

A Graph Theoretical Approach

R. Maqbool Qadir, PE
Enpro Solutions, Inc.
6500 Dublin Blvd., Suite 215, Dublin, CA 94568
Qadir@enprosolutions.com

Keywords: PHA, HAZOP, FMEA, FTA, graph theory, hazard

Abstract

Process Hazard Analysis (PHA) methods and techniques are constantly being developed. Yet, PHAs often fail to recognize discoverable, credible hazard scenarios. Unrecognized hazards were implicated in the BP Texas City disaster, Formosa Plastics explosion, Chevron Richmond Refinery fire and other high profile incidents. Unrecognized scenarios represent a gap in understanding of process hazards and keep us from preventing accidents and disasters. This paper focuses on improving the completeness of the hazard identification. The PHA process is visualized by mapping it in the form of graphs. Ammonia tank hazard and operability study (HAZOP) and Failure Modes and Effects Analysis (FMEA) examples are mapped as graphs to reveal tree structures. These tree graphs or inference maps directly allow us to visualize causal connections and inferences. Chains of cause and effect are mapped as unique pathways that can be represented as both graphs and matrices. For example, one can graphically trace the pathways to hazardous events such as loss of containment.

This paper investigates the role of errors, omissions and constraints in reducing the discovery of process hazards. What are the sources of errors and omissions (E/O)? What are the effects of constraints like budget and time? How can PHAs be audited efficiently to identify and correct E/O? A fault tree analysis (FTA) of the generic HAZOP process is performed to trace the sources of the E/O and how they may be minimized. The FTA identifies seventeen distinct sources of E/O. The sources of E/O are mapped to the issues they affect and a quality assurance checklist is derived from the FTA of the HAZOP process. CSB case studies are cited showing the impact of the E/O on hazard discovery in real incidents.

1. Introduction

Process Hazard Analysis (PHA) is an analytical discipline whose purpose is to discover process hazards and initiating causes that if uncorrected could lead to process upsets, accidents and disasters. CSB and OSHA have found that a large number of incidents could have been prevented if the hazards implicated in the incidents had been recognized and mitigated. PHA is a fundamental part of any Process Safety Management (PSM) program and is one of the keys to preventing accidents. Furthermore, PHAs are required for facilities with hazardous materials subject to the OSHA PSM Standard [1] and/or the EPA Risk Management Program or RMP [2] as well as state and local regulations.

In the past decades improvements have been made to existing PHA methods and new techniques are constantly being developed. Yet, PHAs conducted often fail to recognize discoverable, credible hazard scenarios. Unrecognized hazards and other PHA deficiencies were implicated in the BP Texas City disaster [3], Formosa Plastics explosion [4], Chevron Richmond Refinery fire [5] and other high profile incidents. Unrecognized scenarios represent a gap in understanding of process hazards and limit our ability to prevent accidents and disasters. There is a need therefore to improve the effectiveness of PHA to maximize the discovery of credible hazard scenarios. This would improve our ability to identify measures that can reduce risk and increase safety performance. It is also essential to understand the limitations of analysis - that all hazards cannot necessarily be identified even with the best teams and methods. Constant vigilance and periodic re-evaluation of existing PHAs are required to operate safely.

This paper investigates the role of errors, omissions and constraints in reducing or limiting the discovery of process hazards by PHA. Specifically I explore the following questions – Why do PHAs fail to identify credible hazard scenarios? What are the sources of errors and omissions (E/O)? What is the effect of constraints like budget and time? How can PHAs be audited efficiently to identify and correct errors and omissions? I perform a fault tree analysis (FTA) of the PHA process to trace the sources of the E/O, show how they arise, and how they may be minimized. I cite CSB reports where the E/O identified in the FTA occurred in real incidents. The paper will also discuss regulatory requirements, standards/ guidelines and best practices that apply to PHA. This article will focus mainly on the hazard and operability (HAZOP) method but the arguments presented are applicable to other techniques as well.

2. PHA Methods and discoverability of hazards

There are many structured brainstorming methods used by multidisciplinary teams of analysts to conduct hazard analyses on complex systems such as process plants, aviation systems and manufactured components. The HAZOP method is arguably the most popular in the chemical process industries. Another popular method is the failure modes and effects analysis (FMEA). Both methods are recognized in OSHA and EPA regulations [1, 2]. Each method was developed for a specialized purpose and its evolutionary history makes one method more suited to a particular situation than another. For instance the HAZOP method is particularly well suited to the analysis of process plants while the FMEA was evolved to study the failure modes of aircraft components. Because of the variations in approach and methodology among the various PHA methods, discovery of hazard scenarios depends to an extent on the method selected. A HAZOP and FMEA conducted on the same system will tend to identify a similar but not identical set of hazard scenarios. Furthermore, the combinatorial possibilities in complex systems are far too many for a team applying a single PHA method to identify, much less evaluate *all* potential hazard scenarios in any reasonable timeframe. This is one reason why PHAs can fail to detect certain hazards. This dependence of hazard scenario discovery on PHA method is illustrated in Figure 1.

Figure 1:Hazards discovery using various PHA methods

3. Visualizing PHA - A Simple HAZOP and FMEA example

HAZOP teams study complete processes or plant sections within a process to predict how hazards may arise and how they might be mitigated. For this purpose they examine drawings and schematics, especially piping and instrumentation diagrams (P&IDs) to understand how individual plant equipment and control systems work together as complex systems. The use of HAZOP and FMEA methods is illustrated below to analyze hazards associated with an anhydrous ammonia storage tank depicted in the simplified P&ID in Figure 2.

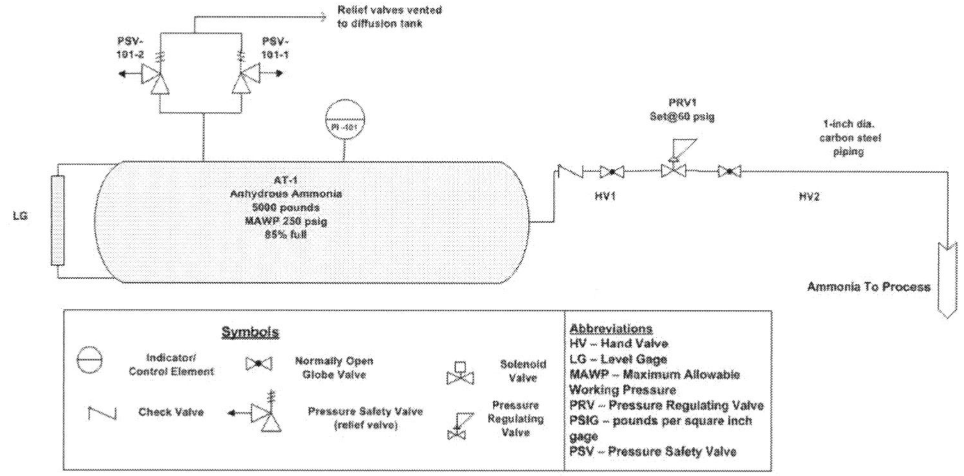

Figure 2: Simplied P&ID for an Anhydrous Ammonia Storage Tank

Any PHA examines a process step by step and comes up with a series of inferences that lead to identifying and evaluating hazard scenarios, initiating causes and safeguards. PHA team findings are typically documented in spreadsheet or tabular format. However, this mode of presentation obscures an underlying structure to the PHA. Therefore, this section introduces the use of graphs for visualizing the sequential steps of a PHA. The graphs shown in Figures 3 and 4 illustrate the HAZOP and FMEA analyses as applied to the anhydrous ammonia tank system of Figure 2.

The illustrations are not meant to be complete analyses or full explanations of the use of HAZOP/FMEA methods. Rather the main point of these graphs is to reveal a tree structure that will allow us to gain some useful insights into the analytical process. In the language of graph theory the HAZOP analysis can be modeled as connected, acyclic graphs, with the "root vertices" at the nodes and "leaves" at the terminal ends ("recommendations" in Figure 3). These tree graphs or inference maps directly allow us to visualize causal connections and inferences. For example, one can graphically trace the pathways to hazardous events such as loss of containment (LOC) as discussed below.

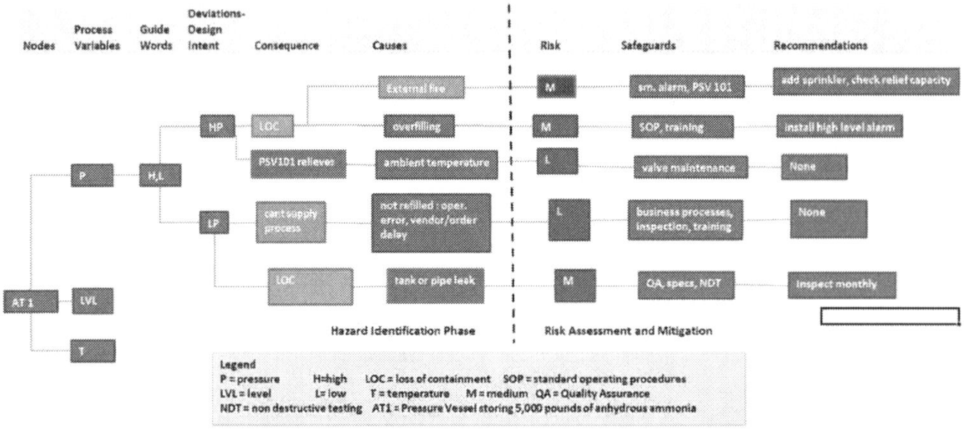

Figure 3: HAZOP GRAPH

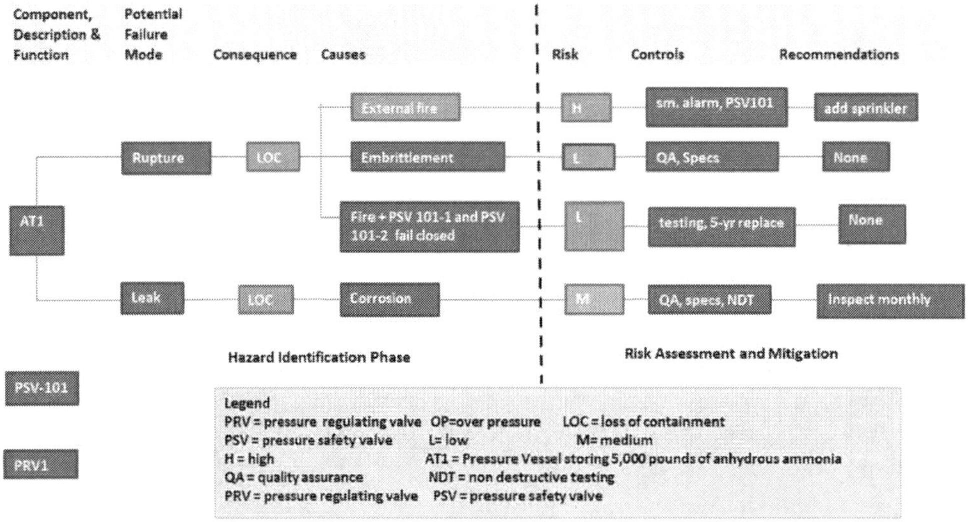

Figure 4: FMEA GRAPH

4. Hazard Identification and Consequence Assessment

The steps "nodes" through "causes" shown in the area to the left of the dotted line in Figure 3 and "components" through "causes" in Figure 4 might be termed the "hazard identification" or "discovery" phase of the HAZOP or FMEA analysis. The HAZOP method [6] combines key process variables such as pressure, flow and level with guide words such as less, more, high and low to explore deviations from design intent such as high pressure, high level, no flow etc. For example in Figure 3 the deviations high pressure (HP) and low pressure (LP) were identified for the ammonia tank AT1 of Figure 2.

The team then brainstorms what could cause the deviations and how the system would behave when perturbed from steady state design conditions. Hazards may arise from the effects that would occur as a result of these initiating causes. The team documents the plant section or equipment studied (nodes); process deviations, potential causes and their effects or consequences - credible hazard scenarios are identified as cause-consequence pairs. For example the hazard identified in the topmost branch of figure 3 is LOC from tank-AT1 caused by fire. The complete pathway is - fire causes high pressure in Tank AT1 resulting in LOC. This pathway is easily seen in the color coded graph. A detailed consequence assessment could be required for the LOC such as estimating exposure of employees to a toxic release and offsite consequences to nearby receptors.

FMEA is similar to HAZOP in some ways but focuses on failure modes of major components [6] or equipment and their causes rather than deviations in process variables and the causes of these deviations. For example, the FMEA in Figure 4 identifies rupture and leakage as modes by which AT1 and the associated pipeline might fail. Like the HAZOP the FMEA also identifies LOC pathways for tank AT1. However, the causes identified are corrosion, embrittlement, and stress from high pressure sustained due to fire. Corrosion and embrittlement were not identified as potential causes of a LOC in this example HAZOP (although the tank or pipe leak mentioned as a cause should lead one to consider corrosion). This illustrates the effect of differences in PHA methods on hazard scenario discovery.

5. Hazard Scenarios - Understanding Cause and Effect

Investigating deviations and understanding their causes and effects is at the heart of the hazard discovery process. This is where much of the team brainstorming effort is directed. To further investigate this portion of the analysis one can abstract the chain of cause and effect shown as graphs in Figure 5 below with the following notations: D for deviations, C for causes and E for effects or consequences. Some common cause and effect patterns signifying hazard discovery pathways are depicted as a series of graphs in Figure 5. These pathways can also be represented as matrices - which we will call Deviation Matrices - as shown in Figure 5. Four common patterns are demonstrated with examples that are not necessarily related to the system shown in Figure 2.

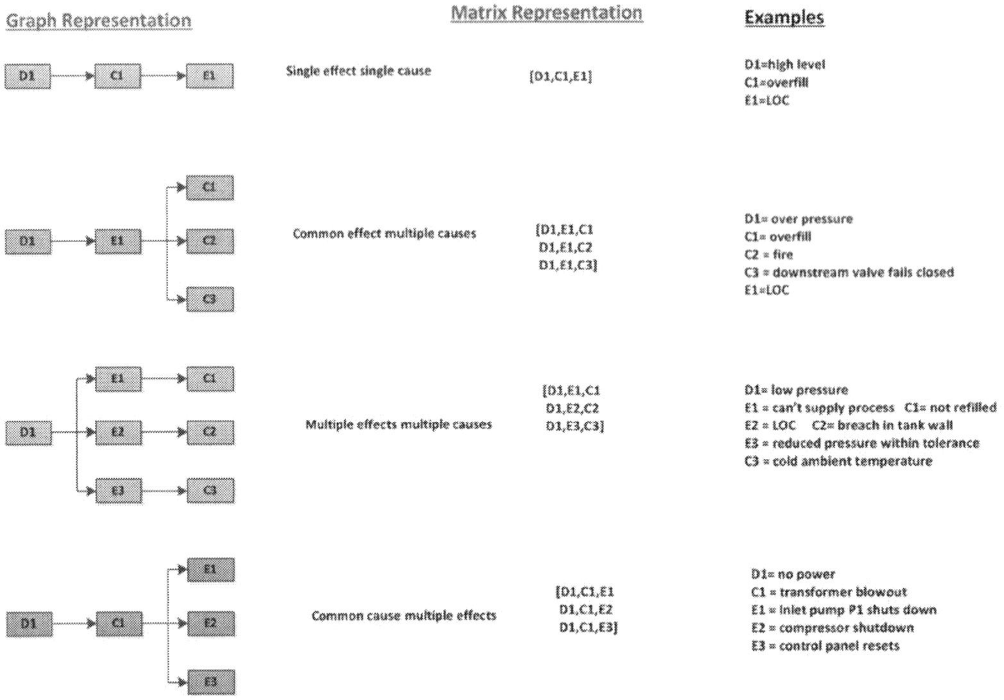

Figure 5: Cause and Effect Patterns

The first case "deviation, single effect single cause" depicts the general case of cause and effect and may be regarded as the fundamental unit of hazard scenarios. Logically, the cause always precedes the effect. The deviations can be regarded as intermediate effects, thus the logical sequence would be C-D-E e.g. an initiating cause such as overfilling leads to a deviation such as high level which if continued leads to LOC as the ultimate consequence or effect . However, the HAZOP process uses deviations from design intent to examine the behavior of a system perturbed from its design or steady state. The team reasons backward to find causes and forward to estimate effects. Therefore the brainstorming sequence is usually D-C-E or D-E-C. The two sequences are logically equivalent but there are advantages to each. It is easier to think about the cause leading to the deviation leading to the effect. However, for reasons of economy it can be advantageous to list effects first so that deviations leading to moderate or benign effects can be eliminated from further consideration. Thus the team's brainstorming efforts in search of causes can be reserved for the more serious or harmful effects.

The HAZOP process may first identify an immediate or proximate cause but the complete pathway may require going to a more primitive or root cause. For example in Figure 3, a LOC caused by a tank or pipe leak begs the question – what caused the leak? So a complete pathway would be for example, a LOC caused by a leak which is caused by corrosion. This may seem obvious but in the recent Chevron Richmond Refinery catastrophic pipe failure and fire of 2012 the CSB found [5] that a HAZOP team had looked at the deviation "leak/rupture" associated with piping from a crude distillation tower. However, they did not identify corrosion as a potential cause – and ultimately the tubes failed due to corrosion releasing hot hydrocarbons that caused the explosion! The CSB further determined that this apparent oversight occurred because the team had not identified a damage mechanism for the type of corrosion that occurred, perhaps because the team had no corrosion experts on it.

Similarly, in Figure 3 the LOC is identified as a consequence of a leak. But a complete consequence assessment would flesh out the scenario to look at the impacts of the LOC such as worker exposure or offsite consequences. In the case of BP Texas City explosion of 2005, the CSB found [3] that consequences for the deviations "high level" and "high pressure" in the raffinate splitter tower where the explosion took place were not fully identified. Specifically that safety relief valves would be over-pressurized and liquid would overflow from the tower into a blow down drum and then overflow from the blow down drum were not identified.

Looking at the graphs in Figure 5, there are three ways of failing to identify hazard pathways for a given process variable – miss a deviation, miss one or more causes, and miss one or more effects.

6. Risk Assessment and Risk Reduction

The steps to the right of the dotted line in Figures 3 and 4 are devoted to risk assessment and mitigation of the hazards identified in the first phase. The risk of an event is estimated from the combination of its estimated frequency of occurrence and severity of consequence. Existing safeguards are identified such as process controls, alarms and trips that would prevent a process deviation from turning into a hazardous event. If adequate safeguards are lacking the team comes up with recommendations for additional engineering and/or administrative controls to reduce the frequency or consequence of a potential event. This paper does not address risk assessment - risks can be estimated by various methods: qualitative risk assessment such as a Hazard Matrix, semi-quantitative methods such as Layer of Protection Analysis (LOPA) and quantitative risk assessment (QRA) methods. Risk reduction includes the study of inherently safe design, process controls and alarm systems, passive and administrative safeguards and other preventive measures and is also not the focus of this paper.

7. Mapping the Hazard Identification Phase

The more complete the hazard identification the more successfully the associated risks, safeguards and mitigation measures can be evaluated. This paper focuses on improving the completeness of the hazard identification phase. In order to do this we map and analyze its structure as discussed below. Figure 6 shows an abstract representation of a hypothetical PHA where the process contains three nodes (N), three process variables of interest (V) in each node, three potential guide words (G) for each variable and three causes (C) for each deviation (D) identified by combining variables and guide words. The example is constructed to show that consideration of a complex system can easily lead to a very large number of deviation-cause-consequence pathways to be evaluated. Referring to Figure 6, an example of a complete pathway is N1-V1-G1/D1-C1-E1, where the hazard scenario would consist of the cause-consequence pair C1-E1.

This formulation of the problem would lead to N*V*G*C potential pathways i.e. 3^4 or 81 hazard scenarios. Of course in a real case many pathways would not be of interest and would not lead to "credible" scenarios. But the question is - what is the impact of not identifying a relevant guideword? In this example nine cause-consequence pairs or hazard scenarios are obtained for a single process variable from one node. Failure to recognize a cause- consequence pair at the terminal branches results in missing one scenario. Not recognizing a single valid guide word results in the failure to recognize three potential hazard scenarios. Not recognizing an important process variable would lead to a loss of information about nine potential hazard scenarios. Thus an omission at the front end or "upstream" in the analysis is more consequential than at the terminal end.

The above analysis has implications for auditing a PHA and performing completeness checks on a PHA. Suppose that a review of process variables indicated that loss of containment had not been analyzed. Then referring to Figure 6, one would add a variable "containment" as V4 for example. The guide word could be "no" and the deviation would be loss of containment and one would proceed to investigate causes and consequences. Our analysis shows that it is easier to review process variables and guide words to successfully detect an omission (Figure 6) rather than trying to find a missing cause or consequence among the many that might have been generated. Therefore, a gap analysis should first check process variables and guide words as control points when completing or revalidating a PHA.

The PHA tree or inference map allows us to perceive the structure of the hazard discovery process and perform the following:

- See paths leading to hazard scenarios for each node and mode of operation.
- Recognize and remove unimportant branches – saving effort for analysis.
- Add a branch back in if re-considered.
- Do Quality Assurance (QA) on PHA to do "completeness checks".
- More easily audit for errors and omissions in analysis.

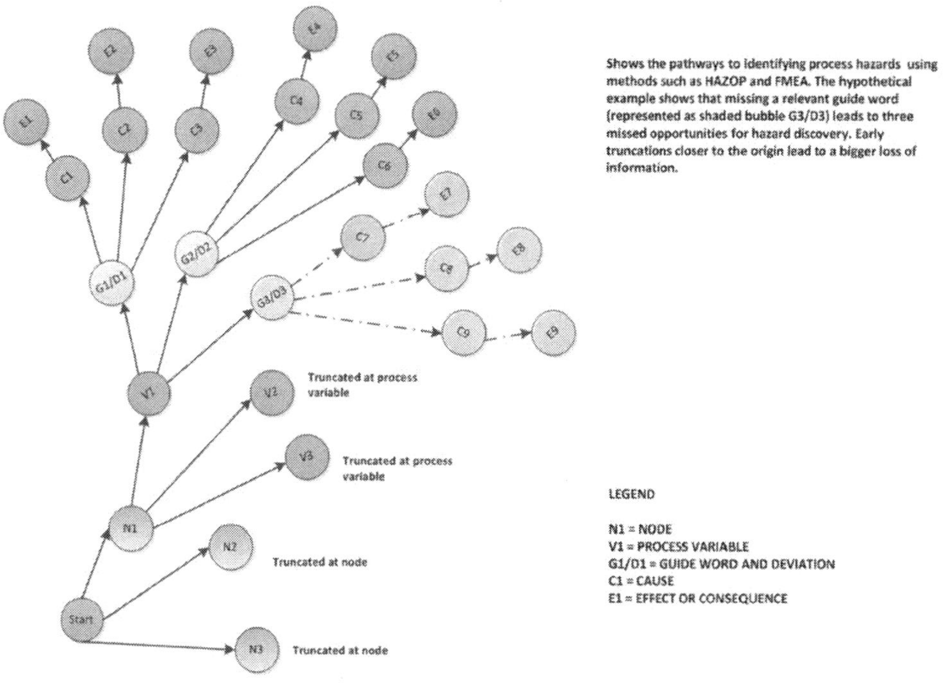

Figure 6: PHA Tree

8. Sources of Errors and Omissions in PHA - A Fault Tree Analysis

We mapped the PHA process using HAZOP as a template and illustrated how deviations from design intent are generated from a consideration of nodes, key process variables and guide words. At the next stage a consideration of causes and consequences of deviations results in identifying hazards as cause-consequence pairs. We demonstrated the loss of information that can occur at each stage of the analysis such as when a PHA team fails to identify a process variable or guide word. We also discussed the limitations of individual methods. Next we bring together all these factors along with a consideration of PHA input data to identify additional sources of errors and omissions that lead to non-recognition of hazard scenarios. For this purpose a FTA of the PHA process itself was performed as shown in Figure 7 to discover the root causes of errors and omissions.

For purposes of this analysis an error is a mistake in applying an analytical method, a calculation error, obsolete data, incorrect data entry etc. An omission is a failure to include any valid consideration or data in the analysis. A constraint is a limitation imposed by time, budget, scope, technical expertise, or industry state of knowledge. Through sheer reasoning alone the FTA identifies 17 distinct sources of errors, omissions and constraints that could result in a failure to identify a credible hazard. These findings are summarized in as a checklist in Table 1 below. To demonstrate the real world relevance of these findings, cases are cited where CSB found the E/O to have occurred. A further insight provided by the FTA is that the sources of E/O are mapped to the issue they affect e.g. missing process safety information (PSI) are mapped to missing deviations from design intent and ignoring incident history is mapped to failure to recognize hazards and incorrect evaluation of consequences.

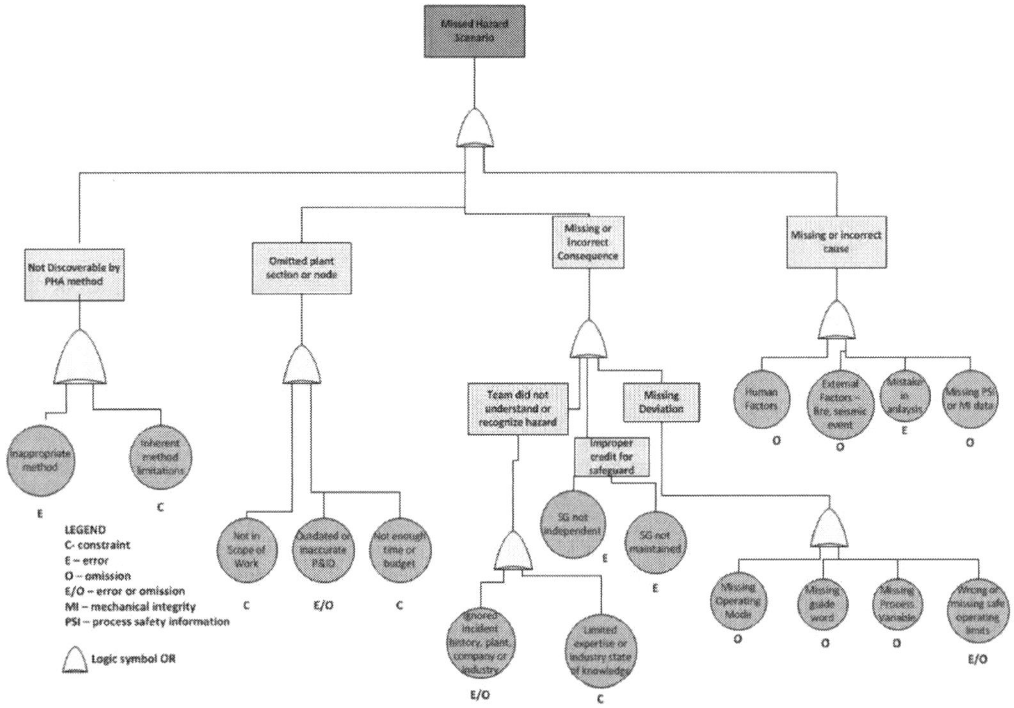

Figure 7: Fault Tree, Errors and Omissions in Process Hazard Analysis

Table 1 – A Checklist for Reducing Errors and Omissions in PHA (Sources of E/O Identified by Fault Tree Analysis)

Item No.	FTA Finding	Constraint, Error, or Omission	CSB case study documented examples of occurrence
1.	Inherent PHA method limitation	Constraint	None provided
2.	Not in scope of work	Constraint	None provided
3.	Not enough time or budget	Constraint	None provided
4.	Limited Expertise or industry state of knowledge	Constraint	T2 labs [7], BP Amoco [9], process chemistry - reaction hazards not understood or poorly understood.
5.	Inappropriate PHA method	Error	None provided
6.	Outdated or inaccurate P&ID	Error	Oleum release [8], emergency power supply not included on P&ID
7.	Ignored incident history – plant, company or industry	Error	BP Texas City [3], BP Amoco [9], Formosa Plastics [4], Chevron [5] –

			PHAs and revalidations did not take into account documented plant incidents or company incidents, and/or ignored recommendations from incidents.
8.	Took credit for safeguard that is not independent	Error	Oleum [8], All safeguards required operator attention and action, no engineering controls independent of operator. Chevron [5] PHA only listed qualitative safeguards such as inspection and corrosion allowance.
9.	Took credit for safeguard that is not maintained	Error	BP Texas City [3] – raffinate splitter alarm and blow down drum high level alarm did not work. Chevron [5] PHA only listed qualitative safeguards such as inspection and corrosion allowance.
10.	Mistake in analysis	Error	Chevron Richmond [5], Did not identify failure mechanism for sulfidation corrosion
11.	Incorrect safe operating Limits	Error	BP Amoco [9] Safe operating limits not identified.
12.	Did not consider human factors	Omission	BP Texas City (3), Oleum [8], Formosa Plastics [4] – Human error not taken into account in evaluating safeguards.
13.	Did not consider external factors like fire or earthquake	Omission	None provided
14.	Did not have process safety information or mechanical integrity (MI) data	Omission	Oleum – PHA team lacked information on hazards of transferring oleum using emergency power supply.
15.	Did not consider non-routine operating modes such as startup and shutdown	Omission	BP Texas City and BP Amoco did not consider startup and shutdown.
16.	Did not consider valid guide word	Omission	Formosa cited by OSHA [10] for leaving out key guide words for batch processes.
17.	Did not consider key process variable	Omission	CSB Sterigenics [11] No explosion scenarios evaluated for explosive concentrations.

9. Deficiencies in PHA Data Sources

Incorrect or missing data are implicated in a number of sources of errors and omissions identified in Table 1 above. Table 2 maps data required for conducting a HAZOP to the output that is affected by deficiencies in those data. For example deficiencies in process chemistry data might cause reactive hazards to go unidentified. Similarly, a team working with old and outdated operating procedures might fail to identify hazards associated with starting up a process unit.

Table 2 - Impact of Data Deficiencies on PHA Findings

INPUT		OUTPUT
P&ID	HAZOP	Nodes, upstream, downstream, loops, system effects
Equipment Specs, materials and safe operating limits		Deviations and consequences of deviations
Instrumentation, alarms and interlocks		Safeguards, BPCS, ERP
Process Chemistry		Reactive Hazards
Utilities		Deviations, common cause failures, consequences
Labeling, siting, HMI		Human Factors
Standard Operating Procedures		Task design, batch operations, startup, shutdown, maintenance

10. Regulations, Standards and Best Practices

Applicable regulations such as OSHA PSM [1] and EPA RMP [2] contain requirements and guidelines for conducting PHAs. OSHA has detailed requirements and guidelines including team composition, team expertise, recognized PHA methods, human factors, siting, evaluation of hazards and safeguards. There are also OSHA letters that address the interpretation of specific issues. CCPS has many guidelines for hazard evaluation [6] that include best practices to minimize deficiencies and produce robust PHAs.

10. Conclusion

What does the analysis presented in this paper show us about controlling or reducing errors and omissions in PHA to maximize the discovery of hazards based on?

10.1 Team: All of the errors and omissions identified in conducting a PHA are human errors. Therefore, picking the right team with the appropriate training, expertise and leadership is essential. PHA is a team exercise requiring a properly balanced and knowledgeable team with the appropriate composition. This is an OSHA PSM requirement [1]. A good team leader can help the team maintain cohesion and apply the proper discipline, skill, and awareness to the analysis. The leader and the team must be confident, curious, be willingness to question assumptions and have the ability to express disagreement without being disruptive.

10.2 Errors: Upfront preparation and a thorough review of the data sources should be performed prior to conducting the PHA to prevent errors from incorrect or outdated data. A review of the completed PHA documentation with appropriate checklists can aid in detecting errors. Tables 1 and 2 serve as checklists in addition to the many excellent literature references available to aid in successfully conducting PHAs.

10.3 Omissions: Omissions can be minimized with appropriate quality assurance procedures such as the checklists in Tables 1 and 2. A review of data sources prior to conducting the analysis should be conducted to ensure that PSI and MI data are current and available to the team. An incident history of the plant, company and industry should be compiled and reviewed to ensure that known problems are not ignored. Process incident databases such as the CCPS database can help to identify industry incidents and lessons learned. LOC should always be considered as a deviation. Our analysis shows that it is easier to review process variables and guide words to successfully detect an omission (Figure 6) rather than trying to find a missing cause or consequence among the many that might have been generated. Therefore, a gap analysis should first check variables and guide words as control points when completing or revalidating a PHA.

10.4 Constraints: Proper scoping is essential to correctly prioritize unit selection so that the most hazardous units and the most hazardous equipment are selected for analysis. For proper scoping the PHA team and the consumers of the PHA must take into account the limits imposed by time, budget, PHA method, and company or regulatory priorities.

10.5 Insights Gained from graph theoretical analysis and FTA

1. The logical steps of a PHA can be modeled as a connected, acyclic graph or tree. The tree structure provides useful insights into the chain of inference.
2. The tree structure shows a logical precedence of positions in variables for generating hazard scenarios or pathways - from nodes outward to cause-consequence pairs (Figure 6).
3. A PHA completeness check or audit should follow this precedence from node outward to branches. This means that omissions or errors are more likely to be found by first looking at the process variables and guide words and then proceeding to examine causes and consequences.

4. An FTA of the generic HAZOP process (abstracted for logical form) identifies seventeen distinct sources of E/O – examples of real world incidents are cited. The findings serve as a QA checklist of the PHA process and an aid to maximize hazard discovery.
5. Use checklists, best practices and incident databases to minimize E/O.
6. Recognize constraints that impose limits on analysis. Proper project scoping will ensure that safety is enhanced by prioritizing plant areas of concern for PHA analysis.

11. References

[1] OSHA. "Standard for Hazardous Materials - Process Safety Management of Highly Hazardous Chemicals", 29 CFR, 1910.119, U.S. Occupational Safety and Health Administration

[2] USEPA. "Chemical Accident Prevention Provisions", 40 CFR, 68, Subparts C and D. U.S. Environmental Protection Agency

[3] CSB. "Refinery Explosion and Fire, (15 Killed, 180 Injured), BP Texas City, Texas, March 23, 2005", Investigation Report No. 2005-04-I-TX, U.S. Chemical Safety and Hazard Investigation Board, Washington DC (March 2007).

[4] CSB. "Vinyl Chloride Monomer Explosion, (5 Dead, 3 Injured, and Community Evacuated), Formosa Plastics Corporation, Illiopolis, Illinois, April 23, 2004", Investigation Report No. 2004-10-I-IL, U.S. Chemical Safety and Hazard Investigation Board, Washington DC (March 2007)

[5] CSB. "Chevron Richmond Refinery Fire, Chevron Richmond Refinery, Richmond, California, August 6, 2012", Interim Investigation Report , U.S. Chemical Safety and Hazard Investigation Board, Washington DC (April 2013)

[6] CCPS. *Guidelines for Hazard Evaluation Procedures.* 2nd ed., Center for Chemical Process Safety, American Institute of Chemical Engineers, New York, NY, 1992.

[7] CSB. "T2 Labs, Inc., Runaway Reaction, Jacksonville, Florida, December 19, 2007", Investigation Report No. 2008-3-I-FL, U.S. Chemical Safety and Hazard Investigation Board, Washington DC (September 2009)

[8] CSB. "Uncontrolled Oleum Release, Petrolia, Pennsylvania, (Three towns evacuated)", Case Study No. 2009-01-I-PA, U.S. Chemical Safety and Hazard Investigation Board, Washington DC (September 2009)

[9] CSB. "Thermal Decomposition Incident, (5 Killed), BP Amoco Polymers, Inc., Augusta, Georgia, March 13, 2001", Investigation Report No. 2001-03-I-GA, U.S. Chemical Safety and Hazard Investigation Board, Washington DC (June 2002)

[10] OSHA. "Citation and Notification of Penalty – Formosa Plastics Corporation, 19800 Old Route- 36, Illiopolis, IL 62539." U.S. Occupational Safety and Health Administration, Inspection No. 305893679, Issuance Date 10/22/2004

[11] CSB. "Sterigenics, (4 employees hurt), Sterigenics, Ontario, California, August 19, 2004", Investigation Report No. 2004-11-I-CA, U.S. Chemical Safety and Hazard Investigation Board, Washington DC (March 2006)

349055 Deflagration Incident Case Review

Abstract:

A deflagration incident occurred during the cleaning procedure of a drum that was used in a coating process. The pressure generated from the incident damaged the drum as well as several pieces of equipment that were connected through ductwork for the dust collection system. Thankfully, no personnel were injured during the incident. This case study will describe the coating process and the incident, provide an incident analysis, review the process safety information and evaluate results from additional testing that was performed. Lessons learned will be summarized including hazards related to material compatibility, thermal stability and combustible dust.

349382 Enhanced Lessons Learned Approach from the Bscat Investigation Approach

Abstract:

A key aspect of process safety management is the effective transmission of lessons learned both around the organization and specifically to those who most need to know. This often occurs poorly.

Most investigation methods have been optimized for occupational accident investigation. The BSCAT approach – Barrier-based SCAT – builds on an earlier root cause method, but integrates bow tie barrier structure into the investigation and provides a better focus for major accidents and their pre-cursor near-miss events. All incidents can be converted into a bow tie sequence of barriers that were challenged and which either failed (and gave the incident) or some succeeded (terminating or mitigating the incident).

The BSCAT approach provides many benefits for lessons learned. Its structure matches the API754 Tier 3 Process Safety Indicator as it records clearly challenges to safety barriers – and whether they worked or failed. A feature of bow ties is that barriers are frequently repeated between different Threats and Outcomes and different bow tie Top Events. Thus a barrier that was identified as failed or degraded from one investigation will warn that the same barrier used for other threats or appearing in different bow ties affecting other plants are also likely to be degraded. Software that records all incidents into a database, can show barrier failures on all bow ties in which that barrier appears and thus automatically warn the barrier or bow tie owners throughout the orgnaization. This does not require proactive communication from the plant where the incident occurred or from the Safety Department – as this often is incomplete.

This paper also compares the application of BSCAT to other traditional root cause approaches and demonstrates the clarity of communication possible with the bow tie structure. It has the added advantage that the focus on barrier failures tends to screen out "motherhood" recommendations that may be good practice but did not contribute to the incident.

349552 Death of "Landlord" or Collapse of "Tomb", Which Matters More? - Some Perspectives of Engineering Ethics and Engineering Philosophy on Enterprise Global Risk Management

Abstract:

Within the paper presented at 9th GCPS, the author named CCPS RBPS Temple the house for a "landlord" i.e. core value to live within and also stated that the 'landlord" could even buy a new house with new style of design to live within. In other words, a "landlord" could own a couple of, even many a few, houses at same time to enjoy life. Should "landlord" die withnin, however, the house may seem still a house outwardly, but substantively with a purpose of tomb.

And the author argues further that inhabited house could be timely repaired and maintained to good condition, otherwise uninhabited house whose "landlord" died within is more likely to degrade neglectedly, and eventually collapse. In context of enterprise global risk management, it seems that the structural soundness of the houses has been attracting more attention than should deserve, whereas the death of "landlord" neglected to some extent within process industry and beyond. And for the fact that the lack of correct core value and healthy process safety cultural communication was arguemented by the author to be the ultimate root cause for all the catastrophic process safety accidents, some attempts are made to explore and prospect if the absence of alive correct core value should be inherently of organizational and episodic conflicts of interest i.e. 'natural' multi-chain of command in light of Engineering Ethics.

Then the author goes one step further to characterise RBPS Core Value and Process Safety Cultural Communication with new implications in light of the emerging subject Engineering Philosophy, so that it could be prospected that how tradeoffs between different functions, stakeholders, regions and businesses could be negotiated, potential failure diagnosed and prevented, design style of house modelled, and ultimately how more critical "landlord" enjoying life could be than tomb's structural soundness, which is in fact of new implications for addressing some of the biggest challenges from enterprise global risk management, especially of merger and acquisition.

Finally, upon all of above, a few of important recommendations would be given to process industry and beyond so that a brighter future of enterprise global risk management could be prospected.

349837 New Tools to Aggregate Operational Risk Across an Enterprise of Assets and to Help Govern This Risk through Policy Which Can be Directly Linked to Front Line Decision Making

Abstract:

Major catastrophic incidents typically result in a whole spectrum of consequences, the highest being the loss of life and serious injury. Beyond the harm to people, the financial cost can be extreme to the extent that it can jeapordize the future of the corporation responsible. Following recent headline grabbing incidents, board members who are responsible for the secure and prosperous stewardship of corporations are asking the question, how do we properly govern such risks in our organisations.

This is the challenge most organisations now face, how to make sure that the major Operational Risks with our plant, equipment and people are being properly managed and how within a large global organisation a picture of this aggregated risk can be presented. This paper will explore the nature of operational risk and its contribution not just from Process Safety Hazards but also from ongoing work hazards. It will propose that the combination of compromised protection layers and the risk from new hazards introduced as a result of ongoing work represent the cumulative Operational risk which needs to be managed. The day to day status of these protection layers and the ongoing work schedule represent the dynamic involved in Operational risk Management. Currently we lack proper tools to effectively measure this risk in pace with the dynamic of operations where every day is different. Similarly we lack the ability to aggregate this risk across many units in a large plant or across many plants in different regions to provide an overall picture of operational risk exposure to the organisation. There are, however, new technology solutions emerging which will allow this to be

achieved. This new category of application platform helps manage ongoing work from a risk perspective taking into account real time data on the status of protection layers in the plant. Working in concert with the maintenance management system, a fuller picture of protection layer status can be presented and for the first time a picture showing maintenance and repair work efficiency along with risk status can be achieved. These tools allow better decisions to be made at the frontline. Furthermore, they are designed to be deployed across large organisations and with the ability to aggregate risk data across the corporation effectively in real time. They also provide tools to allow organisations to better manage the common operational risks helping directly drive policy in front line decision making.

The paper will present some realistic scenarios and explain how future operations can be not only be manged more safely but that governance of the enterprise organisation in terms of operational risk management is now more achievable.

Inherently Safer Design of Stirred Reactors and VisiMix® Modeling Software

Y. Nekhamkin
VisiMix Ltd.
8 Hamarpe Street
7th Floor, Bynet Bldg.
Jerusalem 97774 Israel
email: info@visimix.com

L. Braginsky, Y. Kokotov
VisiMix Ltd.
8 Hamarpe Street
7th Floor, Bynet Bldg.
Jerusalem 97774 Israel
email: info@visimix.com

Prepared for Presentation at
American Institute of Chemical Engineers
2014 Spring Meeting
10th Global Congress on Process Safety
New Orleans, LA
March 30 – April 2, 2014

UNPUBLISHED

Inherently Safer Design of Stirred Reactors and VisiMix® Modeling Software

Y. Nekhamkin
VisiMix Ltd.
8 Hamarpe Street
7th Floor, Bynet Bldg.
Jerusalem 97774 Israel
email: info@visimix.com

L. Braginsky, Y. Kokotov
VisiMix Ltd.
8 Hamarpe Street
7th Floor, Bynet Bldg.
Jerusalem 97774 Israel
email: info@visimix.com

Keywords: stirred tanks/reactors, inherently safer design, process safety technology, process simulation, VisiMix

Abstract

Stirred reactors belong to the key type of equipment used in all branches of the chemical process industry (CPI). Risk of accidents during various chemical processes is caused by deviation from normal course of the technological process. It means that any calculation technique (software) permitting to design an apparatus (chemical reactor) for a required chemical process and to define a range of normal operations of this process can be considered as an ultimate element of the "inherently safer design" (ISD) tool of the corresponding complex "chemical reactor – chemical process". This paper demonstrates an approach based on application of VisiMix® software intended for analysis and calculations of various chemical processes in stirred tanks/reactors and to provide their enhanced safety.

1. Introduction

Stirred reactors belong to the key types of the process equipment used in various technological processes in practically all branches of the chemical process industry (CPI) – chemical, pharmaceutical, biotechnological, manufacturing cosmetics, food products, etc.

The current status of CPI features considerable number of various accidents occurring during process operation [1…7] caused by different reasons. As regards to the stirred reactors, accidents in such equipment are associated with inadequate process analysis of heat transfer (23%), reaction problems (23%) and process contamination (16%) [2].

Providing process safety technology becomes now one of the basic requirements of the day and it resulted in the Inherently Safer Design (ISD) concept [8]. This concept asserts that the process can be "described as inherently safer if it reduces one or more hazards... associated with the materials and operations used in the process when compared to some alternative process, and this reduction or elimination is accomplished by characteristics which are permanent and inseparable parts of the process" [9]. Since risk of incident/accident is caused by any deviation or consequence of deviations from normal course of the technological process, adequate process analysis is a fundamental issue of the reactor safety design.

One of the main requirements to calculation tools used for the ISD is its potential to perform calculations linking equipment characteristics and process parameters. However the current state of the art in the calculation tools used for the stirred reactors design is mainly focused upon the reactive nature of the chemical processes [10, 11] aiming to predict thermal hazards (runaway reactions) and to diminish their impact ([12], [13]). This approach is based on the Well Stirred Reactor concept that eliminates mixing related processes from consideration.

2. Stirred Reactors

Stirred reactors, extensively used in the CPI, allow implementing various technological processes, including, in particular:

- homogeneous blending;
- liquid-solid mixing (among them suspension, dissolution of solid, liquid-solid extraction, etc.);
- liquid-liquid mixing (among them emulsification, liquid extraction, etc.);
- liquid-gas mixing;
- homogenization of multi-component mixtures;
- single phase chemical reactions in batch, semibatch and continuous flow reactors;
- heterogeneous reactions in liquid-liquid and liquid-solid mixing media;
- temperature-dependent chemical reactions in batch, semibatch and continuous flow reactors)

Stirred reactors' productivity and their quality should be based on thorough calculations, capable of covering all the unit operations implemented in the equipment.

Despite the complexity of processes in stirred tanks/reactors, the VisiMix® software [14] is capable of simulating these processes. This software has gained recognition because it provides adequate and complete description of process and equipment configurations based on reliable models verified in practice.

This paper describes the use of VisiMix® for ISD of various chemical processes in stirred reactors. Main features of the VisiMix® software are briefly described below.

3. VisiMix® Software. Main Features.

VisiMix® software was developed for process engineers as a universal tool for solving a wide range of technological problems. VisiMix® includes VisiMix Turbulent® for turbulent flow regime, VisiMix Laminar® for laminar flow regime, VisiMix Different Impellers® for simulating mixing devices with different impellers (to be used with VisiMix Turbulent®), VisiMix RSD® for simulation of rotor stator dispersers, VisiMix Excel® that integrates VisiMix reports in a standard Excel worksheet, and VisiMix Pipe-Line® for hydraulic calculations for low and high viscous and non-Newtonian liquids in plant pipe lines. These components perform simulation of different technological processes with respect to real equipment design and process regime parameters for both turbulent and laminar flow regimes. VisiMix® GUI allows for selecting the equipment types and to enter the equipment parameters and media properties (Fig.1).

Figure 1. Examples of the VisiMix® GUI Elements

In addition to powerful calculation capabilities, VisiMix® has functions of an expert system: firstly, it allows for entering properties of materials from built-in databases, and secondly (and most importantly), it analyses the initial data and calculation results, and issues warning messages whenever the input results in unacceptable process course. These messages serve as a means for defining the safety range of the basic process parameters.

4. VisiMix® Application in ISD of Chemical Processes

This example demonstrates the use of VisiMix® for checking the safety of a process in a $5m^3$ stirred jacketed reactors. The device has variable impeller rotational speed in the range 60-90 RPM, the motor rated power is 7500 W. The process is based on exothermal catalytic reaction with a solid catalyst.

The main risk for such a process is a runaway reaction that may start in the reactor because of a number of factors that we need to evaluate and prevent. First, we should check that there is no sedimentation of solid particles on the tank bottom, and that catalyst distribution inside the tank is uniform. Second, we need to check if the reactor's cooling system is capable of preventing the 2nd order irreversible reaction from running away.

As input data for all calculations, Visimix® uses basic dimensions of the tank and baffles, type, dimensions, positions and rotation speed of the impeller, characteristics of the shaft, motor power and relevant properties of the media (both liquid and solid phases), as well as the jacket parameters and characteristics of cooling and heating agents.

4.1 Evaluation of Just Suspension Speed and Uniformity of the Catalyst Distribution

In order to ensure absence of sedimentation, we will calculate Just Suspension Speed (JSS), which is the minimum rotational velocity of the impeller at which there are no stagnant zones at the bottom.

Although VisiMix® does not calculate JSS directly, it enables the user to easily determine this parameter.

When calculating any of the parameters of the Liquid-Solid Mixing, VisiMix® issues warning messages whenever settling of solid particles may occur. Therefore, to find JSS, it is enough to enter any reasonable value of rotational speed, and then to gradually increase it until VisiMix stops issuing the warnings. The first RPM value, which does not result in the warning message, will be the desired JSS value.

1. Before starting the calculations, enter the equipment parameters, including tank dimensions and impeller power and rotational speed, to display the diagram of your mixing system (Fig.2), and then enter the properties of solid and liquid phases (Fig. 3). Start with **60 RPM** as impeller rotational speed.

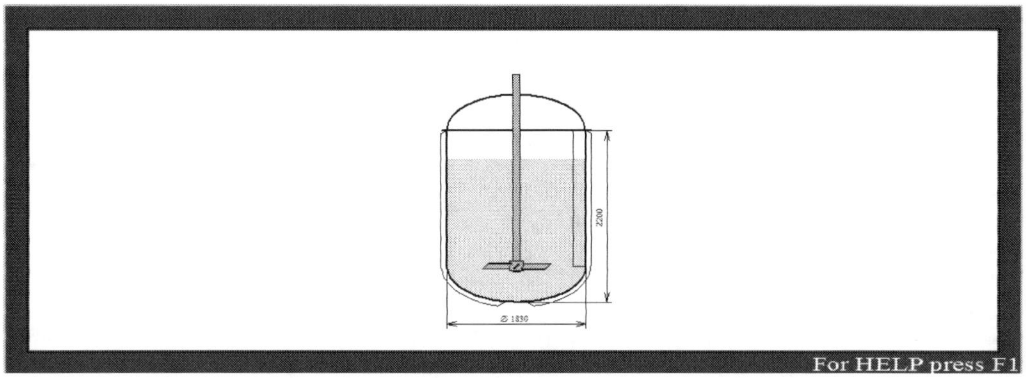

Figure 2. The diagram of the mixing system

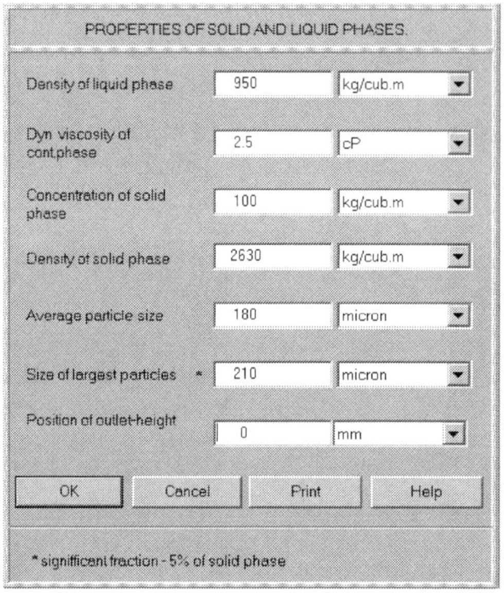

Figure 3. Properties of solid and liquid phases

2. To proceed to calculations, select **Calculate - Liquid-Solid Mixing**, then select any parameter, for instance, **Axial distribution of solid phase**.
3. For the data you have entered, VisiMix® issues a warning message (Fig.4).

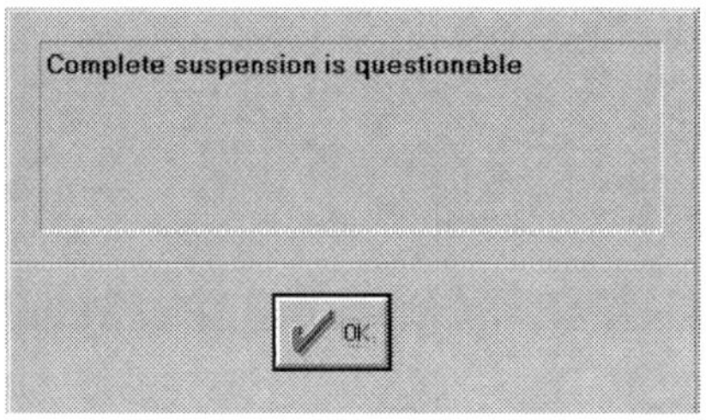

Figure 4. The message warning of possible settling of solid phase

This means that the entered value of rotational speed (60 RPM) is lower than JSS and does not ensure absence of sedimentation.

4. VisiMix® also checks uniformity of the catalyst distribution inside the tank. The resulting graph of axial distribution of solid phase is shown in Figure 5.

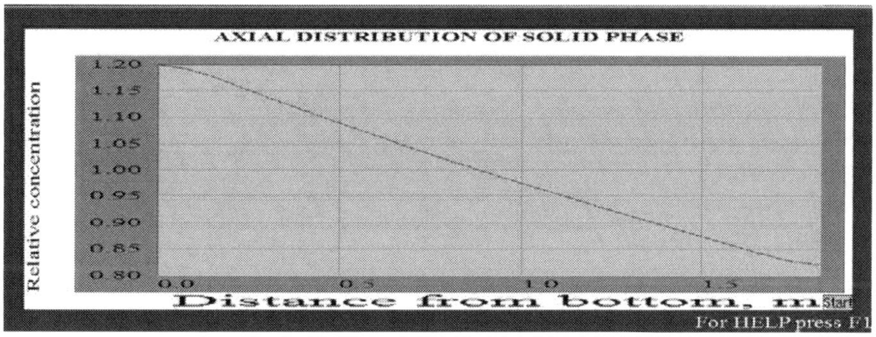

Figure 5. Axial distribution of solid phase (60 RPM).

This graph shows that at 60 RPM, the non-uniformity is about 20%. Catalyst non-uniformity may cause local overheating, which may also result in a runaway reaction.

5. Gradually increasing the impeller rotational speed, we see that at 80 RPM the message warning of possible settling of solid phase message appears no longer, so we can assume this RPM value as JSS. However, because we also want to improve our process configuration with regard to non-uniformity, we continue increasing the rotational speed, until, at 85 RPM, the axial distribution of the solid phase becomes more uniform, with maximum degree of non-uniformity of about 15% (Fig.6).

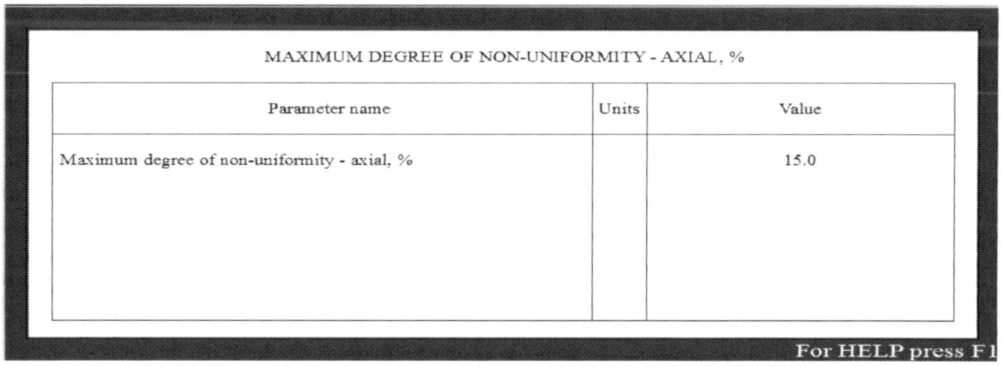
MAXIMUM DEGREE OF NON-UNIFORMITY - AXIAL, %

Parameter name	Units	Value
Maximum degree of non-uniformity - axial, %		15.0

For HELP press F1

Figure 6. Maximum degree of non-uniformity of catalyst axial distribution (85 RPM)

Based on the above calculations, we will select **85 RPM** as safe rotational speed for our device and will now check the reaction thermal parameters.

4.2 Simulation 2nd Order Exothermal Reaction Carried Out in a Stirred Batch Reactor

VisiMix® performs simulation of exothermal reaction based on the analysis of the equipment and process parameters to prevent runaway reaction that may take place when the energy generated by the reaction is greater than the energy removed from the reactor.

The reaction is run according to the stoichiometry

$$\boldsymbol{A} + \boldsymbol{B} \rightarrow \boldsymbol{C} \quad \text{[Eq. 1]}$$

The equation for the reaction rate is:

$$r = kC_A\, C_B \quad \text{[Eq. 2]}$$

where r is reaction rate, moles of A/(l·sec);
k is reaction rate constant, l/(mole·sec);
C_A is concentration of reactant A, mole/l;
C_B is concentration of reactant B, mole/l.

The reaction rate constant is a function of the system temperature and is given by

$$k = k_0 \cdot e^{-E/RT} \quad \text{[Eq. 3]}$$

where k_0 is Arrhenius constant, l/(mole·sec);
E is energy of activation, kJ/mole;
R is gas constant, J/(mole·K);
T is absolute temperature, K.

Kinetic reaction parameters were partially borrowed from [15].

For thermal calculations, additional input data are necessary.

The initial temperature inside the reactor is 20°C. Initial concentration of reactants A and B is the same and is equal to 5 g-mole/l. The actual process is performed in two stages: 1) initial heating which starts the reaction and 2) subsequent cooling required for removing excessive heat. In our case, the heating is with steam at atmospheric pressure supplied into the jacket, and the cooling is with ordinary water circulated through the jacket. The cooling water temperature is 20°C, and its volume flow rate is $40 m^3/h$.

The properties of the reacting mixture are identical to those of ordinary service water.

The maximum allowable temperature of the reactor is equal to the media vapor saturation temperature (143°C). The corresponding maximum working pressure of the reactor is 4 bar. Heat effect of reaction is 90 kJ / mole, energy of activation is 125.5 kJ/mole, and Arrhenius constant is $k_0 = 10^{17}$ l/mole·sec.

To evaluate the suitability of the reactor equipment to the exothermal process described above, we need to first check a number of important general parameters, such as **Mixing power, Macromixing time**, etc. Then we need to simulate the reaction to determine the process time before runaway regime starts, and make sure that efficient mixing is achieved. Thereafter, the cooling stage should be analyzed, and only then it will be possible to conclude whether the process safety is ensured.

1. In the VisiMix® **Hydrodynamics** menu, calculate **Mixing Power** (Fig. 7) to make sure that the consumed power does not exceed the motor rated power (7500 W). The calculated result shows that the power consumption is acceptable.

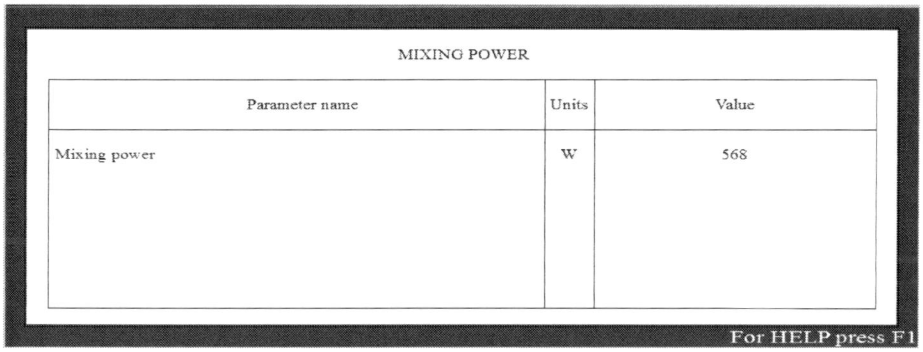

Figure 7. Mixing power

The second parameter, **Macromixing time** needed for the general check-up can be calculated using **Single-phase liquid mixing** submenu, the result is shown in Figure.8.

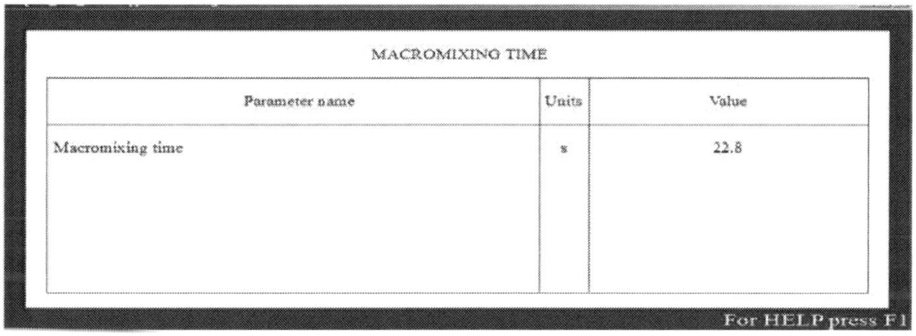

Figure 8. Macromixing time

We see that **Macromixing time** in our reactor is about 23 sec. This result is of great practical importance. Since acceptable mixing can only be achieved if the characteristic process time is at least 10 times greater than **Macromixing time**, the process time in the reactor should be at least 4 minutes.

2. We will now determine the process time before runaway regime starts. Select **Media temperature** in **Heat Transfer. Batch (BH) - Vaporous agent (VA).** At first we receive a warning message about a runaway reaction starting in about 7 minutes (Fig. 9). We can also see it in the graph shown in Figure10.

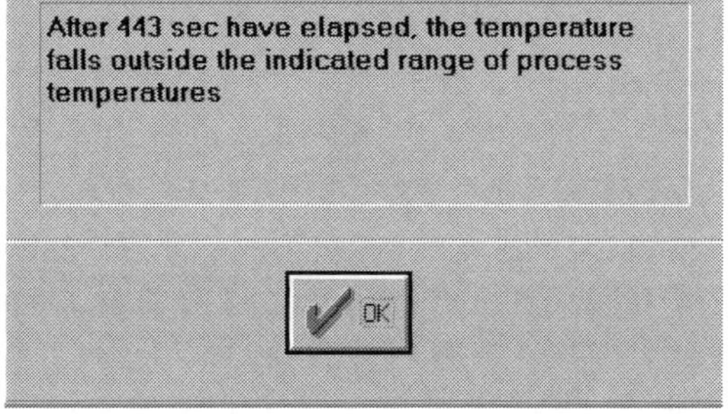

Figure 9. The message warning of the temperature exceeding the prescribed limit

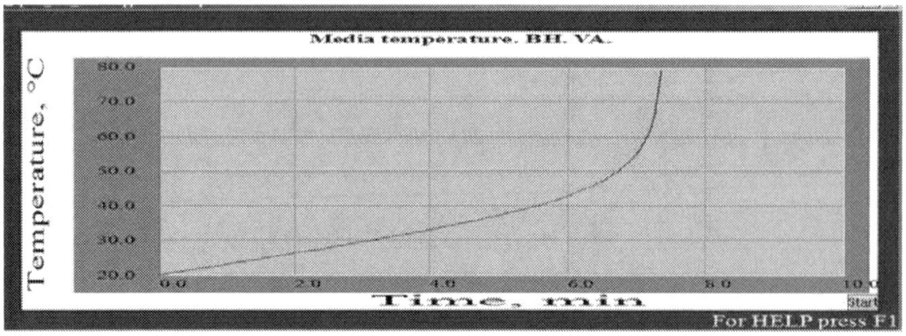

Figure 10. Media temperature. BH. VA.

To see the course of the reaction, select **Concentration of reactant A. BH. VA (**Figure 11).

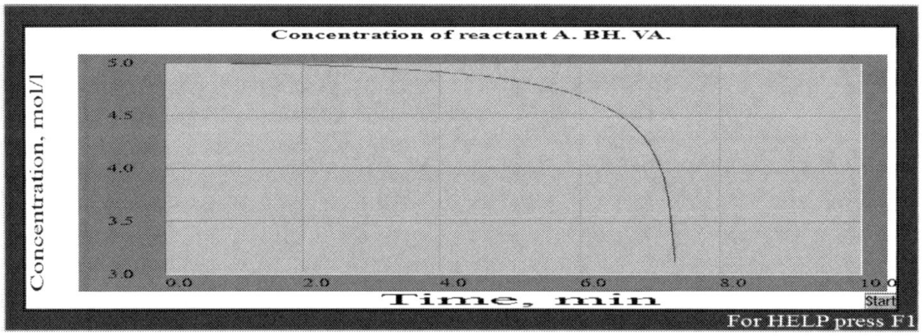

Figure 11. Concentration of reactant A. BH. VA.

Figures 10 and 11 show that runaway regime begins almost at the end of the reaction, when the reactant concentration is dramatically falling down, in about 7 minutes from the start of the process. This time is more than twenty times higher than the **Macromixing time** (23 sec in Figure 8). This corresponds to the basic requirement for the efficient mixing, and the reactor, therefore, may be considered as well stirred.

This information indicates that the second process stage – cooling should begin before the process approaches the runaway regime, i.e. about 5.5 minutes from the start of the process. At this moment the steam supply stops, the cooling water fills the jacket, and the second (cooling) stage starts. The results of the heating stage (temperature and reactant concentration values in Figures 10 and 11 corresponding to 5.5 minutes from the start of the process) will thus serve as the initial data for the cooling stage of the simulation, and the desired values are 40°C and 4.75 mole/liter (for both reactants).

3. Calculations for the cooling stage will be performed by means of the **Heat Transfer. Batch, Liquid agent (LA)** submenu. The results are presented in Figure 12 (**Media temperature. BH. LA**) and Figure 13 (**Concentration of reactant A. BH. LA.**).

The results of the calculations show that we have started the cooling early enough to prevent the process from reaching the upper temperature limit (143°C) and, consequently, the reactor maximum working pressure (4 bar). The figures indicate that heat release stopped for the only reason that both reactants in the reactor had ended (about 2.8 minutes from the start of the cooling). The total duration of the process is about 8.3 minutes.

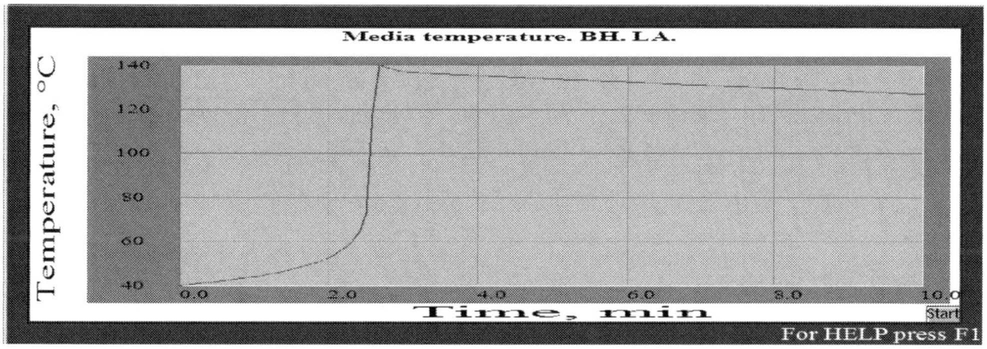

Figure 12. Media temperature. BH. LA. The heating stage is 5.5 minutes.

The results presented above show that if cooling starts 5.5 minutes from the start of the reaction, runaway regime will be prevented, and the reaction completed.

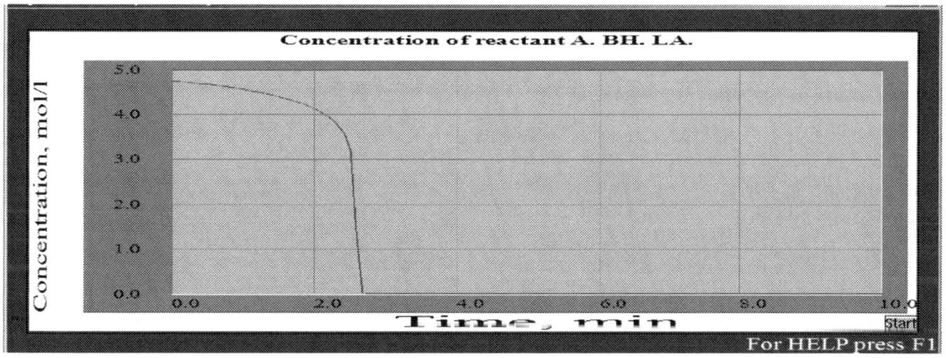

Figure 13. Concentration of reactant A. BH. LA. The heating stage is 5.5 minutes.

5. Conclusion

1. Process analysis based on valid technical calculations and simulation is a mandatory element of the ISD concept. The ISD software should meet the following requirements:
    - Provide adequate and complete description of the process and equipment configuration based on reliable models verified in practice.
    - Adequately detect any deviations from the process course which are generally regarded as main factors causing accidents in the process industries and provide quantitative analysis of the influence of these factors.
    - Be user friendly, accessible to plant engineers and allow quick results.
    - Combine a calculation tool with an expert system that analyses initial data and calculation results, and warns the user whenever the system approaches dangerous operation regimes.
2. Nuclear engineering experience can be used for advancing inherent safety in process industries. In the USA, for example, the Nuclear Regulatory Commission (NRC) grants a license to a plant if it submits, among other materials, the computer simulation results of the plant's expected behavior according to standard scenarios [17]. Computer codes for such simulation have been developed either by NRC or by its contractors. In the process

industries application of approved software tools for process analysis will improve safety status.

3. VisiMix® software which has proven capabilities in simulation of various technological processes in stirred tanks/reactors and estimation of their safety ranges can serve as a prototype for software tools applied for ISD in the process industries.

6. References

[1] Ta-Cheng Ho, Yih-Shing Duh, and J.R.Chen, Case Studies of Incidents in Runaway Reactions and Emergency Relief. Process Safety Progress, Vol.17, No.4, pp.252-262, (1998)

[2] Kamarizan Kidam and Markku Hurme, Analysis of equipment failures as contributors to chemical process accidents. Process Safety and Environmental Protection, 91 (2013), pp.61-78.

[3] C. Kirchsteiger, Trend in accidents, disaster and risk sources in Europe. Journal of Loss Prevention in the Process Industries, Vol.12, (1999), pp. 7…17.

[4] H. - J. Uth, Trend in major industrial accidents in Germany. Journal of Loss Prevention in the Process Industries, Vol.12, (1999), pp. 69…73.

[5] S. J. Kang, Trend in major industrial accidents in Korea. Journal of Loss Prevention in the Process Industries, Vol.12, (1999), pp. 75…77.

[6] S. Wakakura, Trend in chemical hazards in Japan. Journal of Loss Prevention in the Process Industries, Vol.12, (1999), pp. 75…77.

[7] Kamarizan Kidam and Markku Hurme, Origin of equipment design and operation errors. Journal of Loss Prevention in the Process Industries, Vol.25, (2012), pp. 937…949.

[8] T.A. Kletz and P. Amyotte, Process plants. A handbook for inherently safer design. 2nd Ed., Taylor CRC Press, Boca Raton. Fl. (2010).

[9] D.C. Hendershot, Inherently Safer chemical process design. Journal of Loss Prevention in the Process Industries, Vol.10, (1997), pp. 151…157.

[10] David J. Leggett, Safe process development from reaction hazards testing. Thermochimica Acta, 367-368, (2001), pp. 351…365.

[11] Theodor Grewer, David J. Frurip, and B. Keith Harrison, Prediction of thermal hazards of chemical reactions. Journal of Loss Prevention in the Process Industries, Vol.12, (1999), pp. 391…398.

[12] Valeria Gasson, David George Lister, Maria Francesca Milazzo, and Guiseppe Mascio, Comparison of criteria for prediction of runaway reactions in the sulphuric acid catalyzed esterification of acetic anhydride and methanol. Journal of Loss Prevention in the Process Industries, Vol.25, (2012), pp. 209…217.

[13] R. D. McIntosh and P. F. Nolan, Review and experimental evaluation of runaway chemical reactor disposal design methods. Journal of Loss Prevention in the Process Industries, Vol.14, (2012), pp. 17…36.

[14] www.visimix.com

[15] Safety, Health, and Loss Prevention in Chemical Processes. Problems for Undergraduate Engineering Curricula, (The Center for Chemical Process Safety of the American Institute of Chemical Engineers) pp. 162 - 164, Problem No. 81.

[16] A. Mak, S. Yang and N. G. Ozcan-Taskin, The Effect of Scale on the Suspension and Distribution of Solids in Stirred Vessels. In the book: Recents Progres en Genie des Procedes, (MIXING - 97. Multiphase Systems), Vol.11, number 52 (1997), pp. 97 - 104.

[17] L. M. Shotkin, "Development and assessment of U.S. nuclear regulatory commission thermal-hydraulic system computer codes," Nuclear Technology, vol. 116, no. 2, pp. 231–244, 1996

An Easy Design of Inlet Piping Systems for Relief Valves in Gas/Vapor Service

Guibing Zhao, Ph.D.
Chilworth Technology Inc.
113 Campus drive, Princeton, NJ 08540
Guibing.Zhao@dekra.com
www.chilworth.com

Prepared for Presentation at the
American Institute of Chemical Engineers
2014 Spring Meeting
10th Global Congress on Process Safety
New Orleans, LA
March 30 – April 2, 2014

UNPUBLISHED

An Easy Design of Inlet Piping Systems for Relief Valves in Gas/Vapor Service

Guibing Zhao, Ph.D.
Chilworth Technology Inc.
113 Campus drive, Princeton, NJ 08540
Guibing.Zhao@dekra.com

Keywords: Safety relief valve, inlet piping system, 3% pressure loss, flow capacity, allowable maximum velocity head loss

Abstract

Safety relief valves that are used to protect against overpressure require well-designed inlet piping for proper operation. Good engineering practice requires that the un-recoverable pressure loss of the inlet piping of a safety relief valve be less than 3% of its set pressure to avoid chattering, while attaining its desired capacity. Therefore, the design of a proper inlet piping system may require several trial-and-error fluid-mechanical calculations to reach the goals of both the 3% rule and the capacity requirement. This article presents a simplified method for the design of inlet piping systems without trial-and-error and iteration calculations. To accommodate the 3% rule and the required flow capacity, the maximum allowable velocity head loss of an inlet piping system can be easily calculated from classical one-dimensional adiabatic compressible fluid-flow principles. Then the inlet piping system (including pipe length, diameter, elbows, tees, etc.), and the corresponding velocity head loss, can be easily adjusted to satisfy both the 3% rule and the capacity requirement. A case study involving the selection of a safety relief valve for a gas validates the proposed method.

1. Introduction

Safety relief valves are widely and effectively used to protect equipment from excessive overpressure. The proper design of a relief system requires not only determining the correct size for the valve, but also determining the proper size and configuration of the upstream and downstream piping, and providing a proper effluent-handling system. The inlet-line pressure loss in a safety relief valve system can be significant even when the piping is quite short, between the protected pressure vessel and the safety relief valve. These losses are due mostly to the length of the inlet piping, but also to friction losses in valves and other fittings, and the effects of changes in the direction of flow, as in elbows and tees.

A poorly-designed inlet piping system – having a large inlet-piping pressure loss – will cause (1) a reduction in the flow capacity of the valve, resulting from a reduced stagnation pressure at the valve nozzle inlet; thus, the pressure in the vessel might exceed the Maximum Allowable Working Pressure (MAWP), and/or (2) unstable operation of the valve due to a rapid cyclic opening and closing the valve, commonly known as "chattering", when the pressure loss in the inlet piping periodically exceeds the blowdown of the safety relief valve. The blowdown of a safety valve is the difference between the set pressure and the closing pressure of a safety valve, expressed as a percentage of the set pressure. When the inlet loss is low – as during the initial flow toward the valve – the valve will open at its set pressure. However, when the inlet loss is high – as during flow through the valve – the pressure at the inlet could decrease below the valve-closing [re-seating] pressure. Repetition of this cycle can lead to not only a reduction of the flow capacity, but also failure of the safety relief valve itself or the piping system, with possible rupture of the pressure vessel. Therefore, good engineering practices (API 520 [1] and ASME B&PV Code Section VIII [2]) requires that the "un-recoverable" pressure loss of the inlet piping of a safety relief valve be less than 3% of the set pressure, to avoid chattering and reduction in the flow capacity. Although the 3% rule has long been controversial [3] and been thought to be overly conservative [4], a discussion the 3% rule is beyond the scope of this article.

The methods for designing the inlet piping for a safety relief valve – to satisfy both the 3% rule and the required capacity- have been well established and are used in industrial practice [5-9]. However, as discussed below, the conventional method is quite lengthy and complicated, and requires several fluid-mechanic and trial-and-error calculations to reach the goals of the 3% rule and the required capacity. Thus, this article presents an easy and objective-oriented method for the chemical and petrochemical industry.

The method presented in this article directly incorporates the 3% rule and capacity requirement in the design calculations, with an assumption of choked [sonic] flow at the exit nozzle of a relief valve. The maximum allowable velocity head loss of the inlet piping system can be easily calculated from classical one-dimensional adiabatic compressible-fluid-flow principles for gas or vapor relief. Therefore, any inlet piping system (including pipe length, diameter, elbows, tees, etc.) with a velocity head loss that is less than the calculated maximum allowable velocity head loss will automatically satisfy both the 3% rule and capacity requirement.

2. A basis for the design of inlet piping for safety relief valves

A schematic for a typical inlet piping system for a safety relief valve is shown in Figure 1. It is assumed that the characteristics of the protected tank or pressure vessel and its contents – such as the MAWP, and the temperature and pressure of the gas or vapor – in an emergency-venting situation are known. The calculations involve the pressure, temperature, density, and the velocity (Mach number) at the entrance to the piping between the protected vessel and the relief valve, at the entrance to the relief valve, and the nozzle exit of the relief, with the assumption that flow at nozzle exit of the valve is choked.

It is also assumed that the flow throughout the system is turbulent and one-dimensional, and that the flow through the valve-inlet piping is adiabatic, with no heat losses from the gas/vapor stream. Adiabatic flow can be either frictional or frictionless, and reversible adiabatic flow (without friction loss) is isentropic flow. Also, the gas or vapor is assumed to be a perfect gas, such that the ideal gas law applies.

The four variables of pressure, temperature, density, and velocity (Mach number) at locations 1, 2, and 3 can be solved by four equations of continuity (mass balance), momentum balance, energy balance, and "state" (pressure, temperature, and volume) at any given boundary conditions. Such solutions [10] are shown in Table 1, where the stagnation value (subscript o) is used as the reference condition for isentropic/frictionless flow, while the critical value under sonic/choked conditions (Mach number = 1, denoted by asterisks, whether or not it is actually reached in the pipe exit) is used as a reference condition for adiabatic flow with friction. The stagnation state has a macroscopic flow velocity of zero, which can be an actual state in a pressure vessel, or a hypothetical state in frictional flow where given state conditions are isentropically transformed to a stagnation state. The mass flux can be calculated through Equation [Eq.8a] at any location in the venting line for any given Mach number. Equation [Eq.8b] is the critical mass flux when assuming choked flow at the exit from a pipe or from a nozzle.

Figure 1. Inlet piping system of safety relief valve

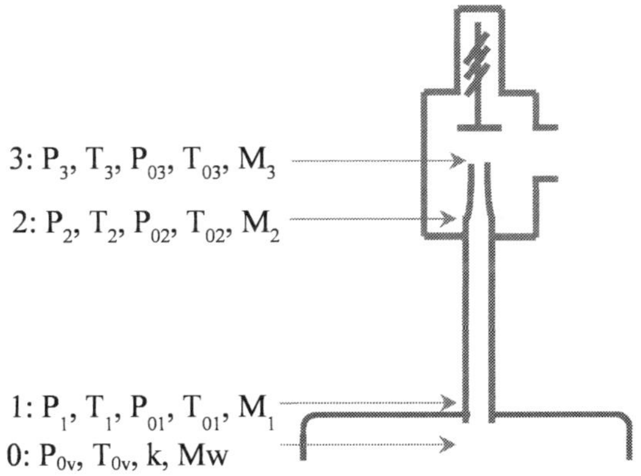

Table 1. Equations for calculating the state conditions and Mach numbers along the inlet piping of a safety relief valve

Variable	Isentropic (frictionless) process	Adiabatic process with friction
Pressure	$\frac{P_0}{P} = \left(1 + \frac{k-1}{2}M^2\right)^{\frac{k}{k-1}}$ [Eq.1]	$\frac{P}{P^*} = \frac{1}{M}\left[\frac{k+1}{2+(k-1)M^2}\right]^{\frac{1}{2}}$ [Eq.4]
Temperature	$\frac{T_0}{T} = 1 + \frac{k-1}{2}M^2$ [Eq.2]	$\frac{T}{T^*} = \frac{k+1}{2+(k-1)M^2}$ [Eq.5]
Density	$\frac{\rho_0}{\rho} = \left(1 + \frac{k-1}{2}M^2\right)^{\frac{1}{k-1}}$ [Eq.3]	$\frac{\rho}{\rho^*} = \frac{1}{M}\left[\frac{2+(k-1)M^2}{k+1}\right]^{\frac{1}{2}}$ [Eq.6]
Mach number relationship		$kN = \frac{1}{M_1^2} - \frac{1}{M_2^2} + \frac{k+1}{2} ln\left[\frac{2+(k-1)M_2^2}{2+(k-1)M_1^2} \cdot \frac{M_1^2}{M_2^2}\right]$ [Eq.7]
Mass flux	$G = \rho u = \frac{PMw}{RT} M\sqrt{\frac{kRT}{Mw}} = PM\sqrt{\frac{kMw}{RT}} = P_0 M\sqrt{\frac{kMw}{RT_0}\left(1 + \frac{k-1}{2}M^2\right)^{-\frac{k+1}{k-1}}}$ [Eq.8a] $G_c = P_0\sqrt{\frac{kMw}{RT_{0v}}\left(\frac{2}{k+1}\right)^{\frac{k+1}{k-1}}}$ (choked flow) [Eq.8b]	

It is worth noting that the actual piping section from location 1 to 2 may be complex – including valves, elbows, tees, and rupture disks, as well as the piping – and not necessarily a straight pipe, as shown in Figure 1. Therefore, the inlet piping velocity head loss, N, is calculated from the actual pipe length and fittings as

$$N = f\frac{L}{d} + \sum K_i \qquad \text{[Eq.9]}$$

The Darcy friction coefficient, f, and resistance coefficient, K_i, of various fittings can be obtained from the often-used Crane Technical Paper 410 [11]. The "state" emergency-relief conditions of pressure and temperature at location 0, i.e., in the vessel, and the required relief capacity, i.e., mass flow rate, W_r, in kg/s, are obtained from tests.

Generally, the following assumptions are widely accepted by the industry:

- For adiabatic flow, $T_{01} = T_{02} = T_{03} = T_{0v}$,
- The fluid flow from location 0 to 1, i.e., vessel nozzle, and from 2 to 3, i.e., safety relief valve nozzle, is isentropic,
- The effect of elevation change on pressure drop is assumed to be negligible,
- Both API 520 and ASME code allow an accumulated pressure in the vessel beyond the MAWP during discharge, and is given as a percentage of the MAWP. Therefore, P_{0v} is calculated as

  $$P_{0v} = (1 + x)MAWP + P_{atm} \qquad \text{[Eq.10]}$$

  where x is accumulation factor; for example, 10% for single-valve installation if the overpressure scenario is not from fire exposure.

The flow capacity calculation for a safety valve according to API 520 [1] is based on isentropic flow through a nozzle with a correction factor, the so-called discharge coefficient K_d, which is the correction factor between the mass flow rate for isentropic flow in a nozzle and the experimentally determined mass flow rate in a safety valve. Section UG131 of the ASME Code [2] requires that the manufacturer-provided coefficient, K, be the product of the actual test-result coefficient, K_d, and a discounting factor of 0.9. The basic safety relief valve nozzle formula for sonic flow in API 520 is

$$A = \frac{W_r}{CKP_0K_bK_c}\sqrt{\frac{T_0 Z}{Mw}} \qquad \text{[Eq.11]}$$

which is the same as [Eq.8b] when Gc = w_r/A; when the value of each of the correction factors K, K_b, K_c is assumed to be 1.0; when the value of the compressibility factor Z is assumed to be 1.0; and when the dimensionless coefficient, C, is given by

$$C = \sqrt{\frac{k}{R}\left(\frac{2}{k+1}\right)^{\frac{k+1}{k-1}}} \qquad \text{[Eq.12]}$$

For conventional and pilot-operated valves, the capacity correction factor, K_b, due to backpressure is 1.0. The combination factor, K_c, is 1.0 when a rupture disk is not installed. When the compressibility factor Z is less than 1, the pressure loss calculation in which ideal-gas behavior is assumed will yield a larger value than that in which the compressibility factor is considered. Thus, for most common industrial gases, the assumption of an ideal gas will generate conservative results.

3. Conventional method to design safety relief valve inlet piping system

The following is the conventional procedure for designing the inlet piping system of a safety relief valve that is summarized from reference [9]:

(1) Determine the required relieving capacity, from all possible sources and causes of overpressure, using a hazard and operability (HAZOP) study, for example. This would include failure of a valve to close, in gas or liquid piping to the protected vessel; pressure or temperature control system failure; fire exposure; failure of a pump in the discharge from the vessel; uncontrolled chemical reaction; and blocked-in vessel or piping subjected to abnormal temperature. The worst case combination of these factors is used to determine the required capacity [12].

(2) Calculate the required area of the internal nozzle or orifice, $A_{3,cal}$, of a safety relief valve from [Eq.11] based on manufacturer-provided discharge coefficient and assuming $P_0 = P_{0v}$. The next-larger standard nozzle size, A_3, of a safety relief valve than the calculated size, $A_{3,cal}$, usually is selected. Accordingly, the standard size of inlet, A_2, and outlet piping of a safety relief valve can be determined based on API 526 [13] or manufacturer-provided safety relief valve manual.

(3) Calculate the ASME rated flow capacity, W_{rated}, based on the actual nozzle size of a safety relief valve from [Eq.11] by assuming $P_0 = P_{0v}$.

(4) Calculate the Mach number in location 2, M_2, by combining [Eq.8a] and [Eq.8b] based on a mass balance.

$$W_{actual} = G_2 \cdot A_2 = G_3 \cdot A_3 \cdot (\frac{K}{0.9}) \quad \text{[Eq.13]}$$

where G_2 and G_3 are from [Eq.8a] and [Eq.8b], respectively. Because the actual capacity is considered, the ASME rated capacity is divided by 0.9, as in the last term of [Eq.13]. With known A_2 and A_3, M_2 can be calculated by combining [Eq.8a] and [Eq.8b]

$$\frac{0.9A_2}{K \cdot A_3} = \frac{1}{M_2}\left[\frac{2+(k-1){M_2}^2}{k+1}\right]^{\frac{k+1}{k-1}} \quad \text{[Eq.14]}$$

(5) With a known inlet piping configuration – length, diameter, and fittings – for the safety relief valve system, the velocity head loss, N, of the inlet piping can be calculated from [Eq.9].

(6) With the values of N and M_2 now known, the Mach number at location 1, M_1, can be calculated by solving [Eq.7].

(7) Since $P_{01} = P_{0v}$ and $T_{01} = T_{0v}$, with a known M_1 from step (6), all other state variables, P_1, T_1, and ρ_1 can be calculated from [Eq.1], [Eq.2], and [Eq.3].

(8) By applying [Eq.4], [Eq.5], and [Eq.6] to both locations 1 and 2, the state variables, P_2, T_2, and ρ_2 can be calculated based on known P_1, T_1, and ρ_1 from step (7).

(9) With the known state variables P_2, T_2, and ρ_2, and the known Mach number M_2 at location 2 from step (8), the stagnation parameters P_{02}, T_{02}, and ρ_{02} can be calculated from [Eq.1], [Eq.2], and [Eq.3].

(10) With a known $P_{02} = P_{03}$ from step (9), the ASME-creditable flow capacity can be calculated from

$$W_{ASME} = 0.9G_2 \cdot A_2 = G_3 \cdot A_3 \cdot K \quad \text{[Eq.15]}$$

(11) Now comparing W_{ASME} to W_r, and P_{01} to P_{02}, if either of the two design criteria

$$\gamma = \frac{W_{ASME}}{W_r} \times 100\% > 100\%, \text{ or} \quad \text{[Eq.16a]}$$

$$\beta = \frac{P_{01}-P_{02}}{P_{sg}} \times 100\% < 3\% \text{ (3\% rule)} \quad \text{[Eq.16b]}$$

is not satisfied, it would be necessary to change the inlet piping configuration, usually to decrease the inlet-piping velocity head loss, *N*. Steps (5) to (11) would then be repeated until both [Eq.16a] and [Eq.16b] are satisfied.

The above design calculation method is very tedious and time-consuming. Even with current computer technology, it may still take time and effort to find an inlet-piping configuration that satisfies both [Eq.16a] and [Eq.16b]. The purpose of this article is to propose a calculation method that avoids this trial-and-error process.

4. An easy method to design a safety relief valve inlet piping system

An increase of stagnation pressure at location 2, P_{02}, would increase flow capacity and reduce unrecoverable pressure loss in the inlet piping system for a safety relief valve. There are two calculation strategies that can result in a minimum stagnation pressure at location 2, $P_{02,min}$, for satisfying either the 3% rule or the flow capacity requirement.

For satisfying the 3% rule, the design of the inlet piping of a safety relief valve system requires that P_{02} cannot be lower than $(P_{0v} - 3\% \cdot P_{sg})$. Therefore,

$$P_{02,min} = P_{0v} - 0.03 \cdot P_{sg} \qquad \text{[Eq.17]}$$

Based on the selected safety relief valve, the ASME-creditable flow capacity, W_{ASME}, can be immediately calculated from [Eq.15] with a known nozzle area and $P_{02} = P_{03}$, which is based on the assumption of isentropic flow in the nozzle. Generally, the calculated W_{ASME} will be greater than the required flow capacity, W_r. Therefore, both the 3% rule and flow capacity requirement can be satisfied.

In addition, $P_{02,min}$ can also be calculated based on the minimum flow capacity, W_r, the required flow,

$$P_{02,min} = P_{03,min} = \frac{W_r}{CKA_3}\sqrt{\frac{T_{03}}{Mw}} \qquad \text{[Eq.18]}$$

Unfortunately, in most cases, substituting the calculated $P_{02,min}$ from [Eq.18] into [Eq.16b] will violate the 3% rule.

Based on the above analysis, an easy method is proposed to design the inlet piping system of a safety relief valve. The procedure involves

(A) The same as step (2) in the conventional method.
(B) The same as step (3) in the conventional method.
(C) The same as step (4) in the conventional method.
(D) Calculating the $P_{02,min}$ from [Eq.17]. With the known $P_{02,min}$, and M_2, all state variables in location 2 can be calculated from [Eq.1] to [Eq.3].
(E) Calculating W_{actual} from [Eq.13] from state variables at either location 2 or 3 and the corresponding flow area.
(F) Calculating M_1 by solving [Eq.8a] through use of trial-and-error or a spreadsheet solver

$$G = \frac{W_{actual}}{A_1} = P_{01}M_1\sqrt{\frac{kMw}{RT_{01}}\left(1 + \frac{k-1}{2}{M_1}^2\right)^{-\frac{k+1}{k-1}}} \qquad \text{[Eq.19]}$$

where W_{actual} is calculated from step (E) and $P_{01} = P_{ov}$, $T_{01} = T_{ov}$, and $A_1 = A_2$.

(G) Calculating the maximum allowable the inlet piping velocity head loss, N_{max}, by solving [Eq.7] through

$$N_{max} = \frac{1}{kM_1^2} - \frac{1}{kM_2^2} + \frac{k+1}{2k} ln \left[\frac{2+(k-1)M_2^2}{2+(k-1)M_1^2} \cdot \frac{M_1^2}{M_2^2}\right] \quad \text{[Eq.20]}$$

(H) Comparing the actual velocity head loss of the inlet piping system, N, (calculated from [Eq.9]) to N_{max}, if N is less than or equal to N_{max}, the inlet piping system satisfies both the 3% rule and capacity requirement. If N is greater than N_{max}, it would be necessary to analyze the inlet piping configuration and determine what practical changes can be made to reduce the velocity head loss of the inlet piping system until N equals or is less than N_{max}.

(I) If the inlet piping size needs to be changed, repeating steps from (E) to (G) to find N_{max} of a larger piping size and adjusting the inlet piping configuration till $N < N_{max}$.

Measures to reduce the velocity head loss include decreasing the length of the piping, increasing the diameter of the piping, avoiding high-loss fittings, and avoiding unnecessary changes in flow direction. It is generally economical to determine the smallest standard diameter capable of sustaining an acceptable pressure loss. If a rupture disk is installed upstream of a safety relief valve – for protection against process fluids – it must burst completely, with all fragments remaining attached to the holder. Installing a shutoff valve in the piping to a relief valve is prohibited – with few closely-controlled exceptions – and, if installed, should be selected to contribute the lowest possible pressure drop to the system.

The following case study demonstrates how the proposed method facilitates the design of the inlet piping system of a safety relief valve, as compared to the conventional method.

5. Case study

Reference [9] illustrates the design of the inlet piping to a safety relief valve by using the conventional design method. The design requirements are used to compare the conventional method to the proposed simplified method.

A hazard analysis has identified that a pressure vessel should be protected by a safety relief valve, since a process upset could result in a flow of 26,000 kg/h of gaseous nitrogen that must be relieved, to prevent overpressurization. The vessel's MAWP is 50 psig (3.45 bar-g), and a 10% pressure accumulation – above the MAWP – is permitted. The safety relief valve to be installed is 25 meters downstream of the vessel outlet nozzle, and its set pressure is equal to the MAWP in accordance with ASME requirement. Also, a 5% blowdown difference between the opening and seating pressure is allowed. Between the vessel nozzle and the safety relief valve, there would be five long-radius pipe bends and one tee, through which the nitrogen must flow, from one end of the run through the branch. The original design model data are shown in Table 2.

Following steps (1), (2), (3) in the conventional design method, the required nozzle area $A_{3,cal}$ of a safety relief valve is calculated from [Eq.11],

$$A_{3,cal} = \frac{26000/3600}{0.2375 \times 0.95 \times 4.827 \times 10^5} \sqrt{\frac{288.71}{0.02801}} = 67.35\ cm^2$$

where the value of C is obtained from [Eq.12]

$$C = \sqrt{\frac{1.4}{8.314}\left(\frac{2}{1.4+1}\right)^{\frac{1.4+1}{1.4-1}}} = 0.2375\ mol^{\frac{1}{2}} \cdot K^{\frac{1}{2}} \cdot J^{-\frac{1}{2}}$$

This requirement can be met by a spring-loaded safety relief valve, "6Q8"[12], with a "Q" (11.0-square-inch, or 71.29 cm^2) orifice and with a 6-inch inlet flange and an 8-inch outlet flange. Therefore, this valve is selected, and $A_3 = 71.29\ cm^2$.

Table 2. Original design model data

Required flow capacity, Wr, kg/hr	26,000
Molecular weight, Mw, kg/mol	0.02801
Specific heat ratio, k	1.4
Safety relief valve set pressure, bar-g	3.45
Safety relief valve nozzle coefficient, K	0.95
Gas stagnation temperature in vessel, T_{0v}, K	288.71
Atmospheric pressure, Patm, bar-a	1.034
Accumulated pressure, $P_{0v} = 1.1 \times$ MAWP + Patm, bar-a	4.83

With the selected valve and with the 6-inch inlet flange reference [9] describes the trial-and-error calculations and adjustments in the configuration of the inlet piping system to satisfy both the 3% rule and the required flow capacity, as shown in [Eq.16a] and [Eq.16b]. Table 3 shows the four cases of trial calculations that were performed by the author of reference [9], and only Case D satisfied both the 3% rule and the required flow capacity. The detailed calculations and results for each case are also shown in reference [9].

Table 3. Vent piping setups tried in Reference [9]

Inlet piping system	**Case A: 5 bends, Flow through branch of tee, 25 m of 6-inch pipe**	**Case B: 2 bends, Flow through run of tee, 25 m of 6-inchpipe**	**Case C: 2 bends, Flow through run of tee, 25 m of 8-inch pipe**	**Case D: 2 bends, Flow through run of tee, 7.9 m of 8-inchpipe**
vessel nozzle exit and vent pipe inner diameter, d, cm	15.41	15.41	20.27	20.27
inlet piping straight length, m	25.00	25.00	25.00	7.90
Darcy friction factor, f	0.0149	0.0149	0.0141	0.0141
velocity head loss, Ki				
vessel nozzle	0.5	0.5	0.5	0.5
Piping	2.42	2.42	1.74	0.548
90-deg long radius elbows	1.04	0.417	0.394	0.393
Tee	0.894	0.298	0.282	0.282
Total velocity head loss, N	4.85	3.62	2.92	1.72

Table 4. Data and calculation results for piping systems that satisfy the 3% rule or the capacity requirement, with the corresponding maximum allowable velocity head losses

Variables	Case D in Reference [9]	Satisfying the 3% rule with 6-inch inlet piping	Satisfying the lowest capacity requirement with 6-inch inlet piping	Satisfying the 3% rule with 8-inch inlet piping	Satisfying the lowest capacity requirement with 8-inch inlet piping
Location 1					
A_1, mm^2	32276.3	18640.1	18640.1	32276.3	32276.3
M_1	0.133	0.236	0.228	0.133	0.129
P_1, bar-a	4.767	4.643	4.656	4.767	4.771
P_{01}, bar-a	4.827	4.827	4.827	4.827	4.827
T_1, K	287.69	285.52	285.75	287.69	287.76
T_{01}, K	288.71	288.71	288.71	288.71	288.71
ρ_1, kg/m^3	5.581	5.476	5.487	5.581	5.584
u_1, m/s	46.03	81.43	78.46	46.15	44.52
W_1, kg/hr	29916	29925	28889	29925	28889
Location 2					
A_2, mm^2	32276.3	18640.1	18640.1	32276.3	32276.3
M_2	0.136	0.242	0.242	0.136	0.136
P_2, bar-a	4.661	4.535	4.378	4.662	4.501
P_{02}, bar-a	4.722	4.723	4.560	4.723	4.560
T_2, K	287.64	285.37	285.37	287.64	287.64
T_{02}, K	288.71	288.71	288.71	288.71	288.71
ρ_2, kg/m3	5.457	5.352	5.167	5.459	5.270
u_2, m/s	47.18	83.33	83.33	47.18	47.18
W_2, kg/hr	29916	29925	28889	29925	28889
Location 3					
A_3, mm^2	7129.0	7129.0	7129.0	7129.0	7129.0
M_3	1	1	1	1	1
P_3, bar-a	2.495	2.495	2.409	2.495	2.409
P_{03}, bar-a	4.722	4.723	4.560	4.723	4.560
T_3, K	240.59	240.59	240.59	240.59	240.59
T_{03}, K	288.71	288.71	288.71	288.71	288.71
ρ_3, kg/m^3	3.492	3.493	3.372	3.493	3.372
u_3, m/s	316.25	316.25	316.25	316.25	316.25
$W_3 = W_{ASME,a}$, kg/hr	26924	26932	26000	26932	26000
Results					
$\Delta P_{0,1\text{-}2}$, Bar	0.105	0.103	0.267	0.103	0.267
β, %	3.04	3.00	7.74	3.00	7.74
γ, %	103.6	103.6	100.0	103.6	100.0
N_{max}		0.54	1.48	1.70	4.63

The second column of the following Table 4 shows the calculated results for Case D of Table 3 by the author of this present article, using the conventional method as described above. These results are fairly close to those in reference [9], with an unrecoverable pressure loss of 3.04%, a little bit larger than 3.00% as reported in reference [9].

By using the proposed method in this article, the third and fourth columns of Table 4 show the results of calculations to determine the maximum allowable velocity head loss through the 6-inch-diameter inlet piping that would satisfy the 3% rule alone or the lowest capacity requirement alone, respectively.

As shown in the third column of Table 4, the calculated ASME-creditable capacity – assuming an unrecoverable pressure loss of 3% of the set pressure of the selected safety relief valve, i.e., $\beta = 3\%$ – is larger than the lowest capacity requirement, W_r, i.e., $\gamma > 100\%$. However, the maximum allowable velocity head loss is only 0.54 and would be difficult to attain with 6-inch piping.

As shown in the fourth column of Table 4, the calculated unrecoverable pressure loss would be 7.74% - and, thus, greater than 3% - if an e ASME-creditable flow capacity, $W_{ASME,}$, is assumed to be equal to the required flow capacity, W_r, i.e., $\gamma = 100\%$. Therefore, both the 3% rule and flow capacity requirement can be satisfied by using $P_{02,min}$ as [Eq.17] to calculate the maximum allowable velocity head loss of inlet piping system. These results are consistent with the analysis in Section 4, above.

If 6-inch-diameter inlet piping is chosen, the maximum allowable velocity head loss of the inlet piping system, N_{max}, is 0.54 as shown in the third column of Table 4. Any design of the inlet piping system with a velocity head loss larger than 0.54 will violate the 3% rule. This immediately eliminates Cases A and B of Table 3 because the total velocity head loss for each Case is far greater than 0.54. With a vessel nozzle velocity loss of 0.5, there is only 0.04 velocity head loss left for adjustment, which includes the length of piping length, the 90-deg elbows, and the tee. Thus, it would not be possible to design a practical inlet piping system to satisfy both the 3% rule and the flow capacity requirement for such a small margin of velocity head loss (0.04).. Therefore, any further design calculation must consider enlarging the inlet piping size.

Therefore, a larger inlet piping size of 8-inch-diameter is selected. The fifth and sixth columns of Table 4 show calculation results with corresponding maximum allowable velocity head loss, while satisfying the 3% rule alone and lowest capacity requirement alone, respectively. Both the 3% rule and the flow capacity requirement can be satisfied by using $P_{02,min}$ as given by [Eq.17] to calculate the maximum allowable velocity head loss of the inlet piping system.

With a larger pipe diameter of 8 inches, the maximum allowable velocity head loss of the inlet piping system, N_{max}, is 1.70, which leaves more room for evaluating various configurations of the inlet piping system. Thus, a practical inlet piping system could be designed$N < N_{max}$to satisfy both the 3% rule and the flow capacity requirement, while maintaining N less than N_{max}. It is not necessary to evaluate Case C in Table 3 because its velocity head loss is 2.92, greater than the maximum allowable velocity head loss of 1.70. Interestingly, Case D in Table 3 as shown in reference [9] has a velocity head loss of 1.72, which is fairly close to the maximum allowable velocity head loss of 1.70. As shown in the second column of Table 4, a velocity head loss of 1.72 gives an unrecoverable pressure loss of the inlet line of 3.04%. To completely comply with the 3% rule and the flow capacity requirement, the straight piping length of Case D in Table 3 should be 7.5 m, rather than 7.9 m.

By comparing the proposed method to the conventional method for the design of the inlet piping system of a safety relief valve, the following advantages are:

- The proposed method is quite straightforward and avoids extensive and time-consuming trial-and-error calculations.
- With a maximum allowable velocity head loss – as calculated by the proposed method – design engineers can quickly identify where or what part of the inlet piping system (including pipe diameter, fittings, piping configuration, etc.) needs to be adjusted or modified to satisfy both the 3% rule and flow capacity requirement.

6. Conclusion

An easy method is proposed to facilitate the design of the inlet piping system to a safety relief valve for gas/vapor service. This proposed method allows calculating the maximum allowable velocity head loss, based on both the 3% rule and the required flow capacity. Any practical inlet piping system with a velocity head loss that is larger than the calculated maximum allowable velocity head loss would violate either the 3% rule or the flow capacity requirement. With the known maximum allowable velocity head loss, design engineers can quickly identify where or what part of the inlet piping system needs to be adjusted or modified. This method is quite straightforward and avoids extensive and time-consuming trial-and-error calculations.

Nomenclature

A	Cross-sectional area, m^2
C	Thermodynamic coefficient for valve sizing, $mol^{1/2} \cdot K^{1/2} \cdot J^{-1/2}$
D	Pipe internal diameter, m
F	Darcy friction factor, dimensionless
G	Mass flux, $kg \cdot m^{-2} \cdot s^{-1}$
k	Specific heat ratio, dimensionless
K	Manufacturer-provided discharge coefficient, $K = 0.9K_d$, dimensionless
K_b	Capacity correction factor due to backpressure, dimensionless
K_c	Capacity correction factor due to combination with a rupture disk, dimensionless
K_d	Tested discharge coefficient, dimensionless
K_i	Resistance coefficient of pipe fitting, dimensionless
L	Pipe straight length, m
M	Mach number, dimensionless
MAWP	Maximum allowable working pressure, bar-g
Mw	Molecular weight, $kg \cdot mol^{-1}$
N	Velocity head loss or pipe resistance coefficient, dimensionless
P	Pressure, bar-a
R	Gas constant, 8.314 $J \cdot mol^{-1} \cdot K^{-1}$
T	Temperature, K
u	Velocity, $m \cdot s^{-1}$
W	Mass flow rate, $kg \cdot hr^{-1}$
x	Accumulation factor in percentage
Z	Compressibility factor, dimensionless
β	Ratio of unrecoverable pressure loss to the valve set pressure, dimensionless
γ	Valve capacity sufficiency ratio, dimensionless
ρ	Density, $kg \cdot m^{-3}$
Subscripts	
0	Stagnation condition
1,2,3	Location of relief line
01,02,03	Stagnation at location 1, 2, and 3
0v	Stagnation in vessel
actual	Actual
ASME	ASME creditable
atm	Ambient conditions
c	Critical or sonic condition
cal	Calculated
max	Maximum
min	Minimum
R	Required
Rated	ASME rated
Sg	Set gauge pressure

References

[1] API STANDARD 520: Sizing, selection and installation of pressure-relieving devices in refineries, 8th Edition, 2008

[2] ASME International, Pressure Vessels – Division I, Boiler and Pressure Vessel Code – Section VIII, ASME, New York, 2010

[3] Smith D., Burgess J., Powers, C., Relief device inlet piping: Beyond the 3 percent rule, Hydrocarbon Processing, November 2011, p 59-66

[4] Cremers J., Friedel L., Pallaks B., Validated sizing rule against chatter of relief valves during gas service, Journal of Loss Prevention in the Process Industries, 2001, V14, p 261-267

[5] Cox O. J., and Weirick M. L., Sizing safety relief valve inlet lines, Chemical Engineering Progress, 1980, V76(11), p 51-54

[6] Kandell P., Program sizes pipe and flare manifolds for compressible flow, Chemical Engineering, 1981, V88 (13), p 89-93

[7] Kern R., How to size piping and components as gas expands at flow conditions, Chemical Engineering, 1975, V82 (22), p 125-132

[8] Van Boskirk, B. A., Sensitivity of relief valves to inlet and outlet line lengths, Chemical Engineering, 1982, V89(17), p 77-82

[9] Westman, M. S., Rapidly design safety relief valve inlet piping system, Chemical Engineering Process, 1997, V93 (3), p 80-88

[10] White F. M., Fluid Mechanics, McGraw-Hill, 2001

[11] Crane Co., Flow of fluids through valves, fittings, and pipe, Technical Paper No. 410, Crane Co., New York, NY., 1982

[12] Crosby Valve Inc., "Crosby Pressure Relief Valve Engineering Handbook", 1997, p 2-6

[13] API STANDARD 526: Flanged steel pressure-relief valves, 6th Edition, 2009

348683 Off-Gas Flammability Control for a High Level Nuclear Waste Glass Melter System Based on Process Modeling and Pilot Testing

Abstract:

Off-Gas Flammability Control for

A High Level Nuclear Waste Glass Melter System

Based on Process Modeling and Pilot Testing

Alexander S. Choi

Savannah River National Laboratory

Aiken, SC 29808

In February 1996, the Defense Waste Processing Facility (DWPF) at the U.S. Department of Energy's Savannah River Site (SRS) commenced its first radioactive operation to convert over 300 million liters of high level nuclear waste into a stable borosilicate glass matrix for a long-term storage in a geological repository. At the heart of the DWPF operation is the joule-heated melter which converts a 50 wt% slurry feed containing pretreated waste and glass-forming frit into glass and the off-gas treatment system which removes volatile contaminants such as cesium and mercury from the melter exhaust. One of the critical safety requirements for the melter off-gas system is to maintain the concentration of flammable gases in the off-gas downstream of the Quencher below 60% of the lower flammability limit (LFL) under all operating conditions, while keeping the entire system under moderate vacuum.

The original DWPF design for the melter off-gas flammability control was to monitor the concentration of flammable gases using the LFL analyzers and to interlock the feeding off when readings exceeded a preset limit. However, it was shown during cold run tests that even after frequent calibration and maintenance these analyzers still were determined to be not reliable enough to function as a Safety Class instrument. Furthermore, in the event of sudden off-gas surges that could lead to a significant spike in the flammable gas concentration, these LFL analyzers could not provide adequate protection against potential off-gas flammability, since the duration of surges is typically much shorter than the inherent time delay in the response of these analyzers.

As a result, all LFL analyzers and associated interlocks and alarms were removed, and a new strategy was adopted to mitigate potential off-gas flammability that is in compliance with the NFPA Code 69; it requires that all operating variables that can contribute to potential off-gas flammability be monitored, and necessary safety interlocks and operating procedures to control those variables be devised. This paper describes a comprehensive method currently in place at the DWPF to monitor and control these variables using two process models; a steady state model that describes the batch-to-glass conversion chemistry and a transient model that describes the melter off-gas system dynamics and combustion kinetics of flammable gases. Some of the pilot test results that were used to develop and validate the model parameters are also covered briefly.

Advancing Process Safety: Operating Risks and Solutions to SOP Compliance

Samuel H. Solomon
Chairman
sam@mobideo.com

Prepared for Presentation at
American Institute of Chemical Engineers
2014 Spring Meeting
10th Global Congress on Process Safety
New Orleans, LA
March 30 – April 2, 2014

UNPUBLISHED

Advancing Process Safety: Operating Risks and Solutions to SOP Compliance

Samuel H. Solomon

Chairman
sam@mobideo.com

Keywords: inspection, compliance, operations risk, mechanical integrity, process safety, EHS, mobile, mobility, software, enterprise software, analytics, procedures, cloud, big data

Abstract

Advancing process safety requires rethinking technologies necessary to ensure adherence to procedures and effectively monitoring of compliance. With the advent of mobility, cloud and big data analytics, new tools are available to directly impact process safety and its implementation. *Key to successful implementation is the ability to capture and monitor activities at the most granular level of procedural detail – knowledge "one step at a time."* The case study in this paper will be on the inspection process with a specific attention to coating detection and failure as we discuss SOP compliance challenges resolved by this approach.

1. Introduction

Significant progress has been made during the past few years to improve the quality of data captured at the point of service in a continuous process facility. This has evolved to improving sensing devices and the data collection from these devices to deliver analytics that reveal the lifecycle of equipment and predicting failures before they occur. This area of machine-to-machine engineering and communications is critical to the long term safety and efficacy of equipment. What has not been addressed is the other side of this issue – the human interaction with this equipment in the form of Standard Operating Procedures, best practices and experience which is, for the most part, still complied with using documents, forms and spreadsheets.

The paper will address these challenges by detailing an information technology approach and its business value impact. The approach will cover just three of the many potential outcomes:

- Increasing process safety while reducing maintenance cycle times
- Increasing asset availability by ensuring latest procedures are followed

- Reducing maintenance costs by aggregating maintenance work and increasing pattern analysis leading to pro-active safety management

As we have said, these procedures are being performed by following checklists, job guides and documentation manuals. And, in every process industry these procedures are typically maintained in hardcopy or an image-equivalent such as PDF. Due to this critical lack of appropriate automation and controls, a number of key business concerns have been uncovered:

- How do we measure risk in people operations?
- How do we ensure compliance with protocols and regulatory standards?
- What do we do as our knowledge and expertise retires at a more rapid rate?
- How do we analyze best practices and evaluate processes?

The answers to these and other important operational questions are hidden in the details of a company's procedures which document exactly "how" a specific activity needs to be executed. These "Standard Operating Procedures" or SOPs, typically cover how every process is supposed to be performed and, yet, for the most part, companies have absolutely no visibility into these activities.

Today, a number of key enablers have come about pointing to new and comprehensive approaches for technology to address issues of operational integrity, and in particular process safety imbedded in these SOPs.

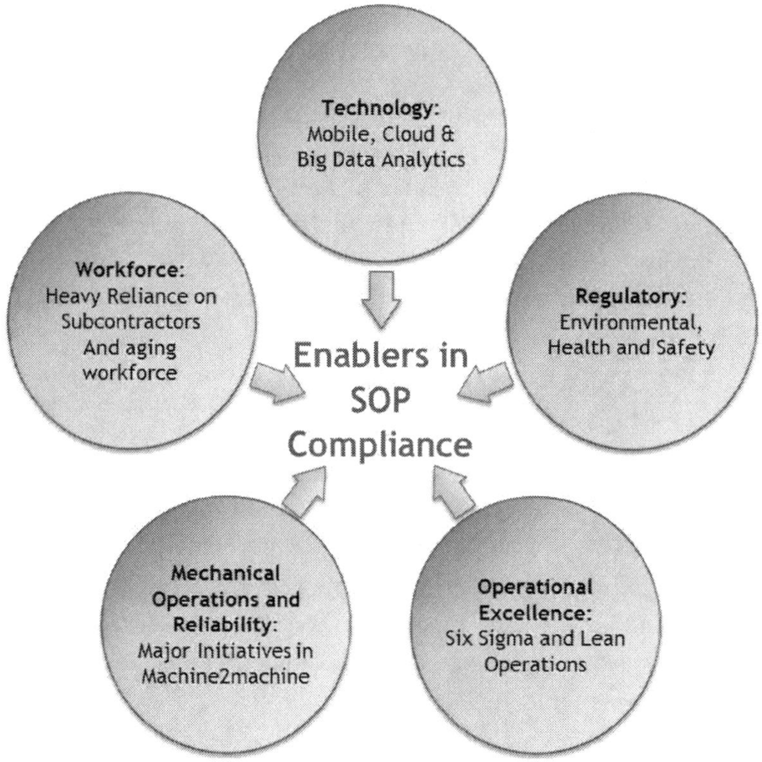

Figure 1 – Enablers in SOP Compliance

Source: Mobideo Technologies LLC

Let's briefly touch on each of these enablers:

Technology: Mobile, Cloud and Big Data Analytics – The availability and maturity of these technologies has lowered the cost of acquisition and precipitated their convergence. In addition, there is a strong desire within companies today to adopt new solutions that are based on Mobile, Cloud and Big Data Analytics.

Regulatory: Environmental, Health and Safety – Process industry companies now have internal and external compliance requirements. This oversight has grown because of significant risks associated with failure to comply entailing both civil and criminal penalties. SEMS (Safety & Environmental Management System) – a new regulation that is now being imposed on Owner/Operators engaged in off-shore Oil & Gas exploration in the Gulf of Mexico – is just one example of new and more comprehensive compliance requirements being adopted by the industry.

Furthermore, EH&S requirements are becoming critical "KPIs" (Key Performance Indicators) as to the sustainability and viability of companies throughout all process industries, with industry experts and analysts paying increased attention to these indicators. Again, failure to ensure compliance may result in loss of revenue, fines and at-risk employees.

Workforce: Heavy Reliance on Subcontractors and an Aging Workforce – We see this as another critical element in enabling the adoption of workforce process tools, since both the risk of non-compliance is significant, while, simultaneously, the traditional resistance to employee "monitoring" is negated.

For example, a company operating an oil rig must monitor and ensure compliance of its service provider subcontractors (who may represent as much as 90% of the workforce on the platform or in the plant). Liability for subcontractor performance is increasingly being applied not only to the service provider, but also to its customer, the Owner/Operator of the facility.

In addition, companies are constantly looking for ways to solve the "aging workforce" problem – experienced and knowledgeable employees who are retiring from the industry. They are being replaced with a new generation of workers who are technologically savvy and are comfortable working with mobile devices such as an iPAD, as opposed to hard copy checklists or solutions found in a printed reference book. For the new generation of workers, using a mobile device is both natural and expected.

Mechanical Operations and Reliability: Major Initiatives in Machine2Machine – There are major business initiatives by companies such as GE and Emerson to invent and deploy "pervasive sensors" on complex systems, thereby enhancing predictive maintenance and operational KPIs. Monitoring and analysis at the granular level of human interaction with complex systems adds a critical and complementary component to industry trends to better performance measurement and equipment reliability.

Operational Excellence: Six Sigma and Lean Operations – Operational Excellence principles, systems, and tools such as Six Sigma, Lean Manufacturing and Scientific Management are being adopted by companies in order to gain sustainable improvement of key performance metrics. The granular data collected by operators and inspectors provides invaluable information to be used by such Operational Excellence programs.

Our paper will examine key criteria for a successful technological solution to this problem as well as challenges and personnel considerations. Examples will be provided from an actual case study.

2. Challenges Facing SOP Compliance

Standard Operating Procedures are a broad description of protocols and procedures which detail the step-by-step guidance necessary to perform a specific task, such as corrosion inspection in a pipe or vessel. They are typically based upon industry standards as are promulgated by organizations including the American Petroleum Institute (API) and would be modified by the company to address actual plant or field conditions or best practices by experienced subject matter experts (SME) at the company. SOPs provide workers with guidance and checklists which detail every step to be taken, whether it is a check, warnings, meter readings and other data gathering requirements. The importance of these procedures in addressing monitoring and compliance is the fact that these SOPs must be followed in a very specific manner – step by step – in a granular fashion in order to adhere to these protocols.

These workflow processes are at the core of a compliance regime, according to the Chevron Richmond Refinery Report on the Aug 6, 2012 refinery rupture and fire, which "were intended to be <u>goal-setting or performance-based</u> (*the emphasis is ours*), in practice they appear to function primarily as reactive and activity-based regulatory schemes that require extensive rulemaking to modify. As a result, the federal and California PSM standards have become static in the face of advancing best practices and technology, with the emphasis placed on the completion of a task or activity rather than achievement of continuous risk reduction."[1] To ensure that goal-setting and performance-based adherence to protocols actually occurs as part of a strategy for continuous risk reduction, one needs to rethink how technology and best practices are implemented to enforce this compliance and to provide the level of monitoring and analytically reporting necessary to comply with emerging standards.

At the center of the issue of adherence to standard procedures and protocols is the question of the origination of errors, and in particular, human errors. The Guidelines for Preventing Human Error in Process Safety by Center for Chemical Process Safety (CCPS)[2] describe opposing approaches to Human Error Reduction and as can be seen from the diagram in Figure 2 below, an appropriate approach requires more than just an understanding of human/organizational/cultural factors – i.e. <u>tools and techniques</u> are on an equal footing. Their suggestion is to focus on a proactive approach that involves five key factors – all of which relate to ensuring compliance to SOP guidelines and procedures. Our focus will be on the tools and techniques related to supporting not just a reactive understanding of what happened but also a proactive approach to error prevention.

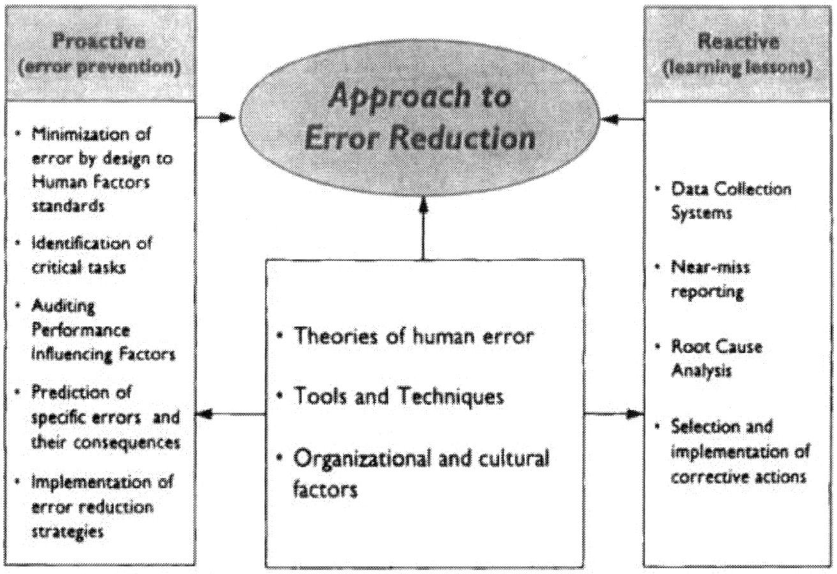

Figure 2 – General Error Management Structure [2]

A report from the Institution of Chemical Engineers (IChemE)[3] declares that 20% of equipment failures in chemical process industries are the result of human and organizational errors. The top 5 reasons are all directly related to the administration and deployment of protocols and procedures: poor contractor control, failure to follow procedures, lack of planning, poor management and supervision, and simple misjudgments. David Brown, chief executive of the Institution of Chemical Engineers (IChemE), said: "Accidents caused by equipment failures have many potential causes including contamination, corrosion, poor installation and mechanical failures. But it is worrying to see that human and organizational failures head the list of contributing factors. Everyone employed in the process industries must play their part in managing risk. The recent explosion in West, Texas, which killed 15 people, is another sad reminder of the consequences."

The textbook Human Error – Causes and Control[4] states that in a recent analysis of chemical plant problems, there were five main sources of these problems:

- Maintenance Practices 18%
- Maintenance Materials 7%
- Raw materials 5%
- Design 5%
- Operational discipline 45%

And we quote their conclusion: "The major human error problem was called operational discipline, that is, the failure to follow procedures and to stay within operational envelope. The second human factors problem was maintenance practices, that is, the inspections, repairs and modifications of equipment. Together, these two human error categories account for 63% of the chemical plant problems." They continue to elaborate: "Very often, mechanics do not consult procedures in printed service manuals or those that are available on computer terminals, since they often use the diagnostic challenge of trial-and-error methods to solve problems. In some situations, written instructions are not conveniently available, practical, reasonable or understandable." (*The emphasis is ours*). [4]

When one considers the human error component in compliance to SOPs, it is clear that this major challenge to accountability needs to be addressed at policy, cultural and procedural levels. We will address the procedural level which is really about finding an effective technological solution to SOP compliance given the convergence of mobility, cloud and big-data.

From a technological standpoint, there are a number of challenges to the effective deployment of SOP solutions including:

- Data necessary to carry out the field work is scattered in multiple locations – silo-ed within Enterprise Application Software, inspection, reliability, MOC and other systems for project/content/process management.
- Important data is embedded within disparate fill-in forms and support documents in word/excel/pdf formats.
- Enterprise application software products (such as ERP, EAM, and PLM) and Enterprise middleware (such as BPM and EAI) enable activity management only at higher levels of data resolution (task and sub-task). They fail to provide fine granularity at the "nano" level, necessary to provide deep, detailed and effective business process management and optimization.
- The difficulty of SOP protocol creating, maintenance, distribution and updating for most companies is an expensive nightmare and is embedded within many groups.

SOPs today are still complied with manually, even if the medium of display is an excel spreadsheet or PDF document. And, in many cases, SOPs are followed by adhering to each written step and completing the form with ink and paper. This approach creates severe challenges in addressing compliance and operational risk. These key challenges include:

- Data quality and consistency
- Costly over-maintenance and risk of under-maintenance
- Supervision of sub-contractors
- Defect identification to defect resolution
- Defect identification across equipment and personnel over time

Let us briefly touch on each of these areas:

Data quality and consistency – GIGO is a well-worn acronym in the computer industry: Garbage-in Garbage-out. Current approaches to managing and monitoring SOPs have no effective data validation and consistency systems where information is checked to ensure it is correct the first time. Typically mistakes are only discovered way after the work has already been completed. And, in many cases, the data can never be corrected or made consistent since it cannot be accomplished – sometimes days, weeks or months – after the fact.

Costly over-maintenance and risk of under-maintenance – This is directly related to data availability which drives metrics and decision making. A key part of this decision making process is to contextualize problems, for example, within the overall scope of the maintenance issues and objectives for this part of the plant. Data will come from gauges, sensor and human observation and following SOPs. If the data is unavailable or inconsistent/error prone, then it will be nearly impossible to accurately predict the best maintenance or replacement to support and extend the life cycle and safety of the equipment.

Supervision of sub-contractors – This is probably the most difficult area to ensure compliance to SOPs. On the other hand, they are the easiest to enforce new technology standards. Understanding what service providers are doing, how they are doing it and whether they are in compliance with company/operator standards cannot be accomplished without some sort of real automation at the point of service.

Defect identification to defect resolution – Similar to our comments regarding over and under maintenance, making the identification to resolution process most efficient requires consistent and accurate data at the most granular level of SOP compliance.

Defect identification across equipment and personnel over time – Ultimately, when one starts to contextualize data on human interaction with equipment via SOPs and other procedures, this granular data starts to reveal patterns way beyond the specific piece of equipment and allows for pattern and pier group analysis of the people, processes and equipment to determine errors and failures. We show an example of this in the case study.

3. Key Criteria for Competent Technology Solution

Over the past 20 years, Information Technology has successfully automated key back office operations such as Enterprise Resource Management (ERP), Enterprise Asset Management (EAM), Manufacturing Execution Systems (MES), Electronic Document Systems (EDS), Project Lifecycle Management (PLM) and more. A significant gap exists between these back office systems and the work actually being performed by workers in the plant or field.

Figure 3 – Gap Between Back Office and Field Force

What are key components of a technological solution to address SOP guidelines and their proper deployment in a company?

- Deployment to the workforce with no disruption in the way they currently follow and utilize SOPs.
- Rapid deployment based on inputting existing hardcopy or electronic technical guidelines, which are then converted into confirmation-dependent conditional steps;
- User-defined step-levels for each task, allowing the highlighting of details with whatever degree of granularity is needed to ensure consistent, error-free task completion;
- Real-time remote monitoring for supervision and/or collaboration;
- Multi-level guidance sensitive to user experience and expertise allows for significant training and employee mentorship and monitoring;
- Event tracking, information-tagging, real-time analytics, and data mining for continuous improvement programs;
- Cross-platform mobile-device support including: iPAD, Android and Windows;
- Architecture for flexibility and functionality across all deployment scenarios (public/private cloud or on-premises);
- Choice of standalone functionality, or seamless integration with other enterprise application software systems, including: Enterprise Resource Planning (ERP); Enterprise Asset Management (EAM); Enterprise Content Management (ECM); Product Lifecycle Management (PLM); project management and other middleware;
- Allow the system to have full functionality when isolated from the network with automatic updating whenever there is connectivity.

The combination of all factors directly impacts customer bottom line results by reducing operation costs, minimizing potential risks, increasing safety, ensuring quality, improving assets performance and ensuring regulatory compliance.

4. Case Study

The solution demonstrated in this case study is based upon Mobideo Technology LLC[5] platform to help companies increase asset capability, performance and situational awareness by ensuring proper execution of document-based activities and best practices across the organization; enabling users to have information where and when they need it. Rapid deployment is made practical by leveraging existing knowledge and expertise in the organization. A key business area impacted by this technology involves adherence and monitoring of SOP procedures and best practices. An important consideration was the need to address the most "granular" or "nano-level" data in an operation.

We have displayed a sample "coating inspection form" within a chemical manufacturing plant. The inspection process occurs on a regular basis as part of a mechanical integrity discipline to ensure the reliability and safety of the plant assets such as vessels, pipes and pressure valves. During the inspection process, an inspector examines the equipment while adhering to the "coating inspection form" and documenting his/her findings. In order to prepare for this pilot we identified the key SOP forms which were incorporated into the Mobideo application. Following this, "metadata" was added, i.e., information useful for validating and making consistent all data entry by the inspector (we used Mobideo's Studio product to accomplish this); we then created links to specific data we will need from the inspection and ERP databases; and finally these forms were distributed to standard (and ruggedized) tablets. We have chosen to implement the coating inspecting report and to focus on just one section being the coating thickness assessment. It took about 3 hours to complete the application presented in this paper.

We found that the manual approach introduced numerous errors by the nature of using a form. Even more important, with automation, we discovered useful and more fine-tuned data that was not possible in a form based approach but become very practical using a mobile solution, such as assessing thickness over a smaller inspection area. In addition, we looked at how mobilization with the right tool impacts the time expended performing the inspection without effecting the trade work itself, such as painting or welding. Finally, we looked at the contextual picture to see how this data, now granular and consistent, would provide a "bigger picture view" of the patterns of the coating failures.

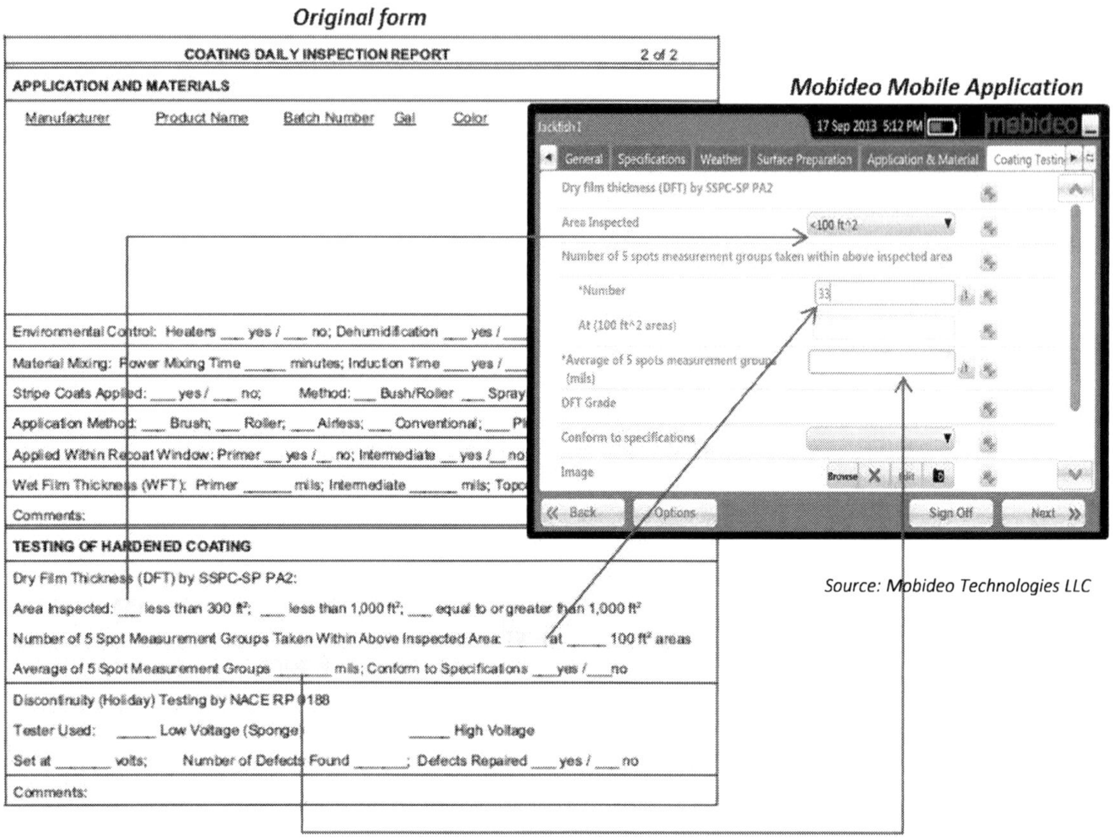

Figure 4 – Chemical Plant Application

In the form depicted above, each individual field is a granular step. For example, as highlighted in yellow, one step is to check-off the area inspected, and the next step is to enter the number of 5 spot measurements and finally to enter the average thickness for each of these measurements.

We created the mobile application where each of these granular steps is validated, time and location stamped and sent to the server in real-time (as well as any linked IT systems). The mobile application is built from all the information sources necessary to ensure proper task execution – documents and forms, as well as dynamic content that is retrieved in real time from third party back office systems. These data sources are integrated into a uniform application to be presented to the user. Along with the real time monitoring of the mobile application, a portal is automatically configured to observe, analyze and interact with the people and the data reflecting what they are doing during the SOP process. This flow is depicted in Figure 5.

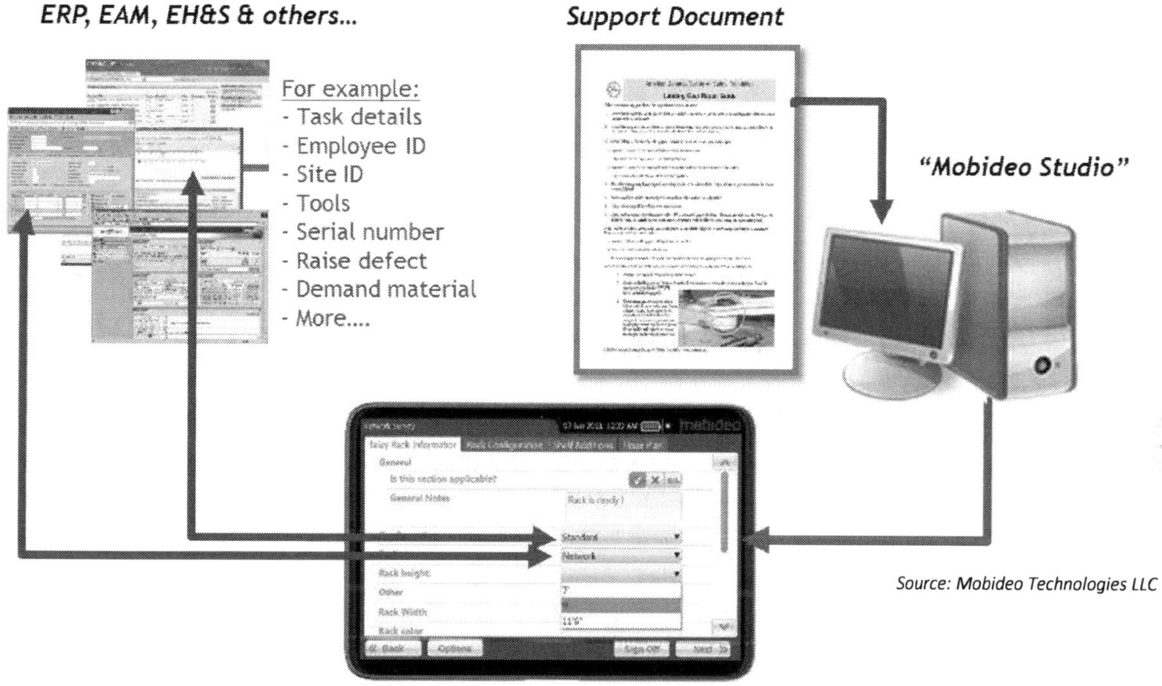

Figure 5 – Integration of Support Documents with Existing IT Databases

With the basic design flow in mind, let's look at a few sample images captured from the tablet running the application. The first screen shot (Figure 6) shows how the resulting SOP was incorporated and arranged with meta-data. Along the top there are tabs for each section of the SOP and then the actual SOP details, whether it is a question, selection, notice or display, they are arranged in a similar manner to the SOP therefore requiring little if any training in terms of content. The application itself is based upon simple "big glove" interaction – back, next and options. Everything else is the data within context of how the SOP presents the options. The meta-data would include validation checks, types of data entry (pull-down box, choices, free text, GPS settings, reading from an external meter or device, etc.), warnings, messages and many other data types.

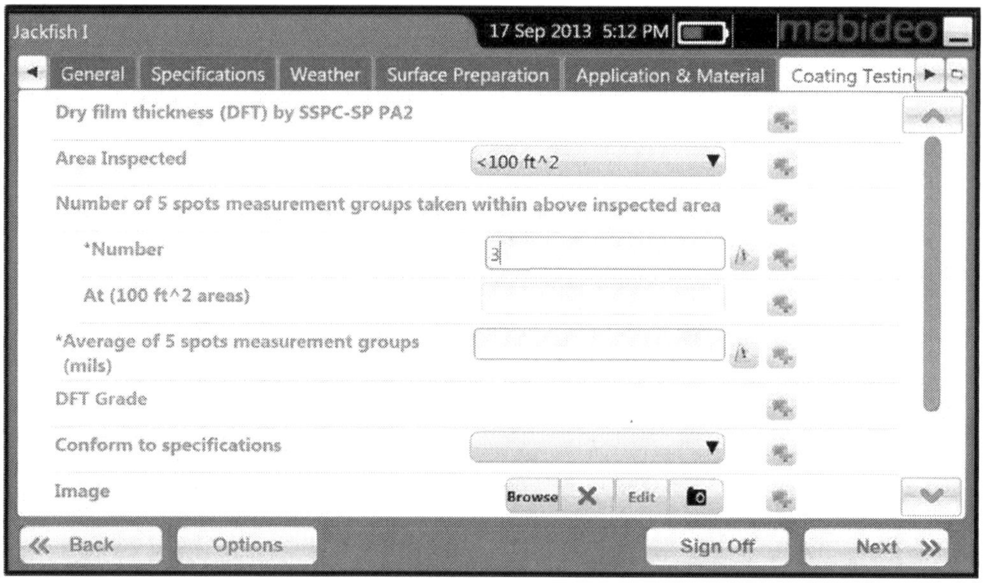

Figure 6 – Mobideo Form Used During Inspection Process

Figure 7 shows another example of an interesting input, in this case being a photo that was drawn upon and integrated into the SOP responses. This is an example of extending the impact of the SOP because it is now a mobile application. The other two changes we made by way of Figure is the ability to easily reduce the size of the inspection area (from 100 ft^2 to 300 ft^2) and allowing a number of spot inspections – something hard to implement and then change and make dynamic in a hardcopy form. In addition, the application makes certain calculations and lookups, such as the DFT Grade, and gives immediate feedback to the operator of the grade assessment.

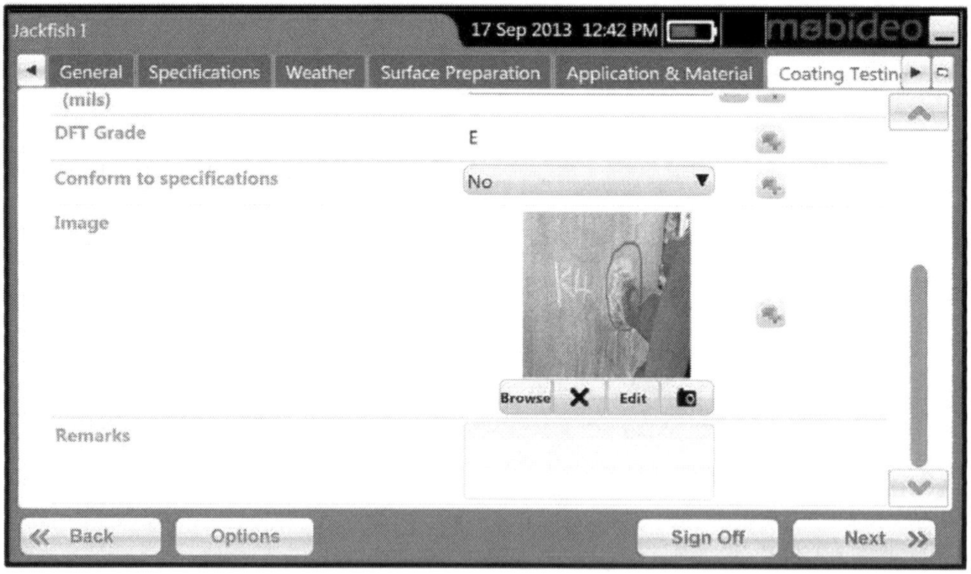

Figure 7 – Mobideo Form with Picture Illustrated During Inspection Process

Figure 8 shows how comments and social collaboration allow for easy capture of issues or best practices at the most granular step level of the process. Each step may be commented upon. Figure 9 shows how metadata enforces validation and consistency.

Figure 8 – Mobideo Form with Comments Added at the Granular Step Level

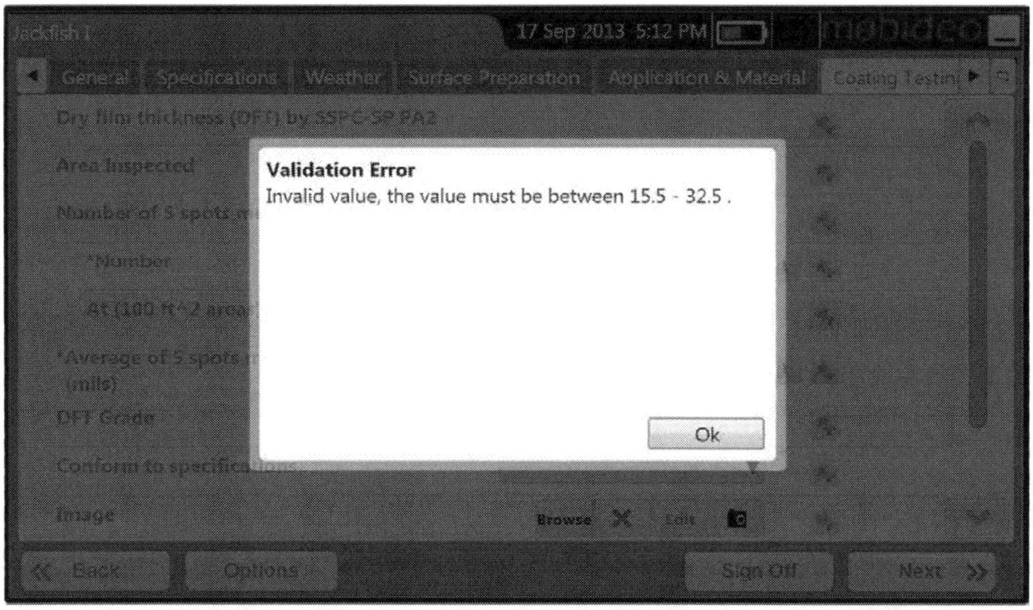

Figure 9 – Mobideo Form Demonstrating Validation of Data Entry

Figure 10 is very important. It shows the signature page which ends up becoming a very time consuming and error prone activity since signers may be in different locations and time frames. The system automatically coordinates proper signature signoff and there by enforces oversight and significantly reduces signature coordination time.

Figure 10 – Mobideo Form Show Authorization and Signature Coordination

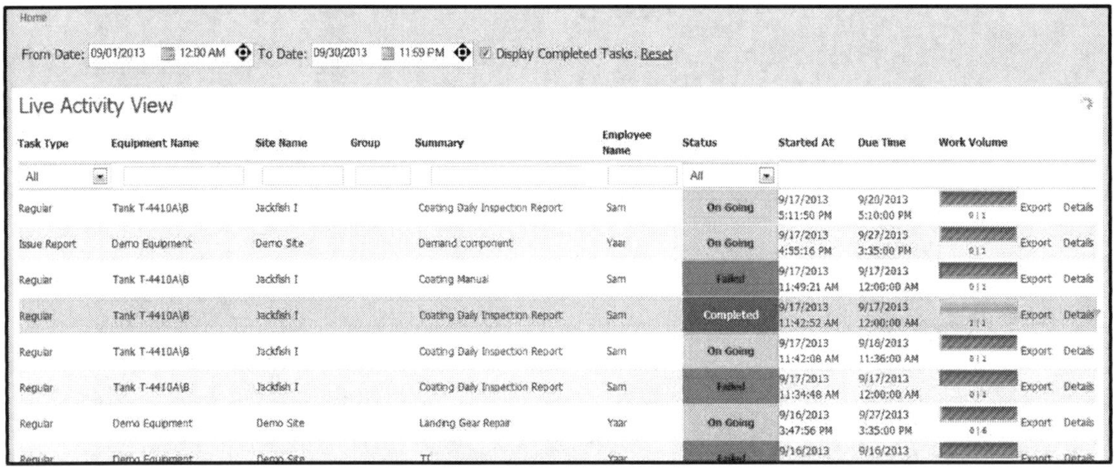

Figure 11 – Mobideo Web Portal Status Screen

Figure 11 is a sample display of the portal. Because every step is being monitored, supervisors have a complete picture of all processes, issues, errors and progress. In addition, traceability of all inputs is assured.

Figures 12 and 13 show an example of the impact the system had on turnaround time of the coating project without attempting to use the system to reduce the time the trades take to do their work. The saving of nearly 30% in time is just related to the logistics of the process, i.e., coordinating sign-offs, generating documentation and facilitating the timing and actions of the various trades. The streamlining of workflow is a direct result of using mobility technology and granular data capture to ensure compliance to the SOP. We now have ways to modify workflow to make it more logical and efficient and we gain productivity just by eliminating the loss due to logistics management.

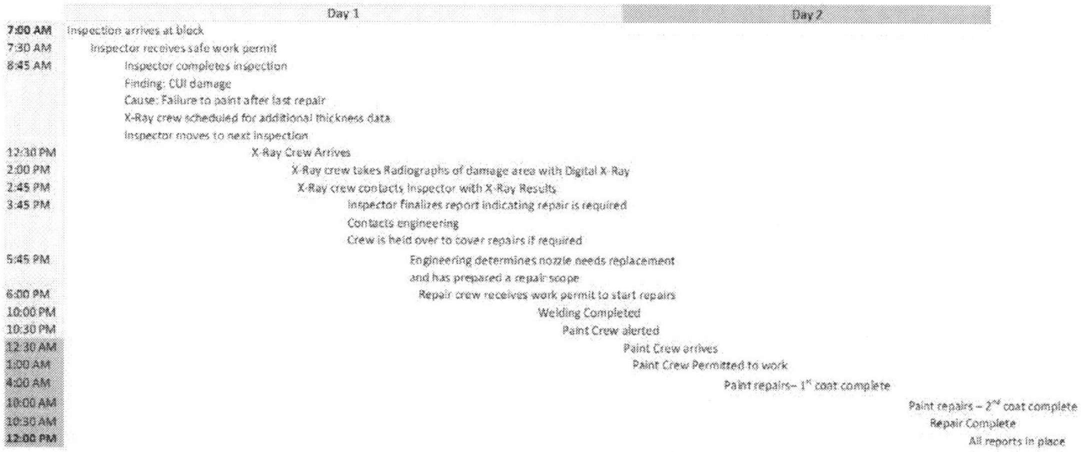

Figure 12 – Workflow Time Duration Using Standard Manual Approach

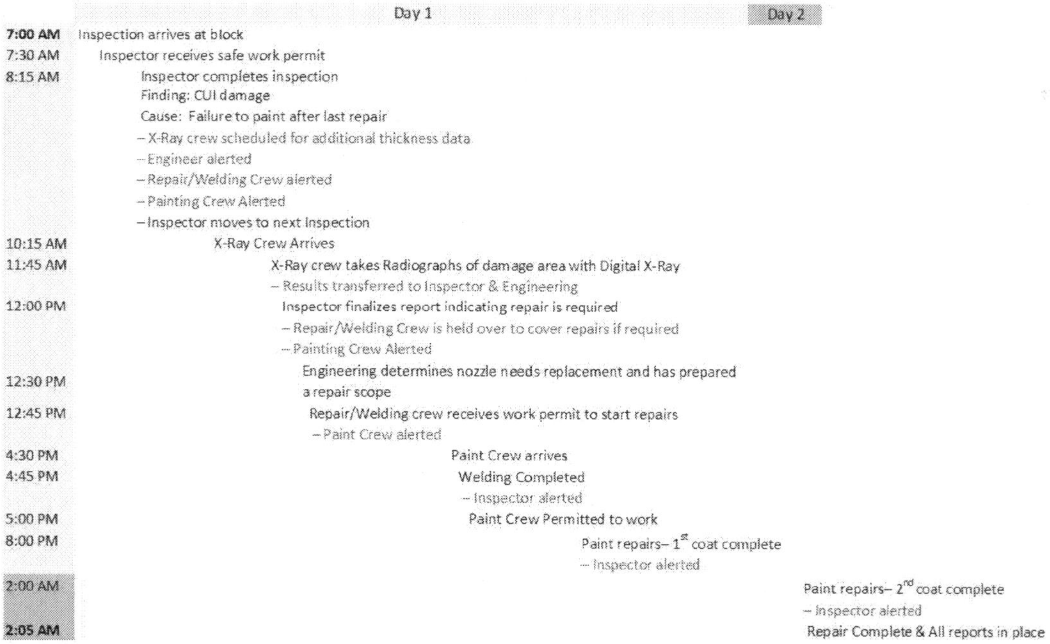

Figure 13 – Workflow Time Duration Reduced with Mobideo Coordination of Tasks and Trades

Source: Mobideo Technologies LLC

This last Figure 14 demonstrates another business value – the ability to now look across equipment and locations to determine appropriate KPI and intervention strategies. In this case, by looking at the inspection pattern of DFT spot measurements, we can quickly map out an area of corrosion that tells us to potentially redo all the painting as part of a more in-depth inspection process of this area.

DFT Spot Measurements of 60% "E"; 20% "D" - Schedule Remediation of All Equipment in this Area and Determine Root Cause of Coating Failures

Lower Gate Area					
Gauge reading	Spot measurement (DFT in mils)				
	A	B	C	D	E
1	18.2	20.6	21.0	16.7	26.4
2	18.6	19.1	21.8	17.1	29.8
3	17.8	20.4	20.8	16.8	29.0
Average	18.2	20.0	21.2	16.9	28.4

Upper Gate Area					
Gauge reading	Spot measurement (DFT in mils)				
	A	B	C	D	E
1	22.6	20.4	18.6	19.5	29.3
2	22.1	21.7	18.3	19.9	31.0
3	21.9	21.3	19.4	19.2	30.4
Average	22.2	21.1	18.8	19.5	30.2

Source: https://mail.google.com/mail/u/0/?ui=2&ik=26ab16b259&view=att&th=14136591b9723ffa&attid=0.1&disp=emb&realattid=51fefff28ef9e23e_0.1&zw&atsh=1

Figure 14 – Analysis of Corrosion Pattern

5. Issues to Consider

We have experienced a number of customer implementations and it is important to highlight critical success factors outside of an excellent software platform.

The first factor to consider is creating an opportunity to "dip one's toes in the water" before considering a full in-house implementation. With the advent of Software as a Service (SaaS) over the cloud, departments can implement a limited and very defined project to determine design, implementation and organization issues. This also allows for a quick jump to pilot stage without the need of a capital investment based ROI analysis, which could come later but is not necessary to start.

A second success factor is related to training and culture. For the emerging younger workforce, they have little fear of mobile devices and tablets as interactive devices. But the culture of accountability and an understanding that systems make one more effective (and not just big brother) still needs to be addressed. This is where training and cultural awareness by management will be very important.

The third success factor we believe is important is that deployments must be very rapid and the implementations non–technical. In that way, one may push accountability to the teams doing the work as close as possible to the field operations and avoid the loss of knowledge and efficiency as well as cost of implementation in staff functions to support SOPs, as is typically done today.

The forth success factor is the use of existing SOP documents to drive the application process. This is so important since we want to avoid errors and resistance to change because the actual inspection process has been re-engineered. We were able to assure operators and inspectors that they would continue to work and interact with the data exactly the way they had the day before roll-out.

6. Conclusion

With the convergence of mobile, cloud and analytics at the point of service, we are ready as an industry to tackle the difficult issues relating to compliance and monitoring of SOP guidelines. This area has essentially remained untouched while the back office and administrative side of the business has been completely automated. ERP, Plant Maintenance System, Inspection database, and Electronic Document Systems, to name a few, have tackled the "what" needs to be done to support operational integrity and safety. The "how" to accomplish this has remained locked in documents, spreadsheets and best practices without any direct way to ensure compliance and to monitor effective deployment from a KPI perspective.

The author believes it is now entirely possible to accomplish this transition of the culture and will be there to make this happen. We also want to keep in mind a number of important ingredients to a successful implementation:

Converge of all task execution data into a unified application – Converge of all task execution information from all sources, whether third party enterprise software such as ERP and EAM, or from documents and forms into a unified, personalized and contextualized mobile application. Integrating ERP, EAM, PLM and other third party enterprise applications enables customers to leverage existing investments in these systems, to extend their reach to the field, to ensure that they are updated in real time, and to increase their value

Ensure compliance – Enable companies to strictly enforce compliance and adherence to standard operating procedures and ensure proper execution of field activities according to company's best practices even in the most remote locations, regardless of whether they are being performed by company's direct employees or sub-contractors.

Track granular interaction and data entry in real time – By offering visibility of critical and complex processes at the most detailed and granular level (nano-step), management has the most comprehensive and detailed insights enabling them to gain situational awareness, perform predictive analytics, manage by exception, assure compliance to protocols and shield employees from mistakes while validating their work.

Enable rapid deployment driven from the procedures themselves – Utilizing the existing knowledge base by incorporating documentation that already exists within the company, as opposed to a top-down approach to a new system, ensures that the deployment is transformative and non-disruptive.

Technological convergence – Take full advantage of mobility with cloud/SaaS for rapid piloting and prototyping, analytics based upon granular data and social collaboration to facilitate the sharing of experience and best practices.

7. References

The author wishes to thank Mr. Keith Lapeyrouse, Reliability Manager at Dow Chemical Corporation for his support, insights and counsel.

[1] U.S. Chemical Safety and Hazard Investigation Board, Regulatory Report, Chevron Richmond Refinery Pipe Rupture and Fire, December 2013.

[2] Guidelines for Preventing Human Error in Process Safety, By Center for Chemical Process Safety (CCPS), Page 357, 1994.

[3] http://www.icheme.org/media_centre/news/2013/human-error-risk-for-process-industries.aspx, May 28, 2013

[4] Human Errors – Causes and Control, George Peters and Barbara Peters, CRC Press, page 14, 2006

[5] For more information on Mobideo Technologies LLC please contact yona@mobideo.com

349674 "Reducing the Frequency and Lowering the Severity of Human Error: Optimize Performance"

Abstract:

"Reducing the Frequency and Lowering the Severity of Human Error: Optimize Performance"

Process safety within the chemical, pharmaceutical, and petroleum industries demands extreme rigor where there is little room for human error. However, people are not perfect, and even the best make mistakes. Embracing this truism allows for real safety improvement strategies that are designed with the understanding that human error is inevitable and caused by complex factors. Failure to accept the inevitability of human error will result in continual frustration over adverse events and poor performance. When reactionary measures focus too heavily on the last person(s) involved in the error rather than focusing on the cumulative effects of systemic organizational breakdowns that allowed the event to occur, improvement opportunities are lost. This undue attention on symptoms instead of causes will yield only small gains, whereas accurate analysis and correction of system failures will return great benefits. Human performance will improve once organizational factors are redesigned to promote success and avoid failure. And, as individuals learn and use a set of well-defined, proven tools, common error-traps are mitigated. Thus, a comprehensive, holistic approach is needed that addresses all components of performance – error traps, organizational factors, and performance tools.

Optimize Performance (OP) is the approach that is brilliantly simple, yet profoundly impactful: Simple enough for everyone to learn and use, and impactful because it works. The practical value of Optimize Performance has been realized by more than 9000 individuals over the past 15 months. People embrace the OP brand of human performance because it works for them and their families, on and off the job. When employees embrace error-reduction principles off the job, employers win because of the inseparable connection between safe lifestyles off-the-job and error avoidance on-the-job. Employers reap tremendous benefits as employees put their training to work because they want to, not because they were told they had to.

Optimize Performance methods include:

- A set of well defined "**Tools**" that act as mental PPE to prevent errors and adverse events.
- Recognition of "**Traps**" that create error-likely situations. Both Situational-Factor Traps and Normalized-Drift Traps are identified and mitigated with OP "Tools".
- An **Error-Review** process that evaluates how OP Tools/Traps can be mined to prevent recurrence and generate concrete and applicable lessons learned.

OP Traps increase the risk of error-likely situations. Situational Factors affect individuals at a given point in time, whereas Normalized Drift is a trap in the form of weakness and breakdowns in organizational and personal defenses, which become accepted over time. Both sets of OP Traps result in sub-standard performance and adverse events.

OP Tools are a set of thought-provoking defenses that will prevent, predict, or reduce the likelihood of errors and events only after the methods and techniques are fully embraced, learned and practiced.

A simple, but highly effective **Error-Review** process (OPER) offers a more complete understanding of adverse events and is a critical component of the Root Cause Analysis. The OPER helps determine OP Traps that created error-likely situations, and OP Tools that could have helped predict or mitigate the circumstances that resulted in the event.

Regulatory agencies are embracing human performance concepts and have aggressively begun incorporating them into regulations. On many fronts, human performance is emerging as a very effective and valuable performance improvement process.

Once learned, these strategies improve all aspects of operational performance. When applied, Optimize Performance changes culture. Improvements will be widespread and widely recognized. The frequency of human error will be reduced, and the severity of the effects of human error will be lowered. Safety will improve, reliability will be increased, productivity will rise, and people's lives will be enriched.

Brochure Description:

People are not perfect, and even the best make mistakes. Organizational factors and values greatly affect performance and human errors. Optimize Performance is a proven strategy that is brilliantly simple, yet profoundly impactful. Recognize the "Traps" that create error-likely situations and learn how to apply well-defined "Tools" to prevent errors.

349820 Management of Process Safety Performance Indicators

Abstract:

Those working in the oil and gas industry are reminded with depressing regularity of the importance of process safety management, as post-incident investigation invariably points to one or more deficiencies related to process safety as root causes. Clearly, recent process safety related incidents and the development in process safety regulations have enabled wider awareness and adoption of process safety performance indicators (PSPI).

However, the primary challenge has been related to the selection, collection, analysis, and management of key process safety performance indicators that will eventually contribute to improving process safety in general. A systematic approach to manage these indicators is still evolving.

In this paper, the authors highlight a methodology for selection and measurement of process safety performance indicators which can help drive continuous process safety improvement and manage the performance of critical process safety barriers such as pressure relief devices, instrumentation systems, and mechanical integrity programs. This methodology considers complex set of intra-related factors and parameters that are available to the process operator; however, lack of systematic utilization of such parameters may hinder the realization of valuable process safety trends. In addition, the benefits and drawbacks of manual and automated systems for tracking near real time PSPI's will be presented.

349897 Improving Process Safety Performance for Mature Asset By Implementing of the Process Safety Key Performance Indicator

Abstract:

A mature asset has its own risk; it especially deals with plant integrity, facility changes, obsolete technology, people culture and stagnant procedure. A mature asset also frequently lay in the area of comfortable zone. Process safety hazard might be forgotten as time goes on and will lead to major accident scenario if it doesn't maintain.

The Offshore North West Java facilities belong to Pertamina Hulu Energy has lots of mature assets. Therefore Pertamina need to measure its process safety performance and be part of the corporate safety performance. The Process Safety Key Performance Indicators (PS KPI) is proposed to be used by Pertamina Hulu Energi Offshore North West Java (PHE ONWJ) to minimize risks from process safety incidents and to improve business performance.

The use of PS KPIs has been championed by the oil and gas industry as follow up lesson learned from several major incidents. It shown eventhough Process Safety incidents have low frequency but it potentially has high consequences. Hence, designing of the PS KPI should be carefully customized, practically implement and provide value to the organization, so that the process safety incident can effectively prevent.

The Pertamina Hulu Energy ONWJ used the combination of published guidance (API 754, CCPS, OGP Report No. 456 and UK HSE HSG254) for designing its PS KPIs. Understanding of the major gaps in facilities integrity, design and personnel culture are the important in designing PHE ONWJ's PS KPI. The effective PS KPI should consider the following criteria:

- The PS KPIs should address operational efficiency, reliability, and avoidance of production disruption.
- At the first phase of deployment, the PS KPIs should leverage the available data from the existing business process.
- The PS KPIs should provide senior management team in the company in the ability to understand the effectiveness of the PS programs
- The leading PS KPIs should be structured and easily tracked.
- Collecting data and tracking PS KPIs should add value to the organization and resources spent on the PS KPIs program should be in line with the value provided.

Pertamina believes that with designing Process Safety KPI that meet the criteria established can improve process safety performance as well as business goals for the company.

350057 An Innovative Work Flow for Performing Overpressure Protection Analysis Incorporating Process Simulation, Pressure Relief Valve Sizing, and Flare System Analysis

Abstract:

An innovative work flow for performing overpressure protection analysis incorporating process simulation, pressure relief valve sizing, and flare system analysis

By Wilfried Mofor, Aspen Technology Inc.; Nicholas Brownrigg, Aspen Technology Inc.; Dinu Ajikutira Aspen Technology Inc. and Ron Beck, Aspen Technology Inc.

Relief valve sizing is a critical element of the early design of any process facility, both from the point of view of ensuring the feasibility of achieving a safe design but also to understand the capital cost scope for both the relief valves themselves and the associated relief system.

The design of the facility itself is usually performed by a process modeler, employing a simulation system such as Aspen HYSYS or equivalent. Traditionally, the relief valve design has had to be performed in a stand-alone fashion, by an individual charged with that particular design task, employing a spreadsheet-type application that incorporates a particular organization's safety and design rules. Because a project cannot progress in the FEED workflow until the safety elements of the design have been verified, this is often a bottleneck in the process. There are also specific examples (and most organizations can attest to them), where incorrect data was transferred from the process model to the relief sizing programs, leading to incorrect results, sometimes only discovered after the relief valve has been procured and possibly even installed. Additionally, it is important for the safety engineer to rigorously consider all pressure source, routes, and failure potentials in enumerating the scenarios that need to be considered in the sizing analysis. These are usually only able to be derived by examining the process flow schema in the process model (or P&ID).

An innovative solution to this design challenge has recently been introduced by AspenTech, involving integration of an accepted, tested and rigorous relief valve sizing program within the process model workflow. This workflow spans from the equipment source of the pressure through the relief devices into the flare network. And additionally, from the process model, to the valve system design, to the overpressure relief network design and incorporating the documentation needed for the OSHA regulatory process and operator SOPs.

This paper will; (1) describe the new overpressure protection analysis workflow that is enabled by this new technology solution, (2) report on a comparative analysis of a complex overpressure protection analysis design problem as executed with the traditional and new workflow, (3) and report on results to date in industry, both on the E&C and owner operator side, from application of this new workflow.

We expect to be able to report on 3-4 industry case studies by the time this paper is presented, which we are currently in the process of gaining permission to present and to cite the names of the organizations involved.

350082 Risk Based Inspection Applied at Aging Chemical Facilities

Abstract:

Integrity safety is important to all companies and understanding the asset risk is the major key indicator in this process. Integrity management is underpinned by inspection and maintenance of equipment. Risk based inspection is becoming the major tool in integrity management of equipment at chemical facilities. Process risk management within Mechanical Integrity applying the challenges of using a risk based inspection programme, using a combination of the principles American Petroleun Institute (API) 580, ASME Post Construction Code (PCC-3), and MIL 882 introduced at older chemical plant mixed use environment with a combination of pressure vessels and Atmospheric Storage Tanks.

350323 Identifying Early Indicators of Incidents through Near-Misses

Abstract:

Post-incident investigations show that there are several near-misses or precursors that occur before undesirable events in different industries (chemical, manufacturing, aerospace, IT, etc.). Near-misses can be categorized into two categories – observable and non-observable near-misses – both of which indicate conditions which had the potential to cause incidents but did not actually result in much impact – due to sheer luck or existence of other protection systems. Observable near-misses refer to events that are readily visible to operating personnel, such as equipment failure, whereas non-observable near-misses refer to events that are not casually visible to the operating team, such as special-causes related to control loops. To improve reliability of systems and processes, it is of utmost importance for the risk management and reliability teams to not only identify and prioritize these near-misses, but also to learn from them. Particularly, cross-industry lessons on near-misses can provide new insights in their identification and management.

The presentation will demonstrate that having a strong near-miss management philosophy, system as well as culture, addressing both observable as well as non-observable events, help companies predict risks at their budding stages and prevent catastrophic events. It will include case studies from different industries on the importance of such events and their timely identification for improvement of reliability, and process safety. New quantitative techniques and methodologies to identify them systematically will also be discussed.

10TH GLOBAL CONGRESS ON PROCESS SAFETY

Methodology to Determine the Minimum Number of PHAs for Projects

Humbert Joseph Howard
Process Risk Specialist
Suncor Energy Inc.
150 6th Avenue SW
Calgary, AB, Canada
hjhoward@suncor.com

Prepared for Presentation at

American Institute of Chemical Engineers

2014 Spring Meeting

10th Global Congress on Process Safety

New Orleans, LA

March 30 – April 2, 2014

UNPUBLISHED

Methodology to Determine the Minimum Number of PHAs for Projects

Humbert Joseph Howard
Process Risk Specialist
Suncor Energy Inc.
150 6th Avenue SW
Calgary, AB, Canada
hjhoward@suncor.com

Keywords: Process Hazard Analysis, PSM, Projects

Abstract

The purpose of this paper is to provide Process Hazard Analysis (PHA) study team leaders and project managers, who drive the need for PHAs during projects, guidance relating to a decision methodology that can be used to assess the minimum number of PHAs to be done for various complexity projects. We will use change complexity analysis to assess the change and use the change complexity factor to determine the number of PHAs required for a project.

Some of the most challenging decisions involve Process Safety risks which often require experience and knowledge in order to be addressed. The use of proper methodology, applied at the right time, can drastically improve the design of new project installations handling hazardous materials.

The determination as to what type of hazard review and the number of PHAs required should be done during the initial engineering phase of the project. This allows recommendations from the PHA(s) to be implemented prior to installation and/or start-up.

1. Introduction

In general, the decisions faced by project engineers of process facilities in the chemical and petroleum industries on the number of Process Hazard Analysis reviews are difficult and complex. Companies do many small projects that may only involve a few pieces of equipment and thousands of dollars. Companies also do large projects involving hundreds of pieces of equipment and millions of dollars.

The purpose of this paper is to provide Process Hazard Analysis (PHA) study team leaders and project managers, who drive the need for PHAs during projects, guidance relating to a decision methodology that can be used to assess the minimum number of PHAs to be done for various complexity projects.

Some of the most challenging decisions involve Process Safety risks which often require experience and knowledge in order to be addressed. The use of proper methodology, applied at the right time, can drastically improve the design of new project installations handling hazardous materials.

The determination as to what type of hazard review and the number of PHAs required should be done during the initial engineering phase or front end engineering design (FEED) of the project. This allows recommendations from the PHA(s) to be implemented prior to installation and/or start-up.

This paper is not intended to give guidance on how to conduct the PHA or what methodology should be used. Individual company procedures and practice should be consulted for specific guidance on how to conduct a PHA.

2. Methodology

We will use a risk assessment process that I am calling Change Complexity Analysis to assess the change and use the Change Complexity Factor to determine the number of PHAs required for a project.

2.1 *Process Boundary for Change Complexity Analysis*

It is very important to determine the process boundary in order to do a proper complexity analysis. The process boundary identifies the area of immediate process impact due to this small project. When the small project change involves equipment that is integral to operation of another piece of equipment the boundary should include both pieces of equipment. This is done to make sure that the true complexity of the change is assessed properly.

Example 1

Replacing a storage tank recirculation pump.

Boundary for the complexity analysis should include the new pump, the related piping, and the storage tank it recirculates.

Example 2

Replacing a tower overhead condenser.

Boundary for the complexity analysis should include the new condenser and the tower that the condenser is attached to with all related piping and controls. Intermediate column take-offs should be included with the tower.

Example 3

Replacing a process unit Programmable Logic Controller (PLC).

Boundary for the complexity analysis should include the new PLC and all the process instrumentation and equipment that is attached to the PLC with all related controls for the equipment.

2.2 *Change Complexity Analysis*

Change Complexity Analysis seeks to identify how difficult a change a project will be. The more complex the project, the more carefully the project will need to be analyzed and managed. A project involving water does not need the same level of review as a project using a toxic chemical. A project to purchase and install a bucket conveyor system in alumina service may not need the same level of review as the installation of a new distillation column system in HF service.

We start by identifying the toxicity and flammability of the chemicals involved in the change *without* considering any interaction. Nh is the health factor of a material as defined in NFPA 704 [2] and given in NFPA 49 [1] or in Appendix A in DOW's Fire and Explosion Index 7th edition [3]. Nf is the flammability factor of a material as defined in NFPA 704 [2] and given in NFPA 49 [1] or in Appendix A in DOW's Fire and Explosion Index 7th edition [3]. We use Table 1 to calculate what is called a ***Hazard Material Factor***.

Table 1 – Hazard Material Factor Multiplier (HMF)

Description	Toxic Material (Nh=0)	Toxic Material (Nh=1)	Toxic Material (Nh=2)	Toxic Material (Nh=3)	Toxic Material (Nh=4)
Flammable material (Nf=0)	**1**	**1**	**2**	**3**	**4**
Flammable material (Nf=1)	**1**	**1**	**2**	**3**	**4**
Flammable material (Nf=2)	**2**	**2**	**3**	**3**	**4**
Flammable material (Nf=3)	**3**	**3**	**3**	**4**	**4**
Flammable material (Nf=4)	**4**	**4**	**4**	**4**	**5**

Nh is the health factor of a material as defined in NFPA 704 [2] and given in NFPA 49 [1] or in Appendix A in DOW's Fire and Explosion Index 7th edition [3].

Nf is the flammability factor of a material as defined in NFPA 704 [2] and given in NFPA 49 [1] or in Appendix A in DOW's Fire and Explosion Index 7th edition [3].

If the fluid you are dealing with is a mixture of chemicals, as is typical in a distillation column, then you should use the chemical (minimum 3% concentration) which results in the highest Hazard Material Factor.

If the fluid you are dealing with is a mixture of mostly water then you should consider the concentration of other chemicals in the water and assess what will be in the vapor space. If the hydrocarbon or toxic will be the dominant material in the vapor space it should be the dominant Hazard Material Factor.

If the fluid in question has special reactivity issues (heat or light sensitivity or not covered in NFPA list), they should be considered but are outside the scope of this paper.

2.3 *Correction Factor for Incomplete Design*

Small projects may not have complete information about design when the complexity analysis is performed. The following correction factors should be used to adjust for incomplete design:

CFID =4	PFD drawings only
CFID = 2.5	PFD drawings and P&ID's 20-30% complete
CFID = 1.5	PFD drawings and P&ID's 31-60% complete
CFID = 1.0	PFD drawings and P&ID's 60-100% complete

Note: All P&ID drawings should comply with ANSI/ISA Standard 5.1 (2009) [4]

2.4 *Change Complexity Factor*

To determine the Change Complexity Factor (CCF) for the project change we use the following equation:

$$CCF = HMF * (ET + IT + LT + RD) * CFID \qquad \text{[Eq. 1]}$$

IT: Instrument Total (Non Control)

LT: Total # of Control Loops

ET: Total # of Equipment Pieces (Pumps, Columns, Tanks, etc)

RD: Total # of Relief Devices

HMF: Hazard Material Factor (See Table 1)

CFID: Correction Factor Incomplete Design

2.5 *Change Complexity Factor Score*

0 to 25	one PHA
25 to 50	two PHAs
51 and above	three PHAs

See Table 2 for type of PHA required for each CCF.

Table 2 - Project PHA Type Required

Change Complexity Factor Score	**# of PHAs Required**	**Type of PHA required**
0 to 25	1 PHA to be Done	Detailed PHA only
26 to 50	2 PHAs to be Done	FEED and Detailed Design PHAs
51 and above	3 PHAs to be Done	FEED, Early in Detailed Design and Before start of Construction PHA 's

3. Conclusion

Using Change Complexity Analysis it is possible to determine the complexity of projects. This technique will allow Process Hazard Analysis Study Team Leader and Project Managers to determine the minimum number of PHAs to be done for various complexity projects. The technique will effectively utilize project resources and provide needed input to project teams designing a safe process.

References

[1] NFPA 49: Hazardous Chemical Data, National Fire Protection Association. Technical Committee on Hazardous Chemicals, American National Standards Institute 1994

[2] NFPA 704: Standard System for the Identification of the Hazards of Materials for Emergency Response, National Fire Protection Association. Technical Committee on Classification and Properties of Hazardous Chemical Data, American National Standards Institute 2012

[3] Dow's Fire & Explosion Index Hazard Classification Guide, 7th Edition ,American Institute of Chemical Engineers (AIChE) 1994

[4] ANSI/ISA 5.1: Instrument Symbols and Identification, International Society of Automation, American National Standards Institute 2009

Additional references not cited:

[5] Robert Johnson, Steven W. Rudy, Stephen D. Unwin, "Essential Practices for Managing Chemical Reactivity Hazards," John Wiley & Sons 2010

"People" Means Leadership, NOT Simply Mean People -- 4 New Dimensions of Process Safety Competency

Steven (Long) Zhang
ConocoPhillips China
Steven.Zhang@conocophillips.com
E4B4, East Wing Binhai Finance Zone, TEDA, P. R. China

Prepared for Presentation at
American Institute of Chemical Engineers
2014 Spring Meeting
10th Global Congress on Process Safety
New Orleans, LA
March 30 – April 2, 2014

UNPUBLISHED

"People" Means Leadership, NOT Simply Mean People
-- 4 New Dimensions of Process Safety Competency

Steven (Long) Zhang
ConocoPhillips China
Steven.Zhang@conocophillips.com
E4B4, East Wing Binhai Finance Zone, TEDA, P. R. China

"Probably the most difficult point to make and make clearly is that not only is culture imposed upon man but it *is* man in a greatly expanded sense."

"By broadening our conception of the forces that make up and control our lives, average people can never again be completely caught in the grip of patterned behavior of which they have no awareness."

-- Anthropologist Edward T. Hall *The Silent Language*

Keywords: Leadership, Process Safety Competency, Plant Procedure People, Core Value

Abstract

Within this paper, some attempts to rewrite the CCPS RBPS element *Process Safety Competency* with 4 new dimensions and to characterize the newly upgraded Pillar *Leadership* and new Pillar *Weather/climate* are made so that a *Reflection Model of RBPS Egg* could be crafted, to shed light on the current dead small cycle where the process industry and beyond still might have been trapped till very recently, and to contribute to prospect a future potential alive big cycle which is in fact a real start point of rationale for sustainable cultural change. And furthermore, the implications of the *Reflection Model* are broadened in light both of new RBPS Socio-Technical Systems and of a new map of process safety culture, so that a few of important recommendations regarding sustainable culture change and other important issues rationalized in a strictly anthropological sense.

1. Introduction

Within the paper presented at 9th GCPS, the author identified 2 new elements/ universes, i.e. RBPS Core Value and Process Safety Cultural Communication (PSCC), for CCPS RBPS Temple or 4 Pillars and recommended an upgraded model, i.e. RBPS Egg. [1]

And at 1st CCPS APCPS, some benchmarking study of RBPS Egg against *The Art of War* was shared: the element *Process Safety Culture* was absorbed into PSCC; 2 underlying potential pillars *Leadership* & *Weather/climate* were identified from within, and instead of, the original CCPS 1st Pillar *Commit to Process Safety*; the author arrived at a RBPS version of *The Art of War,* a new architecture of 7 mega elements (2 universes + 5 pillars), except with a main gap, i.e. lack of correct *Leadership* or the "people" among 3 "P" not meaning people but leadership in fact. The author recommends that the element *Process Safety Competency* should be revised with the 4 new dimensions. [2]

<table>
<tr><td colspan="2">RBPS Egg</td><td>The Art of War</td><td>Philosophy of Process Safety</td><td></td></tr>
<tr><td colspan="4">RBPS Core Value - Tao - RBPS Worldview</td><td>Worldview</td></tr>
<tr><td rowspan="4">Commit to Process Safety</td><td>Process Safety Competency</td><td rowspan="2">Leadership</td><td rowspan="2">People</td><td rowspan="7">Process Safety Cultural Communication (Process Safety Culture) RBPS Methodology</td></tr>
<tr><td>Workforce Involvement</td></tr>
<tr><td>Compliance with Standards</td><td rowspan="2">Weather/Climate</td><td rowspan="3">Plant</td></tr>
<tr><td>Stakeholder Outreach</td></tr>
<tr><td>Understand Hazards & Risk</td><td></td><td>Terrain</td></tr>
<tr><td>Manage Risk</td><td></td><td>Discipline</td><td>Procedure</td></tr>
<tr><td>Learn from Experience</td><td></td><td>Espionage</td><td>Engineering & Humanities</td></tr>
</table>

Upon the previous work, within this paper, the element *Process Safety Competency* is rewritten with the 4 new dimensions so that the new Pillar *Leadership* could be upgraded as strong as the Sun Tzu's one and RBPS Egg becomes real RBPS version of *The Art of War*. The other new Pillar *Weather/climate* and its relationships with *Leadership* and other mega element/ pillars/ universes would be also examined. And the Reflection Model not sheds light on the current dead small cycle, i.e. failure to *Learn from Experience* or *Espionage,* but also is likely to prospect a potential alive big cycle, upon both of which not *Socio-Technical Systems* for RBPS Egg would be characterized but also *a map of Process Safety Culture* presented to process industry and beyond.

2. "People" Means Leadership
— 4 New Dimensions of Process Safety Competency

2.1 "People" Means Leadership in Internal Communication – 4 New Dimensions of Process Safety Competency

One of the recommendations the author made of benchmarking RBPS Egg against *The Art of War* is that *"People" means Leadership, not simply mean people*, which might have broad implications, especially to the sphere of original CCPS 1st Pillar *Commit to Process Safety* or the current Leadership (Process Safety Competency + Workforce Involvement) & Weather/Climate (Compliance with Standards + Stakeholder Outreach).

Leadership in Process Safety Cultural Communication - Within the paper presented at 9th GCPS, the author went through some of the greatest lessons learned and experience within the cultural traditions of ancient East and West and arrived at conclusion with those *4 ethical and highest strategic imperatives* to process safety practitioners, board of directors, executive management, operations manager, and functions of strategy, legal and HR [3], among which some new concepts, which the author borrowed from Historian Arnold J. Toynbee, have also emerged as "creative minority" and "dominant minority" for process safety leadership. Or more correctly, as the author defined, as the ultimate purpose of PSCC is to convey the message of RBPS Core Value, the correct process safety leadership should be a joint leadership of dominant minority & creative minority in PSCC, allied with corporate functions of strategy, legal & HR.

A Shorter Chain of Command & Inherently Safer Operational Discipline - And as the author has commented, RBPS Core Value and PSCC mandate a shorter chain of command directly and primarily from RBPS Core Value rather than from superiors. In another word, the ultimate purpose of the superiors is to support to cultivate and facilitate this new type and shorter chain of command of their subordinates, which is in fact to mentor their subordinates (or their new colleagues) of conveying the message of RBPS Core Value to the latter themselves and others. And some of the subordinates may be likely to become creative minority and ceaselessly remind, communicate or challenge upward the actual status of this desired shorter chain of command, even to the highest level when necessary, till all the superiors' behavior universe sends same messages as tamed as from their verbal universe. And the matches of dominant minority and creative minority might be called inherently safer PSCC leadership, which is in fact in the right direction along PSCC leadership enlightenment roadmap, of which the destination would probably be the *Brotherhood of Bian Que or called "powerful powerless" dominant minority with the "powerless powerful" creative minority*[4].

And within this shorter chain of command and as illustrated within the scheme of the author formally defined Philosophy of Process Safety [5], we also could recognize that RBPS Core Value dictates PSCC, and PSCC dictates PSCC leadership (People), Plant and Procedure. In another word, the role the 2 universes of RBPS Core Value and PSCC jointly play within RBPS Egg is actually of inherently safer operational discipline (ISOD) in a broader sense. And only through RBPS Core Value and PSCC could correct decision making and correct behavior determining be reliably "manufactured" and guiding, organizing and accumulating with other non-ethical building raw materials into the People, Plant & Procedure.

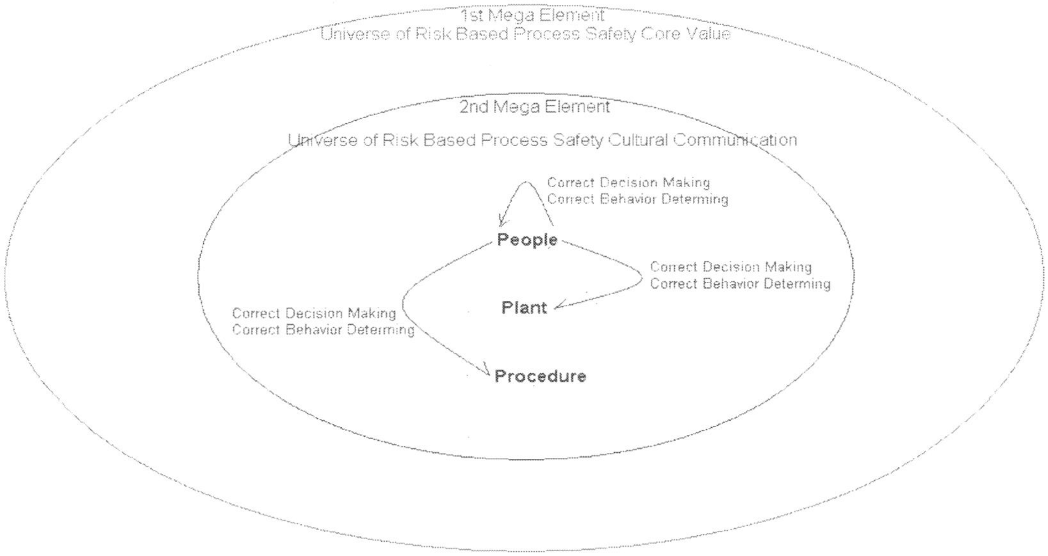

4 Ethical Imperatives as PSCC Contract [6] - In retrospect, we might currently find some content within the Baker Report's Recommendation #1 and its commentary not much correct as before. In author's opinion, the Board of Directors, executive management and other members of the corporate management might hardly be able to provide effective leadership or communication leadership until RBPS Core Value could be resiliently documented throughout their collective verbal and behavior universe in light of the new game code or PSCC contract, i.e. those *4 ethical & highest strategic imperatives*. In another word, the PSCC partner "creative minority" among the workforce may get empowered from RBPS Core Value in personal commitment and professional ethics [7] to challenge upward, even to highest level of the organization, the pitfalls and truthfulness of the verbal and behavioral universes of the "dominant minority", who expectedly to be prepared psychologically and rationally for getting empowered to be "powerless", i.e. to obey RBPS Core Value through completely giving up their personal power and authority in deep integrity throughout their verbal and behavioral universes. So some content of the recommendation #1 and its commentary might be revised or added as below [8]:

Process Safety Leadership in fact means PSCC Leadership, not dominant minority but also creative minority;
Dominant minority's leadership could never be effective except to cooperate with creative minority;
4 ethical and highest strategic imperatives are the new game code or PSCC contract;
Creative minority gets empowered to upward communicate and dominant minority empowered to obey;
Dominant minority set the "tone at the top" and RBPS Core Value set above the top of the tone;
Dominant minority and creative minority are the inherently safer concepts of PSCC leadership;
......

4 New Dimensions of Process Safety Competency or PSCC Competency - Along with the new game code or PSCC contract and the new defined game players, let's examine the required competency of the game players. Sun Tzu defined *Leadership* (Roger T. Ames put it here as *command*) as a matter of wisdom, integrity, humanity, courage and discipline, all of which seems functioning as the military process safety competency of the organization to pursue and convey the message of the Tao. In context of the globalization and global operations, however, Sun Tzu's *Leadership* may be upgraded with the following 4 new dimensions of process safety competency of efficiently facing the challenges of workforce increasing diversity and operational discipline's higher cultural and communication sensitivity [9]:

Competency in embracing RBPS Core Value;
Competency in upward Process Safety Cultural Communication;
Competency in downward Process Safety Cultural Communication; and
Competency in process safety leadership planning, assessment and intervention

Theoretically, these 4 new dimensions of process safety competency or PSCC Competency are very correct but in fact less measureable as these 4 ones inherently relate to inner ourselves, the "natural" resources of the professional ethics and personal commitment across the organization, the former of which seems functioning as soil as well as sunshine and the latter seeds.

"People" means PSCC Leadership & Internal Communication – Based on above analysis, some following conclusions might be arrived: **Process Safety Leadership is in fact the leadership in Process Safety Cultural Communication; those 4 new dimensions of Process Safety Competency are actually of Process Safety Cultural Communication; RBPS Core Value and Process Safety Cultural Communication mandate a shorter chain of command which substantively functions as an Inherent Safer Operational Discipline (ISOD) in a broader sense; and those 4 ethical & highest strategic imperatives make up Process Safety Cultural Communication Contract.**

2.2 Stakeholder Outreach – Extension of Internal Communication

Another important recommendation the author made of benchmarking RBPS Egg against *The Art of War* should be that "Plant" as small Nature consists of Weather/Climate and Terrain, which seems slightly abstract but is in fact meaningful that our understanding of the "Plant" as well as of "Procedure" have been elevated to a philosophical level, where we might not be able to disagree with currently popular sayings of the contemporary philosophers of technology that our "Plant" is an social artifact rather than pure engineering product[10], i.e. both Weather/Climate and Terrain are of an artifact and social process influenced product. So is "Procedure".

Without ISOD, What "People, Plant & Procedure" would look like? – As the scheme within section 2.1 and the form in section 1 jointly illustrate, incorrect Leadership in PSCC would reliably bring about incorrect decision making and incorrect behavior determining across internal communication, external communication, terrain, discipline; and similarly maintaining blindness of humanities espionage[11]. In other word, incorrect Leadership dictates incorrect "People, Plant & Procedure", whereas correct Leadership otherwise with regards "People" (internal communication), Weather/Climate, Terrain, Discipline as well as Espionage.

External PSCC - Weather/Climate Allied with Terrain & Discipline

Correct Leadership in PSCC dictates correct "People, Plant & Procedure" through correct decision making and correct behavior determining. And the correct Terrain and Discipline consistently convey the message of RBPS Core Value to Weather/Climate that *there is no power, authority or governance on process safety except on behalf on Process Safety Expertise.* Without receiving correct message from Terrain & Discipline, however, Weather/Climate only could do external communication but not PSCC. And put another way, Weather/Climate only functions as verbal communication, whereas the performance of Terrain and Discipline as the behavior universe, which might convey another message that is very different from the verbal one. In this sense, incorrect Leadership dictates incorrect decision making and behavior determining of Weather/Climate so that Weather/Climate could not be honest enough and eventually failed to convey the message of RBPS Core Value to the public through its verbal universe as well as the behavior universe of Terrain & Discipline.

And simultaneously, Weather/Climate conveys incorrect message back to Terrain and Discipline so that Terrain and Discipline are permanently exposed within the sea of incorrect messages. Marshall McLuhan, one of the greatest thinkers of 20 century, has a popular saying: the medium is the message [12]. The allied media of incorrect Leadership & incorrect Weather/Climate etc completely surrounds, and conveys incorrect messages to, the Terrain & Discipline who inevitably get corrupted and witnessed with one another within the sea of incorrect messages. And the day to day mild Weather/Climate public communication universe would suddenly become fraudulence when the actual message of the wrong behavior of the Terrain and Discipline is perceived by the community. And for the fact that the organization successfully conveyed the message of the wrong core value through the bad behavior of the operations or performance of Terrain and Discipline, the community especially the regulator would "anthropologically" convey message of the community's correct core value to the "unethical citizen" through force measures, i.e. extreme Weather/Climate suddenly arrives.

So effective management of Weather/Climate should ultimately counted on the inherently safer Leadership (correct Leadership in PSCC) and its media the sea of correct messages rather than on Weather/Climate expertise in isolation such as communication initiatives, tools and skills. And interestingly, the author's opinion seems having resonance with some of the catastrophic incident warning signs of CCPS regarding leadership poor behavior on interrelationships between Leadership (internal communication), Weather/Climate (external communication), Terrain and Discipline [13]. Either way, Weather/Climate really does have something critical to do with the status of Leadership's actual belief in RBPS Core Value and of their trust and rationale in PSCC.

In context of those 4 ethical and highest strategic imperatives, the author recommends: external public communication functions should involve into and learn from the internal communication or internal PSCC; and dominant minority should consult with creative minority for the correct decision making and behavior determining throughout the activity universe of Weather/Climate, i.e. Weather/ Climate, allied with Terrain and Discipline, shall be of PSCC. And the failure in PSCC throughout Weather/Climate is the extension of the failure of Leadership (internal communication). Both of them often triggers extreme weather/climate, i.e. the community might have to do their "tribal" PSCC to the bad "citizens" to force them to be ethic enough again. So in the author's opinion, **the ultimate purpose of stakeholder outreach should be to lead the community to the true enlightenment of process safety, i.e. to embrace, and convey the message of, RBPS Core Value, rather than only tactically maintaining stable relationships with stakeholders.**

3. Socio-Technical Systems of RBPS Egg -- Leadership Media Universe

3.1 RBPS Reflection Model – Dead Small Cycle & Alive Big Cycle

Sun Tzu's Legacy of Espionage – The objects Sun Tzu's Espionage is made to are: Tao (RBPS Core Value), Weather/Climate, Terrain, Leadership and Discipline.[14] And the author added another one i.e. PSCC, which is in fact an advanced form of process safety culture and whose ultimate purpose is to convey the message of RBPS Core Value. So totally, we have 6 strategic assessment criteria/facets, along with Espionage, within the RBPS Egg.

As per Traditional Chinese Medical Methodology, the 6 criteria/facets might be categorized into 2 groups: the group of symptoms of Weather/Climate, Terrain and Discipline, and the group of roots of RBPS Core Value, PSCC and Leadership.

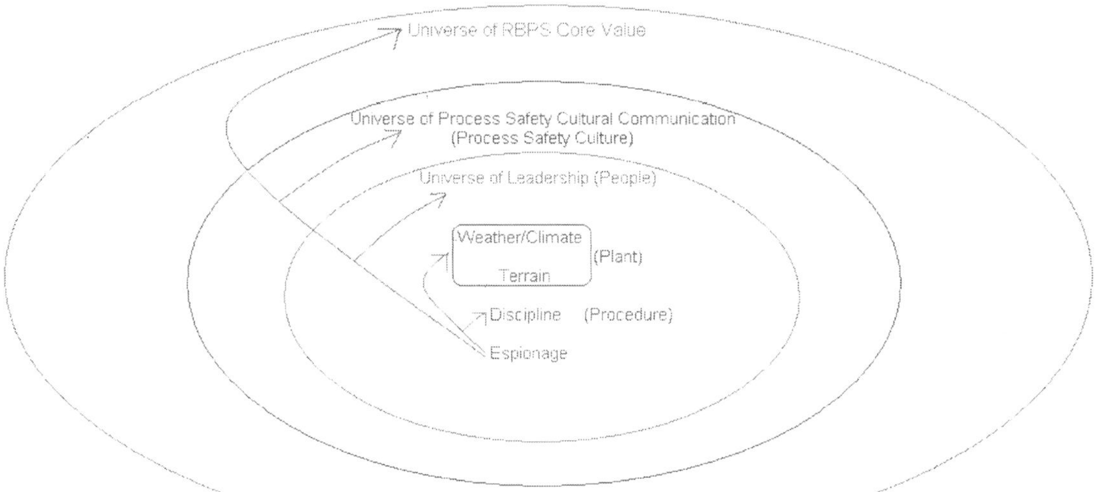

Symptoms (Syndromes) - Small Dead Cycle Locked within Universe of Leadership - Whenever creative minority is absent, the dominant minority might probably become 3rd universe which, however, functions as a prison and isolates whatever within from the other 2 universes of RBPS Core Value and PSCC as well as from Leadership itself. And in this manner only could Weather/Climate, Terrain and Discipline be reflected upon in isolation at the level of symptoms so that Weather/Climate is probably not to be deemed as a contributing factor to the catastrophic accidents, but rather, as a post accident remediation measures for the damaged public image and reputation. In another words, in light of the Reflection Small Dead Cycle, Weather/Climate has nothing to do with Espionage and hence its 4 sub-elements i.e. Incident Investigation, Measurements & Metrics, Auditing and Management Review.[15] And the author captures within the form below the inter-relationships mentioned above between Espionage and the symptoms group of Weather/Climate, Terrain and Discipline.

	Espionage (Learn from Experience) – Reflection Small Dead Cycle			
	Incident Investigation	**Measurements & Metrics**	**Auditing**	**Management Review**
Weather/ Climate	If Weather/Climate is examined as a contributing factor to incident? NO	If Weather/Climate is an item within process safety KPI? NO	If Weather/Climate is an item within auditing? NO	If Weather/Climate is an item within management review? NO
Terrain	If Terrain is examined as a contributing factor to incident? YES	If Terrain is an item within process safety KPI? YES	If Terrain is an item within auditing? YES	If Terrain is an item within management review? YES
Discipline	If Discipline is examined as a contributing factor to incident? YES	If Discipline is an item within process safety KPI? YES	If Discipline is an item within auditing? YES	If Discipline is an item within management review? YES

From the form above, we might recognize that, in context of lack of big alive cycle, the rationale only could be to solely count on the improvement of the current practices and development of new practices within the scope of Terrain and Discipline. In author's opinion, however, Terrain and Discipline function together as "a car", which might get increasingly complicated and complex, whatever extent. But we need to ask other questions: How about the "driver"? Does the driver have a valid "driving license"? Has the driver passed an "alcohol test"?

Roots (Diseases) - Big Alive Cycle as Espionage through 3 Universes - And whenever creative minority is active enough in leading PSCC, the wall of the prison of dominant minority would collapse and the organization would succeed in Espionage into the 3 universes as well as subsequently into Weather/Climate, Terrain and Discipline. And due to the nature of Weather/Climate, it might immigrate from small dead cycle to big alive cycle so that Terrain and discipline make up the symptoms group "the car" and RBPS Core Value, PSCC, Leadership and Weather/Climate come together as roots group "the driver". And in this scenario, as captured within the form below, Weather/Climate does have something important to do with Espionage and hence with its 4 sub-elements, and so do RBPS Core Value, PSCC and Leadership. So the rationale becomes of classic Traditional Chinese Medical Methodology, i.e. "both symptoms and roots should be treated, and symptoms only could be eliminated through treatment of roots." In another word, whenever a car knocks upon some people, we need to ask not "what's wrong with the car?", but also "what's wrong with the driver?"

	Espionage (Learn from Experience) – Reflection Big Alive Cycle			
	Incident Investigation	**Measurements & Metrics**	**Auditing**	**Management Review**
RBPS C.V.	If RBPS Core Value is examined as a contributing factor to incident? YES	If RBPS Core Value is an item within process safety KPI? YES	If RBPS Core Value is an item within auditing? YES	If RBPS Core Value is an item within management review? YES
P.S.C.C. *	If PSCC is examined as a contributing factor to incident? YES	If PSCC is an item within process safety KPI? YES	If PSCC is an item within auditing? YES	If PSCC is an item within management review? YES
Leadership	If Leadership is examined as a contributing factor to incident? YES	If Leadership is an item within process safety KPI? YES	If Leadership is an item within auditing? YES	If Leadership is an item within management review? YES
Weather/Climate	If Weather/Climate is examined as a contributing factor to incident? YES	If Weather/Climate is an item within process safety KPI? YES	If Weather/Climate is an item within auditing? YES	If Weather/Climate is an item within management review? YES

* P.S.C.C. is the author's addition in light of anthropology and intercultural communication

So the author might say to the process industry and beyond that Sun Tzu's legacy of Espionage has been rediscovered completely through the Reflection Model of RBPS Egg, within which there are 2 symptoms i.e. Terrain and Discipline and 4 roots i.e. RBPS Core value, PSCC, Leadership and Weather/Climate; or otherwise, **there are one symptom with 2 dimensions of Terrain and Discipline and one root with 4 dimensions of RBPS Core value, PSCC, Leadership and Weather/Climate. And Espionage might function as of Traditional Chinese Medical Diagnosis.**

3.2 RBPS Socio-Technical Systems – Leadership Media Universe

Hinted by the topic or perspective of socio-technical systems' failure for catastrophic accidents[16], the author crafts the following RBPS Socio-Technical systems based on RBPS Egg of 7 mega elements (22 sub-elements). And due to the critical role, as the author analyzed above, Leadership the 3rd Universe would play in isolating whatever within from the rest 2 universes as well as from Leadership itself and in closing the big cycle and keeping dead small cycle "work", the author prefers to call the Socio-Technical Systems a Leadership Media Universe. In another word, different Leaderships, or more exactly different Leadership in PSCC, inherently embrace different approaches i.e. either "sole symptom approach" or "symptom-root double approaches". Dominant minority usually favorites on the former, whereas creative minority often holdfasts the latter. And accordingly, they might have polarizing behavioral patterns toward the actual relationships among RBPS Core value, PSCC, Leadership and

Weather/Climate. As we all know, however, RBPS Core Value dictates PSCC; PSCC dictates leadership (dominant minority and creative minority); and leadership dictates Weather/Climate as well as Terrain, Discipline & Espionage.

And obviously, the "sole symptom approach" concerns all the interactions among these 3 mega elements of Espionage, Terrain and Discipline. And all these interactions in fact might have been or are being effectively studied, implemented and leveraged to some extent within and outside the corporate, e.g. with regards Centers of Excellence in asset integrity and in other sub-elements across these 3 mega elements.[17] By and large, the "symptom approach" could be called Process Safety Techno Cultural Communication (PSTCC). And also the "symptom-root double approaches" have actually been defined long earlier by the author, and its name is Process Safety Cultural Communication (PSCC), which consists not only of PSTCC but also of Process Safety Socio Cultural Communication (PSSCC).

Unlike PSTCC which could be acted upon and leveraged effectively to some extent within specific area of media such as Centers of Excellence within corporate, the media for PSSCC is still emerging and to evolve. As the author quoted the popular saying of Marshall McLuhan: "the medium is the message", however, in light of those 4 ethical and highest strategic imperatives, dominant minority such as board of directors, executive management, along with functions of strategy, legal & HR etc, shall explicitly determine what the ultimate purpose /message of their PSSCC media is, or if they are determined to commit faithfully to the PSCC contract.

And it is the ultimate purpose /message of the PSSCC media (or PSCC media) that the board of directors, executive management either explicitly or implicitly testify to, through their verbal and behavior universe organizationally and psychologically aligned with functions of strategy, legal & HR etc, would inherently correlate to the future failure or safe of the RBPS Socio-Technical Systems – the Leadership Media Universe. Or the author might borrow from the concept of the name of McLuhan's masterpiece *Understanding Media: The Extensions of Man* and make saying like this: Leadership Media Universe is to be pursued as the "extensions" of not only dominant minority but creative minority. And in the author's opinion, Leadership Media Universe is in fact an acronym for Process Safety Culture. So the author goes further saying: **Process Safety Culture is not only the "extensions" of dominant minority but also of creative minority. And sure, "performance contract system and other incentive-based compensation plans", probably even the functions of strategy, legal, HR as well as of external communication, remain critical part of the extensions of dominant minority, whereas "decisions about corporate level initiatives, operations, financial performance, resource allocation, capital project, personnel changes etc" the main themes of the "media", either of dominant minority alone or of both.**[18]

And we could say: **the "media" of the 7 mega elements is the "message" of the actual core value, either of RBPS or not at all. And the sustainable change of Leadership Media Universe only initiates when the people at the highest level, probably all board members & executive management, one by one, testify verbally, and subsequently through behavioral universe, to the RBPS Core Value and those 4 ethical and highest strategic imperatives are faithfully committed by all the dominant minority, creative minority and functions of strategy, legal & HR etc.**

<table>
<tr><th colspan="15">RBPS Socio-Technical Systems - Leadership Media Universe</th></tr>
<tr><td rowspan="3">22 Sub-elements</td><td colspan="11">PS Socio Cultural Communication (PSSCC)
Sub-Socio System (7 Elements + Espionage 4 Elements)</td><td colspan="3"></td></tr>
<tr><td colspan="7"></td><td colspan="7">PS Techno Cultural Communication (PSTCC)
(Espionage 4 Elements + 11 Elements) Sub-Technical System</td></tr>
<tr><td>RBPS Core Value</td><td>Process Safety Cultural Communication</td><td>(Process Safety Culture)</td><td>Process Safety Competency with 4 new dimensions</td><td>Workforce Involvement</td><td>Compliance with Standards</td><td>Stakeholder Outreach</td><td>Incident Investigation</td><td>Measurement & Metrics</td><td>Auditing</td><td>Management Review & Continuous Improvement</td><td>Process Knowledge Management</td><td>Hazard identification & Risk Analysis</td><td>9 Sub-elements (in light of organizational effectiveness, inherently safer chain of command & logistical support)</td></tr>
<tr><td>7 Mega Elements</td><td>Tao/ PS Worldview</td><td colspan="2">P.S. Methodology</td><td colspan="2">Leadership</td><td colspan="2">Weather/ Climate</td><td colspan="4">Espionage (RBPS 4th Pillar Learn from Experience)</td><td colspan="2">Terrain (RBPS 2nd Pillar Understand Hazards & Risk)</td><td>Discipline (RBPS 3rd Pillar Manage Risk)</td></tr>
<tr><td>CTM*</td><td colspan="7">One Root (4 dimensions)</td><td colspan="4">Traditional Chinese Medical Diagnosis</td><td colspan="3">One Symptom (2 dimensions)</td></tr>
<tr><td>AIC**</td><td colspan="7">Formal Culture</td><td colspan="4">Informal Culture</td><td colspan="3">Technical Culture</td></tr>
</table>

CTM*: Chinese Traditional Medicine
AIC**: Anthropology & Intercultural Communication

3.3 Desert & Oasis in Ourselves – A Map of Process Safety Culture

In order to get more rationale, based on the newly updated pillar Leadership and borrowing anthropologist Edward T. Hall's approach, the author draws a map of process safety culture or called a map of process safety cultural communication, upon which we may be able to estimate how much infra cultural activity of process safety culture still remain desert and how much has already become oasis.

Beside Sun Tzu called "5 strategic assessment criteria" and Clausewitz put it "elements of strategy"[19], the author prefers to treat these 7 mega elements as 7 Primary Message Systems as anthropologist Edward T. Hall did for the culture study so that a similar two-dimensional grid could be constructed as below [20]. And in this manner, we could see there are totally possible 49 infra (strategic) cultural activities resulting from the various combination of the 7 mega elements. And the author defines each infra (strategic) cultural activity as the critical verbal statement & behaviors related to the relationship between mega element A and mega element B. And in special cases, i.e. these 7 ones in the diagonal from upper left to lower right are in fact the critical verbal statement & behaviors related to the definition of the mega element concerned. And for the purpose of simplicity, all the 21 infra (strategic) cultural activities above the diagonal which are exactly the same with the activities collection below the diagonal, along with these 7 ones in the diagonal, would make up the map of process safety culture which the author likes more to call the map of process safety cultural communication.

	RBPS Core Value 0	P.S.C.C. (Process Safety Culture) 1	Leadership 2	Weather/ Climate 3	Espionage 4	Terrain 5	Discipline 6
RBPS Core Value 0	00	01	02	03	04	05	06
P.S.C.C. (Process Safety Culture) 1	10	11	12	13	14	15	16
Leadership 2	20	21	Leadership with 4 new dimensions 22	23	24	25	26
Weather/ Climate 3	30	31	32	33	34	35	36
Espionage 4	40	41	42	43	44	45	46
Terrain 5	50	51	52	53	54	55	56
Discipline 6	60	61	62	63	64	65	66

A Map of Process Safety Cultural Communication (PSCC)

As we analyzed above, whenever creative minority is absent in PSCC leadership, the leadership would probably become a prison isolating whatever within from the 2 universes of RBPS Core Value and PSCC as well as from Leadership itself so that all the 18 infra (strategic) cultural activities in yellow background would remain dry desert area of the process safety culture. And similarly as we discussed earlier, incorrect leadership also dictates strategic failure bonded Weather/Climate so that those 4 infra (strategic) cultural activities in blue background (33, 34, 35 & 36) remain dry desert of the process safety culture as well. And furthermore, we found the Espionage is solely restricted to the One Symptom with 2 dimensions of Terrain and Discipline. So due to lack of Espionage to the One Root with 4 dimensions, the actual infra (strategic) cultural activity (44) is in fact a half desert of the health area unit. **Totally and approximately, the percentage of the desert area of the current actual industry average process safety culture should be: (18+4+0.5)/28 = 80.4%, whereas the actual oasis 19.6%.**

These statistics of the actual situation of our process safety culture seems a shock and hard to be accepted. Anthropologist Edward T. Hall not only already mentioned this kind of shock and resistance as saying "... people as cultural being are bound by hidden rules and not the masters of their fate ..." within his masterpiece *The Silent Language,* but also reminded that people would become really bound if "they remain ignorant of the nature of the hidden pathways culture provided for them" [21]. And similarly, for our process safety culture, there also have been hidden pathways i.e. RBPS Core Value, PSCC and 4 new dimensions of process safety competency, all of which we could not afford to remain ignorant anymore.

3.4 Governance on the Root – Coaching Fishing Better Than Giving Fish

As anthropologist Edward T. Hall defined, culture has three levels, i.e. formal, informal and technical. Formal culture is always learned through mistake and correction, with a clear right-wrong character. Technical one is also learned through mistake and correction, but being offered with reasons. And informal culture, however, is learned without knowledge of occurrence of the learning process at all.[22] The author innovatively matches the 3 levels of culture with One Root, Diagnosis & One Symptom so that it become manifest that those 4 ethical & highest strategic imperatives would supposedly function as the core principles of the formal culture (One Root with 4 dimensions) and shall be faithfully committed by dominant minority, creative minority and functions of strategy, legal & HR etc. And the rest whole organization psychologically holds the same intentionality as well.

In the author's former opinion, PSCC functions as the impulses from the heart RBPS Core Value. And the impulses consistently transport the nourishment of the highest design principle i.e. RBPS Core Value across the tree of process safety culture so that RBPS Core value solely chairs and dictates correct decision making & correct behavior determining which would guide, organize, and accumulate with, other non-ethical raw materials building and maintaining the rest mega elements i.e. leadership, Weather/Climate, Terrain, Discipline and even Espionage. Put another way, without PSCC, some enemy value or ideology may share the exclusiveness of chairing decision making and behavior determining with RBPS Core Value so that incorrect decision making and incorrect behavior determining are produced and accumulated consistently within last 5 mega elements till, and ever after, the collapse of the "tomb" or what within the 2 universes.[23]

There was an ancient Chinese proverb saying: coaching fishing is better than giving fish. And the author would re-examine in light of the fishing-fish relationship the two accidents CCPS formally used as examples in the 2 elements: asset integrity and readiness.

a. The accident investigation of DC-10 crush on May 25, 1979 concluded that "...the original replacement procedure calls for removal of the engine prior to the removal of engine pylon. To save time and money, the airline instructed its mechanics to remove" them together ...[24]

In the author's opinion, giving up upward PSCC equates, even worse than, giving up fishing the sole livelihood, being bound to starve to death, sooner or later.

b. That major leak of flammable gas and the subsequent fire did occur, and we found that "the startup checks for this equipment had not been performed as a part of overall turnaround , because this particular piece of equipment was not go to be used immediately."[25] The author challenges the logic behind that "because", which was probably a natural mindset and patterned behavior after giving up fishing the sole livelihood or even not attempted to it at all. During that moment of startup check, the accident might have been prevented should the organization concerned have been given with some specific outsourced recommendations toward improvement or reconstruction of the organizational effectiveness and the logistical structure. But could being given with some fish be strategically meaningful for the life of the fishing man who has given up permanently his sole livelihood?

		Sun Tzu's 3 Aspects of Discipline			**RBPS Core Value**	**Process Safety Cultural Communication (PSCC)**	**Leadership**	**Weather / Climate**
		Organizational Effectiveness	A Chain of Command	A Structure for Logistical Support	**Formal Culture** (4 ethical imperatives as Shorter Chain of Command)			
Terrain*	Knowledge	Of social construction in context of RBPS Core Value & PSCC (RBPS Core Value dictates correct decision making and correct behavior determining which reliably accumulates within the construction process of Terrain & Discipline)	To update to ***Shorter Chain of Command* [26]**, *as illustrated at right,* primarily and Directly from Tao (RBPS Core Value), rather than from superiors; *4 Ethical & Highest Strategic Imperatives* as PSCC contract	Of social construction in context of RBPS Core Value & PSCC (RBPS Core Value dictates correct decision making and correct behavior determining which reliably accumulates within the construction process of Terrain & Discipline)	Learn, understand, embrace, testify and communicate of RBPS Core Value through verbal and behavioral universes.	Learn, understand and communicate that the ultimate purpose of PSCC is to convey the message of RBPS Core Value; what are upward PSCC & downward PSCC; those *4 ethical and highest strategic imperatives*	Learn, understand & communicate of *creative minority, dominant minority; Bian Que-Duke Cai, Moses-Pharaoh, Sheba Queen, Solomon and brotherhood of Bia Que;* [27]	Weather/Climate consistent honestly with Terrain & Discipline; ultimately to lead stakeholders and community to the enlightenment of RBPS Core Value
	Risk							
Discipline (Manage Risk) - 9 Sub-Elements	Procedure							
	safe work							
	asset integrity							
	Contractors							
	Training							
	MOC							
	Readiness							
	Operations							
	Emergency							

*: Sun Tzu's 3 aspects, which were originally specific to Sun Tzu's Discipline, are extended by the author to Terrain.

3.5 Recommendations for Sustainable Cultural Change

To creative minority, dominant minority, functions of strategy, legal and HR etc, the author makes the following recommendations:

a. PSCC is the advanced form of Process Safety Culture, and Process Safety Culture's future is PSCC. And the Map of Process Safety Culture (PSCC) might become a paradigm for Process Safety Culture assessment & sustainable change;

b. The RBPS Reflection Model might prove to be a profound framework for Espionage (incident investigation, metrics, auditing & management review) within which the critical role Leadership (internal communication) would play as the 3rd universe is manifested for both small cycle and big cycle;

c. As RBPS Socio-Technical Systems shows, PSCC consists of both PSTCC (Process Safety Technical cultural Communication) and PSSCC (Process Safety Socio Cultural Communication). And the ultimate purpose of PSCC would be fulfilled mainly through PSSCC. And PSSCC, PSCC and Leadership Media Universe are guided by those 4 ethical & highest strategic imperatives, which shall be faithfully committed by creative minority, dominant minority, and functions of strategy, legal & HR. And Process Safety leadership shall be the leadership in PSCC or PSSCC, and Leadership Media Universe is in fact an acronym for process safety culture;

d. The ultimate purpose of Weather/Climate as part of Leadership Media Universe is to convey the message of RBPS Core Value and lead stakeholders and community to the enlightenment of RBPS Core Value;

e. The health of RBPS Egg, or the actual percentage of desert of the map of process safety culture, is concerned more with messages of actual value transferring among mega elements rather than network and control system. And this concept may be a fundamental challenge to the current popular ones e.g. Bow-Tie & LOPA which inherently have far less relevance with Process Safety Culture or PSCC or RBPS Core Value;

f. Same as culture, process safety culture (Leadership Media Universe) also consists of Formal Culture (One Root with 4 dimensions), Informal Culture (TCM diagnosis) and Technical Culture (One Symptom with 2 dimensions);

g. Couching fishing (PSCC or PSSCC) is better than giving fish (PSTCC). And the true sustainable cultural change only set out the journey with couching fishing (PSCC or PSSCC). PSCC or PSSCC is a growing sphere of communication context, which is governed and dictated by RBPS Core Value as the highest design principle and in which the overall quality of the correct organizational decision making and correct behavior determining could be reliably achieved; and

h. Anthropologist Edward T. Hall reminded us that "the study of institutions and their structures and the study of individuals and their psychological makeup" should be included in study of culture on a higher organizational level. [28] And Hofstede went further with "institutions" within his masterpiece *Cultural Consequences*. [29] The author has to be honest enough to admit that Process Safety Culture is inherently concerned with institutions rather than otherwise. And the content within right side (Formal Culture) of the form within the section 3.4 is in fact a curriculum or competency map for creative minority, dominant minority, functions of strategy, legal and HR etc, to substantively set out a journey of sustainable culture change.

4. Conclusion

At the beginning of this paper, the new mega element/ pillar Leadership was upgraded with those 4 new dimensions and newly characterized as leadership in internal communication or in PSCC. And then the mega element Weather/Climate, its interaction with other mega elements as well as it ultimate purpose was explored and examined.

And based on the analysis and characterization and other previous studies, a RBPS Reflection Model was constructed so that we could see how critical the role Leadership (in internal communication or PSCC) actually plays is in locking organizational rationale within the small cycle or otherwise in unlocking rationale through the big cycle.

And this mechanism was further interpreted through not only newly crafted RBPS Socio-Technical Systems but also an unprecedented map of process safety culture or PSCC.

At the end, several important recommendations toward true sustainable culture change were made.

5. References

[1] Zhang, Long, *An Upgraded Definition of Risk Based Process Safety Core Value and Its Impact in Light of Anthropology & Intercultural Communication on the Future Evolution of Process Safety Culture & Management*, 2013 AIChE Spring Meeting and 9th Global Congress on Process Safety, San Antonio, April 2013

[2] Zhang, Long, *Benchmarking Risk Based Process Safety Egg (RBPS Temple+2 Universes) against The Art of War*, 1st CCPS Asia-Pacific Conference on Process Safety, Qingdao, Sept 2013

[3] Zhang, *An Upgraded..., ibid.*

[4] Zhang, *An Upgraded..., ibid.*

[5] Zhang, *Benchmarking..., ibid.*

[6] Schramm, Wilbur & Porter, William E., *Men, Women, Messages, and Media: Understanding Human Communication*, Peking University Press, 2007, pp.42-45.

[7] Martin, Mike W., *Meaningful Work: Rethinking Professional Ethics*, Oxford University Press, 2000, pp. vii-xiii.

[8] B.P. U.S. Refineries Independent Safety Review Panel, *The Report of the B.P. U.S. Refineries Independent Safety Review Panel*, 2007, p.244.

[9] Zhang, *Benchmarking..., ibid.*

[10] Bucciarelli, Louis L., *An ethnographic perspective on engineering design*, Design Studies, Volume 9, Issue 3, Jul 1988, pp.159-168.

[11] Zhang, *Benchmarking..., ibid.*

[12] McLuhan, Marshall, *Understanding Media: The Extensions of Man*, MIT Press, 1994

[13] Center for Chemical Process Safety (CCPS), *Recognizing Catastrophic Incident Warning Signs in the Process Industries,* John Wiley & Sons, Inc., 2012, pp.40-41.

[14] Sun-Tzu and Ames, Roger T., *The Art of Warfare,* Ballantine Books, 1993, p.103.

[15] Center for Chemical Process Safety (CCPS), *Guidelines for Risk Based Process Safety,* John Wiley & Sons, Inc., 2007, pp.549-647.

[16] Hurst, Nick W., *Risk Assessment: The Human Dimension*, Royal Society of Chemistry, 1998, pp.25-26.

[17] CCPS, *Guidelines for Risk Based Process Safety,* John Wiley & Sons, Inc., 2007, p.319.

[18] B.P. U.S. Refineries Independent Safety Review Panel, *ibid*, p.244.

[19] Clausewitz, Carl von, *On War*, Wordsworth Editions Limited, 1997, pp.149-150.

[20] Hall, Edward T., *The Silent Language,* Anchor Books, 1990, pp.188-194.

[21] *ibid*, p.117.

[22] *ibid*, pp.67-68.

[23] Zhang, *An Upgraded..., ibid.*

[24] CCPS, *Guidelines for Risk Based Process Safety,* John Wiley & Sons, Inc., 2007, p.317.

[25] CCPS, *Guidelines for Risk Based Process Safety,* John Wiley & Sons, Inc., 2007, p.449.

[26] Zhang, *Benchmarking..., ibid.*

[27] Zhang, *An Upgraded..., ibid.*

[28] Hall, *ibid.* p.186.

[29] Hofstede, Geert, *Culture's Consequences: Comparing Values, Behaviors, Institutions & Organizations Across Nations,* Shanghai Foreign Language Education Press, 2008

346632 Evaluating The Need For Depressuring Systems - A Methodology

Abstract:

Evaluating the Need for Depressuring Systems - A Methodology

A thorough Overpressure Protection Philosophy should consider all options for overpressure protection, and should not be limited to relief devices and rupture disks - they are not the only line of defense.

Another effective means of overpressure protection are depressuring systems. These can be used to mitigate against vessel leaks by reducing inventory or leakage rate; or to mitigate against the effects of external fire.

ioMosaic recently conducted a site-wide overpressure protection study for a large LNG production facility. Part of this scope involved assessing equipment items which did not have a depressuring system installed, and determining if a depressuring system was warranted, in addition to the equipment's other overpressure protection.

This paper outlines the methodology used to determine the need for a depressuring system, and also highlights some of the issues to be aware of when designing a depressuring system.

347261 The Capability-Demand Gap In US Refining and Petrochemical Console Operations

Abstract:

The Capability-Demand Gap in US Refining and Petrochemical Console Operations

By George Dzyacky

Director of Training Services, RSI Simcon, Inc.

Of all the professional, trade and administrative disciplines required to operate a refinery or petrochemical facility, managers, accountants, engineers, electricians, pipefitters, machinists, administrative assistants etc.; console operators who unarguably occupy the most safety-critical position in the plant, are the only ones trained and brought all the way to "qualification" by their coworkers.

Who is best suited to train operators?

"We are", say most of the operation training departments in the U.S. RPI , and this has been true for a very long time. The roots of these training practices go back generations, and pre-date the existence of process training companies that now serve our industry. Over the last twenty-five or so years, professional training consultants have developed extensive expertise in the border region between two previously distinct disciplines: unit-specific process operator training and the science of adult learning.

The Unit Trainer

The responsibility for operator training (both field operators and console operators) in the U.S. RPI typically resides at the process unit level. This is chiefly due to the fact that for many decades there simply didn't exist external training companies that could tackle the complexities of equipment specificity and the apparent idiosyncratic training requirements encountered in a typical process unit. Conventional wisdom correctly led the industry to conclude that no one was better positioned to train operators, than other operators who had mastered the operational nuances of these complex units.

Traditionally and out of necessity, Operator Trainers are pulled from the ranks of operators who, it should be noted, are very often hired into operating positions without prior specialized training or educational achievement beyond a high school diploma.

They train what they know

The operators who become Operator Trainers as I once did, and like the operation supervisor who trained me years before that, train well on the things they know. But their body of unit process knowledge is too often encompassed by the instructions they received from their operator trainer, and the knowledge and skills they acquired as a result of their personal motivation and inquisitiveness. As a result Operator Trainers naturally focus on the essential tasks that made them good at what they do, such as:

- Understanding alarms and how to respond to them
- Knowing where to locate and read instrumentation
- Knowing where to locate and how to make rounds on process equipment, vessels and control valves
- Knowing the purpose and safe use of utilities
- Knowing where to locate and when to utilize procedures
- Describing the basic functionality of technologies in the plant

While these training topics are appropriate for most field operator positions, they fall short of the skills required to be a competent console operator. There still exist in our industry today, many sites where console operator training is accomplished by osmosis. Absent any formal curriculum, operators are assigned to parallel or "shadow" on the console with an experienced, qualified operator for weeks or months until they can replicate the key tasks of their trainer.

The problem with training by osmosis

From the Center for Process Safety, American Institute of Chemical Engineers, "Guidelines for Preventing Human Error in Process Safety, on operator training: *"In general, little use is made in the process industry of more sophisticated approaches such as job and task analysis to define the mental and physical skills required for specific types of work, and to tailor the training program accordingly. Instead, informal on-the-job training is common, even in more complex types of work such as control room tasks. Although the necessary skills will eventually be acquired by this process, its inefficiency leads to the need for extended periods of training. In addition, there is the problem that inappropriate or even dangerous practices may become the norm as they are passed from one generation of workers to the next."*

To achieve the level of competency demanded in this job, Console operators in the U.S. RPI must have formal classroom, simulator, and on-the-job training to achieve full capability. Certainly no less rigorous than that of many regulated trades such as, Journeyman Electrician or Machinist. Console operator training must cover topics such as process theory, process control, safety guidelines, safety procedures, math for technicians, instrumentation, pumps, valves, thermodynamics, chemistry, distillation, boilers, furnaces, compressors, reactors, heat exchanges and troubleshooting to name some.

On human error

From the Center for Process Safety for the American Institute of Chemical Engineers, the book "Guidelines for Preventing Human Error in Process Safety," defines the cause of human error in the U.S. RPI this way; "... *a systems perspective is taken, which views error as a natural consequence of a mismatch between human capabilities and demands, and an inappropriate organizational culture*."

In stark contrast

The State of Texas requires a hair stylist to successfully complete 1500 hours, ten months, of training in a state licensed institution. Only after successful completion of state sanctioned written and demonstration tests, does the Texas State Department of Licensing and Regulation issue a Cosmetology Certificate and allow the graduate to cut hair.

But for console operators, in control of process units that have more instrumentation and greater cost than a Boeing 787 Dreamliner, training and testing is not sanctioned by any state or regulatory body in the U.S.. It is not uncommon for console operators to be in complete, sole control of a process with fast dynamics where hazardous chemicals and hydrocarbons are held under hundreds or thousands of pounds of pressure. Further, it is also not uncommon for these operators to have achieved this level of responsibility with no more than a high school diploma and to have been brought to "qualification" by a coworker over a period of several months.

A cursory review of completed investigations by the US Chemical Safety Board, dramatically illustrates just how dangerous and deadly are these processes. The U.S. RPI conducts business in an era of increased governmental, community and legal scrutiny. Yet despite the known dangers to workers and surrounding communities posed by the variety of refining and chemical processes, there is no regulatory oversight or industry-wide standards that mandate minimum curricula or validate the competency of those in control of these plants.

Can the industries do it?

The majority of operator trainers and operator training departments in the U.S. RPI are not prepared to train console operators against a competency framework to appropriate proficiency levels, for reasons of skill, resources and culture. The training interventions (courses) are highly specialized and tend to exceed the expertise and resources of most in-house training departments. The level of training needed to meet the demands of the job is greater in duration and cost than historically presumed appropriate. Lastly, training is not the industry's core business.

What needs to change?

We must no longer accept the outdated culture and practices of operators-training-operators. console operators in refining and petrochemical industries should be required to successfully complete comprehensive curricula, based on a competency framework, targeted to teach technical fundamentals, process fundamentals and unit-specifics.

One approach is to have the methods and standards for competency identification, curricula development and assessments managed through an Industrial-Academic Consortium, coordinated by an industry-recognized trade association such as the AFPM or API.

The training must include formal classroom and simulator training covering the entire range of topics from fundamental concepts to unit-specific emergency response.

After trainees demonstrate proficiency in the fundamentals they enter the unit-specific portion of training where they learn to operate their unit. At that point the Operator Trainer comes back into the picture. But this time the trainer is not burdened with the responsibility for designing training programs, that task requires the skill sets of training professionals. The Operator Trainer fills the role in the development of a training program for which they are perfectly suited and qualified – that of Subject Matter Expert (SME). The Operator Trainer works with professional training consultants who specialize in the development and execution of fundamental and unit-specific training material and process simulations, to insure the training is customized for their unit.

After successfully completing the fundamental and unit-specific training courses the console operator would serve an apprenticeship, paralleling on the actual console with a previously qualified operator before the trainee is certified to work the console alone.

As a model for rigorous competency assurance we can look to the nuclear industry, where the knowledge base and performance levels of licensed nuclear board operators parallel closely in depth and breadth to those of console operators in the refining and petrochemical industries, along with salaries.

This type of approach to achieving uniform competency assurance in the refining and petrochemical industries, will not only give plant managers and surrounding communities the innate assurance that all is well in the control room, but will pay for itself many times over, in day-to-day operational benefits and in the timely mitigation of emergencies.

348246 Process Safety Management (PSM) In Pilot Plants and Research Laboratories

Abstract:

Process Safety Management (PSM)

In Pilot Plants and Research Laboratories

Abstract

Pilot Plants are considered to be the important tool for tackling the scale up problems. Even though the quantity of the chemicals used is less compared to a commercial plant, they are capable of creating significant hazards and risks. Hence the hazards and risks posed by pilot plants are critical. The increasing number of accidents in research laboratories and pilot plants all over the world is a testimony to the above fact. Many organizations struggle to determine what guidelines they should apply to their own units. It is essential that the risks at pilot plants and laboratories are to be assessed and properly controlled to protect lives, property and environment.

If the elements of Process Safety Management (PSM) are implemented in a research and development facility accordingly safety can significantly improve. The key strategies suggested include properly conducted preconstruction safety reviews, leak tight design and construction, close attention to area classification, well designed ventilation, well planned storage, properly conducted maintenance, integrity management and strictly controlled procedures for the management of change.

Paper/Presentation Details

- Takreer Research Centre (TRC) in brief
- Commit to Process Safety
- Understand Hazards and Risks in Pilot Plants and Research Labs
- Hazard Identification and Risk Analysis
- Manage Risk
- Asset Integrity and Reliability
- Training and Performance Assurance
- Management of Change
- Emergency Management
- Process Safety Performance Measurement Metrics
- Case Study
- PSM implementation in Pilot Plants and Laboratories in TRC

348476 Resolving Inherently Safer Design Conflicts with Decision Analysis

Abstract:

The Chemical Safety Board (CSB) investigated a fatal explosion at a chemical plant in Institute, West Virginia. The CSB noted that the incident placed at risk a large inventory of methyl isocyanate, the toxic and reactive chemical responsible for the 1984 Bhopal tragedy. The CSB commissioned a study by the National Academy of Sciences (NAS) to investigate the application of inherently safer design (ISD) concepts to the design of the subject chemical plant. One of the findings of the NAS report was the absence of guidance documents to assist the process safety professional when evaluating ISD alternatives. The challenge identified in the NAS report is that process technology alternatives must satisfy multiple objectives, some of which may be in conflict. The NAS panel recognized the potential value of multiattribute utility theory as a basis for evaluating ISD alternatives, but they did not illustrate its use in this report.

In this paper we illustrate the use of multiattribute utility theory (MAUT) as a decision analysis tool for evaluating ISD alternatives. Using the NAS report as the foundation of our study, we show how to use MAUT to evaluate ISD alternatives by formulating utility functions and weights for the decision objectives. A sensitivity analysis is also presented to demonstrate how the final ranking of alternatives varies with the weights. This case study demonstrates that the MAUT technique is an effective tool for resolving ISD conflicts.

10TH GLOBAL CONGRESS ON PROCESS SAFETY

Operating Safely Through Integrated Process Safety Management

Srinivasan Ramabhadran
DuPont Sustainable Solutions
#11-01 Harbour Front Tower One, Singapore
Ramabhadran.Srinivasan@dupont.com

Alfonsius Ariawan
DuPont Sustainable Solutions
1919 Minnesota Court, Mississauga, Ontario
Alfonsius.B.Ariawan@dupont.com

Prepared for Presentation at
American Institute of Chemical Engineers
2014 Spring Meeting
10th Global Congress on Process Safety
New Orleans, LA
March 30 – April 2, 2014

UNPUBLISHED

Operating Safely Through Integrated Process Safety Management

Srinivasan Ramabhadran
DuPont Sustainable Solutions
#11-01 Harbour Front Tower One, Singapore
Ramabhadran.Srinivasan@dupont.com

Alfonsius Ariawan
DuPont Sustainable Solutions
1919 Minnesota Court, Mississauga, Ontario
Alfonsius.B.Ariawan@dupont.com

Keywords: Operational Discipline, Felt Leadership, Holistic Approach, Integrated

Abstract

Recent major incidents across the world have brought the need to operate safely as a top of mind issue for most CEOs. This presentation will focus on how effective implementation of an integrated Process Safety Management system can help achieve safe operation and improve productivity. It will address three key areas: operational discipline - focuses on carrying out each task the right way, every time; integrated approach - a holistic approach to implementing process safety elements that is based on risks rather than implementing individual elements of the system; felt leadership – the critical role of leadership in driving a culture of safety.

1. Introduction

Many companies today are beginning or accelerating programs designed to reduce their risks and in particular, prevent potential catastrophic incidents. These efforts are being driven by a number of factors: increased public and regulatory scrutiny of their operational risks, and increased exposure in terms of liability to leaders, and attrition of a competent and skilled workforce. Recent major incidents and the resultant public outcry have also contributed to raising the need to manage operational risks to the top of the agenda. Among other things that are required to effectively implement process safety management, there are three primary factors:

- Operational Discipline
- Integrated Approach
- Felt Leadership

1.1 Operational Discipline

Effective implementation of PSM does not happen only by *designing* good PSM systems; effective implementation requires a continuous, consistent and persistent *application* of these good systems in the day-to-day operations (Figure 1). The principle of "Operational Discipline" is recognized as a key success factor in driving effective PSM implementation. Organizations are increasingly realizing that operational discipline can result in rapid, substantial and sustainable PSM improvements.

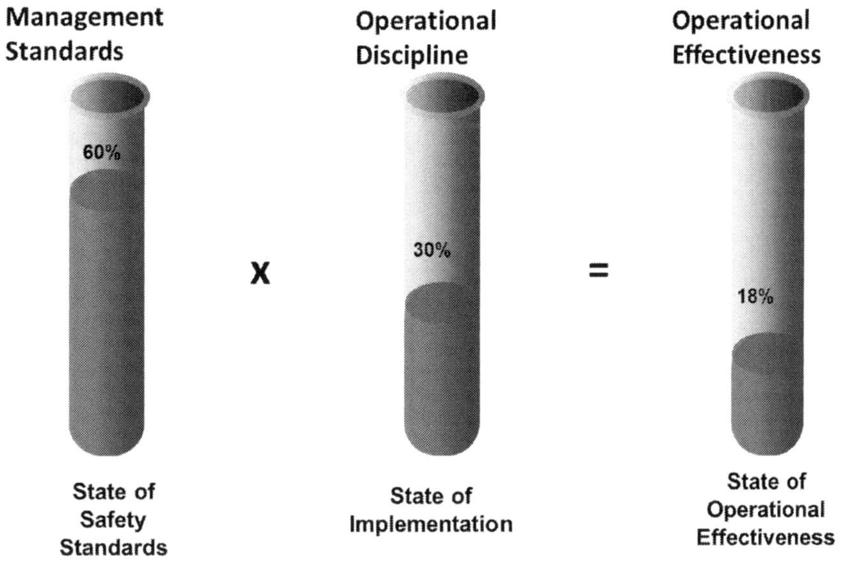

Figure 1. Performance is a function of the management system and level of execution

At DuPont, we define operational discipline as "the deeply rooted dedication and commitment by every member of an organization to carry out each task the right way, every time." We believe operational discipline requires change at both the organizational and individual levels in order to be substantial and sustainable.

Based on many years of cumulative working experience internally within DuPont and with clients, we have identified success characteristics that are the foundation of world-class operational discipline. The characteristics cover leadership, culture, processes and standards. A focused effort on improving these characteristics at the organizational and individual levels can lay the foundation for sustainable improvement and change.

1.2 Integrated Approach

Many companies implement the various elements of process safety management, but those elements must work together to achieve success.

DuPont has realized the value of this integrated approach and has defined the process safety management system as a wheel where the hub is the leadership and culture and the spokes are the critical elements of the management system. The way it works is through operational discipline (Figure 2).

Figure 2. DuPont PSM Model

1.3 Felt Leadership

What does "felt leadership" have to do with Process Safety Management?

A more descriptive term for what process safety is all about might be "Operational Risk Management." Operational Risk Management implies that:

- It is the operation or the materials processing activities that must be controlled.
- It is the reduction of risk to an acceptable level with the ultimate elimination of significant process incidents that is the goal.
- It is the management of these risks that is one of the organization's key responsibilities and challenges.

The degree of success achieved in the management of operational risks relates to the quality of decision-making by and within the operation's organization. And leadership is all about influencing and improving the quality of decision-making (Figure 3).

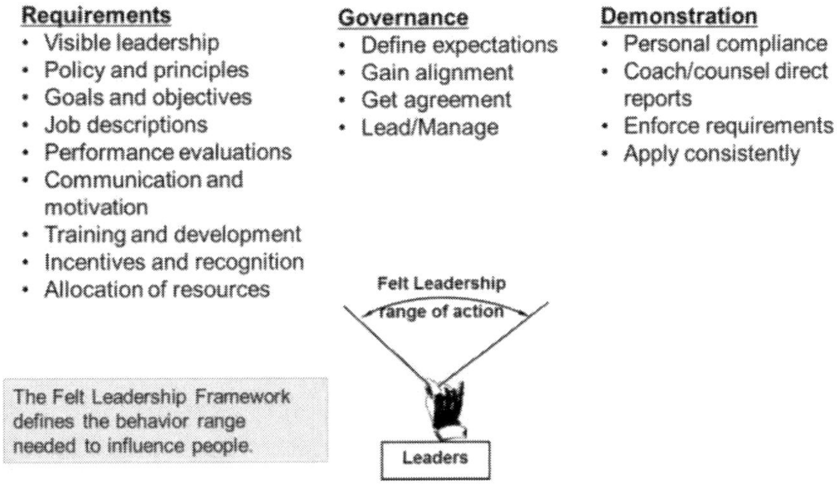

Figure 3. Felt Leadeship defined

2. Conclusion

- Leaders must have a clear vision of the ultimate goal of the company's process safety management efforts.
- Leaders must set the goals and strategic objectives.
- Leaders must ensure that sufficient, quality resources are dedicated to PSM system requirements.
- Leaders must make line management accountable for PSM implementation and results.
- Leaders must recognize and reward PSM accomplishments.
- Leaders must drive continuous improvement through the use of metrics and audits.

349434 Improving PSM Performance through Workforce Culture Assessment

Abstract:

Increasing PSM Performance Through Workforce Culture Assessment

By Peter Baker and Carl Green

This paper describes MetaPower's focus on increasing PSM performance by routinely measuring compliance to regulation-based criteria (structure) and simultaneously measuring key dimensions of work groups (culture). This dual approach is built on extensive work process design and implementation experience which convinced us that workforce culture is a key component to successful implementation. So MetaPower went "back to school" for five years, working with scholars who had organizational theory and culture expertise. Our learning and understanding led to the approach presented in this paper.

This presentation will introduce you to the radical difference between structure and culture, to the unique "logic" of each, and to the tools appropriate to cultural issues that can be secured and mastered. You will discover that it is possible, to measure culture and gather clues on how it is impacting PSM performance. You will learn that, because culture is such a powerful force, it is important to change culture only when necessary. You will also be introduced to a method for addressing cultural issues when change is necessary. You will find that when culture is approached with a new understanding, it is not mysterious and forbidding and it can be successfully navigated to achieve consistent performance. Finally, you will learn how structure and culture, integrated into a holistic Work System, can help you achieve your process safety goals.

349514 Tracking Instrumentation and Controls Reliability

Abstract:

To improve process safety and instrument maintenance decisions, CCPS PERD and the MKOPSC Instrument Reliability Network are developing better reliability tools to provide high quality, valid, and useful data pertaining to the hydrocarbon and petrochemical industries. The documentation required for the validation and maintenance of instrumentation and control devices must be completed properly in accordance with established criteria to demonstrate the appropriateness of credited risk reduction. Maintenance personnel are quite often overwhelmed with the challenge of tracking the testing frequency, inspection procedures, due dates and compliance for the instrumentation and controls in their areas of responsibility.

This presentation will review procedures and work processes that have been incorporated successfully for testing and validation of these devices. Technicians can quickly document compliance, pending due dates, completion dates, corrective actions and comments when an efficient documentation management system is in place. This system can be utilized to track and assess reliability, instrument and compliance data from proof test activities during plant turnarounds, and periodic safety instrumented function validation and routine maintenance.

10TH GLOBAL CONGRESS ON PROCESS SAFETY

PSSR: The Easiest and Sometimes Forgotten PSM Element

Brian D. Rains
DuPont Sustainable Solutions
Brian.D.Rains@dupont.com

Srinivasan Ramabhadran
DuPont Sustainable Solutions
Ramabhadran.Srinivasan@dupont.com

Prepared for Presentation at
American Institute of Chemical Engineers
2014 Spring Meeting
10th Global Congress on Process Safety
New Orleans, LA
March 30 – April 2, 2014

UNPUBLISHED

PSSR: The Easiest and Sometimes Forgotten PSM Element

Brian D. Rains
DuPont Sustainable Solutions
Brian.D.Rains@dupont.com

Srinivasan Ramabhadran
DuPont Sustainable Solutions
Ramabhadran.Srinivasan@dupont.com

Keywords: PSSR. Pre-start up safety review

Abstract

Pre-Startup Safety Reviews (PSSR) is to Process Safety what Personal Protective Equipment is to Personal Safety. Often, the PSSR is the last line of defense to ensure that a change or other major plant activity has not introduced an unintended or unmitigated risk. The principles behind the PSSR process are straightforward. The actual PSSR effort is basic and logical. And everyone says they do them and that they are easy. Is there data to support these statements? In this presentation, not only will PSSR best practices be reviewed but common PSSR weaknesses and improvement opportunities will be exposed along with suggested remedies and solutions.

1. Importance of PSSR

PSSR is one of the simplest elements of PSM. Does simple translate to less important? The simple answer is NO!

Figure 1: Where does PSSR come in to play?

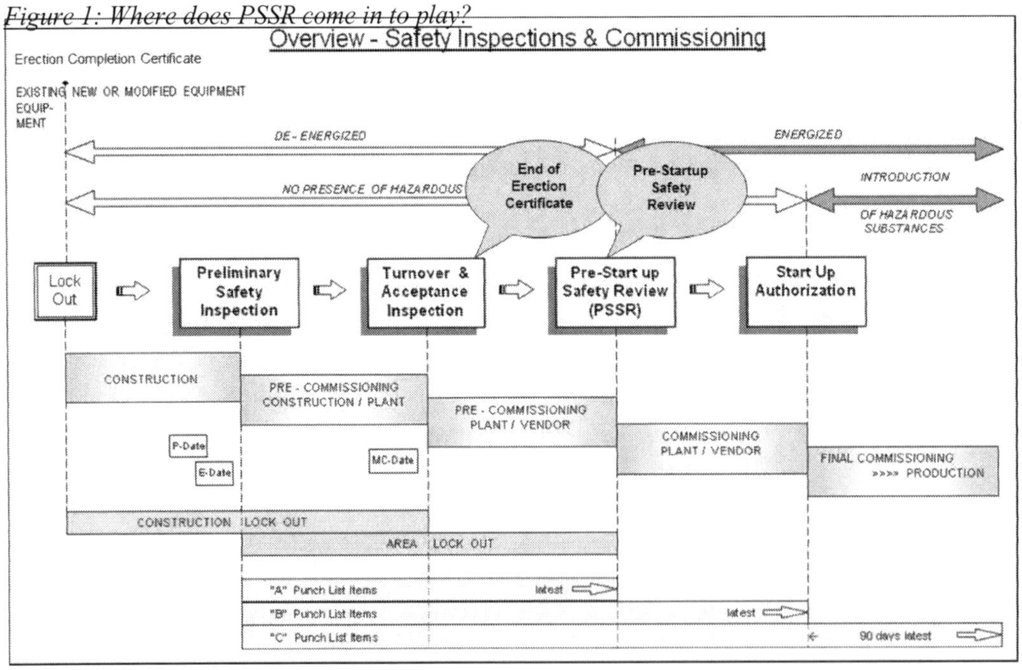

It is usually done after mechanical completion and before commissioning. The "ownership" for PSSR lies with operations as they will be pressing the button to start the operations. The operations leader along with a cross functional team has to satisfy himself that it is safe for him to start up the facility. While mechanical completion will be a responsibility of projects team, PSSR is clearly the responsibility of "operations".

Each element of PSM has a specific purpose and intent and failure to apply even one element in practice can contribute to potential catastrophic incidents. PSSR provides the last line of defense and events that can trigger the need for a PSSR include any change to an existing process or facility or the introduction of a new process or facility.

PSSR is NOT a last minute hazards analysis.

2. About PSSR

2.1 Principle

PSSR provides a FINAL CHECKPOINT for new and modified equipment to confirm that all appropriate elements of PSM have been addressed satisfactorily and the facility is safe to start up. For capital projects, planning for success with the pre-start-up review begins with assembling a competent project team. Effective project management, including careful planning and coordination of all PSM elements, is important to achieve a smooth start-up.

2.2 Essential features

2.2.1 Scope—PSSR

Sites shall perform pre-start-up safety reviews for all new (including equipment on hire or lease) and modified facilities for which the modification necessitates a change in the documented process technology.

PSSRs shall be conducted for significant "replacement in kind" equipment capital projects (compressors, heat exchangers, vessels, centrifuges, incinerators,..) even when no change has been applied, to help ensure that all relevant elements of PSM have been evaluated (e.g., updating of equipment files, performing QA checks for fabrication and installation, and resetting MI test and inspection values for the new components).

PSSRs may also be conducted for the following circumstances to help ensure that all relevant aspects of process safety have been completed prior to returning to full operations:

- Restarting an idled or mothballed process
- After a major shutdown or turnaround where significant or extended maintenance or repair activities occurred (i.e., emphasizing the field portion of the PSSR process).

Sites may alternatively use process-specific start-up procedures and safety checklists for start-up after these activities.

Prior to the introduction of hazardous substances and associated energies to a facility, the pre-start-up safety review team shall confirm that:

- Construction and equipment are in accordance with design specifications.
- All relevant elements of process safety management have been appropriately addressed (i.e. PSM assessment).
- Basic safety, health, and environmental - and fire-protection items have been checked during a physical inspection of the facility. These inspections are usually guided via a previously pre-prepared "good practice" checklist.
- Installation is in compliance with local regulatory requirements and site and unit operating permit(s), including environmental air permits and approvals.
- The facility is safe to start up. This may be contingent on completion of certain recommendations.

2.2.2 Team composition

The pre-start-up safety reviews needs to be conducted by multidisciplinary teams consisting of manufacturing (including operators), technical, design as-needed, maintenance, appropriate safety representatives on an as-needed basis and the Change of Design (COD)/Project originator/leader or the COD Coordinator. Other resources, specialists, or independent or third-party competent persons as specified within various RAGAGEPs (Recognized and Generally Accepted Good Engineering Practices) may be included as applicable (e.g., electrical, fire safety, instrumentation and control, personnel; safety interlock guardians; ergonomics specialists; or equipment manufacturers). When the new facilities will be used by a resident contractor then a member of that organization must take part in the PSSR.

For very minor changes, a smaller pre-start-up safety review team may be used and could consist of two team members. In this situation, the facility owner shall designate the team members. Examples of minor changes include installation of a new drain valve and instrument transmitter bracket modification.

The team composition is the key to maximizing the value and results of a PSSR. The facility owner must take responsibility for assembling a multidisciplinary team with knowledge and experience appropriate to the task. The Team make up will be predetermined by function and indicated on the Copy of the PSSR circulated with the COD for approval. Team members should include personnel who were not directly involved on the applicable COD/project team.

The following points on team makeup are for guidance only. Advice on this issue can be provided by site or regional/external PSM resources.

- Even for the smallest PSSRs, such as those would be conducted prior to starting up a single piece of equipment (e.g., a pump, heat exchanger or vessel), the team should have three or four representatives from different backgrounds (e.g., manufacturing, maintenance, design/technical, and appropriate safety representatives). This helps create a more searching and valuable review. People who operate or maintain the facility (e.g., an operator and a mechanic) should be included.

- Where the change involves introducing hazards (i.e., process, material or environmental) that is new to the site, the PSSR team should consider including off-site personnel who have relevant knowledge and experience. These could be manufacturing, maintenance, technical or appropriate safety representatives.

- For significant capital projects and for larger projects a formal PSM audit may be recommended. In this case the facility owner must ensure that the team has the skills and capability to carry this out. Consideration should be given to having one or more safety and occupational health consultants on the team.

- The leader of the PSSR team should foster an open atmosphere where the project team members look on the review as an opportunity to upgrade the process safety of the facility and do not become defensive of the design or preparations for start-up.

2.3 PSSR Application

The flow chart below represents a recommended process for applying PSSR.

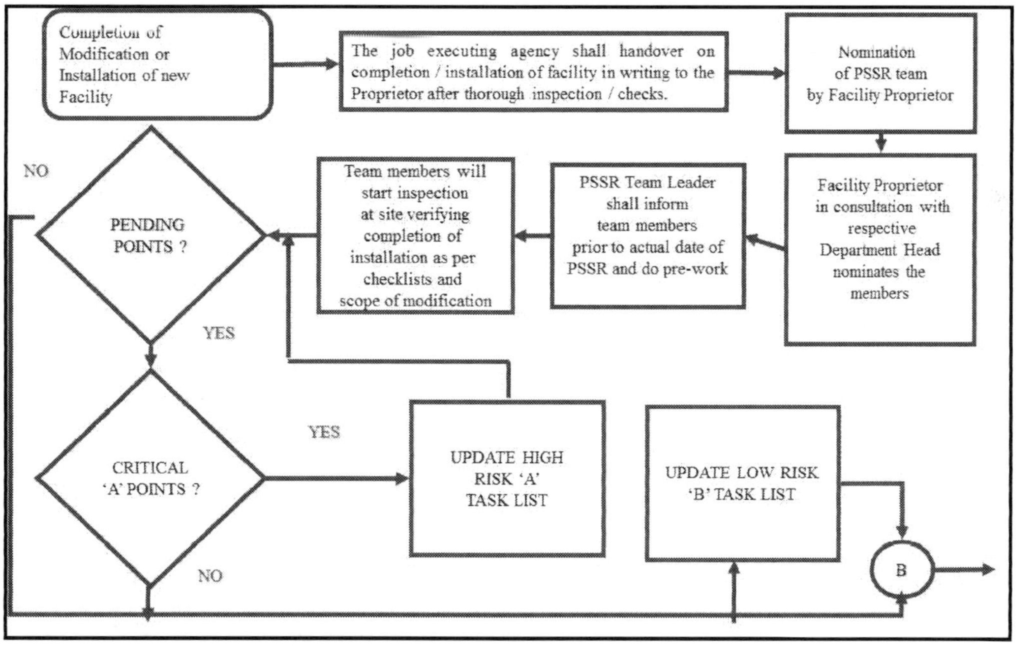

Figure 2a: PSSR Flow Chart

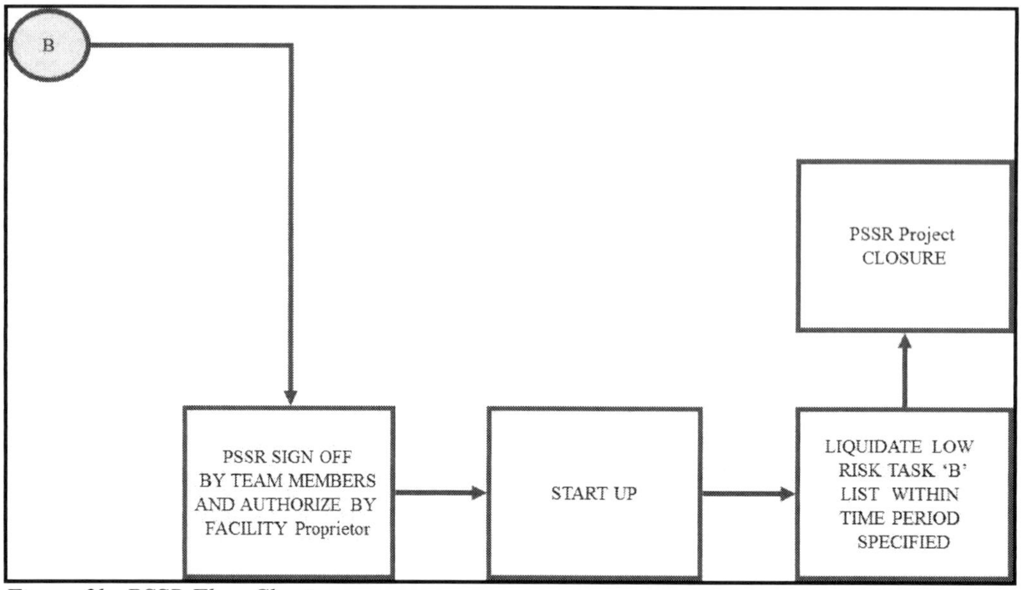

Figure 2b: PSSR Flow Chart

2.3.1 Significant capital expenditures

Where the change involves significant capital expenditures (typically needing off-site authorization) and one or more complex unit operations with hazardous materials and activities, then it is recommended that the PSM element assessment be conducted by either a first-party or a second-party PSM audit using the appropriate corporate PSM audit checklist(s).

For significant capital expenditures, it may be helpful to carry out a two-part review. The first should take place when design is complete and construction is at an early stage or has not yet begun. This allows operating procedures, emergency planning, and training to be upgraded well in advance of start-up.

The second part of the review can then take place just prior to introduction of hazardous substances and focus on checking that construction and equipment are in accordance with design. For example:

- Prior to charging a substation, instrumentation building (if applicable) for the first time.
- Before introducing hazardous substance / feed in a new, modified, hired or temporary facility

2.3.2 Small projects/modifications

For smaller projects/modifications, the PSM element assessment is conducted using a PSSR checklist. The PSSR team should select relevant parts of a PSSR checklist to review, as appropriate, for the project/modification. Although it is not the PSSR team's function to review the process design, its input may affect construction.

- Significant projects involving "replacement in-kind" equipment applications
- Changes to operating facilities
- Temporary changes
- Restarting mothballed process
- Post major turnaround
- Sites may alternatively use specific start-up procedures and safety checklists after these activities

2.4 PSSR - PSM element assessment focus

The pre-start-up safety review team shall ensure that all relevant PSM elements have been appropriately addressed and shall be guided by pre-written checklists. Physical inspections should cover:

- Documentation Items
- Vessels, Piping and Valves
- Rotating and Mechanical Equipment
- Electrical systems
- Instruments and control systems
- Structures
- Safety, Health, Security and Fire Protection
- Ergonomics
- Waste, Stream and Environment
- Operation, Inspection and Maintenance and Others
- Fume Hood/Chemical Storage/Gas Cylinder Storage and other Laboratory Area

It also needs to include appropriate checks to confirm that *local regulations* are being followed. The PSM elements to consider included the following:

PROCESS SAFETY INFORMATION	Assists in • verifying PT for equipment, • material hazards & technology • Updating in timely fashion.
PROCESS HAZARD ANALYSIS (PHA)	• Recommendations have been addressed and appropriate actions associated with s/u have been completed.
OPERATING PROCEDURES	• Has the operating/emergency procedures and safe work practices procedures revised to reflect the change?
TRAINING AND PERFORMANCE	• Has the operating and maintenance personnel required to be trained for the new facility or change been done? Competency of personnel have been evaluated.
CONTRACTORS	• Do contractors require any special training or instructions to be given; are they properly trained?
MECHANICAL INTEGRITY and QUALITY ASSURANCE	• Maintenance task procedures are in place? • Equipment needs to be placed in mechanical integrity program? • Installation as per codes, standards and manufacturer's recommendations?
MANAGEMENT OF CHANGE	• MOC related activity and its documentation. • 2nd level protection to ensure MOC is working. • Influences complexity of PSSR. • Evaluates MOC scope and intent are met. • Establish criteria that maintain minimum levels of knowledge and experience
INCIDENT INVESTIGATION	• PSSR documentation support investigation teams. • LESSONS LEARNED ARE POWERFUL TOOLS OF IMPROVEMENT
EMERGENCY PLANNING & RESPONSE	• Verifies that emergency response plan changes are made & affected people are trained.
AUDITS	• Following start-up, the new facility is included in the site PSM audit system.

2.4.1 PSSR Attributes

The attributes of a "good practice" checklist are as follows.

- A clear front sheet stating what equipment has been inspected, when and by whom.
  - A statement that the installation is consistent with design specifications.
  - Statements that the PSSR team concludes that the facility is safe to start up after certain recommendations have been satisfactorily resolved.
  - A record of recommendations (if any) with timing and responsibility for completion either before or after start-up.

- A check that the following basic safety and occupational health areas have been appropriately addressed:
  - General Safety
  - Machinery safety
  - Ergonomics
  - Occupational health

- A check that the following PSM elements/topics have been appropriately addressed:
  - Process technology
  - Management of change – technology and facilities
  - Process hazards analysis
  - Quality assurance
  - Mechanical integrity
  - Operating procedures and safe work practices
  - Training and performance
  - Interlocks and alarms
  - Highly toxic materials
  - Contractor safety

- A check that other relevant SHE topics have been addressed:
  - Environmental
  - Community awareness and emergency response
  - Electrical safety
  - Firc protection

- Site should include in their pre-start-up review checklist appropriate checks to ensure that local regulations are being followed.

2.5 PSSR documentation, recommendations and closure

The pre-start-up safety reviews shall be documented, signed by each member of the review team, and authorized by the facility owner. The pre-start-up review document shall include a statement as follows:

The PSSR team concludes that the equipment is in compliance with current regulatory requirements and limits stated in the site and/or unit operating permit(s) and is safe to start up after completion of recommendations required before start-up.

A system shall be established and controlled by the owner to ensure review recommendations are resolved (including documentation of resolutions) before hazardous substances and associated energies are introduced to the facility. Start-up authorization shall be issued after verification that all PSSR items to be resolved have been satisfactorily completed. The start-up approval shall be documented on PSSR document. At the site the required for startup recommendations must be signed off as complete before the startup of the facilities is authorized for start up

Similarly, a system shall also be established by the facility owner to ensure resolution of PSSR review recommendations that are to be completed after start-up. At the site those recommendations that are not required for startup are entered in the Site's "Manage IT" system for follow up and tracked to completion.

Periodic reports listing open recommendations should be published at least quarterly and distributed to the site business area management, and all persons assigned to follow up a recommendation, and their supervisors of persons assigned the recommendation. The report should highlight recommendations that are past the due date.

Each item in the Site PSSR Authorization Form is referenced back to this document so that the inspection team may obtain additional clarification as the intent of the items in the Site PSSR Check sheet. It is important to specifically track all and especially punchlist A items for closure and verification before start up. Having a tracking system provides the confidence that all critical recommendations items have been closed and verified and assures that the plant is safe for start-up.

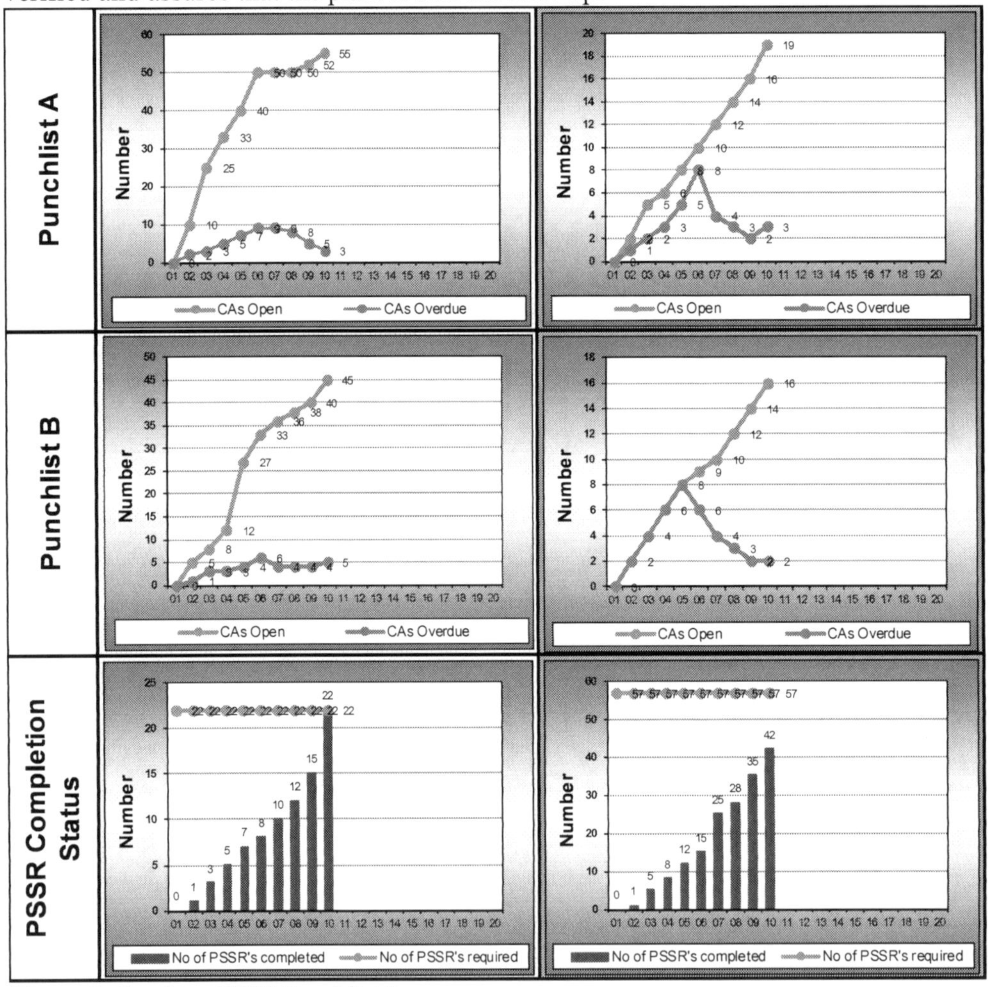

Figure 3: Tracking PSSR Punchlist Items

2.6 PSSR Responsibilities

Senior Line Supervision: Senior Line Supervision shall authorize the checklist after the team members have signed the checklist and shall ensure that all recommendations required prior to startup are completed and documented prior to the introduction of hazardous materials into the facility.

Review Team Leader: The Review Team Leader is responsible for assembling a multi-discipline team and establishing a review schedule for that team. A member of the safety organization may also be invited to take part in the PSSR inspection. The leader shall guide the team through the review, using the checklist to ensure that all relevant questions have been asked and answered. The leader shall document the answers to the checklist questions and note all deficiencies. The leader will ensure that all deficiencies are assigned to a specific person for correction. The leader will forward the authorized PSSR to the COD Technical Assistant for entering Manage It Central database and filing in the COD folder. Upon completion of all recommendations, the leader will document the date the recommendations completed and initial on the PSSR. The leader is responsible for ensuring all recommendations are completed in a timely manner.

All COD's are entered into a tracking system and their status reviewed by the area once a month.

Review Team Members: These personnel are selected based on their individual skills/knowledge relative to the change being made. They shall use their individual skills and knowledge to ensure that all relevant questions from the checklist are correctly answered and those any and all additional concerns are brought forward for discussion by the team. Members shall sign the checklist, stating that the facility is safe to start up after completion of recommendations required before startup.

Recommendation Assignees: These are the personnel assigned to follow-up on recommendations generated during the safety review. They shall complete their assigned recommendations in a timely manner and document the action taken in Manage It Central database.

COD Technical Assistant: The COD Technical Assistant will be responsible for entering a PSSR and recommendations in Manage It Central database and filing it in the COD folder once received from the Review Team Leader.

Verifier: The area person who is responsible for confirming proper documentation related to the new/modified facilities shall be responsible for checking the PSSR Checklist to ensure all recommendations have either been signed off as complete or are listed on the cover sheet with the respective tracking number from Manage IT prior to signing off for closure of the COD which required the change

3 Conclusion

In conclusion, while PSSR is a simple element to understand and deploy, it is an important element and plays a critical role in preventing incidents during start-up. Any process is "more prone" to incidents during start-up due to:

- Involvement of multiple functions and departments
- Understanding of the new technology and/or modified process requirements
- Transfer of ownership from projects to operations

PSSR is a good way to bring all parties together and align on the changes and needs to ensure safe start up.

4 References

[1] Guidelines for Performing Effective Pre-Startup Safety Reviews, Center for Chemical Process Safety (CCPS), April 2007

[2] DuPont Internal Standards

349905 Guidelines for Pressure Relief and Effluent Handling Systems, 2nd Edition

Abstract:

The second edition of Guidelines for Pressure Relief and Effluent Handling Systems will be issued later this year. It is the only book to comprehensively cover both design sizing of emergency relief devices and the systems to safely control the effluent after the relief device has activated. It also includes software for designers to size the emergency relief systems, plus worked examples to show how the software can be used for several design scenarios. It is the only book in this field that covers the topic of 2-Phase flow, based on the most up to date research and studies by the Design Institute for Emergency Relief Systems (DIERS). It is also the only book to cover in detail the safe control of the release after the relief device had activated. An overview of the book will be presented.

349983 Journey to World Class through Capability Development

Abstract:

DuPont has been on a 7 year journey towards world class operational excellence, resulting in significant earnings gains and cost-savings ($6B+) built off a foundation of capability development for individuals and teams.

This strategy has been proven across a wide variety of assets and businesses, both inside and outside of DuPont, deployed and sustained at over 200 DuPont locations.

This framework will be shared/discussed and compared/contrasted with attendee's experiences. Several of the questions to be discussed are:

- How do you intentionally work to build a culture of continuous improvement throughout your organization?
- What specific areas of capability building have you found to be critical (For individuals? For teams? For key leaders?)?
- What cultural elements can be used to engage a multi-generational workforce?

DuPont will share key frameworks and elements of their journey that can be used as criteria for building capability and operational excellence in organizations.

349330 Incident Lessons Learned Portal

Abstract:

Incident Lessons Learned Portal

Marco Vela

Corporate Process Safety Engineer

Celanese

1601 West LBJ Freeway

Dallas, TX 75234

(214) 417-1920

marco.vela@celanese.com.mx

Abstract

The Celanese incident lessons learned portal consist of four levels: company lessons learned, technology specific lessons learned, site specific lessons learned and external incidents of interest. This paper will present how we capture knowledge from previous incidents to retell our story at various levels in the organization. It is designed to engrain learnings from past incidents into the corporate process safety culture to prevent repeats.

Company Lessons Learned

Company lessons learned use CSB type videos to share the key causes of the major incidents that have occurred in company with all Celanese employees. The system includes the initial delivery method and definition of how the messages are reinforced over time. They describe why we have established a process safety system.

Technology Specific Lessons Learned

Technology specific lessons learned are gained either by experience, past incidents, or standard design. These learning's are targeted to hourly workers in each technology, and describe what could go wrong in that technology and the safeguards in place to prevent.

Site Specific Lessons Learned

This is a common area on the process safety portal that gives everyone access to one page that contains the information of the most significant incident lessons learned that have happened at each site. Site specific lessons learned are used for toolbox meetings, operator training, new leader process safety certification, PHA preparation, and other process safety activities.

External Incidents of Interest

This is a place to share with all the employees the lessons learned from external incidents of interest based on relation that they have with our companies manufacturing operations.

349394 A Functional System Approach to Criticality Analysis, Fsca

Abstract:

ABSTRACT

A Functional System Approach to Criticality Analysis, FSCA.

This paper will outline and discuss a new and unique approach to physical asset criticality analysis: Functional System Criticality Analysis (FSCA) and illustrate key results with two sample case studies. This approach is applicable to any type of complex facility or process where process security and continuity are paramount. It has consistently identified previously unknown significant risks that were missed by other methods and analyses such as HAZOPS, FMA, process reviews, and can quantify the risk potential in dollars. A functional system based approach (FSCA) is more efficient than current methods through analyzing and ranking system risks via system failure scenarios rather than failure modes at the asset level. It is an invaluable tool and approach when risk mitigation is important to assure overall success balanced against the reality of limited time, capital, maintenance, and staff budgets.

The two case studies featured involve an Envirex ground water remediation facility, and an industrial RO water treatment system within a refinery. Each study had a unique set of evaluation criteria determined by the client's values and the specific performance, contractual, and regulatory issues affecting the facilities. Importantly, in each case study previously unknown significant system risks were identified, as well as highly effective mitigating steps.

Approach:

Several key principles are foundational to the functional system based approach to criticality analysis, FSCA:

- A deep and well structured hierarchy of the asset register, leading to a rigorous parent/child ratio
- Inclusion of all valves and pipes (strive for 100 percent asset listing) in the hierarchy
- Uniform asset levels (every pump is an asset no matter where it is or how big)
- The grouping of assets into single-purpose functional systems

A very clear description of the mission statement/priorities is very important in this type of an analysis. Whether it is defined as a level of service or a more numerical set of KPIs, the defined mission of the physical assets under review will determine the categories evaluated against: safety, environmental, up-time, product quality, maintenance cost, capital replacement, etc. These in turn must also be clearly and uniquely defined from one another on a severity scale ranging from virtually 'no impact' through to 'worst case' in the client specific nomenclature. The next step is to proportion or allocate a contribution ratio for combining categories into overall risk. This will determine, for example, the relative importance or ranking of a worst case scenario compared against a scenario with 2 or 3 instances of a slightly lesser consequence. These can be adjusted to suit a client philosophical approach.

With these principles in place, a risk analysis is then performed at the functional system level derived from the hierarchy, prior to the asset level. Failure scenarios and their resulting consequences are evaluated per the evaluation criteria defined (not failure modes since those are at the asset level). The failure scenarios for each functional system define a composite risk space for each system, and then each system is ranked against all others in terms of risk.

Important to note is that assets within a parent system initially assume the risk ranking of the parent system. Thus, systems with highest risk ranking (mission critical) are then further evaluated at the asset level, employing a suitable mix of evaluation tools (RCM, FMA, etc) to discern which of the contained assets pose the highest risks to the mission objective. Systems with low risk ranking do not have assets that can significantly impact the mission, are thus not mission critical, and are thus relegated to a lesser priority for further evaluation.

In this approach mission critical assets are more rapidly identified with subsequent evaluations more efficiently utilizing complementary tools saving time, staff resources, and money. We have also found that, due to the unique arrangement of the assets within the system level and overall integration, this approach captures risks that might otherwise be missed.

Base Case 1. Envirex Ground water remediation system

In mid 2004 a sequence of unusual, but not extraordinary, events led to an extended shutdown of a key ground water remediation process consisting of extraction wells, a fluidized bed activated media treatment process with filtration, and internal re-use of the treated water. The result of this was that the owner was at considerable risk of consent decree violations. Excellent engineering, standard process reviews, HAZOPs, and an aggressive maintenance program had not correctly identified risk. The need to find the true risks in the system was paramount.

Response:

The owner requested assistance in the evaluation of the overall process to identify the as yet hidden risks and related potential issues in order to prevent future similar and unforeseen occurrences. The Uberlytics criticality analysis approach and an early version of a software tool were brought in and applied in a workshop environment where the owner's system experts, including engineers and operators, were brought together with a criticality analysis expert facilitator.

Inclusion of 100% of the overall process units in this functional area of the refinery, and inclusion of pipes and valves as a separate group asset again proved essential. Successful discovery depended on the aforementioned and what had previously not been done: leveraging the network system approach rather than a discrete assets approach.

Results:

While the owner's focus was primarily on preventing recurrence of the event that had occurred, a full analysis of the system revealed several key aspects that led to a more comprehensive understanding of the process and associated risks, as well as providing specific improvement opportunities that would immensely reduce risk with remarkable little effort and expense. The networks systems approach to criticality analysis pinpointed the issues behind the major occurrence, and also found a number of other previously unidentified risks, including one previously thought to be insignificant. Commensurate simple and cost effective solutions that immensely reduced the risk profile for the entire process in all risk areas were identified.

Base Case 2: Refinery boiler feed pretreatment RO system

Situation

A boiler feed water pretreatment system was due for an expansion and had undergone several design reviews, two HAZOP reviews, and several construction reviews. However, given the importance of this system within the refinery operation, as well as being a key part in the refinery overall water reuse initiative, the refinery elected to commission an analysis on the plant boiler feed water pretreatment system. The decision to perform a criticality analysis was also based, in part, on the success of the previous review methodology. However, since the system had also undergone some annual refurbishments and, after 18 years of operation, was thought to be fully 'known', it was not expected that any previously unidentified issues would surface. This assumption proved to be false.

Response

As in the previous example, the success of this analysis again hinged on the flexibility of the analysis to include not only areas such as safety, environmental, capacity, and up-time, but also commercial terms, liquidated damages, and contract terms. This proved to be a truly unique analysis in comparison to the many previous ones that had been conducted on the process. The results were reflective of more than just a process view, showing the unique position of this process's risk profile within the larger context of the client operation.

The same steps were conducted as in the previous case, except in here an asset registry had previously been generated as part of a HAZOP review. The database was imported for the analysis. Again, a team comprising various client experts was assembled with a criticality analysis expert facilitator, and again the unique approach proved essential.

Results

While the analysis did clarify and further define the generally known critical areas central to the overall mission, the analysis brought to the surface as yet uncovered critical risks inherent to the system as well as a different than expected risk ranking. One system was a particular surprise to the investigating team as it could shut down the whole operation in a single shift, and two other systems were elevated to a higher criticality ranking than previously thought given the potential for overall system failure. The inclusion of contractual and commercial evaluation parameters highlighted the relationship between process and commercial risks that had previously gone unnoticed. This provided the opportunity to monetize those risks for a more comprehensive business risk evaluation.

Conclusion

A sophisticated yet efficient approach is required to identify all risks, including hidden ones, in a cost effective way that also lends itself to solution identification. Conventional analysis and evaluation often fail to identify some significant mission critical risks because they do not take context into account. The functional systems approach to criticality analysis, FSCA, is a unique and proven approach that consistently meets and exceeds that objective. This approach to criticality analysis reliably identifies previously undiscovered system risks, means of mitigation, does so in a more efficient way than conventional means, and delivers verifiable and reproducible results.

Ignorance of risk it is not absence of risk.

349504 Can Black Swans be Red Herrings?

Abstract:

At present, our industry faces its biggest challenges due to mismanaged risks and the need to be sustainable. Over a period of time, along with the experience of various major accident events, risk management philosophies and techniques have become more sophisticated. However, there still remains much opportunity to extract better value from risk management activities and investments. This presentation will discuss some of the current weaknesses and highlight specific opportunities for improvement. It is intended to provoke change in how we look at existing risk management processes.

Specific topics will include:

- The potential for "black swan" events is a fallacy. Rather, they are a failing of threat identification, risk assessment and/or mitigation investment decisions.
- Rather than truly exploring the specific causes that might be applicable in that special situation, many threat identification exercises are content with simply listing the generic causes of potential failures.
- Current technology and thinking could enable much more value to be extracted from expensive hazard identification activities that have not evolved since their inception in the 1960s.
- Many risk metrics do not have a practical intrinsic meaning. There is a need to monetize risk metrics so that their intrinsic meaning can be better understood and they can be better related to risk mitigation investments.

Historically, much risk management effort has gone into demonstrating that the status quo is OK. We need to truly embrace the concept of continuous improvement and focus risk management activities on helping to determine where we can make our best risk mitigation investments.
This will improve our business.

349717 Emergency Response Plan: It's Above and Beyond Best Practices

Abstract:

Human life is of prime importance in the any industry – offshore, refinery, chemicals, mining, etc., where potential for exposure to hazards can result in harm to people. Although the best way of managing an unexpected event or emergency is through prevention, it is also critical to have measures in place such as a robust Emergency Response Plan (ERP) to manage personnel in emergencies. A well-designed ERP that includes proper escape plans and rescue facilities and procedures can decrease confusion and increases safety of personnel. Among various elements of ERP one of the key decisions to be made is whether or not to follow a Temporary Refuge (TR) philosophy which begins with instructions to assemble in the muster points. Currently, there are regulations and guidelines that require or recommend having a TR and some that do not. This paper explores various escape and rescue philosophies, and the pros and cons of having an evacuation plan. These will be discussed to provide a clearer picture on implications of having one onboard. Chain of command during an emergency is another aspect of ERP that needs clarity. This paper will gather information from major accident investigations and the current ERP regulations and best practices. This information will be combined with expert experience with process safety design and human factors to help determine pros and cons of including a TR in the ERP as well as show the significance of having a specified chain of command for all emergency events.

349885 Process Safety Hazard Management Plan : Help You in Sustaining Production and Preventing Lossess

Abstract:

ABSTRACT

Managing mature assets is a challenge, especially when we have to deal with production target. Sometimes it is not easy to straight on safety as your first priority while in the other hand we need to maintain the production or even increase it. How safety and production can stand together without scarifying each other? Process safety hazard management plan is the key answer to resolve this matter.

Process safety practices and formal safety management systems have been in place in some companies for many years, including Pertamina Hulu Energy ONWJ which brought the standards from former company. Process safety hazard management plan is a part of the implementation during the operation stage. Not only assessing safety as a behavior but also assessing the plant and process for reduction in potential major accident risk and an improved plant performance. What can be done on the plant and process area to ensure that hazards are identified, addressed and the consequences are reduced?

This paper describes how process safety hazard manament plan can help you in sustaining your production, which includes:

- *identifying your process hazard and risk*
- *learning your protection layer (Swiss cheese model)*
- *agreeing your acceptable risk*
- *giving commitment on your actions, who is to do what*

In process safety perspective, all incidents are preventable. The key element is managing risk consistently to face any obstacles, minimize losses and achieve your goals.

349882 Process Risk Assessment and "Safe Area" for a Petro-Chemical Plant in China

Abstract:

Process Risk Assessment and "Safe Area" for a Petro-Chemical Plant in China

Jing Yu, DuPont Engineering, Shanghai, China

S. Dharmavaram, DuPont Engineering, Wilmington, DE, USA

Ji Ming Wang, SINOPEC, Beijing, China

Many process hazards exist and highly hazardous events are possible in a typical petro-chemical plant. Even though the frequencies of such events occurring are low, the community has limited knowledge of consequences, frequencies, and process risk. Process hazards are like wild animals trapped in cages. Understanding hazards and minimizing risk can still lead to safe operation and contribute to society. Severe objections have been raised recently on building a para-xylene plant in China.

Risk assessment for a petro-chemical plant starts with a thorough identification and understanding of process hazards. Such process safety information is important for conducting process hazards analyses. Practical experience on incidents that have occurred in the past in China and elsewhere the world is also very valuable for determining the frequency of hazardous events.

In this paper/presentation, information about the severity and consequences of hazardous events at a para-xylene plant and associated process risk will be presented. Detailed modeling done for simulating the consequences of accidents and a study of domino effects to determine how fires/explosions propagate within a plant will be discussed. Use of process dynamic modeling tools such as Aspen® to calculate process conditions leading to incidents will be described. Deflagration and detonation tests used as a basis for Aspen® modeling to identify the worst-case scenarios for fires/explosions and toxic releases will also be presented. An understanding of off-site and on-site consequences and frequencies of hazardous events is used ultimately to determine a "safe area" concept for a petro-chemical plant, which will help local government make proper decisions and help the community in understanding process risks.

IDENTIFYING AND QUANTIFYING MAJOR HAZARD FOR PLATFORMS DECK RAISING USING SYNCHRONOUS HYDRAULIC JACKING SYSTEM

Authors:

Akhmad Tutus Harmantoro
PT Pertamina Hulu Energy Offshore North West Java (PT PHEONWJ)
Arkadia Office Building, TB Simatupang Street, Kav. 88, South Jakarta, Indonesia, 12740
akhmad.harmantoro@pertamina.com

Margaretha Thaliharjanti
PT Pertamina Hulu Energy Offshore North West Java (PT PHEONWJ)
Arkadia Office Building, TB Simatupang Street, Kav. 88, South Jakarta, Indonesia, 12740
margaretha.thaliharjanti@pertamina.com

Prepared for Presentation at
American Institute of Chemical Engineers
2014 Spring Meeting
10th Global Congress on Process Safety
New Orleans, LA
March 30 – April 2, 2014

UNPUBLISHED

IDENTIFYING AND QUANTIFYING MAJOR HAZARD FOR PLATFORMS DECK RAISING USING SYNCHRONOUS HYDRAULIC JACKING SYSTEM

Akhmad Tutus Harmantoro
PT Pertamina Hulu Energy Offshore North West Java (PT PHEONWJ)
Arkadia Office Building, TB Simatupang Street, Kav. 88, South Jakarta, Indonesia, 12740
akhmad.harmantoro@pertamina.com

Margaretha Thaliharjanti
PT Pertamina Hulu Energy Offshore North West Java (PT PHEONWJ)
Arkadia Office Building, TB Simatupang Street, Kav. 88, South Jakarta, Indonesia, 12740
margaretha.thaliharjanti@pertamina.com

Keywords: Failure Mode, Effect, and Critically Analysis (FMECA), Quantitative Risk Analysis (QRA), subsidence, java sea, deck raising, hydraulic jacking system.

Abstract

Lima Flow Station is one of the gathering manned stations of Pertamina Hulu Energy Offshore North West Java (PHEONWJ) which is located in Java Sea, Indonesia. The Lima Flow Station consists of four bridge linked platforms and a bridge linked flare structure. Lima Flow Station had experienced subsidence for about 3 meters since it was built in 1973 to early 2013. The subsidence caused significant reduction in the platform air gaps (distance between cellar deck bottom of steel and maximum wave height) which caused Lima Flow Station to not be reliable for safe operation. The synchronous hydraulic jacking system was selected to raise the platform to 4 meters elevation and remedy the air gap for safe operation. Three platforms, Flare tripods, and interconnecting bridges were raised while one platform was not raised as it was isolated and utilized as a muster point and for personnel transfer purposes only. The same technology has been successfully applied in Gulf of Mexico in 2006.

Quantitative Risk Assessment (QRA) study was conducted for the design of a synchronous hydraulic jacking system to measure whether the design provided tolerable risk or not and identify mitigation of any intolerable risk. The assessment was carried out in two main parts: hazard identification in the form of failure modes, effects and criticality analysis (FMECA) to determine major accident events to be carried forward into the QRA; and determination of the impact of safety risk to personnel and assessment of financial risks in the form of delays to resumption of production.

FMECA was used for qualitative and semi-quantitative analysis of engineering systems. The reviewed system was divided into system, subsystem, and equipment. Failure mode, cause, safeguard and effect/consequence for equipment are identified. High consequence was considered as major accident and was further studied in QRA.

QRA consisted of Fault Tree Analysis (FTA) to estimate the probability of a major accident event identified by FMECA, Event Tree Analysis (ETA) to estimate the probability of a major accident leading to the ultimate event, Consequence analysis to estimate the impact of the ultimate event, risk estimation by multiplying probability and impact of ultimate event, risk assessment by comparing estimated risk and risk criteria, and identification of mitigation for any major accident having intolerable risk.

FMECA has identified some major accident events caused by misalignment and structural failure during deck rising. The results of the QRA determined that the individual risk per annum for the most exposed individual was still within tolerable risk of company's criteria as the deck raising was conducted during shutdown and minimum manning level. While the financial risk was high for platform damage due to structural failure for lifted platforms as there was no detection method for structural failure and sudden movement in case of structural failure may cause major damage. The financial loss was mostly contributed by production loss. The major contributor for structural failure was weld failure. Recommendation was raised to minimize the risk by providing extra quality check for welding and material of structural parts on fabrication and installation.

By implementing good project and risk management, Lima Flow Station Deck Raising has completed successfully and started up in 2013 without any accident.

1. Introduction

LIMA Flow Station is one of gathering manned station of Pertamina Hulu Energy Offshore North West Java (PHE ONWJ) is located on Java Sea which were installed in 1973. It gathers and produces oil and gas from the LIMA Normally Unmanned Installations (NUIs), with current production rates of 3.000 bopd and 20-30 MMscfd gas and also gathers and produces gas from other stations around 200 MMscfd and oil 4000 bopd. It consists of four bridge linked platforms: LA, LPRO, LCOM, and LSER; and a bridge linked flare structure.

Lima Flow Station has been experiencing subsidence of about 3 meters since it was built in 1973 to early 2013. The subsidence caused significant reduction in the platform air gaps (distance between cellar deck bottom of steel and maximum wave height) which caused Lima Flow Station to not be reliable for safe operation. The synchronous hydraulic jacking system was selected to raise the platform to 4 meters elevation and remedy the air gap for safe operation. LCOM platform, LPRO platform, LSER platform, Flare tripods, and interconnecting bridges were raised while LA platform was not raised as it was isolated and utilized as a muster point and for personnel transfer purposes only. The same technology has been successfully applied in Gulf of Mexico in 2006.

As deck raising activity was a new technology for PHEONWJ, it was decided to conduct a detailed risk assessment on the design of the synchronous hydraulic jacking system proposed by Contractor as part of project risk assessment. It was expected that the risk can be measured quantitatively, whether the design provided tolerable risk or not, the major contributor to risk can be identified, and recommendations can be established to reduce the risk for intolerable risk. The risk assessment will focus on fatality and financial risk.

2. Deck Raising Overview

The platforms to be raised were LCOM platform, LPRO platform, LSER platform, including interconnecting bridges and Flare tripods. LA platform was not raised as it was isolated and utilized as a muster point and for personnel transfer purposes only. The deck raising was conducted in two stages due to minimum remaining air gap. On the first stage, the platforms were raised 1 meter and on the second stage the platforms were raised for another 3 meters to give a total of 4 meters elevation.

The deck raising system consisted of 3 main systems: Hydraulic System, Control System, and Structural Parts. The hydraulic system was used to raise the platform by using hydraulic rams. The control system was to ensure the same deck raising elevation and speed to avoid platform tilting during raising. The structural parts consisted of pad eyes for hydraulic jack support, sleeves as the extended deck legs were also used to protect from lateral load during raising, and sleeve pins for temporary securing after completing deck raising.

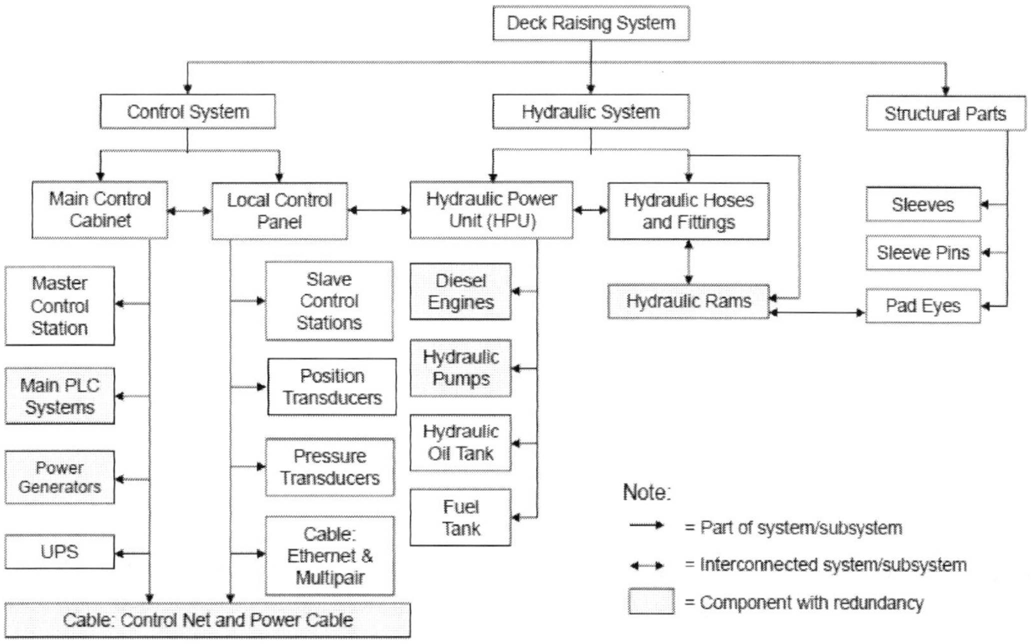

Figure 1 Simplified Function Diagram For Deck Raising System

3. Methodology

The overall methodology for the risk assessment is shown in the flow chart on Figure 2.

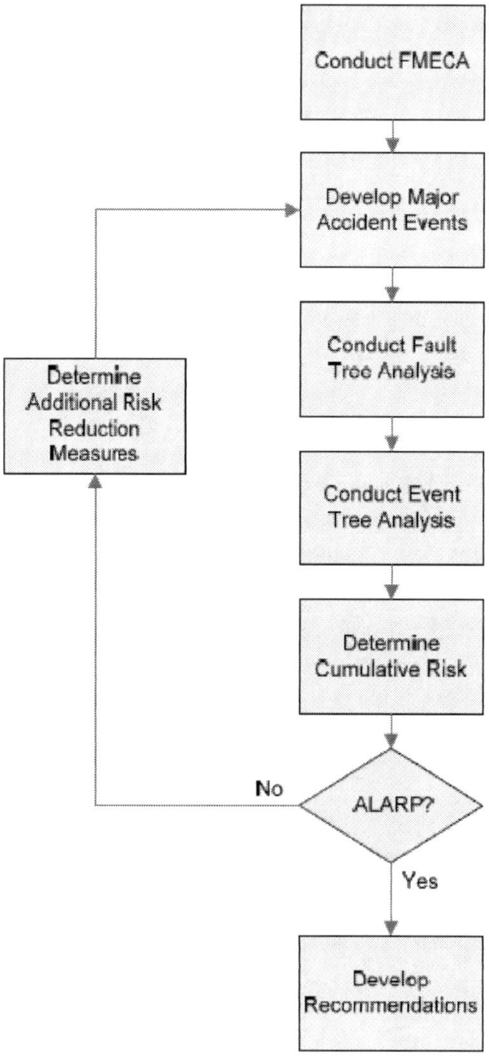

Figure 2 Flow Chart for Risk Assessment Methodology

3.1 Failure Mode, Effect, and Critically Analysis (FMECA)

Failure Mode, Effect, and Critically Analysis (FMECA) was used for qualitative and semi-quantitative analysis of engineering systems. The review was performed in the meeting with personnel that have good knowledge (designers/operators) of the system. The reviewed system was divided into system, subsystem, and equipment. Failure mode, cause, safeguard and effect/consequence for equipment are identified. Critically Analysis was conducted by using PHE ONWJ risk matrix as reference.

3.2 Development of Major Accident Events (MAE)

Major Accident Events (MAE) are any unwanted events with potential to cause injury property damage, business loss, and environmental damage. From FMECA result, any event having major and catastrophic consequence based on PHE ONWJ risk matrix was included as MAE and the risk to be further analyzed.

3.3 Fault Tree Analysis (FTA)

A fault tree analysis is developed based on a fault tree diagram. A fault tree diagram is a structured logic diagram which shows how series and/or combination of failures may lead to MAE. By inputting available failure frequencies into the fault tree diagram, the frequency of MAE can be estimated. Failure frequency was obtained from vendor data, generic failure frequency databases such as OREDA 2009[1] and NPRD 2011[5], and engineering judgments.

3.4 Event Tree Analysis (ETA)

An event tree analysis which is complementary of fault tree analysis is developed based on an event tree diagram. For developing event tree diagrams, layers of protection were identified. Consecutive events from the initiating event can be identified for each layer by developing two branches for its success and failure. The ultimate consequence can be obtained if review of all layers of protection has been completed. By inputting probability of failure or success to the event tree diagram, the frequency of ultimate consequence was estimated. The probability data was obtained from OGP [4] for ignition probability and engineering judgments for other probabilities, for example the probability of successful escape during structural collapse.

3.5 Consequence Analysis

The consequences being considered for this risk assessment are uneven lifting of deck leg leading to structural damage and damage to hydrocarbon piping leading to personnel fatality due to flash fire, personnel fatality due to impact of structure collapse, fatality due to personnel over board, and business loss.

3.6 Risk Estimation

The results of frequency and consequence analysis are combined to obtain the risk for each ultimate consequence. The risk is represented in the form of Individual Risk Per Annum (IRPA) for fatality risk and Probable Loss for Financial Risk. Since the platform raising activity is a short duration activity (3 months), the estimated frequencies will be equivalent 'annualized' frequencies to enable risk comparisons.

3.6.1 Individual Risk Per Annum (IRPA)

Individual Risk Per Annum (IRPA) shows the risk exposure to any individual by considering the duration of a specific person on board. To calculate the IRPA, it is required to obtain the fraction of time spent on board and Location Specific Individual Risk (LSIR) which shows the risk exposure to any individual who is present in a particular area for a full one year.

$$LSIR(/year) = \sum_i f_i \times P_{i,impact} \times P_{i,exposure}$$

$$IRPA(/year) = \sum_i LSIR_i \times f_{time}$$

Where :

fi = event frequency (event/year)

$P_{i,impact}$ = probability of harm from individual incident (injury or fatality)

$P_{i,exposure}$ = probability that an individual may be present within the impairment zone

f_{time} = Fraction of time the specified worker group is present at that point

3.6.2 Probable Loss for Financial Risk

The financial loss from a major accident event may consider the following costs:

- Asset damage and replacement costs
- Loss of production during the period of investigation, repair and decommissioning
- Cost of investigation

The total probable loss for an associated operation is therefore measured as:

$$Probable\ Loss\ (\$/year) = \sum_i L_i \times f_i$$

Where,

Li = Loss from an accident event ($)

fi = accident event frequency (event/year)

3.7 Risk Criteria

The result of risk estimation is then compared with the risk criteria to identify whether the risk of the deck raising system design is within the tolerable region or not. IRPA Criteria is used for fatality risk while PHE ONWJ risk matrix will be used for financial risk.

3.7.1 Individual Risk Per Annum (IRPA) Criteria

The Individual Risk Per Annum (IRPA) Criteria as shown in the Figure 3 is used to define whether the risk is tolerable or not which refers to Health & Safety Executive in the UK. The risk below $1x10^{-6}$ is considered acceptable risk, while risk above $1x10^{-3}$ is considered unacceptable risk. The risk between $1x10^{-6}$ to $1x10^{-3}$ is considered as tolerable risk. For unacceptable risk, implementation of mitigative action is mandatory to reduce the risk to at least the tolerable region. For tolerable risk, basically the risk can be undertaken and recommendation may be considered to be established.

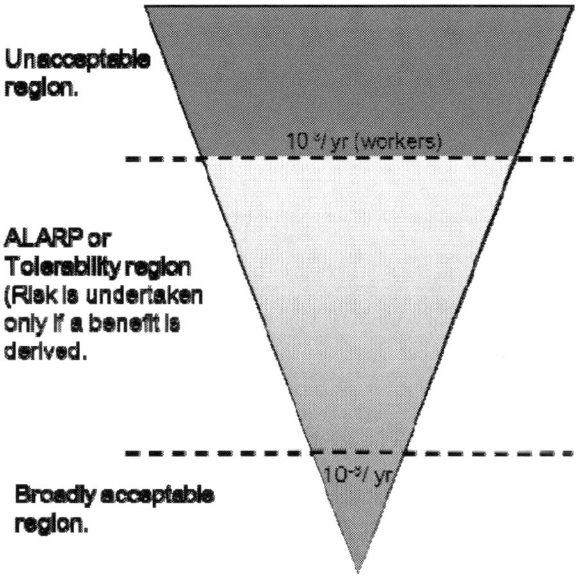

Figure 3 Individual Risk Per Annum (IRPA) Criteria

3.7.2 Financial Risk Criteria

The PHE ONWJ Risk Matrix as shown in the Figure 4 is used to define whether the risk is tolerable or not. The risk that falls on green cells is considered as low / acceptable risk, risk that falls in yellow cells is considered as medium / tolerable risk, the risk that falls in red cells is considered as high / unacceptable risk.

3.8 Develop Recommendation

For high / unacceptable risk, the additional examination is performed to identify the factor that provides major contribution to the risk. The recommendation is then developed based on the identified major risk contributors. For some cases, residual risk after implementing recommendations is not estimated due to data difficulties.

			Likelihood				
			Almost Impossible $<10^{-6}$ per year	**Very Low** $10^{-6} - 10^{-4}$ per year	**Low** $10^{-4} - 10^{-2}$ per year	**Medium** $10^{-2} - 1$ per year	**High** > 1 per year
Severity	A	**Catastrophic** > $ 200 M	6	7	8	9	10
Severity	B	**Major** $ 100 M - $ 200 M	5	6	7	8	9
Severity	C	**Serious** $ 50 M - $ 100 M	4	5	6	7	8
Severity	D	**Moderate** $ 10 M - $ 50 M	3	4	5	6	7
Severity	E	**Minor** $ 1 M - $ 10 M	2	3	4	5	6
Severity	F	**Slight** ≤ 10 M	1	2	3	4	5

Figure 4 Financial Risk Criteria

4. Result

4.1 Major Accident Event (MAE)

Based on Failure Mode, Effect, and Critically Analysis (FMECA), it was identified that uneven raising where there is a deck raising elevation difference relative to other platforms and/or bridges will cause the platform to tilt. The tilting platform and/or bridge will cause the pipe bridge or bridge structure to break down if the stress exceeds the material allowable stress or shifting of the center of gravity that will cause the load to concentrate on a specific platform leg leading to platform collapse. The uneven raising may be caused by misalignment of hydraulic rams during raising or failure of a structural part to withstand the topside load. Table 1 summarizes the identified Major Accident Event (MAE) including its frequency based on Fault Tree Analysis (FTA).

4.2 Individual Risk Per Annum (IRPA)

Based on frequency and consequence analysis, Table 2 and 3 show the resulting Location Individual Specific Risk (LSIR) and Individual Risk Per Annum (IRPA). The risk for misalignment and structural failure is shown to be able to identify the contribution of failure to the risk.

Table 1 Summary of Major Accident Event (MAE) and Its Probability

MAE No.	Description	Event probability (per activity)
1	Potential for damage to LCOM – LSER bridge (misalignment)	5.40×10^{-5}
2	Potential for damage to LPRO – LSER bridge (misalignment)	5.40×10^{-5}
3	Potential for damage to LPRO – LA bridge and LA lift frame (misalignment)	6.14×10^{-5}
4	Potential for damage to LPRO – Flare support bridge A and Flare lift frame A (misalignment)	6.14×10^{-5}
5	Potential for damage to Flare support bridge A – Flare support bridge B and Flare lift frame B (misalignment)	6.97×10^{-5}
6	Potential for damage to LCOM platform (misalignment)	8.57×10^{-5}
7	Potential for damage to LSER platform (misalignment)	8.57×10^{-5}
8	Potential for damage to LPRO platform (misalignment)	8.57×10^{-5}
9	Potential for damage to LCOM platform (structural)	7.59×10^{-4}
10	Potential for damage to LSER platform (structural)	7.59×10^{-4}
11	Potential for damage to LPRO platform (structural)	7.59×10^{-4}
12	Potential for damage to LPRO – LA bridge and LA lift frame (structural)	7.22×10^{-4}
13	Potential for damage to LPRO – Flare support bridge A – Flare support bridge B and Flare lift frame A (structural)	1.23×10^{-4}
14	Potential for damage to Flare support bridge A – Flare support bridge B and Flare lift frame B (structural)	1.23×10^{-4}

The maximum IRPA is 4.6×10^{-6} located on LPRO platform. The risk is considered tolerable as the value is within the range of 1×10^{-6} and 1×10^{-3}. The tolerable fatality risk can be achieved as deck raising was conducted during shutdown and minimum manning level. Structural failure contributes more than 50 % for LSIR and more than 90% for IRPA, except for failure on LPRO-LSER bridge where structural failure is not considered as Major Accident Event. The structural failure provides major contributor as there was not detection method and sudden movement in case of failure may cause major damage.

Table 2 Location Specific Individual Risk (LSIR) Result

Location	LSIR (risk / year)		
	Misalignment	Structural	Total
LCOM	1.1E-05	7.6E-05	8.7E-05
LSER	1.4E-05	7.6E-05	9.0E-05
LPRO	1.4E-05	8.8E-05	1.0E-04
HPU (LPRO-LSER)	2.7E-06	0.0E+00	2.7E-06
HPU (FSA)	1.3E-05	2.5E-05	3.8E-05
HPU (FSB)	7.0E-06	2.5E-05	3.2E-05
HPU (LPRO-LA)	6.1E-06	7.2E-05	7.8E-05
Total	**6.9E-05**	**3.6E-04**	**4.3E-04**

Table 3 Individual Risk per Annum (IRPA) Result

Location	IRPA		
	Misalignment	Structural	Total
LCOM	6.2E-08	3.9E-06	3.9E-06
LSER	7.7E-08	3.9E-06	3.9E-06
LPRO	7.9E-08	4.5E-06	4.6E-06
HPU (LPRO-LSER)	1.5E-08	0.0E+00	1.5E-08
HPU (FSA)	7.2E-08	1.3E-06	1.3E-06
HPU (FSB)	3.8E-08	1.3E-06	1.3E-06
HPU (LPRO-LA)	3.4E-08	3.7E-06	3.7E-06

4.3 Financial Risk

The result of financial risk is shown in Table 4 which shows the frequency of event, potential financial loss which is detailed to production loss and repair cost, and the risk figure which refers to the PHE ONWJ risk matrix.

Structural failure in LCOM, LSER, and LPRO Platform is considered as high / unacceptable risk as it causes major damage and requires a long period for repair. More than 50% of the potential loss is caused by production loss.

Table 4 Financial Risk Result

	MAE No.	Event frequency (per year)	Potential Loss (USD $M per year)			Risk Ranking
			Production	Repair	Total	
Damage due to misalignment during raising	1	5.56×10^{-5}	41.4	3.0	44.4	Low
	2	5.56×10^{-5}	41.4	3.0	44.4	Low
	3	6.28×10^{-5}	22.3	2.5	24.8	Low
	4	6.28×10^{-5}	41.4	3.9	45.3	Low
	5	7.08×10^{-5}	41.4	4.3	45.7	Low
	6	8.57×10^{-5}	44.6	8.8	53.4	Low
	7	8.57×10^{-5}	44.6	5.8	50.4	Low
	8	8.57×10^{-5}	44.6	7.3	51.9	Low
Damage due to structural failure during raising	9	7.59×10^{-4}	69.5	54.3	123.7	High
	10	7.59×10^{-4}	69.5	30.3	99.7	High
	11	7.59×10^{-4}	69.5	42.3	111.7	High
	12	7.22×10^{-4}	41.4	3.2	44.6	Medium
	13	1.23×10^{-4}	41.4	5.9	47.3	Medium
	14	1.23×10^{-4}	41.4	5.9	47.3	Medium

4.4 Recommendation

The study revealed that structural failure is the major contributor to both fatality and financial risk while production loss is the major contributor to financial risk. Further examinations found that welding failure contributes more than 90 % to structural failure. Therefore, the following recommendations are developed to reduce the risk:

- Conduct robust quality check for welding and material of structural parts on fabrication and installation
- Set-up contingency plans during an accident event to minimized downtime to reduce financial loss, such as installation of a by-pass line.

5. Project Implementation

The Lima Flow Station Deck Raising was successfully completed in September 2013. Now the production is operated under normal condition, and is expected to stay in operation until 2026. During the project execution, the recommendations from the risk assessment were consistently implemented. The structural part design was carefully reviewed by a third party and a robust quality check was performed during purchasing, fabrication, and installation. A series of production contingency plans were also set up to minimize loss in case of failure.

6. Conclusion

The potential for overlooked hazards is high for implementation of new technology, such as implementation of the synchronous hydraulic jacking system for Lima Flow Station Deck Raising. Whenever qualitative risk assessment is considered inadequate, , quantitative risk assessment can be conducted by adopting methodology as described on this paper.

The quantitative risk assessment study for Lima Flow Station Deck Raising provided these conclusions:

- Major Accident Events include uneven raising whether it is caused by misalignment of hydraulic rams during raising or failure of a structural part to withstand the topside load. This uneven raising will cause bridge and pipe bridge failure due to excessive stress or platform collapse due to shifting of the center of gravity leading to a concentrated load on a specific leg.
- The maximum IRPA is 4.6x10-6 located on LPRO platform. The risk is considered tolerable as the value is within range of 1x10-6 and 1x10-3.
- Structural failure in LCOM, LSER, and LPRO Platform is considered as high / unacceptable risk as it causes major damage and requires a long period to repair.
- Structural failure contributes more than 50% to fatatility and financial risk while production loss contributes more than 50% to financial risk.
- It was recommended to provide robust quality check for welding and material of structural parts on fabrication and installation and set-up contingency plans during an accident event to minimize downtime to reduce the risk.
- By implementing good project and risk management, Lima Flow Station Deck Raising has completed successfully and started up without any accident.

7. References

[1] OREDA, Offshore Reliability Data Handbook, 2009, Vol.I, Topside Equipment, 5th Edition, Oreda participants, Distributed by DNV, Hovik, Norway.

[2] Mannan,S. (Editor), Lees' Loss Prevention in the Process Industries, 3rd Edition, Elsevier Butterworth-Heinemann, 2005

[3] HSE Books, HMSO, London, Reducing Risks, Protecting People: HSE's Decision Making Process, 2001

[4] Ignition Probabilities, OGP Risk Assessment Data Directory Report No. 434 – 6.1, March 2010.

[5] NPRD, Nonelectronic Parts Reliability Data, Reliability Information Analysis Center (RIAC), Utica, NY, 2011.

[6] Safety Equipment Reliability Handbook, Exida, 2005, Exida.com L.L.C, Sellerville, PA, USA.

348096 A Creative & Strategic Initiative (Champions Model) In Managing Process Hazard: Process Safety Culture

Abstract:

A Creative & Strategic Initiative (Champions Model) In Managing Process Hazard: Process Safety Culture

Rathore, Sharad

Health Safety & Environment

Honeywell India Technology Centre, UOP IPL, Gurgaon

Abstract

A robust Process Safety Culture within an organization is as important as having a reliable process safety management system; it makes process safety a true part of the day-to-day operations. It develops trust and a sense of importance for process safety, allowing process safety management systems to be effective

Culture is the tendency in all of us - our organization - to want to do the right thing in the right way at the right time, ALL the time – even when no one is looking. Culture is the result of all the actions - and inactions - in institutional/workforce memory. These shape/influence individual behaviors and tendencies. The culture transcends the members of the organization, passing on to new members and enduring over time .Good cultures evolve from common values and attitudes developed as the group seeks to properly solve common problems. It is hard to measure and more difficult to change. Culture will be the "root cause" of the decade.

The significance of Safety Culture is becoming increasingly recognized as a key factor to achieve the safe operation not only of hazardous processes but also in Experimental Pilot plants. This paper describes the systematic approaches that are being practices at HITC R&D facilities to go through in their journey to a world class safety culture and applies this concept to Process Safety Management in experimental pilot plant

The foundation for this journey will be illustrated in a model that describes the strategy for HSE culture strength. The model (called Champions Model) has been calibrated against performance and is based on years of data derived from HSE metrics HLS- ACS facilities. In addition, useful tools, employed by facilities to improve their safety culture and performance, will be presented

Key words: Process Safety Management, Safety Culture

348700 Defining Dust Hazard Areas

Abstract:

Defining Dust Hazard Areas

Michelle Murphy, Partner

ioMosaic Corporation

As part of the 2013 update, NFPA 654 provides clarification on how to determine if a flash fire or explosion hazard exists, using a hazard assessment. This hazard assessment incorporates dust characteristics, design of the facility and localized accumulations in the analysis. This presentation will present a multi-step approach to conducting these hazard assessments, simplifying the process. Once complete, the hazard assessments define the areas to evaluate during the process hazard analysis and assure that valuable resources are spent appropriately.

Challenges and achievements in implementing Management of Change system at Binh Son Refinery, Vietnam

Bong Nguyen Thanh
Binh Son Refining and Petrochemical Company Limited
208 Hung Vuong Street, Quang Ngai City, Quang Ngai, Vietnam

Thuc Do Van
Binh Son Refining and Petrochemical Company Limited
208 Hung Vuong Street, Quang Ngai City, Quang Ngai, Vietnam

Vu Vo Hoang
Binh Son Refining and Petrochemical Company Limited
208 Hung Vuong Street, Quang Ngai City, Quang Ngai, Vietnam

Prepared for Presentation at
American Institute of Chemical Engineers
2014 Spring Meeting
10th Global Congress on Process Safety
New Orleans, LA
March 30 – April 2, 2014

UNPUBLISHED

Challenges and achievements in implementing Management of Change system at Binh Son Refinery, Vietnam

Bong Nguyen Thanh
Binh Son Refining and Petrochemical Company Limited
208 Hung Vuong Street, Quang Ngai City, Quang Ngai, Vietnam
bongnt@bsr.com.vn

Keywords: Management of Change, Process Safety, Management, MOC, Refinery, implementation

Abstract

This paper describes the challenges and achievements in implementing a MOC system in BSR, Vietnam, where there is no regulations such as OSHA PSM rule. Binh Son Refining and Petrochemical Co. Ltd, a subsidiary of PetroVietnam commissioned its refinery and polypropylene units located at Dung Quat bay, Quang Ngai Province in the year 2009. This plant was designed and built by a consortium of international process licensors and detailed engineering contractors and has a capacity to handle 6.5 million tons per year of Bach Ho (80 %) and some other imported crude oils. This is the first and only refinery in Vietnam.

This paper describes the evolution of Management of Change system at BSR, the challenges faced and the achievements so far.

1. Introduction

Binh Son Refining and Petrochemical Co. Ltd, (BSR) a subsidiary of PetroVietnam commissioned its refinery and polypropylene units located at Dung Quat bay, Quang Ngai Province in the year 2009. This plant was designed and built by a consortium of international process licensors and detailed engineering contractors and has a capacity to handle 6.5 million tons per year of Bach Ho (80 %) and some other imported crude oils. This first and only refinery in Vietnam is a project of national importance, is responsible for the production of petroleum products, and is contributing to energy security and stabilization of oil prices in the domestic market of Vietnam.

The refinery configuration includes 14 technology process units to produce the following products: Polymer Grade Propylene, LPG, Domestic Gasoline, Kerosene and Jet A1, Auto Diesel and Fuel Oil.

Table 1. Products of Binh Son Refinery

Product Name	Tons / year
LPG	400000-420000
A92 MOGAS	1400000-1800000
A95 MOGAS	600000-700000

Kerosene / Jet A1	80000-400000
Auto Diesel	2900000-3200000
Fuel Oil (FO)	60000-10000
Polypropylene	130000-150000

The refinery also has its own utility and power generation systems. Crude feedstock is unloaded via a single point mooring system and stored in the crude tank farm.An intermediate refinery tankage is provided within the refinery fence, however, finished products are stored in a separate product tank farm. Product loading facilities are provided for both Ocean and Coastal ships. Truck loading is also provided for supplying products locally.

The number of employees at the refinery is around 1100. A very systematic training of employees by the Licensors and EPC contractors, including training in overseas refineries was done before the commissioning of refinery and contributed to the successful commissioning and safe and stable operations thereafter.

2. QHSE Management and Process Safety Management

The refinery is certified to Quality Management Standard: ISO 9001:2008, Environmental Management System ISO 14001: 2004, and Management System for Occupational Health and Safety: OHSAS 18001: 2007. Under the QHSE management system, ensuring safe, stable and efficient production is given prime importance. The Process Safety Management (PSM) is part of the HSE Management system and enhances the effectiveness of HSE management system to deal with low frequency high consequence events.

The PSM system of BSR is adopted from the US OSHA PSM elements. References were also made to the UK COMAH Regulations, EU SEVESO Directives. DOE-HDBK-1100-2004, DOE-HDBK-1101-2004; OSHA 3132-2000, OSHA 3133-1994; CCPS; 2007, and Australian Major Hazard Facilities Regulations. The PSM system consists of 14 elements. Process Safety Management implementation started in the year 2010 and one of the initial elements that was put into practice is Management of Change.

Under OSHA PSM rule, Management of Change requires documented procedures aimed at insuring safe system operation as changes to the system occur. The management of change process guides updates or development to process safety information, process hazard analyses, operating procedures, and training to insure all elements of the PSM program are current.

3. Evolution of Management of Change system

The objective of Management of Change is to provide a system of administrative controls that ensures:

1. The proper consideration of direct and indirect potential hazards before the change is implemented;
2. That procedures are modified and essential training and communication occurs before the modification is put into service;
3. Enhanced emphasis on the preparation and retention of records of each change; and

4. The ability to audit the changes.

The procedure to follow Management of Change was first issued in 2009. Since then it has undergone many revisions to improve the Management of Change system and its practice. The improvements to the procedure were done based on periodical audits and addressing the gaps in the MOC system. The current MOC procedure is the sixth version from the beginning and put into practice since July 2013.

Key aspects of improvement to MOC procedure and the reason:

Version	Key aspects of improvement	Reason
Version 2	Develop supporting discipline and hazards review checklists for hazard evaluation and design reviews of change	- To assist people to address the important discipline concerns on safety, reliability, operability - Improve the MOC documentation and tracking to ensure it complies with QHSE management system requirements
Version 3	Adopt HAZOP and PSSR procedure to the implementation of MOC	Make links of the MOC procedure with other PSM elements
Version 4	Integrate the MOC procedure with project implementation	To ensure it is in line with the organization and consistency in roles and responsibilities
Version 5	Adopt the Safety Interlock and Protective Systems (SIPS) bypass procedure into MOC system	To manage all bypasses to safety interlocks (e.g. Maintenance Override Switch-MOS) and protective systems (RV, ESD, etc.)
Version 6	Adopt some advanced PHA methods in hazard evaluation of changes such as SIL study, QRA	Improve quality and effectiveness of the process hazard analysis of changes

The current MOC workflow consists of the following key steps:

1. Change initiation
2. Change screening and hazards evaluation
3. Design review and approval
4. Verification of modification and ready for start-up
5. Post change review

The key role in the success of MOC system implementation is played by MOC Element custodian and the Change Responsible Persons who are responsible for implementing each change.

4. Challenges in implementing of MOC

Apart from the most common problems in implementing of MOC that have been seen at our facility as of in worldwide industry, the company has faced some significant challenges which adhere to our local safety and management culture and experiences, which can be addressed as below:

4.1. Process Safety Cultural Issues

In the initial years, a culture that encouraged raising process safety concerns regarding any change proposal was not existing. Corrective actions were not promptly taken when unsafe process safety conditions were brought to management's attention. The awareness of employee about safety was still limited in boundary of occupational safety such as slip, trip or fall. Beside that commitment to Process Safety Management was limited and inconsistent in the context of the production was the first interest.

4.2. Employee professionalism and experiences

As mentioned earlier, Binh Son Refinery is the first refinery in Vietnam. Therefore BSR employees lacked experiences, not only in operating and maintenance but also in management of refinery. It gave a challenge in identification and consideration of suitable and sufficient information to support the MOC process. In addition, the Process Safety Information documents were not sufficient available, e.g., project engineering drawings had not been verified as as-built.

4.3. Organizational Issues

In the early period of the refinery operating, there were several gaps and inconsistencies in defining roles, responsibilities and authorities among facility owner and disciplines. The designing of MOC system therefore became more complex. These gaps in the organization caused a tendency to pass the buck to others whenever a change was proposed.

The employee participation was one of our big challenges during the implementation of MOC and PSM as well. Less people are responsible for process safety task force. Workers at all level of the refinery did not actively participate in hazards evaluations and design review of changes.

One of the other organizational issues is that training and performance systems were not systematically established for effectively providing the knowledge and skill to the MOC implementers.

4.4. Other challenges

The other elements of PSM system had not been implemented fully leading to gaps in MOC activities such as HAZOP study, PSSR, Operating Procedure and PSI availability and updating, etc.

The lack of a Vietnam regulatory in PSM like the OSHA PSM regulation also limited the application of system like MOC in BSR.

5. Achievements

As a result of a continuous improvement process, we have accomplished some encouraging achievements in implementation of the MOC system in our refinery which is in line with the implementation of the Process Safety Management in compliance with OSHA's PSM regulation:

1. Good commitment to adhere to MOC procedure at all levels of management for all changes other than replacement in kind.
2. The MOC system covers all disciplines. MOC is even raised for a variety of changes, not having process safety concerns. Examples are: changes such as format change in bill of loading or software upgrade also follow the MOC route and approval process and updating relevant documents.
3. A good training system is in place for maintaining the proper process safety awareness and competency of MOC implementers.
4. An effective compliance audit system which consists of 2 levels (Tier 1 and Tier 2) and continual improvement.
5. Competency of MOC implementers has been much improved and key performance indicators of MOC process have become the job performance evaluation indicators.

6. Way forward

The recent PSM internal high level audit has brought out the following gaps, not only with MOC but also with other PSM elements that make links to the MOC implementation:

- The compliance of capital project implementation with MOC requirements.
- Lack of editable engineering documents leading to some difficulties in updating of modification in affected drawings.
- Paper-based MOC document handling leading to cost and schedule over run and gaps on communication and tracking during MOC approval processing and execution

It is proposed to address the above gaps through short-term and long-term action plans for enhancement and consolidation of the MOC implementation within PSM system.

BSR is also developing an electronic-based MOC system to track MOC right from initiation and integrate MOC with work order and purchasing system using our current CMMS (Computerized Maintenance Management System).

Excellence is a never ending journey and BSR is moving towards a MOC system covering all kinds of change and having a simple robust process which is commensurate with the risk introduced by the changes.

7. References

[1] Process Safety Management Guidelines for Compliance, https://www.osha.gov/Publications/osha3133.html (accessed Feb. 20, 2014)

[2] Center for Chemical Process Safety, *Guidelines for Management of Change for Process Safety,* 2008

349936 Process Safety Culture Applied in Latin America Oil and Gas Industry: Experiences in Ecopetrol Colombia

Abstract:

In some installations of oil and gas industry a wide range of risks and hazards are likely to come up together, since there is usually a space shortage. Additionally the complexity of processes has been increased dramatically in petroleum and petrochemical industry, especially in Latin American oil and gas installations. Therefore, there must be a continuous effort to implement strategies that optimize the preventive treatment when hazards like space shortage and process complexity combined by a lack of discipline all mixed with a irresponsible cost reduction policies. This paper will show experiences in Colombia, focused in ECOPETROL improvement in that way.

348497 Ammonium Nitrate Condition-Dependent Thermal Decomposition

Abstract:

Runaway reactions present a potentially serious threat to the chemical process industry. In general, this research topic will investigate the mechanisms that drive these reactions, develop predictive models, and create appropriate and effective mitigation strategies. Specifically, this project will look at identifying the root causes associated with ammonium nitrate (AN) explosions, so that the risk can be reduced, while maintaining its agricultural benefit. This work focuses on developing reaction stoichiometry under different conditions, thermodynamic parameters and kinetic parameters related to AN thermal decomposition. In this research, Reactive Systems Screening Tool (RSST), Automated Pressure Tracking Adiabatic Calorimeter (APTAC), and Differential Scanning Calorimetry (DSC) will be used.

Ammonium nitrate can explode under favorable conditions, such as the existence of specific additives. Some materials change the chemical reaction conditions, inhibiting the explosion; others act as promoters, increasing the explosion risk. This study will look into the behavior of AN with different types of additives including promoters and inhibitors, combined and in isolation. Apart from that, weathering effect, humidity effect, and volume effect will be examined to find out how the storage condition will affect decomposition of AN. Those results will be compared with the work done in literature and some regulatory issues associated with it.

348758 CFD Modeling for Prediction and Prevention of Runaway Reaction

Abstract:

Thermal runaway reactions have caused many industrial incidents with dire consequences, including injury and death to humans, negative effects on the environment, and economic losses for companies. The U.S. Chemical Safety and Hazard Investigation Board (CSB) released a report on 167 reactive chemical incidents in the U.S. between 1980 and 2001. About 35% of the incidents were caused by runaway reactions. With the aim to prevent the occurrence of these incidents, it is necessary to have a good understanding of the events that can lead to a runaway reaction.

Mixing operations have an important effect not only in the adequate contact of reactants which affects the performance of the reaction, but also in removing the excess heat generated by the reaction. For exothermic reactions this removal of heat is a crucial operation because when the heat generated exceeds the heat removal, the excess of temperature will increase the heat generation rate, which would produce an auto-acceleration behavior triggering uncontrolled conditions of temperature and pressure inside the reactor.

The mixing inefficiencies lead to non-uniform mass flow distribution which causes inhomogeneous temperature profiles across the reactor. Therefore, it is important to predict this temperature distribution to locate and understand the characteristics of local hot spots that ultimately may provoke a thermal runaway.

Computational Fluid Dynamics (CFD) techniques have been used to identify local and instantaneous values of reactor temperatures and reactant concentrations in mixing tanks to indicate in advance the local hot spots in the reaction mixture. This kind of information is of particular interest to determine situations that can produce a thermal runaway in order to make engineering decisions to avoid them. Specifically, in some cases this information has been used to create an Early Warning Detection System (EWDS) for runaway detection.

The main objective of this work is to discuss the effect of mixing by paddle on the heat accumulation rate, considering the difference between the heat of reaction and the loss of heat by transfer. The results can provide a good understanding of the reaction system using an anchor impeller.

In this work, a time dependent simulation was carried out using a Multiple Reference Frame (MRF) technique; in this method a region away from the impeller is defined as a stationary frame and a region adjacent to the impeller is defined as a rotating frame, but the impeller itself remains stationary with respect to the MRF. Geometry of the tank was created similarly to the Mettler Toledo RC1*e* reaction calorimeter stirred by an anchor impeller. The simulations were carried out with a three dimensional CFD model using the ANSYS FLUENT 14.0 software.

A moderate to high exothermic reaction, the hydrolysis of acetic anhydride, is used as a model reaction in this work. This is a pseudo first-order reaction, ideal for validating the simulation by experimental results on the dynamic response of a calorimetric reactor. This reaction model is a simple mechanism that has been well studied with much experimental work reported in the literature.

The conclusions drawn from this work will have important implications to elaborate future research by developing a method for early warning detection of runaway reaction and extending this idea to more complex mechanisms of highly exothermic reactions.

348859 Sensitivity Analysis of Variables Affecting the Runaway Decomposition of Dicumyl Peroxide

Abstract:

Dycumil Peroxide (DCP) is widely used as an initiator, hardener and as a cross linking agent in the petrochemical industry. The use of this peroxide poses an intrinsic risk due to its reactivity and the possibility of its highly exothermic self-decomposition and subsequent runaway reaction, which is accompanied by a rapid rate of temperature and pressure increase.

The hazards associated with the process, transport and storage of DCP are clearly reflected by the frequent incidents that have occurred in Asia in the last two decades: Taiwan (2008) [1], Taiwan (2003) [2], Japan (1999) and Taiwan (1988)[1]. Though the accidents associated to DCP have not led to any fatality, it is important to be aware of its decomposition properties so as to prevent further incidents and major consequences.

Emergency relief systems are the most commonly selected measure to protect reactors from thermal explosions (CCPS, 1998[3]). The main advantages of their use are the independence from the control system and passivity (and therefore reliability and robustness). However, the exiting relief sizing methods are consolidated mainly in the case of vapor systems (in which the relief operation is usually enough to cool down the reacting mixture), leading to oversizing in the case of untempered gassy systems (such as the decomposition reaction of DCP[4]).

In the present work, the results of adiabatic experiments using two different calorimeters (PHI TEC I and PHI TEC II) are presented. A sensitivity analysis around the equipment configuration (open and close cell), the concentration of the peroxide (diluted in2,2,4-Trimethyl-1,3-pentanediol diisobutyrate), initial fill level, starting pressure and phi factor on the different runaway parameters (maximum pressure, gas generation, maximum rise of temperature and pressure rates, time to maximum rate and onset temperature) is discussed. Finally vent sizing calculations are performed using different methodologies available in the open literature. Based on the calorimetric data and the vent sizing calculations the limitations of these methodologies are discussed and recommendations for the proper vent sizing and safe scale up of the system studied are made.

References

1. Shen S-J, Wu S-H, Chi J-H, Wang Y-W, Shu C-M. Thermal explosion simulation and incompatible reaction of dicumyl peroxide by calorimetric technique. *Journal of Thermal Analysis and Calorimetry*. 2010;102(2):569–577. doi:10.1007/s10973-010-0916-4
2. Wu KW, Hou HY, Shu CM. Thermal phenomena studies for dicumyl peroxide at various concentrations by DSC. *Journal of Thermal Analysis and Calorimetry*. 2006;83(1):41–44. doi:10.1007/s10973-005-6983-2.
3. Center for Chemical Process Safety, 1998, Guidelines for Pressure Relief and Effluent Handling Systems, American Institute of Chemical Engineers
4. Round robin vent sizing exercise on a gassy system: 40% dicumyl peroxide in butyrate solvent. Health and Safety Laboratory ; INERIS Collaboration, 2011. (s)
5. Luc Véchot, Jake Kay, Jill Wilday, Douglas Carson Jean-Pierre Bigot, Runaway reaction of non-tempered chemical systems: Development of a similarity vent-sizing tool at laboratory scale, Journal of Loss Prevention in the Process Industries, Volume 21, Issue 4, July 2008, Pages 359-366, ISSN 0950-4230, http://dx.doi.org/10.1016/j.jlp.2008.01.005

Beyond the Phi Factor: Correction of Experimental Data for Vaporization in Tempered Reactions, for Pressure-Relief System Design

Guibing Zhao, Ph.D.
Chilworth Technology Inc.
113 Campus drive, Princeton, NJ 08540
Guibing.Zhao@dekra.com
www.chilworth.com

Prepared for Presentation at
American Institute of Chemical Engineers
2014 Spring Meeting
10th Global Congress on Process Safety
New Orleans, LA
March 30 – April 2, 2014

UNPUBLISHED

Beyond the Phi Factor: Correction of Experimental Data for Vaporization in Tempered Reactions, for Pressure-Relief System Design

Guibing Zhao, Ph.D.
Chilworth Technology Inc.
113 Campus drive, Princeton, NJ 08540
Guibing.Zhao@dekra.com

Keywords: Emergency relief system, fill ratio, vaporization, calorimeter, thermo-kinetics

Abstract

Design of emergency relief systems to accommodate runaway reactions requires using bench-scale adiabatic calorimeters to evaluate thermo-kinetic data of a full-scale reactor. However, the well-known method to correct bench-scale experimental data for "thermal inertia" – the Phi factor – does not consider the effect of vaporization on thermo-kinetic data. Calorimeter volumetric fill ratio is an important parameter directly relating to the vaporization effect and is discussed in this paper. For a 60% fill ratio, a case study of an aqueous reaction system shows that the measured reaction heat would be 7% less than the actual value for a reaction with a heat release of 800 J/g, 5% less than the actual value with a heat release of 600 J/g, and 2% less than the actual value with a heat release of 300 J/g. The self-heating rate is even more sensitive to the fill ratio. A high fill ratio is desirable, to get more-accurate self-heating data for a tempered exothermic reaction. A low fill ratio requires a significant Phi correction and could also give a misleading thermo-kinetic interpretation of a plant-scale reactor. Applying the same fill ratio of a plant-scale reactor to tests in an adiabatic calorimeter – to get thermo-kinetic data for emergency relief sizing – is recommended.

1. Introduction

The design of an emergency relief system (that is, a pressure-relief or safety valve or a rupture disk, and the associated piping) for process vessels – in which runaway reactions might occur – requires knowledge of the thermo-kinetics of the reactions involved. The use of adiabatic calorimetry for the quantification of the hazards of exothermic runaway reactions has considerably increased over the last 30 years or so. To a great extent, this has been the result of process hazards analyses that have identified the need for emergency relief systems, and the design of such systems involves calculation methods which require adiabatic data input.

A plant-scale reactor has a smaller surface-to-volume ratio as compared to a laboratory-scale reactor (or to an adiabatic-test cell), since this ratio is inversely proportional to radius for a spherical vessel. As a result, the heat loss to the surrounding environment (which is proportional to the heat transfer/surface area) and the heat loss to the body of the reactor (proportional to the mass of the reactor) per unit mass of reactants loaded are smaller than those characteristics in laboratory equipment. Thus, great efforts have been made throughout the industry to develop an adiabatic calorimeter that can simulate exothermic runaway reactions in a large plant reactor, using a small-scale cell that has similar heat losses to both the environment and to the mass of the cell. Such widely-accepted and commercially-available adiabatic calorimeters include, but are not limited to, the VSP II (Venting Sizing Package II) and the APTAC (Automatic Pressure Tracking Adiabatic Calorimeter). Both of them have adopted automatic pressure-tracking technology to balance the pressures inside the test cell and outside the test cell, and this allows the use of a thin-wall test cell at high pressures. This is critical to reducing the cell mass and, therefore, reducing the heat loss to cell wall. Adiabaticity is accomplished by the surrounding guard heaters which track the cell temperature and thus maintain a near-zero temperature difference between cell wall and surroundings. Three key thermo-kinetic parameters – heat of reaction, temperature-increase rate, and pressure-increase rate – can be measured accurately under the operating conditions of a full-scale reactor. Also, the activation energy, the frequency or pre-exponential factor, and the reaction order can be extracted from the data if a single reaction is involved [1].

The thermal inertia, or Phi factor, is a well-known parameter to describe the heat loss to the test-vessel wall. A higher Phi factor means more energy is absorbed by the testing cell, while a lower Phi factor means more energy is retained in the sample being tested. A Phi factor of unity means that all of the reaction energy is retained in the tested material. The results of testing with a low Phi factor (close to 1) are considered to be representative of the results that would be obtained in a plant-scale reactor and, therefore, could be directly used for designing an emergency relief system. However, the results of testing with a higher Phi factor can be corrected to that with a Phi factor close to 1 if a single reaction is involved [2]. This allows experimentalists to load a smaller quantity of reactants in the test cell (thus, a higher Phi factor, but less likely to rupture the cell) and later correct the results of the test to a Phi factor close to 1. Thus, the results would represent the thermo-kinetic behavior of a plant-scale reactor. The HSE workbook recommends [3] that it may not be necessary to correct self-heat rate data for thermal inertia if the thermal inertia of the calorimeter is less than or equal to that of the full-scale reactor. However, DIERS [2] recommends filling the test cell to 60% or more to obtain the best results, but it does not explain why and what the effect would be with a lower fill ratio (fill ratio means the ratio of sample volume to cell volume). Changes in fill ratio, with corresponding differences in Phi factors, would affect the test results. The real problem is that the corrected results – to obtain the behavior for a Phi factor that would be close to unity – from tests with different fill ratios (with different Phi factors) may not give the same result. For example, a low fill ratio will lead to a high Phi factor and also a high head-space volume, and this would allow more vaporization of volatile materials, such as solvents. The well-known basis for the Phi factor correction [2]Error! Bookmark not defined. only considers heat loss to the cell wall, without considering heat loss due to the endothermic process of vaporization. During the past 30 years of practice with adiabatic test equipment, no reports or publications are found to explore this issue.

For a gassy reaction system, there is no vaporization issue, because little energy is carried from the sample with the generated gas. For a tempered (vapor-generating) system or a hybrid (gas and vapor generating) system, vaporization plays an important role during the testing of exothermic reactions if a low fill ratio is used. The goal of this article is to understand the effects of vaporization on the tested thermo-kinetics of exothermic runaway reactions.

2. Modeling of Vaporization in a Closed Test Cell

2.1 Reaction Heat

As shown in Figure 1, consider a closed test cell with a total volume V_0 with a mass of m_0 of reactants, with the following assumptions:

(1) Liquid specific volume, v_l, is far less than vapor specific volume, v_v,

(2) Vapor-liquid phase equilibrium is established instantaneously at any temperature, and the relationship between vapor pressure and temperature follows the Antoine equation,

(3) Vapors obey the ideal gas law,

(4) Constant liquid density and constant heat capacity of reaction system,

(5) Constant solvent vaporization latent heat and $\Delta H_{lv} \approx \Delta U_{lv}$.

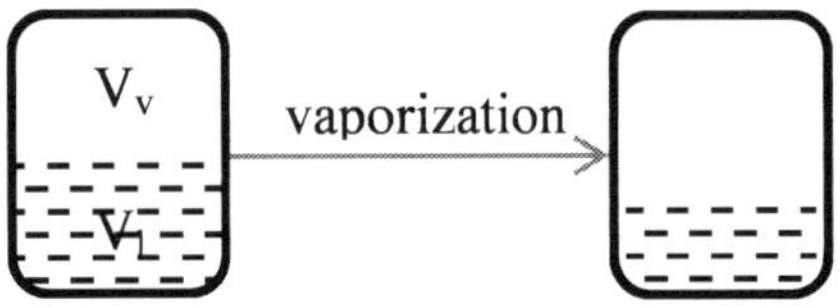

Figure 1. Liquid vaporization process

Vapor mass, m_v, at temperature T, is

$$m_v = \frac{P_v V_v M}{RT} = \frac{P_v M}{RT}\left(V_0 - \frac{m_l}{\rho_l}\right) = \frac{P_v M}{RT}\left(V_0 - \frac{m_0 - m_v}{\rho_l}\right) = \frac{P_v M}{RT}\left(V_0 - \frac{m_0}{\rho_l}\right) + \frac{P_v M}{RT}\frac{m_v}{\rho_l} \quad \text{[Eq.1]}$$

Solving [Eq.1] gives

$$m_v = \frac{V_0 - \frac{m_0}{\rho_l}}{\frac{RT}{P_v M} - \frac{1}{\rho_l}} \quad \text{[Eq.2]}$$

Applying Antoine Equation to [Eq.2] gives

$$m_v = \frac{V_0 - \frac{m_0}{\rho_l}}{\frac{RT}{M \cdot \exp\left(A - \frac{B}{T}\right)} - \frac{1}{\rho_l}} \quad \text{[Eq.3]}$$

From onset temperature, T_0, to any temperature, T, of a runaway reaction, the vapor mass ratio in the head space is

$$\frac{m_v}{m_{v0}} = \frac{T_0}{T} \cdot exp\left[B\left(\frac{1}{T_0} - \frac{1}{T}\right)\right] \quad \text{[Eq.4]}$$

From the onset temperature to the maximum temperature of a runaway reaction, the vapor mass change is

$$\Delta m_v = m_{v0}\left\{\frac{T_0}{T_f} \cdot exp\left[B\left(\frac{1}{T_0} - \frac{1}{T_f}\right)\right] - 1\right\} \quad \text{[Eq.5]}$$

The total energy balance is

$$Q = \phi m_0 C_{vl}\left(T_f - T_0\right) + \Delta m_v \cdot \Delta H_{lv}$$
$$= \phi m_0 C_{vl}\left(T_f - T_0\right) + m_{v0}\Delta H_{lv}\left\{\frac{T_0}{T_f} \cdot exp\left[B\left(\frac{1}{T_0} - \frac{1}{T_f}\right)\right] - 1\right\} \quad \text{[Eq.6]}$$

where ϕ is the Phi factor. If the heat loss to vaporization is not considered, the last term in [Eq.6] is omitted, and the final temperature would be calculated from

$$T_f = T_0 + \frac{Q}{\phi m_0 C_{vl}}$$ [Eq.7]

However, if vaporization is considered, the final temperature, T_f, can be solved by "trial-and-error" iteration from [Eq.6].

2.2 *Self-heat rate: dT/dt*

From [Eq.7], the heat release rate without considering vaporization is

$$q_{nv} = \frac{dQ}{dt} = \phi m_0 C_{vl} \left(\frac{dT}{dt}\right)_{nv}$$ [Eq.8]

The heat release rate with considering vaporization is, from [Eq.6]

$$q_v = \frac{dQ}{dt} = \frac{dm_v}{dt} \Delta H_{lv} + \phi m_0 C_{vl} \left(\frac{dT}{dt}\right)_v$$ [Eq.9]

From [Eq.2],

$$\frac{dm_v}{dt} = -\frac{V_0 - \frac{m_0}{\rho_l}}{\left(\frac{RT}{P_v M} - \frac{1}{\rho_l}\right)^2} \frac{R}{M} \left(\frac{1}{P_v} - \frac{T}{P_v^2} \frac{dP_v}{dT}\right) \left(\frac{dT}{dt}\right)_v$$ [Eq.10]

Applying the Clapeyron equation

$$\frac{dP_v}{dT} = \frac{\Delta H_{lv}}{T \Delta v}$$ [Eq.11]

to [Eq.10]:

$$\frac{dm_v}{dt} = \frac{V_0 - \frac{m_0}{\rho_l}}{\left(\frac{RT}{P_v M} - \frac{1}{\rho_l}\right)^2} \frac{R}{M P_v^2} \left(\frac{\Delta H_{lv}}{\Delta v} - P_v\right) \left(\frac{dT}{dt}\right)_v$$ [Eq.12]

Applying [Eq.12] to [Eq.9], the heat release rate with considering vaporization can be obtained as

$$q_v = \left[\frac{V_0 - \frac{m_0}{\rho_l}}{\left(\frac{RT}{P_v M} - \frac{1}{\rho_l}\right)^2} \frac{R}{M P_v^2} \left(\frac{\Delta H_{lv}}{\Delta v} - P_v\right) \Delta U_{lv} + \phi m_0 C_{vl}\right] \left(\frac{dT}{dt}\right)_v$$ [Eq.13]

The heat release rate from an exothermic runaway reaction is, using the Arrhenius equation

$$q = r \cdot V_l \cdot \Delta H_r = k \cdot f(C) \cdot V_l \cdot \Delta H_r = A e^{-\frac{E}{RT}} \cdot f(C) \cdot V_l \cdot \Delta H_r$$ [Eq.14]

where f(C) is an "unknown" function of concentration in the expression of reaction rate. Therefore, the heat release rate increases exponentially with an increase in temperature as shown in [Eq.14]. If vaporization is significant, the endothermic vaporization process would exponentially affect the reaction rate, thus the rate of temperature increase, dT/dt. For an easy comparison, the ratio of temperature rise rate without considering vaporization and with vaporization considered at the same heat release rate can be deducted from [Eq.8] and [Eq.13],

$$\frac{\left(\frac{dT}{dt}\right)_{nv}}{\left(\frac{dT}{dt}\right)_v} = \frac{V_0 - \frac{m_0}{\rho_l}}{\phi m_0 C_{vl} (\Delta v)^2 M P_v^2} \frac{R}{} \left(\frac{\Delta H_{lv}}{\Delta v} - P_v\right) \Delta H_{lv} + 1$$ [Eq.15]

3. Case Study

A case study was conducted to understand how significantly the fill ratio would affect the heat of reaction and the temperature rise rate. A lower fill ratio – corresponding to a higher Phi factor – will have more head space, which allows more solvent to vaporize. Therefore, the effect of fill ratio on the thermo-kinetics of a runaway reaction would be the same as the effect of vaporization on the thermo-kinetics. For an aqueous reaction system, it is not uncommon to find the following physical properties:

$\rho_l = 1200\ kg \cdot m^{-3}$,

$Cv_l = 3000\ J \cdot kg^{-1} \cdot K^{-1}$,

$M = 0.018\ kg \cdot mol^{-1}$.

The Antoine constants for water vapor, with pressures in Pascals and absolute temperatures

$A = 24.24$

$B = 4736$

$\Delta H_{lv} = 2188\ kJ \cdot kg^{-1}$

The error of the calculated vapor pressure based on the above Antoine constants and the reported water vapor pressure data [4] is about ±3%. An adiabatic VSP cell material made of Hastelloy C has a $Cp = 420\ J \cdot kg^{-1} \cdot K^{-1}$, an empty cell weight of 41.5 g (including stirrer), and a volume of 110 ml.

Assuming the onset temperature of a runaway reaction to be $T_0 = 100°C$, i.e., the normal boiling temperature of water, the final temperature, T_f, can be calculated based on the reaction heat, ΔH_r, in J/g by solving [Eq.6] or [Eq.7]. The total energy released, Q, can be calculated by the multiplication of mass of reactant loaded and the reaction heat,

$$Q = m_0 \cdot \Delta H_r \qquad \text{[Eq.16]}$$

The result of the energy release is: increased temperature of the sample, increased temperature of the cell wall, and increased vaporization of water from the sample. The heat loss to cell and sample is expressed in the first term of [Eq.6]. The heat loss to vaporization is expressed in the second term of [Eq.6]. The numerical iteration is conducted to calculate the final temperature, T_f, for different fill ratios from [Eq.6]. The fraction of heat loss to vaporization (x_v), test cell (x_c), and sample (x_s) can further be calculated as follows

$$x_v = \frac{m_{v0}\, \Delta H_{lv} \left\{ \frac{T_0}{T_f} \cdot exp\left[B\left(\frac{1}{T_0} - \frac{1}{T_f} \right) \right] - 1 \right\}}{m_0 \cdot \Delta H_r} \qquad \text{[Eq.17a]}$$

$$x_c = \frac{(\phi - 1) C_{vl} (T_f - T_0)}{\Delta H_r} \qquad \text{[Eq.17b]}$$

$$x_s = \frac{C_{vl} (T_f - T_0)}{\Delta H_r} \qquad \text{[Eq.17c]}$$

As an example, the calculated results for a reaction heat of 800 J/g are shown in Table 1, for a 110-cubic-centimeter test vessel. For different scenarios, similar calculations can be made.

Table 1. The calculated results for a reaction heat of 800 J/g for an aqueous system

m_0, g	V_l, ml	fill ratio	Phi factor	$(T_f)_{nv}$, C	m_{v0}, g	$(T_f)_v$, C	Pv, bar	m_v at T_f, g	Δm_v,g	x_v	x_c	x_s
1.00	0.830	0.00758	6.81	139	6.56E-02	130	2.63	1.55E-01	8.91E-02	0.243	0.646	0.111
5.00	4.17	0.0379	2.16	223	6.36E-02	188	11.7	5.86E-01	5.22E-01	0.284	0.385	0.331
10.0	8.33	0.0758	1.58	269	6.11E-02	223	23.9	1.07E+00	1.01E+00	0.273	0.267	0.460
20.0	16. 7	0.152	1.29	307	5.61E-02	259	45.9	1.77E+00	1.71E+00	0.231	0.173	0.596
30.0	25.0	0.227	1.19	323	5.11E-02	280	64.9	2.20E+00	2.15E+00	0.192	0.131	0.677
40.0	33.3	0.303	1.15	333	4.61E-02	296	81.6	2.44E+00	2.40E+00	0.160	0.107	0.734
50.0	41.7	0.379	1.12	339	4.11E-02	307	96.4	2.53E+00	2.49E+00	0.132	0.090	0.777
60.0	50.0	0.455	1.10	343	3.61E-02	317	110	2.50E+00	2.47E+00	0.109	0.079	0.813
70.0	58.3	0.530	1.08	346	3.11E-02	325	122	2.37E+00	2.34E+00	0.088	0.070	0.842
80.0	66.7	0.606	1.07	349	2.61E-02	331	133	2.15E+00	2.13E+00	0.070	0.063	0.867
90.0	75.0	0.682	1.06	350	2.10E-02	337	143	1.86E+00	1.84E+00	0.053	0.057	0.889
100	83.3	0.758	1.06	352	1.60E-02	342	153	1.50E+00	1.49E+00	0.039	0.053	0.908
110	91.7	0.833	1.05	353	1.10E-02	347	162	1.09E+00	1.08E+00	0.026	0.049	0.926
120	100	0.909	1.05	354	6.01E-03	351	170	6.22E-01	6.16E-01	0.013	0.046	0.941
130	108	0.985	1.04	355	1.00E-03	355	178	1.08E-01	1.07E-01	0.002	0.043	0.955

3.1 Reaction Heat

Figure 2 shows the calculated results with vaporization considered, by solving [Eq.6], and without considering vaporization, by solving [Eq.7], at reaction heats of 300 J/g, 600 J/g, and 800 J/g. For the same fill ratio, the higher reaction heats lead to higher final temperatures, whether vaporization is considered or not. With increasing fill ratios, the Phi factor decreases more than proportionally, as shown in Table 1. For the same reaction heat, ΔH_r, the final temperature, T_f, increases with an increase in fill ratio. An increase in fill ratio leads to (1) a lower Phi factor, with a smaller fraction of energy absorbed by the cell wall; (2) a smaller head-space volume, and less energy absorbed by vaporization. Therefore, an increase in fill ratio leads to an increase in the final temperature. For the same reaction heat, the temperature difference between vaporization considered and without considering vaporization is small at a low fill ratio, gradually increasing to a maximum, and then decreasing at further increases in fill ratio. This type of trend is further shown in Figure 3.

Figure 2. The final temperature and Phi factor as a function of fill ratio

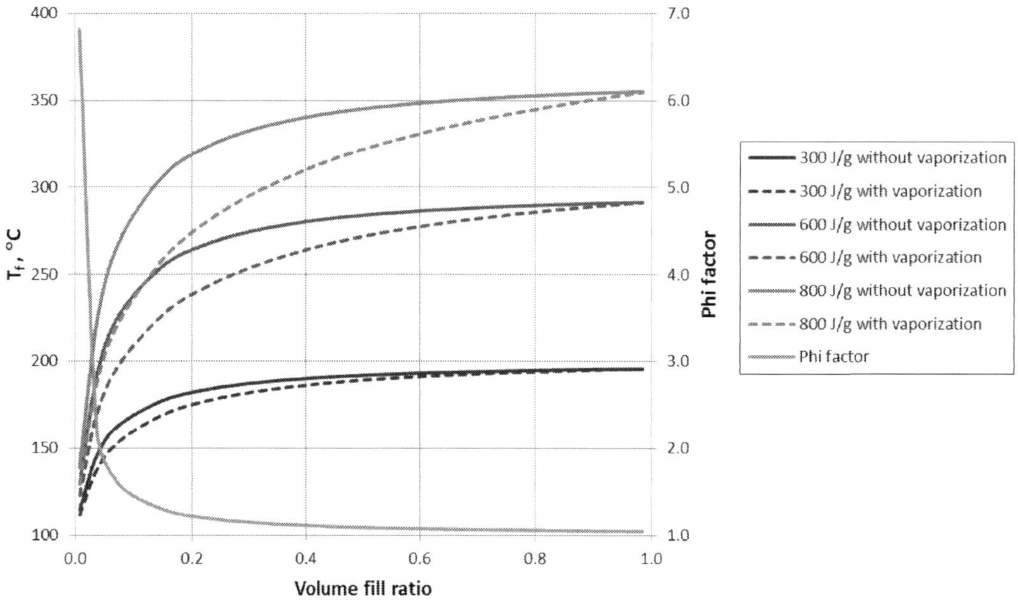

At very low fill ratios, the Phi factor is high. Most of energy that is released from the reaction contributes to heating the cell wall, and the energy contributing to vaporization is low. That is the reason that the temperature difference between vaporization considered and without considering vaporization is small at very low fill ratios, as shown in Figure 3.

With slight increases in fill ratio, more energy is released. Although the head space is reduced (which leads to less volume available for vapor), more energy is contributed to vaporization, since more energy is available from a larger reaction mass. This behavior continues until a maximum difference is attained.

With further increases in fill ratio, more energy is available for heating both the cell wall and the sample mass, while less volume is available for vaporization. This leads to a decrease in the temperature difference between vaporization considered and without considering vaporization. For a fill ratio of unity (100% full), there would be no head space available for vaporization.

As shown in Figure 3, the maximum temperature difference – when vaporization considered and when vaporization is not considered – is attained at higher fill ratios with higher reaction mass and energy. The temperature difference is smaller for a lower reaction mass (fill ratio) and energy. This is consistent with [Eq.17b] and [Eq.17c].

Figure 3. Final temperature difference between considering vaporization and without considering vaporization, and Phi factor, as a function of fill ratio

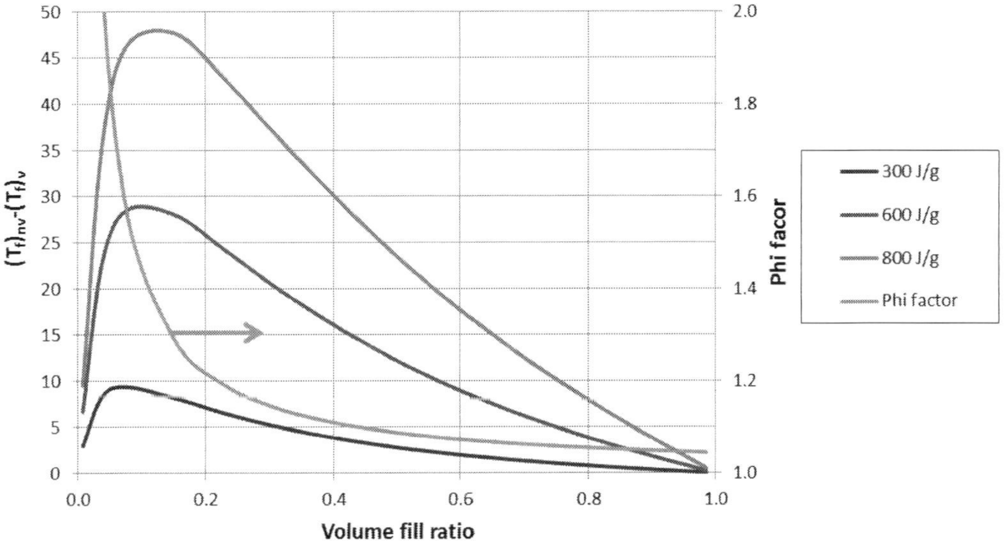

With a known final temperature, the fractions of heat losses to vaporization, the cell, and the sample can be calculated from [Eq.17a-c]. Figure 4 shows the calculated fractions of these heat losses as a function of fill ratio, for reaction energies of 300 J/g, 600 J/g, and 800 J/g. For the same fill ratio, the heat losses to the test cell and the sample decrease, while heat losses to vaporization increase, with an increase in the reaction heat. As would be expected, more-energetic reactions have more energy to expend in vaporization. For a lower reaction heat – such as 300 J/g – the fraction of heat loss to vaporization monotonously drops with an increase in fill ratio. However, for higher reaction energies – such as 600 J/g and 800 J/g, the fraction of heat loss to vaporization initially increases, then arrives at a maximum, and then decreases with fill ratio. These results are consistent with the results as shown in Figure 3. At a lower reaction heat – such as 300 J/g – the heat losses to both the test cell and the sample is higher, and this limits the heat loss to vaporization in the whole range of fill ratios. However, at a higher reaction heat – such as 600 J/g and 800 J/g – the heat losses to vaporization would increase with the fill ratio at lower ranges – due to more energy expended in vaporization, but the heat loss to vaporization decreases at the higher fill ratios, due to less space available for vaporization, and more energy contributes to increasing the temperature of the sample.

Figure 4. The fraction of heat loss to vaporization, cell, and sample as a function of fill ratio.

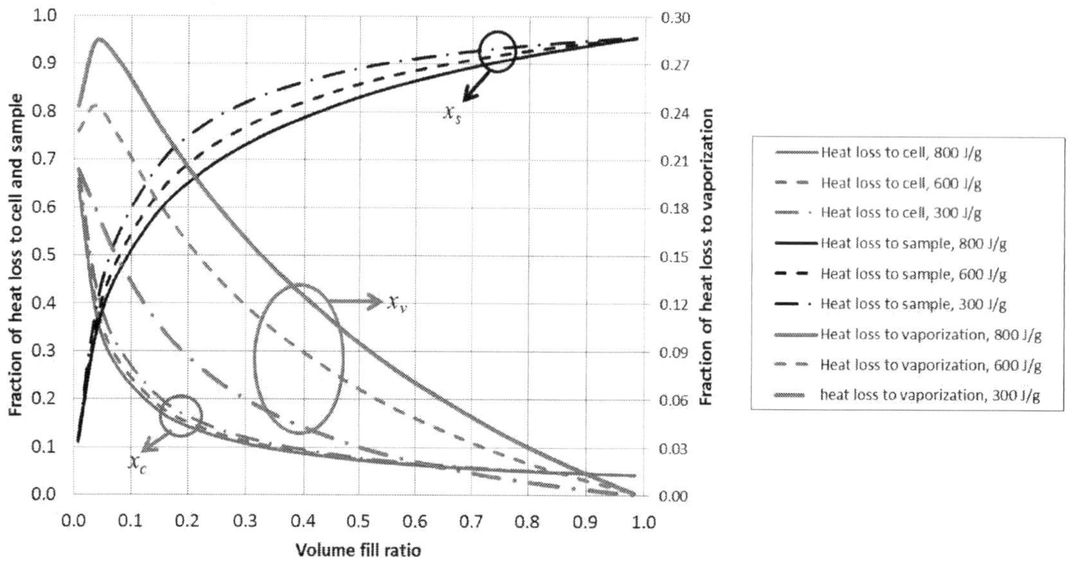

The results in Figure 4 indicate that, for a fill ratio of 60% – as recommended as a low fill-ratio limit by DIERS [2] – the measured reaction heat would be 7% less than the actual value for a reaction with a reaction heat of 800 J/g, 5% less than the actual value with a reaction heat of 600 J/g, and 2% less than the actual value with a reaction heat of 300 J/g, if vaporization is not considered in the energy calculations.

3.2 Self-heating Rate: dT/dt

Figure 5 shows that the effect of vaporization on the temperature-increase rate, dT/dt, as a function of fill ratio, at temperatures of 150 °C, 200 °C, 250 °C, and 300 °C. The ratios of temperature-increase rate without considering vaporization and with vaporization considered are calculated from [Eq.15]. The reaction heats calculated from [Eq.6] and [Eq.16] are also shown in Figure 5. At the same fill ratio, a higher temperature indicates that more energy is generated by the reaction system, which would cause more vaporization. Therefore, the ratio of temperature-increase rate without considering vaporization to the temperature-increase rate with vaporization considered increases with an increase in temperature. At a given temperature, combining [Eq.2], [Eq.6], and [Eq.16] gives the reaction energy

$$\Delta H_r = \phi C_{vl}\left(T_f - T_0\right) + \frac{\frac{V_0}{m_0} - \frac{1}{\rho_l}}{\frac{RT_0}{P_{v0}M} - \frac{1}{\rho_l}} \Delta H_{lv}\left\{\frac{T_0}{T_f} \cdot exp\left[B\left(\frac{1}{T_0} - \frac{1}{T_f}\right)\right] - 1\right\} \qquad \text{[Eq.18]}$$

Therefore, as the fill ratio increases – for a given volume – the Phi factor decreases and the sample mass increases, and this leads to decreases in both terms in [Eq.18]. Therefore, the measured reaction heat decreases, as shown in Figure 5. At the same reaction temperature, the ratio of the temperature-increase rate without considering vaporization to the temperature-increase rate with vaporization considered decreases with an increase in fill ratio. As the fill ratio approaches 1.0 – where no head space available for vaporization – the temperature-increase rates become equal. As would be expected, the effect of vaporization on the temperature-increase rate becomes more significant with lower fill ratios and more energetic reactions.

Figure 5. Effect of vaporization on temperature-increase rate, dT/dt, as a function of fill ratio and the ratio of dT/dt without consideration of vaporization to dT/dt with vaporization considered.

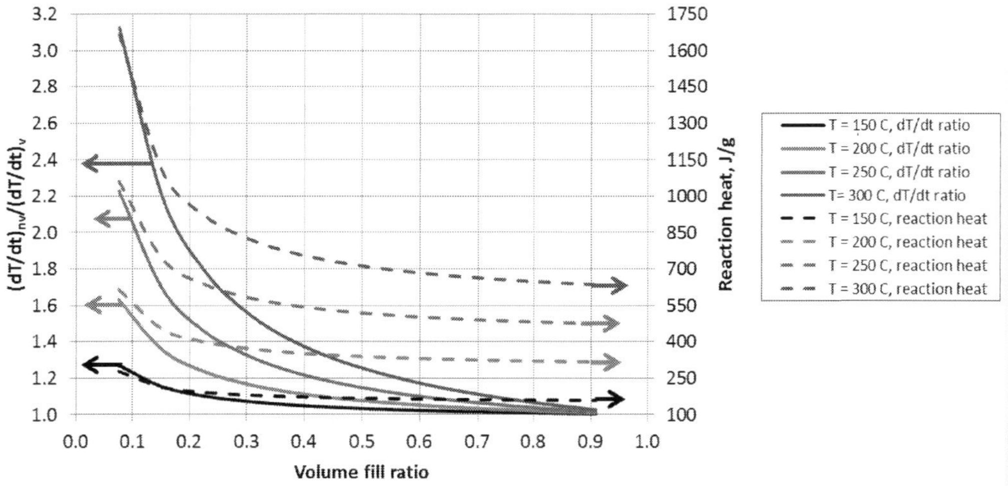

Figure 6 shows the effect of vaporization on temperature-increase rate, dT/dt, as a function of temperature at different fill ratios and Phi factors. At a given fill ratio, the ratio of temperature-increase rate without considering vaporization and with vaporization considered increases with an increase of temperature. A lower fill ratio causes a higher ratio of temperature-increase rate without considering vaporization to the temperature-increase with vaporization considered.

Figure 6. Effect of vaporization on temperature rise rate, dT/dt, as a function of temperature at different fill ratios and Phi factors.

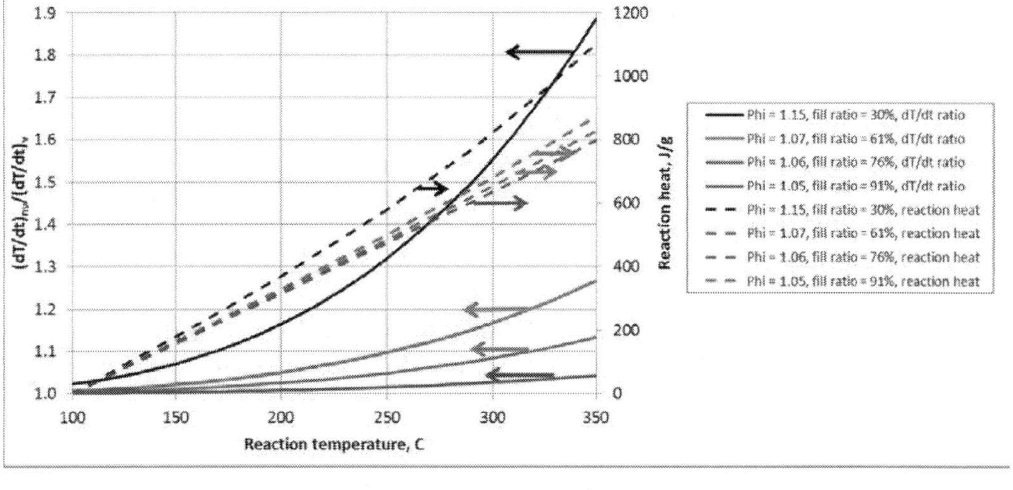

Figure 7 shows the ratio of temperature-increase rate without considering vaporization to the temperature-increase rate with vaporization considered, as a function of reaction heat at different fill ratios and Phi factors. For a given fill ratio, the ratio of these temperature-increase rates increases as the reaction energy increases. For a given heat of reaction, the ratio of temperature-increase rates increases as the fill ratio decreases. For a fill ratio of 60% – which is the lower limit of fill ratio that is recommended by DIERS [2] – the measured temperature-increase rate would be 22% lower than the actual value at a reaction energy of 800 J/g, 13% less than the actual value at a reaction energy of 600 J/g, and 5% less than the actual value at a reaction energy of 300 J/g, if vaporization is not considered in the calculations. For a lower fill ratio of 30%, the measured temperature-increase rate would be 52% less than the actual value at a reaction energy of 800 J/g, 34% less than the actual value at a reaction energy of 600 J/g, and 13% less than the actual value at a reaction energy of 300 J/g, if vaporization is not considered in the calculations.

Figure 7. Ratio of temperature-increase rate without considering vaporization to the temperature-increase rate with vaporization considered, as a function of reaction heat at different fill ratios and Phi factors.

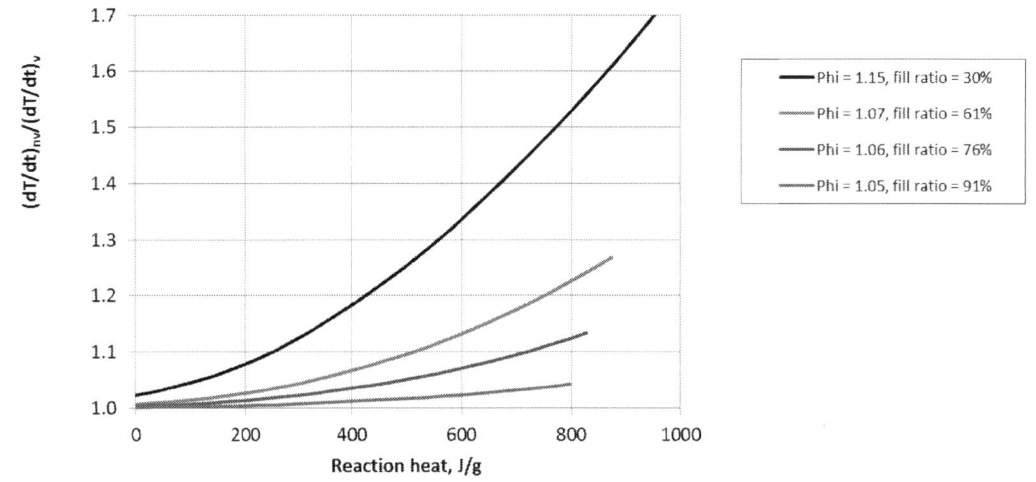

4. Conclusion

(1) A case study of an aqueous reaction system shows that reaction heat with a fill ratio of 60% would be 7% less than the theoretical value for a reaction heat of 800 J/g, 5% less than the theoretical value for a reaction heat of 600 J/g, and 2% less than the theoretical value for a reaction heat of 300 J/g, if vaporization is not considered in the evaluation of the test results.

(2) The self-heating rate is sensitive to the fill ratio. For a fill ratio of 60%, the measured temperature-increase rate would be 22% less than the actual value at a reaction heat of 800 J/g, 13% less than the actual value at a reaction heat of 600 J/g, and 5% less than the actual value at a reaction heat of 300 J/g, if vaporization is not considered in the evaluation of the test results.

(3) A high fill ratio is required to obtain more-reliable data for a high-energy exothermic reaction that involves a volatile component in a tempered system.
(4) A low fill ratio would give misleading thermo-kinetic data for the design of an overpressure-protection system on a full-scale reactor that involves one or more volatile components.

5. Recommendations.

This investigation of adiabatic calorimetry results in the following recommendations:
(1) For emergency relief sizing, use the same fill ratio of the full-scale reactor in the test vessel of the adiabatic calorimeter, to obtain valid thermo-kinetic data, rather than just considering thermal inertias.
(2) Generally speaking, the fill ratio of 60% as recommended by DIERS should be a minimum charging ratio for adiabatic calorimeters, for obtaining the valid thermo-kinetic data for emergency relief sizing.
(3) Considering the heat transfer limitations of full-scale reactors, Phi correction of the measured self-heat rates from adiabatic calorimeters is always necessary.

Nomenclature

ΔH_{lv}	Vaporization latent heat at constant pressure, $J \cdot kg^{-1}$
ΔU_{lv}	Vaporization latent heat at constant volume, $J \cdot kg^{-1}$
ΔH_r	Reaction heat, $J \cdot kg^{-1}$
Δv	Specific volume difference between vapor phase and liquid phase, $m^3 \cdot kg^{-1}$
A	Antoine constant [natural log], or frequency factor
B	Antoine constant, for temperatures in °K
C	Concentration, $mol \cdot m^{-3}$
C_p, C_v	Heat capacity, $J \cdot kg^{-1} \cdot K^{-1}$
E	Reaction activation energy, $J \cdot mol^{-1}$
k	Reaction rate constant
m	mass, kg
M	Molecular weight, $kg \cdot mol^{-1}$
P	Pressure, Pa
Q	Total energy released, J
q	Heat release rate, $J \cdot s^{-1}$
R	Gas constant, 8.314 $J \cdot mol^{-1} \cdot K^{-1}$
r	Reaction rate, $mol \cdot m^{-3} \cdot s^{-1}$
T	Temperature, K
t	Time, s
V	Volume, m^3
ρ	Density, $kg \cdot m^{-3}$
ϕ	Phi factor, $\phi = 1 + \frac{m_c C_{pc}}{m_0 C_{pl}}$
Subscripts	
0	Initial, or onset

c	Cell
f	Final or end
l	Liquid
nv	Without considering vaporization
s	Sample
v	Vapor, or with considering vaporization

References

[1] Townsend D. I. and Tou J. C., "Thermal Hazard Evaluation by an Accelerating Rate Calorimeter", Thermochimica Acta, Vol. 37, Pages 1 to 30 (1980).

[2] Fisher H. G., et al., "Emergency Relief System Design Using DIERS Technology", Experimental Emergency Relief System Design – Some Do and Do Not Recommendations, American Institute of Chemical Engineers, New York, Pages 382 to 391 (1992)

[3] Etchells J., Wilday J., "Workbook for Chemical Reactor Relief System Sizing", Contract Research Report 136, HSE books, Page 142 (1998).

[4] Green D. W., and Perry R. H., "Perry's Chemical Engineers' Handbook", 8th Edition, The McGraw-Hill Company, Inc., Page 2-413 (2008).

349001 The Effect of Non-Uniform Distribution of Obstacles on Deflagration-to-Detonation Transition (DDT)

Abstract:

The effect of non-uniform distribution of obstacles on

Deflagration-to-Detonation Transition (DDT)

Camilo Rosas[1, 2], Hao Chen[1, 2], Eric Petersen [3], and M. Sam Mannan[1, 2]*

[1] Mary Kay O'Connor Process Safety Center

[2] Artie McFerrin Department of Chemical Engineering

[3] Department of Mechanical Engineering

Texas A&M University, College Station, Texas 77843-3122

*Corresponding author: mannan@tamu.edu

The mechanisms for explosions are usually classified as two different types. The first mechanism is deflagration, a sub-sonic combustion wave with respect to the unburned gas ahead of the flame, during which the flame front and the shock front can be distinguished. The second type is detonation, which is a supersonic combustion wave propagating at 1500-2000m/s in fuel-air and can produce overpressures up to around 2 MPa. To obtain a detonation there are two different approaches. The first way is by direct initiation with a high energy source, and the second one is by inducing a transition from deflagration to detonation (DDT) with a low energy source and turbulence enhancers, which are also known as obstacles.

From previous research, the obstacles' shape, distribution, and blockage ratio along the tube were always designed to be uniform. However, these designs deviate from real processes. For instance, different layouts of obstacles (*e.g.* pipe racks, turbines, and vessels) are substantially present in any facility, which could either accelerate and lead to DDT or quench the combustion process. Thus, in this work, the obstacles were designed to be distributed non-uniformly along the detonation tube to determine the effect of the randomness in the obstacles on the distance necessary to obtain DDT. Details of these experiments and the results are presented in the paper.

Computational Fluid Dynamics (CFD) models were also applied in this work to better understand the mechanisms. Even though CFD models do not represent DDT, various models are still able to predict high pressure values which are typically in the range of detonation pressures. Moreover, a CFD model capable of predicting the explosion behavior is of particular interest for the situation that the flame front captures the pressure front by the spatial pressure gradients across the flame front, which is one of the main mechanisms that may initiate a DDT.

Dispersion Modeling of a Cloud Generated by the Depressurization of a Flashing Multi-Component Liquid System

Laurent Nouailhetas
ERM, America Risk Practice,
1776 I Street Northwest, Suite 200
Washington D.C., 20036, USA
laurent.nouailhetas@erm.com

Ralph Mancik
ERM, America Risk Practice,
840 West Sam Houston Parkway North, Suite 600
Houston, 77024, USA
ralph.mancik@erm.com

Prepared for Presentation at
American Institute of Chemical Engineers
2014 Spring Meeting
10th Global Congress on Process Safety
New Orleans, LA
March 30 – April 2, 2014

UNPUBLISHED

Dispersion Modeling of a Cloud Generated by the Depressurization of a Flashing Multi-Component Liquid System

Laurent Nouailhetas
ERM
1776 I Street Northwest, Suite 200
Washington D.C., 20036, USA
laurent.nouailhetas@erm.com

Ralph Mancik
ERM
ralph.mancik@erm.com

Keywords: Depressurization, multi-component mixture, flashing liquid, dispersion, modeling

Abstract

Depressurization is usually a process operation during which the gas contained in a pressurized system is sent either to a flare or to a cold vent. However, depressurization can also be the consequence of an unexpected event as, for example, the accidental opening of a vent or an unintended leak of a pressurized system. One of the immediate outcomes is the release and dispersion of a cloud into the atmosphere that creates a hazard when the cloud is flammable and/or toxic. The prediction of the harmful extent of such dispersing clouds is a common task when assessing the risk associated with such events.

Depressurization events are of several types which differ in the number and complexity of the mechanisms involved. In general, there is either a system containing a single compressed vapor phase (referred to in this paper as a "gas system"), or a system containing a flashing liquid phase that is in thermodynamic equilibrium with a vapor phase (referred to in this paper as a "flashing liquid system"). The flashing liquid is either a pure component or a multi-component mixture.

With a multi-component flashing liquid system, the temperature and the composition of the released vapor will change over time. The resulting dispersing cloud could transition from a buoyant vapor mixture early in the depressurization event to a dense vapor mixture later in the depressurization event. One immediate consequence is that in stable atmospheric conditions, a dispersing cloud could ascend initially, but after a few minutes it could descend and disperse at ground level. Not accounting for the changing characteristics of the system could result in a misleading assessment.

With a good consequence modeling program, modeling the time-varying discharge and dispersion of a depressurizing gas system is relatively easy to handle. However, the same problem for a system containing a flashing liquid under pressure (LNG, LPG, etc.) is more complex even with a good commercial consequence modeling program. Adding an appropriate accounting of the changing characteristics of a multi-component system further complicates predicting the harmful extent of a dispersing cloud.

The objective of this paper is to present a few methods to appropriately model such a complex dispersion problem. After comparing the merits and demerits of each method, ERM proposes the best method for the modeling of a time-varying multi-component flashing liquid release.

1. Introduction and context

Depressurization is a common process operation during which the gas contained in a pressurized system is sent to a flare or to a cold vent. However, depressurization can also be the consequence of an unexpected event as in the accidental opening of a vent or an unintended leak of a pressurized system. One of the immediate outcomes is the release and dispersion of a cloud into the atmosphere. This creates a hazard when the cloud is flammable and/or toxic. A depressurization through a flare would also generate a hazardous cloud in the event of a flame out. Figure 1-1 illustrates common situation for controlled and uncontrolled depressurizations.

Figure 1-1: Vessel depressurization and gas discharge generating a hazardous cloud

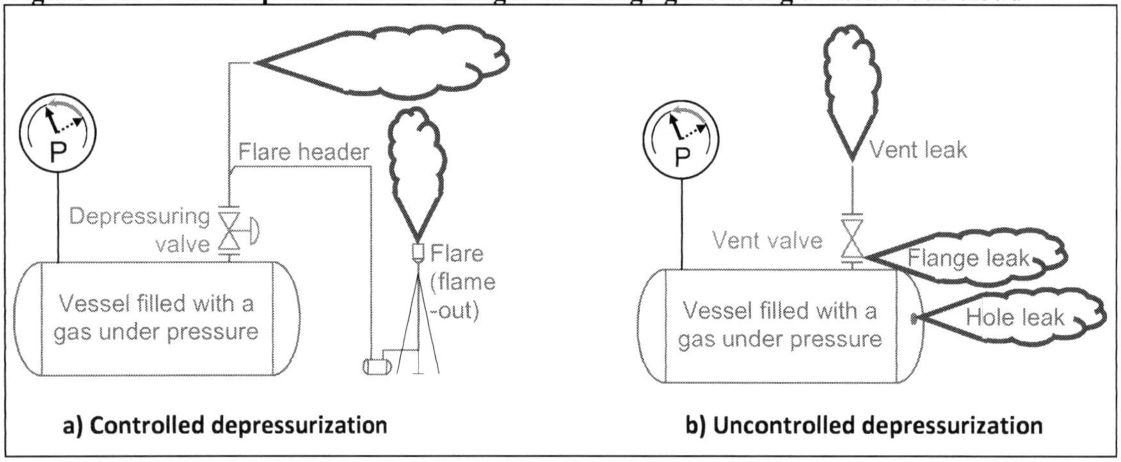

Depressurization events are of several types which differ in the number and complexity of the mechanisms involved. In general, there is either a system containing a single compressed vapor phase (referred to in this paper as a "gas system"), or a system containing a flashing liquid phase that is in thermodynamic equilibrium with a vapor phase (referred to in this paper as a "flashing liquid system"). The gas or the flashing liquid is either a pure component or a multi-component mixture. Figure 1-2 illustrates a typical depressurization of a gas system and a flashing liquid system.

Figure 1-2: Depressurization through a vent: two accidental configurations

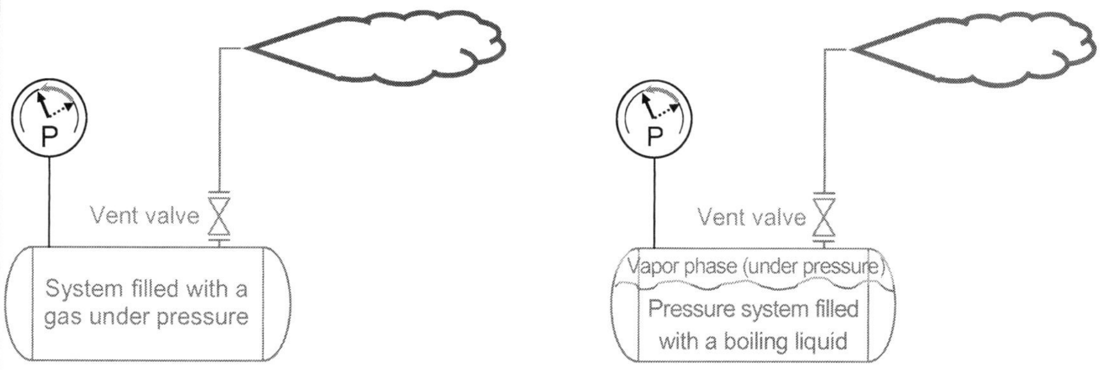

a) Depressurization of a gas system **b) Depressurization of a flashing liquid system**

Other variations of these include systems containing a liquid and a vapor phase that are not in equilibrium, systems with a vapor phase in presence with multiple liquid phases, and others. However, this paper will focus on the "flashing liquid system" with a multi-component composition, and will make references to the "gas system" for comparison purposes.

2. Characteristics of the discharged gas

2.1 Gas system

During the depressurization of a gas system, the characteristics of the gas will evolve over time as follow:

- The mass inventory will decrease as gas inventory (and gas volume) is withdrawn from the gas system;
- Since the volume of the gas system is fixed (illustrated by curve ① in Figure 2-1), the gas remaining in the system will expand to fill the entire system volume, leading to a reduction of pressure (curve ②) and reduction of temperature (curve ③);
- The average molecular weight and the composition of the discharged gas will remain constant (curves ④, bar charts ⑤ and ⑥).

Figure 2-1: Vapor phase characteristics for the depressurization of a gas system

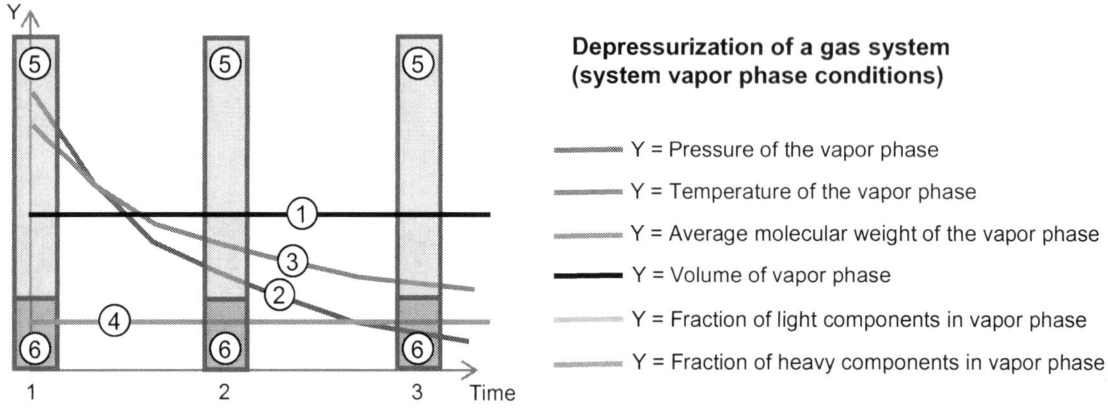

2.2 Multi-component flashing liquid system

With a multi-component flashing liquid system, the gas fills only a fraction of the system volume (the vapor phase), the rest being filled with the boiling liquid phase. As gas is withdrawn from the vapor phase, the gas remaining in the system expands leading to a decrease of system pressure. The presence of the boiling liquid phase however introduces additional phenomena:

- As pressure drops, a fraction of the liquid is vaporized to reach new boiling conditions under the lower system pressure;
- The volume of the gas is not fixed. Since a fraction of the liquid inventory is vaporized, the liquid mass inventory decreases. The liquid being incompressible, a reduction of mass inventory leads in a decrease of the liquid phase volume and an expansion of the vapor phase volume (illustrated by curve ① in Figure 2-2).
- The composition of the vapor phase becomes proportionately leaner in light components and richer in heavy components as the pressure decreases over time (bar charts ⑤ and ⑥).
- Consequently, the average molecular weight of the vapor phase increases over time (curve ④).
- Temperature of the liquid and the vapor phases decrease gradually over time due to the expansion of the gas in the vapor phase and the vaporization of the liquid (curve ③). Note that the temperature decreases slower than for the gas case (Figure 2-1) because of the presence of the liquid.

Figure 2-2: Vapor phase characteristics for the depressurization of a multi-component flashing liquid system

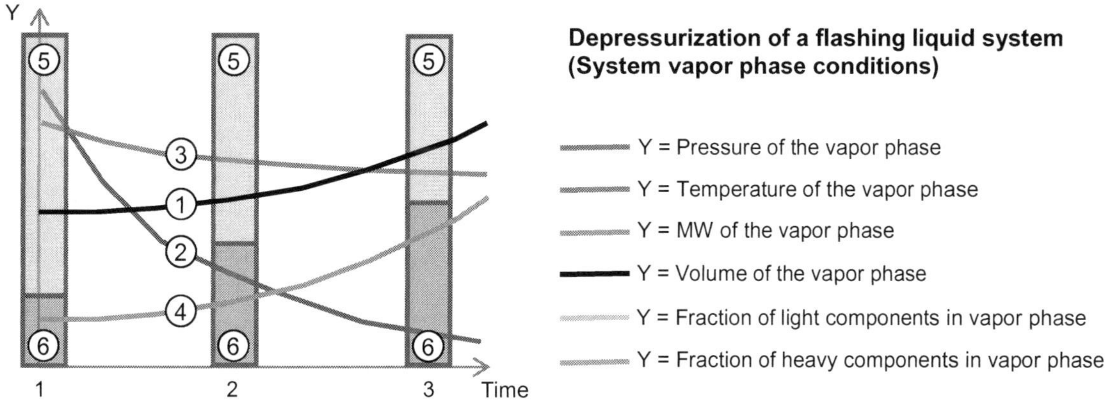

2.3 Important difference

One important difference between these two systems is the increase in density of the vapor phase over time. For the gas system, it is driven only by the drop of temperature over time. For the multi-component flashing liquid system, it is driven by both the drop of temperature over time and the increase of the average molecular weight over time.

For dispersion, the density of the vapor phase is of primary importance since a light gas discharge would disperse like a buoyant cloud (raising in atmosphere) but a heavy discharge would disperse like a heavy cloud (slumping on the ground). Thus, for a multi-component flashing liquid system, both the change in the temperature and the average molecular weight of the vapor over time need to be considered.

The rest of the paper will relate to the multi-component flashing liquid system only. Examples will present toxic cloud dispersions because they emphasize the different behaviors of the cloud. For flammable dispersions, those differences would be more subtle in our example.

3. Dispersion modeling of a discharged gas

To model the dispersion of the cloud following a gas discharge, two approaches are generally used:

1. Dispersion modeling based on a single constant discharge rate and
2. Dispersion modeling based on multiple fixed discharge rates (referred as "time-varying modeling")

In case #1, the single fixed discharge rate is used to represent the entire duration of the discharge. This is a simplification that does not take into account the change of pressure over time.

In case #2, the effects of a pressure decrease over time can be accounted by using a time-varying release model which divides the discharge into multiple segments and uses a separate fixed discharge rate for each segment. Thus the inputs into the dispersion model "varies with time".

3.1 Dispersion modeling based on a single fixed discharge rate

This approach is often used since it presents the advantage of being simple and adequate for many situations: The modeling requires very few input data and outputs are relatively straightforward and easy to interpret.

Two types of fixed rate can be chosen:

1. Case A: A single fixed rate equal to the initial discharge rate
2. Case B: A single fixed rate equal to an intermediate discharge rate

This approach presents a few limitations: the difficulty to justify the choice of an adequate fixed rate and therefore, the difficulty to defend the reasonable conservatism of modeling results.

3.1.1 Case A: single fixed rate equal to the initial discharge rate

The evaluation of the initial discharge rate is based on the initial process conditions in the multi-component flashing liquid system. If the pressure in the system drops slowly, the discharge is nearly constant and therefore using the initial discharge rate would provide accurate results.

In the case where the pressure drops quickly, this will often lead to conservative dispersion results since the initial discharge rate is higher than the average discharge rate over the depressurization duration.

The graph in Figure 3-1 illustrates toxic results (100 ppm H_2S concentration) for a 10 minute depressurization (discharge duration of 10 minutes) of a flashing liquid system containing a light crude oil extracted from a well. In the initial conditions, the vapor phase in equilibrium with the crude oil has an average molecular weight of 26 g/mol (main components ranging from Methane to n-Decane, plus a small amount of H_2S, N_2 and CO_2).

Figure 3-1: Side-view of a 10min-depressurization – Cloud Footprint (F1.5) at initial conditions

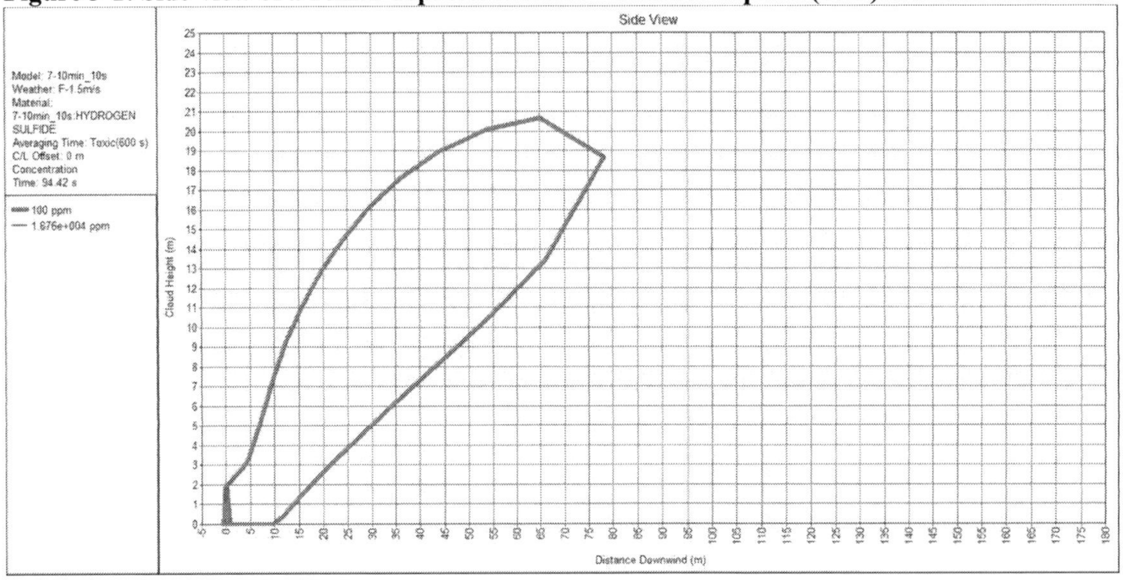

In the initial discharge conditions, the gas is lighter than air hence the dispersing cloud ascends in the atmosphere as illustrated by the side view plot. Based on this modeling, one would conclude that there is no risk of being exposed to a 100 ppm H_2S-cloud at ground level beyond a 10m radius from the source.

3.1.2 Case B: single fixed rate equal to an intermediate discharge rate

Choosing which intermediate rate to use is a challenge: too high intermediate rate may lead to conservative dispersion results and too low may lead to optimistic dispersion results.

Using the same example as for case A) with an intermediate rate corresponding to discharge conditions at 3.5 minutes would lead to results illustrated in Figure 3-2.

Figure 3-2: Side-view of a 10min-depressurization – Cloud Footprint (F1.5) at 3.5min conditions

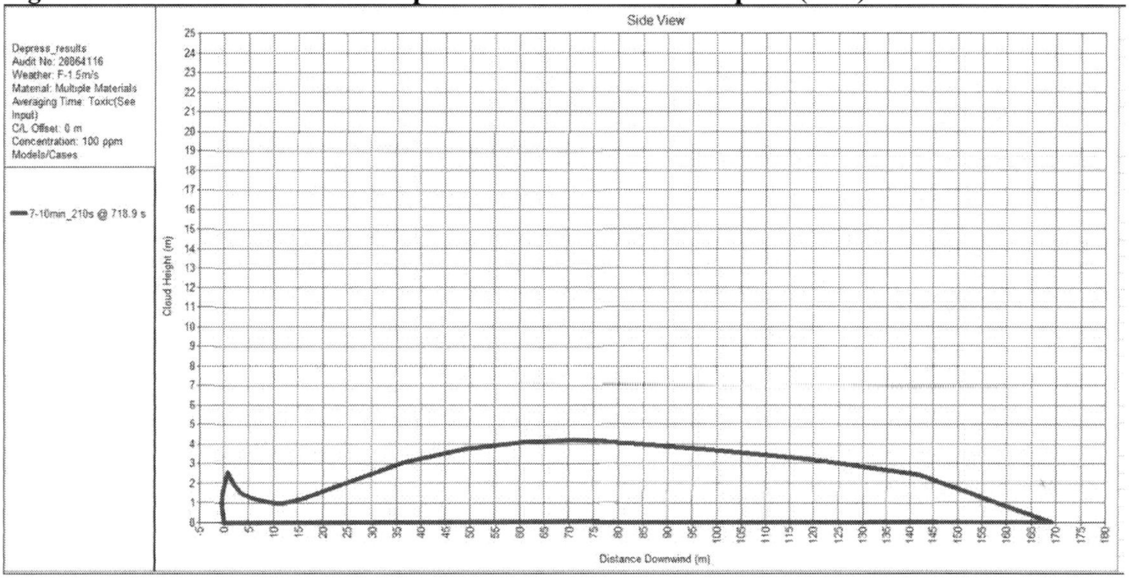

In this intermediate discharge condition, the discharged gas is heavier than air (molecular weight of 33 g/mol and cold discharge temperature) hence the cloud descends and disperses at ground level. Based on this modeling, one would conclude that there is a toxic risk at ground level up to 170m from the source. Consequences would obviously be much worse that for case A).

Therefore, considering that the dispersing cloud transitions from a buoyant vapor mixture early in the depressurization event to a dense vapor mixture later in the depressurization event, what would be the correct fixed discharge to use?

If the location of concern is somewhere above the ground, case A) might provide better results than case B). But if the location of concern is at ground level, case B) would provide better results.

Since toxic risk is dependent not only on concentration but also on exposure duration, accurate results should also account for exposure duration. Assuming that someone can be exposed to the toxic cloud during the whole duration of the depressurization might therefore lead to over-conservative results.

3.2 More rigorous modeling based on time-varying discharge rates

For a multi-component flashing liquid system, both the change in the temperature and the average molecular weight of the vapor over time need to be considered. This is accomplished by modeling the time-varying discharge as multiple discrete steps. Each step will be represented by a different discharge rate, gas temperature and gas composition. Figure 3-3 illustrates an example of a time-varying discharge model using six discrete steps.

Figure 3-3: TV dispersion by six steady-states dispersions – 10s, 2min, 3.5min, 5min, 7min, 9 min

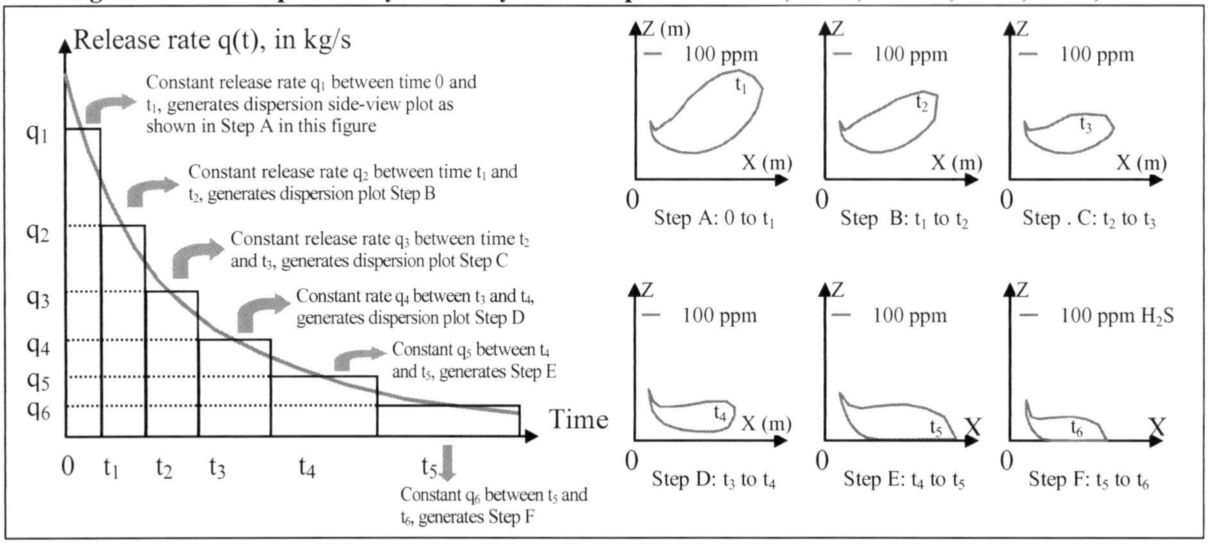

For each step, the dispersion of the cloud is modeled using as inputs, the fixed discharge rate, composition and temperature values from the corresponding quasi-steady state discharge. Dispersion results from each step can be illustrated by the side-view concentration plots, as shown in Figure 3-3 (Step A, Step B, Step C, Step D, Step E and Step F).

To facilitate the analysis, results for all steps can be superimposed on a single figure as presented in Figure 3-4.

Figure 3-4: TV dispersion summary with six discrete steps

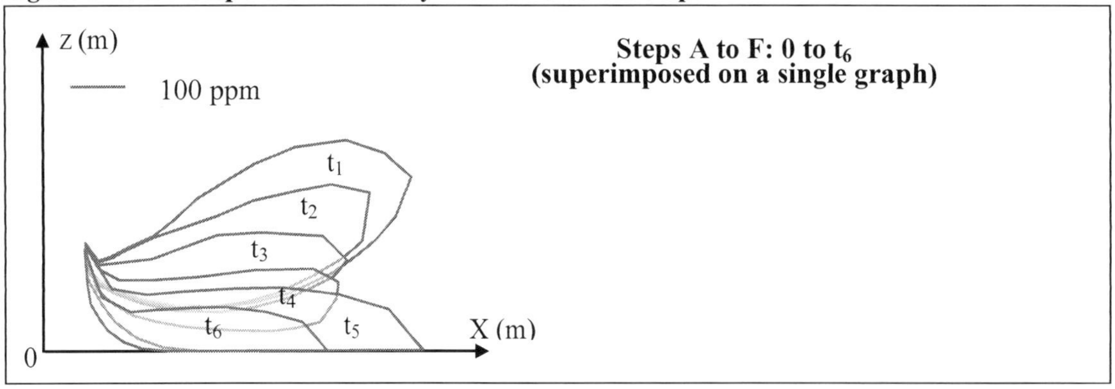

This figure reflects the dynamic nature of the dispersing cloud which in this example starts as a buoyant cloud and transitions to a slumping cloud. The use of a single discharge dispersion modeling could only reflect one of these steps. Depending on the location of interest, attempting to rely on a single fixed discharge release could result in misleading conclusions.

4. ERM recommended approach to the modeling of the multi-component flashing liquid depressurization and cloud dispersion

To model properly this complex release and dispersion case, one shall:

- Model the time-varying discharge using multiple steps. Each step will be represented by a different discharge rate, gas temperature and gas composition;
- Model the dispersion individually for each step.

Commercial dispersion programs like PHAST [1] are currently unable to model the depressurization of a vessel filled with a multi-component flashing liquid. Therefore, the multi-component time-varying discharge shall be performed with a chemical process simulation software like HYSYS [2], PRO/II [3], ASPEN PLUS [4], or through the use of a specific modeling tool developed internally for this purpose. Once modelled, the depressurization shall provide the following outputs:

- Gas pressure inside the vessel over time
- Gas discharge rate over time
- Gas temperature inside the vessel over time
- Gas composition over time

These outputs can then be used to generate multiple fixed-discharge-rate dispersion cases that will be modelled with a commercial dispersion modeling program like PHAST.

In the example chosen for this paper, the multi-component time-varying discharge is performed with HYSYS specific depressurization module. Then based on HYSYS outputs, six PHAST dispersion cases are defined to represent six steps along the time-varying discharge. These steps correspond to 10s, 2min, 3.5min, 5min, 7min and 9min. The 100 ppm H_2S side-view iso-concentrations for each of the six steady-states' discharge modeling are presented on a single graph in Figure 4-1.

Figure 4-1: Side-view of a 10min-depressurization – Cloud Footprint (F1.5) at 10s, 2min, 3.5min, 5min, 7min

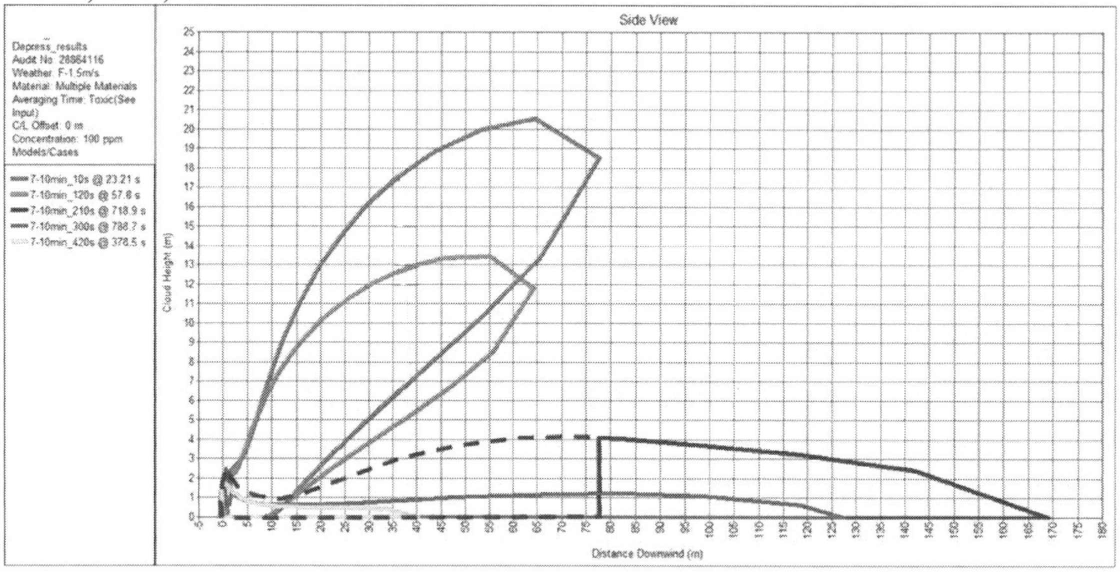

Based on these results, it can be concluded that a receptor located at 100 meters from the leak is exposed to the 100 ppm toxic effect of H_2S only after the first 2 minutes of the depressurization and is no longer exposed after 7 minutes. Hence the exposure duration would be approximately 5 minutes. This result differs radically from the results based on the dispersion modeling performed with a single fixed rate as shown in Figure 3-1 (toxic exposure would be evaluated as 0 minute) and Figure 3-2 (toxic exposure would be evaluated as 10 minutes).

5. Conclusion

When modeling the dispersion following the depressurization of a multi-component flashing liquid system, the use of a single fixed discharge rate might provide results that are either too conservative or too optimistic.

A multi-steps/multiple dispersion source modeling approach as presented in this paper would more rigorously represent the phenomenon involved in the depressurization of a multi-component flashing liquid system. Results would therefore be more representative of the consequences. However, this modeling approach presents the following limitations:

- This approach is relatively complex and requires the use of a specific tool like a chemical process simulation software, to model the depressurization discharge.
- The depressurization module built in the chemical process simulation software do not provide the final velocity and the final temperature of the gas after expansion to atmospheric conditions. These parameters would have to be evaluated separately for each dispersion case before the modeling with a dispersion software like PHAST.

A future paper should investigate whether an intermediately complex approach could be defined, using solely a dispersion modeling software like PHAST.

6. References

[1] PHAST (Process industry Hazard Analysis Software Tool), an integrated consequence modeling software package developed by DNV GL to analyze situations which present hazards to life, property and the environment, and to quantify their severity (http://www.dnv.com/services/software/products/phast_safeti/phast/).

[2] HYSYS, a steady-state process simulator (process simulation) for process design and operational analysis for process engineers in the chemical, petroleum, natural gas, solids processing, and polymer industries, initially developed by Hyprotech and now developed and released by Aspen Technology (http://www.aspentech.com/hysys/).

[3] SimSci PRO/II, a steady state process simulator (process simulation) for process design and operational analysis for process engineers in the chemical, petroleum, natural gas, solids processing, and polymer industries, developed initially by SimSci and now released by Invensys (http://software.invensys.com/products/simsci/design/pro-ii/).

[4] ASPEN Plus, a steady-state process simulator (process simulation) for process design and operational analysis for process engineers in the chemical and specialty chemical organizations and related industries to design and improve process plant. Aspen Plus is developed and released by Aspen Technology. (https://www.aspentech.com/products/aspen-plus.aspx)

349966 Quantitative Analysis of Environmental and Societal Risk for Onshore FUEL Pipelines

Abstract:

Combustible materials transportation by pipeline has been established as a safe and efficient mean. Due to the low accidental frequency of incidents associated, it is socially accepted as favorable. However, like any other industrial practice it can be dangerous and represents risks to the society, environment and infrastructure. Since production sites are often away from consumption centers, increased demand in Colombia and the world has led to increased transportation needs, forcing the growth of pipeline networks.

Carrying out a-priori risk analysis could provide engineering tools to support decision making, related to the compatibility of activities within a territory (Land-use Planning). This allows taking preventive measures with respect to potential accidents. Hence, implementing scenario building enables to explore and clarify chain of possible events, including actions and consequences. Also, it is important to define event probabilities; this is done using Event Tree Analysis (ETA). Subsequent phase consists of consequence analysis, where simulation is performed in order to determine effects intensity and then vulnerability values. Once all this is accomplished, risk assessment is completed.

So, this work aims to develop a methodology for the assessment of societal and environmental risk associated with: loss of containment in onshore fuel liquid pipelines and consequences of critical events, as radiation, overpressure and contamination effects. As risk values vary along pipelines due to differences in environmental and societal conditions, analysis is performed for segments of a pipeline, including analysis of causes and types of events.

For societal risk assessment the methodology includes some guidelines to develop a segmentation process. Risk changes along the pipeline due to differences in environmental and societal conditions in order to define manageable pipeline sections for analysis. It is also necessary to identify the probable events and their probabilities; in this case MIMAH methodology and historic analysis is used. Then consequence analysis is performed for each segment, determining the intensity of overpressure and radiation effects caused by the previously defined events. This determination is done by means of simulations performed on specialized software and then using *probit* equations the vulnerability is estimated. With vulnerability and probability values known, risk can be estimated. As a result, risk values for people are determined and expressed as individual risk contours and FN curves for societal risk.

In addition to the societal impacts, assessment of environmental contamination is also done. It is known that hydrocarbon releases are a real hazard to the environment as they produce contamination to soil and groundwater. In Colombian topography, mountains and slopes are mostly present. This generates scenarios out of the right of way making analysis of onshore hydrocarbon spillages even more relevant in this territory. Therefore, environmental effects on soil are determined considering mass balance of infiltrated and evaporated material in horizontal and inclined surfaces. Consequently, effective volume remaining on soil surface is determined to establish amount of material available for migration pathways. Environmental damage is then estimated taking into account depth of infiltrated material, associated to soil excavation and off-site disposal, soil clean-up and groundwater clean-up costs. As a result, environmental risk values are presented as risk indexes in terms of remediation action costs, defining an environmental local monetary damage. Contamination is analyzed in order to determine average soil volumes which need to be treated, considering intervention, mitigation and remediation times.

This methodology is targeting to support decision-making in Colombia in early stages of analysis, with respect to land-use planning, from a societal and environmental point of view. A total risk index is calculated, in order to be used as a comparative tool for ranking different projects or pipeline sections. This total risk index is calculated turning both societal and environmental risk in equivalent terms, meaning monetary terms. It is represented as an FM curve (frequency vs. millions of dollars, euros, etc.). Simultaneous assessment of both human and environmental risk is remarkable, as it integrates and supports decision making.

This methodology is applied on a case study to illustrate how it can support decision making (related to Land-use planning and to alternatives choice).Taking into account international societal risk criteria, it will be determined whether these values can be accepted or should be reduced. Therefore, coupling risk assessment of environmental contamination and damage to exposed population allows quantifying risk and thus, supporting decision making within the territory.

Small Scale Experimental Study of Vaporization Rate of Liquid Nitrogen Released on Ice

Nirupama Gopalaswami[1]
M. Sam Mannan[1]
[1]Mary Kay O'Connor Process Safety Center
Artie McFerrin Department of Chemical Engineering Department
Texas A&M University System
College Station, Texas 77843-3122, USA
nirupama89@tamu.edu
mannan@tamu.edu

Luc Vechot[2]
Tomasz Olewski[2]
[2]Mary Kay O'Connor Process Safety Center - Qatar
Texas A&M University at Qatar, PO Box 23874, Education City, Doha, Qatar

Luc.vechot@qatar.tamu.edu
Tomasz.olewski@qatar.tamu.edu

Prepared for Presentation at
American Institute of Chemical Engineers
2014 Spring Meeting
10th Global Congress on Process Safety
New Orleans, LA
March 30 – April 2, 2014

UNPUBLISHED

Small scale experimental study of vaporization rates of liquid nitrogen released on ice

Nirupama Gopalaswami[1], Luc Vechot[2], Tomasz Olewski[2] and M. Sam Mannan[1]
[1]Mary Kay O'Connor Process Safety Center
Artie McFerrin Department of Chemical Engineering Department
Texas A&M University System
College Station, Texas 77843-3122, USA
[2]Mary Kay O'Connor Process Safety Center - Qatar
Texas A&M University at Qatar, PO Box 23874, Education City, Doha, Qatar
nirupama89@tamu.edu
mannan@tamu.edu
Luc.vechot@qatar.tamu.edu
Tomasz.olewski@qatar.tamu.edu

Keywords: Source term, liquid nitrogen, conduction, ice, vaporization

Abstract

One of the LNG accident scenarios is the collision of an LNG carrier on an iceberg while transporting on the sea. A collision can result in damages to the vessel and lead to the leakage of contents on ice or an ice-water mixture. When cryogenic liquid comes in contact with ice, it undergoes rapid vaporization and boiling due to the difference in temperature between the ice and cryogenic liquid. This process is different from the heat transfer between water and cryogenic liquid as ice is a solid and thus heat transport to the pool occurs through the conduction. In this paper, the heat transfer ~~phenomena between ice and cryogenic liquid has~~phenomena between ice and cryogenic liquid have been studied and the boil-off rates have been quantified. A small scale experiment was conducted to investigate the heat transfer phenomenon of cryogenics released on ice. The amount of cryogenic liquid spilled was varied to determine its effect on vaporization rates and the temperature of ice was monitored. Additionally the vaporization rates were determined by direct measurement of the mass loss during the experiment. As expected, the vaporization rates were found to be a function of ice temperature and were found to be independent of time for long duration releases. The one dimensional conduction model was validated against experimental results. The predicted temperatures and heat flux were found to be in good agreement with the experimental data.

1. Introduction

One of the commonly considered LNG accident scenarios is the collision of an LNG ship with an iceberg while transporting on the sea. Detailed consequences analysis considering various scenarios and hypotheses through calculations are required for LNG carriers. It is also important to take into account the fatigue effect of vibrations in risk analysis when LNG ship traverses through ice. Consequence analysis in such cases depends on features of the ice and the structural strength of the LNG carrier. Two of the common types of collision that can occur are the side collision between vessel and iceberg and bow collision [1]. A collision can result in damage of the vessel and allow leakage of contents on ice or ice-water mixture. Such a scenario is common in areas where ice is predominantly present throughout the year like the Arctic Circle. Upon spillage of contents, one of the most important factors that affect cryogenic liquid release on ice is the heat flux. This process is different from the heat transfer between water and cryogenic liquid as ice is a solid and conduction phenomena is more dominating than convection due to boiling.

Currently, very few experiments have been performed by releasing cryogenic liquids on ice. In 1971, Nakanishi observed that when cryogenic liquids like Condensed Pipeline Gas (CPG, with composition CH_4 92.7%, N_2 7.3%; boiling point about -111.6K) and Liquid Nitrogen (LN_2, boiling point -77 K) are poured on an ice layer maintained at -150 °C, there is no vapor film formation under the cryogen's layer and both liquids are boiling in nucleate regime [2]. When the experiment was repeated by pouring onto an ice layer maintained at -5 °C, a vapor film was observed to occur for liquid nitrogen alone, but not for CPG. It was concluded when CPG was spilled on ice, it stays in film boiling for a brief period. This might have been due to the fact that the temperature difference between ice and CPG would have been small to sustain film boiling for a long period. This phenomenon was not addressed in detail in the research work.

In another study, small scale experiments were performed with liquid nitrogen and LNG (Burgess, Murphy and Zabetakis, 1970a). The primary objective of this study was to measure the heat flux and the vaporization rates. The experiments were performed at laboratory scale, (spill area - 0.18 m^2) where cryogenic liquid was poured onto a flat tray (30.5 x 61 cm) of ice. The ice was made 4 cm thick and covered the entire surface of the tray such that the cryogenic liquid contacted only ice. LNG and LN2 were released on ice. An array of thermocouples was placed below the ice to measure the time dependent temperature and associated heat transfer. However, wrinkling of the ice-cryogen interface disturbed the position of thermocouples and a spatial variation of heat transfer was not obtained in the test. The amount of cryogenic liquid poured was varied and the vaporization rate was reported for the experiment. The boil-off rates of nitrogen on ice varied from 0.08-0.12 kg $m^{-2}s^{-1}$. The small scale tests showed a lot of variation in the boil-off rates reported for nitrogen. The vaporization rates were reported for discrete values of time.

A heat transfer model for growing ice thickness involving conduction and convection phenomena for release of cryogenic liquids on shallow water was undertaken by Vesovic [3].However, the model was not validated with any experimental data. In all these experiments the source term, the heat flux from the substrate (ice) to the pool was calculated by measuring the temperature difference between the substrate and cryogenic liquid pool with time. This was possible by measuring the temperature difference between the substrate and cryogenic pool and using analytical expressions to determine the heat flux. The conclusions of small scale experiments had limited effects due to the small confined space of the boiling cell. Most of the tests were also qualitative in nature and currently there are very few quantitative results for cryogenic liquids released on ice.

An experimental investigation is undertaken here to improve the understanding of cryogenic liquid releases on ice and to provide a benchmark qualitative and quantitative results. The small scale experimental study is performed to verify the heat transfer mode of cryogenic liquids released on ice and independently validate the one dimensional conduction for cryogenic liquid released on ice.

2. Experimental Setup and Procedure

The experimental setup designed for small scale spill of liquid nitrogen on ice is shown in Figure 1. The experiments were performed in a wind tunnel present at the Fire Station 2 of the Ras Laffan Industrial City, Qatar. The wind tunnel was 12 m long, 2 m wide and 0.855 m tall offering a convenient space for safe dispersal of vapor. The setup consists of a metallic cylinder of inner diameter 58.5 cm and height 87.5 cm made of carbon steel. A stainless steel chimney with dimensions of 35 x 35 x 80 cm was placed inside the metal cylinder to protect the cylinder from fractures resulting from a cryogenic liquid spill. The stainless steel chimney was square in cross-section of 0.1225 m^2 and devoid of top and bottom walls, which offered a well-defined area for the spill on water. The material was robust in cryogenic liquid temperature and offered additional protection for the external cylinder. The stainless steel chimney was lowered into the metallic cylinder using a steel support in such a way that it was half filled with water at any point of time during the experiment. Such a setup made possible for the liquid nitrogen to interact only with water and stainless steel walls rather than the cylinder, which was susceptible to fractures at cryogenic liquid temperatures. An ice slab of dimensions 25 x 25 x 10 cm was placed inside the stainless steel chimney on top of water. The thermocouple to measure the temperature of the ice was embedded into the ice slab. Additional ice cubes were added around the ice slab to create an ice-water mixture and to limit contact of LN_2 with water.

The temperature of water, liquid nitrogen and nitrogen vapor was monitored by distributed N-type thermocouples. The thermocouples were mounted on two polycarbonate boards and each included sixteen thermocouples distributed vertically. One board was mounted near the wall of stainless steel box and another was held in the center of the box. The dimensions of the thermocouples relative to water level are given in Table 1. The polycarbonate board was suspended into the metallic cylinder in such a way that 11 thermocouples measured the temperature below water, three measured liquid nitrogen temperature and two measured liquid nitrogen vapor temperature. Each of these thermocouples was calibrated for measuring liquid nitrogen and ice temperature and the temperature was measured every second. The thermocouples were distributed with smaller distances (<0.5 cm) near the water-liquid nitrogen surface. The position of each thermocouple and cryogenic discharge hose nozzle is shown in

Table 1. The cryogenic hose was placed in the middle of the setup to ensure smooth discharge of liquid nitrogen on ice. The width of the tunnel was considered X axis and length of the wind tunnel was considered as the Y axis (and the positive value indicates downwind direction). The entire setup was placed on a balance (maximum load – 300 kg and sensitivity – 100 g) to monitor the change of mass when nitrogen vaporizes on water. The sensors' output was recorded every second (1 Hz) by a Data Acquisition system (DAQ), and the recorded data were sent to a computer for processing.

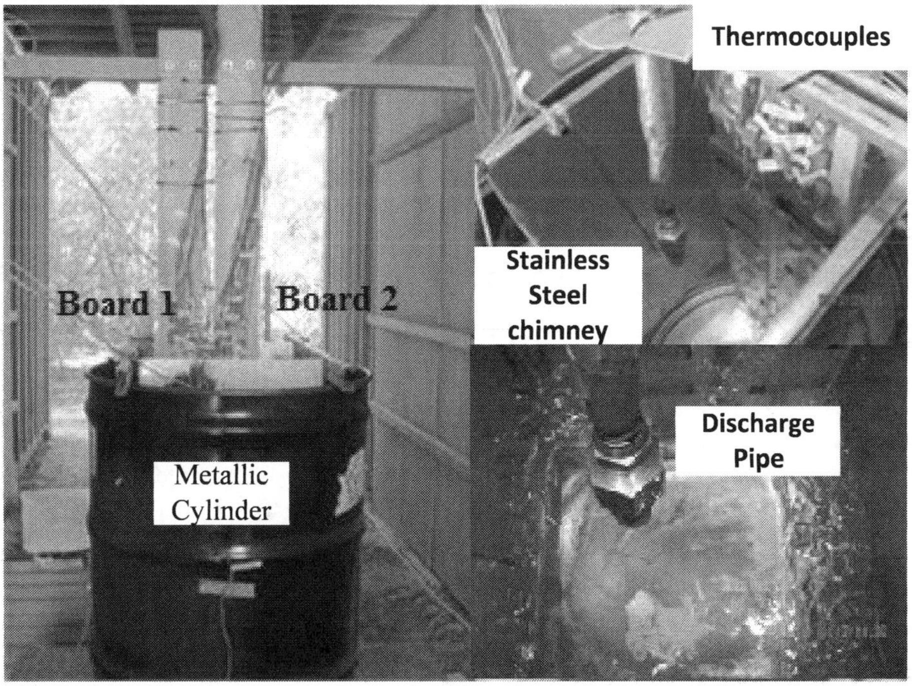

Figure 1: Top and lateral views of experiment setup

Table 1: Thermocouple positions

Thermocouple number	Position with respect to cylinder's corner (cm)			Thermocouple number	Position with respect to cylinder's corner (cm)		
	X	Y	Z*		X	Y	Z*
TC-201	14	26.6	-10.25	TC-204	26	26	-9.85
TC-202	14	26.6	-8.15	TC-205	26	26	-7.7
TC-203	14	26.6	-6.25	TC-216	26	26	-5.67
TC-206	14	26.6	-4.25	TC-218	26	26	-3.65
TC-207	14	26.6	-3.25	TC-219	26	26	-2.65
TC-208	14	26.6	-2.35	TC-222	26	26	-2.25
TC-209	14	26.6	-1.25	TC-224	26	26	-0.75
TC-210	14	26.6	-0.8	TC-225	26	26	-0.25
TC-211	14	26.6	-0.25	TC-226	26	26	0.35
TC-212	14	26.6	0.25	TC-227	26	26	0.75
TC-213	14	26.6	0.65	TC-228	26	26	1.25
TC-214	14	26.6	1.65	TC-229	26	26	2.35
TC-215	14	26.6	2.65	TC-230	26	26	3.25
TC-217	14	26.6	4.75	TC-231	26	26	5.35
TC-220	14	26.6	7.75	TC-232	26	26	8.25
TC-221	14	26.6	22.25	TC-233	26	26	23.45
Discharge Pipe	14	14.5	13.75				

*Distance from top ice surface; positive value means above the ice surface, negative one below the ice surface (inside ice).

The liquid nitrogen was discharged continuously from a tank of 180 liter capacity through a cryogenic liquid hose of length 15 m and internal diameter 0.0127 m, until desired amount of LN_2 was reached. The mass of the box was recorded every 1 second and used to measure the LN_2 vaporization rate. The time of opening the cryogenic liquid valve and the time when liquid nitrogen contacted with the water was noted. The filling rate of LN_2 was then calculated from the time and mass data (where filling rate referred to discharge rate ~~substracted~~subtracted by vaporization rate during the spill). The time to complete vaporization of liquid nitrogen was also noted. The discharges were repeated several times and the spilled amount of liquid nitrogen varied. The temperature of ice decreased after each experimental run due to release of liquid nitrogen on it. A summary of the six experimental runs with initial temperature of water bulk, initial temperature of ice and the amount of liquid nitrogen is given in Table 2.

Table 2: Summary of experiments

Run Number	Total Amount Spilled, (kg)	Release rate, ($kg.s^{-1}$)	Initial temperature of ice, (°C)	Water Temperature below ice, (°C)
Run 1	4.3	0.02	-11	8
Run 2	10.5	0.08	-5	13
Run 3	5.3	0.05	-66	10
Run 4	2.6	0.02	-51	9
Run 5	8.0	0.03	-44	8.3
Run 6	13.6	0.01	-119	7.4

3. Theory

When a cryogenic liquid contacts the ice surface, a large temperature difference exists between ice and liquid nitrogen which results in rapid vaporization of liquid nitrogen. The heat transfer from ice to liquid nitrogen is predominantly controlled by conduction phenomenon. In this study, one dimensional semi-infinite unsteady heat conduction equation with ideal contact of liquid nitrogen and ice is used to describe the heat transfer process [4]. The heat flux q from ice to liquid nitrogen is given by

$$q = \left(\frac{k}{\sqrt{(\pi\alpha t)}}\right)(T_i - T_b)\exp\left(-\frac{z^2}{4\alpha t}\right) \quad (1)$$

The heat flux is also proportional to inverted square root of time and is given by

$$q = \left(\frac{k}{\sqrt{(\pi\alpha)}}\right)(T_i - T_b)\exp\left(-\frac{z^2}{4\alpha t}\right) * t^{-0.5} \quad (2)$$

The temperature T inside the ice at the z depth is given by

$$T = T_i - (T_i - T_b).erfc\left(\frac{z}{2\sqrt{\alpha t}}\right) \quad (3)$$

$$q = \left(\frac{k}{\sqrt{(\pi\alpha t)}}\right)(T_i - T_b)\exp\left(-\frac{z^2}{4\alpha t}\right) \quad (1)$$

~~The temperature T inside the ice at the z depth is given by~~

$$T = T_i - (T_i - T_b).erfc\left(\frac{z}{2\sqrt{\alpha t}}\right) \quad (2)$$

Where the T_i represents the initial temperature of ice, T_b is the temperature of liquid nitrogen, k is the thermal conductivity and α is the thermal diffusivity of ice. A perfect thermal contact between liquid nitrogen and ice is assumed in calculating the temperature variation of ice.

4. Results and discussion

The results of small scale experiments are discussed in subsequent sections. Section 4.1 discusses the vaporization flux determined from mass loss data obtained during the experiment. Section 4.2 and 4.3 discusses the validation of model with experimental data.

4.1. Vaporization Mass Flux

The mass flux of vapor (vaporization rate) was obtained from direct measurement of mass loss data obtained during the experiment (Figure 2). An averaging time of 60 seconds was used to compute the vaporization rates of the liquid nitrogen on ice. The selection of averaging time was based on a sensitivity analysis shown in Figure 3. Four averaging times were chosen to calculate the mass vaporization flux of LN2 vapor. An averaging time of 10 and 20 seconds resulted in large fluctuations in vaporization flux due to low resolution in mass measurement (maximum capacity of balance was 300 kg and 100 g resolution). An averaging time of 30 seconds resulted in fairly smooth curve but still with periodic fluctuations. An averaging time of 60 seconds was found to give low variations and thus sufficiently describing the vaporization mass flux without much loss in the actual physics. The spill area included the constant cross sectional area of the stainless steel chimney and a variable area based on the height of the liquid nitrogen interacting with the stainless steel. The chimney cross sectional area of 0.1225 m^2 was used to determine the vaporization rates.

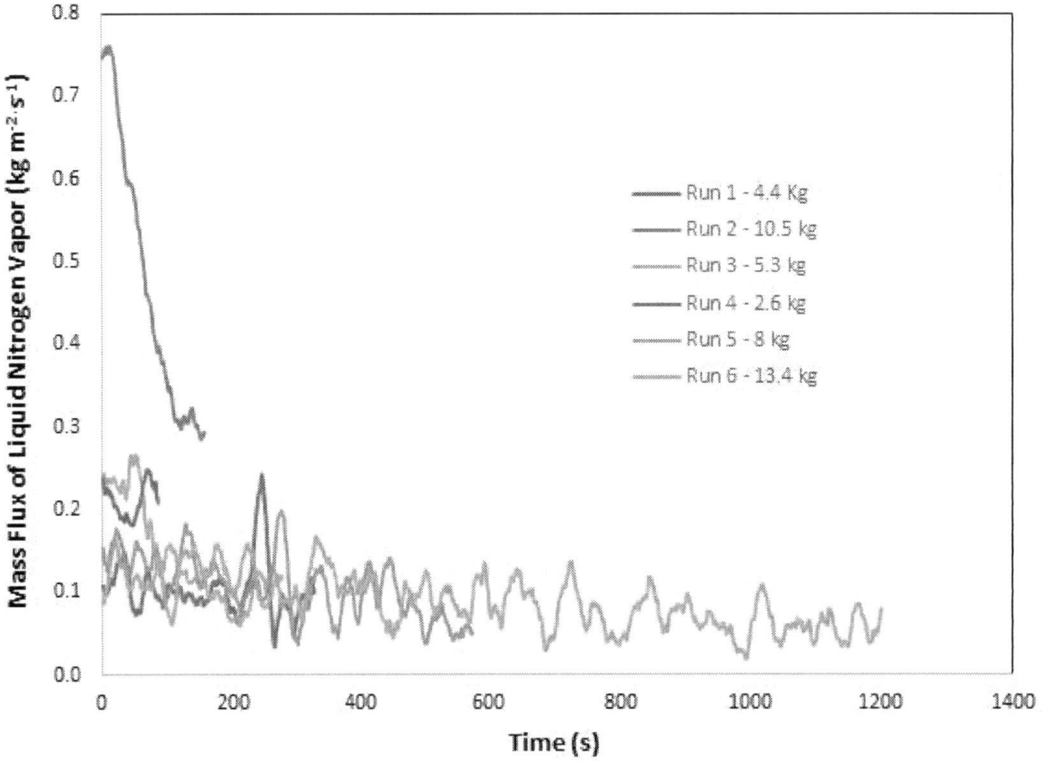

Figure 2: Vaporization Mass Flux

Atypical for conduction, a decreasing trend of vaporization rate was observed when the ice temperature was relatively high (-5 °C at run 2). However, the mass flux became comparatively constant, when the ice temperature dropped down below -11°C, (run 1 and runs 3 to 6). However, small peak in vaporization rate was observed during run 1 (red line in Figure 2). This is with the agreement to the result obtained at the Bureau of Mines [5], where a decreasing trend was observed at the initial stage of the spill on ice.

Initially during the start of experimental runs, the ice block was transparent in nature. With the release of liquid nitrogen in every test, the thickness and opacity of ice increased. Small ice balls were formed during each experiment on the surface of ice, which might be attributed to freezing of condensed water from air and their aggregation. The ice balls and frost on the walls of the equipment are shown in Figure 3. It has been observed that the growing ice thickness in the pool created resistance to heat transfer and thus controlled the heat flux supplied to the cryogenic pool. An average heat flux of 36.6 kW/m^2 was measured in the tests, which represents an average vaporization flux of 0.18 kgm^{-2}s^{-1}. The values of vaporization rates were compared with experiments conducted by Bureau of Mines (1970). The average heat flux obtained in their experiment was around 25.85 kW/m^2 with corresponding vaporization flux determined to be 0.13 kg m^{-2}s^{-1}. The variability can be attributed to the lack of water and ice mixture that was not present in the previous experiments. A combination of ice and water mixture results in heat fluxes that are intermediate between heat flux obtained due to convection and conduction alone.

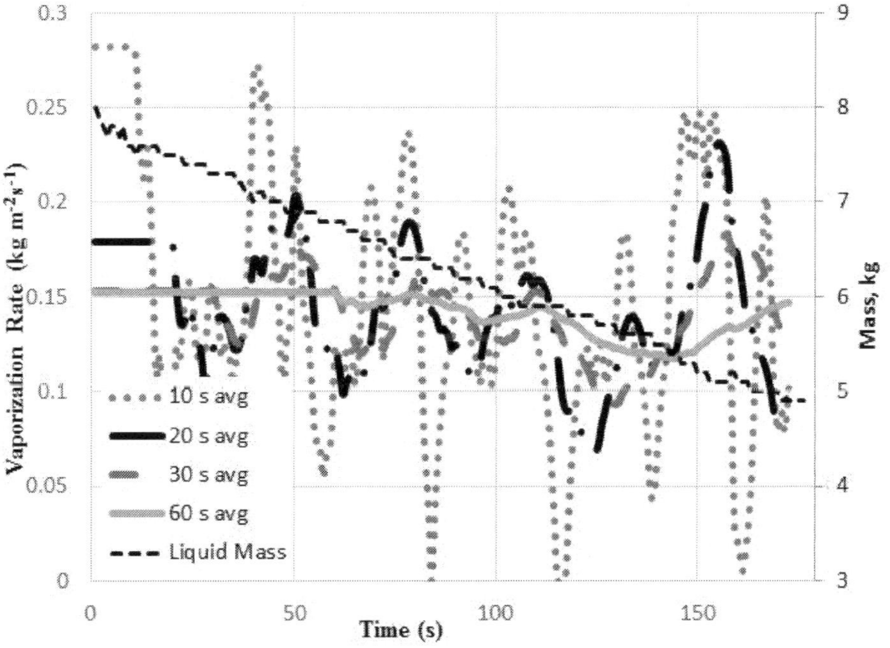

Figure 3: Vaporization flux for different averaging times

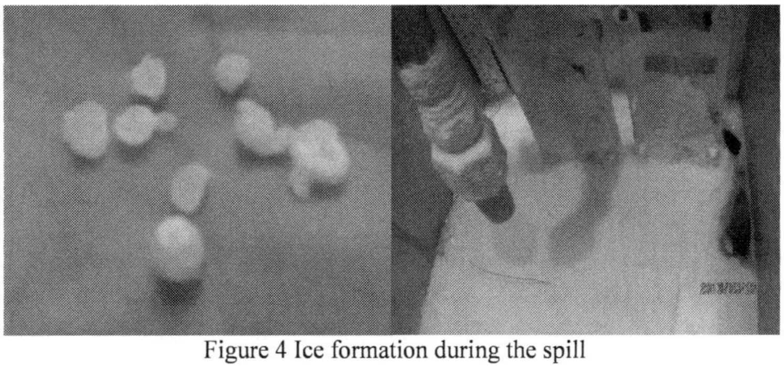
Figure 4 Ice formation during the spill

4.2. Temperature profile of ice

The variation of ice temperature during the spill was analytically calculated from equation (2) and compared with experimental values at seven different depths, and the result obtained for experimental run 5 is shown in Figure 5. The solid lines represent the temperature measured during the experiment and the dashed lines represent the predicted temperatures. A good agreement was observed between the experimental and predicted temperature. It is to be noted that the initial temperature profile of ice in this run (run 5) was an effect of previous spill and this explains lower temperature near to the ice top surface and higher values for thermocouples placed farther from the ice surface. Also, the thermocouples placed near the top surface are cooled earlier than the thermocouples below, due to the release of liquid nitrogen. A deviation in predicted temperatures is observed after 750 seconds, which might be an effect of heat transfer from warmer water below ice. This might also be due to the variation in thermal properties of ice with temperature, which is not accounted for in the model.

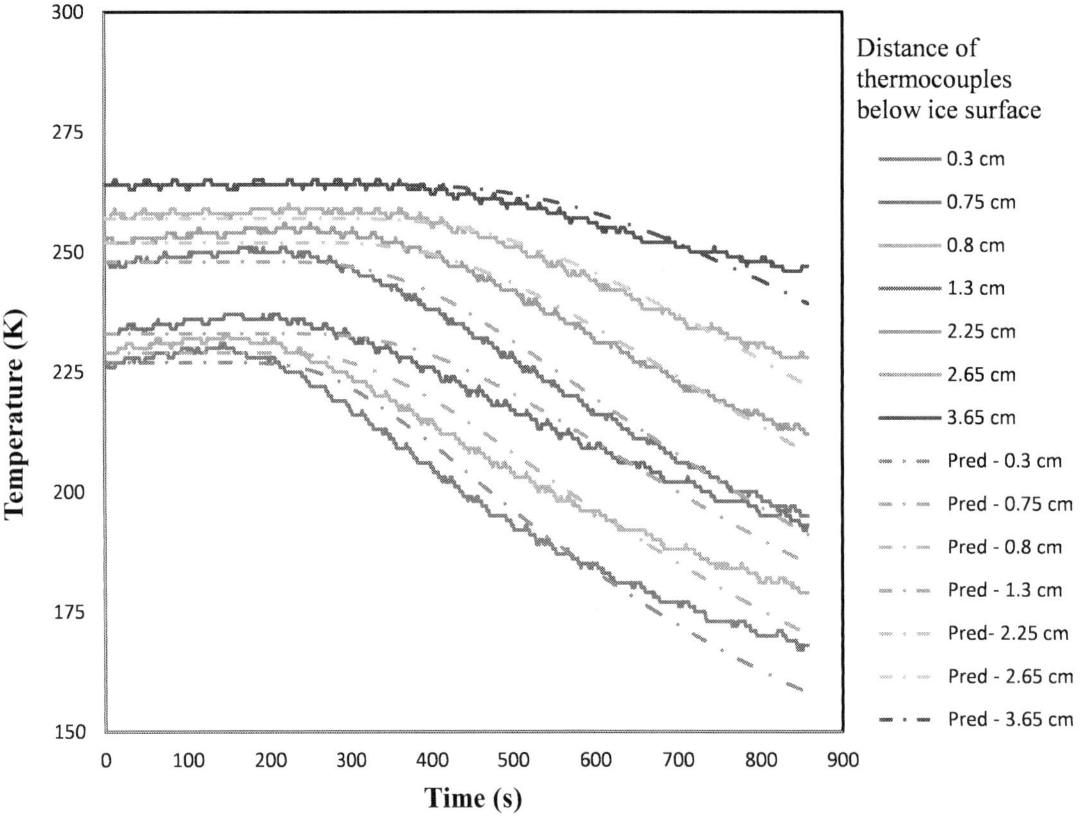

Figure 5: Temperature profiles of thermocouple below water surface

4.3. Comparison of experimental and predicted transient heat flux

A comparison of the predicted heat flux and experiment heat flux as the function of time is given for run 5 is given in Figure 7. The trend for one experimental run is shown here since all the experiments except run 2 followed similar trends in data. Figure 7(a) represents the comparison of predicted heat flux obtained from equation (1) and experimental heat flux with respect to time. The heat flux obtained from experiment was found to be varying above and below the predicted heat flux. However, the variations of experiment were comparatively low. Figure 7 (b) represents the comparison of predicted heat flux obtained from equation (2) and experimental heat flux with respect to inverted square root of time. The results seem to have good agreement except for the run 2. However, run 2 was a case where the liquid nitrogen was spilled on a mixture of ice and water, thus it cannot be well modeled by the conduction and perhaps is closer to the convection as for the spill on water.

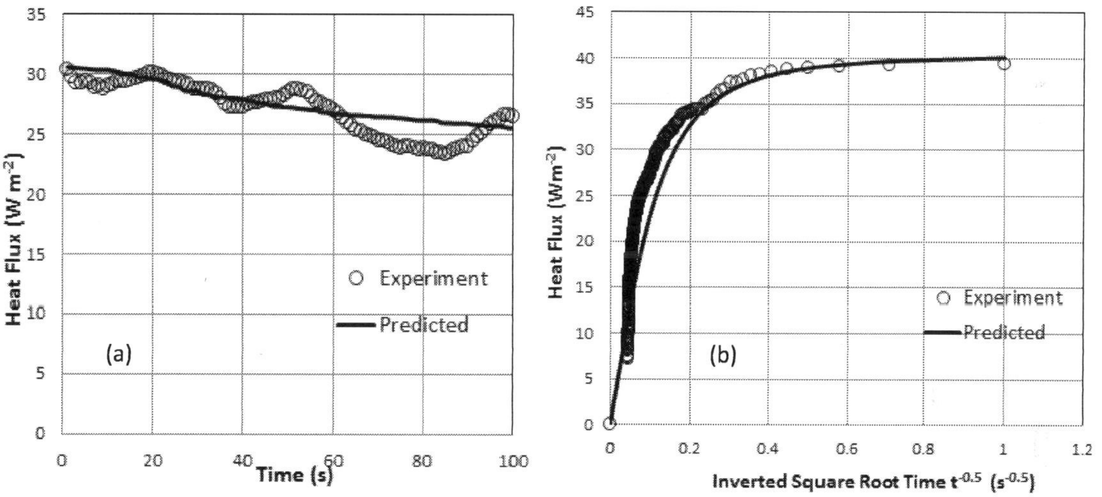

Figure 6 Comparison of Experimental and Predicted Heat Flux for Run 5
(a) Time (b) Inverted Square Root of Time for Run 5

5. Conclusion

An experimental investigation of small scale releases of liquid nitrogen on ice was performed to improve the understanding of the source term modelling of cryogenic liquid spilled on ice. The vaporization rates obtained were independent of the amount of liquid nitrogen spilled, and as expected, were a function of the ice temperature. A very good prediction agreement of the temperature inside the ice obtained by 1D ideal conduction model was observed. The performance of the theoretical model in predicting heat flux for the release on ice was found to be fair, but not for ice-water mixture. The vaporization rates were found to be constant for long duration releases of liquid nitrogen on ice.

6. Acknowledgements

The authors would like to thank BP Global Gas SPU for their financial support and guidance. They would also like to acknowledge the support of Qatar Petroleum (QP) for offering facilities at Ras Laffan Industrial City. The authors personally thank QP's staff who worked with LNG research team of Texas A&M University at Qatar for the experiment.

7. References

[1] K. A. Dahl, "Analysis of accidental iceberg impacts with LNG tank carriers," Norwegian University of Science and Technology, 2012.

[2] E. Nakanishi and R. C. Reid, "Liquefied Natural Gas-Water Reactions," *Chem. Eng. Prog.*, vol. 67, no. 12, pp. 36–41, 1971.

[3] V. Vesovic, "The influence of ice formation on vaporization of LNG on water surfaces," *J. Hazard. Mater.*, vol. 140, no. 3, pp. 518–526, Feb. 2007.

[4] H. S. Carslaw and J. C. Jaeger, *Conduction of Heat in Solids*, 2nd ed. Oxford University Press, USA, 1986, p. 520.

[5] D. Burgess, J. Biordi, and J. Murphy, "Hazards of spillage of LNG into water," Pittsburg, Pennslyvania, 1972.

348750 Consequences Analysis Associated with the Failure of the Safety Interlock System of Methanator (I-351) of the Ammonia Plant of Fertilizantes Nitrogenados De Venezuela, C.E.C. (FertiNitro)

Abstract:

Consequences Analysis associated with the failure of the safety interlock system of methanator (I-351) of the Ammonia Plant of Fertilizantes Nitrogenados de Venezuela, C.E.C. (FertiNitro).

Fertilizantes Nitrogenados de Venezuela, C.E.C. (FertiNitro) is a petrochemical company, dedicated to the production and sale of anhydrous ammonia and granular urea high quality, located within the Oil and Petrochemical Complex General José Antonio Anzoategui, in the town of José, north of the Anzoategui state, adjacent to National Highway José Antonio Anzoategui (troncal 9), between Barcelona and Puerto Píritu Population. 30% of the production of these fertilizers are for the domestic market and 70% is exported to the international commercial market.

Throughout the value chain of Fertinitro, ammonia and urea are produced by transforming raw materials such as natural gas (which comes from the gas extraction plant of PDVSA) and atmospheric air. Fertinitro is formed by two identical trains, which in turn are cut by two anhydrous ammonia plants whose maximum production capacity is 1,800 tons per day each, and two granular urea plants, with maximum production capacity of 2,200 tons per day each. To achieve optimum performance of such plants, they are fed by the Industrial Services Unit and then the final product is sent to the storage unit and dispatch for further marketing.

Ammonia plant is divided into several sections (Desulfurization, Reformation, conversion, carbon dioxide removal, methanation, Synthesis Gas Compression, Synthesis and Ammonia Refrigeration). In the methanation section, the reactor is also known as methanator reactor (12/22-R-301), which introduces the process gas coming from the sections of Conversion and Carbon Dioxide Removal respectively (rich in hydrogen and Nitrogen with small traces of carbon monoxide and carbon dioxide).

Considering that the presence of carbon monoxide (CO) and carbon dioxide (CO_2) into the reactor 12/22-R-501 (located upstream of methanation section) produce adverse effects design intention inside thereof, the catalyst content in the methanator promotes an exothermic reaction converting the traces of CO and CO_2 to methane (CH_4). The resultant process gas methanation section contains primarily hydrogen (H_2), Nitrogen (N_2) with trace amounts of methane (CH_4) and argon (Ar), the latter two being inert in the ammonia synthesis section.

The inlet temperature to the methanation reactor is controlled by the Temperature Indicator Controller 12/22-TIC-3353 that, in split-range, the first controls the "by pass" associated with the exchanger gas / gas 12/22-E-311 A / B generating the opening of the temperatura control valve 12/22-TV-3353[a], and if is necessary additional heating, the 12/22-TIC-3353 sends a signal for the opening of the valve 12/22-TV-3353B and therefore increases the amount of the converted gas that passes through the heater 12/22-E-211 of the methanation reactor.

On the other hand, the temperatures in the catalyst bed is measured at four different levels in order to monitor the temperature profile. If there is a pass of CO and / or CO2 in the methanator due to poor operation of the CO conversion section and / or CO2 removal section, the temperature rise through the methanator may be so high that exceed the design temperature of the methanator. To avoid such an incident, it has been installed in the catalyst bed a high temperature switch to trigger the safety interlock system of the methanator (I-351). If the I-351 is triggered, the flow of process gas that is directed to the methanator is cut, being automatically vented downstream of the CO2 absorber (belonging to the CO2 removal section) through 12/22-PV- 3350 and downstream of the CO converter of low temperature through 12/22-PV-2076 12/22-PV-2076 a and B.

The I-351 activates the manual air vent on the top of methanator, which also can be operated safely from the control room. This action allows the gas to be vented through reverse flow of the methanator to protect the vessel and the catalyst from overheating during the release. When the I-351 is activated, the 12/22-HY-3050 decrease the pressure at 2 kg/cm2g per minute, opening the 12/22-HV-3050.

In that sense, due to hazardous materials that are handled in this equipment, plus the operating conditions, it is considered a high-risk equipment, that is why there is a need to realise an consequences analysis associated with the failure of the safety interlock system of methanator (I-351) of the Ammonia Plant of Fertilizantes Nitrogenados de Venezuela, C.E.C. (FertiNitro).

349822 Building Siting Evaluation: A New Software Tool for the Determination of Blast Loads from Potential Vapour Cloud Explosions (VCEs)

Abstract:

A central element of building siting evaluation—especially where vapour cloud explosion (VCE) hazards are credible—is the calculation of the blast parameters/loads to buildings of interest. This determination is typically approached in two ways. Using simplified models that utilize blast curves (notably the TNO Multi-Energy and Baker Strehlow Tang models), or advanced methodologies using computational fluid dynamics (CFD) technology. Simplified models are more widely used as they offer a good level of characterization potential and are more pragmatic to use. The latter (i.e. CFD) can be resource and time intensive and is typically reserved for instances where the simplified models are limited (e.g. in the determination of near field blast loads).

A rigorous application of the simplified approaches would typically involve dispersion modelling (under specific conditions e.g. weather and wind speed) for the identified loss of containment scenario (or scenarios) to evaluate the size of the resulting flammable cloud and the potential for (and degree of) interaction with any identified regions of congestion and/or confinement. However, such a task is not trivial and to reduce the calculation burden this analytical step is sometimes bypassed and a simplifying assumption that the entire region is full of flammable gas is made. Whilst such an assumption ensures a conservative, cautious approach to the analysis, it does not reflect reality and can result in outcomes that are excessive to requirements – the upshot of which can be sub-optimal resource allocation. Consequently, there is a need for an approach that allows for a more realistic analysis of VCE potential.

This paper describes an advanced implementation of the two most widely used vapour cloud explosion models mentioned above, namely the TNO Multi Energy (ME) Method and the Baker-Strehlow-Tang (BST) model, into Phast, the well-known general purpose consequence analysis software package developed by DNV GL Software. A key aspect of this new addition is the capability to examine the interaction of a dispersing dynamic flammable cloud profile with congested/confined regions in all three dimensions (length, width and height) and in multiple directions. This is in direct response to the aforementioned challenge and the intention is to provide a robust platform by which a more detailed analysis of the VCE hazards can be evaluated without recourse to simplification.

The implementation of both models is extensive and includes all the key parameters that have a bearing on the magnitude of a vapour cloud explosion. A case study of the tool as applied to a hypothetical occupied building siting challenge is presented and the value of using the tool discussed.

Modeling of time-varying dispersion from ground-level liquid pools or vapor area sources

Henk W.M. Witlox
DNV GL Software
3 Cathedral Street, London SE1 9DE, UK
henk.witlox@dnvgl.com

Mike Harper
DNV Software
Exchange Street, Stockport, SK0 3ET
michael.harper@dnvgl.com

Maria Fernandez
DNV Software
3 Cathedral Street, London SE1 9DE, UK
michael.harper@dnvgl.com

Prepared for Presentation at
American Institute of Chemical Engineers
2014 Spring Meeting
10th Global Congress on Process Safety
New Orleans, LA
March 30 – April 2, 2014

UNPUBLISHED

Modeling of time-varying dispersion from ground-level liquid pools or vapor area sources

Henk W.M. Witlox
DNV GL Software
3 Cathedral Street, London SE1 9DE, UK
henk.witlox@dnvgl.com

Mike Harper
DNV GL Software

Maria Fernandez
DNV GL Software

Keywords: dispersion, time-varying, pool, along-wind diffusion

Abstract

The current paper discusses a new enhanced dispersion formulation accounting for time-varying effects resulting from a time-varying ground-level liquid pool or vapor area source. This new formulation is implemented in the Unified Dispersion Model (UDM), and is planned to be included in a future version of Phast. The new model includes a considerably improved model for linking between the pool and the cloud. It furthermore includes along-wind-diffusion effects (due to ambient turbulence) at the upwind and downwind edges of the cloud. This results in reduced concentrations while the cloud travels in the downwind direction which is particularly relevant for toxic releases. The new formulation presumes a number of 'observers' to be released at successive times from the upwind edge of the pool. The UDM first carries out 'pseudo steady-state' calculations for each observer, where the 'release data' correspond to the time at which the observer is released, and where the observer accumulates vapor while travelling over the evaporating pool. Subsequently effects of along-wind diffusion (AWD) are included by means of Gaussian integration over the downwind distance, whereby the downwind dispersion coefficient is chosen to be consistent with the current finite-duration correction (FDC) formulation adopted in the UDM.

First the above new UDM formulation has been verified against the time-varying HGSYSTEM model HEGADAS-T. The UDM provides more accurate results in the near-field because of more accurate observer calculations while observers are moving over the pool. Subsequently the UDM FDC and new UDM models have been validated against experimental data including finite-duration carbon-dioxide vapor ground-level area releases (Kit Fox experiments), and accurate results were obtained using both formulations.

1. Introduction

Many accidents involve releases of hazardous chemicals into the atmosphere. Commonly used atmospheric dispersion models are limited to continuous or instantaneous releases only. They do not accurately account for time-varying effects following time-varying releases i.e. resulting from a pressure drop in a vessel or pipe, and/or from a time-varying pool (following rainout or from direct spill). These effects include along-wind-diffusion (due to ambient turbulence) resulting in reduced concentrations.

The consequence modelling package Phast includes the Unified Dispersion Model UDM (Witlox and Holt [1], Witlox and Harper [2]) for modelling two-phase outdoor releases including potential rainout. The UDM in the current Phast (versions 6.7 and 7.1) only accounts for along-wind-diffusion (AWD) for instantaneous releases or finite-duration continuous releases that do not rain out.

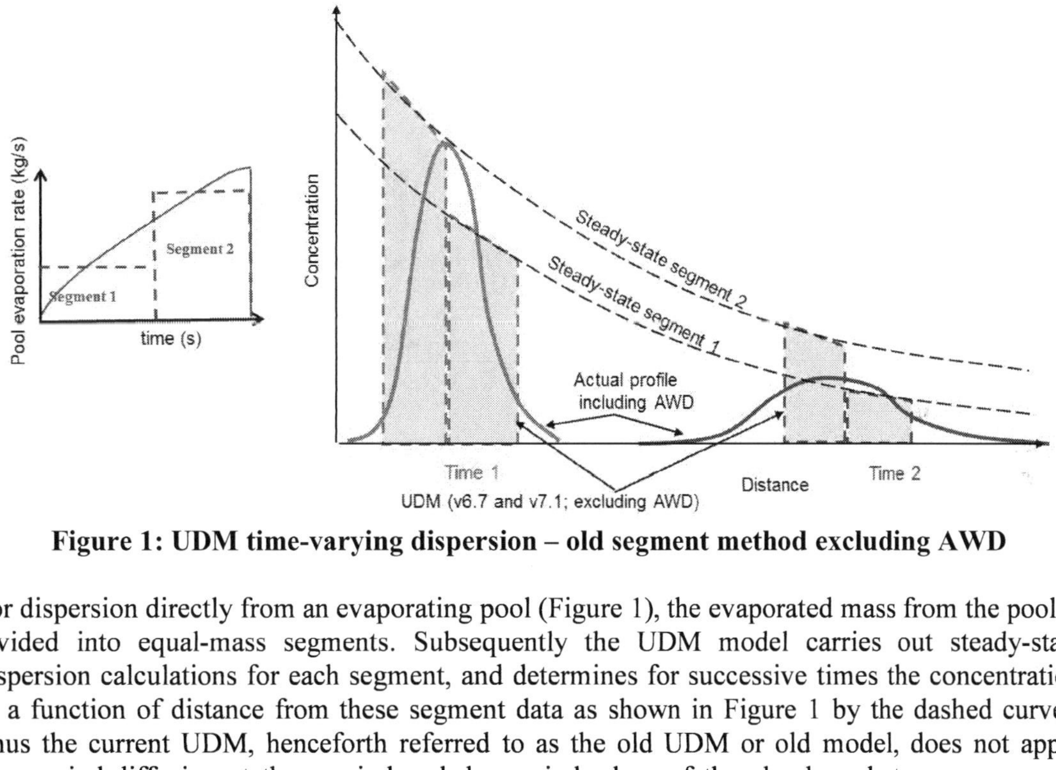

Figure 1: UDM time-varying dispersion – old segment method excluding AWD

For dispersion directly from an evaporating pool (Figure 1), the evaporated mass from the pool is divided into equal-mass segments. Subsequently the UDM model carries out steady-state dispersion calculations for each segment, and determines for successive times the concentration as a function of distance from these segment data as shown in Figure 1 by the dashed curves. Thus the current UDM, henceforth referred to as the old UDM or old model, does not apply along-wind-diffusion at the upwind and downwind edges of the cloud, or between segments, which may lead to significant over-prediction of concentration and under-prediction of duration in the far-field (see Figure 1 at time 2). This is particularly important for toxic releases, where dispersion calculations are required to be carried out to small concentration levels such as ERPG levels. It is less important for flammable releases with calculations to relatively high concentration levels only, such as LFL or 0.5LFL.

The Shell consequence modelling package HGSYSTEM 3.0 (Post [3]) includes the time-dependent dispersion model HEGADAS-T (Witlox [4]) for modelling the time-dependent dispersion of a heavy-gas cloud moving with the ambient wind. It can be used to model the dispersion downwind of a time-dependent ground-level source (unpressurised release). The transient behaviour of the cloud is approximated by a quasi-steady-state description in which so-called "observers" are released upwind of the source at a series of times. These observers travel with the cloud. For each observer, the observed concentration is set from steady-state HEGADAS-S calculations using the observed source/transition data. Thus by calculating the position of each observer at a given time t, the concentration c is set for a number of downwind distances. Subsequently the actual concentration is set from Gaussian integration with respect to the downwind distance x of the above observer concentrations as illustrated in Figure 2. This involves a downwind dispersion coefficient σ_x, which allows along-wind diffusion to be taken into account.

A new enhanced dispersion formulation has been developed to account for time-varying effects resulting from a time-varying release. It generalises the above HEGADAS-T formulation to account for along-wind-diffusion both for ground-level unpressurised releases (e.g. evaporating pools) and for elevated two-phase pressurized releases including potential rainout. This method has been implemented in a new version of the UDM model, which is to be included in a future version of Phast. The current paper considers the case of an unpressurised release from a ground-level liquid or vapor area source only. Previous separate papers (Witlox and Harper [5], [6]) have dealt with the cases of an elevated pressurized release without rainout and an elevated pressurized release with rainout.

2. New UDM along-wind diffusion formulation

In the case of a liquid pool, first pool evaporation calculations are carried out to determine the time-varying pool spreading/evaporation data [radius (m), evaporation rate (kg/s), temperature (K)]; see Fernandez et al. [7] for details.

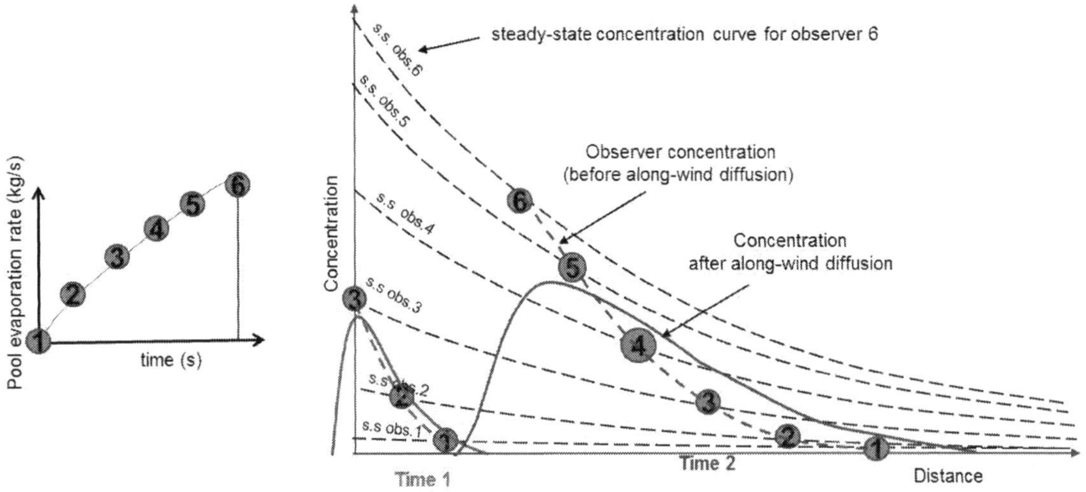

Figure 2: UDM time-varying dispersion –new observer method including AWD

Subsequently 'observers' are released from the upwind edge of the pool at intervals which can be chosen to correspond to equal-mass increments as for the old UDM method (see Figure 2 for the case of 6 release observers). These observers are assumed to move along the cloud centre-line (located at the ground) at the UDM cloud centroid speed. This is illustrated by Figure 3, where the first observer (1) is released at the start of pool evaporation, and subsequent observers (2,3) start from the upwind edge of the spreading pool.

For each observer steady-steady calculations based on the observed source data are carried out to evaluate the observer concentration prior to including effects of along-wind diffusion; see Figure 2. Primary variables and associated differential equations are as follows (see [1], [2] for details):

- Mass of moist air added to the cloud (kg/s) – air entrainment law describing turbulent mixing of air with the cloud, accounting for heavy-gas entrainment and/or passive entrainment
- Excess downwind cloud momentum (kg m/s^2) – conservation of cloud momentum
- Position of downwind edge of cloud (m) – relation between cloud position and speed
- Heat conduction from substrate (J/s) – heat transfer equation between substrate and cloud
- Water evaporated from water substrate (kg/s) – water-vapor transfer equation between water substrate and cloud
- Cross-wind dispersion coefficient (m) - empirical spreading law in case of heavy-gas dispersion and based on formula for ambient crosswind dispersion coefficient in case of passive dispersion

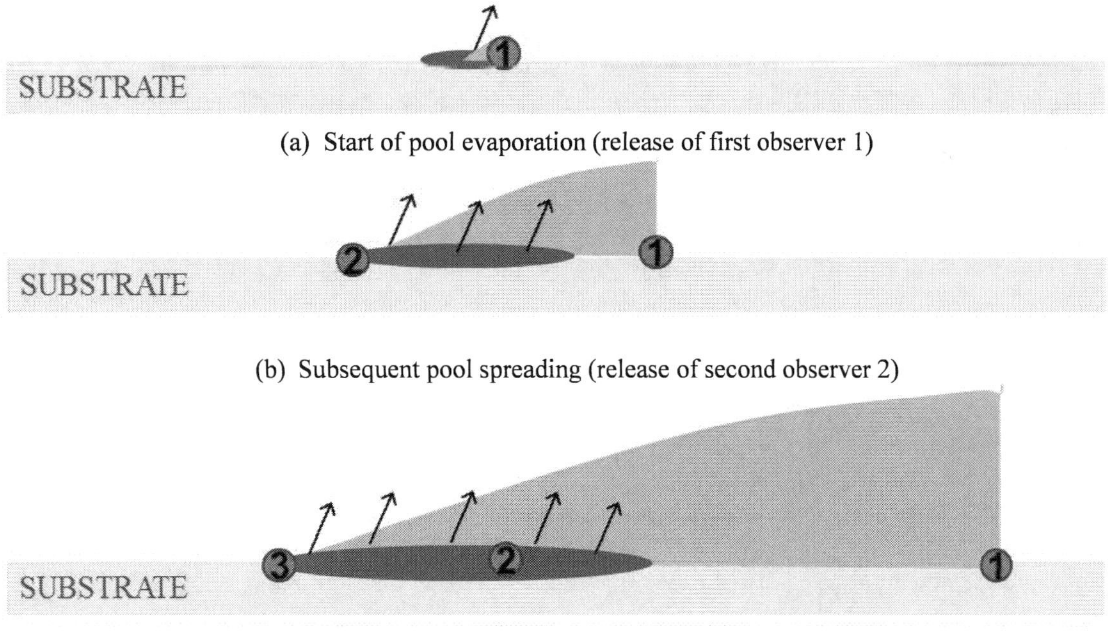

(a) Start of pool evaporation (release of first observer 1)

(b) Subsequent pool spreading (release of second observer 2)

(c) Subsequent pool spreading (release of observer 3)

Figure 3: New UDM dispersion phases for time-varying dispersion from pool

While the observer moves above the pool, the above observer dispersion equation (momentum conservation, cloud crosswind gravity spreading, heat and water-vapor transfer from the substrate) are modified to account for the pool vapor added back to the cloud. Two equations are solved for two additional primary variables: the mass of component present in the cloud (to account for mass added from the pool) and the component enthalpy (to account for component enthalpy added from the pool).

Additional equations are solved to derive the so-called 'secondary' variables from the 'primary' variables, which includes nonlinear equations for cloud thermodynamics (isenthalpic equation to evaluate cloud temperature) and cloud geometry. Furthermore an empirical concentration profile at each calculated downwind distance x is adopted to evaluate the observer concentration C as function of crosswind distance y, and vertical height z.

Inclusion of along-wind diffusion

The actual plume concentration c(x,y,z,t) including effects of along-wind-diffusion is a function of time t, distance x downwind of the release location, crosswind distance y, and vertical height z; this function is calculated by means of Gaussian integration of the observer concentration C(ξ,y,z,t),

$$c(x,y,z,t) = \int_0^{\infty} \frac{C(\xi,y,z,t)}{(2\pi)^{1/2}\sigma_x(\xi)} \exp\left\{-\frac{(x-\xi)^2}{2\sigma_x^{\,2}(\xi)}\right\} d\xi \qquad (1)$$

In the above equation ξ is the downwind distance from the release point at time t of an observer travelling with the cloud in the downwind direction. At this position the observer observes the concentration C(ξ,y,z,t). In Equation (1) along-wind diffusion is taken into account by assuming that the concentration C(ξ,y,z,t) spreads out around ξ according to a Gaussian distribution with a downwind dispersion coefficient $\sigma_x = \sigma_x(\xi)$. Figure 2 depicts the pre-AWD observer concentration C and the post-AWD concentration c at a short time after the release (time 1; limited AWD effects), and at a larger time after the release (time 2; larger AWD effects).

The downwind dispersion coefficient σ_x is evaluated as function of downwind distance ξ according to the formulation by Ermak [8] [9]. This assumption is consistent with the current finite-duration-correction (FDC) formulation adopted in the UDM [2] to apply AWD effects for finite-duration releases; see Witlox and Harper [6] for further details.

Averaging time effect because of time-varying release rate

In addition to the averaging time effect of wind meander, the user can optionally apply additional time-averaging of time-dependent concentrations resulting from time-varying pool evaporation,

$$[c(x,y,z,t)]_{avg} = \frac{1}{t_{av}} \int_{t-t_{av}/2}^{t+t_{av}/2} c(x,y,z,\tau)\,d\tau \qquad (2)$$

3. UDM verification against HGSYSTEM for dispersion from pool

The case of an instantaneous spill of 10000 kg of butane liquid is considered with both cases of a bund (radius 10m) and no bund (pool on water; dispersion over land). Ambient data applied are stability class D, wind speed at 10m height of 5m/s, temperature 300K and humidity 70%. Furthermore a maximum pool duration of 1 minute is applied, and thus no pool evaporation is assumed to occur after 1 minute.

Figure 4 shows the values predicted by the UDM of butane pool radius and evaporation rate for both cases with and without a bund. It is seen that the immediate butane rainout within the bund results in a virtually steady-state pool of 10 meter radius, 12.3 kg/s evaporation rate and a temperature of 271.8K.

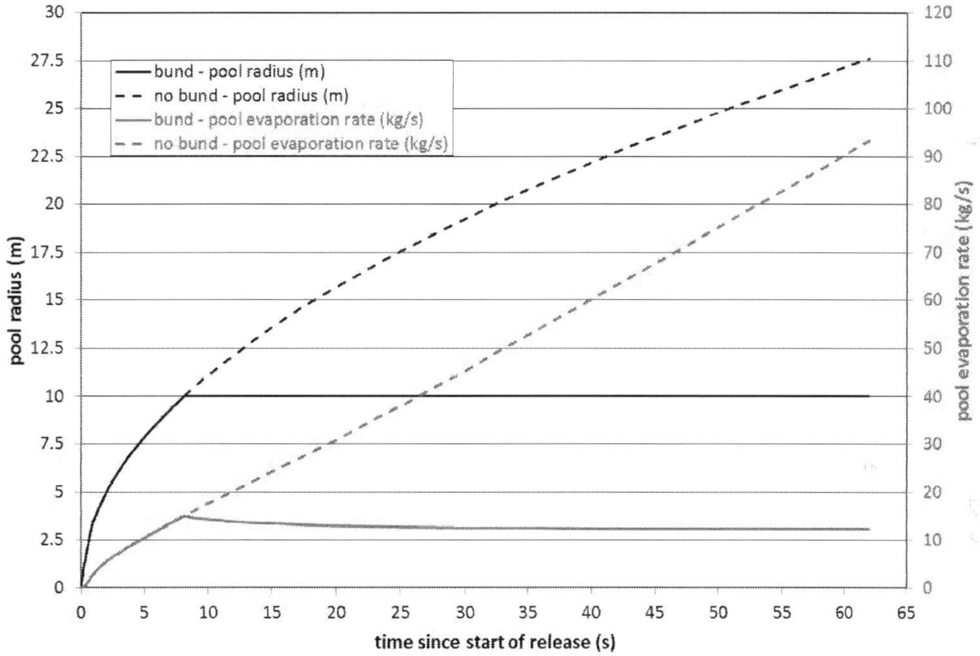

Figure 4: UDM pool predictions (instantaneous 10000kg spill of butane on water)

Figure 5 includes results of the following calculations for the case of a bunded pool:

- HGSYSTEM 3.0 calculations (the rectangular HEGADAS source was chosen to be square with the same area as the UDM circular pool)
    - steady-state HEGADAS-S calculations (ground-level steady-state heavy-gas dispersion from area source) based on the steady-state pool data mentioned above.
    - FDC finite-duration correction (using Ermak's formula) to above HEGADAS-S results (60 seconds duration) using the post-processor program POSTHS
    - HEGADAS-T calculations (ground-level time-varying heavy-gas dispersion from area source). In line with the new UDM model, no gravity-shape correction was

applied and the along-wind-diffusion coefficient was chosen to be based on Ermak's formula (non-default HEGADAS-T options)

- UDM calculations
    - Steady-state UDM calculations based on the above steady-state pool data
    - FDC correction to the above steady-state UDM results (60 seconds duration)
    - UDM AWD calculations, with 30 observers released from the upwind edge of the pool

Figure 5a includes the steady-state HEGADAS-S and UDM results of the centre-line ground-level concentrations in the near-field for the case of a steady-state pool. HEGADAS-S assumes a uniform concentration (top-hat profile) above the area source (about 15% mole fraction), which is envisaged to lead to an over-prediction near the source. On the other hand, the UDM model more rigorously solves the dispersion equations across the pool allowing for a variation of the concentration across the pool. The very rapid decrease predicted by the UDM of the concentration near the upwind edge of the pool is a numerical artefact and may not occur in reality. Further downwind from the pool a closer match is observed between the UDM and HEGADAS-S data.

Figure 5b includes results of all the above calculations. It includes results of the time-varying dispersion calculations by UDM AWD and HEGADAS-T at output times 300s, 600s and 1200s. It is seen that the maximum values of these curves correctly touch the maximum centre-line concentration (maximum value over all times) predicted by UDM FDC and HEGADAS-S FDC, respectively. This confirms that the AWD integration is carried out correctly by both programs and is consistent with the FDC formulation. Also adequate agreement between the UDM and HEGADAS FDC results is observed. It is seen that initially the HEGADAS-T cloud is heavier (larger concentrations) and therefore moves more slowly than the UDM AWD cloud, but this effect reduces in the far-field when the predictions become closer.

Likewise, Figure 6 includes UDM AWD and HEGADAS-T results at times 300s, 600s and 1200s for the case without the bund. Here the UDM calculated pool data were specified as input to HEGADAS-T with a step size of 2 seconds. The observed differences are confirmed to be very similar to that seen for the case with the bund (compare Figures 5b and 6).

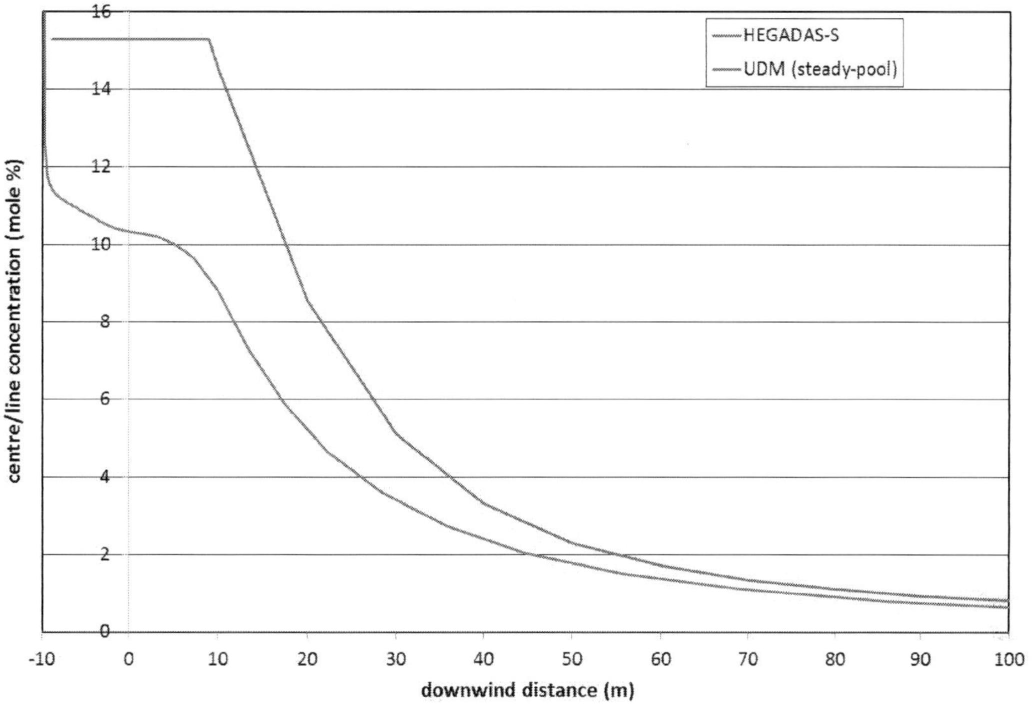

(a) Near-field centre-line concentrations (above pool and downwind of the pool)

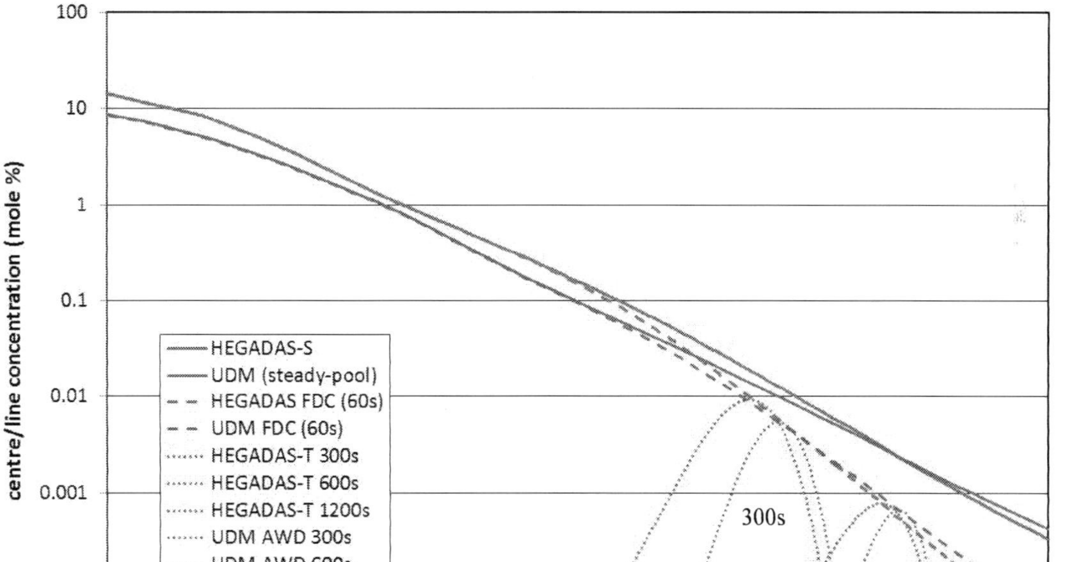

(b) Far-field concentrations

Figure 5: UDM verification against HEGADAS for spill of butane with bund (radius 10m)

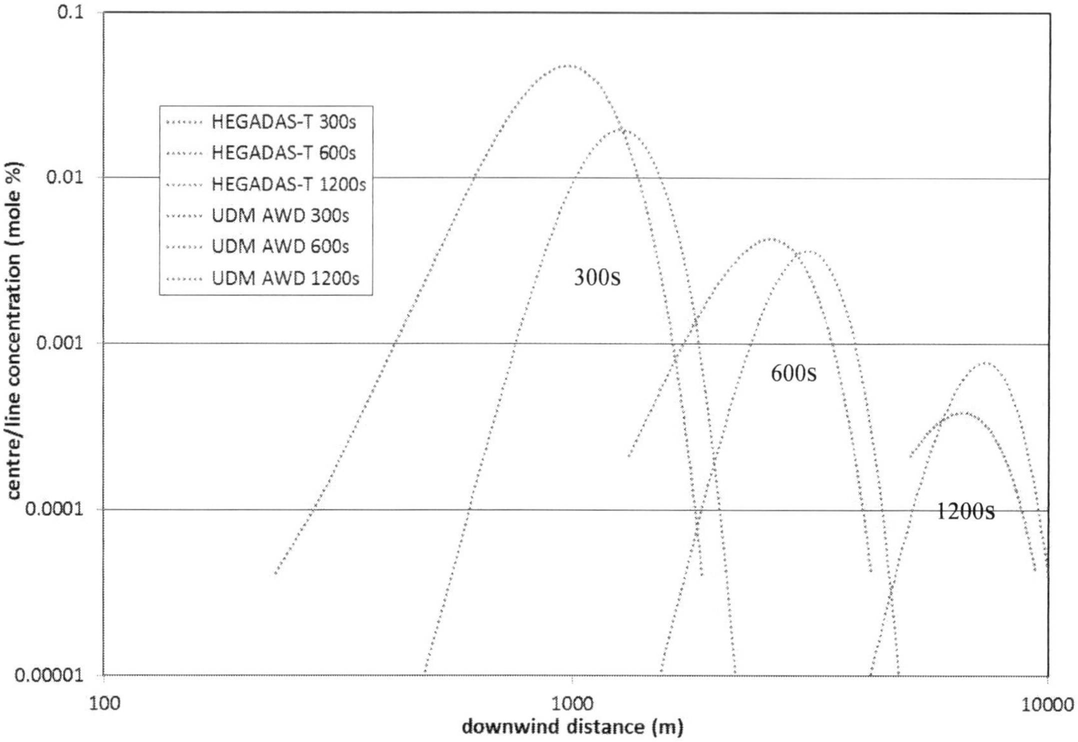

Figure 6: UDM verification against HEGADAS for spill of butane without bund

4. UDM validation against Kit Fox experiments (CO_2 area source)

In these experiments the CO_2 was released as a dense vapor from a 1.5m x 1.5m ground-level area source for continuous and 20-second finite-duration releases, during both neutral and stable conditions. In this paper results are reported of experiments with a uniform surface roughness (URA; roughness estimated between 0.01 and 0.02 m; adopted value 0.01m); see Table 1.

Table 1. List of Kit Fox experiments for UDM validation

Kit Fox Series (URA)	duration (s)	flow rate (kg/s)	release velocity (m/s)	Experiments grouped by stability class		
				D	E	F
continuous	$\geq$ 120	1.5-2.1	0.4-0.6	604, 805	702,808,605,703,705	606,811,709, 609,712
puff	20	1.1-1.8	0.3-0.5	801,601,803,802, 804,602,603,806	807,809,706,810,812	704,708,710,607,711, 608,714

UDM input data have been obtained from the MDA database given by Hanna and Chang [10]. The ground-level area source is modeled as a circular source with diameter 1.69 m and corresponding source area A equal to that of the actual 1.5m x1.5m square source. The release velocity (m/s) is obtained by dividing the release rate (kg/s) by the source area A (m^2) and the CO_2 vapour density (kg/m^3).

In a previous paper (Witlox et al. [11]) results were reported for the above experiments (as well as other experiments), where the velocity was imposed as a horizontal instead of a vertical velocity, and where the plume was modeled as a horizontal jet. In the current paper the above new logic is applied where observers are released upstream of the source and where the ground-level cloud is modeled physically more correctly above the source and in the near-field. Below results are presented using the 'finite-duration-correction' model (UDM FDC) as well as results using the new time-dependent UDM AWD formulation.

Figure 7 compares steady-state, UDM FDC and UDM AWD results for ground-level centre-line concentrations (maximum value over all times) against experimental data for experiment KF0706 (URA puff with 20 second duration, stability class E, wind-speed 2.66m/s). Both approaches of a horizontal jet (Figure 7a) and pool source (Figure 7b) are considered, as well as the effect of additional time-averaging [(see Equation (2)], using an averaging time of 20s as adopted for the experimental values. The following conclusions can be drawn:

- In the near-field the AWD and (un-averaged) FDC results are close to the steady-state results, and increasingly deviate from the steady-state results with increasing downwind distance because of effects of along-wind diffusion.
- In the far-field the UDM AWD results show less effects of along-wind diffusion than the FDC results. The reason for this discrepancy is being investigated.
- Since the release duration of 20s is identical to the averaging time, the effect of time-averaging should be limited in the near-field (no AWD effects), more pronounced in the intermediate distance (more time-variation over the averaging period of 20s because of AWD effects), and less in the far-field (relatively less variation over the averaging period of 20s because of increasing cloud duration due to increasing AWD effects). This is confirmed by the AWD results, but not in the near-field for the FDC results. Thus the analytical FDC results seem to be less correct in the near-field.
- At the experimental data points the results between horizontal jet and pool source are very close.

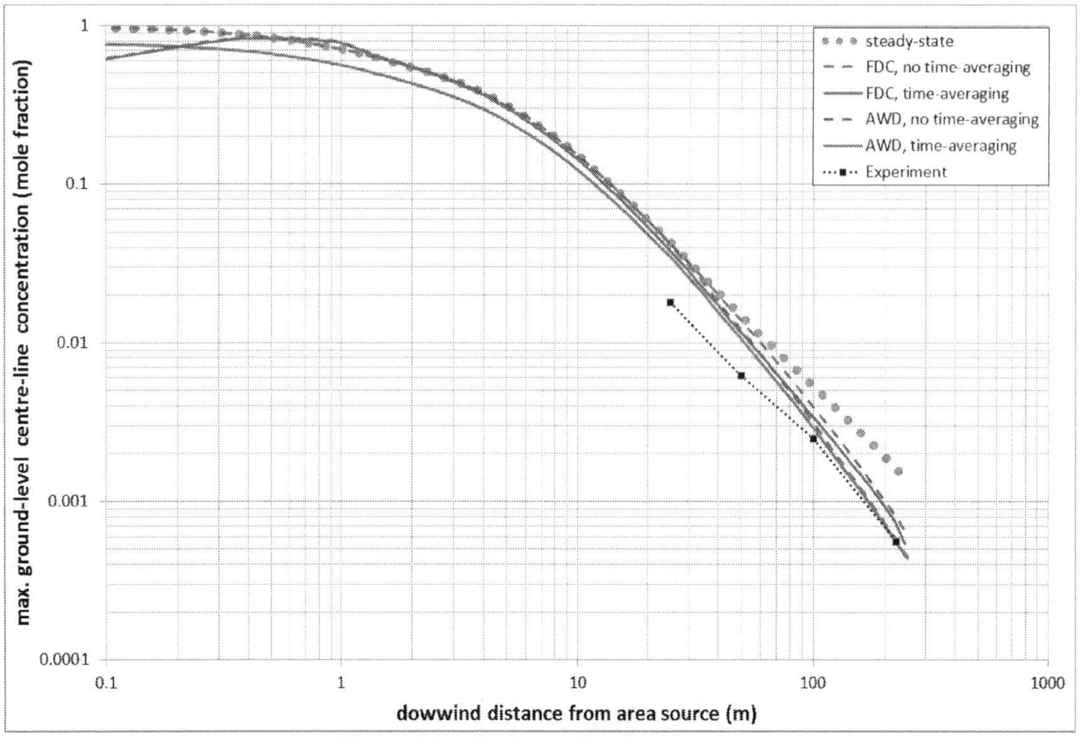

(a) Model as horizontal jet from area source

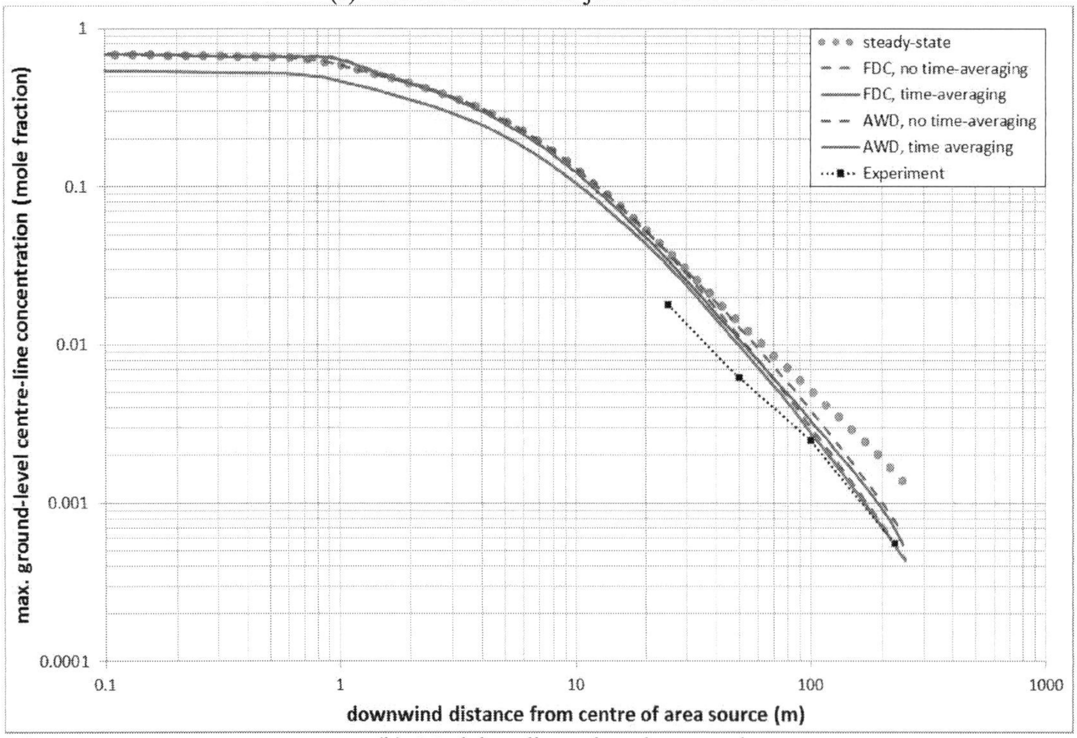

(b) Model as dispersion above pool

Figure 7. UDM predicted maximum concentrations for Kit Fox experiment KF0706 (20s release)

For a given experimental dataset, it is common practice [Hanna et al. [10]] to calculate the geometric mean bias MG (averaged ratio of observed to predicted concentrations; MG<1 over-prediction and MG>1 under-prediction) and the geometric variance VG (variation from mean; minimum value = 1). Ideally, MG and VG would both equal 1.0. Geometric mean bias (MG) values of 0.5 and 2.0 represent a factor of 2 in over-predicting and under-predicting the mean, respectively. Likewise, a geometric variance (VG) of about 1.6 indicates scatter from observed data to predicted data by a factor of 2.

A graphical presentation of the overall MG and VG validation results for arc-wise concentration is shown in Figure 8 for both FDC and AWD models presuming a pool source including additional time averaging. The overall results can be summarized as follows:

- URA continuous (black markers): excellent prediction for both concentration and cloud widths, and very close consistent agreement between FDC and AWD results
- URA puff (red markers): The FDC method (solid markers) results in concentration over-prediction for neutral stability class D, accurate results for stability class E and in under-prediction for stability class F. The AWD method (hollow markers) results in increased over-prediction for D, E and improved results for F (slight under-prediction).

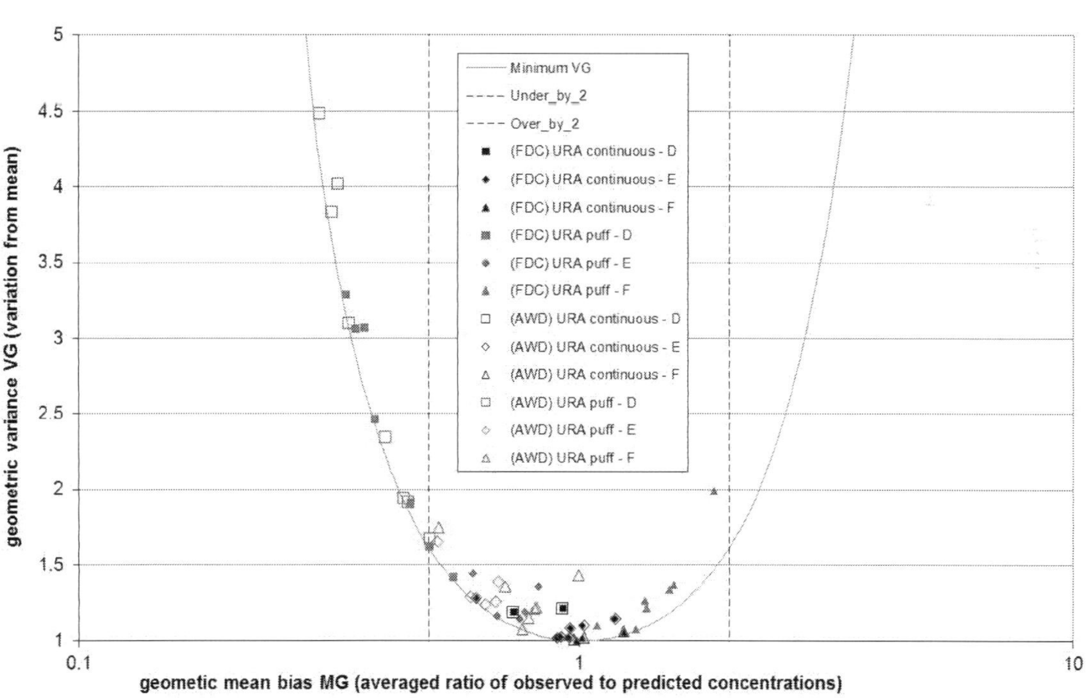

Figure 8. UDM validation statistics for Kit Fox URA experiments
(case of pool source including additional time-averaging; FDC and AWD results)

5. Conclusions and future work

The following main conclusions can be drawn:

- The Phast dispersion model UDM has been extended to improve modeling for time-varying ground-level releases. The new formulation presumes a number of 'observers' to be released at successive times from the upwind edge of the pool. The UDM first carries out pseudo steady-state calculations for each observer, accounting for vapor accumulation over the pool. The effects of along-wind diffusion are included by means of Gaussian integration over the downwind distance, and additional time-averaging can be carried out subsequently.
- In the old model (Figure 1) the equations were discontinuously modified at the original point of rainout to account for the vapor added back from the entire pool, and no along-wind-diffusion was applied. In the new model the cloud adapts continuously to the presence of an evaporating pool.
- The new method is based on a generalization of the HGSYSTEM model HEGADAS-T of dispersion from area sources.
- The new UDM AWD model has been verified against HEGADAS-T for the case of dispersion from a ground-level area source and good agreement was obtained
- The new UDM FDC and UDM AWD models have been validated against the Kit Fox experiments (CO_2 area source). Accurate results were obtained using both the FDC and AWD formulations, with most accurate results by FDC for stability classes D, E and by AWD for F.

The current further work is being planned:

- Extension of UDM instantaneous model to account for along-wind-diffusion after rainout.
- The current new model includes effects of along-wind-dispersion resulting from ambient turbulence corresponding to passive dispersion. For heavy-gas continuous or time-varying releases, it only takes into account gravity spreading in the cross-wind direction and not gravity spreading in the downwind direction. Thus the model should be extended to include effects of along-wind gravity spreading, which are particularly relevant in the near-field in the case of small wind speeds
- Validation against additional experimental data
- Implementation of new UDM AWD model into Phast
- Further refinement of empirical formulas for along-wind-diffusion coefficient σ_x

References

[1] Witlox, H.W.M., and Holt, A., "A unified model for jet, heavy and passive dispersion including droplet rainout and re-evaporation", *International Conference and Workshop on Modelling the Consequences of Accidental Releases of Hazardous Materials*, CCPS, San Francisco, California, September 28 – October 1, pp. 315-344, 1999

[2] Witlox, H.W.M and Harper, M.., "UDM theory manual", Phast 6.7 technical documentation, DNV Software, 2011

[3] Post, L. (editor), Section 7C in "HGSYSTEM 3.0 Technical Reference Manual", Report TNER.94.059, Thornton Research Centre, Shell Research, Chester, England, 1994

[4] Witlox, H.W.M., "The HEGADAS model for ground-level heavy-gas dispersion, II. Time-dependent model", Atmospheric Environment, Vol. 28, No. 18, pp. 2933-2946, 1994

[5] Witlox, H.W.M, and Harper, M., "Modelling of time-varying dispersion for releases including potential rainout", Process Safety Progress, In Press, 2013; also published at Loss Prevention Symposium, 9th Global Congress on Process Safety, San Antonio, Texas, April 28 – May 1, 2013

[6] Witlox, H.W.M., Harper, M., Natarajan, S., and Williamson, A., "Modelling of time-varying dispersion for elevated pressurised releases without rainout", 2013 International Symposium, Mary Kay O'Connor Process Safety Centre, Texas A&M University, College Station, Texas, 22-24 October 2013

[7] Fernandez, M.I., Harper, M., Mahgerefteh, H., and Witlox, H.W.M., "An integral model for pool spreading, vaporisation and dissolution of hydrocarbon mixtures", pp. 466-472,Symposium Series No. 158, Hazards XXIII, Southport, England, 13-15 November 2012, ICHEME

[8] Ermak, D.L., "Unpublished notes on downwind spreading formulation in finite-duration release version of SLAB", Lawrence Livermore National Laboratory, California, 1986

[9] Ermak, D.L., "User's Manual for SLAB: an atmospheric dispersion model for denser-than-air releases, Lawrence Livermore National Laboratory, Livermore, California, February 1989

[10] Hanna, S.R., Chang, J.C., New MDA database version of Kit Fox, Provided as part of official CD together with overall Kit Fox report by Hanna, Chang and Briggs, 1999

[11] Witlox, H.W.M. Harper, M., Oke,A. and Stene, J., "Validation of discharge and atmospheric dispersion for unpressurised and pressurised carbon dioxide releases", Journal of Process Safety and Environmental Protection 92 , pp. 3-16, 2014

Acknowledgment

Financial support of the work reported in this paper was provided by DNV Software, RIVM (Dutch Government), and TOTAL. The contents of this paper including any opinions and/or conclusions expressed, are those of the authors alone and do not necessarily reflect the policy of these organizations.

349889 Nanotechnology on Removing Arsenic Using Modified Carbon NANO- Tubes (MCNTs)

Abstract:

This study explains the percentage removal of arsenic from water by using modified multi- walled carbon nanotubes (MCNTs) as adsorbents. Apart from the raw carbon nanotubes, various metals has been impregnated on the large surface area of the CNT namely iron, titanium, silver, gold, copper and analyzed further to compare the percentile removal of each. The study investigates further the effect of the presence of carboxylic functional group (-COOH) functionalized on the surface of the carbon nanotubes on the removal of arsenic. To study the influence of the arsenic removal from water, the independent variables considered in this part of the paper are pH, running time and agitation speed. The concentration of the arsenic solution used in throughout the experiment was 1000 μg/L i.e. 1 ppm. The presence of Nano materials in the CNTs was characterized by using field emission scanning electron microscopy (FESEM) and transmission electron microscopy (TEM) in order to measure the lengths and the diameters of the CNT. Results of the study showed that iron and titanium impregnated on the surface of the CNT has the highest removal percentage compared to pure CNT and the functionalized CNT with carboxylic group. Almost 70% of the arsenic was removed from the water that contained 1 ppm arsenic contamination. The highest removal was at pH 3, contact time of 24 hours and agitation speed of 200 RPM. Part of the study is also dedicated to investigate the effect of dosage in the removal of arsenic from water.

Possible Two Physical Hazard Scenarios for Polystyrene Foams, Based on Life Cycle Stages

Toyoaki Nakarai, Graduate School of Environment and Informaion Sciences, Yokohama National University, Yokohama, Japan
nakarai@ynu.ac.jp

Atsumi Miyake, Graduate School of Environment and Information Sciences, Yokohama National University, Yokohama, Japan, atsumi@ynu.ac.jp, Satoru Yoshino, College of Industrial Technology, Nihon University, Narashino, Japan, yoshino.satoru@nihon-u.ac.jp

Prepared for Presentation at
American Institute of Chemical Engineers
2014 Spring Meeting
10th Global Congress on Process Safety
New Orleans, LA
March 30 – April 2, 2014

UNPUBLISHED

Possible Two Physical Hazard Scenarios for Polystyrene Foams, Based on Life Cycle Stages

Toyoaki Nakarai, Graduate School of Environment and Informaion Sciences, Yokohama National University, Yokohama, Japan
nakarai@ynu.ac.jp

Atsumi Miyake, Graduate School of Environment and Information Sciences, Yokohama National University, Yokohama, Japan, Satoru Yoshino, College of Industrial Technology, Nihon University, Narashino, Japan

Keywords: Polystyrene, Foam, Life cycle, Physical hazard, Scenario, HBCD

Abstract

Polystyrene foams are good thermal insulators and therefore widely used in building insulation materials, such as roofs, and/or walls, down to foundation. However these foams have significant physical hazards (fire, etc.), because flammable blowing agent has come to be used, and foamed polystyrene itself is combustible. And so, HBCD (Hexabromocyclododecane) is generally used as flame retardant in polystyrene insulation foams, while currently a global ban on HBCD is being considered under the framework of Stockholm Convention on POPs (Persistent Organic Pollutants), due to persistence, toxicity and eco-toxicity.

Under these circumstances, attempts to find environmental-compatible flame retardants are being performed by, such as ENFIRO, a European Commission-funded project of offering a prototypical case study on substitution options for specific brominated flame retardants for plastics, based on flammability, environmental and toxicological risks.

Regarding the toxic potency in fire hazard caused by plastic burning, a model, based on the calculation of a FED (Fractional effective Exposure Dose) of mixture of smoke gases, had been previously developed by Building and Fire Research Laboratory, NIST, where some most toxic chemicals in LC_{50}, played important role.

On the other hand, for chemical process, the incident scenarios are related to the immediate impact of unplanned releases of material and energy, where the effects of fire, explosion or acute toxicity are generally targeted. In case of flammable blowing gas release from polystyrene foam to the atmosphere, for example, in a warehouse of low air exchange rate, it is possible to cause fire by static electricity as the ignition source.

This paper describes the possible two physical scenarios for polystyrene foams, based on life cycle stages: 1) Production/import, 2) Transportation, 3) Industrial use, 4) Private use, 5) Waste & disposal. One is acutely toxic gas release scenario by incomplete combustion case (nearly 400˚C), etc., and the other is flammable blowing gas release scenario directly from the foam itself (nearly 100˚C). The evidence was shown mainly by using gas chromatography.

The acutely toxic gas concentration from newly fabricated EPS, were compared with different flame retardants (HBCD, FP-2200) heated by electric furnace at 400°C, and analyzed by gas chromatography. The detected CO gas concentration was higher for EPS foams with HBCD than foams of tested phosphorous retardant.

1. Introduction

This paper does not describe ecological/environmental hazards related to HBCD flame retardant under Stockholm Convention on POPs, but possible two main physical hazards for insulation foams made by polystyrene.

1.1 Physical Hazards and Risk Assessment Types

The three major types of risk assessment follow essentially the same logic but differ in emphasis. R. Kolluru presents an overview of physical (safety), health, and ecological/environmental risk assessments including 'major process steps', 'typical endpoints', and 'typical applications' in his "Risk Assessment and Management Handbook" [1]. All the assessments start with hazard identification or problem definition. A hazard is a chemical, biological, or physical agent (including electromagnetic radiation) or set of conditions that presents a source of risk but not risk per se.

In physical (safety) hazard analysis, the end points are well defined: fatalities, injuries, and economic losses. The impact is immediate and transparent, and the cause-effect relationship is generally clear. In contrast, considerable uncertainty pervades health risk analyses and also ecological/environmental risk analyses.

1.2 ENFIRO Project

ENFIRO, the EU-funded collaborative research project, offer a prototypical case study on substitution options for specific brominated flame retardants (BFRs) [2]. It started out based on the assumption that not enough environmental and health hazard data were known for the alternatives to the brominated fire retardants such as HBCD. In order to make the evaluation fully comprehensive, it was decided to compare fire performance as well as a life cycle assessment of halogen free flame retardants (FRs) versus BFRs. 17 halogen-free flame retardants as alternatives for three BFRs were selected.

Three FR/product combinations (e.g. metal-based FRs, phosphorous-based and nanoclay-based FRs in printed circuit boards, paints and foams) are studied for: environmental and toxicological risks, viability of industrial implementation, fire safety, application of the FR into products (electronic equipment, specific plastics, paints and textiles). And overall it was found that alternative materials have much lower tendency to bioaccumulate in fatty tissue than the studied brominated flame retardants.

In above EU project, alternative FRs for plastic foams are studied, however, this paper describes about the physical hazards of polystyrene foams from the stand point of hazardous gas release.

1.3 Toxic Data by NIST Fire Hazard Analysis

It has long been realized that most fire victims die from smoke inhalations rather than from burns. However, there has also been long-term controversy over how to characterize the potential for harm from fire-generated smoke and an apparatus from which to obtain accurate data.

A comprehensive methodology has been developed for obtaining and using smoke toxicity data in fire hazard analysis. The methodology comprises: focusing of smoke inhalation deaths: bench scale LC_{50} measurement method, using a minimal number of test animals: interpreting gas measurement data using the N-Gas Model, etc. A small fraction of the chamber atmosphere is removed for chemical analysis of CO, CO_2, O_2, HCN, HCl, HBr, and NOx. The Model has been developed, to enable the use of these data to obtain approximate LC_{50} values, based on the calculation of FED of mixture of these gases. For the LC_{50} measurement, animals are exposed to the smoke for 30 minutes, where the lethality is noted both during this time and for 14 days afterwards, including the delayed effect of those animals who received a mortal insult [3, 4].

The toxic data by the study of NIST indicate that a few toxic gases play important role for toxic fire hazards.

2. Polystyrene Insulation Foams

2.1 Production Flow

EPS and XPS, two foam types for polystyrene, are both rigid, thermoplastic materials. Starting from the polymerization of styrene, additives such as flame retardants, blowing agent are incorporated during or at the end of foam production.

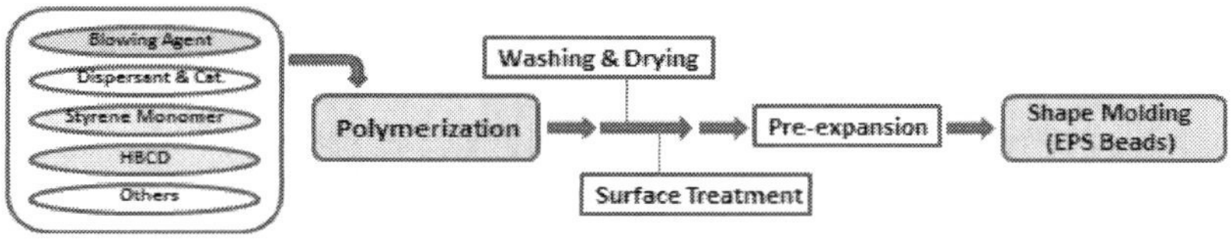

Figure 1. Expanded polystyrene (EPS) manufacturing process

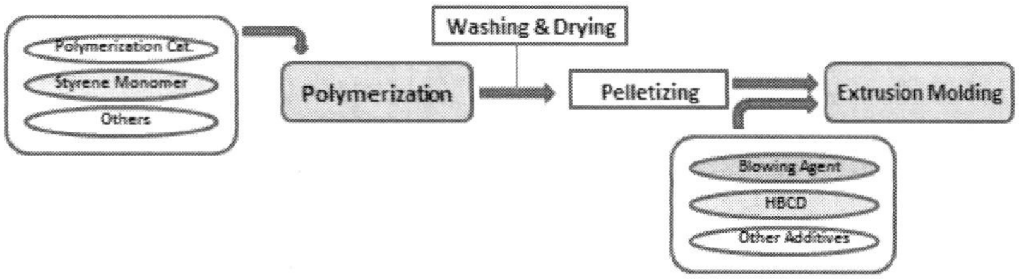

Figure 2. Extruded polystyrene (XPS) manufacturing process

2.1.1 EPS production

Foamable polystyrene beads are generally produced by two basic methods; 1) suspension polymerization of styrene into spherical beads containing a blowing agent, 2) incorporation of a blowing agent during the extrusion process of bulk foaming. Since extruded EPS is normally

processed directly to packaging applications, not to insulation materials [5], then Figure 1 showed suspension polymerization process.
The term of suspension polymerization means a process in which water-insoluble monomers are dispersed as liquid droplets with suspension stabilizer, and vigorous stirring to produce polymer particles as a dispersed solid phase. The polymerization takes place within the styrene monomer droplets, where oil-soluble initiators are used as starting substance. And the blowing agent and flame retardant, HBCD, and some additives are applied in this polymerization step.
EPS particles based on above suspension polymerization, are converted into final foam through pre-expansion, which is achieved due to small amounts of gas contained in the polystyrene bead. The gas expands when heat of steam is set on, and the large EPS blocks beads can be fabricated.

2.1.2 XPS production

To produce XPS foam, polymerized styrene is converted into pellets by extrusion machine. These pellets are, then, fed into extruder along with special additives and a blowing agent, shown as Figure 2. Within the extruder the mixture is combined and melted under controlled conditions of high temperature and pressure into a viscous plastic fluid. The hot, thin liquid is then forced in a continuous process through a die. As it emerges from the die, it expands to foam, is shaped, cooled, and then trimmed to dimension.

2.2 Flame Retardant

Since plastic foam itself is combustible, then flame retardant is essentially necessary for building insulation material. Under the framework of the Stockholm Convention on Persistent Organic Pollutants (POPs), a global ban on HBCD (hexabromocyclododecane), a flame retardant used mainly for polystyrene foams, is currently being considered.
EPS board is reported to contain typically 0.5% HBCD by weight in the final product [6] and the maximum concentration of HBCD in EPS beads is assumed to be 0.7% [7]. In XPS, HBCD is used in up to 3% loading to meet technical and flammability foam requirements [8].
Based on the standards such as JIS A9511, classification of insulation board, combustibility, and thermal conductivity are required. Flame retardants are used to clear the requirements of, for example, "the flame disappears within three seconds, there is no remainder dust, and it does not burn exceeding the combustion limit instruction line". However, this kind of standard does not mention physical hazards based on blowing agents and their requirements for use [9].

3. The Possible Physical Hazards of Polystyrene Insulation Foams

3.1 Physical Hazards and ISO 14040 LCA Frameworks [10, 11]

LCA (Life Cycle Assessment) provides a framework, approach, and methods for identifying and evaluating environmental burdens associated with life cycle of chemicals from cradle to grave. Since the 1970's, there have been several efforts to develop LCA methodologies, mainly in the USA and Europe. As a result of these efforts, consensus has been achieved on an overall LCA framework and a well-defined inventory methodology has been outlined. For example, LCA procedures are a part of the ISO 14000 environmental management standards as ISO14040 and ISO14044. ISO14040: 2006 describes the principles and framework for life cycle assessment including 'the definition of the goal and scope of the LCA', 'the life cycle inventory analysis phase', 'the life cycle impact assessment phase', and 'the life cycle interpretation phase'. However, the LCA standard methodology noted above is 'not a framework to evaluate the effects of physical hazards, but rather to evaluate the environmental effects of chemicals'.

In this manuscript a chemical's life cycle is described in totally five life stages: in fixed facilities, four life stages (raw material collection/storage/import/production, industrial use/distribution, commercial product private use, waste & disposal), and additional one in transport stage (waterways, roads, railways and pipelines).

3.2 The Outline of Life Cycle Physical Hazards of Polystyrene Insulation Foams

Figure 3 shows the outline of life cycle physical hazards of polystyrene insulation foams (EPS, XPS), divided into 5 stages: Production/import, Industrial use, Private use, Waste disposal and Transportation. The main physical hazards are 1) flammable gas release caused by abnormal storage temperature (below 100°C), and 2) carbon monoxide (CO) release, heated at higher temperature (nearly 400°C), as found in normal case of hydrocarbon heat decomposition.

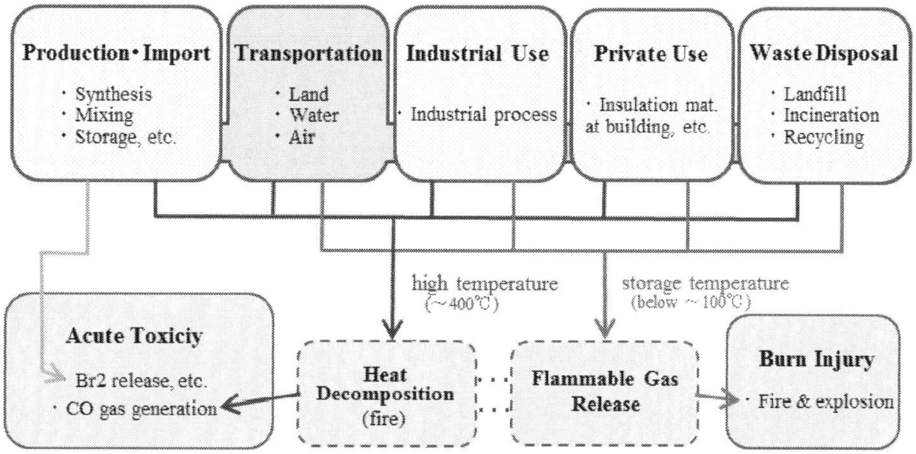

Figure 3: The outline of life cycle physical hazards for polystyrene insulation foams

Raw materials necessary for fabrication of XPS products, such as polystyrene, flame retardant, additives like dyestuffs and blowing agents constitute "Production /import stage".
During the "Industrial use stage", the raw materials above are fed to an extruder continuously under high pressure as shown in Figure 2. After foamable gel is formed, it is then cooled just before the orifice, with foam structure created by evaporation of blowing agent. The introduction of chemically stable blowing agent into the molten polymer, at normal pressure, reduces the foam product density, by forming a myriad of closed cells. The foam is then trimmed to the desired shape: the foam boards are packed into shrink-wrapped bundles on pallets, and then stored for curing over a certain period, until shipment to the next stage [12].
At "Private use stage", XPS boards are mainly applied as insulation materials for office buildings, residential houses, refrigerated warehouses, Tatami floor mats, and as road construction material, where light foam product is necessary to solve soft soil problems.
As for EPS products, blowing agents are generally introduced in the stage of polymerization, and the gases are contained in the polymer beads. At the "Industrial use stage", by heating these

polymer beads, most of EPS foam products are fabricated in shape molds as Figure 1, due to the expansion of polymerized beads through pre-expansion process.

In each stage, physical hazards were identified based on accident/incident information of polymer foams. Hereinafter, we define physical hazards as those caused by mainly high energy events, such as fire, explosion and/or acute toxicity, leading to death by heat, blast overpressure, and acutely toxic chemicals.
Thus, flammable blowing agent, from polymer insulation foam, will become the cause of physical hazards with toxic hazardous gas, because it may catch fire in a certain gas concentration, with the existence of ignition source.

4. Experiments

4.1 FID-Gas Chromatography Analysis

The following experiments of commercially available XPS foams by gas chromatography (flame ionization detector) showed evidence of possible blowing gas release from the polystyrene insulation foam board.

4.1.1 Methodology

Polystyrene foam from Company A, Company B and also Company C, was cut by hot-wire to the dimensions of W8, L8, D8 mm, and then 5 samples were placed in 16 ml test tubes with a septum cap (made of silicone) for gas chromatography analysis. Test tubes were heated at 90˚C for 1 hour in an oil bath. Released gas in the test tubes was injected into a gas chromatography (GC-2010, SHIMAZU), equipped with ULBON capillary column HR-1, 0.25mm in diameter and 30m in length. Column temperature was 35˚C, and helium carrier gas was used for FID detection, together with hydrogen gas. Figure 4 shows the retention time comparison between XPS released gas and standard gas, which was made by Sumitomo Seika Chemicals, Co., Ltd., with n-pentane (Kanto Chemical, Co., Ltd.) added.

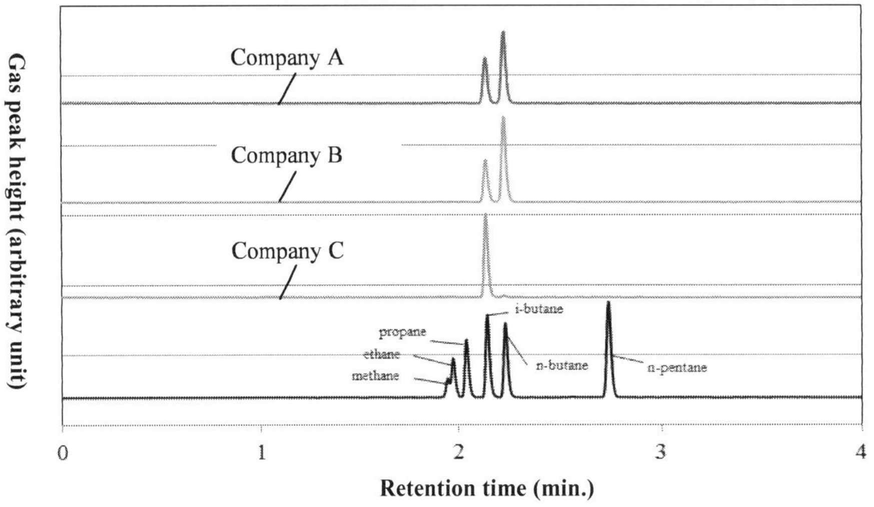

Figure 4: Gas chromatograph comparison between XPS released gas and standard gas

4.1.2 Results

Qualitative analysis by FID-GC of foams, heated at 90°C for 1 hour, detected two main kinds of gases, as shown in Figure 4. Comparison with commercially available standard gas, containing methane, propane, iso-butane, and n-butane, with pentane added, found retention times for released gas peaks to be the same as those of iso-butane and n-butane in foams of Company A and Company B, however only iso-butane for Company C.

4.2 TCD-Gas Chromatography Analysis (1) -commercially available foams-

The following experiments by TCD-GC showed evidence of carbon monoxide generation from the polystyrene foam board, heated at 400°C. The Thermal Conductivity Detector (TCD) is a universal detector and can detect air, hydrogen, carbon monoxide, carbon dioxide, nitrogen, inorganic gases and many other compounds. For most organic molecules, the sensitivity of the TCD is lower compared to the Flame Ionization Detector.

4.2.1 Methodology

Commercially available XPS foams from Company B and Company C, were cut by hot-wire and 100mg each was prepared, together with commercially available EPS bead from Company E. Then, each by each, they are placed in 20ml test tube (outer diameter;11mm ϕ, length;75mm) in Titanium sphere type vessel (63.5mm ϕ) for gas chromatography analysis. The metal vessel, which contained glass test tube in the neck part (inner diameter;11.5mm ϕ), was placed in ceramic tube type electric furnace at 400°C, and was kept for 1 hour.
Gas release from test tube in sphere vessel due to heat decomposition, was caught by micro syringe with silicone cap, and then injected into a gas chromatography (GC-2012, SHIMAZU), equipped with SHINCARBON ST column made by Shinwa Chemical Industries Ltd. Column temperature was 40°C for 3minutes at first stage, and then heated up to 200°C in 10°C/min., and then kept additional 3 minutes at 200°C, with helium carrier gas.

4.2.2 Results

Figure 5 shows the gas chromatograph comparison of decomposition gases from two kinds of XPS foams (Company B,C; 100mg each), one EPS bead (Company E; 100mg) and CO standard gas, made by Sumitomo Seika Chemicals. X axis is retention time in minutes and Y axis is gas peak height in arbitrary unit.
Detected released gases from two foams and EPS bead, in 400°C for 1hour, showed the similar chromatographs.

4.3 TCD-Gas Chromatography Analysis (2) -commercially nonavailable foams-

Three kinds of EPS foams (flame retardants of HBCD, FP-2200, and no retardants) were heated in 400°C, and thereafter released decomposition gas, containing CO, was compared by TCD-GC analysis.

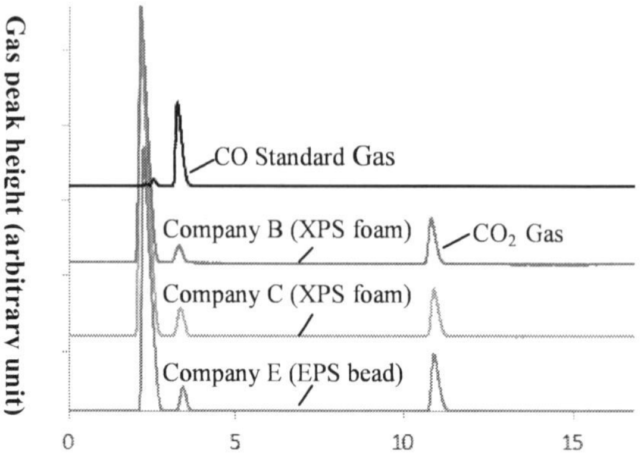

Figure 5: Gas chromatograph comparison of released gas from XPS foams, EPS bead and CO standard gas

4.3.1 Methodology

Suspension polymerized beads and 1.0wt% flame retardant was placed in glass bottle, and pre-expanded (100˚C, 30minutes). And then EPS foams were fabricated by using metal mold, heating in 100˚C for 20minutes. Here, HBCD was from Wako Pure Chemicals Industries, Ltd., and FP-2200 (Phosphorus/intumescent) from ADEKA Corporation.

As mentioned before, released gas of heat decomposition by test tube in sphere vessel, was injected by micro syringe into TCD-GC (GC-2012, SHIMAZU) for analysis. In each test, 140mg of EPS foam materials were used.

4.3.2 Results

Three kinds of EPS foams (flame retardants of HBCD, PF-2200 and no retardants) were shape molded, and then decomposed by heating in 400˚C, for 1 hour. Table 1 shows CO gas concentration, by TCD-GC analysis: Concentration of CO gas from EPS (HBCD, 1.0wt%) was 3.9%, while EPS foam without flame retardant was 2.8%, and EPS (PF-2200, 1.0wt%), phosphorous flame retardant, was 2.2%. The density of fabricated EPS foams was 0.04g/cm^3.

Table 1. Released CO gas concentration analyzed by TCD-GC

Polystyrene foams		Fire Retardant	CO gas conc.
EPS	Fabricated foams of Company D beads	HBCD 1.0wt%	3.9%
		PF-2200 1.0wt%	2.2%
		-	2.8%

4.4 Analysis in Polystyrene Foams

The amount of HBCD flame retardant in polystyrene foams was measured by energy dispersive X-ray fluorescence spectrometer, EDX1800B, Skyray Instrument Inc.

Suspension polymerized beads, with no halogen based retardant, were mixed and expanded with certain amount of HBCD at 100˚C for 30minutes in 100ml glass bottle, and calibration curve was prepared.
Based on above data, HBCD content in polystyrene foams and beads were measured. EPS molded piece from Company C, contained 0.6wt%. HBCD in beads from Company D and E were not detectable. However XPS molded foams contained retardant a little higher: Company A article was 6.1wt%, Company B was 2.4wt%, and Company C was 2.5wt% as shown in Table 2.

Table 2. X-ray fluorescence analysis (XRF) of HBCD in polystyrene foams

Polystyrene foams			HBCD analysis by XRF
XPS	Company A	foam	6.1wt%
	Company B	foam	2.4wt%
	Company C	foam	2.5wt%
EPS	Company C	foam	0.6wt%
	Company D, E	bead	ND

5. Result and Discussion

5.1 HSE Information Sheet and Experimental Results

HSE information sheet titled "Fire and explosion risks from pentane in expandable polystyrene (EPS)" suggests that it is necessary to calculate air exchange rate to reduce risk at the storage site [13].
In Japan, following a warehouse fire and explosion accident (1982), a notification for physical hazard for EPS beads was sent to local prefectural governments by the Fire and Disaster Management Agency (FDMA) [14]. The accident cause was attributed to beads in flexible containers with no explosion-proof structure for cooling equipment switching, in a low temperature (2-3˚C) warehouse [15]. HSE's suggestion will also be useful in reducing physical risks from the flammable gas used in XPS.

5.2 Fire Retardant

HBCD, hexabromocyclododecane, is used as fire retardant for XPS insulation products, because it prevents fire through direct contact between ignition source and polystyrene foam. It is not yet known whether HBCD has capabilities as a flammable gas barrier or not.
Hence, additional countermeasures are necessary for physical risk (fire, explosion), such as sufficient air ventilation for flammable gas just after wire cutting of large quantities of foam, and/or excluding ignition sources, etc., in indoor storage areas or underground construction sites.

HBCD content in EPS was found to be lower than those of commercially available XPS, made by Company B or C. This was below 3.0wt% as the literature shows [6, 7 and 8]. However XPS foam from Company A was found to contain HBCD more than 3.0wt%.

5.3 Gas Chromatography Experiments

In FID-GC analysis, sample heating temperature of 90˚C was selected, supposing much more gas release at abnormal storage temperature. In Figure 4, n-pentane was not found in three kinds of

commercially purchasable XPS foams, because no gas peaks were detected between 2.5minutes and 3.0minutes retention time, as n-pentane in standard gas.
We understand that only two gas peaks were found for foams from Company A and Company B and one gas peak from Company C, because the concentration of these butane gases was high.

In TCD-GC, the detected gases from decomposed XPS at 400˚C, for 1 hour, were N_2, CO, and CO_2. In Figure 5, three commercially available polystyrene materials showed the similar gas chromatographs.
The CO gas concentration from newly molded EPS (HBCD, 1.0wt%) was higher than that of EPS with 1.0wt% of phosphorous flame retardant (FP-2200). Intumescent Char seems to decrease CO gas concentration in the latter EPS foam, compared with bromine flame retardant.

Heating studies for EPS foams were performed at the temperature between 200˚C and 400˚C, and it was reported that CO gas increased much more at 400˚C [16]. So, 400˚C was selected for foam or bead decomposition temperature.

6. Conclusion

Possible two kinds of physical hazard scenarios for polystyrene foams (EPS, XPS) were shown in this paper for their life cycle: 1)Production/import, 2)Transportation, 3)Industrial use, 4)Private use, 5)Waste disposal. One scenario is acutely toxic gas release by incomplete combustion case (nearly 400˚C), etc., and the other is flammable blowing gas release from the foam itself (nearly 100˚C). The evidence was shown mainly by experiments with two kinds of gas chromatography (FID-GC, TCD-GC).

Qualitative analysis by FID-GC of foams heated at 90˚C for 1 hour detected two main kinds of gases. Comparison with commercially available standard gas, containing methane, propane, iso-butane, and n-butane, with pentane added, found retention times for released gas peaks to be the same as those of iso-butane and n-butane for Company A and Company B, however only iso-butane for Company C foam. However, pentane was not found by FID-GC analysis for commercially available 3 kinds of XPS foams.

In TCD-GC, the detected gases from decomposed XPS at 400˚C, for 1 hour, were N_2, CO, and CO_2, where O_2 was not found. The three commercially available polystyrene foams showed the similar gas chromatographs as Figure 5.

The CO gas concentration from newly fabricated EPS foams, with different flame retardant (HBCD, FP-2200) was compared, by gas chromatography analysis. The CO gas concentration was higher for EPS foams with 1.0wt% HBCD than EPS foams of tested phosphorous retardant.

HBCD content in EPS foam was found to be lower than those of commercially available XPS made by Company B or C, which was below 3.0wt% as the literature shows. However XPS foam from Company A was found to contain HBCD more than 3.0wt%.

7. References

[1] R. Kolluru, S. Bartell, R. Pitblado and S. Stricoff, "Risk Assessment and Management Handbook for Environmental, Health, and Safety Professionals", New York: McGraw-Hill, Inc., 1996

[2] http://www.enfiro.eu/ (accessed on February 27, 2014)

[3] V. Babrauskas, B.C. Levin, R.G. Gann, et al., "Toxic Potency Measurement for Fire Hazard Analysis", NIST Special Publication 827, 1991

[4] R.G. Gann, V. Brabrauskas, E. Brawn, B.C. Levin, et al., "Toxic Data for Fire Hazard Analysis", Proceedings of 12th Joint Meeting of the UJNR Panel on Fire Research and Safety, Oct. 27-Nov.2, 1992

[5] J. Scheirs and D. Priddy, "Modern Styrene Polymers: Polystyrene and Styrenic Copolymers ", John Wiley&Sons, Ltd., 2003

[6] G. Morose, "An Overview of Alternatives to Tetrabromobisphenol A (TBBPA) and Hexabromcyclododecane(HBCD)", Lowell Center for Sustainable Production, University of Massachusetts Lowel, 2006

[7] European Chemical Bureau, European Commission. Joint Research Centre. EU Risk Assessment Report, Hexabromocyclododecane, CAS No: 25637-99-4, EINECS No: 247-148-4, 2008

[8] J. Troitzsch, "The relevance of hexabromocyclododecane for polystyrene EPS/XPS foams to meet fire safety requirements as construction products in Europe. Expert opinion for the European HBCD Industry Working Group", 2008.

[9] Y.Ohara, K.Tanaka, T.Hayashi, H.Tomita and S. Motani, "The Development of a Non-Fluorocarbon Based Extruded Polystyrene Foam which Contains a Halogen-Free Blowing Agent", Bull. Chem. Soc. Jpn, 77, 599-605, 2004

[10] T. Nakarai, S. Yoshino, I. Nishimaru, M. Kutsumi, A. Miyake, "Accident/Incident Scenarios based on Life Stages Across the Complete Chemical/Product Lifecycle", 7th Global Congress on Process Safety (GCPS2011), March 2011

[11] M. Finkbeiner, A. Inaba, R.B.H. Tan, K. Christiansen, and H.J. Kueppel, "The New International Standards for Life Cycle Assessment: ISO14040 and ISO14044", Int. J.LCA 11(2) 80-85p, 2006

[12] Fire and Disaster Management Agency, Fire Safety Countermeasures for Expandable Polystyrene Beads, etc., Notification, Dec. 14, 1982

[13] Health and Safety Executive 2001, Fire and Explosion Risks from Pentane in Expandable Polystyrene (EPS), HSE Information Sheet, http://www.hse.gov.uk/pubns/ppis1.pdf

[14] Fire and Disaster Management Agency, Fire Safety Countermeasures for Expandable Polystyrene Beads, etc., Notification, Dec. 14, 1982

[15] H. Koseki, Warehouse Fire at Yokkaichi City, Shoubou Shuuhou, No.36, 25-29p, 1982

[16] S. Yoshino, T. Nakarai, I. Nishimaru, A. Miyake, "Hazard analysis for plastics contained HBCD based on physical risk scenarios", Society of Environmental Sciences, Japan, Annual Meeting, Sep. 14, 2012

Acknowledgements

Part of this study was promoted by the "Effective Cooperation of Science and Technology Policies" based on special coordination funds for promoting science and technology in 2007-2009, by the Ministry of Education, Culture, Sports, Science and Technology, and was also financially supported partly by the Environment Research and Technology Development Fund (c-1003) of the Ministry of the Environment, Japan, for 2010-2012.

349296 Problems Encountered in the Development of a Process Safety Climate Tool

Abstract:

Following on from the successful launch of the Health and Safety Laboratory's (HSL) Safety Climate Tool in 2010, Human factors specialists at HSL have been investigating the feasibility of developing of a process specific safety climate tool. There is an appetite in industry for a safety culture tool aimed specifically at major hazards and market research identified a gap for a tool that would enable the perceptions of workforce on process safety to be collected within an organisation. HSL have identified a number of key topics from the literature and market research to be included in a tool, these will be described. The paper will also discuss the challenges HSL have faced when developing a process specific tool with a valid/ reliable question set, and how this relates to the HSL's SCT.

348745 Upper Explosible Limits for Combustible Dusts

Abstract:

Upper Explosible Limits for Combustible Dusts

by Richard W. Prugh

Principal Process Safety Engineer

Chilworth Technology, Inc., a DEKRA Company

113 Campus Drive, Princeton, New Jersey 08540

609-799-4449; richard.prugh@dekra.com

Abstract

Although Upper Flammable/Explosible Limits are sharply defined for flammable liquids and gases, combustible dusts do not appear to have well-defined Upper Explosible Limits. The causes for this behavior include the typically-wide an explosibility test, the larger particles may act as inert materials, and the smaller particles represent a low During concentration in the surrounding mixture of air and inert particles, and they burn readily. If the composition and heat of combustion of the combustible dust are known, thermodynamic calculations can yield the concentration at which the combustion temperature is insufficient for propagation of flame through a cloud of small particles. Examples are given for a typical carbohydrate dust and a metal dust.

Agglomeration Effect on Combustion and Explosion Properties of Nanoparticles

Jiaqi Zhang
Mary Kay O'Connor Process Safety Center,
Artie McFerrin Department of Chemical Engineering, Texas A&M University
College Station, Texas 77843-3122, USA
jiaqizhang@tamu.edu

Yi Liu, Hao Chen, and M. Sam Mannan
Mary Kay O'Connor Process Safety Center,
Artie McFerrin Department of Chemical Engineering, Texas A&M University
College Station, Texas 77843-3122, USA
mannan@tamu.edu

Prepared for Presentation at
American Institute of Chemical Engineers
2014 Spring Meeting
10th Global Congress on Process Safety
New Orleans, LA
March 30 – April 2, 2014

UNPUBLISHED

Agglomeration Effect on Combustion and Explosion Properties of Nanoparticles

Jiaqi Zhang
Mary Kay O'Connor Process Safety Center,
Artie McFerrin Department of Chemical Engineering, Texas A&M University
College Station, Texas 77843-3122
jiaqizhang@tamu.edu

Yi Liu, Hao Chen, M. Sam Mannan
Mary Kay O'Connor Process Safety Center,
Artie McFerrin Department of Chemical Engineering, Texas A&M University

Keywords: Dust explosion, Agglomeration, Nanoparticles

Abstract

Nanoparticles have been demonstrated to possess significantly high ignition sensitivities compared with micro-particles, *e.g.*, much lower minimum ignition energy (MIE) and feasible spontaneous ignition phenomena, and have been predicted to have a high deflagration index of dust explosion (K_{St}) and a low minimum explosive concentration (MEC). However, these predictions are inconsistent with experimental observations, which were attributed to the agglomerations of nanoparticles. Multi-scale agglomerations, including soft and hard agglomerations, have been observed in nanofiber systems, which may serve as proper platforms to validate this assumption. This study tested the MECs of carbon nanofibers (CNFs), and found that MECs decreased after the milling process. This finding calls for attention to a widely used processing method in nanomaterial industries. Comparing to MECs of micro-sized graphite particles, this study supports the hypothesis that it is the size of agglomerates rather than that of single nanoparticles can be used as an index to estimate CNFs' MEC. Theoretical analyses were performed for carbonaceous particles with different sizes at different surface temperature to explain how specific surface area affects particles' combustion and the limiting step in combustion process was also discussed.

1. Introduction

Particle size plays a significant role in affecting dusts' combustion and explosion behaviors. Minimum ignition energy (MIE) was found to reduce exponentially as particle size decreases; the "plateau effect" of particle size on MEC exists, which means that the minimum explosive concentration (MEC) declines with reduced particle size until reaching a limiting stage where particles are too fine. Thus, the explosion severity can be significantly increased by reducing particle size [1]. This trend was summarized from much experimental data of micro-size particles [1].

However, when it comes to nanoparticles, the situation becomes much more complicated. On one side, some experimental data and observations indicate that nanoparticles possess significantly high ignition sensitivities; most metal nanoparticles' MIE is lower than 1 mJ while corresponding micro-size particles' MIE are ~20 mJ [2]. Furthermore, an explosibility research of several different types of carbon nanotubes (CNTs) reached a conclusion that the explosion severity is indeed higher for shorter CNTs [3]. On the other hand, much research states that nanoparticles' explosibility is limited by limiting dispersibility and high coagulation rates [4]. For example, an explosibility research of aluminum nanoparticles found that explosion of 200 nm particles' cloud was stronger than that of 100 nm particles [5]. One possible explanation is the larger agglomerates of 100 nm particles [6]. Another possible reason is the different oxidation degree [5].

Since it is difficult to exclude the influences of other factors in the aluminum nanoparticle research, this study intended to choose samples with identical properties, except for the different sizes, to study agglomerations' effect on combustion and explosion behaviors. There are both milled and unmilled carbon nanofibers (CNFs). Considering the probability that different combustion and explosion behaviors are dominated by different properties of combustible dusts, this study focuses on agglomerations' effect on MECs.

In this study, MECs were identified according to the ASTM standard E 1515 [7]. A scanning electron microscope (SEM) was used to measure the size gaps between different samples. Theoretical analyses were also applied to explore the mechanisms.

2. Samples and Experiments

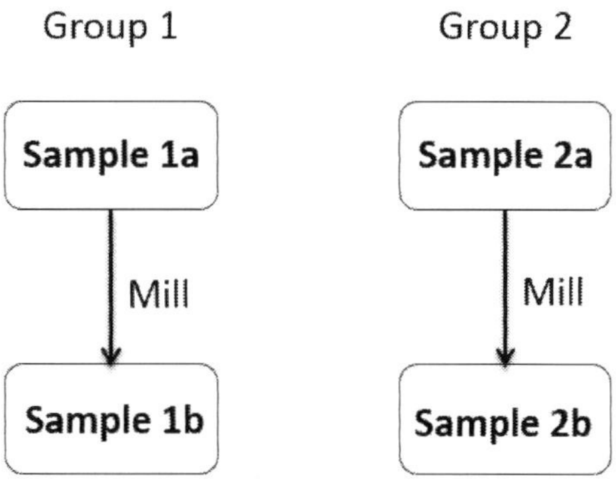

Figure 1. Samples

Two groups of CNFs were used in this study, as shown in Figure 1. There are two samples in each group. Samples 1a and 1b are in group 1; samples 2a and 2b are in group 2. Sample 2a was obtained by annealing sample 1a at 3000 °C. They are CNFs with 100 nm diameter and 10-30 μm length. Samples 1b and 2b were obtained by milling samples 1a and 2a with improved bulk densities up to 12 pounds per cubic feet (pcf). Consequently, samples in the same group are of quite similar properties, except for the particle sizes.

2.1 MEC tests experiments

MEC tests were conducted in a customized 36 L dust explosion vessel, as shown in Figure 2. It consists of a vaccum system, an air reservoir, a fast acting valve, a dust container, a rebound nozzle, an ignition system, and pressure transducers. Compared to the standard 20 L sphere, besides the difference of volume, this equipment's dust container is also located at a different location – the bottom of the vessel. Therefore, the dispersion process may be slightly different. However, after calibrated several essential parameters, it generates very similar data as the 20 L sphere does [8].

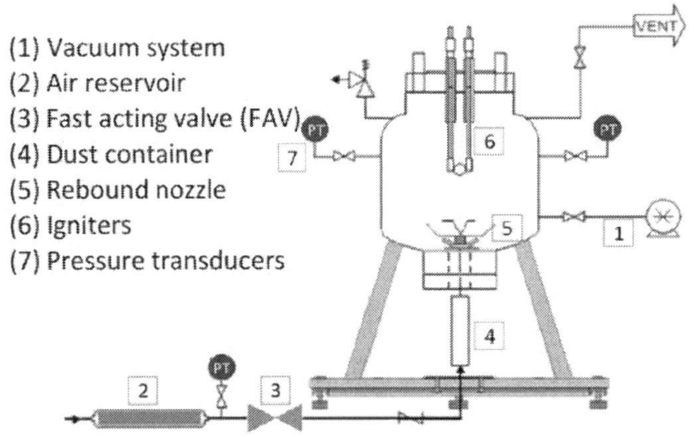

Figure 2. Schematic diagram of 36 L dust explosion vessel [8]

Following the standard ASTM E 1515 [7], this study identified the MECs by increasing the dust concentration until the pressure ratio (PR) ≥ 2.0:

$$PR = \left(P_{ex,a} - \Delta P_{ignitor}\right) / P_{ignition} \qquad \text{[Eq.1]}$$

Where $P_{ex,a}$ is the maximum explosion pressure; $P_{ignition}$ is the absolute pressure when the ignitor is activated; and $\Delta P_{ignitor}$ is the pressure rise in the chamber due to the occupied energy of the ignitor in air at atmospheric pressure.

2.2 SEM characterization

A JOEL JSM-7500F field emission scanning electron microscope (FE-SEM) was used to characterize the nanoparticles. Dry sampling was adopted to study the differences between the agglomeration sizes of the original and milled nanoparticles. Wet sampling was also performed to observe the size and length of individual particles

3. Results and discussion

Figure 3. SEM images for A) unmilled sample (sample 1a), magnification: 400X; B) milled sample (sample 1b) magnification: 400X

Figure 3 shows that CNFs agglomerated together and formed micro-sized spheres, rather than single fibers. It also indicates that the milling process effectively reduced the size of agglomerated spheres from approximately 30 μm to less than 10μm.

Sample 1a and 1b were dispersed in a 90% isopropyl alcohol solution. The resulting samples are shown in Figure 4. From Figure 4, it can be seen that the milling process also reduced the length of a considerable portion of fibers from about 10μm to 5μm or less, although long fibers still existed.

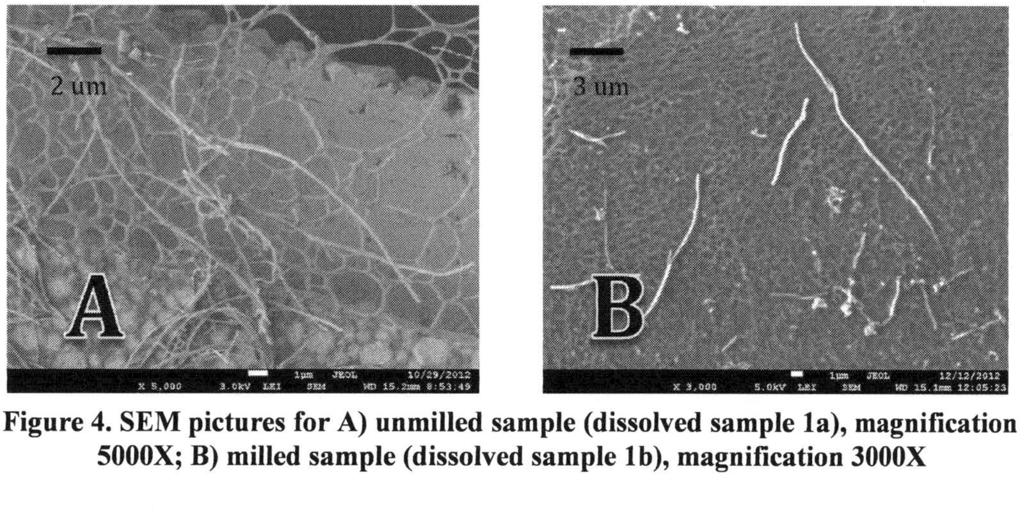

Figure 4. SEM pictures for A) unmilled sample (dissolved sample 1a), magnification 5000X; B) milled sample (dissolved sample 1b), magnification 3000X

MECs of sample 1-4 were also measured and corresponding results are shown in Table 1. It is easy to find that milled samples have much lower MECs, which means higher risks of explosion.

The differences between MECs show that the milling process increases the explosion risks of CNFs. Given that the milling method is widely used in customizing CNFs, and no specific attention is given to those customized samples, this finding calls for attention to this "blind spot".

Table 1. MECs of Samples

	Sample	Description	MEC (g/m^3)
Group 1	1a	Unmilled	>300*
	1b	Milled	105
Group 2	2a	Unmilled	>300*
	2b	Milled	210

* Due to the samples' low density and dust container's limited volume, MECs of samples 1a and 2a were not able to be obtained. The achieved maximum concentration was 300 g/m^3. Therefore, their MECs were identified as larger than 300 g/m^3.

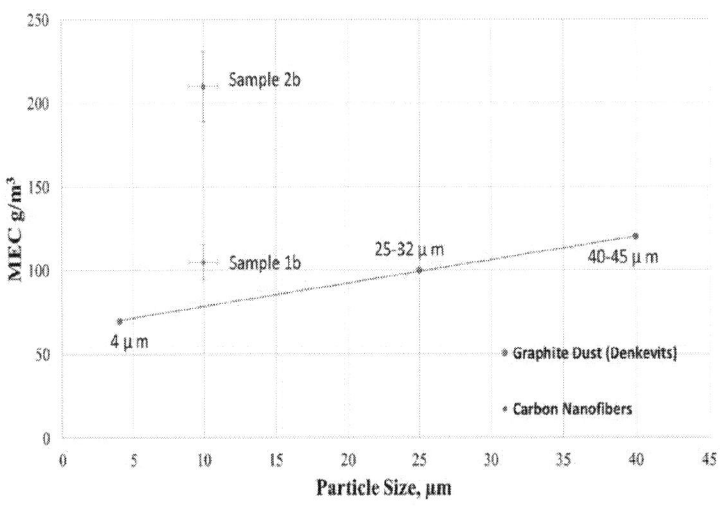

Figure 5. MEC of graphite dust and carbon nanofibers vs. dust particle size

Lower explosive concentration of some graphite dust (4 μm – 45 μm) are between 70 g/m^3 and 120 g/m^3 as shown in Figure 5 [9]. The MECs also exhibit a decreasing trend when particle size goes down. While MECs of carbon nanofibers are even higher than these graphite dust, it supports the hypothesis that the size of agglomerates rather than that of single nanoparticles can be used as an index to estimate CNFs' MEC. Figure 5 also shows that MECs of CNFs of similar agglomerates sizes are higher than their corresponding graphite dust. It might be caused by several other factors. Further study can be done to find them out.

4. Theoretical Analyses

Theoretical analyses were also performed to illustrate how the particle sizes affect carbon particles' combustion behavior at its first stage. In order to simplify the analyses, only single agglomeration's combustion process was considered in this research.

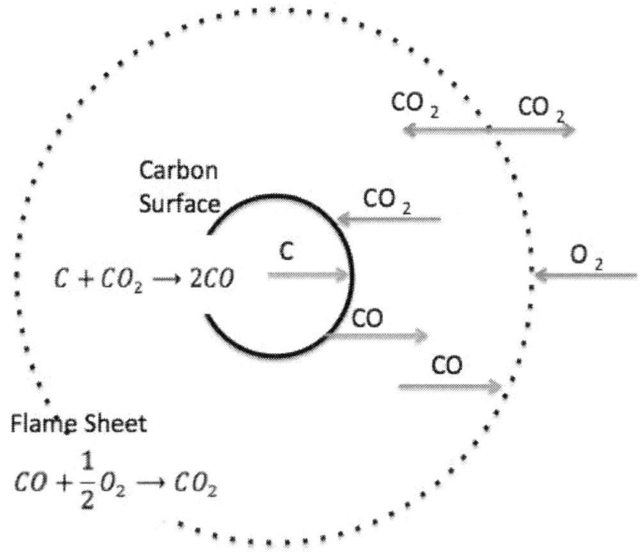

Figure 6. Mass flow at the carbon surface and the flame sheet reproduced based on [10]

In the classic two-film model, proposed by Burke and Schumann [11], developed by Held [12], and supported by experiments involving large particles [13], final combustion takes place in an outer layer, but the formed carbon dioxide attacks the carbon surface according to the reaction:

$$C + CO_2 \rightarrow 2CO \qquad \text{[Eq.2]}$$

The CO generated at the surfaces diffuses outward and is consumed at a flame sheet where it meets an inward-diffusing flow of O_2 and reacts stoichiometric proportions:

$$CO + \frac{1}{2}O_2 \rightarrow CO_2 \qquad \text{[Eq.3]}$$

Figure 5 shows the mass flow at the carbon surface and the flame sheet for this model.

Diffusion of oxygen and reaction on the particle surface are two main parts in the combustion process. Either of them can be the limiting step. The following section will compare the combustion time of single particle to identify the limiting step for the combustion of particles we are interested in. The following section illustrates the calculation example for a 30 μm particle with 1500 K surface temperature. For diffusion, air comes from unburnt zone, therefore, a temperature of 300 K and a pressure of 101.3 kPa is used in calculation. For reaction, since it happens on the hot particle surface, a temperature of 1500 K and a pressure of 516.7 kPa is used.

The collision integral of air, Ω_D is a function of temperature and intermolecular potential field for one molecule of A (Oxygen) and one molecule of B (Nitrogen). It can be approximated by the following expression [14]:

$$\Omega_D = \frac{a}{\left(T^*\right)^b} + \frac{c}{\exp\left(dT^*\right)} + \frac{e}{\exp(fT^*)} + \frac{g}{\exp\left(hT^*\right)} \qquad \text{[Eq.4]}$$

The parameters in this equation are listed in the following table:

Table 2. Parameters for collision integral [13]

Parameter	Value	Parameter	Value
□*	$\kappa T/\varepsilon_{AB}$	**e**	1.03587
a	1.06036	**f**	1.52996
b	0.15610	**g**	1.76474
c	0.19300	**h**	3.89411
d	0.47635	**ε_{AB} /κ, K** (for air)	97

While assuming the temperature of air coming to the particle surface is 300 K:

$$T^* = \frac{T}{\varepsilon_{AB}/K} = \frac{300K}{97K} = 3.09 \qquad \text{[Eq.5]}$$

$$\Omega_D = \frac{1.06036}{(3.09)^{0.15610}} + \frac{0.19300}{\exp(0.47635\times3.09)} + \frac{1.03587}{\exp(1.52996\times3.09)} + \frac{1.76474}{\exp(3.89411\times3.09)} = 0.9424$$

[Eq.6]

Then the diffusivity of air can be determined by using the Wilke and Lee Method [15]:

$$D_{AB} = \frac{\left[3.03 - \left(\frac{0.98}{M_{AB}^{1/2}}\right)\right]\left(10^{-3}\right)T^{3/2}}{PM_{AB}^{1/2}\sigma_{AB}^2\Omega_D} \qquad \text{[Eq.7]}$$

D_{AB} : Diffusion coefficient, cm²/s
M_A, M_B : Molecular weights of A and B, respectively
T: Temperature, K
P: Pressure, bar
σ_{AB} : "Collision diameter," a Lennard-Jones parameter, angstrom
Ω_D : Diffusion collision integral, dimensionless

Then

$$M_{AB} = 2\times\left(\frac{1}{M_A} + \frac{1}{M_B}\right)^{-1} = 29.867 \qquad \text{[Eq.8]}$$

$$\sigma_{AB} = 3.620 \text{ Å} \qquad \text{[Eq.9]}$$

$$D_{AB} = \frac{\left[3.03 - \left(\frac{0.98}{29.867^{0.5}}\right)\right](10^{-3})(300^{3/2})}{1 \times 29.867^{0.5} \times 3.620^2 \times 0.9424} = 0.198 cm^2/s \qquad \text{[Eq.10]}$$

If the reaction is controlled by a diffusion step, then the combustion time, τ_b, can be calculated [6]

$$\tau_b = \frac{\rho_s d_p^2}{8\rho_g D_{O_2}\left(Y_{O_{2,\infty}}\right) \Big/ \left(MW_{O_2} \big/ 2MW_C\right)} \qquad \text{[Eq.11]}$$

ρ_s, ρ_g : Density of solid particle and air, kg/m^3

d_p : Diameter of particle

$Y_{O_{2,\infty}}$: Oxygen mass fraction in air, 0.233

MW_{O_2}, MW_C : Molecular weight of oxygen and carbon

For 30 μm agglomerates

$$\tau_b = \frac{1600 kg/m^3 \times (30\mu m)^2}{8 \times 0.198 cm^2/s \times 1.204 kg/m^3 \times 0.233 \Big/ \left(32 \big/ 2\times12\right)} = 43.126 ms \qquad \text{[Eq.12]}$$

If the reaction is controlled by a surface reaction step, then the combustion time can be calculated [6]

$$\tau_b = \frac{\rho_s d_p}{2k_1 \left(P_{tot} MW_{gas} \big/ RT_s\right)\left(MW_C \big/ MW_{O_2}\right) Y_{O_2,s}} \qquad \text{[Eq.13]}$$

$$k_1 = 3.007 \times 10^5 \times \exp\left(-\frac{17966}{T_s}\right) \qquad \text{[Eq.14]}$$

k_1 : Kinetic constant of combustion, m/s

T_s : Temperature at the particle surface

For 30 μm agglomerates with the surface temperature of 1500 K,

$$k_1 = 3.007 \times 10^5 \times \exp\left(-\frac{17966}{1500}\right) = 1.890 m/s \qquad \text{[Eq.15]}$$

$$\tau_b = \frac{1600kg/m^3 \times 30\mu m}{2\times 1.890m/s \times \left(516.7kPa \times 29.867g/mol \Big/ 8.314J/(mol\cdot K)\times 1500K \right)(12/32)\times 0.233}$$

$$= 156.6ms$$

[Eq.16]

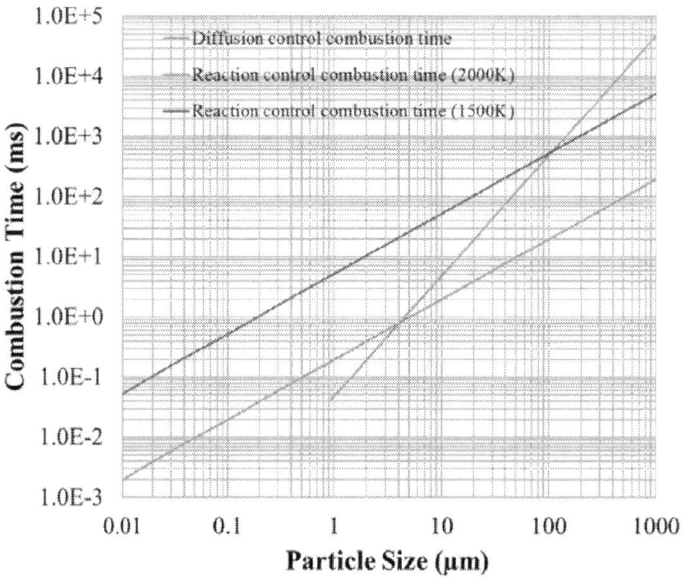

Figure 7. Combustion time for different particle sizes

This study repeated this calculation for a wide range of particle sizes at temperature 1500 K and 2000 K since the temperature significantly impact the surface reaction rate. Results are summarized in Figure 7. These calculation results match the experimental data of carbon particles smaller than 10 μm and larger than 100 μm [6] within a reasonable deviation, which supports the validity of this calculation. Both the experiments and calculation shows that for carbon particles smaller than 10 μm, the combustion process is controlled by the surface reaction. Also with the particle size shrinking, the difference between reaction rates and diffusion rates becomes larger. This is reasonable since the required diffusion path of oxygen is much shorter. The diffusion control becomes the limiting step for carbon particles larger than 100 μm.

For our interested particles – particle size between 10 μm and 100 μm, experimental data is currently not available while the calculation shows that it is difficult to state that there is a dominant step since the differences of combustion time is not obvious. Meanwhile, the exact surface temperature remains unknown, which adds great barrier to identify the key limiting step, the diffusion or the surface reaction.

However, no matter which step is limiting the combustion process, the particle size plays an important role in it. The combustion rates for a single carbon particle, $\dot{m}_c$ can be written as [3]:

$$\dot{m}_c \propto d_p^2 \qquad \text{[Eq.17]}$$

Then, the combustion rate for a certain dust cloud, $\dot{M}_c$, can be expressed as:

$$\dot{M}_c \propto \frac{1}{d_p} \qquad \text{[Eq.18]}$$

For larger particles, the combustion process are limited by diffusion. Then the relationship can be expressed as [3]:

$$\dot{m}_c \propto d_p \qquad \text{[Eq.19]}$$

And,

$$\dot{M}_c \propto \frac{1}{d_p^2} \qquad \text{[Eq.20]}$$

According to [eq. 18] and [eq. 20], further studies on combustion rate and particle size are needed to figure out the mechanism how reaction and diffusion interact in controlling the combustion process.

5. Conclusion

To study agglomeration's influence on particles' combustion and explosion behaviors, this study selected two groups of CNFs, including one milled sample and one unmilled sample in each group. MECs of these four samples were identified in a customized 36 L dust explosion vessel. It was found that MECs decreased after the milling process. This finding calls for attention to widely used processing method in nanomaterial industries. A comparison between the CNFs and micro-sized graphite particles is also discussed in this study, which could confirm the hypothesis that the size of agglomerates rather than that of single nanoparticles can be used as an index to estimate CNFs' MECs.

Theoretical analyses were performed for carbonaceous particles with different sizes at different surface temperature to explain how specific surface area affects particles' combustion and to find out the limiting step in combustion process. Combustion time calculation results using classic equations match well with available experimental data. For interested particles – particle size between 10 μm and 100 μm, experimental data is currently not available while the calculation shows that it is difficult to state that there is a dominant step since the differences of combustion time is not obvious. The unknown surface temperature makes it hard to identify the limiting step, the diffusion or the surface reaction

Acknowledgements

The authors gratefully acknowledge the financial support from the National Science Foundation Eager Program (No.1321581).

6. References

[1] Eckhoff, R.K., *Dust Explosions in the Process Industries: Identification, Assessment and Control of Dust Hazards*. Third ed. 2003: Gulf Professional Publishing.

[2] Wu, H.-C., R.-C. Chang, and H.-C. Hsiao, *Research of minimum ignition energy for nano Titanium powder and nano Iron powder*. Journal of Loss Prevention in the Process Industries, 2009. **22**(1): p. 21-24.

[3] Holbrow, P., et al., *Fire and explosion properties of nanopowders*. Health & Safety Executive: Buxton, UK, 2010.

[4] Eckhoff, R.K., *Influence of dispersibility and coagulation on the dust explosion risk presented by powders consisting of nm-particles*. Powder Technology, 2013.

[5] Wu, H.-C., et al., *Explosion characteristics of aluminum nanopowders*. Aerosol and Air Quality Research, 2010. **10**(1): p. 38-42.

[6] Bouillard, J., et al., *Ignition and explosion risks of nanopowders*. Journal of Hazardous Materials, 2010. **181**(1): p. 873-880.

[7] ASTM, *E1515: Standard Test Method for Minimum Explosible Concentration of Combustible Dusts*. 2007.

[8] Castellanos, D.Y., *The Effects of Particle Size and Crystallinity on the Combustion Behavior of Particulated Solids*, in *Department of Material Science*. 2013, Texas A&M University.

[9] Denkevits, A. and S. Dorofeev, Dust explosion hazard in ITER: Explosion indices of fine graphite and tungsten dusts and their mixtures. Fusion engineering and design, 2005. 75: p. 1135-1139.

[10] Turns, S.R., *An introduction to combustion*. Vol. 499. 1996: McGraw-Hill New York.

[11] Burke, S.P. and T.E.W. Schumann. *The mechanism of combustion of solid fuel*. in *Proc. 3rd Int. Conf. Bituminus Coal*. 1931.

[12] Van der Held, E.F.M., *I3. The reaction between a surface of solid carbon and oxygen*. Chemical Engineering Science, 1961. **14**(1): p. 300-312.

[13] Graaf, D. and J.G.A., Brennst-Warme-Kraft, 1965. **17**: p. 227.

[14] Neufield, P.D., A.R.Jansen, and R.A.Aziz, The Journal of Chmeical Physics, 1972.

[15] Wilke, C.R. and C.Y.Lee, Industrial & Engineering Chemistry Fundamentals, 1955.

348963 Shock Interaction with Dust Layers for Different Mach Numbers and Dust Layer Depths

Abstract:

Secondary dust explosion is a serious industrial issue because it occurs under conditions corresponding to an increased quantity and concentration of dispersed, combustible dust when compared with the primary explosion. The problems of lifting and dispersing of a dust layer behind a propagating shock wave must therefore be understood to ensure safety regarding secondary dust explosion hazards. A new facility was developed recently by the authors for the study of shock waves over dust layers. A shock tube and high-speed (15,000 fps) shadowgraph system were used to film shock waves over limestone dust layers. New data were collected for image analyses over longer periods than found in prior works, where the longer observation time and higher frame rates led to the discovery of trends not previously observed by earlier studies. Further investigation is needed to aid the development of predictive models for shock wave/dust layer interaction.

Data were collected for various Mach numbers and different limestone dust layer depths. Within the current data, it was observed that the rate of dust rise increases linearly with increasing shock Mach number. This trend does not appear to hold true above Mach number 1.32 as the rise rate and overall rise height of Mach 1.32 and 1.40 are nearly identical. However, an increase in rise rate is observed as Mach number increases between 1.10 and 1.32. Further experiments at higher Mach numbers above 1.40 are required to fully observe the effects of shock Mach number on the dust layer rise rate and raise height. When comparing the effects of the dust layer depth, it was observed that increasing the dust layer depth resulted in an increased dust layer rise height but a slight decrease in the rate of the dust layer rise.

348967 Numerical Simulation of Cryogenic Boiling

Abstract:

The main source of hazard associated with cryogen is the loss of primary containment which can lead to vaporization, two-phase spreading and vapor dispersion. Numerous studies have been conducted in modeling vapor dispersion, but much less attention has been given to phenomena such as vaporization and two-phase spreading despite the fact that those are critical to the accurate prediction of the consequences of spill. Vaporization and two-phase spreading models are known as source term model since results of these models are used to perform the vapor dispersion calculations. As a result, LNG vapor cloud dispersion models are highly sensitive to the accuracy of the vapor generation rate estimated from the source term model. The vaporization rate modeling includes boiling and evaporation. During the initial spillage, the surface temperature of the substrate, such as concrete or insulated dyke material, is much higher than the boiling point of the cryogen thus boiling takes place mainly due to the heat conduction from substrate. When the temperature of the substrate falls close or equal to the boiling point, conduction heat transfer becomes insignificant in comparison to convective heat transfer from air to liquid thus the vapor generation is mostly due to the evaporation. In this study, our focus is to model the boiling heat transfer due to conduction for an accurate prediction of the vaporization rate that will subsequently be combined with evaporation model and two-phase spread model. A CFD based approach is considered rather than integral modeling approach for future unification of vaporization, two-phase spreading and vapor dispersion models.

The boiling of a fluid have 3 different regimes, i.e., film boiling, transition boiling and nucleate boiling. Each mode of boiling occurs over a range of excess temperatures (surface temperature exceeding the saturation temperature). Nucleate boiling is characterized by the formation of bubbles from randomly distributed nucleation sites over the surface whereas the film boiling is characterized by the formation of a continuous blanket of vapor film on the heating surface with bubbles leaving the surface at a regular intervals. Transition mode is much less understood but it connects nucleate and film regimes. Liu et al. proposed a volume-of-fluid (VOF) method with a sub-model of phase change to simulate cryogenic pool boiling using the commercial CFD software Fluent. A user defined interphase mass transfer model considering both homogenous and heterogeneous boiling was proposed. Three different boiling regimes were identified, and the boundaries of each regime were well predicted. It was also found that the simulation can reasonably predict the experiments in a film boiling regimes. However, the heat flux for nucleate boiling was under-predicted by Liu's model. In this work, the predictions in the regime of nucleate boiling will be improved by seeking a better sub-model to describe the homogenous model, which is supposed to be critical to predict nucleate boiling.

349002 Homogenous-Gaseous and Particle-Gas-Air Combustion in Turbulent Environment: Analytical Formulation and Experimental Validation

Abstract:

Analytical studies of premixed turbulent flame velocity, namely, the renormalization approach of Bychkov et al [Combust. Sci. Tech. 179 (2007) 137] and the spectral formulation of Chaudhuri et al [Phys. Rev. E 84 (2011) 026322] are reconciled, and validated by Hybrid-Flame Analyzer experiments on turbulent methane-air burning [Combust. Flame 160 (2013) 635]. Both the homogenous gaseous premixtures as well as the particle-gas-air systems containing coal-duct impurities are investigated. The turbulent flame velocity is tabulated as a function of various flame parameters (such as the laminar flame speed, thermal expansion coefficient, etc.) and flow parameters (such as the turbulent intensity, integral and Kolmogorov length scales, etc.) as well as that of the sizes and concentrations of the impurity particles. Specifically, it is shown that the particles typically intensify turbulent combustion. The formulation accounts for the flame-flow feedback due to thermal expansion. Besides, turbulence coupling to the intrinsic combustion instabilities is analyzed by means of the concept of the turbulence-induced Darrieus-Landau cutoff wavelength. It is eventually demonstrated how multi-phase impurities and turbulence-instability interaction modify the conventional turbulent combustion regimes diagram.

349400 Application of Leading and Lagging Indicators to Improve Laboratory Operation Safety

Abstract:

Unlike the chemical manufacturing process with fully automatic equipment, the laboratory operation in the chemical industry will still involve multiple types of manual handling for chemical analysis, reliability test, material characterization and failure analysis. The test operators or lab engineers have different levels of exposure to a lot of chemicals and physical hazards in multiple phases. In order to improve the operation safety and mitigate the potential risk, the study of situation awareness in the laboratory operation aims to investigate the perception of what is happening around the operation environment for both the normal situation and the abnormal situation, which has become an important part in the PSM(Process Safety Management). One of the effective methods to improve individual and team situation awareness and their individual responsibilities to accomplish the team's goals and objectives is the application of leading and lagging indicators, which have been recognized as the effective tools to drive process safety performance improvement. The processes and decisions to develop and implement the process safety leading and lagging indicators within the company involves the integration of the applied criteria, the differences between facility-level implementation and company-wide implementation, the potentially encountered barriers to success and the methodology to overcome, and the demonstrated benefits through the use of leading indicators. This paper will describe the straightforward application of leading and lagging indicators to identify the hazards and assess the risks of laboratory operations. The analysis will be deployed in the ways of manpower, machine, materials, method and environment. The analysis output can be regarded as the aid to safer laboratory design, operator training and real-time fault diagnosis. This will help to debottleneck and achieve operation safety and occupational health in the semiconductor industry.

349855 Graphic Visualization of IPL Status Enables Better Judgement of Inspection, Maintenance and Repair Priorities and Helps Day to Day Assessment of Operational Risk and Work Management Decisions

Abstract:

With hundreds and even thousands of IPL components to be inspected, maintained or repaired at a facility, monitoring and managing their status is a challenge whose difficulty rises exponentially when dealing with ageing assets. It is not uncommon to have a situation where there is a degree of impairment on several prevention or protection systems but it is difficult to judge where the priority lies given constraints on maintenance and inspection resources. With a backlog of safety critical maintenance, dilemmas arise such as whether it is more important to action an overdue callibration on a relief valve or to inspect a set of gas detectors which should have been carried out last month. These practical day to day problems are very common in aging plants or production facilities in mature basin areas. Our maintenance management systems are excellent for setting up and managing an appropriate inspection / maintenance schedule for our critical elements but once a backlog has developed they are not so helpful in prioritising overdue work. Often the right decision is based on proximity of impairments so that you avoid a situation where multiple systems are impaired in the same location, however, the data is not configured to present in this manner.

This paper will describe these challenges and talk about new technology platforms which are now available to help in this area. They are expansions of existing work management technology solutions where work activities are shown graphically with icons on plant plot plans. These systems can now be used to track IPL impairments graphically showing the geographical location of the impairment allowing better decisions to be made on the

cumulative local effect of impairment. They also enable better decisions to be made on day to day work whose activities may have an impact on overall Operational risk based on the status of IPL's in the vicinity of the work. For example if you were about to do maintenance work offshore on a main oil pump which required breaking containment and the tools you were going to use introduced a spark potential. Added to this the deluge system in the area is 3 months beyond its inspection due date. Showing this information graphically rather than in lists and schedules enables better assimilation of the facts and more informed decision making on whether to do the work or what additional precautions to take. The paper will look at how such information is gathered and presented and we will describe some realistic scenarios showing how these systems help with prioritising inspection and maintenance work and how they help with day to day work decisions. While this expanded technology platform is new, the original work management technologies have been around for over a decade and in common use today. We will include some comments on how these systems have evolved, who uses them and how successful they have been.

The growth in this area represents a broadening of the technology tool set which can be used for a risk based approach to inspection, maintenance and repair and helps to combine and share information and data from three disparate functional groups: Operations, Maintenance and Safety whose collaboration is essential for safe and efficient running of the plant.

349881 2. Successes in Implementing PHA/HAZOP/LOPA in Major Capital Projects in a Steel Company

Abstract:

HADEED, the steel making affiliate of SABIC, began implementing a multiple phase approach for major capital projects 8 years ago. For complex processes, such as direct iron reduction, 4 sequential PHAs throughout the project have been required, including a PHA of the operating procedures before commissioning. For simpler processes, such as rolling mills for rod and sheet and plate, 2 to 3 PHAs have been required. This paper describes the approach and successes to date achieved by HADEED using this approach, including discussing the gains in startup efficiency and operational cost savings that more than pay for the PHA efforts during the project. The paper also illustrates the versatility of the phased approach to performing PHAs in non-chemical processes, such as steel making.

<<NOTE: If HADEED cannot get approval from their management to take the lead on this paper, then we will withdraw this submittal >>

349909 Liquid Fuels Release Rate Calculation in Transport Pipelines with Complex Topographical Conditions

Abstract:

Pipelines are the most efficient and economical mean of hazardous materials (HazMat) transportation through long distances. In the particular case of hydrocarbons, production sites are often far away from processing facilities and/or final users, therefore, this industry uses the greatest amount of transmission pipelines. The main concern related to the operation of these systems is the loss of containment events (LoC), due to the fact that the fluid itself constitutes a source of hazard. Such events, despite their low frequency, might have severe consequences on the surrounding communities, the environment and the assets of the operating company or any other company in the vicinity of the pipeline.

In order to ensure a safe and reliable operation of these systems, and to support decision-making processes in a risk management framework, it is required to perform risk analysis. This analysis will allow establishing possible consequences of the accidental events and the required measures to reduce the frequency of events and/or the effects magnitude. One of the main inputs to risk analysis is the release rate, i.e., the rate at which the material is released to the surroundings once the loss containment takes place. Modeling the release rate is particularly important in countries with complex topographic conditions, in which a pipeline undergoes several geographical changes along its route. The amount and manner in which a HazMat is released to the surroundings will define the possible accidental events and therefore the consequences on the elements of interest, such as population, environment or infrastructure. Therefore, it is necessary to properly quantify the available volume for release at each segment of the pipeline and estimate the related release rate profile. As a result, several models have been developed to model hydrocarbon release scenarios, including phenomenological approaches that consider one (axial) and two-dimensional (axial and radial) flow patterns, with or without phase change, transient effects, and more. Although the phenomenological models offer the best representation of reality, their computational cost is extremely high, with computing times on the order of days.

This work presents a tool that supports decision making processes during the design, operation and/or revamping of hydrocarbons transmission pipelines. The tool allows a simple, non-phenomenological calculation for a quick estimation of available volume for release, and the related release profile. The tool considers the presence of active-automatic safety barriers (such as emergency shutdown valves and check valves) in pipelines with complex topographical conditions, given a failure in any point of the pipeline. These calculations are performed at a very low computational cost, with computing times on the order of minutes. The results provided by this tool become a key input to perform *a priori* risk analyses of pipeline systems to identify its critical segments (those with the highest release rates and times). Such information is crucial to define the layers of protection to be implemented (e.g., determine segmentation requirements, application of other safety barriers) and for the strategic design of emergency response plans in case a failure occurs (e.g., establish the critical areas of the pipeline, location of sensing devices, emergency response brigade).

349929 Decision Tree to Optimize NFPA 30 Criteria in Fire Protection Systems Applied in Oil and Gas Industry

Abstract:

In accordance with risks and hazards associated to oil and gas industry operations, criteria must definitely be developed in order to use NFPA 30 code (Flammable and Combustible Liquids Code) in an adequate way. To meet this purpose, ECOPETROL has developed a decision tree so that both engineers and technicians can use NFPA 30 criteria more easily and more efficiently tailoring the decision making process to the most outstanding and recurrent hazards in west.

349931 Sil Determination of High Integrity Pressure Protection System (HIPPS)

Abstract:

High Integrity Pressure Protection Systems (HIPPS) are Safety Instrumented Systems (SIS) implemented to address overpressure scenarios in lieu of a pressure relief valve (PRV). HIPPS essentially replaces the PRV for those scenarios that the SIS is designed to prevent. HIPPS applications are generally pipeline and pressure vessel overpressure protection.

The fourth edition of API 521 allows credit for a favorable response of the instrumented systems that prevent over-pressure and/or over-temperature. The recommended practice refers to these systems as high integrity protection systems and states that they should be at least as reliable as a pressure relief device.

The Pertamina Hulu Energy Offshore North West Java (PHE ONWJ) intents to install new YYA NUI (Normally Unmanned Installation) platform will be sent to the existing unmanned KLB platform via 12 inch three phase pipeline. The KLB platform has fix flare system and has limited space to increase its capacity due to the additional production from the YYA.

The overpressure scenario has been identified in the HAZOP study and need to be mitigated by either release the pressure through PSV to flare or installing another layer of protection to prevent overpressure scenario. In this case of the releasing pressure to existing KLB flare is not sufficient to handle the gas relief from YYA wells. Therefore HIPPS was selected to replace the PSV in preventing the potential overpressure at the KLB.

The LOPA workshop was conducted for SIL determination of the HIPPS system and attended by discipline engineer and operator to identify initiating cause, the protection layer and severity level. It was identified 5 initiating events related to the overpressure scenario above. The independent protection layer also identified the existing pressure switch high which can be assumed to have a credit for SIL 0. The consequences or severity level for safety, environment, cost and business image was determined as impact of the scenario happened. The modifier factor was also included for occupancy (NUI) while the ignition probability was not credited as the ultimate consequence ultimate scenario is rupture.

The assessment of SIL determination resulted IL target IL1 for safety, IL2 for environment and IL0 for commercial. From this LOPA method, maximum IL2 with PFD 6.1E-03 was selected for IL target. The next step is to design HIPPS and verify the SIL level after gathering data of PFD from the vendor.

350028 Comparison of Different Methods to Determine the Activation Energy of Flammable Dusts Mixed with Inert and Inhibitory Materials

Abstract:

Comparison of different methods to determine of the activation energy of flammable dusts mixed with inert and inhibitory materials

C. Wanke, B.Binkau, P.Rettig, U.Krause

Porous flammable media often undergo spontaneous combustion and the spread of smoldering flame fronts. Adding inert or inhibitory substances, the self-ignition and combustion behavior can be influenced. In a series of experiments the activation energy of flammable dusts (lignite, carbon) mixed with inhibitors (ammonium sulfate, ammonium phosphate) was determined and compared to the pure flammable dusts.

To determine the activation energy self-heating methods (isoperibolic and adiabatic hot storage tests) and temperature programmed methods (thervogravimetric (TG) and differential scanning calorimeter (DSC) tests) can be used. The common method in Europe is the isoperibolic hot storage test according to DIN EN 15881. This method is very time and material consuming, due to the fact, that the self-ignition temperature of three to four different sample sizes must be analyzed. An alternative is the adiabatic hot storage test where the activation energy can be determined from just one single experiment. The disadvantage is the difficult analysis of the resulting Arrhenius plot, where the activation energy is the steepest slope of the logarithmic temperature rise. To find this value a Matlab tool was developed.

The advantage, using TG and DSC is the small amount of sample material. The analysis of the mass loss rate from the

TG experiment results in an Arrhenius plot as well, which can be evaluated by the mentioned Matlab tool. Using the DSC, three experiments with different heating rates have to be executed. In spite of the slow heating rates, this method is still faster than the isoperibolic test.

Summing up, the activation energy of pure substances, as well as mixtures, can be calculated by the methods above. The isoperibolic test lead to the highest activation energies, the DSC and TG experiments lead to the lowest values. This can be traced back to the different experimental procedures. Isoperibolic tests start at the self-ignition temperature (SIT), adiabatic tests at the onset temperature, which is below the SIT, and DSC and TG start at ambient temperature. The reactions below the starting temperature can't be taken into account for the determination of the activation energies, therefore the results of DSC and TG are lower.

350059 Economical Approach Quantification of Impacts in Major Accidents

Abstract:

In the industry field, the nature of certain processes has always represented an inherent threat not only to humans but also to the environment in which they are executed. Although the frequency of scenarios that lead to the occurrence of major accidents in facilities is considerably low, their consequences are usually catastrophic. Experience has demonstrated that high level impacts occurred during the development of industrial activities imply significant financial losses. Therefore, one of the main interests for risk assessment results in the economical estimation of these consequences.

Usually, damage costs are estimated once the potential consequences have been already materialized. Nonetheless, in order to generate mechanisms for prevention, mitigation and control over these negative impacts, the implementation of a pre-emptive quantification of potential leakages allows the estimation of useful indicators as incomes for decision making analysis. For this purpose, qualitative, semi quantitative and quantitative methodologies have been developed to support risk analysis.

Within this context the present work pretends to propose an *a priori* quantitative methodology that allows carrying out the economical conversion of impacts originated by major accidents through the integration of two methods. The first method is defined to estimate the impact level over a specific system while the second establishes the transformation of impacts into monetary terms. The suggested methodology is applied later to a particular case of study.

A quantitative vulnerability analysis has been defined for the estimation of impact level. The model used is a classical approach that implements *probit* analysis which gives the level of impact. The fact that the result corresponds to a quantitative estimation, makes easier its conversion into monetary terms. Even so, as a limitation of the method, it must be considered that each type of major accident, and subsequently their specific cases, has a unique equation that represents the system of study. Accordingly, the application of this method for a particular situation requires the definition of the appropriate expression that relates the *probit* variable with the dose associated to the specific event.

Once the level of impact is estimated for a system of interest, the second method is applied in order to express these impacts in monetary terms. In this case, the methodology of CEPAL is used. It defines the types of effects produced by an accident: Direct, indirect and macroeconomic. Since the main interest of this work is not the estimation of impacts effects of an accident at socioeconomic level, only the direct and indirect effects over the system are considered. For direct costs, this method defines a set of parameters for each type of system affected (infrastructure, humans, services). These must be considered for the breakdown of costs including sometimes the appraisal over a period according to the item evaluated. In the case of indirect costs, the execution of two analyses is necessary. As a first step, a projection of each item is done to estimate certain economic parameters dependant of the system type. Then a comparison of the values that will be reached in the supposed accidental scenario, with respect to the ones that would be reached if the accident did not happen is developed. With these values, finally the net incomes for each case are estimated.

An explosion in an energy supply facility is considered as a case study. The correspondent vulnerability analysis with probit method is done in order to estimate the damages caused over the structure. As a result, costs of reparation / reconstruction for damaged infrastructure taking into account the appraisal over the period of time that reparation / reconstruction of the facility takes, costs of equipments and costs of raw materials are determined. Also, operation and capital costs are estimated taking into account the time required for reparation / reconstruction of facility. Capital costs include equipment acquisition, operation, labor force and materials. Finally the net incomes for the accidental and no accidental scenarios are estimated in order to obtain the approximate net indirect costs.

The appropriate use of methodologies, as the one proposed in the present work, defines non conventional paths for generation of indicators that support decision making analysis. Due to the different considerations and approaches used within the methodology, the indicators that could be generated contain valuable information different from the one given by indicators that do not consider an economic perspective of effects. The information provided by these indicators offers new perspectives and more accurate points of view for project evaluation.

350080 Application of Consequence Analysis in the Development of Emergency Response Plans for Accidental Events in Liquid Fuels Transport Pipelines

Abstract:

Pipelines are the world's most common operation for hazardous materials transportation due to its low cost. Oil and gas industry (upstream, mid-stream and downstream) concentrates the greatest amount of pipelines of the world, due to the fact that production fields are normally too far away from final users, export facilities or processing plants. The main concern regarding the operation of these systems is the events of loss of containment (LoC), which might take place for several reasons such as corrosion, external impacts, etc. Such events, despite their low occurrence frequency, could have severe consequences to surrounding populations, environment and assets of the operating company or any other company. The consequences of accidental events might be reduced by application of different protection layers (e.g., construction materials, alarms, shutdown valves). Emergency response corresponds to the last layer of protection, and its objective is to control, if possible, an accidental event, or to reduce its consequences in the case the event develops completely.

In a risk management framework, it is imperative to perform risk analysis to determine possible consequences of accidental events and required actions to reduce the frequency of events and/or the magnitude of effects, and thus, support the decision-making process. As a result, many risk analysis techniques have been developed, which can be qualitative, semi-quantitative or quantitative. The selection of a specific technique depends on the particular characteristics of the system under study (e.g., complexity of the operation, neighboring population, sensitive environment). Typically, emergency response plans have been developed from the application of qualitative risk analysis techniques such as Hazid and Hazop. This approach produces "generic" emergency response plans, which do not differentiate most likely scenarios or possible consequences of an accidental event.

This work presents the application of a quantitative approach of risk analysis, through consequence analysis , to determine major factors in attention of emergencies such as preparedness activities (e.g., drills), critical attention points, potential safety barriers required, shelters, evacuation routes and times, etc., in loss of containment events during pipeline transmission of liquid fuels in complex topographical conditions. These factors will allow the development of emergency response plans that guarantee effective attention of accidental events in pipelines. The methodology developed in this work, given the characteristics of the material being transported and the pipeline, guides the user in the construction of analysis scenarios, the selection of source, effects and consequence models. To illustrate this, a hypothetical pipeline is studied and the most important results are presented.

Real-Time Risk Assessment and Decision Support

Andy Bolsover
DNV GL
Aberdeen, UK
andy.bolsover@dnv.com

Prepared for Presentation at
American Institute of Chemical Engineers
2014 Spring Meeting
10th Global Congress on Process Safety
New Orleans, LA
March 30 – April 2, 2014

UNPUBLISHED

Real-Time Risk Assessment and Decision Support

Andy Bolsover
DNV GL
Aberdeen, UK
andy.bolsover@dnv.com

Keywords: Bayesian networks, major accident, risk assessment

Abstract

Managers and operators of major hazard facilities make complex decisions as a part of their daily work activity. These decisions are made against the background potential for a major accident.

Such decisions may be required to account for daily changes in a large number of factors including plant condition and performance, operational status, knowledge and experience of personnel, interactions with other activities, and the effectiveness of processes. The information involved in the decision comes from multiple sources and may be difficult to assess.

Technical risk assessments provide a useful picture of major accident risk but some widely accepted approaches suffer from some significant problems which limit their value as tools for operational decision making.

The paper describes investigations into an approach that addresses how these difficulties may be addressed in day-to-day assessments. It describes a method and tool in which risks can be monitored in real-time and so enable safer decision making. The method is applicable to the assessment of a wide range of major accident hazard scenarios. The paper will describe how the tool addresses problems in some alternative approaches. A significant feature of the approach is its ability to identify the most probable causes of risk. The speed of the assessment points to its potential use in real-time detection and control systems.

The method employs a Bayesian net to perform the risk assessment. Bayesian nets have been used to aid decision making in many different situations and industries, but have received relatively little attention as risk assessment and decision tools in major hazard industries. The paper will include a description of the benefits offered by this technology as well as a view of its limitations.

1. Introduction

Operators of offshore oil and gas exploration and production installations need to be aware of the hazards associated with high-pressure flammable materials and the potential for a major accident. Successful risk management requires a good understanding of the status of equipment, the effects of human behavior and the effectiveness of safe operating procedures.

Changes in process conditions and the performance of inspection, test and maintenance mean that the risk profile of an installation will undergo frequent change. The offshore installation manager (OIM) is responsible for safe operation and needs to understand where the risks are coming from. Examples of changes to the risk profile may be identified through inspections that identify degraded performance in safety critical systems; variability in the competence of new crew; and potential procedural weakness as identified through verification and audits of the management system. The OIM also needs to know how the risks are influenced by changes in operating conditions and activities on the installation. He must then understand how the risks can be effectively managed.

Day-to-day risk management will often involve operational risk assessments (ORA). An ORA may for example, assess the implications of a temporary loss of redundancy while part of a safety system is repaired or replaced. The OIM must identify potential risk reducing measures and decide what must be done to maintain a tolerable level of risk. The OIM's assessment and decisions will often be difficult because of the multiplicity of relevant factors that must be considered. Such assessments are typically qualitative and identify risk management measures without any clear indication of the relative value or prioritization of these measures.

The work described in this paper arises from a desire to improve the toolkit for operational risk assessments and identify tools that can handle the very wide range of factors that affect the assessment, while allowing the assessment to be performed in real time, accounting for the changes in operating conditions / parameters.

The paper describes a tool that performs a risk assessment that takes account of factors such as hardware reliability, current status of safety critical systems, human reliability and competence. The tool also shows how findings from management system audits and independent verification can be included in the assessment.

The tool, "SOUL" (Safer Operations Upstream Landscaping), has been developed as a demonstration of principles, but enough work has been done to illustrate the feasibility of a full scale tool. The demonstration tool makes an assessment of blowout risk on a drilling rig, but the methods used to develop the tool are applicable to the management of other types of major accident hazard in other industries.

The paper describes some requirements for an operational risk assessment tool. It then gives a simple example of the tool's underlying method (a Bayesian network) before describing the tool itself. The paper describes the benefits and limitations of a Bayesian network as a basis for a practical operational risk assessment tool.

The situation and challenge for improving operational risks assessments is described. Next, Bayesian methods are introduced through simple examples that describe how these methods can assist current widely-used QRA methods. A description is given of a tool, SOUL, that has been developed to demonstrate how Bayesian methods can be used to solve some of the difficulties with widely used risk assessment methods. All risk assessment methods have strengths and limitations, and the main discussion of these is presented in the conclusions.

DNV acknowledges the work of Decision Systems Laboratory at the University of Pittsburgh, in developing the SMILE / GeNIe software [1] that is used to construct and solve the Bayesian network described in this paper.

2. Situation and challenge

The offshore oil and gas industry is a high-hazard industry, where changes in the operating environment may alter the risk profile of an installation. These changes need to be accurately assessed and lead to good decision-making.

The offshore oil and gas industry is also a very dynamic industry, such that operating conditions, activities and the status of safety systems (e.g. override) are likely to be changing daily (or faster for some types of activities, such as drilling). Any assessment of risk should therefore also be rapid. An ORA tool / method should permit reassessment in real time (i.e. it should take account of new information as it becomes available). It should also take account of:

- Current known equipment condition
- Competence, experience and other human factors
- Incident/ accident history
- Inspection, test and maintenance findings
- Day-to-day changes in operating conditions and operational activities (e.g. number of people, hot work, activity)
- Information from audits, safety critical system performance verification, and Key Performance Indicators (KPIs)

The assessment should also take account of the uncertainty that is inherent to each of these listed factors.

Operational Risk Assessments are performed when a problem is identified in the function of a safety critical item of equipment. If the item must be taken out of service or must perform in a degraded state, it is necessary to determine the changed risk and decide what must be done to manage that risk.

A typical Operational Risk Assessment aims to provide a realistic assessment of risk that accounts for current and expected conditions, but it is largely based on a professional judgment, and does not take full account the very large number of parameters that might affect risk.

An ORA must also be a *cumulative* risk assessment - the assessment must account for multiple changes in the operating environment. The risk implication of each change may be small when considered in isolation, but the cumulative effect of small changes may be profound.

A further significant longstanding challenge for persons in professional risk management is the development of risk assessment tools, such as Quantitative Risk Assessment (QRA), that can successfully integrate technical, human and organizational risk-influencing factors in a single assessment. It is also noted that the current approaches to offshore QRA are often focused on the design phase and make an assessment of the average risk that is representative of different types of operating conditions such as normal production and drilling. During the operating phase, the updates to such QRAs tend to account for macro changes in the installation – such as changes to the process conditions and process equipment, revised average manning, and a new bridge linked compression module. In QRAs it is difficult to reflect factors that arise from day-to-day changes. Updates to QRAs are also time-consuming. Issues such as these mean that the current approach to QRA makes it unsuitable as basis for a "real time risk assessment".

3. Bayesian methods

The operational risk assessment method described in this paper is Bayesian. The term "Bayesian" here means that the status of the offshore installation is uncertain, but the probabilities describing this uncertainty can be systematically updated as the result of observations of the installation.

Bayesian methods are named after Thomas Bayes (1701-1761) [2] who was the first to formulate the mathematical theorem that bears his name; Bayes' theorem shows how new information can be used to make better decisions under conditions of uncertainty.

It is useful to understand the principles at work behind Bayesian methods. A comprehensive account of Bayesian methods is beyond the scope of this paper, but the methods can be illustrated by considering a simple system (much simpler than an offshore installation). As an example, Bayes theorem is applied to a very simple system: a coin that may fair (of the standard type) or may have two heads. We are uncertain about the status of the coin because we are not allowed to examine it closely, but want to draw conclusions about it based on observations; is it a fair coin or does it have two heads? For purposes of this thought experiment we learn about the coin by seeing the results of the coin being tossed. How do these observations affect the likelihood that the coin is two-headed?

In a conventional statistical approach we ask that the coin be tossed a number of times and the results of this simple experiment are presented as a p-statistic. For example, if we observe 5 successive heads we might say that the hypothesis that the coin is two headed is confirmed at the 5% level – that is the chance that the coin is not two-headed is less than 1 in 20. (The probability of tossing five successive heads with a fair coin is actually about 3%).

The OIM of an offshore oil and gas installation does not have the luxury of running many "experiments" to determine the status of his installation when the outcome of such an experiment could be a major accident. Instead he must draw conclusions based on limited information.

A follower of Bayesian statistics takes a different approach. He starts with a level of belief about the hypothesis that the coin is two headed. For example he might start with a low level of belief that the coin is two headed, assigning a probability of say 0.01 to this hypothesis. The alternative hypothesis that the coin is fair is assigned a probability of 0.99. Bayes theorem then explains how these hypothesized probabilities should be updated with each toss of the coin.

If the coin is tossed once and a head is observed, it strengthens the two-head hypothesis. So how does the hypothesis probability change? If the coin is two-headed, there is only one possible outcome from tossing the coin – it must be a head, the probability is 1.0. However, if the coin is fair the probability of getting a head is only 0.5.

According to Bayes rule, the updated probability of the two-head hypothesis is now about 0.0198, which is calculated as:

$$\frac{1.0 \times 0.01}{1.0 \times 0.01 + 0.5 \times 0.99}$$

Each successive head tends to confirm the hypothesis as shown in Figure 1.

Figure 1: Bayesian belief in two-headed coin hypothesis

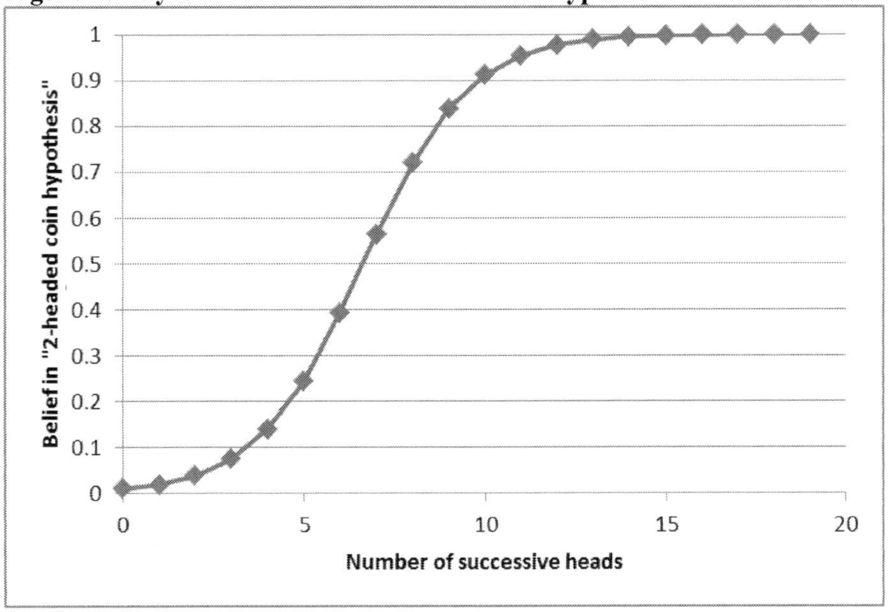

The two-headed coin problem illustrates a very simple application of Bayes' rule. Bayes rule can be written as:

$$P(H|E) = \frac{P(E|H) \times P(H)}{P(E|H) \times P(H) + P(E|\sim H) \times P(\sim H)}$$

Where the meaning of each term is as shown in Table 1.

Table 1. Definition of terms in Bayes Rule

Term	Meaning	Example
P(H|E)	probability of hypothesis, given evidence	probability that coin is two-headed, given that a head has been tossed
P(H)	probability of the hypothesis	probability that coin is two-headed
P(E|H)	probability of the evidence given the hypothesis	probability of head, given that coin is two-headed
P(~H)	probability of alternative hypothesis	probability that coin is fair
P(E|~H)	probability of the evidence given the alternative hypothesis	probability of head, given that coin is fair

More complex forms of Bayes' rule deal with multiple alternative hypotheses and continuous probability distributions, and explain how evidence from different sources should be combined. Bayes rule explains how conditional probabilities should be handled, and for a true Bayesian, all probabilities are conditional.

It is evident that the mathematics underlying the Bayesian approach is more complex than the traditional approach to solving this simple problem. Strict application of the Bayesian approach using hand-calculation methods can make many problems unmanageable, and this accounts for the historical preference for the conventional 'frequentist' approach such as p-statistics etc.

With the advent of modern computers Bayesian methods do, however, offer several advantages over the traditional approaches. Bayesian statistics have been studied in academic circles and with the availability of modern computers this has resulted in sophisticated algorithms being developed which can solve many statistical problems that cannot be addressed by conventional methods in any practical way. [Other advantages and disadvantages of Bayesian networks (BN) are outlined in the Benefits and Limitations.] These solution algorithms are implemented in so-called Bayesian networks that describe, and then solve, complex probability problems. The networks are drawn as *'nodes'* with connecting arrows. Each node represents a factor that is relevant to the problem and the arrows (or *'arcs'*) show how the node factors influence each other. Each node contains probability data that are used in the calculation of probabilities. For the Bayesian network described in this paper, each node can be in a number of discrete *states* and the probability data is presented as a table showing how each of the 'input' nodes affect the node being in each state.

Some have argued that Bayesian networks cannot do anything more than can be done using QRA tools such as fault trees and event trees. This is not correct. A Bayesian network can handle complicated cross linkages between nodes that cannot be handled in any normal event tree. An event tree can be converted to a Bayesian network, but not vice-versa. And while fault trees *can* handle cross linkages that represent common cause failures, most fault trees confine common cause modeling to failures within specific safety systems. Where common cause factors are relevant across the full scope of a risk assessment, Bayesian networks offer an approach that can be simpler than the approach offered by fault trees and event trees. This can be illustrated by considering how two parameters (wind speed and wind direction) affect several factors that must be modeled in a typical QRA.

These factors include:

- Dispersion of gas
- Ignition (affected by gas cloud size, and direction of dispersion)
- Explosion overpressure (affected by gas cloud size)
- Dispersion of smoke
- Impairment of Temporary Refuge (TR) (affected by gas/smoke concentration and direction of dispersion)
- Impairment of lifeboat stations
- Drift of lifeboats / people in water (affected by wind direction)

Bayesian Networks (BNs) can take account of the way that wind speed and direction act as common factors throughout the assessment and also take account of the complex interdependencies between the factors noted above. A typical QRA may make simplifications and assume that some of these factors are independent of each other, when in practice they are not independent. A typical QRA may also assign simple probabilities of occurrence to these factors, when there may be considerable uncertainty about the value of that probability. The BN can take account of that uncertainty. Monte Carlo (MC) methods in QRA can be used to account for uncertainty, but MC methods are usually limited to studies of parts of the QRA problem, such as wind speed and direction, dispersion, ignition and explosion overpressure.

Figure 2 shows a very simple BN which describes a situation while a drilling a well. In particular, it describes how the competence of the mud engineer might affect the probabilities of:

- A critical error in the mud mix
- The mud engineer's appropriate response to a kick

In this example, the level of competence of the mud engineer is uncertain as are the probabilities of an error in the mud mix and a failure to respond to a kick in an appropriate manner. According to the data in this Bayesian net, the competence of the mud engineer is either 'good' (probability = 0.8), or 'poor' (probability = 0.2), and the probabilities of a critical mud mix error depend on the engineer's competence as shown by the probabilities in Figure 2.

The probabilities of failure to respond to a kick are also shown, depending on both the level of competence and whether a mud mix error has occurred.

BNs are able to solve probability problems such as this scenario to evaluate the probabilities of different outcomes. The "mud engineer" problem is sufficiently small that it can be easily solved by someone with a basic understanding of probability calculations.

In this case, a simple calculation (details not presented) using the numbers presented in Figure 2 will show that the probability of a mud mix error is 0.0044, and the probability of a failure to appropriately respond to a kick is 0.0018.

BNs can also perform the reverse calculation. For example if we observe that the mud engineer does indeed fail to respond to a well kick, then the BN can use the probabilities in Figure 2 to infer something about the mud engineer's competence. In this case the revised probabilities of competence will be 'good' probability = 0.443, and 'poor' probability = 0.557. This calculation is not straightforward and is most-easily performed using specific software tools for solving Bayesian nets.

All Bayesian networks are built in a similar way, but BNs are typically much larger than the mud engineer problem. They may contain hundreds of nodes and hundreds of arcs. Practical BN problems may also involve nodes with many more states. It is necessary to use validated software tools to solve large Bayesian nets.

Figure 2: Simple Bayesian network showing nodes, arcs, states and probability tables

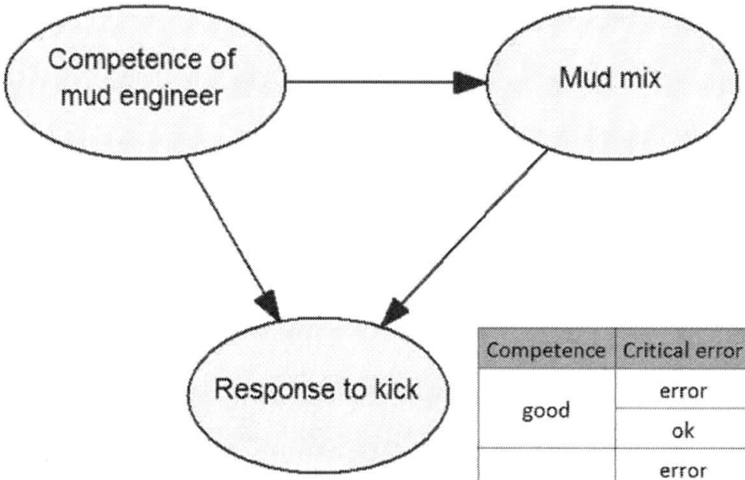

Competence	
good	poor
0.8	0.2

	Critical error	
Competence	error	ok
good	0.003	0.997
poor	0.01	0.99

		Failure to respond	
Competence	Critical error	failure	ok
good	error	0.002	0.998
	ok	0.001	0.999
poor	error	0.01	0.99
	ok	0.005	0.995

The nodes in this simple example each have only two states. Bayesian networks allow the use of larger number of states. For example, the competence states could have been extended to ‘excellent’, ‘good’, ‘average’, and ‘poor’, or further refinement.

4. SOUL (Safer Operations Upstream Landscaping)

Work on the SOUL project was triggered by the recognition that Bayesian networks could offer significant advantages over existing tools for operational risk assessments - potentially addressing each of the problems identified in the challenges noted previously.

The SOUL development project has developed a pilot system (demonstration tool) that illustrates how risks can be monitored, enabling safer decision making for offshore blowout scenarios.

The SOUL tool addresses some of the operating risk assessment challenges identified, through:

- use of location-specific data for plant status
- inclusion of human factors issues including competence and training
- inclusion of findings from safety management system audits
- inclusion of findings from verification of safety critical elements
- update of the risk profile by referring to a database containing relevant plant-specific data

Operational factors include: aspects of hardware condition, procedural elements, information on the competence of personnel and organizational culture. A risk assessment tool should, therefore, take account of this very wide range of factors if it is to identify where the main risks are to be found. The vision has been to provide a tool which shows how these disparate elements can be combined to provide a more complete picture of risk.

Initial development of the SOUL Bayesian network was based on a brainstorm of the factors that influence the likelihood of a blowout while drilling. Causal relationships between these factors were then identified. Additional factors were added as necessary during the construction process while other factors were omitted (either because of low relevance, or due to the demonstration nature of the model).

Initial data for the nodes came from several sources which included: equipment failure rate data from existing studies (adapted to account for uncertainty), blowout probability statistics derived from the Sintef database [3], and human error data from the HEART methodology [4]. Some aspects of the model were omitted, in line with the intended demonstration purpose of the tool. For example, the model includes only three safety critical roles and simplifies the action of the manual BlowOut Preventer (BOP) closure buttons at the driller's control panel and the tool pusher's control panel to a single button. The casing shear ram is also omitted. It can be noted, that the use of a BN instead an event tree does not improve the data, but allows the uncertainty in that data to be easily included.

The tool is also connected to an external MySQL™ database which stores relevant probability data describing the current status of the drilling rig. Again, in line with the demonstration purpose of the tool, only some of the nodes take their probability data from the database.

The Bayesian network in the SOUL tool explicitly includes nodes to model:

1. Kicks due to:
   - Pull-out-of-Hole (POOH) / Run in Hole (RIH)
   - Mud mix errors
   - Inadequate formation modeling / seismic data
   - Loss of down-hole mechanical integrity (e.g. casing, cement failures)
2. Marine risers and subsea equipment failures
3. Topside equipment failures
4. Failure of BOP components: The BOP and Lower Marine Riser Package are assumed to include
   - Upper annular preventer (UAP)
   - Lower annular preventer (LAP)
   - Blind shear rams (BSR)

- Upper pipe ram (UPR)
- Variable bore ram (VBR)
- Lower pipe ram (LPR)
- Hydraulic system
- Power system

The model accounts for various modes of failure in each type of component. For example, the UAP function is represented by seven nodes, each representing a different failure mode. For the BSR, the model includes failure due to the tool joint in the ram (one of the failure modes at the Macondo blowout incident).

5. Emergency push button operation at the driller's control panel and the toolpusher's control panel
6. Human performance in the following safety critical areas
   - Well design
   - Well verification
   - Mud engineering
7. Safety management system audits of:
   - Human Resources (HR)
   - Training
   - Contractor selection
8. Independent verification of safety critical elements
   - Primary well control
   - Emergency well control
   - Marine riser
   - Well monitoring

The complete Bayesian network contains 135 nodes and 158 arcs. It is estimated that a network containing the previously mentioned omitted elements might be double this size. Figure 3 shows the network and indicates groups of nodes with specific modeling functions:

- Drill floor blowout
- Primary well control
- Competence of personnel in critical roles
- Management system audit
- Secondary well control
- BOP push button

Figure 3: Schematic of Bayesian network

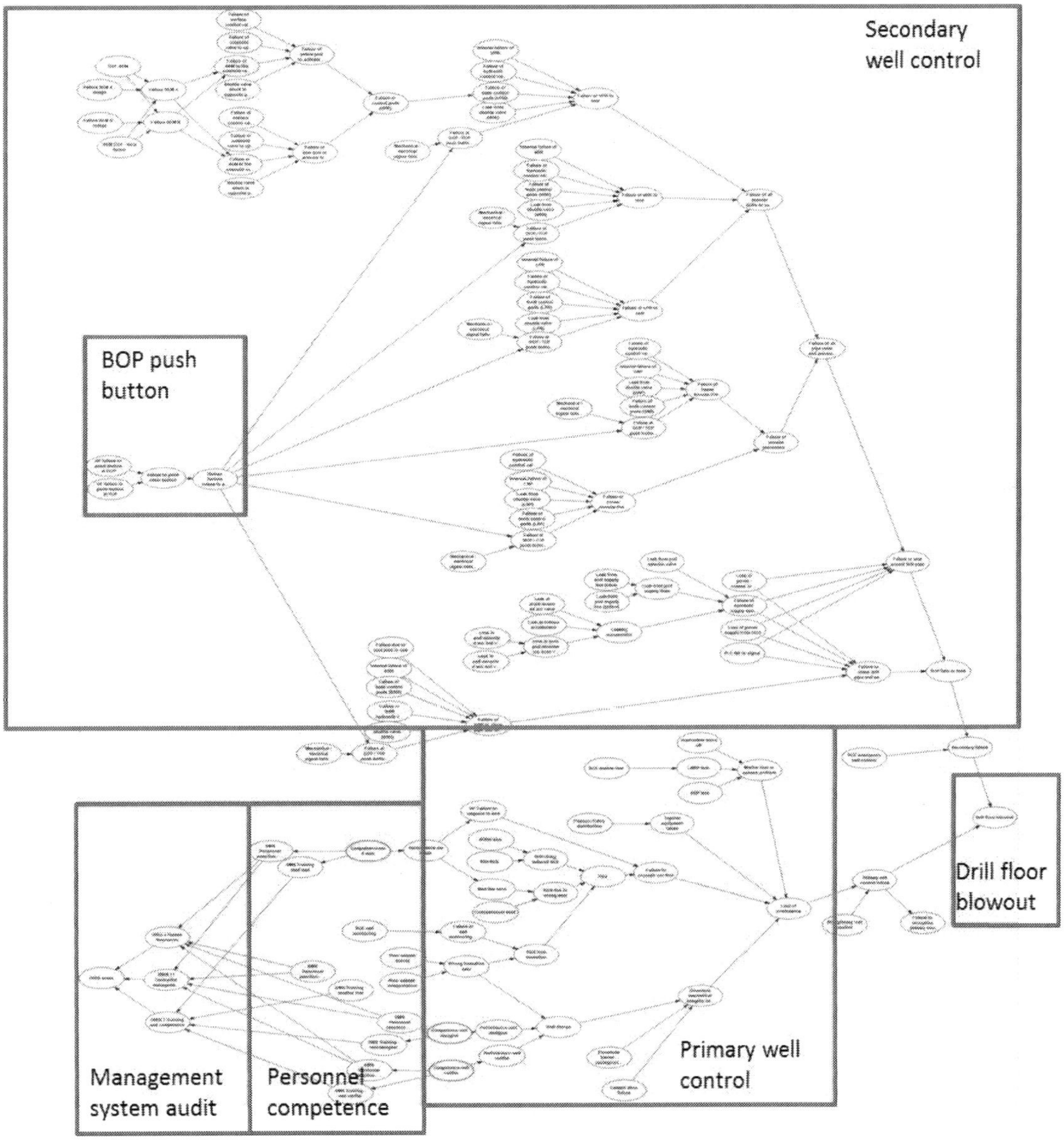

The BN was constructed using GeNIe software [1], but the GeNIe view of the network is not presented to the user. One of the challenges in the development of this tool was the need to provide a user interface that addressed relevant user needs. The GeNIe interface is suitable for BN development, but most of this functionality is unnecessary for a user who wants to make perform a risk assessment on an existing BN. Indeed, it is important that the BN structure and data are not inadvertently altered by the user. The user must however be allowed to enter new data that is relevant to the assessment. A separate viewer has been developed (written in Visual C#) which prevents inadvertent changes to network data and simplifies the display of results. Necessary functionality at each node is provided through buttons on the nodes. Button-clicks allow the user to drill down into the model to view additional nodes or view the state probabilities. The alternative interface allows the tool to be embedded in a standalone tool which gives additional user functions that are only indirectly related to the BN functions. Figure 4 shows a screenshot of the Graphical User Interface (GUI).

Figure 4: Screenshot of SOUL

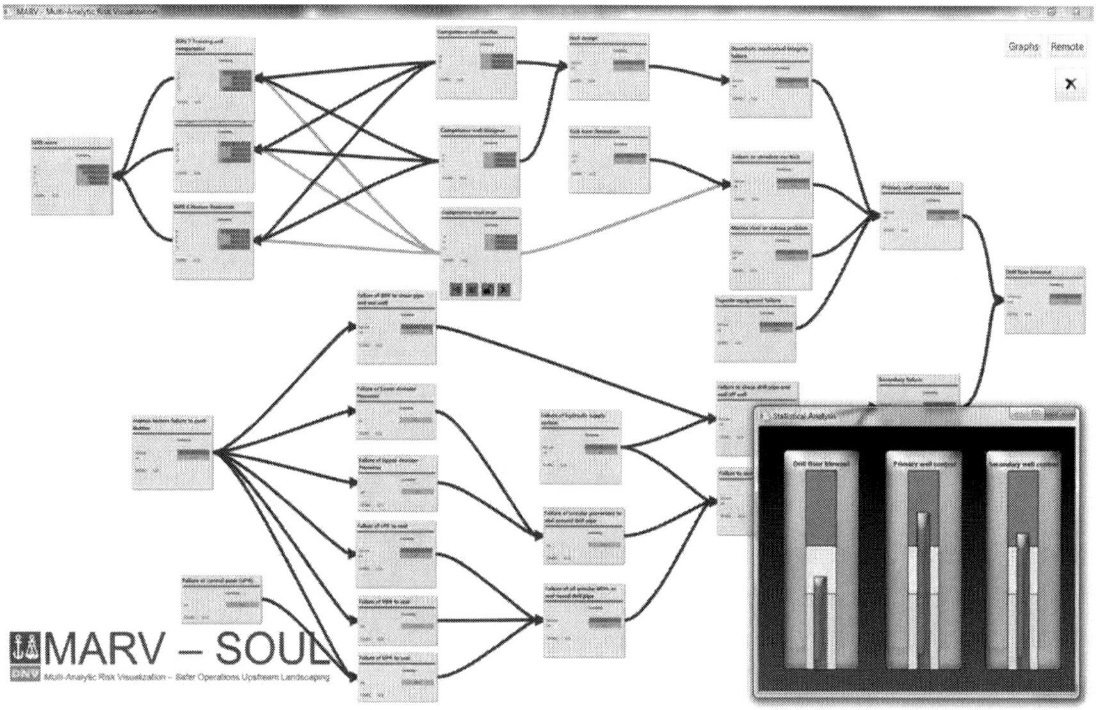

Use of the tool follows a typical pattern. On receipt of new information, the probability data in the relevant nodes is modified. For example, if an item in the BOP is known to have failed then the probability of failure for that item is set to 100%. If the competence of a person in a safety critical role is believed to be below average the probabilities for the relevant safety critical competence node can be appropriately adjusted.

Model run time is very fast (a fraction of a second) such that full results are generated when any item of input data is altered. Basic results are presented as a simple bar chart (Figure 4) which shows the probability of a blowout occurring while drilling the well. In addition to estimating the risk of blowout associated with the well, the tool also identifies the principal contributors (causes) of the risk facilitating prioritization of potential risk management activities.

5. Benefits and limitations

Bayesian methods usually involve more complex mathematics than alternative conventional methods. This difficulty has been addressed by the development of sophisticated computer algorithms, but this does not explain why Bayesian methods should ever be preferred over other approaches. Bayesian methods offer unique capabilities with several advantages over conventional methods, but like all methods they have their own special limitations. Any application of BNs needs to understand the relative strengths and weaknesses of BNs and the alternative methods.

5.1 Design

Bayesian networks are most effective when they are designed according to specific design principles. A comprehensive description of these principles is beyond the scope of this paper, but they include:

- Limiting the number of incoming arcs to each node
- Limiting the number of states within each node
- Trading off memory requirements, speed and accuracy vs. choice of the solution algorithm
- Ensuring that the identified nodes reflect all the concerns of the stakeholders
- Simplifying the network structure (e.g. removing irrelevant factors)
- Eliminating loops (loops are prohibited)
- Implementing expert judgment data (e.g. avoiding bias and unwarranted precision)

Analysis of a well-designed Bayesian network can be accomplished very rapidly. The run-time for the SOUL model is less than one second, but achievement of fast run times requires care in the design. Designs that include nodes with very large numbers of states, or nodes that receive influences from a large number of 'upstream' nodes will generally run more slowly than designs where the numbers of states and incoming links are restricted. Most of the nodes in the SOUL model have two or three states and none have more than four. None of the nodes have more than six parent nodes. The quick run times suggest that Bayesian networks methods can be used for real-time control of hazard management systems. The CODIO (Collaborative Decision Making in Integrated Operations) project at NTNU (Norwegian University of Science and Technology) has investigated the possibility of real-time control of drilling systems responding to changes in well data [5].

5.2 Consequences and frequencies

Bayesian networks analyze only quantified probabilities. But risk assessments involve quantification of both probability and consequence. In practice Bayesian net development for risk assessment will often require supporting consequence studies to assist the design of the node states. The range of consequences will typically be represented in the Bayesian network by the node states, with each state representing a different consequence category.

In stating that Bayesian networks deal with probabilities, it should be specifically noted that they do not deal with frequencies. If the BN is being used to model a specific scenario then this presents no difficulties, because the probabilities can be reinterpreted as frequencies in simple post processing. This post processing is straightforward in our experience.

5.3 Versatility

Bayesian networks have been applied to a wide range of probability problems as part of risk assessments and optimizing decisions. As examples, a brief survey of literature identified applications in areas such as: social policy, investment and strategic planning, assessment of emerging risks, structural reliability, pipeline condition monitoring, and petroleum exploration.

5.4 Breadth and depth

Bayesian networks may be a natural choice for an assessment tool where breadth of assessment is more important than depth. They can address the combined effects of a very wide range of influencing factors, such as technical, human and organizational, in a way that may not be achieved by any alternative tools. This breadth may come at the price of depth; other tools may be preferred if a detailed analysis of a narrow aspect is required. Fault trees and event trees may be relevant for studies of detailed, narrow aspects. A Bayesian approach is expected to more-easily achieve a good balance in addressing the wide range of factors that affect risk, but at the cost of losing fine detail that may be irrelevant to the bigger picture. The Bayesian approach naturally allows for the inclusion of factors that have high inherent uncertainty, such as human error and competence, organizational behavior and factors from the management system.

5.5 Tailor-made solutions

Bayesian networks are generally tailor-made to address specific situations. Bayesian network software is not domain-specific, meaning that most commercial tools are acquired as frameworks that allow the user to build a model that is intended to address a problem in the user's domain of interest. A suite of BN applications is gradually being built that address different types of problems, but domain-specific tools are not available "off-the-shelf". Even in situations where a degree of reuse is possible, it will be necessary to modify the model to address the specific problem situation.

5.6 Prediction and diagnosis

Bayesian networks perform inference in all directions, meaning that the probability distributions in both 'upstream' and 'downstream' nodes will be updated when any node is modified by the user. This means that the probabilities of node states representing both the causes and the consequences of the modified node are updated; risks can, therefore, be evaluated and important causal factors can also be identified. The probability data can be interpreted for the user and presented as risk (by simple auxiliary calculations) in a management dashboard. This allows users to make risk assessments and decide whether corrective actions are necessary. Knowledge of the prioritized causal factors allows identification of those corrective actions that will be most effective in reducing risk. Knowledge of the causal factors also provides an explanation of the changes in assessed risk when new information is obtained by the network; these explanations give the user a level of confidence in the reasoning being performed by the network.

5.7 Cost-benefit assessments

Bayesian networks are the focus of this work. Extensions to Bayesian networks are provided in all of the most well-known commercial Bayesian systems, extending the basic functionality. When the extensions are applied, the tool is usually referred to as an "Influence Diagram". An influence diagram is a Bayesian network with two additional node types: one modeling decisions and one modeling costs. These allow the user to specify decision options (e.g. risk-reduction options), and also define the costs of those options and the monetized cost of the consequences. The influence diagram will then calculate the cost implications of each risk reduction option.

5.8 Handling of incomplete information

Like all tools, Bayesian networks are useful in some situations and not in others. They excel in situations with high levels of inherent uncertainty, such as emerging risk topics where the relevant factors are not yet well defined and the way in which these factors affect each other are only partially understood. Subject areas where the strengths of influencing factors are only partially understood will also benefit. This means that risk assessments where human factors or organizational factors are important can also benefit. This capability extends to handling of incomplete information such that the state probability data can be absent in some nodes, but the analysis will still run. A Bayesian network cannot, of course, deal with "unknown unknowns"; if a relevant factor is omitted from the model then it cannot be expected to affect the solution.

5.9 Learning

Bayes' rule explains how probabilities should be updated on receipt of new information. Networks can therefore start with probability data derived from guesses and gradually improve the probability distribution data on the basis of the new information. It is important to note here, that it will always be better to start with a good initial guess; Bayesian networks do not give something for nothing. Learning also requires that new relevant data is actually available; learning cannot occur in scenarios that are so rare that they are essentially "one-off" events. The SOUL model does not currently include any learning capability.

6. Conclusions

The demonstration tool omits some sources of risk that would be included in a fully functional assessment tool, and the tool also requires calibration. The work presented here is, however, sufficient to show that a Bayesian network can be used as the basis of a tool to assess variations in major accident risk on an offshore drilling rig. The tool integrates normally disparate data into the assessment; data is drawn from sources that include hardware condition and failure characteristics, aspects of human factors (such as error and competence) and the effectiveness of organizational processes as measured in management system audits and formal verification.

The demonstration tool is applicable to the offshore drilling industry, but the work has shown that this approach to risk assessment can also be used for risk assessment and diagnosis across other high hazard industries.

The work has also provided an opportunity to understand the special characteristics of Bayesian networks that make them suitable for risk assessment. Bayesian networks offer a unique set of features that are not readily available in other assessment methods. In particular, they offer: a combination of risk prediction and diagnostic function, speed of reassessment, and ability to perform assessment in the presence of uncertainty and incomplete information.

7. References

[1] GeNIe and SMILE, http://genie.sis.pitt.edu/. (accessed 9 December 2013)

[2] Bayes, T., "An Essay towards solving a Problem in the Doctrine of Chances. By the late Rev. Mr. Bayes, communicated by Mr. Price, in a letter to John Canton, M. A. and F. R. S.," Phil. Trans. Royal Society London 53: 269–71, 1763.

[3] Blowout and well release frequencies based on SINTEF offshore blowout database 2010 (revised), Scandpower AS, 28 March 2011.

[4] Williams, J.C. (1992). Toward an improved evaluation tool for users of HEART. Proceedings of the International Conference on Hazard Identification, Risk Analysis, Human Factors and Human Reliability in Process Safety, Orlando, February, Chemical Centre for Process Studies.

[5] CODIO (Collaborative Decision Making in Integrated Operations, http://www.computas.com/en/Services/Integrated-Operations1/Codio (accessed 9 December 2013)

10TH GLOBAL CONGRESS ON PROCESS SAFETY

The Role of Sequential Automation in Improving Process Safety

David A. Huffman
ABB INC.
29801 Euclid Avenue, Wickliffe, OH 44092 USA
David.Huffman@us.abb.com

Prepared for Presentation at
American Institute of Chemical Engineers
2014 Spring Meeting
10th Global Congress on Process Safety
New Orleans, LA
March 30 – April 2, 2014

UNPUBLISHED

The Role of Sequential Automation in Improving Process Safety

David A. Huffman
ABB INC.
29801 Euclid Avenue, Wickliffe, OH 44092 USA
David.Huffman@us.abb.com

Keywords: Process safety, State Based Control, ISA106

Abstract

All processes operate based on procedures, usually written. Batch based processes more and more run on automated procedures. But even continuous processes constantly use procedures. Among these are equipment and unit startup and shutdown procedures, product or grade change procedures, spare equipment in and out of service procedures, quality control procedures, alarm response procedures, and many more. Frequently there can be significant benefits to automating a high percentage of these procedures.

Safety is one of those benefits and can be derived from opportunities ranging from consistent and mistake proof operations to simply removing the need for an operator to be exposed to potential process or physical hazards. Additionally, significant impacts can be realized on alarm management based on procedural control..

1. Introduction

Generally, process safety automation has been considered to be the responsibility of a specialized safety system and that remains true for Safety Integrity Level (SIL) safety automation requirements. But what if the basic automation of the process could be constructed in such a way that it significantly improved the ability to maintain the process within acceptable operating limits and not reach the conditions that would trigger a response from the safety system?

All processes operate based on procedures, usually written. Batch based processes more and more run on automated procedures. But even continuous processes constantly use procedures. Among these are equipment and unit startup and shutdown procedures, product or grade change procedures, spare equipment in and out of service procedures, quality control procedures, alarm response procedures, and many more. Frequently there can be significant benefits to automating a high percentage of these procedures.

Safety is one of those benefits and can be derived from opportunities ranging from consistent and mistake proof operations to simply removing the need for an operator to be exposed to potential process or physical hazards. Additionally, significant impacts can be realized on alarm management based on procedural control.

2. What is Procedural Control?

Procedural control is historically identified with batch processes, but with the recent efforts of the ISA106 (*Procedure Automation for Continuous Process Operations*) Committee and the release of ISA-TR106.00.01-2013 Technical Report *Procedure Automation for Continuous Process Operations – Models and Terminology*, there is a growing interest in the application of procedural control to continuous processes. The realization is that continuous processes are not truly continuous. In the simplest of cases, a continuous process has at least four (4) process states:

1. Shutdown or idle
2. Starting up
3. Normal running
4. Shutting down

When most continuous processes are analyzed, it is usually found that the Shutdown state is frequently a set of different shutdown states ranging from a complete shutdown, usually for a significant unit turnaround for maintenance, to any number of clearly definable states that brings the unit to a hot, idle condition prior to an actual unit start up. It is also likely that the normal running state is a set of definable operating states that can be classified either by products being manufactured, product performance measurements, unit throughput conditions or other identifiable conditions. Switching between these states readily lends itself to the application of procedural control techniques to execute most if not all of the steps normally performed by operators. For the remainder of this paper, procedural control will be referred to as State Based Control or SBC since the procedural automation actually controls the actions for maintaining and switching between process states.

3. Discussion

State Based Control is not new to process automation. But as valuable a tool as it is, wide spread application in the continuous and hybrid process industries has been very limited except within a few progressive companies. SBC offers a wide range of opportunities for financial benefits, but also creates a number of opportunities to improve process and personnel safety.
There are four key areas where SBC can contribute to safety:

- Consistent adherence to best practices in operation; proper procedures every time removing chances for human error.
- Systematically park the process into safe operating states that keep the process running during abnormal situations that do not demand a unit shutdown; avoids what are typically the most dangerous times for operators.
- Process state management of alarms; activate and deactivate process alarms according to equipment and operating states to prevent unnecessary alarms, standing alarms, alarm overload.

- Manage operator access to control devices during conditions when they should not be able to operate a device.

3.1 Consistent adherence to best practices

Consistent adherence to best practices avoids people making mistakes. Regardless of what condition the process may be in, following proper procedures to the letter is important to maintaining a safe operation. Whether in startup, shutdown, or any normal running state, following proper procedures keeps people safe, maintains equipment integrity, and avoids environmental impact. Frequently there are well documented procedures provided for every aspect of a plant's operation and these procedures are maintained in the plant's operating manuals. But how often do we see plant operators walking around with the procedure book at their side?

During training, operators use the procedures, and the sage guidance of more experienced operators, to learn how to work with the process. Generally to qualify for a position, the individual is tested on their knowledge by demonstrating their ability to work according to procedures and by written or oral testing. But human nature of daily application creates a self confidence that leads to taking shortcuts or sometime performing activities out of order for convenience. Everyone learns and remembers differently. So over time, operators working the same position on different shifts or teams are operating the plant differently and almost always in a manner inconsistent with the written procedures in some way.

Using SBC to embed the proper and best practices for operation into the control logic of the process avoids the inconsistencies and removes the shortcuts that often develop as part of each individual's routine. Even when manual actions are required, the exact instructions and their proper placement within the activities can be generated from the automation system in most cases. Completion of the manual actions can be recorded and the procedure continues along its proper path.

Very few processes remain static throughout their entire lifecycle. When process changes or improvements occur, the procedural control logic can be modified accordingly and all operators are able to run the process immediately according to new best practices. Another thing that fails to remain static is the condition of control equipment. One part of the lifecycle of using procedural control is the need to recognize that not all requested control actions will be carried out. Instances will occur when operators will need to override a stuck valve or bad position feedback information and advance the procedure themselves. The good part about these bad situations is that the actions will be recorded and permanently logged into the operator actions so there is traceability to the actions and overrides that were executed.

3.2 Keeping within safe operating parameters

Keeping the process running within safe operating parameters as an alternative to a process shutdown during certain abnormal process conditions can be a significant contributor to process safety. Operator exposure to dangerous conditions, either physical or environmental, can increase exponentially during process shutdown and startup operations. Additionally, avoiding equipment stress conditions, such as thermal cycles, can increase the useful life of the equipment and reduce the chances of physical failure that can lead to unpredictable failures and the associated risk to people and the environment.

Physical risks may come in the form of exertion working with equipment, risk of falling from heights to access equipment, electrical hazard exposure, or other similar sources. Environmental risks include hazardous atmosphere exposure, chemical exposure or other similar conditions. Additionally, shutdowns may lead to requirements for outside contractors to be brought into the facility. Having outsiders in the facility that are not intimately familiar with physical and process hazards can lead to significant risk opportunities to those individuals.

The alternative is to safe-park the process into a running state that, although not necessarily optimal for production and quality, keeps the equipment in service and allows a quick and controlled return to normal conditions to avoid the risks of process shutdown and restart. For example, a distillation column may be moved to total reflux. Or a unit may be moved to a significantly reduced throughput condition.

Often, operators are trained or encouraged to choose a process shutdown in critical, abnormal situations. The alternative is that the operators try to "save" the process, and our manufacturing history has documented far too many incidents where those attempts have failed, resulting in disastrous consequences. Selecting a process shutdown is often the easy way out of trouble for the operator. Or a Safety Instrumented System (SIS) process safety shutdown results because operators fail to react quickly enough to prevent it from happening due to an inability to immediately recall the proper sequence of actions. In other cases the proper actions cannot be executed fast enough by the operator to prevent the shutdown. In this latter condition, perhaps a safe park option has never been seriously considered since the operator cannot manage the actions fast enough to prevent the shutdown.

In those instances where safe-park conditions can be identified and automated actions can move the processes there safely, the basic plant automation system (BPCS) can be programmed to accurately and quickly move a process into that state, generally much faster than the operator could accomplish the same sequence of steps. If the condition requiring correction does not occur frequently, an operator would typically need to find the procedure even before starting the proper action sequence. But even if the operator does know the complete set of actions required, having the BPCS execute the procedure removes the chance for human error in what are almost always stressful situations that lead to increased chances for mistakes. When the dangerous condition clears, either by corrections made by operations or maintenance or by making other process changes that allow for returning to a more normal processing state, the BPCS will make appropriate changes to return the process to normal.

We can draw an analogy with the latest safety features now available on some automobiles for cruise control where the system detects that the vehicle is approaching another vehicle ahead in the same lane and slows the car to keep a safe distance. The automobile is still in cruise control (process is in automatic), but avoids the increasing potential for an accident by keeping a safe driving distance. If the vehicle ahead increases its speed, or if the driver of the in control automobile changes lanes to where there is clear space for the higher speed, the car automatically increases speed back to the cruise control setpoint.

3.3 Process state management of alarms

Process state management of alarms is rapidly becoming a key aspect of a good alarm management philosophy. One of the key items tied to significantly improving operator performance during both normal and abnormal situations is to be sure that the operator is receiving alarms appropriate to the situation. Having too many alarms occurring, having inappropriate alarms occurring, or having too many standing alarms in the alarm list can all contribute to overloading the operator's ability to identify critical alarms they need to react to for accurate and safe operation of the process. Managing alarms according to the state of the process falls into the category of *alarms suppressed by design* according to ANSI/ISA-18.2-2009.

State Based Control is particularly well suited for managing alarms based on the states of the process and equipment. This can be demonstrated with a simple equipment grouping of a primary and spare pump (standby pump station). In most facilities today, we find that when one pump is shutdown, there tends to be standing alarms related to the out of service pump for motor amps, discharge pressure, low flow, and/or valve positions. When the pumps are switched even though the proper action is occurring for a pump shutdown, the alarms still occur for the pump going out of service and the operator needs to take action to acknowledge them. The alarms were configured to detect the failure of a pump and are valid under that condition, but the alarms are completely inappropriate when the pump is not running by design. The sequential logic of a SBC design provides the structure to place alarms in and out of service according to the desired process or equipment state. As the sequences move in and out of steps, the proper alarm conditions can be established at the same time.

Now consider applying this concept to a full process unit and what are typically hundreds or thousands of alarms. SBC provides for having only the proper alarms configured as being in service for the current state of the process. The overwhelming numbers of standing alarms are eliminated because alarms for out of service equipment are eliminated. SBC can also appropriately change alarm setpoints for the state of the process. Based on product grade, throughput conditions, or other factors, it is generally appropriate that alarm setpoints need to be adjusted or placed in or out of service accordingly.

Shutdown conditions can frequently be the largest offenders and are usually one of the most hazardous times for operators and maintenance personnel. With so much equipment out of service, there are usually an excessively large number of standing alarms. But even during unit shutdowns, some equipment remains in an in-service state. Tanks or storage vessels may still have liquid levels that need monitoring as inventories are moved. Equipment may be under pressure or vacuum. When equipment cleaning cycles are active, they usually require some degree of monitoring for levels, temperature, pressure, and circulating flows that typically have alarm values that are either not used during normal processing states or have significantly different setpoints from their normal operating states. Again, SBC is ideally suited to manage these conditions as individual pieces of equipment or portions of the unit are moved between any number of targeted state conditions as the shutdown progresses.

3.4 *Managing operator access to control devices*

Managing operator access to control devices can also be an important area for personnel and process safety. There can be any number of instances where it is appropriate that operators be prevented from operating motors or valves to prevent a manual action from creating a dangerous situation.

At times when rotating equipment is being prepared for maintenance, and prior to actual electrical lockout, the motor control within the BPCS can be commanded to off and access to the manual changes prevented. Using a state control configuration where a switch to a maintenance setting creates this BPCS lockout can prevent someone from inadvertently starting the motor and causing harm to other operators or maintenance personnel.

A similar control lockout may be appropriate when several different pumps from various material supply tanks are connected to a common material supply header. In these cases, when one pump is in service, it may be appropriate to lockout manual actions to open block valves and start pumps on the other tanks. This can easily be done with SBC for the header, using each individual material supply selection as separate states for the header. Such a design prevents contamination of materials and also can prevent damage to equipment by starting a dead-headed pump.

There can also be instances where valves or pumps should be locked into automatic or running states and the operator prevented from making manual changes. For instance, it may be appropriate to lockout operator ability to switch the controls to manual for cooling systems or heat recovery systems on exothermic reactors while the reaction is active. Another example may be to prevent an agitator motor from being shut off with a reaction running the vessel or to prevent settling of material, either of which can lead directly to a dangerous situation or the need to shutdown the equipment resulting in potential exposure hazards as discuss in previous items.

4. Summary

State Based Control can play a key role in adding significant safety capability into the normal process operations of most continuous process facilities. Although Safety Instrumented Systems will still be required to manage critical safety conditions, using SBC to manage strict adherence to critical procedures, to safe-park a process rather than resorting to shutdown, to manage alarms according to all states of the process, and/or to manage operator access to control devices can increase plant safety compared to traditional control methodologies.

The concepts for SBC are being addressed and promoted by the International Society for Automation (ISA) with the ISA106 committee. The committee has been operating for about three years and has achieved the release of its first technical report ISA-TR106.00.01-2013 Technical Report *Procedure Automation for Continuous Process Operations – Models and Terminology*. Work is continuing on further technical reports for Work Processes and Industry

Examples to help those companies interested in making SBC part of their culture to be able to quick start on a path to increased safety and financial benefits.

5. References

[1] ISA-TR106.00.01-2013 Technical Report Procedure Automation for Continuous Process Operations – Models and Terminology.

Enabling Better Day to Day Process Safety Risk Decision Making by Linking Asset Integrity Risk and Work Management Risk.

Mike Neill
Petrotechnics
10479 Town & Country Way
Houston, Texas
Mike.neill@petrotechnics.com

Prepared for Presentation at
American Institute of Chemical Engineers
2014 Spring Meeting
10th Global Congress on Process Safety
New Orleans, LA
March 30 – April 2, 2014

UNPUBLISHED

Enabling Better Day to Day Process Safety Risk Decision Making by Linking Asset Integrity Risk and Work Management Risk.

Mike Neill
Petrotechnics
10497 Town & County Way
Houston, TX
Mike.Neill@Petrotechnics

Keywords: Operational Risk Management, Production Efficiency, Asset Integrity, Decision Making, Independent Protection Layers (IPL)

Abstract

Lowering and more effectively managing operational risk is a big challenge in process industries. Often organizations struggle with the tension of maintaining production whilst being able to properly maintain and repair their equipment. Daily judgment calls are made on production and risk which can result in an unnecessary sacrifice of one or the other (e.g. increased safety risks in pursuit of higher production attainment or lower productivity levels for a safer environment). Lower productivity over time compromises our ability to properly inspect and maintain our safety critical equipment resulting in increased risk factors. The relationships are complex yet our workforce have few tools to help make these calls, instead relying on their experience and judgment.

Risks are often managed in separate "silos" and while many organizations have expended significant effort to improve understanding of overall risks of plants, a number of key challenges remain.

- The ability to understand that Asset Integrity Risk and Work Management Risk impact each other
- The ability to consistently and automatically generate reliable operational risk data
- Tools to move an organization from an understanding of risk to proactively managing risk
- An understanding of the impact of the "work risk" upon impairments to our independent protection layers
- The ability to consistently manage the compounded risk from the combination of Asset Integrity Risk and Task Hazard Risk
- Bringing order to the complexity of managing operational risk to improve frontline operational decision making

This paper will discuss how technology can be used to dynamically capture, visualize and manage operational risk in real time on our plants in order to improve decision making on daily work activities. By understanding the status of safety critical equipment/ Independent Protection Layers (IPLs) impairments on our plants and the impact of the current or planned workload on those impairments, we can link asset integrity risk with work risk. Consistent data generation provides us with the risk profiles of the compounded risk of doing work against a backdrop of dynamic plant conditions in specific areas of the plant thus improving frontline decision making.

With a holistic understanding of total risk (integrity risk, work risk and other forms of risk such as environmental, drilling etc.), frontline management teams can better prioritize and optimize work and mitigate operational risk.

1. Introduction

High profile events such as Macondo and other significant events have continued to demonstrate the enormous cost of mismanaging operational risk and how the consequences of getting it wrong are significant. It has also served as a key reminder of the importance of having a major accident hazards program that encompasses robust asset integrity strategy and independent protection layers (IPLs)

With the myriad challenges that the industry faces on a regular basis trying to align daily decisions with strategic intent and managing the complexities of the operational environment, it is no wonder a number of unwanted incidents or consequences can occur. They can range from an increased number of incidents - a result of shortcuts in pursuit of achieving production targets, increased Lost Time Injuries (LTIs), increased maintenance backlog, a balance struggle between maintenance and operations on independent protection layers (IPLs) and its impact on production.

The crux of the problem is ultimately lowering and more effectively managing operational risk and the safety of the plant through IPLs, while maintaining or improving production efficiency.

While the industry continues to invest significant time, effort and resources into improving its capacity to manage operational risk, tangible progress for many organizations remains illusory. Often these efforts are embodied in improved policies, procedures, better rulebooks and further training. While they are positive steps, they do little to help frontline management make better day-in-day-out judgment calls regarding the seemingly diametrically opposed objectives of reducing risk and achieving production targets.

Leading organizations recognize the need for a paradigm shift and are using new software technology to better optimize how they manage workload against risk in order to both reduce operational risk and improve production efficiency. By better linking often disparate functional business processes and systematizing operational decision support tools and data collection, frontline operations management has a more holistic understanding of the operational environment and the risk profile. Improved visibility and understanding drives better decision-making, allowing these organizations to both reduce operational risk and optimize production efficiency.

Frontline workers are empowered and more confident to proactively manage risks by their ability to view risks in time, space and geographically. Senior managers can even begin to understand the drivers of risk, capturing key trends from the intelligent data that will enable better decision making, continuous improvements and more sustainable performance.

Technology can empower and enable the workforce and management to proactively identify, track, monitor and manage the health of their IPLs and the impact of workload in real-time to ensure and improve overall plant health, integrity, safety and efficiency.

2. The Operational Environment and Single Biggest Area of Operational Risk

Those ultimately responsible for operational risk and production efficiency and making those daily judgement calls are the frontline, so it's useful to understand the frontline operational context.

At its most basic level, the frontline operational environment can be characterised by three things: the plant, the people and the interaction between the people and the plant. The interaction between the people and the plant represents millions of man hours per year for the typical organization and thousands of people every day routinely intervening in plant operations to do repair, maintenance, inspection and construction work. Understanding the effect this interaction can have on plan accuracy, maintenance and IPLs is critical in understanding why production efficiency or safety can be jeopardized.

Not only does this type of work expose people to the surrounding process hazards but the work they do and the equipment being used can introduce new hazards. These new task hazards can affect those doing the work, other workers and the plant itself. Process Safety Management (PSM) approaches recommend creating multiple IPLs to prevent unplanned releases but when we examine typical Work Execution (WE) operations, we see that often these IPLs intended to keep the plant safe are deliberately removed or inhibited and new hazards are introduced to the workplace in the course of carrying out work. For example, to access, replace or repair equipment it is often necessary to:

- Break containment of hydrocarbon bearing systems,
- Inhibit or take out of service, safety related devices such as alarms and shutdown systems in order to perform work without shutting down the plant.
- Access equipment which often involves working at height or having to perform heavy lifts over live plant

Such scenarios are daily occurrences on most plants.

Figure 1. The Interaction of People with the Plant = Operating Risk

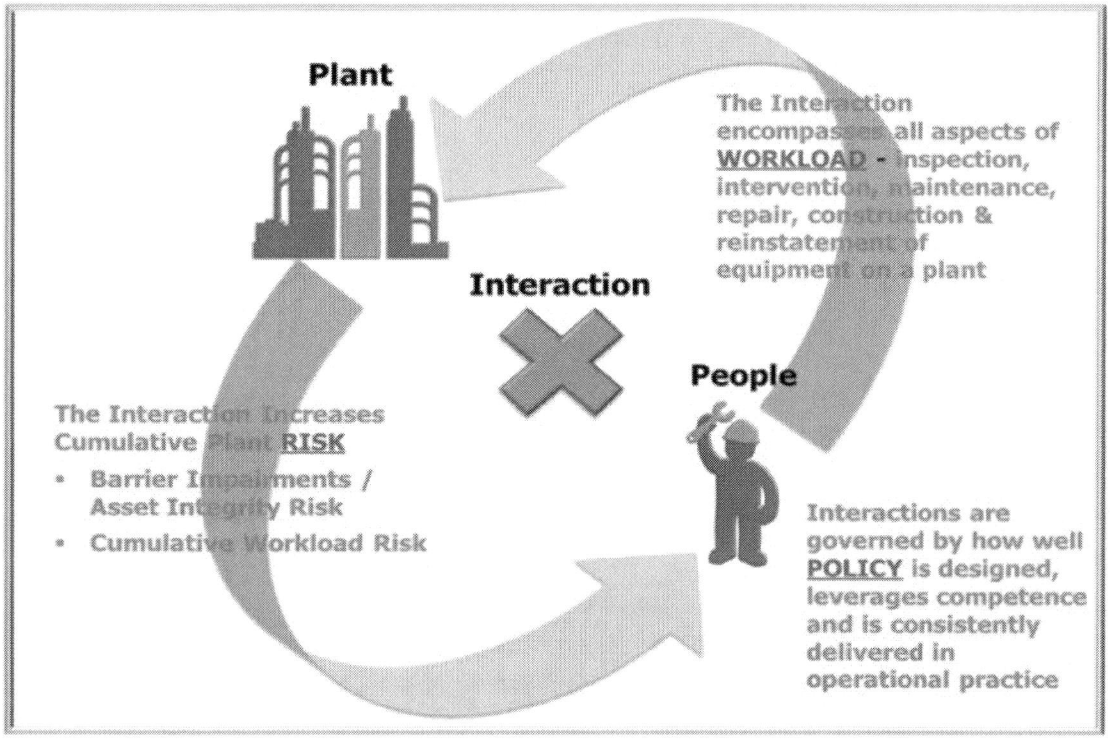

However, the status of the IPLs (equipment, people and processes) put as safeguards and controls to prevent the realization of a major accident hazard is typically managed in isolation from the cumulative workload risk. The lack of a link between asset integrity risk and cumulative workload risk especially within a given area of the plant can be critical. Nancy Leveson, professor of Aeronautics and Astronautics at MIT in her testimony before the United States Senate Committee on Energy and Natural Resources on May 17, 2011 said in reference to confusion between occupational and system safety:

> *Most industries separate these very different problems. Occupational safety focuses on controlling injuries to employees at work by changing individual behaviour. System safety puts an emphasis on designing the system, including the engineered and operational components, to prevent hazardous system states and thus losses. Confusion between these two very different problems and solutions can lead to overemphasis on only one type of safety, usually occupational or personal safety, while thinking that the other types of accidents or losses will also be prevented—which they will not.*[2]

IPL impairments do not exist in isolation of the on-going intervention work that is being performed around them and, in fact, are sometimes caused by the intervention work itself. Without an understanding of how the cumulative workload in a given area of the plant is directly impacting the IPL status, is what we currently understand about the health of our IPLs correct?

And from the workload side, do we understand in simple terms what the impact of the IPL impairments should have on the scheduled work within a specific area of the plant? Are we simply managing a part of the operational risk equation? Or are we managing two critical parts separately?

Figure 2. Asset Integrity Risk, Environmental Risk and Cumulative Workload Risk Impact Each Other

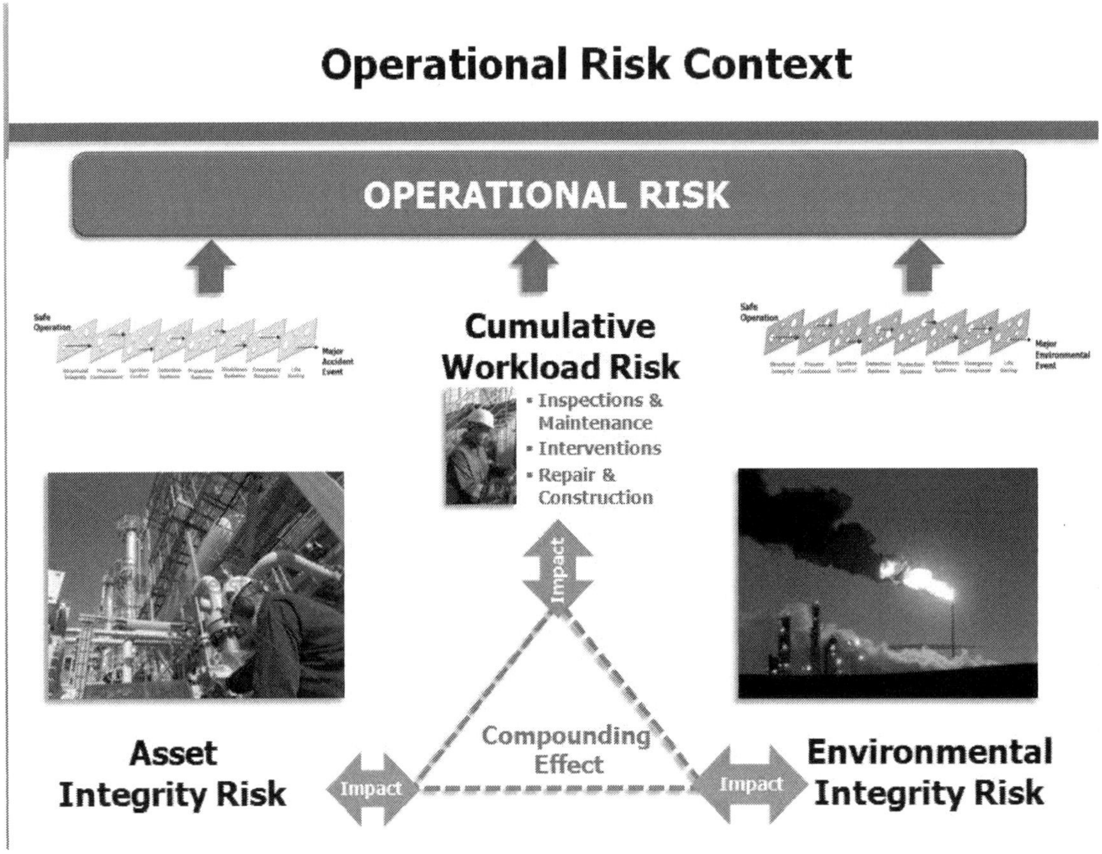

With each day typically presenting different hazards and risks and with inadequate tools to support the decision making process, the quantity and frequency of incidents is not surprising. In the first steps to achieving a holistic understanding of risk and taking appropriate measures to mitigate risks and optimize work schedules, it is also useful to understand how this work is initially planned and the gaps that exist within the planning environment.

3. The Importance of Planning with Frontline Operations and Risk Profiles in Mind

Work on a plant is typically planned across 90-day, 28-, 14- and 7-day time horizons involving a number of different functions and individuals. The dynamic nature of the operational space impacts the "stability" of the original plan. The plans, generated in absence of, or with limited insight of operations, will be issued in a final state as the 14 day plan to be executed on-site. However the dynamic variables such as the state of the plant, IPL impairments, isolation requirements, physical location data, and resource constraints are often missing at the point of plan creation. These plans are then deconstructed by frontline operations to include all of their activities, balanced with overall workload, state of plant and to account for a safe work management perspective (breakdown of all of the hazards, risks and control measures). It is this deconstruction, and the changes to the plan, which affect the planning quality metrics of plan accuracy and plan attainment. The plan always becomes broken 1 week before the execution phase simply because the operations teams have a better understanding and expertise regarding breakdown of the jobs, requirements for preparation, isolation, executing, de-isolation, reinstatements and all of the associated risks. Quite often, when work is initially planned, they seldom account for these factors and, therefore, there is a large disconnect between planners, maintenance and operations.

To address these gaps, the introduction of dynamic operational considerations will lead to the creation of a Safe Work Plan. This new version of the plan will be far more executable, with the known constraints on the plant in the yesterday, today and tomorrow time frame. Ultimately, this Safe Work Plan is what gets done. This Safe Work Plan reflects the operations leadership decisions to manage work around the competing interests of workload, risk, time and resources in the limited space of the plant. New technology allows for a dynamic, integrated system capturing real-time data such as the capacity to do the job, the risks directly associated with completing one job, and also the risks occurring within the greater production environment.

Vizualizing risk views and intelligent profiles on what is driving impairments or risks in IPLs enables the maintenance and asset integrity teams to prioritize safety critical work and most pressing maintenance items or rehabilitation needs within the plans. This level of advanced planning and intricacy of detail provides the platform for holistically managing the risk associated with IPLs alongside workload risk, ensuring optimum production efficiencies and a safer work environment.

4. Improving the Way We Dynamically Manage Operational Risk is Critical

It's no use having the best plans in the world when there are few tools that can help you effectively manage the work execution and ensure it is aligned with the plan, and in a safe work management mode.

Effective operational risk management is an important and highly specialized skill that demands careful balance between risk appetites, risk assessment, risk response decisions and risk awareness whenever operational work is executed. All too often it is a daunting task to provide the safest integrated response to the cumulative risks arising across multiple tasks in multiple locations as well as addressing conflicting production demands and dynamic plant conditions where IPLs may be impaired. It is inevitable that the necessary skills for effective risk management are constrained by entrenched local pressures and cultures, as well as the misleading bias that arises from an absence of incidents.

With the additional pressure of getting more work done, we often leave it up to individual competency to interpret how operational risk is identified and managed, leading to poor quality risk assessments being approved. Furthermore, frontline supervisors and workers lack robust tools to guide them to make the appropriate decisions. In other words, decisions are often being made without the full information, decisions which could lead to catastrophic failure. As is always the case when there is no systematized process with data, the variance in how workload is managed against risk doesn't just vary from facility to facility but from frontline supervisor to frontline supervisor.

As we systematize our frontline work processes in a consistent manner across our organization, we automatically generate data from scheduled and executed frontline operational work and processes. This provides us with visibility on how work is to be carried out, how it is actually carried out, how the current status of IPL impairments may impact it and the degree to which the frontline is executing work processes in line with the organization's policies, procedures and improvement initiatives.

Operational Performance and Predictive risk platforms provide an organization with Operational Decision Support to enable the management of the cumulative effect of multiple jobs happening in the same proximity as impaired IPLs along with other factors and other issues. They provide the following capabilities that can be used to lower operational risk:

- Automated capture of critical IPLs and workload related operational risk data
- Real-time Visualization of IPL Impairments and Simultaneous Work Activity across your plant(s) - in Time, in Space (Area, Plant, etc.)
- Common view of Risk to all Disciplines in time and space
- Operational Decision Support Tools to proactively manage human performance and activity workload against risk
    - Prioritise work to minimise risk
    - Better utilize resources to complete work efficiently

It is not unusual for 20 or more independent jobs to be in progress simultaneously within a defined area. While these individual tasks may have been independently risk assessed and authorized, there are dependencies and interactions between the jobs that need to be taken into consideration in order to reduce the potential overall risk of the combined jobs. The cumulative risk effect of multiple jobs coupled with any human factor or environmental hazards and how they impact or are impacted by the IPL impairments in a specific location of the plant, are seldom if ever assessed on a day to day basis. Work schedules are assessed for clashes where two or more jobs may interfere with each another but we do not equip our plant supervisors with the right tools and techniques to properly assess these types of risks nor consider the cumulative impact.

In the Figure 3 below, the IPL impairments and the concurrent tasks are visualized to provide a view of the risk in time by area and location of the plant. For example, if there are two impaired barriers, a gas detection system and deluge system down, and an individual is about to perform hot work nearby, he may not have knowledge of these impairments and proceeds with the job at high risk. A system like this, providing visual risk profiles, can highlight the impairments and impact on each other enabling better decision-making.

In Figure 4, frontline supervisors can quickly ascertain the status of the IPLs, the cumulative risk of the scheduled work and proactively reschedule work to improve the risk profile in a location in the plant.

While visualization of risk is important, the ability to understand what is driving the risk is critical to improving Operational Decision Support and ensuring compliance with corporate standards. In Figure 5, we can see the status of the IPLs within a specific location of the plant and what jobs or tasks are impacting the risk profile of the IPLs and what specific plant conditions are impacting it as well. With this knowledge, work can either be rescheduled, more effective controls put into place or at a minimum frontline management can more effectively communicate the hazards and the risks associated with executing the work.

Figure 3. Visualization of the IPL Impairments and Concurrent Tasks by Location of the Plant.

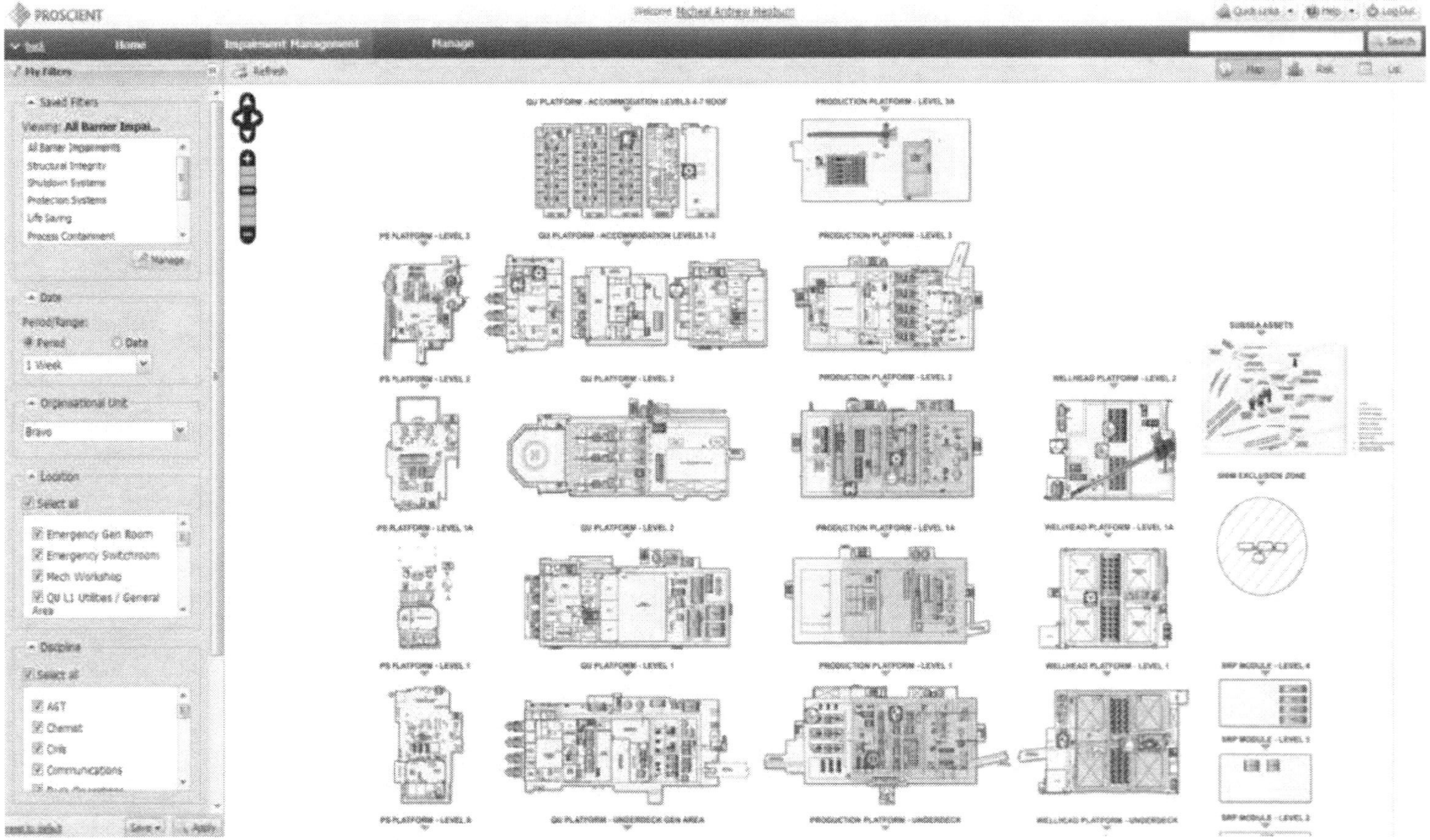

Figure 4. Visualization of the Status and Plant Cumulative Risk Enables Work to be Rescheduled to Improve the Risk Profile of an Asset

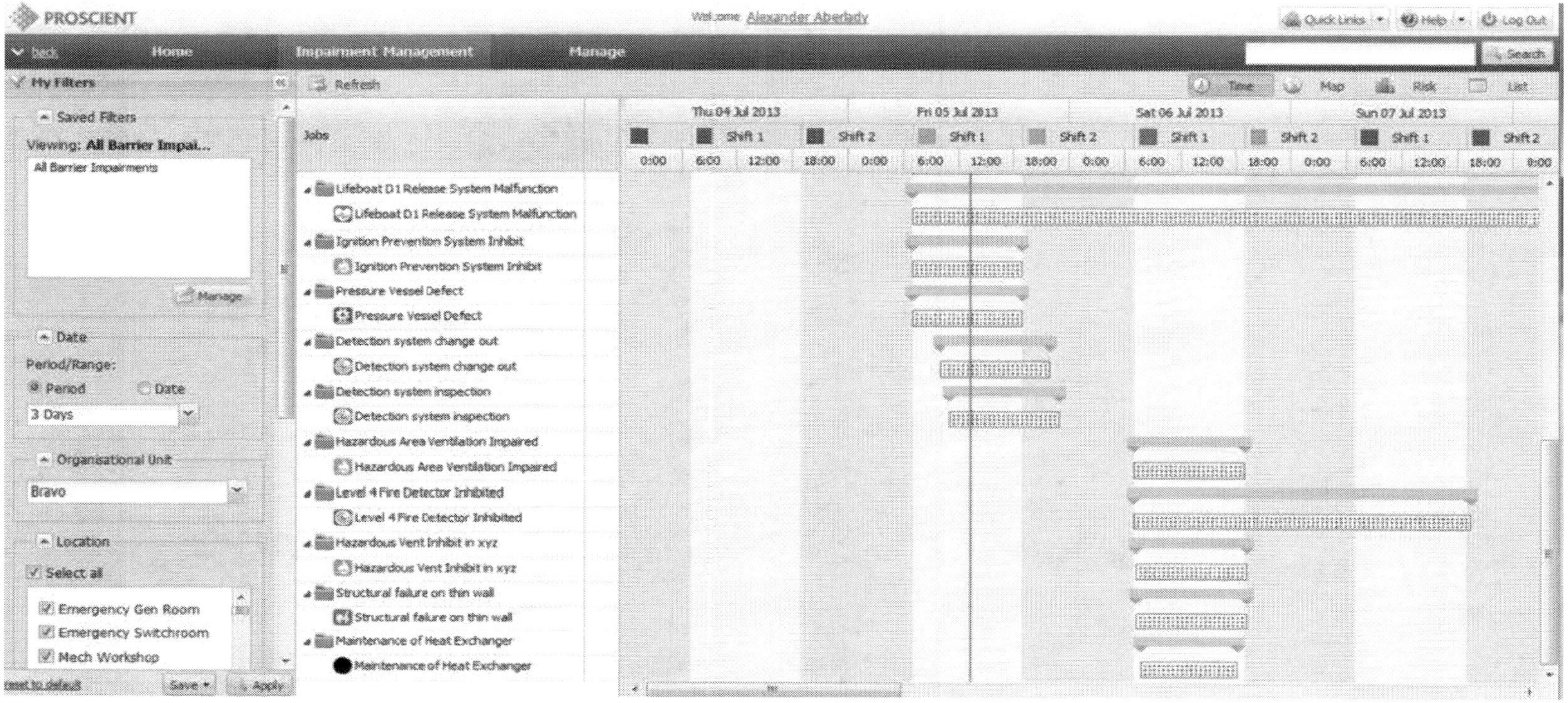

Figure 5. Drill-Down: How do Specific Tasks Impact Barriers and Risk?

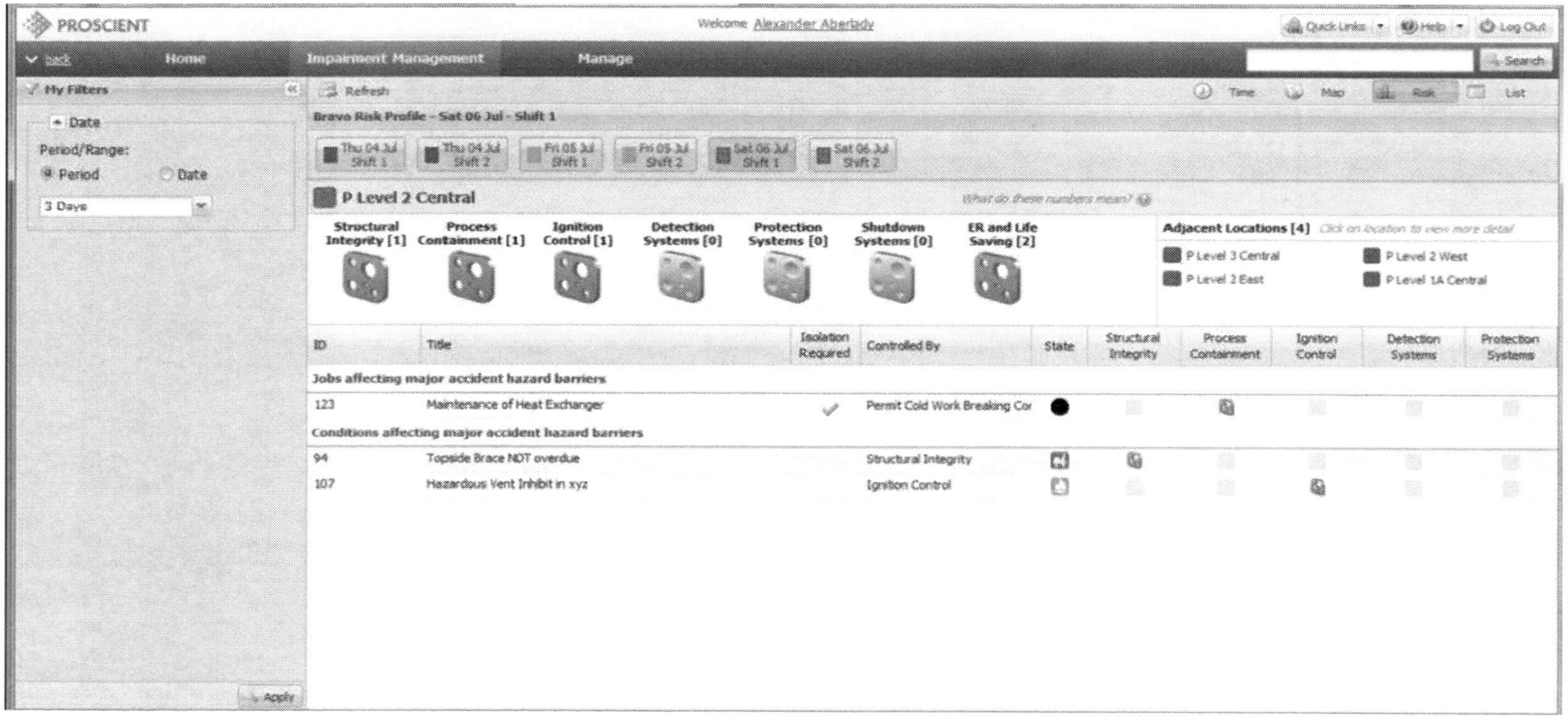

5. Using New Technology to Systemize Work Management Processes Can Bring Significant Gains in Plan Accuracy and Production Efficiency

With planning and decision making tools in place, business processes and execution are more resilient and efficient if it can be underpinned by consistent and standardized ways across the plant(s) or enterprise. In making decisions, people often work accordingly to rules, policies and procedures. However, Hopkins, a world renowned expert in the analysis and causes of major industrial accidents and leading consultant, points out that rule making itself isn't sufficient.

However, new technology can ensure that rules are intelligently embedded within operational workflows. In this way rules are placed within the most pragmatic and effective place to facilitate successful compliance and discipline. It sets the foundation for more effective planning and work execution since it is carried out in a consistent manner.

Following the plan and ensuring that the execution of work is carried out consistently across the plant or organization is critical to the safety and success of operations. Typically organizations have assurance over the PSM approaches and plant design that include policies, standards, practices, systems, data, Key Performance Indicators (KPIs)/metrics and feedback. In this way, they have confidence and feedback that their policies are being followed and they have the means through the use of the right technology and the collection of data, to measure status and progress of their performance.

However, as work execution takes place and the IPLs meant to keep the plant operationally safe are temporarily disabled to carry out the critical maintenance work, there is often ambiguity around the compliance of work being carried out according to policy. Although there are some policies and standards to guide the frontline workers, they are limited often to paper based systems with no data, no metrics and as a result, no feedback loop of any consequence. There is little line of sight to the application of practices intended for work execution, when these IPLs may be weakened temporarily. The rules and guiding principles in these limited systems are provided in a one way and static format that absolutely ensures divergence in the way they are delivered in operational practice even within the same plant. Without the new technology available, organizations do not have data and without data or feedback it is not possible to measure performance, thus severely limiting management's effectiveness in driving improvements in operational performance and operational discipline.

By capitalizing on new technologies such as, Operational Performance and Predictive Risk software platforms, organizations can capture real-time data and workflows and automatically generate consistent and reliable operational and risk related data. Personnel involved in the work management business process are able to ensure all the task risk assessments take into account process hazards relevant to the location and equipment being worked on and are communicated to work crews. Frontline workers can be prompted by the system to be aware of process and hazard safety considerations and control mechanisms relevant to the area of work.

For the first time, this new technology allows operations personnel to document their decisions based on workload, asset integrity risk as well as task risk, schedule and resources. They can also add or import "jobs" created by Computerized Maintenance Management System (CMMS) systems and all of the pre- and post-work operational activities, intelligently building a more complete picture of whatthe original job actually required. As a result of integrating new technologies into the data infrastructure of the organization, data from the live job execution can be captured and fed back into the planning systems immediately, thus improving the overall stability of future plans. Better, more executable plans mean better quality metrics, better resource utilization, increased tool time, and reduction in unplanned jobs.

Through the system's scalable nature, these systems can be deployed across a single plant, multiple assets, and even the enterprise. The standardized and harmonized implementation cultivates a realistic and pragmatic approach to working across the plant and yet retains the flexibility to support localizations to accommodate regional, local and cultural contexts. The frontline workforce can flourish under the guidance of a tool that provides operational decision support, and raises risk management capability and competencies. Over time this can manage out different ways and wide variances of working. It will initially identify and eventually nullify practices that can foster subconscious bias toward inappropriate behaviors leading to unpredictable work execution, compliance failure notices, and inaccurate planning. It can reduce unplanned shutdowns and the number of incidents, while generating and sustaining improved culture, discipline and behavior for operational excellence and enhanced safety.

Figure 6 shows how an enterprise deployment of the platform would allow rules to be set in a cascading hierarchy. Under such a hierarchy, the rules set at a corporate level would be common for the whole hierarchy. Those set at a regional level would contain all the common corporate rules but allow each region to add their own specific rules for flexibility, maximum efficiency and effectiveness. At the plant level, rules can be set which are specific to the plant but include all corporate and regional rules. This hierarchy can even continue down to separate parts of the plant. Ensuring strategic intent is delivered into practice is guaranteed through the hazardous task template and hazard control libraries, where frontline workers are intelligently guided to risk assess in a pragmatic manner, reinforcing what they have been trained in (e.g hazard awareness). This is achieved through visibility of a 'library' of hazards and control mechanisms associated with the task being risk-assessed (e.g confined space entry).

Once a template is created at the corporate level it is locked down and cannot be changed - so for example, if the template is for working in a confined space, one corporate control could be that positive isolations must be in place for any man-entry to a confined space. This rule would be enforced for all confined space work in every region and at every plant. At the regional level, these corporate templates can be modified to create regional templates but only additions can be made. None of the existing corporate hazards and controls can be eliminated. An example might be for a desert region where radiant heat is a problem most of the year. The corporate template for confined space work could be modified for the region adding the need for temperature monitoring and cooling. At the plant level there may be a local Fire Service which provides emergency rescue so the confined space entry template could be modified at that level for the local fire service to be on standby for rescue during any confined space entry event.

Figure 6. Embedding Policy into Practice

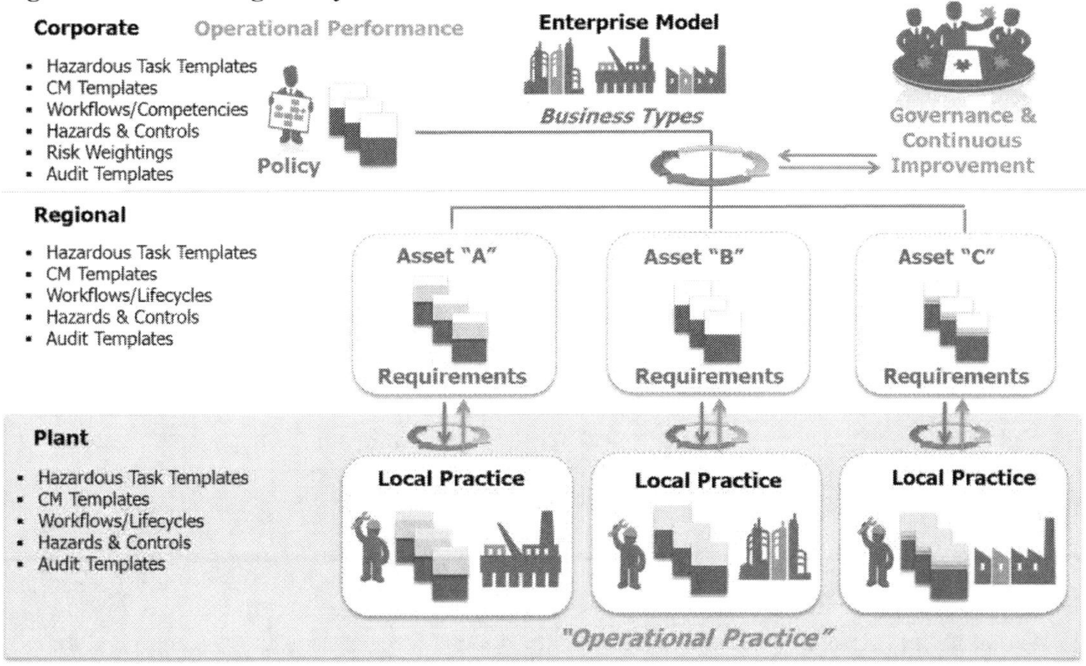

To prevent a 'checklist' mentality, these templates represent the minimum requirements for the task. They will be used to guide the task activities ensuring that the standards at all levels are applied. The expectation is that the person planning a task will also identify additional hazards and specify controls as appropriate for the specifics of the task, specific equipment, local conditions and tools being used. Hence, a comprehensive risk assessment is carried out to ensure maximum safety, but with the flexibility and efficiency to ensure productive operations.

Organizations and senior management can be confident and assured that its strategic intent is translated efficiently and effectively into operational practice. This level of consistency and standardization also provides a streamlined approach, eradicating wide-variances and potential for non-compliances and, therefore, reduction in incidents and accidents as well as efficiency gains.

The technology is underpinned by the systematization of policies and procedures, serving as a foundation and, therefore, improving the way work is planned, risk assessed and executed in work management and operations.

6. Summary and Conclusions

New technology in the form of Operational Performance and Predictive Risk software platforms can support the systematization of a standard approach of safe working practices across the organization, thus leading to improved production efficiency and lowering operational risk.

Over time this can start to improve practices and reduce inappropriate behaviors which lead to unpredictable work execution, compliance failure notices, and inaccurate planning. It can also have tangible benefits across the organization in terms of reducing unplanned shutdowns and incidents, whilst generating and sustaining improved culture, discipline and behavior for operational excellence and enhanced safety.

Furthermore, this standardization drives consistent and reliable data generation including critical operational risk related data. With the tools to dynamically capture, visualize and manage operational risk in our plants in order to get more work done safely. By understanding the status of IPL impairments on our plants and the impact of the current or planned workload on those IPL impairments, we can link asset integrity risk with work risk. Consistent data generation can provide us with the risk profiles of the compounded risk of doing work against a backdrop of dynamic plant conditions in specific areas of the plant, thus improving frontline decision-making.

The ability to provide a real-time view of frontline activities across an organization's global operations is no longer a "pipe dream". Technology can empower and enable the workforce and management to proactively identify, track, monitor and manage the health of their IPLs and the impact of workload in real-time to ensure and improve overall plant health integrity, safety and efficiency. Organizations can start to move from a "find and fix" mode of operating to one of "predict and prevent". With line-of-sight provided by new technologies, organisations now have the ability to see and understand the relationships between workload and risk and can manage the organization to a more optimized and less risky performance.

Improving Plant Safety – An Operator Centric View on Process Safety

Dr.-Ing. Gregor Fernholz
Invensys Systems GmbH
Hermann-Klammt-Str. 1-3, 41460 Neuss, Germany
gregor.fernholz@invensys.com

Dr. Ian Willetts
Invensys Systems Inc.
5760 Fleet Street, Suite 100, Carlsbad, CA 92008, US
Ian.willetts@invensys.com

Prepared for Presentation at
American Institute of Chemical Engineers
2014 Spring Meeting
10th Global Congress on Process Safety
New Orleans, LA
March 30 – April 2, 2014

UNPUBLISHED

Improving Plant Safety – An Operator Centric View on Process Safety

Dr.-Ing. Gregor Fernholz
Invensys Systems GmbH
Hermann-Klammt-Str. 1-3, 41460 Neuss, Germany
gregor.fernholz@invensys.com

Dr. Ian Willetts, Invensys Systems Inc.
5760 Fleet Street, Suite 100, Carlsbad, CA 92008, US
Ian.Willetts@invensys.com

Keywords: plant safety, accident, training, operator training simulator, human error, PHA

Abstract

Reports and published information on plant incidents clearly shows that "human error" is a key cause of abnormal situations and accidents in plants. Numbers on the percentage of "human error" causing abnormal situations vary significantly depending on the industry, scope of the analysis and the definition of "abnormal situation" or "human error". However, typical numbers range between 28% [1] and 40% [2]. In many cases "human error" is related to lack of operator's skills or careless operation [3], [4]. While "human error" is one cause of loss, these statistics do not provide a solution to the problem – how can we reduce this particular cause of accidents or losses? Therefore, this paper focuses on the question how plant safety can be improved through enhancing the process safety critical skills of an operator.

Reviewing the literature on process safety and accidents gives insight into how companies have understood the root cause of issues and addressed specific aspects of improving operator skills in order to improve process safety. The first section of the paper analyzes published data to derive general best-practice recommendations for operator-centric safety improvements from understanding what "human error" means in the context of an operator to how this impacts plant safety and what needs to be done to improve plant safety. The second part of the paper investigates case studies to demonstrate how the operator is impacting plant safety and what the benefits of improved operator readiness to handle critical situations are. The third section of the paper provides an outlook on trends in the industry impacting plant safety and how this will affect immediate and future training requirements and how recent technological advances can help to address these issues.

1. Introduction

Improving plant safety is an objective of every operating company. However, plant safety is not a task that can be considered to be accomplished at a certain point in time. Assuming that plant safety has been accomplished and doesn't need further attention is already an attitude putting safety at risk. Accidents don't only happen in plants that would be considered to be unsafe by safety experts. They happen in supposedly safe plants, too. As we are learning from past accidents, we are continuously improving our safety standards as an industry. We cannot, of course, avoid all accidents but we do need to keep identifying our gaps and how we can close them. A focus on engineering issues alone may not be sufficient to prevent future incidents and better understanding of the psychological basis of human performance is critical to future improvements [5]. The so called "human factor" continues to be a key contributor to accidents [6] as many accidents happen because people in charge did not see the hazard or underestimate the risk [7]. Plant operators have to continuously assess the status of the plant, identify potential risks and decide which actions to take. Thus, improving operator readiness to identify and handle safety critical situations provides significant opportunity to prevent accidents and to improve plant operation. However, identifying operators to be a risk to plant safety misses a very important point: Operators are the last line of defense in preventing an accident: If all other measures fail, it's the operator who can prevent the accident. Thus, the operator is impacting plant safety twice: As a potential cause of the error and as a means to prevent an accident.

If operators can have such an important impact on plant safety, then it is beholden on all operating companies to identify what can be done to ensure that all operators always operate the plant safely and to maximize the operator's readiness to be able to identify critical conditions in the plant and to take the right actions to prevent an accident.

Understanding the factors governing success or failure of an operator in plant operation and in critical situations is one important consideration when designing a successful program to improve plant safety. We intuitively know that it is preferable to have highly skilled and experienced operators. However, we need to understand what skills are needed and what we can do to help operators gain that experience. This paper presents different elements that should be included within a training program to help drive the skills and experience of operators and to enhance the safety concept.

Once these elements are understood, the paper then provides in more detail how operator training simulators can be a critical enabling tool in the training arsenal and the testing and validation of safety relevant operational procedures. Operator training simulators provide comprehensive test and training environments for plant operators, very much like flight simulators do for commercial aircraft pilots. However, an operator training simulator doesn't necessarily improve plant safety unless the training regimen specifically considers emergency preparedness for handling untold operating conditions and includes the validation of procedures and their impact on the plant safety.

The different ways that operator training simulators can be used to accelerate the time to competence for operators is presented, as well as how training can target plant safety as well as how to execute standard operating procedures - and how to ensure that these procedures improve plant safety.

An outlook on trends in the industry impacting plant safety is provided, as well as connections to how this will affect training requirements and how recent technologies can help to address these issues.

2. Understanding the impact of the operator on plant safety

2.1 What is "Human Error"?

"Human error" is a category frequently used to point out that "somebody made a mistake" in deference to "a technical system failed". Better understanding why "human errors" occur will help to identify how to prevent "human errors" in the future. As human errors are too diverse to be treated as a single group, classifying them will provide guidance how to address them [7]:

- Mistakes – They occur because someone does not know what to do or accidently does the wrong thing.
- Violation or noncompliance – They occur because someone decides not to follow instructions or recognized good practice.
- Mismatches – The job is beyond the mental or physical ability of the person asked to do it, perhaps beyond anyone's ability.
- Slips and lapses of attention. These are inevitable from time to time.

Investigating specific incidents might not result in a clear classification according to the list above. However, it will help to analyze what went wrong and how to prevent similar failures in the future.

2.2 The operator perspective.

Obviously, operators don't make errors deliberately. The four error classes will help to understand from an operator's perspective why accidents happen and how to prevent them:

2.2.1 Mistakes

Many accidents start with a minor problem that wasn't addressed properly. This could be that measurements or alarms were ignored or the operator didn't fully understand the meaning and impact of the measurement or alarm. Thus, the operator needs to be capable of diagnosing the problem fully understanding the underlying scientific principles of the process. Taking the right action will prevent an accident, taking a wrong action will either not prevent the accident or will be the cause of the accident in the first place.

Thus, better education and training of operators is a key element in preventing mistakes as only education and training will provide the operator with the required knowledge to understand the meaning of alarms and measurements and to derive the right action from them.

2.2.2 Violations or non-compliance of instructions

Operators, and in some cases even supervisors, turn a blind eye on violations or non-compliance with operation procedures or best-practice operation because rules aren't properly understood, the risk related to non-compliance not understood or simply because existing rules are considered to be too cumbersome to follow. In the worst case instructions are indeed wrong and violating them will prevent accidents.

Thus, an operator-centric approach needs to address:

- Understanding instructions and rules – Operators need to be properly trained and educated to fully understand why rules and instructions need to be followed and what the potential consequences in the plant are, if instructions and rules aren't followed. An operator understanding the instructions and rules will be less likely to violate them.
- Investigating rules and procedures – In many cases a problem has more than one solution. If procedures are too complex, simpler procedures might be feasible. Thus, a system for developing and testing new procedures, or verifying existing procedures, will help to develop better, safer procedures and to increase adoption of rules, procedures and instructions.

2.2.3 Mismatches

If the job is beyond the mental or physical ability of the operator, or maybe even of anyone's ability, the design or method of working needs to be changed. From an operator's point of view it's important to identify these tasks prior to failing to execute them. Thus, we need a system that enables us to test all operators on all tasks. We might identify specific training needs for certain operators. We might even identify tasks that cannot be performed by most or all operators in which case we need to change the design of the plant or need to develop and test a new way to perform a job.

2.2.4 Slips and lapses of attention

These might be considered to be inevitable from time to time. However, slips and lapses of attention can be a result of:

- Confusion – The observation of trainees showed that errors often occur in a confused state of the operator's mental state [8].
- Lack of information – The information required to make a decision is not available, hidden or difficult to find, so that the attention is diverted from addressing the emergency situation to retrieving the information.
- Routine – While routine is required to avoid confusion during a crisis, it might result in overlooking early indicators of an emergency situation emerging, especially if the emergency situation develops over a longer period of time.
- Miscommunication – During an emergency situation, the mind will focus on processing information and taking decisions. This requires that information is exchanged between the different protagonists clearly, unambiguously and quickly.

Thus, an operator-centric approach to improve plant safety needs to address all of the above issues, if operators are to be fully prepared for every situation, to have the required information easily available at all times, to be able to communicate properly and to avoid falling into the trap of overlooking early indicators.

2.3 Requirements

After analyzing the impact of the operator's "human error" on plant safety by investigating the four categories, the following three areas need to be addressed to increase plant safety:

1. Operator education:
    a. Understanding the fundamentals of the process and the operation.
    b. Understanding why accidents happen.
    c. Identification of ways to prevent accidents in the future.
2. Validating and improving procedures and rules.
    a. Identification of best-practices and including them in the operator education.
    b. Understanding why rules and procedures are violated:
        i. Develop better rules and procedures that will be adopted by operators.
        ii. Enhance operator understanding of rules and procedures to increase adoption.
    c. Analyzing new procedures and rules and validate them against plant safety.
3. Improving communication skills and critical situation readiness.

Plant safety improvement is a continuous process and not a task that can be considered to be accomplished at any point in time. Indeed it is of the utmost importance to establish a mechanism that will ensure that lessons learned are not forgotten. Accidents may occur if lessons from past accidents are forgotten. We need to keep in mind, that organizations don't have memory, only individuals have memory. Thus, any operator-centric approach to improve plant safety must become part of the standard operating practices. If we've improved plant safety by addressing the areas above today, we need to address, at the same time, the question how these improvements can be sustained in the long term.

2.4 Consequences

If the areas that need to be addressed are clearly identified and it is understood, that a long-term, sustainable approach needs to be adopted in operations, the question is, what are the best means to achieve this? What's the best way to provide training and education, a test-bed to validate and improve rules and instructions, and to improve communication skills and critical situational readiness at the same time? The answer is twofold:

1. Using the right technology: Operator Training Simulators (OTS)
2. Embedding this technology in design, operation and corporate training practice

2.4.1 Operator Training Simulators

There are numerous ways to provide training to operators and a holistic training approach will encompass classroom training, training on the job and various other means. However, state-of-the-art training with a clear focus on enhancing plant safety through an operator-centric approach will include an operator training simulator as a key element, as this technology best matches all the requirements identified in the previous sections:

- Understanding process fundamentals – An OTS is a replica of the plant. Indeed, the process model of the OTS provides more information about the process than the measurements in the plant. Physical, chemical and thermodynamic principles can be explained to the operator in the context of the operation of the plant.
- Understanding why accidents happen – As the OTS reflects the full range of operation, accidents and near-accidents can be demonstrated in the OTS and the operator can learn to identify indicators for critical situations.
- Preventing accidents – Operators can be trained on how to handle safety critical situations using the OTS. Moreover, the OTS can be used to test different means to prevent critical situations occurring in the plant by changing the process design, safety equipment, and safety relevant controls.
- Best-practice – Different operators or shifts will develop different strategies to identify and handle critical situations. During the training sessions using the OTS these strategies can be developed and assessed. Best practices can be identified and rolled-out to all operators and shifts during future training sessions.
- Adoption of rules and procedures - The adoption of best-practice rules can be monitored by monitoring the performance of all operators in the training. During an OTS training session the reasoning for the rules and procedures can be explained in detail and can be experienced by the operator in a reality-like environment. Operators can explain in the training why rules and procedures are perceived as inappropriate or difficult to follow. Concerns can be addressed and alternatives can be evaluated. In some cases better procedures (best practice) will be developed, in other cases operators will understand the risk related to violating rules and procedures.
- Lack of information/information overkill – Re-running critical situations in an OTS and discussing the actions taken or not taken by the operator will identify if the operator will have the right information in the right context at the right time. Thus, the OTS is not only a means to improve the operator's skills, but also a way to identify if the Human Machine Interface (HMI) of the operator stations in the control room and the alarm systems are properly designed from a process safety point of view.
- Analyzing the impact on plant safety – The OTS is an accurate representation of the plant providing a test bed to analyze all changes in the design, operation, control system and procedures with respect to their impact on the plant safety.
- Improving communication skills – Training sessions for safety-relevant tasks in the OTS typically involve two or more operators. As the training is executed in real-time with the option to include additional failures and to communicate with field operators, training of

communication and teamwork will become part of the training. If a request is not clear, a wrong action might be taken.

- Critical situational readiness – If operators are properly trained, they will be ready to act fast, taking the right actions with determination. Confusion will be avoided and communication will be precise.
- Routine – Re-training of critical situations will keep operators alert ensuring that early indicators of an evolving critical situation will not be overlooked. Training sessions can be slightly modified to ensure that operators will pay full attention at all times.
- Don't forget the lessons-learnt – The combination of establishing best-practice for safety-relevant operations and retraining and certification of operators using the OTS will help to prevent accidents, as safe operation will be less reliant on operators gaining experience in the real plant in critical situations and recalling that a similar situation might have happened many years ago. Lessons-learned need to become part of the training curriculum.

2.4.2 Operation and Corporate Training Practices

The discussion about the fit between the requirements to improve plant safety and the capabilities of an OTS clearly shows the potential to improve plant safety by better training using an OTS [9]. However, the pure existence of a technology or a technical system doesn't necessarily mean, that it is used in the most effective and efficient way. It's a pitfall to assume that the task of better training is accomplished by purchasing a training simulator. If the OTS isn't embedded in the corporate training concept and if no specific OTS-based training program, with a specific focus on safety is developed, and the regular training of operators doesn't become part of the operations plan and schedule, the investment in the OTS will fall short of leveraging the full potential of increasing plant safety. The job of training is only completed, when the behavior is changed, not when people are told to change their behavior.

Glaser [10] discusses the importance of putting in place a systematic training program to complement the investment in the OTS platform, in order to ensure that the simulator remains in use beyond the initial deployment period. He makes a very good point that the relative investment in the accompanying training program is often much less than the cost of building the actual training simulator.

Once the commitment is made to the simulator, there are many opportunities to leverage the simulator as part of an integrated training program in order to maximize the benefits.

Within a process facility there are many different levels of safety training and different functions that require different types of training. Figure 1 below is taken from the John M Campbell Company [11] and it shows a very useful segmentation of the different types of safety training required in an operating company. While an Operator Training Simulator is obviously targeted toward control room operators, the competency pyramid shown in Figure 1 has been enhanced with the red text to highlight the different ways that an OTS could be used in the different safety training areas within an overall safety management program.

Figure 1: Competency pyramid and the impact of OTS

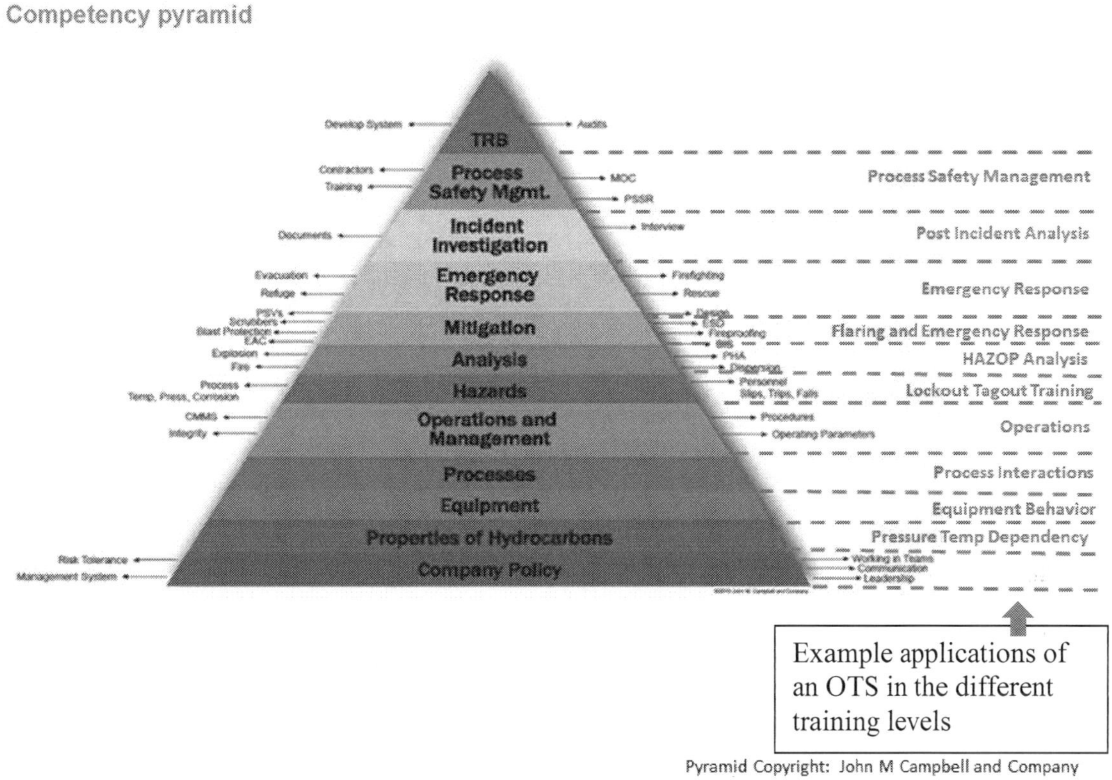

3. The impact of operator skills on plant safety

The analysis of past accidents and the discussion about the impact of an OTS-based training being embedded in a corporate training strategy and operational practice makes it apparent that operators have a significant impact on plant safety. It's difficult to prove that training can avoid an accident. However, there are numerous cases where a clear indication between accident rates and operator experience is seen. This article discusses different examples to address the objection that there is no clear, hard evidence that better training will increase plant safety or that the lack of training doesn't necessarily impact plant safety.

It is certainly worth mentioning that the investigation of the case studies is not meant to pillory the companies or individuals involved in these accidents but to be a reminder of the lessons we should learn from these accidents, as many people have dearly paid for gaining these lessons.

3.1 Case Studies

Although plant safety is a top priority for many operating companies, safety cannot be taken for granted. Indeed, comparing safety records across the global industry clearly shows that plant safety differentiates significantly. According to the "Report of the Interagency Working Group on Refinery Safety" [12] a report by the insurer Swiss Re found accident-related losses four times higher at U.S. refineries than the rest of the world. Inadequate training of workers and contractors has been identified as one of the challenges the Californian refineries are facing in improving plant safety.

3.1.1 Brunner Mond chemical plant, UK

A bulk-chemical manufacturing plant of Brunner Mond in Northwich, Cheshire used to have a higher than average accident rate. A "Susa – safe and unsafe acts" training initiative was established. As a result, the plant operated without a single lost-time injury (LTI) for more than a million hours for the first time in ten years following this training initiative. The achievement gives the plant one of the best safety records in the UK chemical industry. [13]

3.1.2 Chevron Richmond Refinery, US

On August 6, 2012, the Chevron refinery in Richmond, California, experienced a catastrophic pipe failure, releasing flammable fluid that partially vaporized into a large cloud. The vapor cloud engulfed 19 employees, including one fire-fighting employee who ran directly through the ensuing fire. All employees fled quickly, barely escaping serious injury or death. The ignition and burning of the vaporized fluid created a large plume that spread well beyond the refinery confines, causing approximately 15,000 people in the surrounding communities to seek medical attention. [12] The incident was investigated by several agencies, including the California Division of Occupational Safety and Health (Cal/OSHA) and the U.S. Chemical Safety and Hazard Investigation Board (CSB). The Cal/OSHA and CSB findings included that Chevron failed to implement its own emergency procedures to shut down the No. 4 Crude Unit when the leak occurred [12], [14]. Although following emergency procedures to shut down would not have prevented the accident in the first case, it is a strong indicator, that workers aren't properly trained for emergency situations.

3.1.3 BP Texas City Refinery, US

On March 23, 2005, the BP Texas City refinery experienced a catastrophic process accident. It was one of the most serious U.S. workplace disasters of the past two decades, resulting in 15 deaths and more than 170 injuries. As in many cases, the reasons leading to the accident are numerous. However, the accident could have still been prevented even on the 23rd of March 2005. This incident is a good example where a small issue turned into a catastrophic event:

> "...The Day Shift Board Operator overfilled the Raffinate Splitter and overheated its contents without understanding the true status of the unit. ... Intervention steps were inadequate and late, and may have actually made the situation worse. ... Supervisory staff did not verify the correct procedure was being used or followed... Although the startup procedure was not fully up to date, if the procedure had been followed, or if different intervention had been made earlier, this incident would not have happened." [15]

The report of the BP U.S. refineries independent safety review panel that investigated the incident and concluded:

"BP has not adequately ensured that its U.S. refinery personnel and contractors have sufficient process safety knowledge and competence. The information that the Panel reviewed indicated that process safety education and training needs to be more rigorous, comprehensive, and integrated." and recommends: "BP should develop and implement a system to ensure that its executive management, its refining line management above the refinery level, and all U.S. refining personnel, including managers, supervisors, workers, and contractors, possess an appropriate level of process safety knowledge and expertise." [16]

3.1.4 Engro fertilizer Plant, Pakistan

Engro Chemical Pakistan Limited is the second largest manufacturer of Urea in Pakistan. The morning shift identified an issue with the synthesis gas compressor on the 3rd of August 2007 [17]. The operators identified the issue by analyzing the alarms and measurements and immediately tripped the compressor from the control room. After inspection and repairing the compressor was restarted, but the machine still showed abnormal behavior during start-up. The compressor was shut down and fully inspected. The inspection showed that the compressor was severely damaged requiring a rotor replacement. After 10 days the replacement was completed and the system was started up again. An analysis showed that continuing to run the plant with the compressor compromised might have resulted in an even longer shut-down and might have led to a safety incident. A well-trained shift properly analyzed the situation and took the right action, despite the fact that the decision included a 10 days shut-down of the plant.

3.1.5 Bhopal, India

The Bhopal accident in India in 1984 has been investigated in much detail. Various, significant breaches of safety practices had been found. "The plant released a cloud of toxic methyl isocyanate. The leak was caused by a discharge of water into a storage tank. This was a result of a combination of operator error, poor maintenance, failed safety systems and poor safety management."[18] The lack of adequately trained operators was a key contributor to the disaster: "…the work force had been reduced from 450 to 150 and made to do work they had not been found qualified to do during their original job interviews. The production team on the MIC facility was cut from 12 to 6" [19], and "The management never bothered to employ good professionals to implement the safety measures" [19].

3.2 Benefits

The buyers of Operator Training Simulators are interested in safe and incident free operation of, multi-billion dollar assets. Operator Training Simulators provide benefits that far outweigh the cost. Examples of these benefits include avoiding unplanned shutdowns (impacting production and profits) or plant accidents, as highlighted above.

Fiske [20] cites the National Institute for Standards and Technology (NIST) and the Abnormal Situation Management (ASM) Consortium who claim that the inability of control systems and operating personnel to contain abnormal situations costs the US economy at least \$20 billion a year.

O'Brien [21] cites an ASM Consortium Report that indicates that operator error results in 42 percent of unscheduled plant shutdowns. Well trained operators are an important first line of defense against unplanned shutdowns or accidents and yet there are, as discussed below, no regulatory requirements in the process industries to deploy this form of training as a standard best practice.

Further evidence of the benefits of Operator Training Simulators is provided by Komulainen et al. [22] who carried out a comprehensive survey of simulator users in the Norwegian oil and gas industry. One of the most interesting outcomes of the survey was the difference in perceived benefits by the different beneficiaries. The operators reported the three most important benefits to be:

- Improved process understanding increasing confidence of operating the process safely
- Improved ability to handle process upsets
- Increased confidence in being able to execute day to day operations

These benefits are difficult to quantify but as this paper highlights, safety and improved operator competence go hand in hand.

The instructors, managers and engineers claimed the most important benefits to be:

- Faster production start-up
- Reduced operational risk and enhanced facility integrity
- High production performance.

Perhaps it is not too surprising that management focuses on the more easily quantifiable benefits but it is good to see that safety is front and foremost on the minds of the operators.

The paper [22] also reports that over 80% of the respondents reported noticeable to remarkable improvement of operator effectiveness from the use of Operator Training Simulators. The average reported operator effectiveness improvement was 31% but in approximately 10% of users, the improvement was reported to be over 75%.

5. Outlook

There are two key issues the authors wish to highlight that could guide companies to more widely embrace simulator based operator training.

5.1 Increasing Levels of Automation

Today plants are operating with higher levels of automation with fewer staff. Glaser [10] presented the following Figure 2 in his paper in Hydrocarbon Processing. This graphic shows an accelerating trend of more plant control loops under the control of each operator, while a worrying counter trend of the operator making fewer and fewer operational moves yet being responsible for more and more control loops.

This observation is echoed by Gerke [23]. He cites an ARC Advisory Group report that claims that today "operators, on average, are responsible for 300 to 400 control loops." Given this large number of loops are likely in automatic mode the majority of the time, it is quite easy to understand how operators can be challenged when asked to step up and manage a plant that, for whatever reason, has entered an abnormal situation.

Figure 2: Evolution of a Console Operator's Job

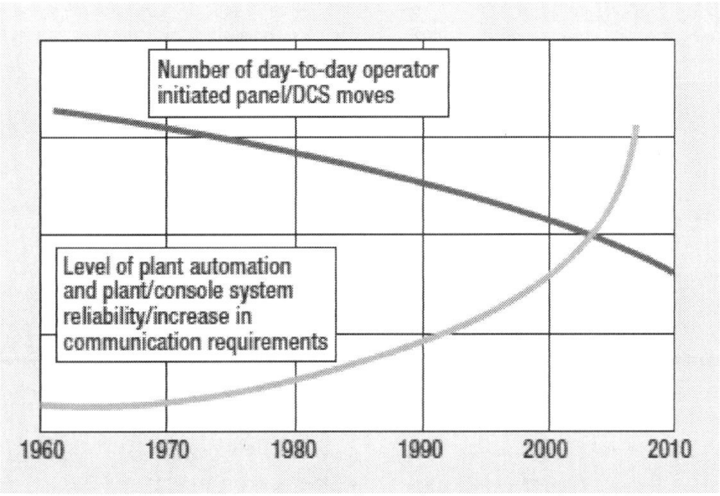

(Copyright June 2011 Hydrocarbon Processing [10])

This type of concern is actually being echoed in the airline industry today. A recent report by the Federal Aviation Industry [24] refers to "Pilots' "automation addiction" eroding their flying skills "to the point that they sometimes don't know how to recover from stalls and other mid-flight problems." The report claims that "the weakened skills have contributed to hundreds of deaths in airline crashes in the last five years." In the article, Rory Kay, an airline captain and co-chair of a Federal Aviation Administration advisory committee on pilot training says "We're seeing a new breed of accident with these state-of-the art planes...We're forgetting how to fly."

This trend of demanding more work from fewer employees is likely to continue and companies should embrace the responsibility to put in place the necessary training programs to prevent a heavy and unsafe reliance on automation. Operator Training Simulators provide a safe and effective way to train plant operators how to respond to a wide range of abnormal situations given their area of responsibility.

5.2 Industry Demographics

In the developed countries, the process industries are characterized by an aging workforce. Figure 3 shows the disproportionately high number of births per 1000 US citizens from 1946 to 1964, creating the "baby boomer" generation.

Assuming a retirement age of 65, the first of the baby boomers started retiring in 2011 and the members of the boomer generation will continue to retire, on average, at a rate of 11,500 every day for the next 16 years. In the developing countries there is a different demographic profile, often referred to as the "youth bulge." Both these demographics are creating a demand for ways to effectively train a new workforce.

Figure 3: US Demographics [25] – the Baby Boomers are retiring

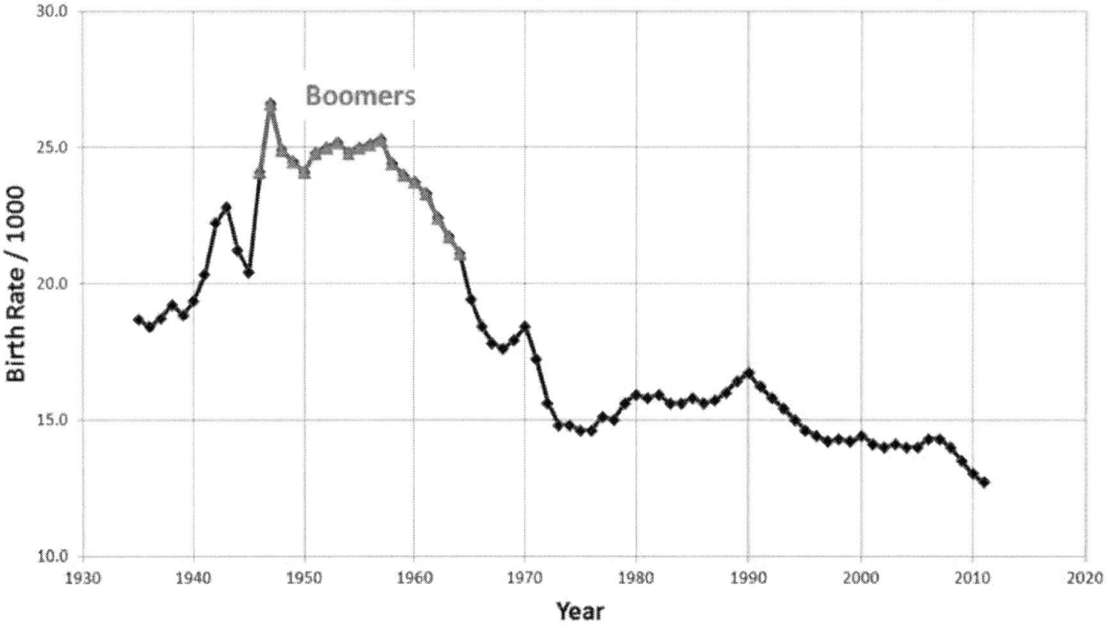

Using Operator Training Simulators the next generation of workers could learn to start up, shut down and operate the plant in a completely safe environment, before they are entrusted with the real plant. This great potential could hopefully help add more drive to companies to embrace Operator Training Simulators as a way to accelerate the learning curve of these workers.

5.3 Changing Regulations?

The United States Nuclear Regulatory Commission Regulations 10 CFR Part 55 requires a nuclear power plant operating company to have and to use an operator training simulator in order to maintain its license to operate. The standard "requires the applicant to demonstrate an understanding of and the ability to perform the actions necessary to accomplish a range of 13 tasks including some of the following [26]:

- Perform pre-startup procedures for the facility, including operating of those controls associated with plant equipment that could affect reactivity.
- Manipulate the console controls as required to operate the facility between shutdown and designated power levels.
- Perform control manipulations required to obtain desired operating results during normal, abnormal, and emergency situations.

These tasks relate to a nuclear power plant but the descriptions are not that dissimilar to the expectations placed upon operators in today's process industries. However there are no mandatory regulations requiring owner/operators in the process industries to have an OTS in order to maintain a license to operate.

The American Society of Safety Engineers publishes ANSI Z490.1 as a national voluntary consensus standard which specifies criteria for best practices in Safety, Health and Environmental training. The standard emphasizes the importance of comprehensive and integrated training programs including development, delivery, evaluation and program management. This is certainly not mandated and certainly the guidance in this standard would be applicable to OTS as well as to on-the-job training, lecture, computer-based training, discussion, classroom exercises, demonstrations, guided practice, activity-based interactive group(s), and distance learning.

In the past few years there have been two significant incidents in refineries, already discussed in this paper, that have led to independent investigations.

The Mogford Report (2005) [27] on the Texas City tragedy likewise observes the lack of simulator training for operators at Texas City and recommends that process simulators be provided:

> "While it is very difficult to quantify "incidents that won't occur" and "improved operator performance" the [investigation] team believes that the advantages to the site from the use of simulator training would be significant. In order to leverage the use of simulators to their best value, a clear simulator strategy is needed, outlining most likely-to-be-realized benefits, and critical success factors. This needs to be backed by resources to develop meaningful ongoing training and optimization programs."

In December 2013, the US Chemical Safety and Hazard Investigation Board (USCSB) issued its draft regulatory report (for public comment) on the Chevron Richmond Refinery Pipe Rupture and Fire, a major incident that occurred August 2006. Although the incident occurred as a result of a mechanical failure and not operator error, the recommendations of the report are still relevant to the topic of safety management and risk minimization.

The report states:

> "In the last decade, the CSB has made a number of process-safety related recommendations to OSHA and EPA in its investigation reports and studies (e.g. Motiva, BP Texas City, and Reactive Hazards). However, none of the regulatory recommendations have been implemented, and there have been no substantive changes made to the PSM and RMP regulations to improve the prevention of major accidents"

The report goes on to recommend that the Governor of California:

> "Develop and implement a step-by-step plan to establish a more rigorous safety management regulatory framework for petroleum refineries in the state of California based on the principles of the "safety case" framework in use in regulatory regimes such

as those in the United Kingdom, Australia, and Norway, and as described in this report, with the following minimum components:

a. A case for safety written by the duty holder that includes a systematic analysis and documentation of all major hazards and effective control methods implemented to reduce those risks as low as reasonably practicable (ALARP);"

The Center for Chemical Process Safety (CCPS) defines ALARP as "a risk reduction goal, where risk reduction efforts are continued until the incremental effort to further reduce risk becomes grossly disproportionate to the level of additional risk reduction."

Given ALARP considers cost, schedule and performance, the authors believe that the cost, schedule and performance of an operator training simulator can easily be implemented within a reasonable cost and schedule to make it a key contributor to enhanced safety in the process industries.

In the same way Operator Training Simulators are mandated by the American Nuclear Society, perhaps the time is right for the process industries to consider mandating OTS solutions given the cost benefit analysis and the recommendations coming out of the USCSB Richmond Refinery report. Per the report, now is the time to update regulations and to establish safeguards and procedures to drive the risk of major accidents in the process industries to levels as low as reasonably practical.

5.4 New technology trends for Operator Training Simulators

Operator Training Simulators are a well established technology in the process industry. The primary focus of the training has been on the control room operators using high-fidelity process simulation models coupled with an emulation of the control system enabling the control room operator to operate a virtual plant for training purposes. Although this approach has been proven to be very effective for control room operator training, it misses out on two safety-relevant aspects:

1. Only about 50% of all activities related to the plant operation are executed in the control room. The remaining tasks have to be done by field operators. However, standard Operator Training Simulators reflect field activities only in a very superficial manner. This is neither suitable for training of field operators nor does it properly reflect the behavior of field operators in a real-life scenario for safety-critical or emergency situations.
2. As the activities of the field operators aren't properly captured in the standard OTS, the communication and interaction between the field and control room operators cannot be included in the training. However, we identified in the previous sections that communication is a critical aspect in handling safety-critical and emergency situations.

Recent developments are addressing the question of how to fully include field activities and field operators in Operator Training Simulator systems. The most state-of-the-art approach connects immersive 3D virtual reality systems fully representing the field with the Operator Training

Simulator [28]. In an integrated system the field operator executes his tasks in the 3D representation of the field while the control room operator operates the plant from the control room communicating with the field operator via radio. With the increasing computational power of computer hardware 3D training simulators have been successfully developed and are already in use for education and training, e.g. the AVESTAR 3D IGCC (integrated gasification combined cycle) power plant [29].

6. Conclusion

Improving plant safety is an important aspect of every operating business. As the "human factor" continues to be a key contributor to plant accidents improving operator process understanding and safety skills will significantly impact process safety. With higher levels of plant automation and the baby boomer generation retiring, training of operators will become more and more critical for improving safety or even to keep plants running according to the current safety level.

The investigation of various case studies has highlighted that properly understanding the process and the operation of the plant is critical to avoid accidents. Operator Training Simulator based training provides the best means to deliver operator-centric training for all operation modes of the plant, especially safety and emergency situation handling, if it is embedded in a corporate learning and training concept that ensures that training and re-training will be delivered to new and existing operators.

Although training and the investment in technologies like Operator Training Simulators might be perceived as costs by many managers, the economic benefits of improved start-ups, higher plant utilization, reduced maintenance costs and, last but not least, increased plant safety in many cases by far outweigh the costs for OTS-based training. With more companies systematically approaching safety through an OTS-based training embedded in a systematic, corporate learning concept, improvements in plant safety are expected.

7. References

[1] Yang, S.H.; Yang, L.; He, C.H., *"Improve safety of industrial processes using dynamic operator training simulators."* Process Safety and Environmental Protection, 79, 329-338, 2001.

[2] Sebzalli, Y. M.; Li, R. F.; Chen, F. Z.; Wang, X. Z., *"Knowledge discovery from process operational data for assessment and monitoring of operator's performance."* Computers & Chemical Engineering, 24, 409-414, 2000.

[3] Goh, S.; Chang, B.; Jeong, I.; Kwon, H.-T.; Moon, I., *"Safety improvement by a multimedia operator education system."* Computers & Chemical Engineering, 22, S531-S536, 1998.

[4] Brambilla, S.; Manca, D., *"Dynamic process and accident simulations as tools to prevent industrial accidents."* Chemical Product and Process Modeling, Vol. 4 : Iss. 2, Article 7, 2009.

[5] *"Cognitive issues associated with process safety and environmental incidents."*, International Association of Oil & Gas Producers, OGP Publications, Report No. 460, July 2012.

[6] Dastous, P.-A.; Nikiema, J.; Marecal, D.; Racine, L.; Lacoursiere, J.P.; *"Risk management: All stakeholders must do their part."*, Journal of Loss Prevention in the Process Industries, 21, 367-373, 2008.

[7] Kletz, T., *"What Went Wrong? Case histories of process plant disasters and how they could have been avoided."*, 5th edition, Elsevier, Oxford, 2009.

[8] Sakata, K.; Ishii, H.; Kurooka, T.; Nishitani, H., *"The characteristics of human operator's behavior during malfunctions in simulator training",* Journal of the Society for Industrial Plant Human Factors of Japan, 4(2), 112–120, 1999.

[9] Ahmad, A.L.; Low, E.M.; Abd Shukor, S.R.; *"Safety Improvement and Operational Enhancement via Dynamic Process Simulator: A Review"*, Chemical Product and Process Modeling: Vol. 5: Iss. 1, Article 25, 2010.

[10] Glaser, D.C.; *"Put training back into operator training simulators"*, Hydrocarbon Processing, 41-43, June 2011.

[11] John M Campbell Company; *"Analyzing Engineering Mistakes to Minimize or Eliminate Mistakes"* http://www.jmcampbell.com/tip-of-the-month/2011/09/analyzing-engineering-tasks-to-minimize-or-eliminate-mistakes, 2011, accessed September 21st, 2013.

[12] *"Improving Public and Worker Safety at Oil Refineries"*, Draft Report of the Interagency Working Group on Refinery Safety Governor Jerry Brown, p. ii & p. 7, 2013.

[13] *"Training transforms Brunner Mond chemical plant",* Industrial and Commercial Training, Vol. 40 Iss: 3, 2008.

[14] USCSB (2013) - US Chemical Safety and Hazard Investigation Board Regulatory Report *"Chevron Richmond Refinery Pipe Rupture and Fire Report"* Number 2012-03-I-CA, p. 8, January 2014.

[15] Mogford, J.; *"Fatal Accident Investigation Report. Isomerization Unit Explosion Final Report. Texas City, Texas, USA."*, pp. 161 ff., B.P. p.l.c.; December 9th 2005.

[16] *"The Report of the BP U.S. Refineries Independent Safety Review Panel"*, p xiii ff, January 2007.

[17] Idrees, M.; Aslam, S.; *"Prevent Major Failures and Accidents with Well-Trained and Empowered Operators",* Process Safety Progress, Vol. 29, No.2, pp. 174-178, 2010.

[18] *"Reducing error and influencing behavior"*, United Kingdom Health and Safety Executive Report HSG 48, p. 9, HSE Books, 2009. (available at http://www.hseni.gov.uk/hsg_48_reducing_error_and_influencing_behaviour.pdf, accessed Dec 29th, 2013)

[19] Gupta, J.P.; *"The Bhopal gas tragedy: Could it have happened in a developed country?"*, Journal of Loss Prevention in the Process industries, 15, 1-4, 2002.

[20] Fiske, T.; *"Uses and Benefits of Dynamic Simulation for Operator Training Systems"* ARC Insights #2007-38MPH, August 9th, 2007, ARC Advisory Group, Dedham, MA

[21] O'Brien, L.; *"Why we Need a Better Approach to Procedural Automation", ARC Strategies Report*, ARC Advisory Group, Dedham, MA 02026, 2010.

[22] Komulainen, T.M., Sannerud, R., Nordsteien, B. and Nordhus, H.; *"Economic Benefits of Training Simulators"*, World Oil Magazine, December 2012.

[23] Gerke, G., *"Changing Workforce Drives Control Room Modernization"*, http://www.automationworld.com/control/changing-workforce-drives-control-room-modernization , AutomationWorld Website, January 2013, accessed December 31st, 2013

[24] Nj.com; *"Some Pilots' Additction to Automation has left them forgetting how to fly";* 2011; http://www.nj.com/news/index.ssf/2011/08/some_pilots_addiction_to_autom.html accessed December 30th, 2013

[25] Martin, J.A; Hamilton, B.E; Ventura, S.J.; Osterman, M.J.K.; Wilson, E.C; Mathews, T.J.; *"Births: Final Data for 2010"*; National Vital Statistics Report, Vol. 61, No. 1; p. 3, August 28, 2012.

[26] U.S. NRC Regulations; *"Part 55 – Operators' Licenses"*, http://www.nrc.gov/reading-rm/doc-collections/cfr/part055/full-text.html#part055-0046 accessed Dec 30th, 2013

[27] Mogford, J; *"Fatal Accident Investigation Report, Isomerization Unit Explosion Final Report. Texas City, Texas, USA."*; p. 88.; BP p.l.c., December 9, 2005.

[28] Richmond, P.; *"How 3-D Visualization is Revolutionising Industrial Training"*, 3D Visualization World, published 22nd of January 2013, http://www.3dvisworld.com/features/articles/3730-how-3-d-visualisation-is-revolutionising-industrial-training.html , accessed 16th of January 2014.

[29] Zitney, S.E., G. Provost, E.A. Liese, P. Mahapatra, R. Turton, and D. Bhattacharyya; *"AVESTAR Center for Operational Excellence of Clean Energy Plants and Invensys DYNSIM OTS / EYESIM ITS Integration"*, SimSci-Esscor North American User Group & Technical Support Symposium, Houston, TX, October 15-16 (2012).

Design Options for Overfill Prevention for Aboveground Storage Tanks

Satyajit Verma Ph.D.
Bechtel OG&C
3000 Post Oak Blvd. Houston, Texas 77056
sverma4@bechtel.com

Freeman Self
Bechtel OG&C
3000 Post Oak Blvd. Houston, Texas 77056
fself@bechtel.com

Prepared for Presentation at
American Institute of Chemical Engineers
2014 Spring Meeting
10th Global Congress on Process Safety
New Orleans, LA
March 30 – April 2, 2014

UNPUBLISHED

Design Options for Overfill Protection for Aboveground Storage Tanks

Satyajit Verma Ph.D.
Bechtel OG&C
3000 Post Oak Blvd. Houston, Texas 77056
sverma4@bechtel.com

Freeman Self
Bechtel OG&C

Keywords: Atmospheric Tanks, Storage Tanks, Overflow Protection, Overfill Prevention, API 2350, Liquid Seal

Abstract

Overfilling of a tank is an important safety hazard. It may result in loss of tank fluid and potentially severe consequences if the fluid is flammable or environmentally sensitive. Additionally, it is necessary to preserve the mechanical integrity of a tank. This article describes different design configurations where overfilling is a possibility.

1. Introduction

Overfill is a major reason for industrial accidents. Since the Buncefield fire and explosion at the Buncefield UK oil storage facility on 11 December 2005, many organizations have updated or produced new standards based on information from the incident. In addition to overflow of hazardous fluids, overfill may encompass fluids that are safe to discharge to the atmospheric drains. A risk assessment is one of the important elements in an overfill management system. One goal of the risk assessment is to develop the tank control configuration. There are many ways that a tank may be configured and this paper presents an overview of possible configurations.

The configurations discussed in this paper are as follows:

1. Fluids Safe to Release to Atmosphere
    1.A. Non-blanketed
    1.B. Blanketed or non-blanketed (3 options)

2. Fluids Not Safe to Release to Atmosphere
    2.A. Categories per API-2350
    2.B. Other alternative categories

2. Overfill Protection for Fluids Safe for Atmospheric Release

Overfill is typically mitigated by an overfill line or instrumentation, depending on the nature of fluid. Following is guidance on overfill lines. Types A, B, C, D are used for discussion purposes only, and have no code/regulation association. It is presumed that all designs discussed have appropriate and redundant (if needed) level transmitters and alarms installed and that periodic tank gaging, inspection and monitoring procedures are in place.

2.1 Type A: Tanks without Inert Gas Blanket

The first, and the simplest, option is that of providing a discharge line that is adequately sized for the incoming fluid flow rate and located at an elevation commensurate with the maximum allowable capacity of the tank. The outlet of this drain pipe should be directed to a sump to prevent the discharging fluid from splashing on the ground and exposure to the personnel in the vicinity. The tank content for such simple overfill protection must be safe for direct environmental discharge and have no exposure concerns to humans or animals. In these tanks, there is no requirement for the tank contents to be protected from long term air or oxygen contact; therefore, such tanks do not require any blanketing by inert gas. Since there is no requirement to maintain a pressure inside tank, the overflow line does not need a seal leg. No instrumentation in addition to those already mentioned above is required. See Figure 1.

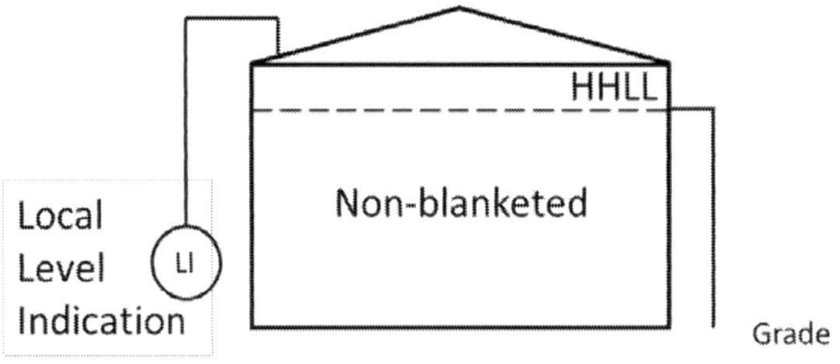

Type A Overfill Protection Option
Figure 1

2.2 Type B: Tanks Blanketed with Inert Gas – Inverted U-Loop

The second type of overfill protection is for fluids that require blanketing with inert gas such as nitrogen because the tank fluid is sensitive to or reactive to oxygen or air. The liquid itself may degrade in quality with prolonged contact with air. Replacing air/oxygen with an inert gas mitigates such a possibility. The tank is operated under a pressure of blanket inert gas higher than atmospheric pressure to ensure that the inert blanket is maintained inside the tank. The pressure-vacuum breathers and emergency vents are still installed on the tank as the last line of defense.

Unlike the Type A design, the overfill line from a blanketed tank cannot be open directly to atmosphere because the blanket gas would be continually escaping to the atmosphere through it. The tank must be kept sealed by a liquid seal leg. A liquid seal can be maintained by two methods.

In one, an inverted seal, the discharge line is connected near the bottom side of the tank at or below the low-low liquid level, and forms an inverted U. The top of inverted U is at an elevation that corresponds to the maximum allowable liquid level in the tank. A siphon break is provided at the top of the seal to avoid siphoning the whole tank content to the drain. In this design, the seal is inherently maintained by a liquid level in the tank, and therefore, no instrumentation is required to monitor and maintain the seal. Only if the tank is essentially empty, is there a chance for the blanket gas to blow-by to the atmosphere protected by lower explosive limit (LEL). The location and elevation of the final discharge point of the siphon break must still meet the 'discharge at a safe location' requirements. See Figure 2.

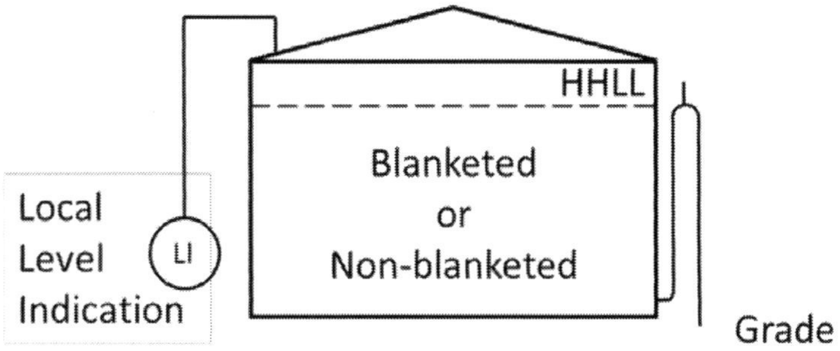

Type B Overfill Protection Option
Figure 2

2.3 Type C: Tanks Blanketed with Inert Gas: Inverted Loop inside Tank

In a slight variation of Type B design, where overfill line cannot be directly released to atmosphere, part of the discharge pipe may be located physically inside the tank, i.e. extending from near the bottom of the tank, and exiting at the side of the tank, with the pipe ultimately draining to a safe disposal location. Here too a siphon break should be provided at the top of the inverted U, and located outside of the tank. See Figure 3.

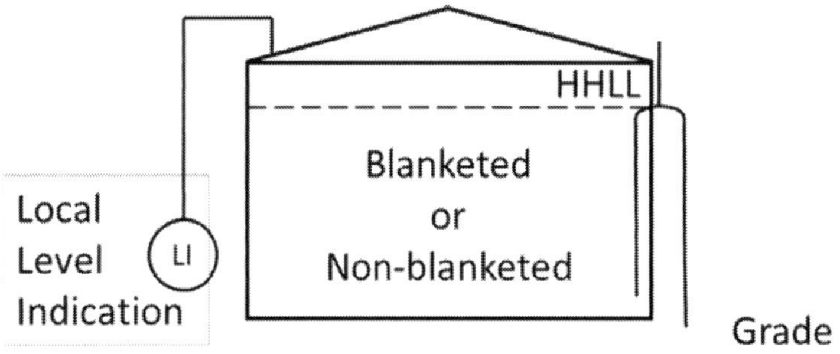

Type C Overfill Protection Option
Figure 3

Tanks that may have a slight layer of lighter material on the top but not blanketed may also use this inverted U-pipe design, to prevent the material from reaching the drain, since the discharge is from the bottom water layer. The material layer will continue to accumulate in the tank and would have to be vacuumed out periodically and disposed of safely. In this design, appropriate consideration must be applied to ensure that the liquid level inside the tank does not exceed the elevation for critical overfill height. If the tank is essentially empty, is there a chance for the blanket gas to blow-by to the atmosphere protected by lower explosive limit (LEL). Therefore, the location and elevation of the final discharge point and the siphon break must still meet the 'discharge at a safe location' requirements.

While calculating the height of the inverted U-loop, it is important to take into account the tank operating pressure because the height of the liquid in the discharge leg will be higher than the liquid level inside the tank by the height of the hydrostatic head. Similarly, appropriate consideration must be taken into account while determining the height of the inverted U loop, in cases where immiscible liquid of different densities exist inside the tank. The liquid in the loop would be the liquid in the bottom layer and, therefore, of higher density than the liquid in the top layer inside the tank. From hydrostatic considerations, the liquid level inside the tank could be different from the height of the inverted U loop at the start/stop of the overflow discharge from the tank.
Note: The hydrostatic equation useful in determining the heights of liquids of two different densities at the same applied pressure is:

$H_2 = \rho_2/\rho_1 * H_1$

Where:
H_1= Height of a column of liquid, 1 m (ft)
H_2= Height of a column of liquid, 2 m (ft)
ρ_1= density of liquid, 1 kg/m^3 (lb/ft^3)

ρ_2= density of liquid, 2 kg/m^3 (lb/ft^3)
$H_T = H_1 + H_2$

2.4 Type D: Tanks Blanketed with Inert Gas – Liquid Seal Leg

In this design, the take-off point for the overfill drain is at the location of the maximum allowable fill level for the tank. A liquid seal leg is placed at the end of the discharge line located at or below grade. This type of overflow line is suitable for the case where it is acceptable to dispose of the top layer-the lighter fluid- to drain. The liquid in the bottom layer would then be periodically vacuumed-out for waste processing.

A level gage is installed at the leg to visibly monitor the presence of liquid seal and a low-level alarm is also provided to alert the operator in case of loss of liquid seal. The seal leg is designed to match the tank blanket design pressure. The disposal location is carefully selected in the event that the liquid seal is lost and personnel exposure may happen because of nitrogen blowing through the seal leg. As mentioned above, additional instrumentation is required in this overfill protection design.

Again, if phase separation or stratification were to occur in the tank because of immiscible liquids, appropriate consideration must be made to account for the different densities of the liquids while determining the liquid seal height, because the exiting liquid will be the one at the top layer and, therefore, would be the fluid that would create the liquid seal. See Figure 4.

Note: The height of the liquid seal can be determined by the hydrostatic head formula:

In metric units:
$H = P / \rho * g$

Where:
H= Height of a liquid seal (m)
ρ= density of liquid (kg/m^3)
P = Tank design pressure (kPa)
g = acceleration due to gravity (9.81 m/s^2)

In US Customary units
$H = 1728 * P / \rho$

Where:
H= Height of a liquid seal (in)
ρ= density of liquid (lbf/ft^3)
P = Tank design pressure (psig)

It should be noted that the seal is still effective even when the liquid level in the seal leg is at least at [½ x H] or higher. This is so because even with half the liquid seal leg, the liquid leg can be pushed to a height H before gas blow-by can happen.

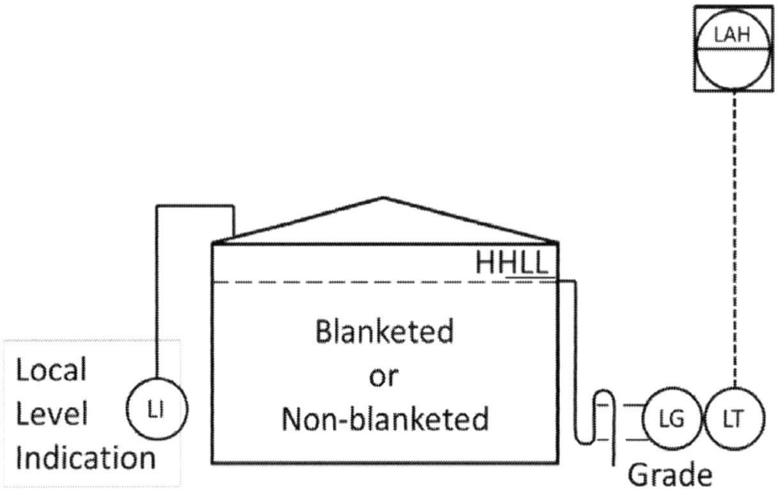

Type D Overfill Protection Option
Figure 4

3. Overfill Protection for Fluids not Safe for Atmospheric Release

In case of hazardous, toxic, or flammable liquid, it is not acceptable to discharge the fluid to the atmosphere for reasons of fire, danger to personnel safety and health or its environmental impact. Such tanks, therefore, do not have any overfill lines as depicted in the previous sections. Tanks containing Class I (flammable) or Class II (combustible) organic fluids fall into this category. These tanks are generally characterized by the presence of inert gas blanketing.

Broadly speaking these tanks can be divided into two types – those that are covered under API 2350, and those that are outside of it. A brief description follows.

According to API 2350, tanks are classified into three main categories based on the features for overfill prevention.

3.1 Category 1 Systems

These preventive systems are characterized by manual operation for tank gaging and stopping/diversion of filling operation. Level instrumentation, if any, is local and no data is transmitted to the control room. On-site personnel surveillance is required throughout the filling operation. See Figure 5.

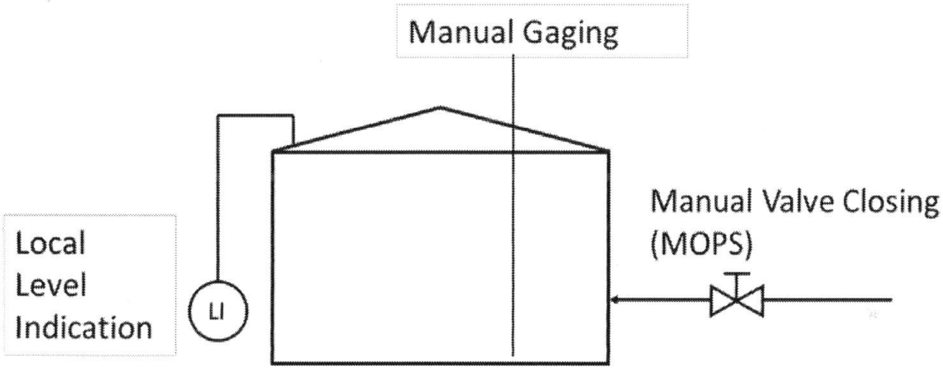

API Category 1 System
Figure 5

3.2 Category 2 Systems

At this level of overfill prevention, automatic and continuous tank level information (also known as ATG – automatic tank gaging) is transmitted to the control room. The level sensor is also used to generate a high-high level alarm. Transmission of alarm data to the control room i.e. remote sensing is the main difference between Category 1 and Category 2 Systems. However, the termination/diversion of filling operation can still be accomplished through manual intervention. See Figure 6.

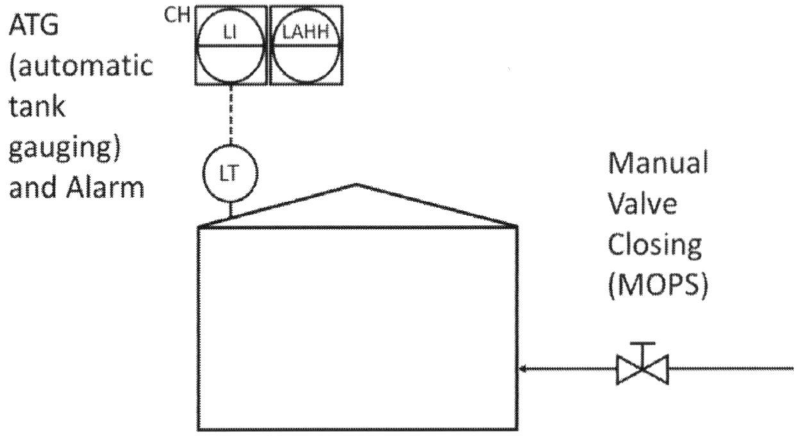

API Category 2 System
Figure 6

3.3 Category 3 Systems

A Category 3 overfill prevention system differs from Category 2 System in that it uses one independent sensor for automatic continuous tank gaging (ATG). It uses a separate level sensor for high-high liquid level detection for Overfill Protection System (OPS) alarm. The level sensor for alarm activation can be a point level device or a continuous level device. The termination/diversion of filling operation upon alarm is but can be local or through remote hand switch electrical, pneumatic or other operator on the valve. See Figure 7.

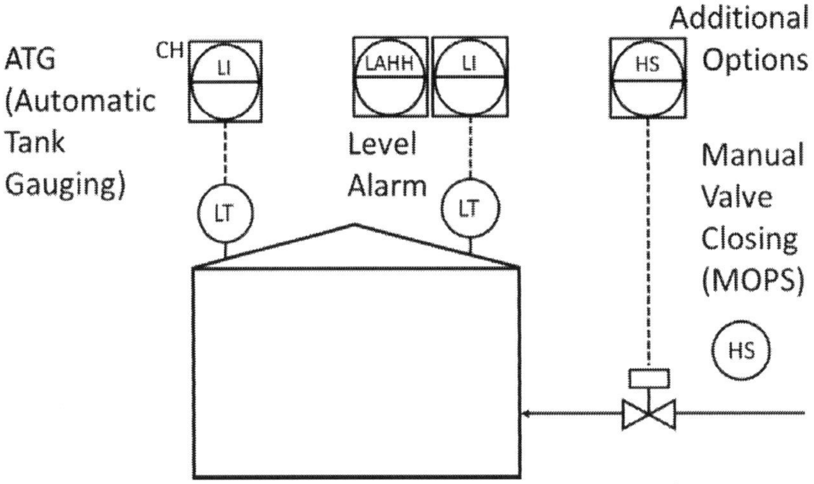

API Category 3 System
Figure 7

3.4 Automated Overfill Prevention Systems (AOPS)

The defining characteristic of an AOPS is that the termination of filling operation is automatic and activated by an independent high level signal. The AOPS is in addition to both the ATG and the independent OPS alarm. Tanks in fully automated operation in unattended facilities are usually equipped with AOPS. Codes [1], [4], [5], [17] recommend an AOPS for tanks with flammables and complex operations. The essential elements of AOPS are shown in Figure 8.

The termination of the filling operation can be achieved by automatic closure of the receipt valve. This is shown in Figure 8. Alternatively, the automatic tripping of the feed pump, or closure of the process valve, or both can be also be accomplished, however, a valve independent of the process control system is preferred. Risk analysis is performed to determine the safety integrity level (SIL) required to ensure that the tank is adequately protected against the severity consequence of overfilling. Generally a SIL of 1 or 2 is required. With the safety interlock and SIL rating, the automated valve closure instrumentation is located in the SIS. Additional redundancy (Independent Protection Layers) may be provided to increase the SIL protection level.

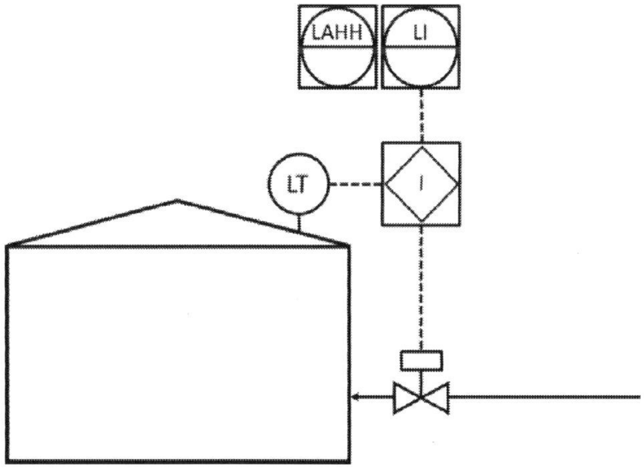

API Automatic Overfill Protection System (AOPS)
Figure 8

In practice, the automatic termination of filling operation feature can be incorporated in any of the other three categories but typically it is used with Category 2 or Category 3 System. See Figures 9 and 10 as examples of added to API Category 2 System. Also, note that although either a high-high level switch or a continuous level transmitter can be used, the latter is preferred because it is easier to detect an error or malfunction in the instrument [5],[10].

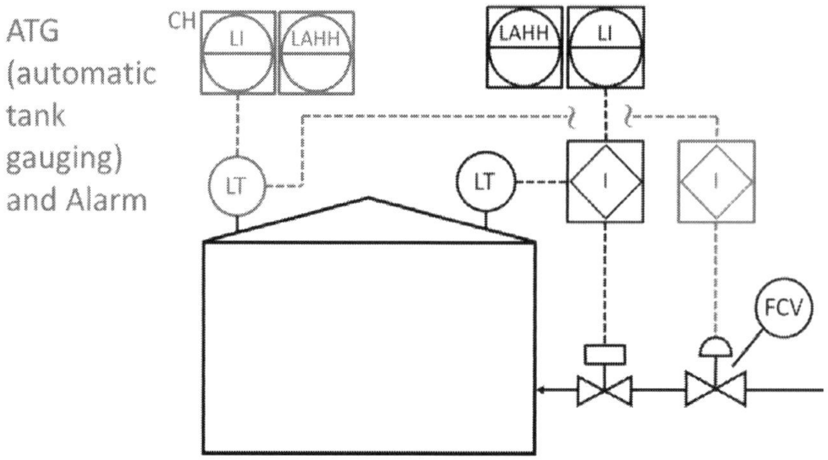

API Automatic Overfill Protection System (AOPS) Added to Category 2
Figure 9

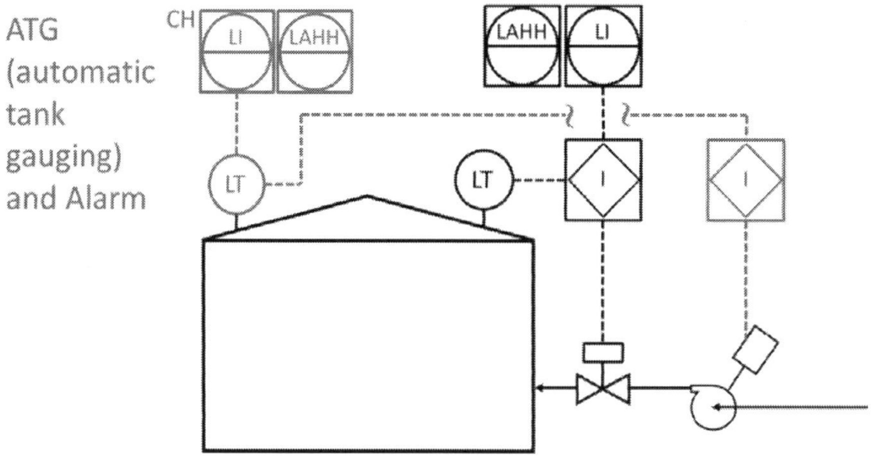

API Automatic Overfill Protection System (AOPS) Added to Category 2
Figure 10

3.5 Shop Fabricated Tanks

Shop fabricated tanks fall outside of scope of API-2350 and are covered under PEI Recommended Practices 600 for overfill protection. Such tanks are used in storage and supply of liquid petroleum products and alternative fuels such as motor fuels dispensing, emergency generator systems, residential and commercial heating oil supply systems and used oil storage and supply. Overfill prevention in these tanks are typically mechanical or electrical flow shut-off devices. Such devices are typically installed as a component of the fill connection such as a mechanical float, or an electrical switch to stop the inflow of the liquid. Electric switches can also be used to cut-off power to the pump. An additional feature of the overfill prevention is the presence of an overfill alarm. See Figure 11.

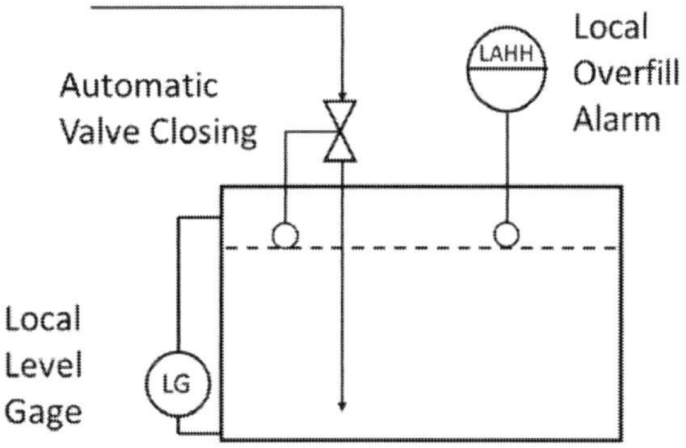

PEI-600 Tanks
Figure 11

4. References

[01] ANSI/API Standard-2350-2012, *Overfill Protection for Storage Tanks in Petroleum Facilities,* 4th Edition 2012, American Petroleum Institute

[02] NFPA-30 *Flammable and Combustible Liquids Code*, 2012 Edition (includes Errata 1-2012, Errata 2-2012, and Formal Interpretation-2012), National Fire Protection Association

[03] Benedetti, Robert and Paul May, *Flammable and Combustible Liquids Code Handbook*, 8th Edition, 2012, National Fire Protection Association

[04] *Model Code of Safe Practice Part 2 - Design, Construction and Operation of Petroleum Distribution Installations*, 4th Ed 2013, Energy Institute

[05] *Recommendations on the Design and Operation of Fuel Storage Sites,* Buncefield Major Incident Investigation Board, 2007

[06] *Identification of Instrumented Level Detection and Measurement Systems Used with Buncefield In-Scope Substances,* RR872, Health and Safety Executive (HSE), 2011

[07] *Safety and Environmental Standards for Fuel Storage Sites*, Process Safety Leadership Group PSLG, Final Report-Superseded, July 2007

[08] *Safety and Environmental Standards for Fuel Storage Sites*, Process Safety Leadership Group PSLG, Final Report, 2009

[09] EN 13616, *Overfill Prevention Devices for Static Tanks for Liquid Petroleum Fuels, 2010,* European Committee for Standardization, www.cen.eu.

[10] *Recommended Practice for Overfill Prevention for Shop-Fabricated Aboveground Tanks*, PEI/RP600-12, 2012, Petroleum Equipment Institute

Further Reading and References

[11] *Buncefield- Why Did it Happen*, Competent Authority for the Control of Major Accident Hazards (COMAH), 2011

[12] IEC 61511 *Functional Safety - Safety Instrumented Systems for the Process Industry Sector"*, 1st Ed 2003, IEC

[13] ANSI/ISA 84.00.01-2004 Part 1 (IEC 61511-1 Mod) *Functional Safety: Safety Instrumented Systems for the Process Industry Sector, International Society of Automation*

[14] *Layer of Protection Analysis: Simplified Process Risk Assessment*, AIChE Center for Chemical Process Safety (CCPS), 2001, American Institute of Chemical Engineers

[15] *Guidelines for Enabling Conditions and Conditional Modifiers in Layers of Protection Analysis*, AIChE Center for Chemical Process Safety (CCPS), November 2013, American Institute of Chemical Engineers

[16] *Guidelines for Initiating Events and Independent Protection Layers in Layer of Protection Analysis*, AIChE Center for Chemical Process Safety (CCPS), 2014, American Institute of Chemical Engineers

[17] *A Review of Layers of Protection Analysis (LOPA) Analyses of Overfill of Fuel Storage Tanks*, RR716, Health and Safety Executive (HSE) 2009

[18] *The storage of flammable liquids in tanks,* HSG176, HSE Books, 1998 <www.hse.gov.uk/pubns/books/hsg176.htm>

[19] API Publication 340, *Liquid Release Prevention and Detection Measures for above Ground Storage Tanks*, 1997

[20] Summers, Angela E., *Overfill Protective Systems-Complex Problem, Simple Solution*, 2009 <www.SIS-Tech.com/papername/papers-in-english/page/2>

[21] Summers, Angela E., *Automatic Tank Overfill Protection: Latest Safety Standards*, 2012 <www.SIS-Tech.com/papername/papers-in-english>

[22] *Complete Guide to API-2350 [4th Ed 2012]*, 2013, Rosemount-Emerson <www2.emersonprocess.com/en-US/brands/rosemounttankgauging/safety/Documents/api2350/index.html>

[23] AS-1940; *Storage and Handling of Flammable and Combustible Liquids*, Australian Standard, 2004

[24] Myers, P.E., *Consider Robust Tank Overfill Prevention and Independent Alarm and Gauging Systems*, February 2014, pp 71-74, Hydrocarbon Processing.

5. Appendices

Appendix 1
Additional Considerations

The configurations discussed in this paper are a starting point for the design of an overfill protection system for an atmospheric tank. There are other considerations, some noted below, that are not discussed in this paper. They are, however, extensively covered in the open literature.

- Risk Analysis and establishment of safety integrity levels (SIL)
- Sensors and Controllers: One important subject is the use of switches. Several codes [5], [1] do not recommend switches but instead recommend sensors with capabilities such as continuous sensing, on-line diagnostics, fully testable without requiring entry into the tank and without raising tank levels, and SIS certified.
- Valve characteristics, including type of valves, fire-safe valves, type of operators, location of initiation on and fail-safe position
- Testing of sensors, logic solver and final elements and testing intervals/frequency.
- Response times
- Fire protection

Appendix 2
Comparisons of Level Definitions between Various Codes

Although most level definitions are consistent with API-2350-2012, 4th Edition (2012), there are differences between other standards and earlier versions of API-2350. Please refer to the standards for guidance on levels and actions initiated.

Action for each Level (Note 1)	**Definitions of Levels or Tank Capacity**			
	API-2350-2012(4th Edition) [1]	**Energy Institute [4] & PSLG [8]**	**PEI-600 [10]**	**API-2350 (3rd Edition)**
Tank overflow and damage (Note 2)	Critical High (CH)	Maximum Capacity (overfill level)	Overfill Level (maximum design capacity)	Overfill Level (design capacity)
Alarm to initiate AOPS stoppage (Note 3)	High-High level (HH for AOPS)	Not specified	Not specified	Not specified
Alarm to initiate manual stoppage (Note 3)	High-High level (HH for OPS)	Tank Rated Capacity (HH)	Shutoff Level	Safe-fill Height (tank rated capacity)
Alarm that routine working level has been exceeded	Not specified	High level (H)	Alarm level (H)	High level (H)
Operational procedures	Maximum Working (MR)	Normal Capacity (normal fill level)	Maximum Operating Level	Normal Fill Level (normal capacity)
Alarm for minimum level	Minimum Working (low level)	Not specified	Not specified	Not specified

Note 1 – Refer to codes for recommended and optional levels, response times, actions and alarms. The level may differ between fixed roof, internal floating roof or external floating roof.
Note 2 – Depending on the code, an alarm or notification may be recommended at CIH
-- API-2350 says should have notification- pages 12 and 27
-- BSLG [8] and EI [4] says no alarm, only action
-- PEI [10] implies alarm – page 4 and definition page 26
Note 3 – If an AOPS is used, which is additional and independent of the OPS and ATG, then the HH level for AOPS will be higher than the HH for the OPS.

Appendix 3

Pressure Relief Protection Used as Overfill Protection

Pressure relief devices should not be used for overfill protection for the following reasons. Even though API-650 12th Edition 2013 suggests that a liquid relief valve may be used, API-650 and the other codes referenced specify overfill protection using the options in this paper.

- **Tank Design Pressure**

The tank may not be designed for the pressure that occurs when the liquid level is at the relief valve height. The mechanical integrity of the tank could thus be comprised.

- **Relief through pressure-vacuum breather vent or fire relief vent**

Weight loaded vapor breather and fire vents are not designed to relieve liquids. They may not open at the set pressure, may require high overpressure to be fully open and may not have the capacity to allow the full flow of incoming liquid. Additionally, the backpressure may affect the pressure at which the valve will open.

- **Relief through pilot operated liquid relief valves**

Pilot operated breather relief valves are available for liquid relief. For one vendor the minimum set pressure is 0.07 barg (1 psig, 28" WC) which would require a high tank design pressure resulting in a more expensive tank. Some codes (e.g. Australian) may not allow pilot operated valves in liquid service for tanks.

- **Relief through frangible roof**

Frangible roof tank with a weak seam would lift as the liquid level rises in the tank beyond the seam such that the hydrostatic pressure exceeds the seam design pressure. API-650 12th Edition 2013, paragraph 5.10.2.6 states that "frangible roofs are not intended to provide emergency venting for ...overfill." However, the roof may not open at the design point. More importantly, case histories demonstrate that cases exist where the roof has lifted and landed by the side of the tank. When an overfill situation occurs, the liquid would overflow to grade uncontrollably, creating a hazard if the liquid is flammable. Refer to reference [3] for more information on frangible roofs.

Appendix 4 -Buncefield Fire and Explosion – December 11, 2005

Early on Sunday morning of December 11, 2005, Tank 912 at the Buncefield oil storage facility was being filled from a pipeline. The tank was equipped with an automatic level gauge and transmitter / controller which displayed the level and initiated three levels of alarms in the control room (user-high; high-level; high-high-level). The tank was also fitted with a high-level switch (set above the gauge high-high-level), which was independent of the level gauge. The switch was intended to alarm and close the valve on the tank feed line from the pipeline.

Courtesy of Royal Chiltern Air Support Unit, from <http://www.buncefieldinvestigation.gov.uk/images/index.htm>

The level gauge was stuck, as it had intermittently on previous occasions. The high-level switch was not operating since the lever used for testing the system was positioned in a mode that disabled the switch. The emergency shutdown button to close the feed valves had never been wired so it was not activated. Gasoline ("petrol") filled the tank and overflowed the top. Subsequently, the bund retaining wall surrounding the tank failed at the joints. The tertiary drainage system was overwhelmed with hydrocarbon and fire water. Over 66,000 gallons (250,000 liters) of gasoline overflowed the tank and retaining walls, producing a vapor cloud, which ignited and exploded. The resulting fire burned for five days, destroying 20 fuel tanks and adjacent buildings and producing widespread environmental damage. Although there were no fatalities, about 40 people were injured and approximately 2,000 people were evacuated. Five companies were convicted of charges relating to the incident, including the facility operators, the pipeline company, and two vendors of the instrumentation.

Day 1 of the fire. Courtesy of HSE, *The Buncefield Investigation Progress Report 1*"

The conclusion outlined in the COMAH Report, 2011 [20] was: "In the Buncefield incident, the story of the sticking gauges and the inoperable high-level switch tells us about the immediate (technical) causes of the incident. However, the underlying managerial failures by others were equally important and have wider implications across all major hazard industries. These managerial failures encompass the cause of an incident and the mitigation processes."

Appendix 5

API 2350 Tank Levels of Concern and Corresponding Instrumentation

(ATG: Automatic Tank Gage, AOPS: Automated Overfill Prevention System)

	Category 1	Category 2	Category 3	AOPS
Personnel Attendance & Monitoring	Local operation & local monitoring	Local operation. Local monitoring with sometimes redundant remote monitoring	May be unattended remote operation & monitoring. Alternative, may be local	May be unattended with remote operation & monitoring
Gauging System	manual	ATG	ATG	ATG
• HH level (High-High)	manual indication	Automatic indication with alarm (remote & local)	Automatic indication with alarm (remote & local)	Automatic indication with alarm (remote & local)
Overfill Protection Systems (OPS)	Manual	Use ATG	Independent OPS	Independent AOPS
• CH level (Critical High)	Use manual gauging	• ATG indication with alarm (remote & local) • Initiate spill management (manual)	• Independent OPS indication with alarm (remote & local) • Initiate spill management (manual)	• Independent AOPS indication with alarm (remote &local) • Initiate spill management (manual)
• AOPS level	N/A	N/A	N/A	• Independent AOPS indication & alarm (remote & local) • Automatically actuated stoppage (remote)
• HH level (High-High)	Use manual gauging	• ATG indication with alarm (remote & local) • Initiate stoppage (local manually actuated)	• Independent OPS indication with alarm (remote &local) • Initiate stoppage (remote manually actuated & local manually actuated)	N/A
• MW level (Maximum Working)	Use manual gauging	ATG indication (remote & local)	Independent OPS indication (remote & local)	Independent AOPS indication (remote & local)
• H level (High)	Not generally noted	Not generally indicated or alarmed	Not generally indicated or alarmed	Not generally indicated or alarmed

Note 1-Remote means at control center not at equipment

Note 2- Local indication, alarm and action is included in all categories

Note 3- A higher level of control may be applied to any category

Note 4- Alarms are to be both audible and visual

Process modeling requirements for the safe design of blowdown systems – changes to industry guidelines and how this impacts current practice

James Marriott
Process Systems Enterprise Ltd
26-28 Hammersmith Grove
London W6 7HA, England
j.marriott@psenterprise.com

Apostolos Giovanoglou
Process Systems Enterprise Ltd

Prepared for Presentation at
American Institute of Chemical Engineers
2014 Spring Meeting
10th Global Congress on Process Safety
New Orleans, LA
March 30 – April 2, 2014

UNPUBLISHED

Process modeling requirements for the safe design of blowdown systems – changes to industry guidelines and how this impacts current practice

James Marriott
Process Systems Enterprise Ltd
26-28 Hammersmith Grove
London W6 7HA, England
j.marriott@psenterprise.com

Apostolos Giovanoglou
Process Systems Enterprise Ltd

Keywords: dynamic flare, blowdown, depressurization, high-fidelity safety analysis

Abstract

Safe design of pressure relief and blowdown systems requires that safety valves, relief orifices, flare piping and knockout (KO) drums are all sufficiently sized, to ensure that the process is protected from over-pressurization and that emergency depressurization can be executed rapidly and effectively. An essential design step is to ensure that temperatures resulting from auto-refrigeration are sufficient to avoid the risk of brittle fracture in both the process equipment and flare system.

However conventional calculation methods rely on a number of assumptions and approximations. In particular, calculations of safe design temperatures and relieving rates for complex blowdown segments are typically based on analysis of a single pseudo-vessel with simplified representations of the fluid's thermodynamics and heat transfer.

Best-practice in engineering and operating companies when assessing risks in pressure relief and blowdown systems now increasingly involves the application of high-fidelity dynamic simulation techniques. For example:

- rigorous, distributed blowdown system models are used to assess the risks of cold temperature brittle fracture during process depressurization
- analytical methods are used when performing pool and jet fire evaluations.

Evolving industry guidelines necessitate a change in the way all oil and gas companies approach process safety, in particular the design and analysis of pressure relief and blowdown systems. This presentation discusses how industry practice, and the tools and methods that support it, are beginning to evolve due to recent guideline changes.

1. Introduction

A blowdown operation is designed:

- to remove, in a controlled manner, all combustible hydrocarbons during an emergency situation or for a planned shutdown
- to reduce the risk of vessel or pipe rupture in the event of a fire
- to decrease the propensity for leaks by reducing process pressures

Typically prior to system depressurization a process plant is isolated into a number of independent blowdown segments. A full plant blowdown operation involves the simultaneous or staggered blowdown (depressurization) of all the pressurized gas (and/or in some cases liquid) in each segment by routing it to one or more flare tips for controlled combustion

The blowdown operation may itself be hazardous and a number of factors must be considered. Of particular importance are the following items reviewed.

1.1 Flare system capacity

A large amount of material must be disposed of simultaneously. It is necessary to ensure that there is sufficient flare discharge capacity to do this without creating excessive pressure in the flare system or excessive velocities and vibration, over-filling knock-out (KO) drums, and exceeding radiation limits from the flare tip.

Flare system capacity limitations and the importance of accurate dynamic modeling are not discussed further in this paper but the reader is directed to an alternate paper presented at this conference "Using dynamic analysis for accurate assessment of pressure relief and blowdown system performance" [1].

1.2. Process equipment mechanical construction

Similarly, potential hazardous situations may also arise in the process (blowdown segment) that is being depressurized. In particular, auto-refrigeration chilling during depressurization can lead to very low fluid, and potentially low metal, temperatures. The requirement to consider the mechanical design and in particular the risk of brittle failure of the process equipment and piping during depressurization is highlighted in the sixth edition of API 521 Guidelines on Pressure Relieving and Depressuring Systems.

For high pressure gas systems where condensate liquid forms, particular attention must be given to the geometry of the blowdown segment and in particular low-points where liquid may accumulate. The importance of these aspects is also identified in the sixth edition of API 521 Guidelines.

1.3. Flare system mechanical construction

Of equal importance is the mechanical construction of the flare system where the effects of low temperature can be addressed by using suitable materials of construction. However materials suitable to very cold temperatures are expensive and more difficult to procure, so it is highly desirable to minimize their use without compromising safety considerations.

The requirement to conduct a brittle fracture assessment of the flare header is again highlighted in the 6th edition of the API 521. Low temperature threats are most predominant in blowdown scenarios (but by no means limited to such cases) and the requirement to assess such scenarios rigorously, taking account of the duration of flaring events is also recommended.

2. Brittle fracture assessments

Brittle fracture is generally defined as the sudden rapid fracture under stress where the fractured material exhibits little or no evidence of ductility or plastic deformation.

Whereas ductile materials are characterized by high toughness, brittle materials have low fracture toughness. Most steels undergo a ductile to brittle transition in fracture toughness as temperature is reduced.

In a hydrocarbon processing facility, a brittle fracture of a pressure vessel or a pipe segment could lead to the release of flammable hydrocarbons with a significant risk of harm to people (for example, as occurred in the Longford Gas Plant Explosion of 1998).

If a pipe or pressure vessel is subjected to a temperature below its minimum design temperature it is considered at risk of being embrittled. Depending on the mechanical stresses the process equipment is subjected to, this can result in brittle failure. Examples of significant mechanical stresses of pipes and vessels are:

1. High internal pressure
2. Shock chilling, resulting in sudden changes in temperature
3. Vibrations, such as those caused by choking in tailpipes
4. Thermal expansion or contraction stresses in long pipes
5. External impact.

To assess integrity risks and avoid brittle fracture, the minimum temperature that process equipment can experience is assessed against its specified mechanical design limits, in particular the Lower Design Temperature (LDT) and /or alternately the Minimum Allowable Temperature (MAT). Abnormal (usually transient) occurrences on the facility most commonly lead to low temperatures for operational events such as start-up, shutdown and emergency situations, of specific concern are:

1. System depressurization ("blowdown")
2. (Re-)pressurization of the process (e.g. following blowdown)
3. Flare piping thermal response during pressure relief and plant blowdown events
4. Transient response following loss of heating.

3. Accurate blowdown analysis

Analysis of the rapid depressurization ("blowdown") of high-pressure vessels is an essential element of the safety analysis of oil and gas facilities and other high-pressure installations.

It is used in order to determine:

- the blowdown times and the relief loads entering the flare network. For new designs, accurate information is needed in order to achieve an optimal design that minimizes the piping diameters required to meet Mach number and back-pressure constraints
- the temperature throughout the process and pipework metal walls, in order to identify the minimum temperature that process equipment will experience for comparison against the lower design temperature for the selected material of construction
- the temperature of the fluid entering the flare system (which may contain evaporating entrained liquids). This provides essential information for choosing the appropriate material of construction for the flare system tailpipes, sub-headers and headers.

Calculation and simulation tools are invariably used as the basis of this analysis; this is regardless of whether the analysis is for the design of new facilities or as part of an asset integrity review. However, the fidelity of the approaches applied vary considerably across the oil and gas industry. With the changes to the API 521 guidelines mentioned previously, a more consistent application of best practice analysis is anticipated.

This paper focuses on the following key aspects of fidelity, all of which are essential for accurate calculation of process blowdown fluid and metal temperatures:

1. The importance of rigorous non-equilibrium modeling to accurately represent the thermodynamic behavior of fluids within the depressuring system – this is considered in the single vessel blowdown example.
2. The use of a distributed 'flowsheet' process model that represents the spatial distribution of volumes in the blowdown segment, rather than the conventional practice of representing ('lumping') the entire segment as a 'vessel' – this is considered in the Blowdown segment representation.
3. The coupling of process and flare network for accurate dynamic flare system analysis – this is considered in the section on flame network metal temperatures.

The minimum metal temperature that occurs during blowdown can be compared against the Lower Design Temperature to determine the risk of brittle fracture. More detailed mechanical analysis allows for an envelope of temperatures and pressures (Minimum Allowable Temperature); so a key outcome of the high fidelity methodology advocated in this paper is the calculated temperature as a function of pressure. However, in such cases it may be necessary to allow the system time to heat-up before a re-pressurization. It is noted that, where possible, such operational procedures should be avoided.

3.1 Single vessel blowdown

The depressurization of a vessel involves a complex set of coupled physical phenomena which need to be characterized accurately in order to understand behavior and provide suitable design values for process equipment and flare network.

Consider the blowdown of a gas-filled vessel. The pressure decrease results in a rapid change in the thermodynamic state of the gas within the vessel. This can result in nucleation of liquids within the gas bulk to form a 'droplet phase' (Figure 1). Once liquid droplets are large enough they will generally fall to the bottom of the vessel. Initially when the liquid hits the bottom it evaporates because of the relatively warm temperature of the metal it encounters; the effect is similar to a drop of water falling on a hotplate. However, once the metal has cooled sufficiently, liquid begins to pool (Figure 1) to form a continuous liquid phase.

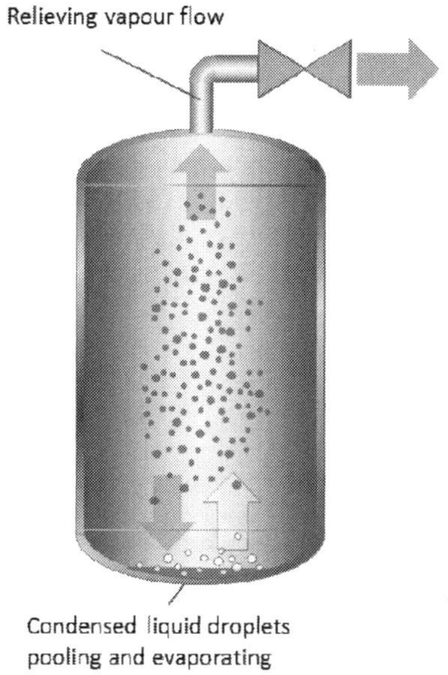

Figure 1. Single vessel blowdown

As pressures decreases the pool boils vigorously and cools, and in turn reduces the temperature of the metal beneath it. This can lead to significant temperature differences between the metal immediately below the pool and its surrounds, presenting a very real threat of brittle fracture and rupture from the base of the vessel. Cold liquids can also drain to lower parts of the pipework, presenting a threat of brittle fracture there.

The blowdown operation is fast and so there is no time for the vapor and liquid phases to reach equilibrium during depressurization; i.e. they show quite different temperature trajectories – see for example Haque *et al.* [1].

3.1.1 Conventional analysis

Traditional practice when performing blowdown analysis of high-pressure vessels is to use off-the-shelf process flowsheeting simulators that employ equilibrium thermodynamic methods and simplistic representations of the vessel walls.

Figure 2 shows the results from a standard equilibrium based calculation against actual measured *metal* temperature data for an approximately 100ft3 vessel [1].

- Figure 2(a) compares the metal temperatures for metal in contact with gas. The calculation under-predicts the temperature (i.e. it predicts lower temperatures than are likely to be observed), which, though inaccurate, is *in this case* conservative.

- Figure 2(b) compares the metal temperatures for metal in contact with liquid. Here the equilibrium approach gives quantitatively and qualitatively incorrect predictions for the wall temperature in contact with the liquid. The results in this case are not conservative, and give rise to a real danger of under-designing the vessel and pipework.

In general, as commented, vapor and liquid are not in fact in equilibrium and so the approach gives poor, and potentially unsafe, predictions. It is noted that many simulation tools enable the user to adjust a number of tuning parameters but this is only useful if the results are known – which by definition is never the case. Essentially, equilibrium based depressurization modeling approaches can never be predictive.

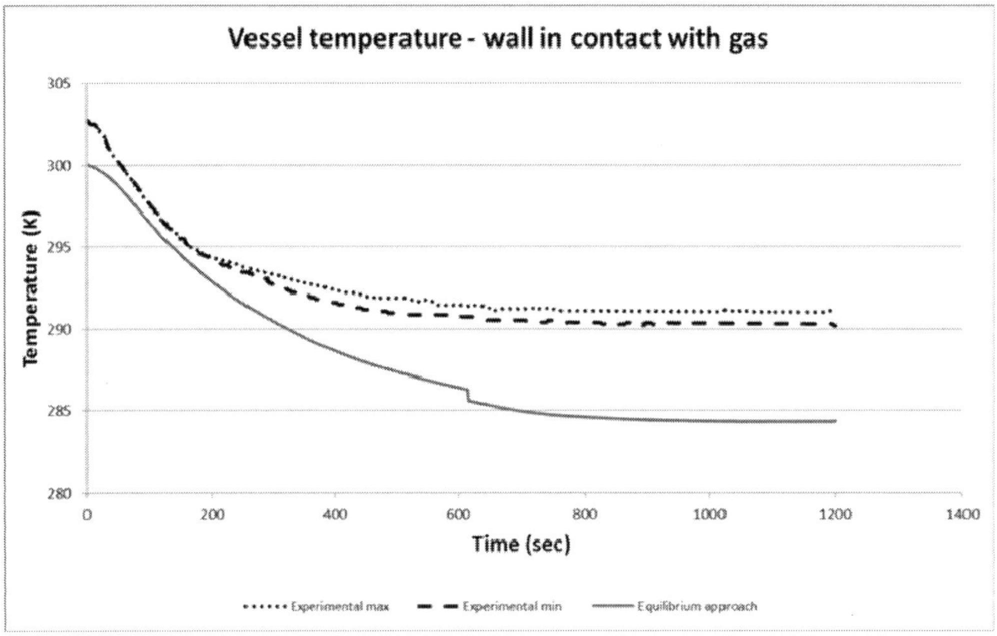

Figure 2(a). Temperatures for metal in contact with gas (equilibrium)

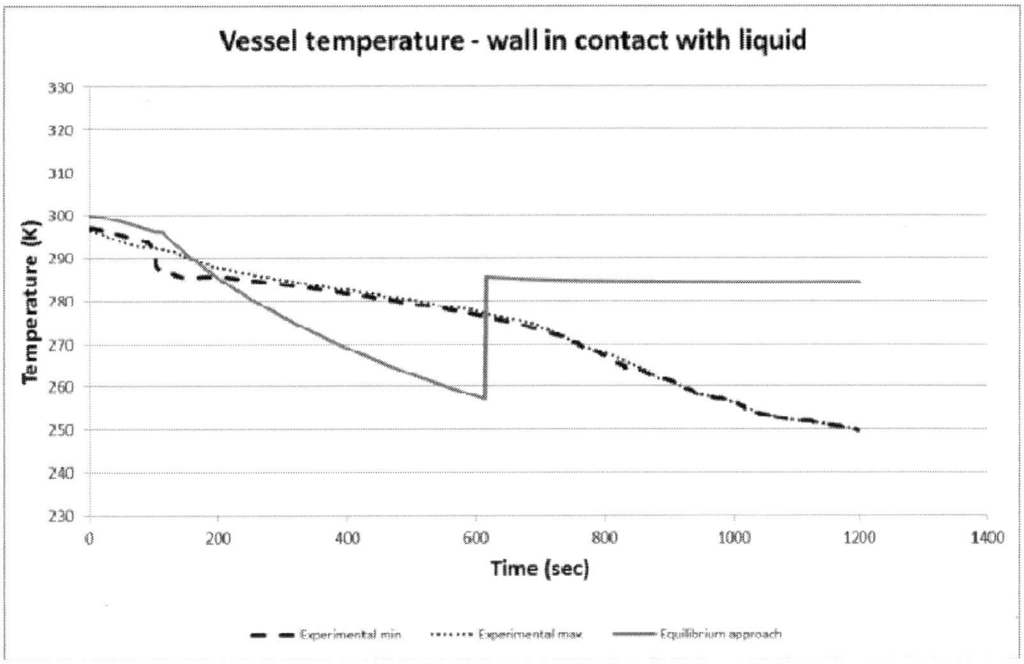

Figure 2(b). Temperatures for metal in contact with liquid

3.1.2 Accurate dynamic analysis

It is clear from Figure 2 that a rigorous rate-based modeling approach is required to accurately model the blowdown operation. This should:

- not assume equilibrium between the liquid and vapor phases; instead a rate-based thermodynamic approach should be applied
- account for the localized thermal mass of the vessel or pipe metal and the heat transfer rate between the adjacent phase (liquid, gas, aqueous) and the metalwork
- be predictive: i.e. have no adjustable or tuning parameters
- be validated for a wide-range of blowdown conditions.

Figure 3 shows the results from such a rate-based calculation against the same measured *metal* temperature data[1].

- Figure 3(a) compares the metal temperatures for metal in contact with gas. The calculation shows a good agreement with experimental data, compared with the results from the equilibrium calculation in Figure 2(a).

[1] Solved in Process Systems Enterprise's gFLARE® Advanced Process Modeling software.

- Figure 3(b) compares the metal temperatures for metal in contact with liquid. Again, the rate-based approach gives good quantitative and qualitative predictions for the wall temperature in contact with the liquid when compared with Figure 2(b).

Finally it is noted that the requirement for such approaches has been known of for a number of years but are still not widely adopted; see for example HSE Offshore: Fire and Explosion Strategy - Issue 1 [3].

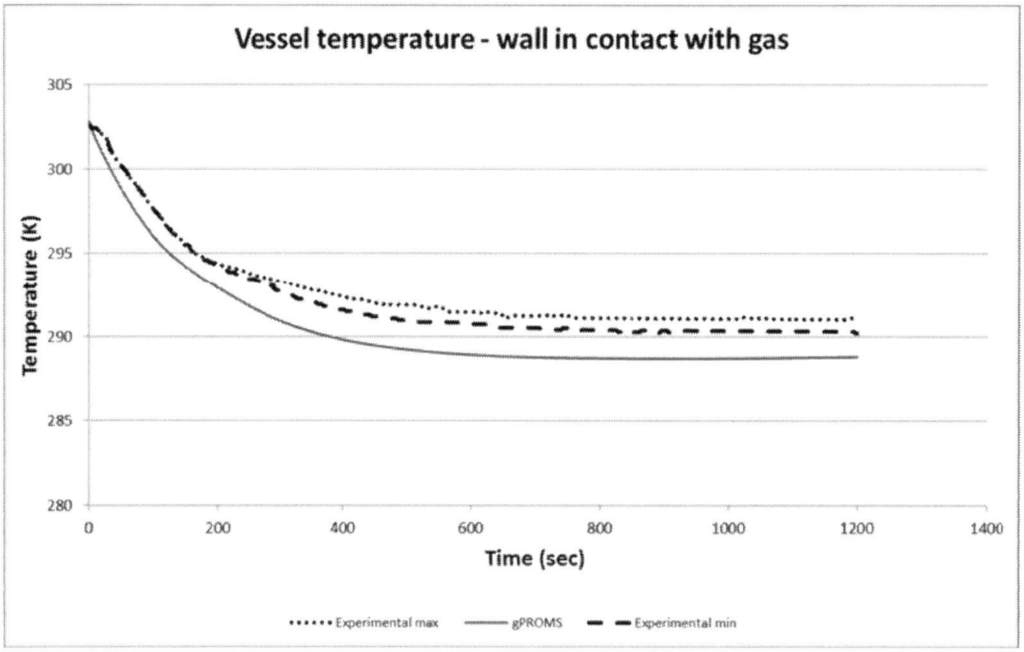

Figure 3(a). Temperatures for metal in contact with gas (rate-based)

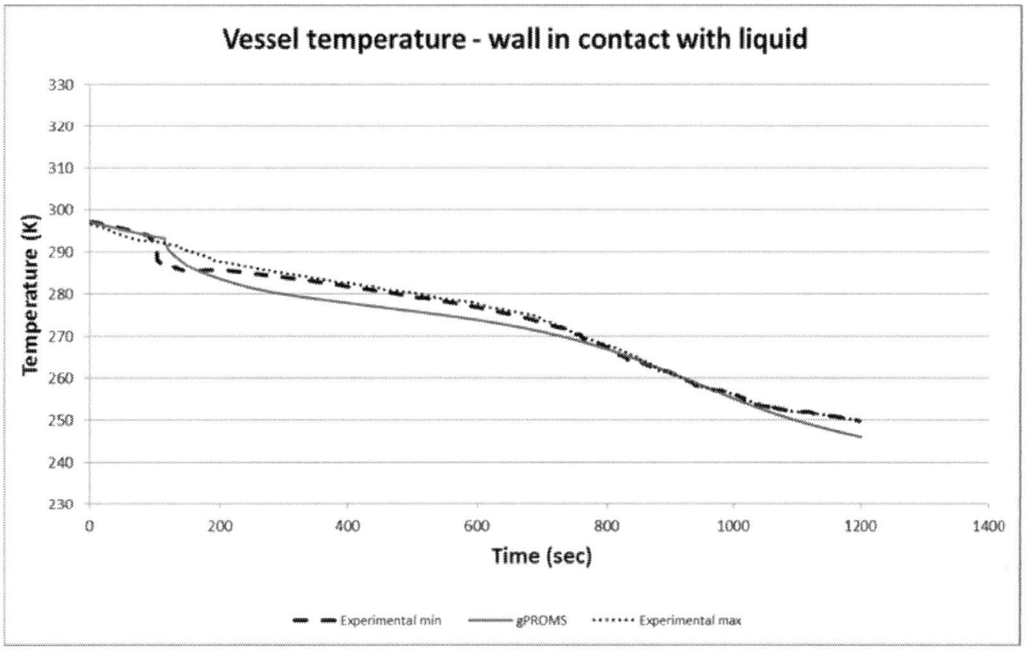

Figure 3(b). Temperatures for metal in contact with liquid (rate-based)

3.2 Blowdown segment representation

Most blowdown segments are not single units, but process sections involving multiple units or vessels (tanks, columns, heat exchangers, etc.) and their interconnecting pipework. Indeed a number of important blowdown systems are simply piping manifolds. During blowdown, each part of the segment undergoes some type of non-equilibrium phase behavior similar to that described in the previous section for fluid in a vessel that is being blowdown.

3.2.1 Conventional analysis

A typical, traditional 'engineering' approach is to represent entire system of each blowdown segment by a single vessel (the so-called 'lumping' approach), as shown in Figure 4. The dimensions of this notional vessel are adjusted so that its volume matches the total sub-system volume. An appropriate aspect ratio is chosen to provide either a suitable heat transfer area or a representative diameter for heat transfer calculations. Vessel wall thickness is chosen to approximate the entire metal mass of the original blowdown segment. The approach is subjective, and the precise approach employed varies from engineer to engineer.

It is noted that a conservative screening calculation takes no credit for metal temperatures (the extreme case); this is equivalent to isentropic expansion to atmospheric pressure. It is noted that for a typical natural gas mixture starting for example at 60 bar and 5°C results in a minimum calculated fluid temperature of -140°C (assuming phase equilibrium).

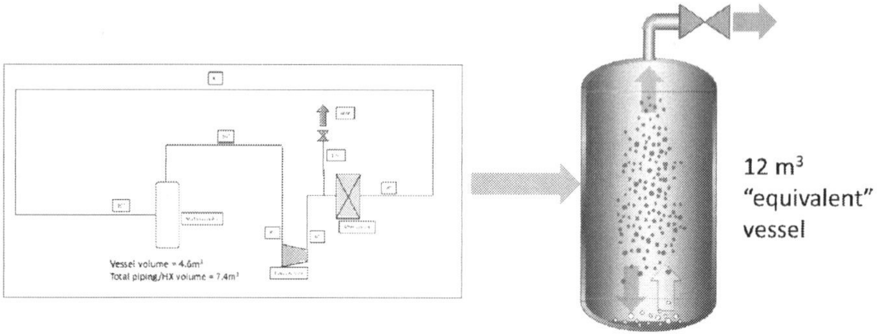

Figure 4. 'Lumping' of process flowsheet to represent it as a single vessel

3.2.2 Accurate dynamic analysis – compressor segment blowdown

By contrast, a rigorous, system-level analysis approach acknowledges that elevations and aspect ratios vary hugely in many systems and it is essential to capture this in the model. The approach thus involves creating a process flowsheet that represents the geometry of the blowdown segment much more realistically. This involves reviewing Piping and Instrumentation Diagrams (P&IDs), equipment data sheets and isometrics to determine pipe and equipment dimensions, and noting whether there are low spots where condensate liquid can form and accumulate.

Figures 5 – 9 show some example results for a typical compressor blowdown segment with a blowdown-point on the compressor discharge line. The calculations were performed with a model constructed in gFLARE® Advanced Process Modeling software for a blowdown starting from a minimum ambient temperature of -6°C.

- Figure 5 shows the rate of blowdown and the pressure upstream of the blowdown valve.
- Figures 6 and 7 show the fluid and inner wall metal temperature in the 1½" blowdown line and 8" by-pass line during the depressurization operation. Two key differences are observed
    - In the blowdown line, heat transfer coefficients are high and so the fluid and metal temperatures show good tracking. By contrast in the bigger 8" line there is a significant difference between the fluid and metal temperatures.
    - The blowdown line has the warmest temperatures in the system; by the time the fluid reaches this location it has been in contact with a significant amount of metal mass.
- Figure 8 shows the fluid and inner wall metal temperature in the 162ft^3 suction scrubber.
    - Profiles are given for the gas, liquid and aqueous phases.
    - It is noted that as in the single vessel blowdown case previously presented, the gas and liquid phases are a long way from equilibrium and though the gas drops to the

lowest temperature it is the metal that is in contact with the *liquid* that gets coldest.

- Figure 9 shows the fluid and inner wall metal temperature in part of the 8" compressor inlet line. This pipe section corresponds to a low point in the system.
  - o This is the coldest point in the system.
  - o As above, the gas and liquid phases are a long way from equilibrium. However, in this case the liquid drops to the lowest temperature. The metal that is in contact with the liquid gets the coldest temperature across the system.

This key observation is that such systems are very distributed in nature. Different parts of the system metalwork experience significantly different temperature profiles. The fluid temperatures in the system also vary greatly, so the temperature upstream of the blowdown point (see the next section on flare network temperatures) is significantly different to that in the vessel.

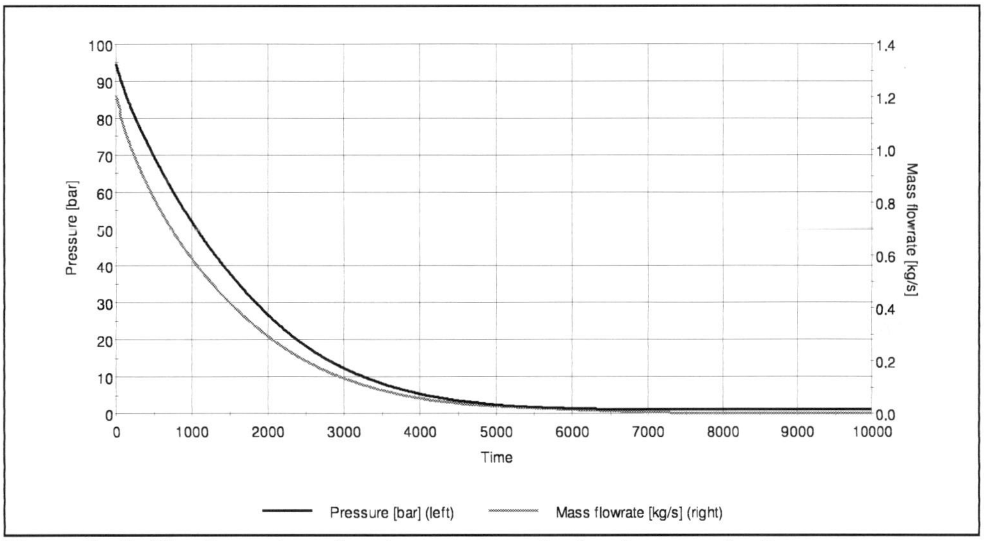

Figure 5. Orifice flowrate and system pressure during depressurization

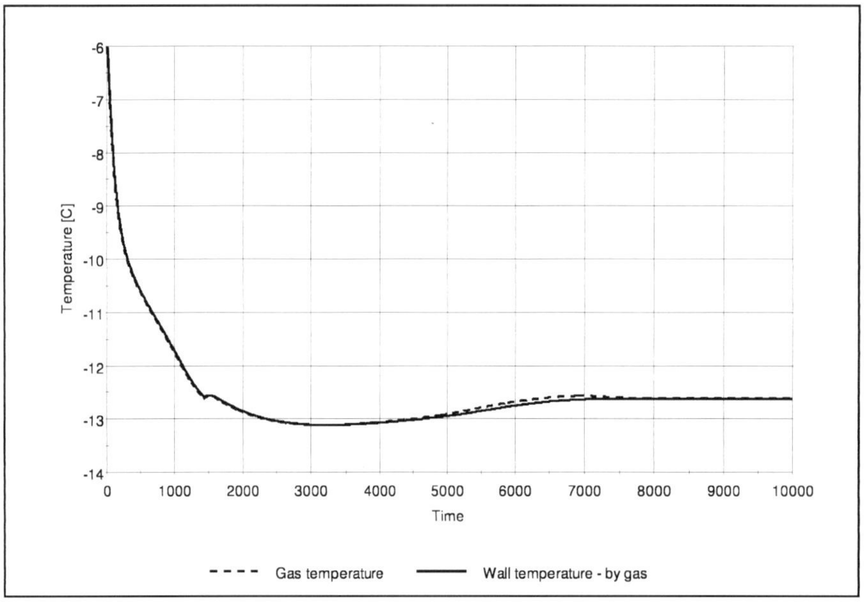

Figure 6. Fluid and inner wall metal temperature in the 1½" blowdown line

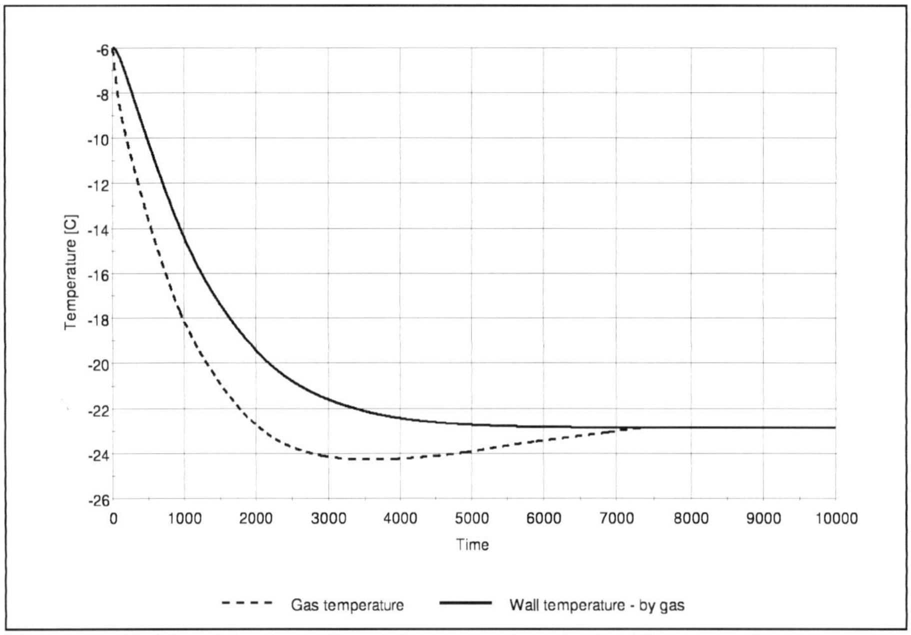

Figure 7. Fluid and inner wall metal temperature in the 8" bypass line

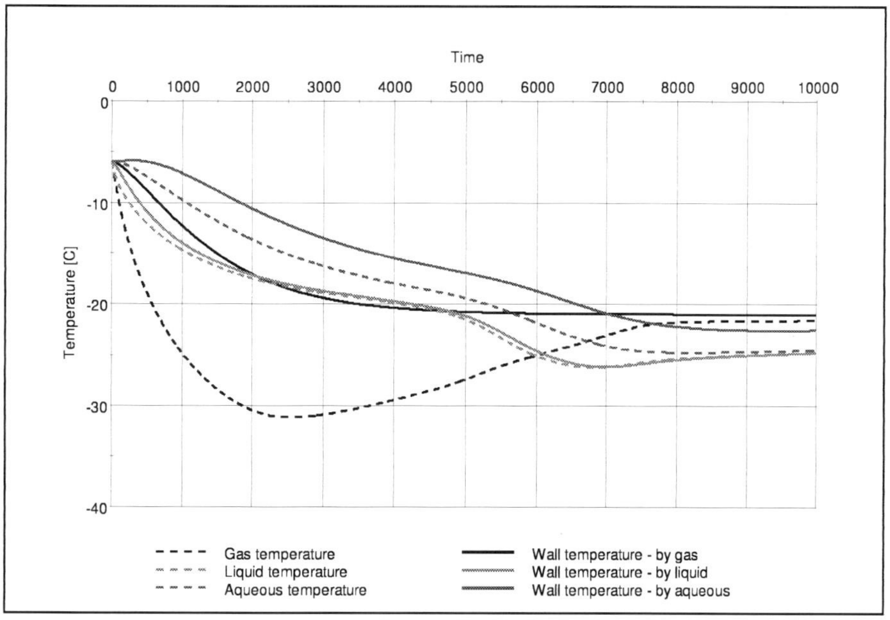

Figure 8: Fluid and inner wall metal temperature in the 140ft3 suction scrubber

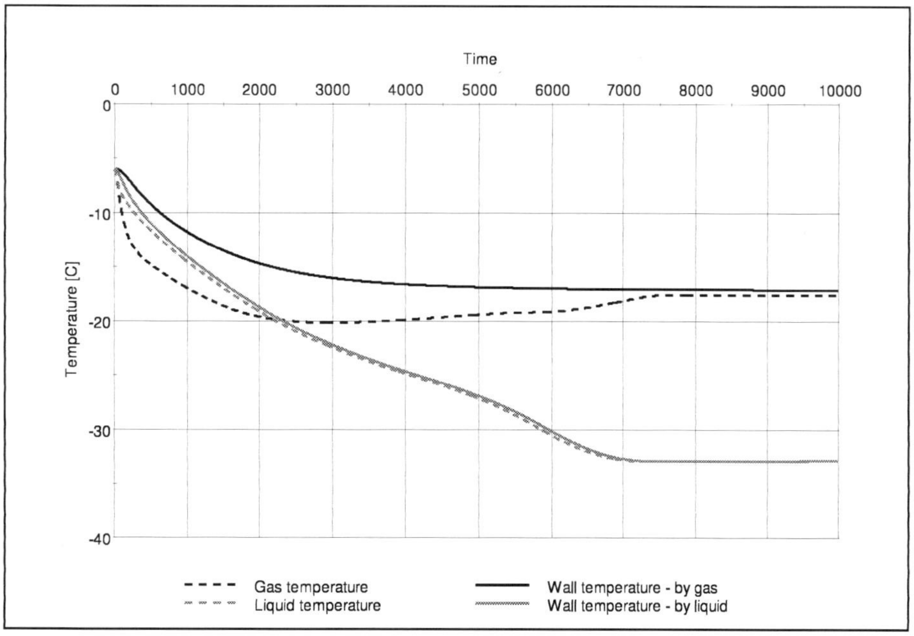

Figure 9. Fluid and Inner Wall metal temperature in 8" compressor inlet line

In most cases[2] an equilibrium and lumped approach will not introduce significant error in the calculation of blowdown rates and valve sizes. However, given the variation in temperatures that occur during blowdown, any attempt to lump units and consider some 'average' behavior (as in the traditional approach) is missing key information and is fundamentally flawed when used to specify minimum metal temperatures.

3.3 Flare network metal temperature assessment

The combination of Joule-Thomson expansion across the valve and the increase in kinetic energy causes the fluid temperature to drop significantly as it enters the flare network. As discussed previously, this can lead to potentially very cold temperatures and a risk of metal embrittlement.

Typically system blowdown determines the minimum temperatures that will be observed in the flare system. This is primarily because

- The process conditions prior to a blowdown are usually considered to be based on minimum ambient conditions (the process having been assumed to settle to ambient conditions prior to blowdown)
- As observed in the previous section the process-side temperature drops further during the depressurization operation

3.3.1 Conventional analysis

As can be seen when comparing Figure 6 (which has a minimum gas temperature of -13°C) and Figure 8 (which has a gas temperature -31°C) neglecting the geometry of the system could lead to significant errors in the fluid temperature upstream of the blowdown valve; in this case there is an 18°C difference though temperature differences of over 30°C are not uncommon.

Furthermore, as seen throughout Figures 5 to 9, the upstream flowrate, temperature and pressure all vary with time. Conventionally, however, flare network analysis is done using steady-state calculations with a fixed source conditions (as opposed to the true scenario of varying upstream conditions). In such a context there are a number of approaches taken to select representative values from dynamic blowdown calculations

1. Minimum upstream temperature, maximum upstream pressure and maximum flowrate
2. Minimum upstream temperature, maximum upstream pressure and flowrate at coincident conditions to the minimum downstream temperature

There is little justification for Approach 2, and Approach 1 is only conservative in the case of a single segment blowdown.

If we consider the blowdown of a single high-pressure segment, then clearly steady-state analysis will predict that the entire flare system will reach the entry fluid temperature[3].

---

[2] There are exceptions where care must be taken; e.g. systems with a significant amount of volatile liquid

[3] Give or take some small heat gain from the environment and the change in pressure and kinetic energy as the fluid flows through the flare network

In reality, as the fluid flows through the system it gains energy by heat transfer from the thermal mass of the pipe metal walls:

- Low temperature fluid will enter the flare network. As the blowdown proceeds the upstream vessel's pressure and temperature decrease and so the fluid temperature entering the flare network will change
- A cold front (fluid and metal) is observed to move down the flare system, with the tailpipes cooling very quickly to the fluid temperature, followed by subsequent cooling of the sub-headers and then the main header
- Each location in the flare network will see a different temperature–time profile.

The actual fluid behavior as it flows through the network is more complex: the fluid's kinetic energy changes as it flows through the network and its pressure will drop further, both of these also affect the fluid temperature.

3.3.2 Accurate dynamic analysis – full plant blowdown example

It is clear from the previous discussion that a rigorous dynamic representation of a flare system is required; this should accurately account for:

- the fluid pressure, composition, velocity and kinetic energy change
    - as it flows through pipes (of increasing size) and pipe fittings
    - as different streams combine at mixing tees
- the complex composition-based thermodynamics of the fluid
    - to determine temperature changes due to the Joule-Thomson effect as it expands
    - to determine the liquid fraction in the flowing fluid and thc possibility of segregated flow
- the thermal mass of all the metal work and the heat transfer rate between the fluid and metal.

However, in addition, it is essential to note that the transient behavior of the flare system is largely determined by the dynamics of the actual process that is undergoing depressurization during the emergency scenario that is being analyzed. To assess the minimum temperatures in the flare system a fully coupled model of the blowdown sources and flare system is required. It is noted that, in principle, for high pressure choked systems the dynamic process blowdown and flare models could be solved separately. The P, T, composition and flow inputs from the blowdown sources would then be fed in via a filed-backed time look-up approach. However, this is not generally advised as:

- the pressure in the flare system can also determine the blowdown rate from a particular blowdown segment:
    - For a blowdown operation the depressurization rate for low pressure sources is greatly affected by the flare network pressure
    - similarly the flare system pressure determines depressurization rates when flare back-pressure controllers are installed
    - in a sequenced blowdown, the blowdown of a particular segment is initiated when the segment depressurized previously reaches a specified time

- given the number of scenarios that need to be assessed, file-backed approaches are laborious and error prone; particularly if cases need to be re-run due to changes, for example, in the assumption basis common during the design cycle
- it allows for different thermodynamic and physical basis between the two models (if for example they are implemented in different software tools); this should be avoided.

Figure 10 shows an example of the fluid and metal temperature profile in a pipe segment that forms part of a flare system main flare header for a full plant blowdown case on a relatively large 25 blowdown source system. The results were again created with a model constructed in gFLARE® Advanced Process Modeling software.

It can be seen that:

- the inlet of the pipe segment experiences colder temperatures than the outlet for the first 4 minutes and its minimum temperature occurs at an earlier time. This is the cold-front moving through the system.
- the coldest fluid only comes through this part of the flare system for the first couple of minutes. After approximately 4 minutes the pipe is actually being warmed by the flowing fluid because the fluid entering flare system is itself warmer due to a less pronounced Joule-Thomson effect across the blowdown valves

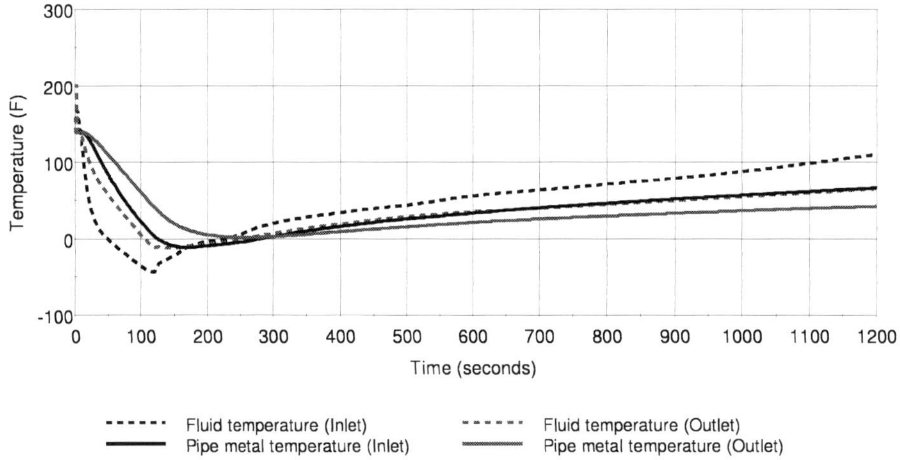

Figure 10: Flare header section fluid and pipe metal temperatures at pipe inlet and outlet

4. Conclusion

In the latest revision of the widely adopted API 521 standard, a number of new recommendations are made for brittle fracture risk assessment during process blowdown operations and in the flare network. To make proper assessment of these risks more rigorous modeling approaches than often employed are required. Such approaches, whilst used in a number of best practice organizations, are by no means yet standard practice.

Conventional approaches usually provide a good determination of blowdown times and required depressurization rates. However, conventional analysis approaches for determining the minimum metal temperatures during blowdown can introduce errors as much as 30°C; with the results obtained very dependent on the subjective approach that has been followed.

The application of detailed modeling approaches affords considerable safety and economic benefits. As an example, the application of the methodology advocated in this paper to a full facility blowdown study has led to design phase projects savings of $100Ms of dollars by avoiding the need to engineer an entire processing facility out of stainless steel. This is a not uncommon case, where the detailed methodology predicts significantly warmer temperatures than conventional analysis. However, it is emphasized that in many other examples colder temperatures are found; in several recent projects this has required thicker vessel end wall thicknesses and modified pipe schedules to cope with liquid condensate drop out and subsequent re-evaporation.

5. References

[1] Lawrence and Marriott, Using dynamic analysis for accurate assessment of pressure relief and blowdown system performance – *For presentation* at the Global Congress on Process Safety (2014).

[2] Haque, Richardson and Saville, Blowdown of Pressure Vessels. II – Experimental validation of computer model and case studies, Transactions of the Institute of Chemical Engineers Part B: Process Safety Environmental Protection, 70(BI), 1 (1992).

[3] HSE Offshore: Fire and Explosion Strategy - Issue 1 (2003).

PHA Analysis of Loss of Containment Events Involving Fixed Assets

F. Russ Davis, CSP
Mistras Group
4000 Underwood, LaPorte, TX 77571
russ.davis@mistrasgroup.com

Prepared for Presentation at
American Institute of Chemical Engineers
2014 Spring Meeting
10th Global Congress on Process Safety
New Orleans, LA
March 30 – April 2, 2014

UNPUBLISHED

PHA Analysis of Loss of Containment Events Involving Fixed Assets

F. Russ Davis, CSP
Mistras Group
4000 Underwood, LaPorte, TX 77571
russ.davis@mistrasgroup.com

Keywords: Process hazard analysis, Mechanical integrity, damage mechanism, fitness for service.

Abstract

In many cases the PHA analysis of fixed equipment risk of failure has only focused on the question "Is it included in the mechanical integrity program?" Very little if any analysis was performed by the PHA team to determine how effective the Mechanical Integrity program was in preventing failures associated with the fixed equipment. Little if any analysis was performed by the PHA team to identify those assets that have a high risk of releasing a toxic or volatile vapor cloud that can result in fires and explosions.

This paper will discuss the methodologies for evaluation of fixed equipment in highly hazardous chemical service. The effectiveness of the Mechanical Integrity (MI) program and techniques for bringing this data into the PHA will be discussed. This paper will further examine current recognized and generally accepted good engineering practices (RAGAGEP) available to assist a PHA team where questions may occur. Questions may include: what damage mechanisms affect the equipment and how will that damage mechanism result in a loss of containment event and what are the most effective methods for detecting the damage being used by the MI group. Methodology for evaluating the consequences of failure e.g., will the equipment leak or rupture, will the release result in fire or personnel injury, will be examined and discussed with a focus on the leak before rupture analysis and how to evaluate this information in the PHA setting.

Loss of containment from fixed assets has contributed to major refinery and chemical plant events in the recent years. This paper discusses the possible failures within the PHA that may have contributed to the events. All fixed equipment will eventually fail at some point. Proper analysis of the risk associated with this equipment is critical to safe, reliable operations.

1. Background

Process hazard analysis (PHA) has been extremely successful in identifying failures that can lead to loss of containment events with respect to instrumentation failures, operational failures and human errors. It is uncertain whether the PHA has been as effective in identifying and assessing failures in fixed equipment, i.e., pressure vessels and process piping. Failures associated with fixed equipment, in many cases, are not always fully analyzed by the PHA team. The PHA team assumes that the MI group has fully analyzed the vessels and piping and that this equipment is safe to operate because it is in the MI program. The PHA team does not assume the same for instrumentation; nor is this the same assumption for operating personnel and procedures.

A Chemical Safety Board (CSB) investigation reported that on December 7, 2009 a high pressure vessel at NDK Crystal manufacturing company ruptured due to stress corrosion cracking. A truck drive at a nearby gas station was fatally injured and 1 other person was injured.

On November 4' 2009 a 10-inch pipe failed on the outlet of a pressure vessel at the Silver Eagle refinery in Woods Cross, Utah resulting in a massive explosion that severely damaged houses in the surrounding neighborhoods. Per the CSB report, no one was injured but the property damage was massive and the refinery was shut down for an extended amount of time.

A recent CSB video presentation notes that on August 6, 2012 the Chevron Refinery in Richmond, CA had a catastrophic failure of an 8-inch pipeline resulting in release of a massive vapor cloud. Nineteen Chevron employees were in the vapor cloud, eighteen were able to make it out of the cloud before ignition. The last individual was in a fire truck that had responded to the release. The responder was able to make it to safety because he was wearing full firefighting gear. He exited the truck through a wall of flame. The Chemical Safety Board's interim report indicates that corrosion was identified as a potential cause of failure for the piping. Approximately fifteen thousand people in the surrounding communities sought medical attention at area hospitals.

The PHA team may not always have a subject matter expert on corrosion and potential failure mechanisms of pressure vessels and piping. As we can see from the incidents above, it is extremely important for the team to understand and analyze all potential failure mechanisms so that the PHA team recommendations can be made when necessary to ensure risk of vessel and piping failure is as low as reasonably practicable.

2. Current HAZOP practice

The dominant method of evaluating hazards during a process hazard analysis is a Hazard and Operability (HAZOP) study. Several other methodologies are recognized by OSHA such as, What-If, Checklist, What-If/Checklist, Failure Mode and Effects Analysis (FMEA), Fault Tree Analysis or other equivalent methodology. In most cases and failure modes, the HAZOP has proven highly successful in identifying causes of failures and in providing a proven tool for a full analysis of the process. There are, however, inherent problems in the use of the guideword HAZOP methodology when we consider fixed equipment failures. Guidewords do not always lead a team to consider all of the potential damage mechanisms that can affect the equipment. The typical guidewords are included in table 1.

Table 1

Guide Word	Meaning
NO OR NOT	Complete negation of the design intent
MORE	Quantitative increase
LESS	Quantitative decrease
AS WELL AS	Qualitative modification/increase
PART OF	Qualitative modification/decrease
REVERSE	Logical opposite of the design intent
OTHER THAN	Complete substitution

Guidewords are then attached to the various operating parameters where the results identify possible deviations. The results can result in a matrix similar to the one below in table 2:

Table 2

Parameter / Guide Word	More	Less	None	Reverse	As well as	Part of	Other than
Flow	high flow	low flow	no flow	rcvcrsc flow	deviating concentra tion	contamination	deviating material
Pressure	high pressure	low pressure	vacuum		delta-p		explosion
Temperature	high temperature	low temperature					
Level	high level	low level	no level		different level		
Time	too long / too late	too short / too soon	sequence step skipped		missing actions	extra actions	wrong time
Agitation	fast mixing	slow mixing	no mixing				
Reaction	fast reaction / runaway	slow reaction	no reaction				unwanted reaction
Start-up / Shut-down	too fast	too slow			actions missed		wrong recipe
Draining /	too long	too short	none		deviating	wrong timing	

Venting					pressure		
Inerting	high pressure	low pressure	none			contamination	wrong material
Utility failure (instrument air, power)			failure				
DCS failure			failure				
Maintenance			none				
Vibrations	too low	too high	none				wrong frequency

It is also observes by the author that the terms leak/rupture used for guiding the team to consider loss of containment (LOC) events. The above guidewords have not fully led the PHA team to an analysis of how fixed equipment fails. Without a full understanding of the failure mechanisms a team cannot determine likelihood of a LOC occurrence.

The OSHA PSM standard, 1910.119, requires that a PHA team be composed of at least individuals with expertise in engineering and process operations, a person with experience and knowledge specific to the process and one individual knowledgeable in the PHA methodology being utilized. Since there is no specific guidance to have a team member familiar with fixed equipment failure mechanisms and potential damage mechanism associated with the process under analysis teams do not always contain these members. [2] In the Chevron Richmond Refinery Interim Investigation Report dated April 2013 the CSB noted that:

Chevron personnel analyze numerous deviations for each portion of a process when conducting a Process Hazard Analysis (PHA). These include conditions such as changes in flow and temperature and pressure extremes. One of the deviations analyzed was "leak/rupture" of the particular vessel or pipe. The team's responsibility was to identify causes, consequences, safeguards, and recommendations. The 4-sidecut line was analyzed in the most recent crude unit PHA. Corrosion was not identified as a potential cause of a leak/rupture in the piping. [1]

The CSB also noted:

"Sometimes referred to as a corrosion review, a damage mechanism hazard review analyzes risks presented by all process failure mechanisms such as corrosion and cracking. Common process failure mechanisms are described in API 571: *Damage Mechanisms Affection Fixed Equipment in the Refining and Petrochemical Industries.* Such a review ensures that potential hazards caused by process conditions, process materials, and external mechanisms are properly identified, analyzed and systems are put in place to control or eliminate the hazard. Despite Chevron knowledge and expertise of potential damage mechanisms (such as sulfidation corrosion), the CSB found these hazards are only identified in a PHA if the participants conduction the PHA happen to have personal knowledge of the relevant mechanism. The Chevron PHA teams do not typically seek assistance from corrosion experts. The inclusion of a damage mechanism hazard review as part of a PHA is not required by the state of California, the California Division of Occupational Safety and Health (Cal/OSHA), Contra Cost County, the City of Richmond, or Chevron standards. Because Chevron does not conduct, and is not required to conduct, a formal damage mechanism hazard review, damage mechanisms are only identified when the PHA team happens to have some knowledge of the mechanism. As a result, many damage mechanisms which occur in various processes are not properly addressed. "[1]

Chevron has a very rigorous PHA methodology and teams are made up of informed and experienced members. The failure to adequately evaluate fixed equipment is an industry problem and is not specific to any one company. The CSB has identified an industry problem.

3. How to Determine Fixed Equipment Fitness for Service

In order to determine if a fixed process equipment is fit for service or safe to operate, the PHA team should first evaluate if the vessel or piping was adequately designed to operate in the current operating parameters. Process safety information (PSI) for pressure vessels include the U-1 form information, manufactures drawings, engineering calculations, material of construction and any other engineering considerations that were used to design the vessels. Piping PSI may consist of the piping specifications that were used to design the piping system under evaluation. The pipe specification should include all of the relevant information about the system including material of construction piping thickness, system components such as valves and gasket materials along with the bolting material and specification, temperature control on the pipes and insulation thickness. If the original design information is not available, the PHA team cannot validate that existing fixed equipment is suitable for any service.

The MI /reliability program will have the responsibility of determining that the fixed equipment is maintained in a fit for service condition. The PHA team will require access to the MI data in order to analyze the equipment likelihood of a loss of containment event. It would be preferable for the team to have at least one member who fully understands the potential damage mechanisms that affect the equipment and who can understand and fully analyze the inspection data. The PHA team must have access to information that fully defines any damage that may be occurring to the equipment. Each damage mechanism such as generalized corrosion, localized corrosion, cracking or fatigue must be evaluated. The remaining life assessments for all of the equipment affected by the various damage mechanisms must be available so a team can determine the likelihood that the piping or vessel will fail. It is very good practice to include the Asset-Life cycle management person as a PHA team member. The consequences of failure can then be evaluated.

In many cases the PHA team evaluation of fixed equipment consists of one safeguard. The team is informed that the equipment is covered by the MI program. If the analysis stops here, then the assumption exists that the MI program is without flaws. In the opinion of the author, this is not a valid assumption for every program or for every facility. Just because inspection and testing is performed on piping and vessels, it does not in any way assure that the correct inspection and testing is being performed. Therefore, in good practice, while it is the function of the MI/Reliability group to perform effective testing, the PHA team should not assume that the correct inspection and testing was performed unless a full damage mechanism review has been performed and that the specific equipment plans are implemented to address each damage mechanism Taking a spot Ultrasonic thickness reading on a pipe subject to stress corrosion cracking is not effective and can give the team a false sense that the proper safeguard is in place to reduce the likelihood of failure.

PHA teams must have access to corrosion rates, remaining life calculations and next inspection dates. This information is essential to determining if a pipeline or pressure vessel is fit for continued service and that adequate safeguards are in place to lower the risk to the company acceptable level.

The leak before rupture philosophy has been utilized in some hazard analysis for fixed equipment. The premise is that the piping or vessel will develop a small leak which will be detected by operations or by area monitors. The leak is the hazard and periodic operations personnel inspections and area monitors are designated as the safeguards. Once again, the author observes that there are a few flaws in this philosophy. The leak before rupture analysis of piping and vessels is again very dependent upon the type of damage that may be occurring. Localized corrosion very well may result in a small leak while the rest of the equipment remains intact and fit for service. Generalized corrosion such as sulfidation corrosion may result in a failure of a large section of piping or vessel as a single event. Cracking corrosion can result in rapid severe rupture when the crack has propagated through the equipment wall to the point where the remaining wall thickness is no longer adequate to contain the internal pressure or to the point where external stresses exceed the strength of the remaining wall thickness. Leak before rupture analysis is a rigorous assessment and must be performed by a qualified engineer. Without this, it is best not selected as an unqualified safeguard by the PHA team members. There are some chemicals that should never leak, such as Hydrofluoric acid. Where small exposures to hazardous chemicals cannot be tolerated the leak before rupture philosophy is entirely not acceptable.

4. Hazard Analysis of Fixed Equipment (Pressure Vessel or Piping System)

The PHA team evaluation of pressure vessels or piping systems must determine if there are adequate safeguards in place to reduce the risk of failure to as low as reasonably practicable (ALARP). While the specific guideword that focuses the team on this evaluation is not critical, the author's opinion is that the leak/rupture guideword does not work as well as intended. The following is a simplified logic flow that can assist a team with the basic analysis:

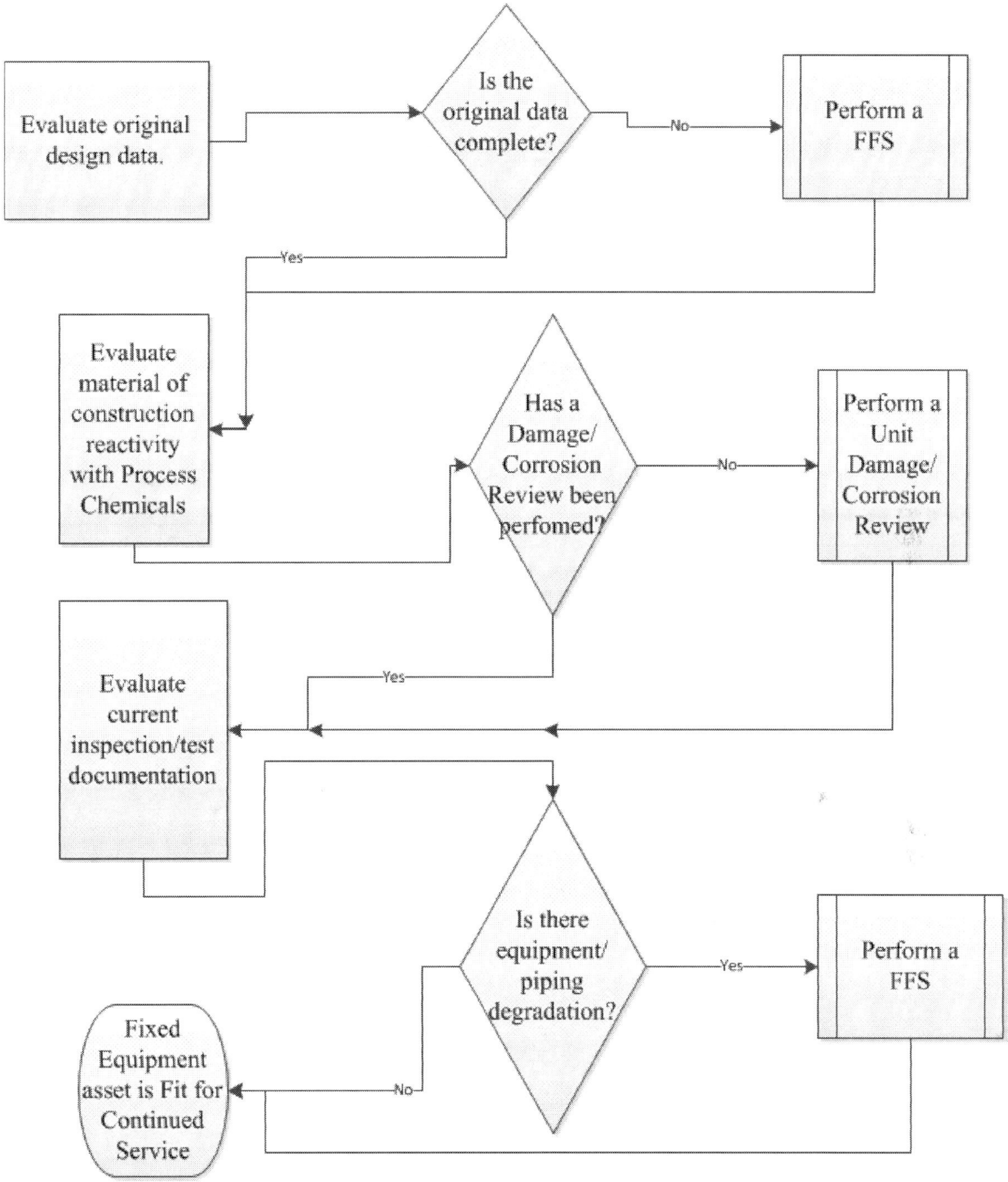

By working through the work/logic flow the PHA has a defined process usable for asking questions that will lead to a determination that is fit for service or to the action item that can result in a PHA recommendation.

The PHA team can evaluate the original fitness for service by verifying the equipment is suitable for the pressure, temperature and the corrosive environment of the process being evaluated. If data is missing an action item to perform a fitness for service (FFS) engineering analysis can be performed. In order for the PHA team to adequately evaluate the current inspection and test plan for the equipment it is necessary to understand what damage mechanisms may be anticipated. It can then be determined if the inspection and testing plan is effective in locating and quantifying the damage. If damage is identified the team will need to determine if the equipment is suitable for continued service. Determining if a fitness for service evaluation has been completed is necessary for the team. If a fitness for service evaluation was not performed; was the equipment repaired to the original design. If no repairs have been performed the team must determine if an FFS evaluation must be done. When the analysis is complete the team will issue action items or document the various safeguards such as the FFS that have been completed.

5. Conclusion

Fixed equipment failures and root cause investigations by agencies such as the Chemical Safety Board indicate that PHA teams may not be fully analyzing fixed equipment during PHA's. It is just as necessary that the likelihood of failure for vessels and piping be understood as it is for the team to understand, for example, the likelihood a pressure transmitter will fail. The team must understand all of the damage mechanisms that can have a detrimental effect on the equipment in order to effectively analyze the primary safeguard which is the MI program. Corrosion and cracking are the two major results from damage mechanisms. Crude feeds containing sulfur may lead to sulfidation corrosion. Water and oxygen in chemical streams or as a by-product of processing also result in corrosion to susceptible materials of construction. Chemical stream containing chlorides can lead to chloride induced stress corrosion cracking. There should be at least one member of the PHA team that understands these damage mechanisms. Where data is missing or incomplete the PHA team should make the recommendation to perform the analysis. Resolution of the PHA action items should either confirm fitness for continued service or result in the necessity to perform additional inspections and testing specific to any potential damage mechanisms. The risk of failures that result in release of hazardous chemical that can lead to injury, loss of life and property damage as well as loss of business opportunities is then lowered and the safeguards such as the Mechanical Integrity program are in place.

It is vitally important for the PHA team to analyze fixed equipment failures to the same degree as the consequences of failure of a control valve. The team must be completely convinced that all potential damage mechanisms that could result in a loss of containment failure have been analyzed. The team must also have an understanding of what potential damage can occur to the fixed equipment. By comprehensive analysis by the PHA team fixed equipment can be determined fit for service or appropriate action items can be identified.

6. References (examples given to show format)

(1) U.S. Chemical Safety and Hazard Investigation Board, "Chevron Richmond Refinery Interim Investigation Report" April 2013, http://www.csb.gov/chevron-refinery-fire/

(2) OSHA 1910.119, *Process Safety Management of Highly Hazardous Chemical*

REQUIREMENTS FOR IMPROVED PROCESS HAZARD ANALYSIS (PHA) METHODS: ADDRESSING WEAKNESSES IN THE HAZARD AND OPERABILITY (HAZOP) STUDY AND OTHER TRADITIONAL PHA METHODS

Paul Baybutt
Primatech Inc.
Columbus, Ohio, USA
paulb@primatech.com

Prepared for Presentation at
American Institute of Chemical Engineers
2014 Spring Meeting
10th Global Congress on Process Safety
New Orleans, LA
March 30 – April 2, 2014

UNPUBLISHED

REQUIREMENTS FOR IMPROVED PROCESS HAZARD ANALYSIS (PHA) METHODS: ADDRESSING WEAKNESSES IN THE HAZARD AND OPERABILITY (HAZOP) STUDY AND OTHER TRADITIONAL PHA METHODS

Paul Baybutt
Primatech Inc.
Columbus, Ohio, USA
paulb@primatech.com

Keywords: Process hazard analysis; hazard and operability (HAZOP) study; process safety.

Abstract

In order to develop better process hazard analysis (PHA) approaches, weaknesses in current approaches first must be identified and understood. Criteria can then be developed that new and improved approaches must meet. Current PHA methods share common weaknesses such as their inability specifically to address multiple failures, their identification of worst-consequence rather than worst-risk scenarios, and their focus on individual parts of a process. These weaknesses are identified and described in this paper to assist in the development of improved approaches. Knowledge of the weaknesses also allows PHA teams to compensate for them when performing studies.

Conventional wisdom holds that the Hazard and Operability (HAZOP) study is the most thorough and complete PHA method. Arguably, it is the most commonly-used PHA method in the world today. However, the method is not without its own weaknesses, some of which are not generally recognized. Therefore, a critique of the method is provided to assist in developing improved methods. Among the weaknesses discussed for the HAZOP study method are the extent to which design intent is defined and considered, completeness of parameter selection, efficiency in identifying hazard scenarios, the counter-intuitive inductive / deductive starting point, and the difficulty in focusing on specific hazard types.

Key criteria to guide the development of improved methods are proposed and discussed in the context of newer PHA methods, such as Major Hazards Analysis, that have been developed to address the limitations of traditional methods such as the HAZOP study. These key criteria include a structure that facilitates meaningful brainstorming of scenarios, ease of understanding and application of the method by participants, ability to identify scenarios efficiently, completeness of scenario identification, exclusion of extraneous scenarios, ease of updating and revalidating studies, and ease of meeting regulatory requirements.

Proposals are made for moving forward with the development of improved methods including the semi-automation of studies and improvements in the training of team members.

1. Introduction

Over the years, hazard studies for processes have evolved from a simple identification of hazards and control measures to detailed evaluation of hazard scenarios and the determination of the need for risk reduction. Hazard studies now involve both the identification of hazards in a process and the determination of hazard scenarios through which they may be realized. Initial hazard identification is often addressed using checklist methods [1]. Hazard scenarios are identified using process hazard analysis (PHA) methods, such as the Hazard and Operability (HAZOP) study. Although not intended specifically for the identification of all the hazards present in a process, PHA methods do provide information indirectly about hazards that are present. However, the focus in this paper is on the identification of hazard scenarios. Current PHA methods address the questions shown in Table 1. However, traditional methods are reaching their practical limits in their current form.

There are both inherent weaknesses in current PHA methods as well as weaknesses in how PHA is practiced, such as inadequate team composition. This paper focuses on the former. Individual PHA methods offer different advantages and disadvantages [1, 2]. However, current PHA methods share a number of weaknesses. In order to develop improved PHA approaches, these weaknesses must be addressed. Also, knowledge of weaknesses allows PHA teams to compensate for them. This paper identifies and describes key weaknesses in PHA methods.

Arguably, the HAZOP study method is the most commonly-used PHA method in the world today and is viewed by many practitioners as the most thorough and complete. However, the method is not without its own weaknesses, some of which are not generally recognized. Therefore, a critique of the method is provided to assist in developing improved methods.

Using the assessment of current methods, criteria are proposed that should be met by new and improved approaches. The paper concludes by offering some suggestions on possible PHA improvements.

2. Weaknesses in Current PHA Methods

This section identifies and describes numerous weaknesses in current PHA methods.

2.1 Role of subjective judgment

PHA studies depend on people to use their imagination and creativity to identify hazard scenarios. Study teams also use judgment in evaluating hazard scenarios. Consequently, PHA studies are inherently subjective. Notably, a US Occupational Safety and Health Administration (OSHA) publication stated [3], "PHA is dependent on good judgment". Subjectivity influences various aspects of studies including:

- Identification of initiating events - Credible initiating events are included in a PHA study while incredible ones are not. However, "credible" may mean different things to different people and subjective judgment is involved in its interpretation. Team judgment must be used to decide whether events are sufficiently probable to be considered credible. It is prudent to include events with a low probability of occurring because catastrophic events will be in a probability domain that is naturally low and such events are of most concern in PHA. They do not occur with a frequency that would cause them to be observed within the typical lifetime of a process but that does not mean that a plant will not experience one. While some data may be available to assist in decision making on credibility, data are often sparse or may not apply to the particular circumstances.

Human perception influences estimates of likelihood. A person's experience is reflected in their estimates. Usually, people underestimate the probability of an event they have not experienced and overestimate the probability of an event they have experienced. Thus, perception plays an important role in estimating the likelihood of events [4]. For example, some external events, such as a 500-year flood, likely will be outside the experience of facility personnel.

- Completeness of scenario identification - Typically, PHA teams brainstorm the identification of hazard scenarios until they can think of no more. Thus, scenario identification is subject to the vagaries of human judgment and the stamina of PHA teams.

- Scenario risk ranking - Risk ranking is performed by qualitatively estimating the severity and likelihood of hazard scenarios based on the collective knowledge and experience of the team. Estimates may not be consistent from team to team within a company or even within the same study owing to the subjective judgment involved. The estimation of scenario severities is easier than for scenario likelihoods. Most team members will have some appreciation of the possible spectrum of consequence severities, from the benign to the catastrophic, that derives from their personal experience or awareness of industry events. Estimating likelihoods for events with frequencies less once per 100 years, the nominal human lifetime, is more difficult for PHA team members. In order to eliminate some subjectivity, some companies provide specific guidelines for the assignment of scenario severities and consequences according to the type of scenario.

- Credit for safeguards - The assessment of scenario likelihoods includes the likelihood of failure of safeguards. Usually, such estimates are made using engineering judgment rather than actual data. PHA teams may optimistically assume the best possible performance rather than the worst-case. This is particularly true when teams credit the actions of people as safeguards. Usually, human failures rates are higher than many people realize.

2.2 Only departures from design intent are addressed

Many practitioners use PHA during the design of a process to improve the design. However, PHA focuses on looking for ways the process may deviate from the design intent and does not evaluate the adequacy of the design intent itself. Verification of the design intent is part of a formal design review and is outside the scope of PHA studies. Notably, the Center for Chemical

Process Safety (CCPS) states [1], "PHA should never be considered a substitute for an organization's customary design review activities. PHA is best used to supplement project design reviews."

Certainly, PHA cannot replace design reviews but it can augment them. Thus, a design review must address matters such as the operability of the process, application of inherently safer design principles, value engineering, and compliance with codes, standards and regulations. Of course, new designs may be hazardous even within the envelope of the design intent and design reviews must address such hazards.

This issue is of greater concern during the design of a process than for an existing process for which the design effectively has been proven in use. However, the issue does arise again when changes are made to an existing process.

2.3 Ability to address all aspects of design intent

Usually, hazard scenarios occur when there is a deviation from the design intent for a process. Unfortunately, there are many aspects of design intent (see Table 2) and it is a significant challenge to identify those aspects for which deviations will result in scenarios of concern. Various design representations are needed to portray design intent in order to address the different aspects. They may be drawings, procedures, etc. For example, traditional HAZOP studies use piping and instrumentation diagrams as the central reference document making such studies equipment-oriented. However, operating and maintenance procedures are important design representations when addressing human failures. Similarly, cause and effect diagrams must be consulted when addressing control system failures. The more aspects of design intent that are included in a study, the more design representations must be consulted by the team and the study becomes more onerous and frustrating for team members.

Not all aspects of design intent need to be considered for each part of a process and, indeed, the important aspects likely will vary for each part. Although this reduces the work needed by the team, the identification and consideration of varying intents from one part of the process to another contributes to the intellectual effort required of the team. PHA teams should not be overburdened by including unnecessary aspects of design intent. For example, construction intent should be addressed in a separate review to confirm compliance with the design.

2.4 Scenario detail

Traditional PHA methods, such as the HAZOP study, usually document hazard scenarios in worksheets that record scenario causes and consequences in columns but do not employ a column to capture intermediate events that may occur between a cause and its consequences. The operation of safeguards is part of the intermediate events for scenarios and, typically, safeguards are captured in a separate worksheet column, although this practice did not become common until the 1990's. Other aspects of the intermediate events may be recorded with the scenario consequences but scenario details that may be important for assessing the importance of the scenario and that may be needed for more detailed studies, such as layers of protection analysis (LOPA), are largely omitted. Also, the omission of details on intermediate events may make consequence entries harder to understand. Furthermore, the inclusion of intermediate events in the consequences column produces consequence entries that can be confusing.

A solution for this issue is to add an Events column to the worksheet to provide important scenario details, such as the specific location of a release, although this practice is not yet common. Such a column also can be used to capture process states, such as hazardous events, that occur as part of a hazard scenario.

Usually, enablers have not been addressed in PHA unless they are perceived to play a critical role for a scenario, such as the possibility of a disabled safety system. However, as more companies perform LOPA studies, there is an increased incentive to identify enablers during PHA studies [5]. Of course, enablers play key roles in real-world accidents and their inclusion in PHA studies is desirable [6]. A solution for this issue is to add an Enablers column to the worksheet and record important enablers and contributing causes for scenarios in the column.

Some hazardous materials pose multiple hazards. For example, the release of a flammable material may result in either a fire or an explosion. If both scenario variants are credible, they should be included as distinct scenarios in the PHA worksheet. Furthermore, there may be multiple scenarios for each hazard type. These are sometimes called incident outcomes. For example, a leak of a flammable material from a process may result in a jet fire, a flash fire, or a pool fire. All or some of these outcomes may be possible. Similarly, several different explosion scenarios may be possible including a boiling liquid expanding vapor explosion (BLEVE), an unconfined vapor cloud explosion, or a confined vapor cloud explosion. Other possibilities include delayed ignition as a flammable vapor cloud drifts before it ignites or simply disperses harmlessly. Delays before ignition of a released flammable material can affect the physical effects that occur and the scenario consequences. For example, immediate ignition of a flammable material within a few seconds of a release precludes the formation of a flammable cloud and flash fires and explosions. Options for the treatment of delayed ignition scenarios in PHA include recording the highest risk and/or the worst-case consequence scenario, or recording both the immediate and delayed ignition scenarios.

Two logical options for addressing incident outcomes are to identify all credible scenarios or choose the worst-case consequence scenario. However, the worst-case consequence may vary according to the receptors at risk, for example, for employees in the immediate area of a release it may be a jet fire but for employees further away it may be a flash fire.

Some incident outcomes can be further subdivided into incident outcome cases which are differentiated by the meteorological conditions that exist at the time of the release, such as wind direction and atmospheric stability class, if they affect the potential scenario consequences. Incident outcome cases are not usually addressed in PHA. Usually, the worst-case consequence is assumed. However, the worst-case consequence may occur so infrequently that its risk could be lower than the scenario that occurs under prevailing conditions. Partly, these are issues with weaknesses in the practice of PHA but partly they are also inherent limitations in the ability of PHA to address all conceivable scenario variations within a reasonable time period.

A related issue is that safeguards that operate successfully or partially may produce undesirable secondary effects. For example, the closing of a control valve that is part of a shutdown system to prevent overfill of a tank blocks the discharge of the feed pump which may lead to pump seal failure and a release if the pump is not successfully shut down. Such scenarios need to be addressed in PHA.

2.5 Identification of human failures

Human failures involving errors of omission are relatively straightforward to identify and to a lesser extent so are errors of commission and violations [2]. However, credible extraneous acts that may result in significant consequences are more difficult to identify owing to the multitude of possibilities and the imponderable nature of many of them.

2.6 Root causes of hazard scenarios

Usually, PHA does not address the fundamental root causes of scenarios such as human and organizational factors [7]. Typically, practitioners identify immediate or basic causes. Currently, there are no consistent practices on the level of causality that should be used in the hierarchy of causes. The deeper that PHA teams probe the hierarchy, the more detailed and time-consuming the study becomes, and the less feasible it becomes to complete the study within a reasonable time period. The key issue is how deep should teams go in order to identify needed risk reduction measures. A related issue is the role that other elements of process safety, such as audits and reviews, should play in eliminating root causes. The onus should not be on PHA or the team to address deficiencies in the implementation of process safety management programs.

2.7 Ability to identify multiple failures

Multiple failures involve two or more events occurring together, either at the same time or separated in time [8]. In the latter case, early failures are usually considered to be latent conditions or events. Multiple failures may involve the initiating event; other elements of the scenario, such as safeguards; and enablers with initiating events or other scenario events. Multiple failures may occur independently of one another or dependently. In the latter case of dependent failures, they can be as likely as single failures. Multiple failure scenarios may have more severe consequences than scenarios involving any one of their contributors. Thus, if multiple failures are not addressed in PHA, important scenarios may be missed and the risks of scenarios may be underestimated. Unfortunately, there is no simple or formal approach for identifying multiple failure initiating events using inductive PHA methods. The identification of multiple failures for other scenario elements is less difficult since they are defined as part of the scenario. For example, the consideration of the multiple failure of redundant safeguards, such as relief valves, is not as difficult as for multiple failure initiating events since scenario safeguards are identified in the PHA.

Unfortunately, since single failures are more easily addressed than multiple failures in current PHA methods, teams tend to focus on them and neglect multiple failures.

2.8 Consideration of dependent failures

Some apparently independent multiple failures may be dependent and the likelihood of the multiple failure scenario will then be higher than otherwise would be estimated. Dependent failures involve two or more failure events that are not independent of each other where the events are related causally [8].

Dependent failures can occur for both initiating events and events within hazard scenarios. Dependent failures within scenarios may occur between the initiating event and other scenario events, such as safeguard responses, and between other events within a scenario, such as the responses of redundant safeguards or sequential safeguard responses.

Common cause failures are a specific type of dependent failure where simultaneous multiple failures result from a single shared cause. Cascade, sequential or propagating failures are another type of dependent failure which can occur in a system of interconnected components where the failure of a preceding component triggers the failure of successive components in a type of chain reaction. One event leads to another in a sequence.

Dependent failures are potentially very important. They can be as likely as single failures since effectively they reduce to a single failure. For initiating events, they may lead to higher estimates of the risk of multiple failure scenarios and the inclusion of multiple failure scenarios that otherwise would be excluded on the grounds of credibility.
For multiple failure events within hazard scenarios, such as the dependent failure of multiple safeguards, they impact the assessment of scenario risk.

Current inductive PHA methods do not incorporate a formal consideration of dependent failures and must rely on the team to recognize their potential occurrence and importance.

2.9 Consideration of domino effects

Domino effects produce hazard scenarios that involve escalating consequences as a chain of linked events propagates throughout and beyond the process where they originate. Such scenarios have been the subject of accident survey studies and quantitative risk analyses but their identification in process hazard analysis (PHA) has received less attention [9].

Historically, to the extent that domino effects have been addressed in PHA, usually they have been considered as external events [2].This approach is limited. PHA teams may not recognize that the domino effect is part of a chain and that further domino effects may be possible. An improved treatment of domino effects in PHA has been proposed recently [9].

2.10 Identification of worst-consequence rather than worst-risk scenarios

Usually, scenario severity values are estimated assuming the worst-case consequence in which all safeguards fail. However, the worst-consequence scenario may not be the worst-risk scenario for the same initiating event, although some practitioners implicitly assume the two are the same. Thus, a team may judge a worst-case consequence scenario to be of tolerable risk when a less-severe consequence scenario originating from the same initiating event poses intolerable risk. For example, a small-bore pipe rupture leading to a vapor cloud explosion may have higher risk than a full-bore rupture. Even though the severity of the former is less than the latter, it may have a sufficiently higher likelihood to produce a higher risk. Furthermore, the worst-case consequence scenario can depend on the type of consequence. For example, If the pressure control system for a hydrocarbon vessel fails, the worst-consequence scenario for people is the failure of the pressure relief system which would result in an explosion. However, the worst-consequence scenario for the environment is the successful operation of the pressure relief system that routes the discharge to a flare which has environmental consequences if the flare fails. Also, scenarios involving the successful operation of a mitigation safeguard may have a higher risk than the scenario in which the safeguard fails.

The obvious solution in such cases is to capture both worst-consequence and worst-risk scenarios in the PHA worksheet. The greater concern will be with the worst-risk scenario but interested parties will also want to be assured that the worst-consequence scenario has tolerable risk. Unfortunately, identifying both scenarios is easier said than done as current qualitative PHA methods do not allow an easy determination of whether a less-severe consequence scenario originating from the same initiating event may have a higher risk than the worst-case consequence scenario.

2.11 Focus on individual parts of a process

In conducting PHA studies, usually the process is divided into sections, called nodes, or systems and subsystems, in order to focus the analysis and make it manageable. Unfortunately, such process subdivision may result in missing scenarios that involve multiple or all parts of the process. For example, the traditional HAZOP study method uses a line-by-line approach for noding in which nodes typically are defined as either individual lines or individual vessels. The analysis focuses on each node in turn and, thus, hazard scenarios that originate within two or more nodes may be missed. For example, if the failure of an operator to open the correct valve within one part of the process is being considered and the team does not consider the possibility of the operator also opening an incorrect valve in a different part of the process, an important scenario may be missed. Similarly, when considering a scenario involving the failure of a valve that allows ingress of a mixture of hydrocarbon and water into an uninsulated line, the enabler of low ambient temperature that allows freezing of water and rupture of the line may not be identified and, therefore, the scenario may be missed. The enabler is a factor external to the node.

Process subdivision leads to a de-emphasis on the identification of system incidents. Lines and vessels could be combined in super-nodes to minimize this issue. However, some hazard scenarios may then be missed for individual lines and vessels since they are no longer considered separately. Clearly, there is an inherent conflict between the desire to focus attention on specific parts of the process to be sure of capturing scenarios that originate within them and the need to address scenarios with contributors internal and external to the specific part of the process.

Most current PHA methods do not incorporate a formal search for such scenarios. As described earlier, it is more difficult to identify multiple failures than single failures but this is true especially when the multiple failures originate in different parts of the process.

Furthermore, some initiating events may affect the entire process or multiple parts of it, producing a global or system scenario that is not specific to any one part of the process. For example, some external events, such as flooding and the complete loss of utilities, e.g. electric power, can produce global scenarios. They should be identified in a global node.

2.12 Uniqueness of process subdivision

There is no unique way of subdividing a process into nodes, or systems and subsystems. It is likely that equivalent PHA results can be expected with alternative subdivisions of a comparable level of detail but this has not been proven. Consequently, it is an open question as to the extent to which process subdivision influences the results of a study. Certainly, if super-nodes are used that combine lines and vessels, it is likely that some scenarios may be missed owing to the difficulty of their identification in a more complex section of the process. Establishing a detailed set of guidelines for subdividing a process can help to address this issue. The goal should be to choose the optimal number of nodes, or systems and subsystems, in order to perform the study efficiently and effectively.

2.13 Utility and support system failures

PHA studies must address the failure of utility systems. Indeed, OSHA has stated in a written clarification that utilities connected to a process covered under the process safety management (PSM) regulations must be addressed in PHA [10]. Some examples of systems that can be viewed as utilities are provided in Table 3. Utility failures are considered to be external events and should be included in PHA routinely.

In addition to utility systems, other support systems are needed to operate a process (see Table 4). There are no rigid distinctions between utilities and support systems. However, the effects of some support system failures can be more challenging to address than for utilities and some PHA practitioners omit them from consideration. However, their failure can play an important role in the safety of a process and they should be addressed.

2.14 Treatment of modes of operation

PHA studies should be performed for all modes of process operation before the process experiences those modes. However, there is no consensus on how this should be accomplished and there are various approaches used by different practitioners [11]. Existing PHA approaches do not require any particular form of consideration of modes of operation, nor, indeed, do they encourage their treatment. Some practitioners focus on normal operation and conduct a single PHA. However, any claims that such a PHA adequately addresses other modes of operation are highly questionable owing to the focus on normal operation. Other modes of process operation should be studied at a comparable level of detail to normal operation.

2.15 Treatment of non-steady-state processes

Traditional PHA methods have been adapted to address non-steady-state processes, such as batch processes, by performing PHA for each step or stage in the process. However, deviations in the timing, duration and sequence of steps and stages, interactions between steps and stages, the operation of interlocks and permissives, and the occurrence of simultaneous operations within the process can be challenging to address. Improvements in their treatment are needed.

2.16 Addressing human factors issues

Regulations such as OSHA's PSM standard require that human factors be addressed as part of PHA. However, there are limitations on what can be covered in PHA for this important aspect of process safety.

Human failures as a cause of hazard scenarios and the factors that impact them must be addressed. The former is a natural part of PHA and the latter can be included by elaborating on the reasons for the identified human failures. However, the performance of a separate human factors study that precedes the PHA is beneficial as it allows a complete focus on this critical aspect of process safety and helps to ensure that human factors issues are identified that might not arise if human factors are addressed only within PHA. The results of such studies can be utilized when PHA is performed. This approach better accommodates OSHA's expectation that human factors issues for a process be identified, evaluated and controlled and the results of the analysis justified [12].

2.17 Addressing facility siting issues

As for human factors, regulations such as OSHA's PSM standard require that facility siting be addressed as part of PHA. However, there are limitations on what can be covered in PHA for this important aspect of process safety.

Traditionally, facility siting has a broad interpretation and includes consideration of the location of a facility, spacing of process units, spacing between equipment, spacing between equipment and potential ignition sources, and domino effects. It can also address some emergency response issues, such as accessibility of fire hydrants and monitors, and the adequacy of hazardous area classifications. However, from a regulatory perspective, a key concern is with the spatial relationship between the hazards of a process and the locations of people in the process and facility, particularly in occupied buildings, such as control rooms. As for human factors, some facility siting issues can be addressed readily within PHA, but others are better examined separately in a study that precedes the PHA to provide the opportunity to address issues that might not arise if facility siting were addressed only within PHA. Moreover, quantitative analyses are often needed to assess facility siting issues and these necessarily must be conducted outside a qualitative PHA study. The results of a separate facility siting study can be utilized when PHA is performed. This approach better accommodates OSHA's expectation that facility siting issues for a process be identified, evaluated and controlled and the results of the analysis justified [12].

2.18 Interactions between processes

Causes of failures in one process may originate within other processes. For example, failures in processes that feed the process being studied may cause failures within it. Thus, the PHA studies for the other processes may need to be consulted to identify failures that may be important for the process being studied. However, it is possible that failures of importance for the process being studied were not addressed in the PHA studies for the other processes because they were not important for those processes. In such cases, those PHA studies would need to be updated. This is an issue whenever a facility is divided for the purposes of PHA into individual processes that are linked.

2.19 Conservative assumptions

Owing to the subjective nature of PHA and the uncertainties involved, conservative assumptions often are made throughout the performance of PHA. The end result can be considerably conservative results that may lead to the inappropriate expenditure of resources that would be better invested elsewhere or in a different way. Managing the subjectivity that enters PHA studies can help to address this issue, for example, by challenging teams to justify what appear to be overly severe estimates of scenario severities.

2.20 Prediction of real-world accidents

Often, accidents that occur in processes involve combinations of many events and it is highly unlikely that a PHA team would ever have suggested the particular sequence of events in a PHA study. At best, a PHA team can identify the most critical initial events of the sequence so that they can determine suitable risk reduction measures.

Furthermore, real-world accidents can involve the realization of one hazard followed by another when a hazardous material poses multiple hazards, for example, a fire followed by an explosion. In order to make PHA studies manageable, usually the hazards are treated separately. There can be an infinite number of possible combinations of events and it is beyond the capablity of PHA to identify them all. Generally, it is not necessary to do so, if addressing each hazard individually will help manage the risk of all such sequences of events.

3. Weaknesses in the HAZOP Study Method

The HAZOP study method is used to identify hazard scenarios that impact receptors such as people, the environment and property as well as operability scenarios where the concern is the capacity of the process to function. Originally, the HAZOP study method was developed for fluid processes found in the process industries but it has also been applied to non-fluid systems such as materials handling, drilling operations, aerospace systems, etc. The HAZOP study method focuses on investigating deviations from design intent, such as "no flow" at a location in the process where flow is intended or "high pressure" in a vessel which should not exceed a certain limit. By definition, deviations are potential problems, e.g., lack of flow in a transfer line or overpressuring a storage tank. Deviations from design intent are generated by applying guide words to process parameters at different locations, called nodes, throughout the process, e.g., for an inlet line to a vessel, no + flow = no flow, or for a vessel, high + pressure = high pressure. The process parameters represent aspects of the design intent for the node.

A standard list of seven guide words is used: No, More, Less, As Well As, Part Of, Reverse, and Other Than. The HAZOP team chooses appropriate parameters for each node, e.g., flow, pressure, temperature, composition, level, addition, cooling, etc. The use of guide words with parameters provides the opportunity to explore deviations from design intent in every conceivable way, thus helping to ensure completeness of scenario identification.

The HAZOP team brainstorms causes of each deviation within each node and identifies the sequence of events that results for each cause including safeguards that may fail and the consequences. The sequence of events represents a hazard scenario. Usually, a qualitative estimate is made of the scenario consequence severity and likelihood so that a risk ranking may be produced.

The HAZOP study method has been the topic of books [13, 14], guides [15, 16] and an international standard [17] and it is designated as an acceptable PHA method by regulations around the world, for example, OSHA's PSM standard. It boasts a long history of application. However, the method shares common weaknesses with other PHA methods described in the previous section and it has weaknesses of its own which are described below.

3.1 Incomplete consideration of design intent

The HAZOP study method shares with other PHA methods the difficulty of addressing all key aspects of design intent. However, the procedure that commonly is employed to generate deviations exacerbates the problem. Often, practitioners select parameters to consider for a node from a relatively short checklist. Indeed, company PHA procedures may provide the checklist to be used. Unfortunately, sometimes practitioners do not consider process parameters that may be important but that are not on the checklist. Thus, the procedure used to conduct the HAZOP study itself acts to inhibit consideration of other aspects of node intention that may be important. As a consequence, important scenarios can be missed. Even if practitioners attempt to address other parameters representing important aspects of design intent, the process of doing so is like gazing into a crystal ball and trying to predict the future or which aspects of design intent, if addressed, will generate scenarios of concern. This is the Achilles heel of the HAZOP study method and it has not been widely recognized. The formal structure of the HAZOP study method can provide a false sense of security. However, the team can never be sure when all key aspects of design intent have been addressed. A large number of parameters may need to be considered and it is not clear when a full set has been addressed.

3.2 Compound deviations

Compound deviations, for example, the occurrence of the contamination of material in a vessel along with the independent occurrence of high temperature of the contents that might cause an explosion, usually are not considered in the HAZOP study method owing to the difficulty of doing so. While it can be argued that the occurrence of such compound events may be of low likelihood, they cannot be dismissed entirely on those grounds. It is possible that some compound deviations could be addressed using enablers but it is not common practice.

3.3 Inductive / deductive starting point is counter-intuitive

The HAZOP study method begins with the generation and consideration of deviations from design intent. This aspect of the HAZOP study method implies that deviations are the central element in PHA. However, deviations are not an end in themselves but are used as an aid in identifying initiating events or causes of hazard scenarios. A deviation from design intent is actually part of the hazard scenario. It immediately follows the initiating event or cause and can be considered the first intermediate event (see Figure 1). This aspect of the HAZOP study method, which involves jumping to a deviation and then reversing to consider causes before procedeeding forward again, can be confusing for novice practitioners and can lead to confounding deviations with causes or consequences.

3.4 Operability issues are included with hazard scenarios

The HAZOP study method is intended to identify both safety and operability scenarios. Operability means the capacity of the process to function. Some practitioners want to include both types of scenarios in a study. However, other practitioners desire only to address safety scenarios. Unfortunately. it is difficult to exclude operabilty scenarios and address hazard scenarios exclusively because the consequences of scenarios may not be apparent until causes have been identified. At that point, much of the work on scenario identification has been performed. Typically, operability scenarios constitute 50% or more of all scenarios identified in a HAZOP study and, consequently, their inclusion can more than double the time required for a study. Obviously, this is a problem when practitioners would prefer to exclude them and militates against the use of the HAZOP study method in such cases.

3.5 Difficulty in focusing on specific hazard types

PHA studies often focus on the major hazards of toxicity, flammability, explosivity and reactivity. These hazards are the only specific ones required to be addressed under OSHA's PSM standard but companies may choose to address additional hazards. It would be valuable if PHA methods could be tailored to address only those hazards of concern to permit more efficient studies. While the HAZOP study method can accommodate various hazard types, it does so by addressing those aspects of design intent for which deviations might result in the realization of the hazards of concern. This indirect method of addressing hazards of concern can lead to the consideration of deviations which do not produce such scenarios and the consequent waste of time spent addressing them.

3.6 Equipment focus

The conventional HAZOP study method focuses on equipment. Separate extensions have been developed to address procedures [18] and control systems [19, 20]. It would be preferable if all pertinent scenarios could be identified in a single study.

3.7 Technical vocabulary

The HAZOP study method employs technical terms that are unfamiliar to facility personnel who have not used the method. Terms such as design intention, node, deviation, guide word, parameter, parameter intention, and design representation can cause confusion. It is not unusual for people, even practitioners, to confound some of the terms.

3.8 Studies are lengthy

Usually, PHA studies performed using the HAZOP study method take more time than other methods. Lengthy studies can have an adverse impact on team performance as team members can become fatigued and jaded [21]. HAZOP studies take longer because they involve a detailed search for hazard scenarios but also because of inefficiencies in the search process. These inefficiencies include the identification of operability scenarios when they are not desired and the consideration of process deviations that do not result in scenarios of concern. The HAZOP study method is theoretically attractive but practically limited.

3.9 Presentation of results

Usually, hazard scenarios in HAZOP studies are captured in individual worksheets for a particular combination of a node and parameter. Thus, scenarios for each node are spread across several worksheets. It is preferable for a PHA method to organize all scenarios for a node in a single worksheet for ease of use and reference.

4. Proposed Criteria for New and Improved PHA Methods

Several attempts have been made to develop improved PHA methods including Structured What-If (SWIFT) [22], Major Hazard Analysis (MHA) [23], and Process Hazard Review (PHR) [24]. In developing new and improved methods, the overall objectives of PHA must be defined together with criteria that should be met in accomplishing the objectives.

PHA methods should have as their principal objective the identification and evaluation of hazard scenarios that are possible for a process for the purpose of determining if the risks of the scenarios are tolerable. Perhaps a better generic term for such methods would be Process Hazards and Risk Analysis (PHRA). Such methods identify and evaluate controls for process hazards. They allow the determination of the need for additional controls and response actions in the event that scenarios were to occur.

There is little guidance in the literature or regulations on technical requirements for hazard analysis. Of course, the weaknesses in existing methods that were described in the previous sections should be addressed to the extent feasible. The following criteria are suggested for new and improved PHA methods for the process industries. For convenience, they are divided into administrative and technical categories.

4.1 Administrative Criteria

4.1.1 Appropriate for the process industries

A number of the traditional PHA methods, such as Failure Modes and Effects Analysis (FMEA), were developed for use in other industries. While they have been applied in the process industries, they do not always meet the needs of the process industries. For example, FMEA does not do a very good job of addressing human failures and external events. Although, generally, it should not be a technique of choice as the principal means of identifying hazard scenarios for a process, it does find specialty application in the process industries, for example, in reliability-centered maintenance programs. The HAZOP study method was developed specifically for the process industries which is one of the reasons it found rapid acceptance in those industries. New and improved PHA methods for the process industries must address the needs of those industries.

4.1.2 Meet regulatory requirements and industry practices, codes and standards

Requirements for PHA have been established in regulations such as OSHA's PSM standard. Also, industry publications have established expectations for PHA practices [1, 16, 17]. An example of a PHA regulatory requirement from OSHA's PSM standard is the need to identify previous incidents that had a likely potential for catastrophic consequences.

Practitioners continue to evolve the practice of PHA. Their work is published in the research literature and in books and articles. As new and improved practices become accepted and are adopted, they become de facto, if not de jure, industry standards.

Obviously, a PHA method must be capable of meeting applicable regulatory requirements, as well as recognized and generally accepted good engineering practices.

4.1.3 Non-proprietary

A PHA method should be accessible to all who wish to use it so that it is transparent to users and interested parties and a wide base of users will be encouraged who will help to further develop and refine the method.

4.1.4 Ease of understanding and application by participants

The easier it is for participants to understand a PHA method, the more effectively they will be able to use it. The easier a method is to apply, the more likely participants will embrace it and be willing to work with it.

4.1.5 Team approach

The combined skills of a multi-disciplinary team have been found to be necessary to identify hazard scenarios properly. This established practice is based on the premise that a group effort is better than the sum of individual efforts. It is highly unlikely that any one individual will know and understand everything that is needed to be able to identify fully the hazard scenarios for a process. Also, the interactions and discussions that occur within a PHA team are essential for a full identification of scenarios.

4.1.6 Facilitator

PHA studies must be managed to ensure they are performed efficiently and effectively. A facilitator, or leader, must guide the team systematically through the facility design so the chosen PHA method can be applied appropriately. The brainstorming that occurs within PHA sessions can easily get out of hand. The facilitator must manage the process carefully.

4.1.7 Brainstorming scenarios

Brainstorming should be an essential aspect of PHA in order to stimulate creativity, generate new ideas, and encourage team members to look at the process in a different way. It helps to overcome mindsets and it encourages the PHA team to identify novel scenarios rather than just referring to a checklist of common scenarios.

4.1.8 Consistency

A PHA method should help to ensure that the analysis is performed in the same way within all parts of a study and across multiple studies conducted by a company. For example, usually the level of detail identified for initiating events should be the same for all studies conducted by a company and the PHA method should help to ensure this is the case. The results of a PHA study should not depend on how the method is applied or which team conducts the study.

4.1.9 Structure

An orderly and systematic approach to scenario identification should be followed to avoid missing scenarios. This is particularly true given the brainstorming involved in PHA which easily can lead to digressions and sidetracking the team. However, there can be a conflict between structuring a PHA method and brainstorming scenarios. A suitable balance must be achieved. The structure of a study includes a road map for the team to follow as the study progresses. Structure also facilitates consistency.

4.1.10 Logical

The approach used in a PHA method must make sense to team members so that they will embrace and not oppose use of the method. Also, they must be able to relate to the technical vocabulary used in applying the method. Ideally, it should use technical terms with which they are familiar already.

4.1.11 Presentation of results

PHA methods should produce and organize study results in a form that can be understood readily by interested parties such as management, facility personnel, emergency planners, regulators, and others. Critical equipment and critical actions should be flagged so that controls for them can be addressed by the company's safety management system.

4.1.12 Ease of updating and revalidating studies

Process safety regulations usually require the periodic updating and revalidation of PHA studies for a process, typically every five years, to ensure that the PHA is consistent with current process at that point in time. Updating is the process of making corrections, edits, and/or adding new content. Revalidation means declaring the PHA valid again. Periodic updating and revalidation provides an opportunity to perform an integrated evaluation of the cumulative, and possibly synergistic, impacts of all changes. Additional items may be addressed during PHA revalidations including omissions and deficiencies in the previous PHA, new process information or technology, and new technical and regulatory requirements. Although PHA studies may not be required to be updated for process changes by regulations, some companies do so for certain types of changes, such as major or extensive changes, and high risk processes or changes.

PHA methods need to provide a systematic and logical editing process to accommodate accurate revisions of PHA worksheets.

4.1.13 Ease of use for other process safety purposes

Information contained in PHA worksheets is valuable for use with other aspects of process safety. For example, it can be used for the identification of critical equipment for a mechanical integrity program and critical actions for writing standard operating procedures, training process personnel, and the development of emergency response plans. Such uses of PHA study results help to justify the investment in conducting PHA studies. PHA methods should record studies in ways that facilitate the various uses of study results. For example, equipment should be identified by tag number and name, and operator actions should be associated with specific steps in procedures.

4.1.14 Conversion of previous studies

In developing new PHA methods, it is important that a smooth transition be possible from the previous methods used and the form in which PHA studies may have been recorded. Many companies have large numbers of studies performed using traditional methods, such as the HAZOP study method, which represent a significant investment. Thus, in choosing to use an improved method, ideally it should be possible to preserve and convert the earlier studies. Typically, conversion is accomplished when studies are revalidated.

4.1.15 Continuous improvement

A PHA method should allow for improvements over time and the incorporation of new technology.

4.2 Technical Criteria

4.2.1 Able to address all types of hazards

Processes pose many types of hazards. Those of principal concern in process safety are the major hazards of toxicity, flammability, explosivity and reactivity. A PHA method must address these hazards equally well. Ideally, the method should be able to handle other hazard types as needed, for example, asphyxia from inert materials such as nitrogen.

4.2.2 Tailored to hazards of interest

It would be exceptionally time consuming, verging on the infeasible, to address every hazard that is posed by a process using PHA. Indeed, many occupational health and safety hazards, such as those of slips, trips, and falls, are better addressed using simpler methods such as checklists and job safety analysis. Consequently, it is desirable if a PHA method can be used to address only those hazards of interest for a particular study and to do so directly.

4.2.3 Exclusion of extraneous scenarios

PHA practitioners may want to exclude certain types of scenarios from a study, for example, operability scenarios or scenarios from lesser process hazards. A PHA method should be capable of excluding undesired scenarios so as to contribute to the efficiency of the study.

4.2.4 Adjustable to the complexity and circumstances of the process

Processes vary in their complexity. They differ in chemistry, equipment, controls, safety systems, sequence of operations, numbers and types of hazards, storage volumes, production volumes, etc. They also differ in size, operating life and history, magnitude of their hazards, and in other ways. Ideally, a PHA method should be able to accommodate such variations without the need to switch between methods.

4.2.5 Reliance on subjective judgment

Engineering judgment has been, and likely will continue to be, a key aspect of PHA studies. Attempts have been made to automate the HAZOP study process using computer software but no completely successful approach has yet been devised. It is questionable whether the creativity of people can be replaced by computer algorithms, at least at this point in time. Furthermore, the regulatory acceptance of PHA studies performed automatically by computer is likely to be an issue. However, the dependence of PHA results on the subjective judgment of team members is not a satisfactory situation. Therefore, a PHA method should minimize the need for subjective judgment to the extent that it does not interfere with scenario identification.

Reducing subjectivity in PHA can reduce conservatism in the analysis and help to avoid overly-conservative conclusions.

4.2.6 Completeness of scenario identification.

PHA methods must be capable, at least in principle, of identifying a full set of scenarios. Techniques such as fault tree analysis and event tree analysis focus on a specific and restricted set of scenarios. They are valuable techniques to augment the identification of scenarios but, by themselves, they do not identify a full set of scenarios.

A PHA method should be capable of addressing multiple failure scenarios, domino effects, all types of human failures, utilities and support system failures, and system and global scenarios.

4.2.7 Coverage of design intent.

A PHA method must be able to address causes of any type of process deviation. As currently practiced, many PHA studies focus on equipment, although some current methods, such as the HAZOP study, have been extended to address procedures [18] and control systems [19, 20] explicitly. However, usually these extensions are made in separate studies. It would be preferable to identify all hazard scenarios within the same study for the sake of efficiency and ease of practice by PHA teams.

4.2.8 Level of detail

A PHA method should generate and record scenarios in sufficient detail for the development of recommendations for risk reduction, and for other purposes such as performing LOPA studies. Studies should address intermediate events, enablers, dependent failures, and other relevant aspects of scenarios. Both worst-case consequence and worst-case risk scenarios should be addressed.

PHA methods must address the impact of human factors and facility siting on scenarios. Also, they should address scenarios that involve interactions between connected processes.

4.2.9 Sequential order of events

People intuitively understand the progression of hazard scenarios from cause to consequence. Therefore, it is valuable if PHA methods follow this progression so that teams identify the elements of scenarios in sequential order.

4.2.10 Robust to team inadequacies

PHA study results are subject to team member bias, motivation, experience, knowledge and creativity. Study success also depends on the interactions of the team members. Team members may be discouraged by the repetitive nature of the analysis which may adversely affect the study quality. It is unrealistic to expect that a PHA team will function in a perfect way, or even in the same way from session to session, owing the nature of people. Thus, PHA study methods should try to minimize the impact of these issues, for example, by minimizing repetition of scenarios, and avoiding time spent identifying obvious scenarios.

4.2.11 Efficiency of scenario identification

Scenarios should be identified in the most efficient way with the minimum time and effort necessary. PHA studies are time consuming and intellectually demanding for participants. Every effort must be made to avoid burnout of team members. The more time consuming a study, the less willing team members will be to spend the time needed to identify scenarios fully. They will convince themselves they have done an adequate job even though more work may be needed. Efficient studies likely lead to better results through the positive impact on team morale. Also, more time is provided for the identification of additional scenarios.

4.2.12 Robust and meaningful scenario risk estimation

The only meaningful basis for making decisions on the need for risk reduction measures for hazard scenarios is to estimate scenario risk and use comparisons of the estimates with risk tolerance criteria to determine if there is a risk gap. Commonly, risk matrices are used for this purpose in PHA to provide a risk ranking of scenarios as a means to distinguish between and prioritize them. However, issues have been identified with the use of risk matrices [25] and improved approaches for risk estimation in PHA are needed. Also, the subjectivity in current risk estimation practices needs to be minimized.

4.2.13 Specific measures for risk reduction

The application of PHA methods should lead to the development of specific risk reduction measures that may be needed to meet tolerable risk criteria. Scenarios must be recorded with sufficient detail to ensure that appropriate remedial actions can be developed. Recommendations for needed risk reduction measures may be developed by the PHA team as the study proceeds or they may be determined once a full set of hazard scenarios has been developed, possibly by a different team.

4.2.14 Use throughout the process life cycle

PHA should begin at the earliest stages of process design in order to avoid expensive design changes later on. As the design of a process evolves, its hazards should be assessed repeatedly. Commonly, simple PHA methods are used early in the design process and more advanced methods are used later on. Similarly, PHA should be applied to each mode of process operation throughout its life cycle. A preferred approach is to use a single PHA method that can accommodate each stage of design and each mode of operation. Focusing the training and experience of practitioners and participants on a single method has the advantage of more fully developing their skills with the method than if several different methods were used.

PHA methods should be capable of adequately addressing issues in the identification of scenarios for non-steady-state processes.

4.2.15 Analysis of process changes

PHA studies on changes can be conducted as a new study or the existing PHA for the process can be updated. It is desirable that the PHA method used should easily accommodate addressing the change or updating the PHA. For example, it is harder to perform a PHA on a change using the HAZOP study method rather than the What-If method as it is more difficult to use the former than the latter method to focus precisely on the change owing to the particular structure of the HAZOP study method.

Certainly, developing a PHA method that meets all of these administrative and technical criteria is a challenge. However, articulation of such criteria is the first step in moving forward with the development of improved PHA methods.

5. Possible Improvements in PHA

The purpose of this paper was to provide a critique of current PHA methods and to propose criteria for new and improved methods which has not been done previously. Although the intent was not to propose a specific new method that meets the criteria, some suggestions are provided here on how to move forward.

While full automation of PHA using computer software has not yet proven successful, software can function to prompt team members with possible worksheet entries and as a quality control tool to help ensure items are not overlooked. A great deal of information has been accumulated over the years about typical entries in PHA worksheets. Much of this information has been peer reviewed. The information can be assembled into data bases that can suggest possible worksheet entries to team members. Such partial automation of the more routine aspects of PHA can reduce the cognitive load on team members and provide them with more time to consider less-obvious aspects of scenarios. Also, it avoids the issue of regulatory acceptance of such PHA results as the team still decides on the entries to be made. Data bases also can be used to provide checks on worksheet entries. Such approaches help to reduce subjectivity in the study results and they can compensate for less-than-adequate performance by teams. Furthermore, software can be programmed to check worksheet entries for compliance with study guidelines leading to more consistent studies.

Regardless of what PHA method is used, training teams in scenario recognition, not just the mechanics of PHA, is valuable. PHA team members must be able to relate what they know of incidents they have experienced to the format in which hazard scenarios are identified and recorded. Real-world examples should be used, preferably from the facility where participants work. Videos from the Chemical Safety Board can be used to bring scenarios to life. Participants must understand important concepts for hazard scenarios such as multiple failures, domino effects, dependent and common cause failures, and latent failures and enablers in the context of real-life incidents, otherwise PHA studies can seem like theoretical exercises whose value may be questioned.

In those parts of PHA where subjective judgment is critical, such as the assessment of the credibility of scenario causes and the assessment of the scenario likelihoods, team members should practice making the assessments using scenarios that have already been assessed by experts so that they can make a comparison with their own assessments and calibrate their judgment against that of experts. The process provides the benefit of sensitizing team members to the difficulty of making objective judgments in PHA.

6. Conclusions

Current PHA methods suffer from a variety of weaknesses which almost certainly result in incomplete studies with hazard scenarios being missed. Most practitioners view the HAZOP study method as the most thorough and comprehensive of the traditional PHA methods. However, the HAZOP study method shares weaknesses with other PHA methods and also has weaknesses of its own. Weaknesses in current PHA methods, including the HAZOP study method, were used to propose criteria that should be met by new and improved PHA methods. Knowledge of weaknesses also allows PHA teams to compensate for them when performing studies. Some suggestions were made for moving forward with the development of improved methods including the semi-automation of studies and improvements in the training of team members.

7. References

[1] "Guidelines for Hazard Evaluation Procedures", 3rd Edition, Center for Chemical Process Safety / American Institute of Chemical Engineers, 2008.

[2] Baybutt, P., Analytical Methods in Process Safety Management and System Safety Engineering – Process Hazards Analysis, in Handbook of Loss Prevention Engineering, Haight, J. M. (ed), Wiley-VCH, 2013.

[3] OSHA Publication 3133, Process Safety Management Guidelines for Compliance, 1994.

[4] Baybutt, P., The Role Of People And Human Factors In Performing Process Hazard Analysis And Layers Of Protection Analysis, J. of Loss Prevention in the Process Industries, Vol. 26, pps.1352-1365, 2013.

[5] Baybutt, P., Analytical Methods in Process Safety Management and System Safety Engineering – Layers of Protection Analysis, in Handbook of Loss Prevention Engineering, Haight, J. M. (ed), Wiley-VCH, 2013.

[6] Baybutt, P., Addressing Enablers in Layers of Protection Analysis, Process Safety Progress, in press.

[7] Baybutt, P., Initiating Events, Levels of Causality, and Process Hazard Analysis, Process Safety Progress, in press.

[8] Baybutt, P., Treatment of multiple failures in process hazard analysis, Process Safety Progress, Vol. 32, Issue 4, pages 361–364, December 2013.

[9] Baybutt, P., The Treatment of Domino Effects in Process Hazard Analysis, 1st CCPS Asia-Pacific Conference on Process Safety, Qingdao, China, September 4 - 5, 2013.

[10] OSHA letter to Mr. Lynton D. Barnett, American Cyanamid Company, March 10, 1994.

[11] Baybutt, P., Process hazard analysis for phases of operation in the process life cycle , Process Safety Progress, Vol. 31, No. 3, pps 279–281, September, 2012.

[12] Petroleum Refinery Process Safety Management National Emphasis Program, OSHA CPL 03-00-004, June 7, 2007.

[13] Kletz, T. A., Hazop & Hazan: Identifying and Assessing Process Industry Hazards, Fourth Edition, CRC Press, 1999.

[14] Redmill, F., Chudleigh, M. and Catmur, J., System Safety: HAZOP and Software HAZOP, Wiley, 1999.

[15] A Guide to Hazard and Operability Studies, Chemical Industries Association (1977).

[16] Tyler, B., Crawley, F., Preston, M., HAZOP: Guide to Best Practice, 2nd Edition, Institution of Chemical Engineers, 2008.

[17] Hazard and Operability Studies (HAZOP Studies) – Application Guide, IEC 61882:2001.

[18] Bridges, W. G., Lorenzo, D. K., Kirkman, J. Q., Addressing Human Errors During Process Hazard Analyses, Chemical Engineering Progress, May 1994.

[19] Andow, P., Guidance on Hazop Procedures for Computer-Controlled Plants, HSE Books, 1991.

[20] Kletz, T. A., Computer Control and Human Error, Gulf Professional Publishing, 1995.

[21] Baybutt, P., On the Ability of Process Hazard Analysis to Identify Accidents, Process Safety Progress, Vol. 22, No. 3, pps. 191–194, September, 2003.

[22] Rausand, M., Risk Assessment: Theory, Methods, and Applications, John Wiley & Sons, 2011.

[23] Baybutt, P., Major Hazards Analysis: An Improved Method for Process Hazard Analysis, Process Safety Progress, Vol. 22, Issue 1, pps. 21–26, March 2003.

[24] Ellis, G. R., Process Hazard Review: The Efficient Risk Assessment of Existing Plants, Loss Prevention and Safety Promotion in the Process Industries, 11th International Symposium, 2004.

[25] Cox, L.A., What's wrong with hazard-ranking systems? An expository note, Risk Analysis, Vol. 29, pps. 940 - 948, 2009.

Table 1. Questions Addressed by PHA

Question	Item
What can go wrong?	Initiating event
What happens?	Scenario events including consequence
How bad could it be?	Scenario severity
How often could it happen?	Scenario likelihood
What is the risk?	Combination of severity and likelihood
Is the risk tolerable?	Comparison of estimated risk with a risk tolerance criterion, considering existing safeguards
If not, what actions are needed to reduce the risk? (If included within the scope of the study)	Recommendations for corrective action

Table 2. Aspects of Design Intent for a Process

Equipment
Process materials
Materials of construction
Conditions
Properties
Operations
Actions
Reactions
Functions
Specifications
Environment
Locations
Positions
Elevations
Measurements
Controls
Software
Maintenance
Calibration
Testing
Sampling
Services / utilities
Communications
Timing
Sequence and order
Etc.

Table 3. Examples of Utilities

Water (process / service, cooling, refrigerated, boiler feed water, fire)
Steam
Nitrogen
Utility gases
Natural gas
Instrument air
Electric power
Uninterruptible power supply
Standby power supply
Cooling / heating medium
Fuel
Lube oil

Table 4. Examples of Support Systems

Communication system
Fire prevention and protection systems
Fire fighting system
Flare
Incinerator
Scrubber
Ventilation system
Medical services
Breathing air
Sewer system
Waste water treatment
Weather station
Access control
Lighting (normal, emergency)
Portable combustible gas and toxic meters
Evacuation alarms

Figure 1. The Role of Deviations in the HAZOP Study Method

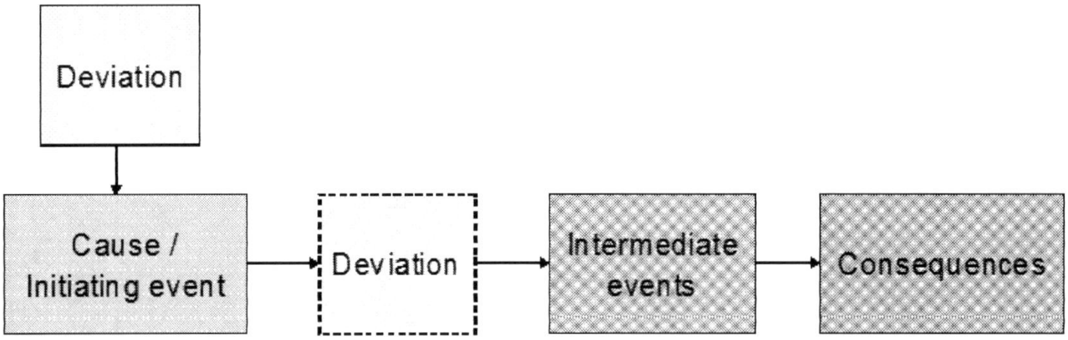

Driving Consistency in the Estimation of Severity Levels in PHA Studies

Marc Guindon
Suncor Energy
1900 River Road, Sarnia, ON, Canada
mguindon@suncor.com

Christopher Wells, Suncor Energy
Rich Piette, Suncor Energy
Greg Knight, ABS Consulting
Gary Fitzgerald, ABS Consulting

Prepared for Presentation at
American Institute of Chemical Engineers
2014 Spring Meeting
10th Global Congress on Process Safety
New Orleans, LA
March 30 – April 2, 2014

UNPUBLISHED

Driving Consistency in the Estimation of Severity Levels in PHA Studies

Marc Guindon
Suncor Energy
1900 River Road, Sarnia, ON, N7T 7J3, Canada
mguindon@suncor.com

Chris Wells, Suncor Energy
Rich Piette, Suncor Energy
Greg Knight, ABS Consulting
Gary Fitzgerald, ABS Consulting

Keywords: PHA, Estimation, Severity, Likelihood, Consistency, Releases

Abstract

During the course of PHA studies, team members are required to categorize the severity level for releases of hazardous materials associated with various cause/consequence pairings. Typically the assessment of severity level is performed in a "qualitative" fashion such that there is a potential that the team may either overstate or understate the impact of a release which will result in inaccuracies in the risk assessment. This may be due to differing perceptions or experience levels of PHA study teams.

In the absence of detailed consequence analysis modeling for a given release scenario, PHA study teams may utilize a table similar to the one provided in Table 3.1 – Example Consequence Categorization contained in the CCPS concept book "*Layers of Protection Analysis*". Alternatively tools such the Dow F&EI or CEI tools can be utilized to estimate the severity level. However all of these tools have limitations in their ability to properly assess severity levels for operating units with differing levels of process congestion or population density, etc... which may result in unconservative categorizations of severity levels.

This paper outlines a methodology and associated tool developed to estimate the severity level for flash fire, vapour cloud explosion, jet fire, pool fire, and toxic exposure consequences for various hazardous materials.

Using some basic Process Safety Information associated with the release scenario and the process unit under study, the PHA study team can use the tool to obtain the severity category as well as the associated inherent risk (prior to the application of safeguards) for a given release. The study team can then proceed with the application of safeguards to determine the residual risk (following the application of safeguards) associated with the scenario.

1. Introduction

When analyzing cause/consequences pairings as part of a What-if, HAZOP, or other hazard analysis technique, PHA study teams are required to assign severity levels for release scenarios. The determination of the severity level is a key input in the assessment of overall risk level, and may help guide the PHA study teams in determining whether or not additional safeguards should be recommended.

However assigning the severity level can at times be a difficult task. As an example if the team has determined that a scenario can result in a release of 80°F gasoline in a tank farm area through a ½ inch equivalent hole at 50 psi, the team may have to decide if the health & safety severity level is a fatality, permanent disability, temporary, disability, medical aid, etc... Without proper information or experience some teams make take a conservative approach and may select a fatality as the category. On the other hand a team may look at experiences within the facility or the company and conclude that a medical aid is the category.

It quickly becomes apparent that PHA study teams studying the same issue at different facilities often have very different views of the risk level associated with a given release scenario. In order to assist PHA study teams in analyzing scenarios in a consistent fashion, a tool has been developed focusing on the health & safety impacts associated with releases that can result in fires, explosions, or toxic exposures.

2. Scenario Selection

The PHA Assessment Tool is a precompiled database of chemical release scenarios and their resulting explosion, fire and toxic hazards. Since the tool is a prototype, only 11 chemicals were modeled and are shown in Table 1. Multiple pressure, temperature and hole sizes were modeled for each chemical. Pressure and temperature ranges were chosen based on the chemical and its use within the manufacturing locations within the company. All chemicals were modeled with hole sizes of 6 mm (0.25 inch), 13 mm (0.5 inch), 25 mm (1 inch) and 51 mm (2 inches). The combination of chemicals and process conditions were obtained by surveying different business areas across the organization to determine which combinations would be of most value.

3. Methodology

In order to understand the magnitude of hazardous contours for releases resulting in fire, explosion or toxic releases, dispersion and fire modeling for each of the release scenarios were performed by using PHAST software. Parameters illustrated in Table 2 were selected as inputs. Fire modeling was performed using the cone model.

Chemical specific explosion and toxic properties are shown in Table 3. The auto ignition temperature was compared with the discharge expanded temperature to determine if a delayed ignition explosion was possible. Reactivity and explosion energy are as defined by the Baker Strehlow Tang (BST) vapour cloud explosion model. AEGL-3 concentrations are provided at 10 minutes and 60 minutes reference times. The toxic Probit values N, A and B are used in vulnerability the calculations associated with toxic releases.

Table 1 Chemicals Included in the PHA Assessment Tool

Chemical	Temperature Range (°C)	Pressure Range (kPa)
Hydrogen	27 – 345	1034 - 16000
Propane (C_3)	0 – 260	345 - 4826
Butane (C_4)	0 – 260	345 - 4826
Hexane (C_6)	0 – 260	345 - 4826
Octane (C_8)	0 – 500	345 - 1724
Decane (C_{10})	27 – 425	345 - 16000
Icosane (C_{20})	66 – 425	345 - 3447
Hydrogen Sulfide (H_2S)	0 – 149	345 - 1724
Sour Water (2% H_2S)	27 – 149	345 - 1724
Hydrogen Fluoride (HF)	0 – 149	345 - 1724
Natural Gas / Fuel Gas (90% C_2, 10% C_1)	0 – 66	345 - 4826

4. Vulnerability Determination

Once dispersion modeling information for selected scenarios has been obtained, the next step in the process is to determine the vulnerability associated to individuals who may be impacted by the release. In the case of flash fires and vapor cloud explosions, it is assumed that individuals will not be able to attempt an escape due to the relatively short duration of these events. Since jet fires, pool fires and toxic releases are longer duration events, individuals may be able to attempt to escape from the hazard. A retreating speed of 1.5 m/s will be assumed.

Unlike buildings impacted by process events which are fixed objects where occupant vulnerability can be easily determined, individuals outside buildings in the process unit could be at any location when the hazardous event occurs, and therefore their vulnerability will differ based on their location. Therefore this methodology will calculate an average vulnerability within the hazardous process unit, and will assume that individuals are evenly distributed throughout the process unit.

Table 1 Dispersion Modeling Weather Conditions and Release Parameters

Weather Conditions
2.2 m/s wind speed
E stability
27° C ambient temperature
50% relative humidity
0 kW/m^2 solar flux
1 m surface roughness
27° C surface temperature
Release Parameters
1 meter release elevation
Horizontal, non-impinging
No bund
Infinite leak inventory

Table 2 Chemical Properties

Chemical	Auto Ignition Temp, °C	BST Reactivity	Explosion Energy in-lbf/ft^3	AEGL at 10 min. (ppm)	AEGL at 60 min. (ppm)	Toxic N	Toxic A	Toxic B
Hydrogen	574	High	732924	None	None	None	None	None
Propane (C_3)	450	Medium	842952	None	None	None	None	None
Butane (C_4)	288	Medium	852291	None	None	None	None	None
Hexane (C_6)	225	Medium	854453	None	None	None	None	None
Octane (C_8)	220	Medium	859190	None	None	None	None	None
Decane (C_{10})	210	Medium	862262	None	None	None	None	None
Icosane (C_{20})	202	Medium	868345	None	None	None	None	None
Hydrogen Sulfide (H_2S)	260	High	651907	76	50	4.55	-8.53	0.44
Sour Water (2% H_2S)	Not Flammable	None	0	76	50	4.55	-8.53	0.44
Hydrogen Fluoride (HF)	Not Flammable	None	0	170	44	1.5	-8.62	1
Natural Gas / Fuel Gas (90% C_2, 10% C_1)	504	Medium	788988	None	None	None	None	None

4.1 Average Vulnerability Determination for Flash Fires

When the release of a flammable gas occurs, the area inside the LFL contour and the area between the LFL and the ½ LFL contours are compared with the total unit area to determine the average occupant vulnerability. Occupants inside the LFL contours are assumed to have a vulnerability of 1 (100% fatality). Occupants between the LFL and ½ LFL contours are assumed to have a vulnerability of 0.5 (50% fatality). The vulnerability of individuals outside the ½ LFL contour is expected to be zero. Therefore the average vulnerability for individuals inside a process unit impacted by a flash fire can be calculated as follows:

$$\mathbf{V_{AVG} = [A_{LFL} + 0.5 \times (A_{HLFL} - A_{LFL})] \div A_{TOT}} \qquad \text{[Eq. 1]}$$

Where:
V_{AVG} = Average occupant vulnerability within the process unit
A_{LFL} = Area within the LFL contour
A_{HLFL} = Area inside the ½ LFL contour
A_{TOT} = Total area of the process unit

4.2 Average Vulnerability Determination for Vapour Cloud Explosions

Explosion calculations start by determining the explosion volume as either the minimum of either the unit volume (Unit Area × LFL Height) or the LFL volume. Explosion energy equals the explosion volume multiplied by the explosion energy per unit volume for the dispersed material. The BST explosion flame speed is calculated using the congestion and confinement (assessed by the user) and the database chemical reactivity. Finally, the BST solver is invoked to calculate the maximum pressure in the cloud as well as the distance from the epicenter to the 10 psi and 1.4 psi contours.
The average vulnerability for individuals inside the process unit in the event of a vapour cloud explosion is determined by calculating vulnerabilities at equidistant points from the epicenter of the vapour cloud explosion to either the 1.4 psi contour or the equivalent unit radius. The vulnerability at each point will be weighted based on the area between segments. Inside the 10 psi contour the vulnerability is assumed to be 1. Outside the 1.4 psi contour the vulnerability is assumed to be 0. In-between the 1.4 psi and 10 psi contours the following Probit equation should be used:

$$\mathbf{Y = A + B \times \ln(P)} \qquad \text{[Eq. 2]}$$

Where:
Y = Scaled probability of fatality
P = Free field overpressure (psig)
A = Probit constant for vapor cloud explosion = 1.47 [3]
B = Probit constant for vapor cloud explosion = 1.37 [3]

The scaled probability of fatality (Y) needs to be converted to a vulnerability using the following equation:

$$\mathbf{V = 0.5 \times [1 + (Y\text{-}5) / (|Y\text{-}5|) \times erf (|Y\text{-}5|/\sqrt{2})]} \qquad \text{[Eq. 3]}$$

Where:
V = Occupant vulnerability due to overpressure at a given point (0 – 1)

The average vulnerability of individuals inside a process unit impacted by a vapour cloud explosion can be calculated as follows:

$$\mathbf{V_{AVG} = [(A_1 \times V_1) + (A_2 \times V_2) + ... + (A_n \times V_n)] \div A_{TOT}} \qquad \text{[Eq. 4]}$$

Where:
V_{AVG} = Average occupant vulnerability within the process unit
$A_{1..n}$ = Area within the segment under consideration
$V_{1..n}$ = Vulnerability within the area under consideration
A_{TOT} = Area of the 1.4 psi contour or area of the process unit (whichever is smallest)

4.3 Average Vulnerability Determination for Jet Fires

Vulnerability is determined by integrating in a radial manner from the center of the jet fire to the minimum of either the 1.6 kW/m^2 contour or the equivalent unit radius. If the distance from the center of the jet fire is less than half the flame length of the fire, it is assumed that 27% of the population will be flame impinged and that 73% will not, due to the aspect ratio of the typical jet fire as shown in Figure 1. It is assumed that for individuals inside the impingement zone that the vulnerability will be 1. For individuals outside of the impingement zone, the following Probit equation should be used:

$$\mathbf{Y = A + B \times ln(L)} \qquad \text{[Eq. 5]}$$

Where:
Y = Scaled probability of fatality
L = Accumulated thermal dose $s(kW/m^2)^{4/3}$
A = Probit constant for thermal radiation = -10.7 [3]
B = Probit constant for thermal radiation = 1.99 [3]

The scaled probability of fatality can then be converted to vulnerability by using Eq. 3.
At each differential annulus in the integration, the vulnerability is calculated using a linear fit model via the following equation:

$$\mathbf{Vi = MAX(0,MIN(1,slope \times distance + intercept))} \qquad \text{[Eq. 6]}$$

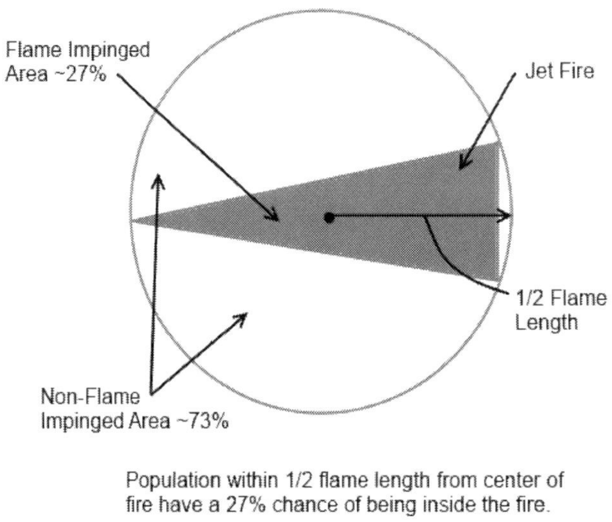

Figure 1 Typical Aspect Ratio for a Jet Fire

The linear fit model is used in order to mathematically express the vulnerability as a function of the distance from the source. An illustration of the linear fit model is shown in Figure 2. The slope and intercept represent a linear fit over the region of the vulnerability curve that transitions from 0% to 100%. The vulnerability curve is calculated based on an evacuation directly away from the fire during which a thermal dose is accumulated.

The integrated vulnerability is expressed as an average vulnerability over the exposed area.

4.4 Average Vulnerability Determination for Pool Fires

Pool fire calculations for the determination of vulnerability are the same as jet fire calculations except that the linear fit model for vulnerability is modified not by a jet fire flame length but by the pool fire radius, inside which the vulnerability is 1.

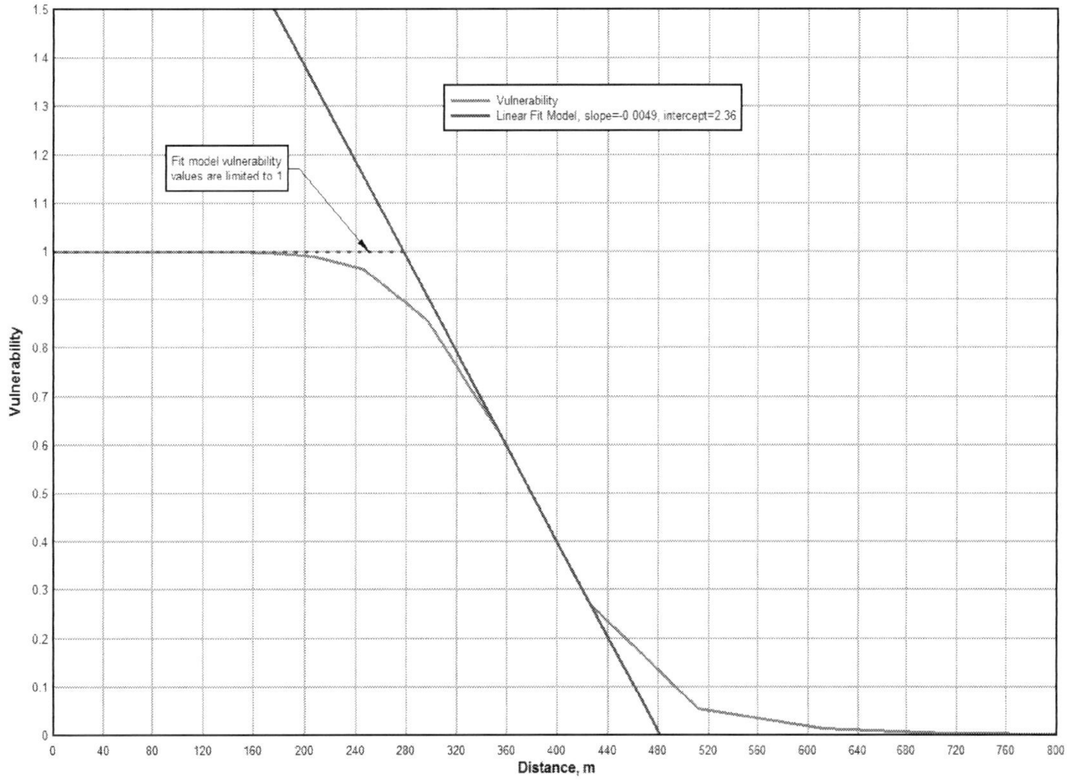

Figure 2 Example of a Linear Fit Model to Complex Vulnerability Curve

4.5 Average Vulnerability Determination for Toxic Releases

Vulnerability for toxic releases is determined by integrating in a radial manner from the release point to the minimum of either the AEGL-3 contour (either 10 minutes for an established evacuation strategy or 60 minutes for no clear evacuation plan) or the equivalent unit radius.

For toxic releases, the scaled probability of fatality can be determined by applying the following equation:

$$\mathbf{Y = A + B \times ln(C^n \times t)} \qquad \text{[Eq. 7]}$$

Where:
Y = Scaled probability of fatality
A, B, n = Probit constants located in Table 3
t = Exposure duration in minutes
C = Concentration at downwind location in ppm

At each differential annulus in the integration, the vulnerability is calculated using Eq. 3. This allows the construction of a linear fit model similar to the one shown in Eq. 6.

5. Severity Determination

The following section will outline the steps that allow PHA Study teams to utilize the vulnerability values obtained in Section 3 in order to estimate the severity level that can be used in a PHA Study.

To help with the assessment of risk levels associated with hazardous scenarios identified by PHA study teams during the HAZOP or "what-if" portions of the study, many companies utilize a risk matrix or equivalent. As an example consider the sample Risk Matrix as illustrated in Figure 3 utilized by Company X. This sample Risk Matrix is limited to Health & Safety Receptors and contains 5 distinct categories of severity which are described on the bottom horizontal axis. The vertical axis provides information relating to the expected frequency for the defined severity category. For this risk matrix, the frequency has been divided into 6 distinct categories. Company X has also established 4 risk categories ranging from low risk to extreme risk.

Company X has also defined the concept of ***inherent frequency*** as the interval at which a given severity level will occur without considering any possible safeguards. The inherent frequency will take into account the initiating event frequency, the average vulnerability as calculated in Section 3, as well as modifiers such as the probability of occupancy and the probability of ignition.

For each of the applicable hazardous release scenario impacts (flash fire, vapour cloud explosion, jet fire, pool fire and toxic exposure) an initial evaluation will be made to determine the inherent frequency of a single fatality. If the inherent frequency of a fatality is found to be outside the lower limit of the frequency range (e.g.: less than 0.00001 events per year), it will be assumed that although a fatality is possible it will not be retained as the severity category, and the team will then evaluate the inherent frequency of the next lower severity level – permanent disability. The team will continue with this approach until the inherent frequency of a defined severity level is within the limits of the risk matrix.

1 or more events per year	F6	High Risk	Extreme Risk	Extreme Risk	Extreme Risk	Extreme Risk
Between 1 and 0.1 events per year	F5	Moderate Risk	High Risk	Extreme Risk	Extreme Risk	Extreme Risk
Between 0.1 and 0.01 events per year	F4	Moderate Risk	Moderate Risk	High Risk	Extreme Risk	Extreme Risk
Between 0.01 and 0.001 events per year	F3	Low Risk	Moderate Risk	Moderate Risk	High Risk	Extreme Risk
Between 0.001 and 0.0001 events per year	F2	Low Risk	Low Risk	Moderate Risk	Moderate Risk	High Risk
Between 0.0001 and 0.00001 events per year	F1	Low Risk	Low Risk	Low Risk	Moderate Risk	Moderate Risk
		C1	C2	C3	C4	C5
		First aid Injury	Medical aid / Restricted duties injury	Temporary Disability / Lost time injury	Permanent disability injury	Fatality

Figure 3 Sample Risk Matrix

5.1 Severity Determination for Flash Fires, Vapour Cloud Explosions, Jet Fires and Pool Fires

In order to determine if a fatality should be the selected severity level, the frequency of a fatality associated with a flash fire, vapour cloud explosion, jet fire or pool fire will be determined by applying the following equation:

$$\mathbf{F_f = IEF \times V_{avg} \times OF \times P_{ign}} \qquad \text{[Eq. 8]}$$

Where:
F_f = Frequency of a fatality associated with a fire or explosion event
IEF = Initiating Event Frequency – see Table 4 for sample values
V_{avg} = Average vulnerability for fires and explosions as calculated in section 3
OF = Occupancy Factor
P_{ign} = Probability of ignition

Sample values of initiating event frequencies for selected events can be found in Table 4[1].

Table 3 Initiating Event Frequency – Sample Values

Description	Frequency (events/year)
Instrument control loop failure	0.1
Rotating equipment failure (pump, compressor)	0.1
Regulator failure	0.1
Pressure vessel – residual failure	0.00001
Atmospheric tank – residual failure	0.001
Pump seal failure	0.1

The occupancy factor is obtained by aggregating the hours spent by any personnel in the operating unit and operating mode under study. This includes operators, tradespeople, contractors, supervisory personnel, etc… This aggregated value is then divided by 168 hours per week to obtain the occupancy factor. It is worth noting that for areas with a large number of individuals occupying a process unit for extended periods of time that the occupancy factor may be greater than 1. Caution must be exercised by the PHA study team if they are considering applying a safeguard involving an operator field response, as the occupancy factor that is applicable under normal operating conditions may become irrelevant if individuals are sent into the process area to respond to an alarm associated with a safeguard. In such a case the occupancy factor should be set at 1.

Another consideration for the PHA study team lies in the fact that hazardous contours may extend beyond the limits of the operating unit. If this occurs the team needs to determine if there is a significant difference in the population density inside and outside the boundaries of the process unit. Significant differences may impact the conclusions made by the PHA study team.

The probability of ignition was based on the UKOAA model. If the discharge liquid fraction (1-flash fraction) is greater than 0.9, the UKOAA ignition curve 09 - Large Plant Liquid is used along with the mass leak rate to determine the ignition probability; otherwise the curve 08 – Large Plant Gas LPG is used. The PHA study team must also determine if the extents of the LFL contour falls within a high energy source such as a fired heater or ground level flare. In such a case the value obtained by the UKOAA methodology is voided and the probability of ignition is set at 1.

In the event that the result obtained from Eq. 8 is equal or greater than 0.00001 fatalities (C5 category on the risk matrix) per year, then a fatality will be retained as the consequence by the PHA study team. The value obtained from Eq. 8 can then be compared with the frequency ranges shown in the risk matrix, allowing the determination of the ***inherent risk*** associated with a fatality, and the team can then proceed with the identification and validation of any applicable safeguards. On the other hand if the result from Eq. 8 is less than 0.00001 fatalities per year, a fatality will not be selected as the severity level. Therefore a severity level one order of magnitude lower – permanent disability – will need to be evaluated by applying the following equation:

$$\mathbf{F_{pd} = IEF \times V_{avg} \times OF \times P_{ign} \times 6.8} \qquad \text{[Eq. 9]}$$

Where:
F_{pd} = Frequency of a permanent disability

The multiplication factor of 6.8 in Eq. 9 was derived from research performed by Wells [2] in which it was discovered that in past process safety accidents involving fires and explosions that on average there are 6.8 permanent disability injuries for every fatality.

Similar to the methodology described above, if the result obtained from Eq. 9 is equal to or greater than 0.00001 permanent disabilities (C4 category on the risk matrix) per year, a permanent disability will be selected as the severity level. The value obtained from Eq. 9 can then be compared with the frequency ranges shown in the risk matrix, allowing for the determination of the **inherent risk** associated with a permanent disability. If the result from Eq. 9 is less than 0.00001 permanent disabilities per year, a permanent disability will not be selected as the consequence level by the PHA team. Therefore a severity level one order of magnitude lower – temporary disability – will need to be evaluated by applying the following equation:

$$\mathbf{F_{td} = IEF \times V_{avg} \times OF \times P_{ign} \times 109} \qquad \text{[Eq. 10]}$$

Where:
F_{td} = Frequency of a temporary disability

The multiplication factor of 109 in Eq. 10 was derived from research performed by Wells [2] in which it was discovered that in past process safety accidents involving fires and explosions that on average there are 109 temporary disability injuries for every fatality.

If the result obtained in Eq. 10 is equal to or greater than 0.00001 temporary disabilities (C3 category on the risk matrix) per year, a temporary disability will be selected as the severity level. The value obtained from Eq. 10 can be compared with the frequency ranges shown in the risk matrix allowing for the determination of the inherent risk associated with a temporary disability. If the result obtained in Eq. 10 is less than 0.00001 temporary disabilities per year, then a restricted duties / medical aid will be the selected severity level, and a likelihood range of 0.0001 to 0.00001 events per year will be assigned. The restricted duties / medical aid (C2 on the sample Risk Matrix) will be the lowest health & safety severity level option for fire and explosion scenarios.

5.2 Severity Determination for Toxic Releases

The determination of the severity level associated with toxic releases is performed in a similar fashion to the methodology for fires and explosions, however the equations used to perform the evaluation will be slightly different. In order to determine the severity level, the frequency of a fatality associated with a toxic release will be determined by applying the following equation:

$$\mathbf{F_f = IEF \times V_{avg} \times OF} \qquad \text{[Eq. 11]}$$

Where:
F_f = Frequency of a fatality associated with a toxic release
IEF = Initiating Event Frequency – see Table 4 for sample values
V_{avg} = Average vulnerability for toxic releases as calculated in section 3
OF = Occupancy Factor

The guidance relating to occupancy factors described in section 4.1 applies to toxic releases as well.

In the event that the result obtained from Eq. 11 is equal to or greater than 0.00001 fatalities (C5 category on the risk matrix) per year, then it may be assumed that a fatality will be selected as the severity level. The value obtained from Eq. 11 can then be compared with the frequency ranges shown in the risk matrix, allowing the determination of the ***inherent risk*** of a fatality, and the study team can then proceed with the identification and validation of any applicable safeguards. On the other hand if the result from Eq. 11 is less than 0.00001 fatalities per year, a fatality will not be selected as the severity level. Therefore a severity level one order of magnitude lower – permanent disability – will need to be evaluated by applying the following equation:

$$\mathbf{F_{pd} = IEF \times V_{avg} \times OF \times 25} \qquad \text{[Eq. 12]}$$

Where:
Fpd = Frequency of a permanent disability

The multiplication factor of 25 in Eq. 12 was derived from research performed by Wells [2] in which it was discovered that in past process safety accidents involving toxic releases that on average there are 25 permanent disability injuries for every fatality.

Similar to the above methodology, if the result obtained from Eq. 12 is equal to or greater than 0.00001 permanent disabilities (C4 category on the risk matrix) per year, a permanent disability will be selected as the severity level. The value obtained from Eq. 12 can then be compared with the frequency ranges shown in the risk matrix, allowing for the determination of the **inherent risk** associated with a permanent disability. If the result from Eq. 12 is less than 0.00001 permanent disabilities per year, then a permanent disability will not be selected as the severity level. Therefore a severity level one order of magnitude lower – temporary disability – will need to be evaluated by applying the following equation:

$$\mathbf{F_{td} = IEF \times V_{avg} \times OF \times 397} \qquad \text{[Eq. 13]}$$

Where:
Ftd = Frequency of a temporary disability

The multiplication factor of 109 in Eq. 13 was derived from research performed by Wells [2] in which it was discovered that in past process safety accidents involving a toxic release that on average there are 397 temporary disability injuries for every fatality.

If the result obtained in Eq. 13 is equal to or greater than 0.00001 temporary disabilities (C3 category on the risk matrix) per year, a temporary disability will be selected as the severity level. The value obtained from Eq. 13 can be compared with the frequency ranges shown in the risk matrix allowing for the determination of the inherent risk associated with a temporary disability. If the result obtained in Eq. 13 is less than 0.00001 temporary disabilities per year, then restricted duties / medical aid will be selected as the severity level, and a likelihood range of 0.0001 to 0.00001 events per year will be assigned. The restricted duties / medical aid (C2 on the sample risk matrix) will be the lowest health & safety severity level option for fire and explosion scenarios.

6. PHA Severity Determination Tool

In order to expedite the evaluation process and reduce the likelihood of a computational error by the PHA study teams, a PHA Severity Determination Tool has been developed that incorporates the methodologies outlined in sections 3 and 4 of this paper. The tool is Excel-based and users primarily interact with a single interface worksheet.

The PHA Severity Determination Tool users should initially select the temperature and pressure closest to the actual process conditions. In addition information on the total surface area of the process unit under study as well as the aggregate occupancy hours is entered. The phase selected should match the actual material phase at the process conditions. Phase is only used as a check on the results returned from the database.

After entering the basic information described above, the user will then obtain preliminary results based on information from the pre-populated database. Users should ensure that the phase returned matches the phase they entered. If they do not match an error message will be displayed, and the user should consider changing the process conditions initially entered. The phase of the release has a significant effect on the overall results. Typically, liquid phase releases will result in higher severity levels due to the much higher discharge mass flow rate.

Based on the flammable cloud parameters obtained as part of the preliminary results, the user can then enter confinement information within the LFL extents. A "yes" answer will be selected if a substantial roof exists above congestion that can be reached by the flammable vapour cloud (i.e. distance from the release is less than LFL extents). Confinement will increase the explosion severity. If there is no confinement or the confinement elevation is greater than two times the height of the flammable cloud (LFL height), then "no" should be selected.

Congestion within the LFL extents is selected by the user as being "none", "low", "medium" or "high". Congestion is defined as piping or equipment located within the process that increases explosion severity.

- Low congestion can generally be walked through without path deviation.
- Medium congestion requires a serpentine path (the majority of process areas are medium congestion).
- High congestion cannot be traversed.

Fired heater within the LFL extents should be selected as 'yes' if any open flame ignition sources such as heaters or ground level flares or any other high energy sources are within the LFL extents, and less than or equal to the LFL height. An affirmative answer will result in an ignition probability of 1; otherwise the ignition probability is set based on the UKOAA methodology.

The final selection will be the 10 minute or 60 minute exposure duration which affects the toxic dose and resulting vulnerability. A 60 minute exposure should be chosen when no clear evacuation plan is defined. A 10 minute exposure is appropriate for personnel who are expected to evacuate the site.

The user can then prompt the PHA Severity Determination Tool to perform the necessary calculations for each of the potential consequences (flash fire, vapour cloud explosion, jet fire, pool fire and toxic exposure), and where applicable will provide an estimate on the severity level as well as the inherent likelihood of each of these consequences. The PHA study team can then proceed with the identification and validation of safeguards for each of the consequences allowing for the determination of the risk level.

An illustration of the interface page through which the study teams enter the inputs and obtain the severity level and inherent likelihood is shown in Figure 4.

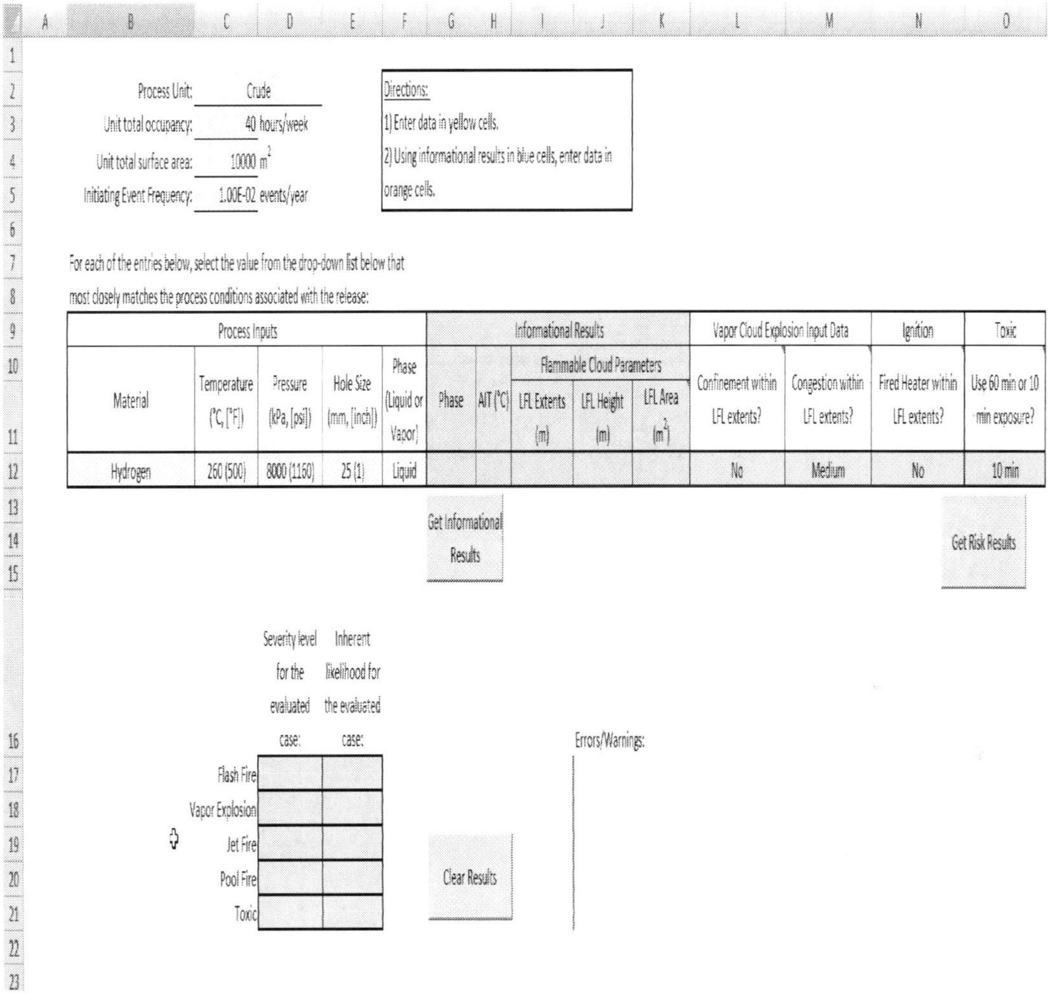

Figure 4 PHA Severity Determination Tool – Interface Page

7. Summary

PHA study teams can select from a variety of techniques to assign severity levels to release scenarios being considered. This paper has outlined a structured approach with the goal of providing greater consistency in PHA studies across the company.

This paper is meant to illustrate the steps considered necessary to evaluate potential impacts of flash fires, vapour cloud explosions, jet fires, pool fires and toxic releases, by understanding the average vulnerability associated with such releases and other factors such as the initiating event frequency, occupancy, the probability of ignition, and the relationship between fatalities and varying degrees of injury.

Although the tool developed is currently limited in scope due to the limited number of scenarios modeled, it is anticipated that over time additional chemicals and process conditions will be added in order to allow for the assessment of additional scenarios as part of PHA studies which should further enhance the usefulness of the tool.

It is understood that different companies have their own unique approaches evaluating the above factors that go into evaluating potential severity levels. In addition, companies have their own risk matrices and risk tolerance which may differ from what is presented in this paper. However it is possible to adapt some of the concepts from this paper to fit the guidelines established as various companies across the chemical process industries.

References

[1] Layers of Protection Analysis, Simplified Process Risk Assessment, American Institute of Chemical Engineers – Center for Chemical Process Safety, New York, 2001

[2] Wells, Christopher J., "Cumulative Probability of Injury to Fatality Ratio" *Suncor Energy*, Unpublished

[3] Lees, Frank P., "Loss Prevention in the Process Industries – Second Edition" – Butterworth-Heinemann, Oxford, 2001

Accident Rates VS. Culture - National and Organizational Cultures;

Suggested Interventions to Modify Organizational Safety Cultures

Fred Infortunio, PhD, PE, CSP
Principal Author
Culture Safety Intervention
Houston, TX 77377
Dr.Fred@Comcast.net

Jiun-Yin Jian, PhD, CPE
Contributing Author
ATKINS
920 Memorial City Way
Houston, TX 77024
Irene.Jian@AtkinsGlobal.com

Prepared for Presentation at
American Institute of Chemical Engineers
2014 Spring Meeting
10th Global Congress on Process Safety
New Orleans, LA
March 30 – April 2, 2014

UNPUBLISHED

Accident Rates VS. Culture - National and Organizational Cultures; Suggested Interventions to Modify Organizational Safety Cultures

Fred Infortunio, PhD, PE, CSP
Principal Author
Culture Safety Intervention
Houston, TX 77377
Dr.Fred@Comcast.net

Jiun-Yin Jian, PhD, CPE
Contributing Author
ATKINS

Keywords: Safety Culture Intervention

Abstract

This paper looks at the accident rate relationship to the anthropological cultural dimensions that are common across all societies, as initially defined by Geer Hofstede that resulted from the factor analysis the IBM multi country study that was run from 1967 to 1972 including approximately 116,000 people. Further the paper looks into accident relationship to the organizational cultural dimensions "1. Process oriented vs. results oriented 2. Employee oriented vs. job oriented. 3. Parochial vs. professional. 4. Open system vs. closed system. 5. Loose control vs. tight control. 6. Normative vs. pragmatic. In addition, the relationships to language primacy and structure along with levels of organizational stress are touched upon in an attempt to encompass the cultural "gestalt" affecting accident rates in industry." This paper will identify some of the variables associated with accident rates and suggest interventions. Hopefully, this paper will provide some guidance in identifying important aspects affecting safety in organizations.

1. The Search for a Healthy Safety Attitude / Culture.

The search for a cultural relationship to accident rates has been ephemeral and fleeting. The problem is that there are many variables and each organization, national history, and industry has a different personality, DNA. Each company will require a different prescription to help it maintain a healthy safety attitude.

2. The search for relationships, modifiers, and causes of accident rates.

The search for a cultural relationship to accident rates has been ephemeral / fleeting. Let us look at the available evidences

2.1 National Differences in Accident Rates

The vehicle death rate and the general levels of occupational accident rates have been tracked by various authors, individual countries, and by the United Nations for many years [16, 17, 21, 22, 29]. Hofstede [16, 17], Lynn [29] and others have done ranking studies that show that these accident rates do vary according to factors that are related to the national cultures.

Lynn [28], using stress-ranking levels of countries, was able to relate national anxiety levels and vehicle deaths in 18 countries. The rank correlation between vehicle deaths (per 1,000 vehicles) and national anxiety level was found to have a Spearman rank correlation coefficient (rs) = .62 at a $p < .01$. Lynn, in his citing of [5], suggests that there is a general association between anxiety and accident proneness. In his citing of Potter (1947) and [24] Lynn states:

> This [impaired performance] is partially due to the well-known effect of anxiety impairing the performance of tasks. It will be recalled that this is the principal of gamesmanship, whose object is to raise the opponent's anxiety level by derogatory criticisms and the thereby impair his performance (Potter, 1947) . . . . Kellogg [24] has also demonstrated a relationship between emotional excitement, and poor muscular control. (p. 66)

Hofstede [16] was able to relate Lynn's anxiety and accident rankings in a rank comparison analysis. Hofstede (in citing [29]) showed that, in the 18 nation comparison to their "neuroticism" or "anxiety" factor his "Uncertainty Avoidance Index" (UAI) had a Spearman rs = .73 ($p < .05$). The UAI index was developed with a large stress component. Their "neuroticism" or "anxiety" factor was developed by a 12 variable factor-analysis across 18 countries which includes high accident death rates (factor loading of .66 and high alcoholism rates (factor loading .68) [28] (p. 124). Warshaw [40] points out that "misuse of alcohol is a factor in at least half of our motor vehicle accidents . . . . ever since it was first discovered, alcohol has been used as a drug to relieve the effects of stress and anxiety and to change mood".

Hofstede [16] (p. 125-126) cites [32] who showed that the national levels of suicides and accidents were associated with the national levels of stress. Again, the rank correlation was high at rs =. 69. ($p < .05$). It was further noted by Hofstede (p. 36) that those countries that had high uncertainty avoidance scores (in a 14 country sample), the auto accident death rate and the speed limits were higher (rs = .59, $p < .05$).

He states that, the countries with high UAI scores are those with societies that are more rigid. The living by strict rules which do not cover all possible circumstances or which cause the inhabitants to skirt excessive rules which might be the cause higher levels of accidental deaths.

Hofstede [16] also finds, in a 14-nation comparison, that his "Individuality-Collectivism Index" (IDV) is negatively related to traffic accidents (rs = -0.61). The more individualistic the country, the lower the accident rate. In his citing of [7], he surmises "individualistic countries show a more calculative involvement in traffic, and that leads to safer driving" (p. 164). The IDV index is also highly correlated to the levels of prosperity:

> In more individualistic countries, traffic is safer. These countries tend to be wealthier, which is more likely to increase the number of vehicles available and to decrease the number of kilometers per vehicle; also, roads are probably better adapted to heavier traffic . . . . the Spearman rank relation between wealth (GNP/capita) and traffic deaths is only -.51,so there is more to the correlation between IDV and traffic safety than just greater wealth. (p. 164)

Hofstede [16] further relates accidents to a second-order correlation to his masculinity index. "In more masculine countries, drivers are allowed to drive faster, and more people die in vehicle accidents per 1,000 vehicles" (p. 205).

Hofstede's Cultural Dimensions

Hofstede [16, 17, 18] provides the following definitions and examples for the four initial cultural dimensions, and a fifth dimension, Long Term Outlook, which was developed later, see [19].

2.1.1 Power Distance (PDI)

Power Distance "indicates the extent to which a society accepts the fact that power in institutions and organizations is distributed unequally" [18] (p. 195). Hofstede gives the following examples in Table 1.

2.1.1.1 Relating power distance factors to accidents.

Several of the above examples could be related to the potential for accidents. High power distance separates management from the work that has to been done. The employees tend to be powerless. There tends to be an underlying conflict between management and workers. Management would tend to be more task oriented (Fiedler's - Contingency Theory - high situational power = high task orientation and low personnel consideration as cited by [3]). Management would have lower consideration for the workers. The management tends to blame people rather than the system when something goes wrong [2] (p. 205); [3] (p. 385).

Table 1. The Power Distance Dimension [18]

Small Power Distance	Large Power Distance
Inequality in society should be minimized.	There should be an order of inequality in this world in which everybody has a rightful place; high and low are protected by this order.
All people should be interdependent.	A few people should be independent; most should be dependent.
Hierarchy means an inequality of roles, established for convenience.	Hierarchy means existential inequality.
Superiors consider subordinates to be "people like me".	Superiors consider subordinates to be a different kind of people.
Subordinates consider superiors to be "people like me".	Subordinates consider superiors as a different kind of people.
Superiors are accessible.	Superiors are inaccessible.
The use of power should be legitimate and is subject to judgment as to whether it is good or evil.	Power is a basic fact of society that antedates good or evil. Its legitimacy is irrelevant
Those in power should try to look less powerful than they are.	Those in power should try to look as powerful as possible.
All should have equal rights.	Power-holders are entitled to privileges.
The way to change a social system is to redistribute power.	The way to change a social system is to dethrone those in power.
The system is to blame.	The underdog is to blame.
Latent harmony exists between the powerful and the powerless.	Latent conflict exists between the powerful and the powerless.
People at the various power levels feel less threatened and more prepared to trust people.	Other people are a potential threat to one's power and can rarely be trusted.
Cooperation among the powerless can be based on solidarity.	Cooperation among the powerless is difficult to attain because of their low faith in people.

2.1.2 Uncertainty Avoidance (UAI)

Uncertainty Avoidance indicates the extent to which a society feels threatened by uncertain and ambiguous situations and tries to avoid these situations by providing greater career stability, establishing more formal rules, not tolerating deviant ideas and behaviors, and believing in absolute truths, and the attainment of expertise. Nevertheless, societies in which uncertainty avoidance is strong are also characterized by a higher level of anxiety and aggressiveness that creates, among other things, a strong inner urge in people to work hard. The people high on this dimension appear to be very rule oriented. [18] (p. 195)

Hofstede [18] gives the following examples in Table 2.

2.1.2.1 Relating uncertainty avoidance to accident rates.

Several of the cultural attributes associated with high uncertainty avoidance could be causal or in other ways related to accidents in society. Certainly, the breaking of rules could be the cause of an accident. As stated by Warshaw [40], stress is certainly related to accidents:

Table 2. The Uncertainty Avoidance Dimension [18]

Weak Uncertainty Avoidance	Strong Uncertainty Avoidance
The uncertainty inherent in life is more easily accepted and each day is taken as it comes.	The uncertainty inherent in life is felt as a continuous threat that must be fought.
Ease and lower stress are experienced.	Higher anxiety and stress are experienced.
Time is free.	Time is money.
Hard work, as such, is not a virtue.	There is an inner urge to work hard.
Aggressive behavior is frowned upon.	Aggressive behavior of self and others is accepted.
Less showing of emotions is preferred.	More showing of emotions is preferred.
Conflict and competition can be contained on the level of fair play and used constructively.	A strong need for consensus is involved.
More acceptance of dissent is entailed.	Conflict and competition can unleash aggression and should therefore be avoided
Deviation is not considered threatening; greater tolerance is shown.	Deviant persons and ideas are dangerous; intolerance holds sway.
The ambiance is one of less nationalism.	Nationalism is pervasive.
More positive feelings toward younger people are seen.	Younger people are suspect.
There is more willingness to take risks in life.	There is great concern with security in life
The accent is on relativism, empiricism.	The search is for ultimate, absolute truths and values.
There should be as few rules as possible.	There is a need for written rules and regulations. If rules cannot be kept, we are sinners and should repent.
If rules cannot be kept, we should change them.	Belief is placed in experts and their knowledge.
Belief is placed in generalists and common sense. The authorities are there to serve the citizens.	Ordinary citizens are incompetent compared with the authorities.

> Stress is known to cause accidents. Stress affects the lives of all people everywhere. It is a cause of illness and accidents . . . . stress affects personality, modifying our perceptions, feelings, attitudes, and behavior . . . . it reaches beyond its immediate victims to affect the political, social, and work organizations (p. 3).

The time is money concept has caused people to overlook safety rules. Overwork is one of the six causes of burnout, which is related to stress [9, 10, 25]. Production pressure, in general, has lead to numerous accidents [35, 37, 38]. Aggression has been related to fatality rates [29].

2.1.3 Individualism – Collectivism (IDV)

Individualism encompasses Individualism and its opposite, collectivism. Individualism is the condition of the social framework in which people are supposed to take care of themselves and of their immediate families only, while collectivism is characterized by a tight social framework in which people distinguish between in-groups and out-groups; they expect their in-group (relatives, clan, organizations) to look after them, and in exchange they feel they owe absolute loyalty to it [18] (p. 196).

Hofstede [18] gives the following examples in Table 3.

Table 3. The Individualism Dimension [18]

Collectivist	Individualist
In society, people are born into extended families or clans who protect them in exchange or loyalty.	In society, everybody is supposed to take care of himself/herself and his/her immediate family.
"We" consciousness holds sway.	"I" consciousness holds sway.
Identity is based in the social system.	Identity is based in the individual.
There is emotional dependence of individual on organizations and institutions.	There is emotional independence of individual from organizations or institutions.
The involvement with organizations is moral.	The involvement with organizations is calculative.
The emphasis is on belonging to organizations; membership is the ideal.	The emphasis is on individual initiative and achievement; leadership is the ideal.
Private life is invaded by organizations and clans to which one belongs; opinions are predetermined.	Everybody has a right to a private life and opinion.
Expertise, order, duty, and security are provided by organization or clan.	Autonomy, variety, pleasure, and individual financial security are sought in the system.
Friendships are predetermined by stable social relationships, but there is need for prestige within these relationships.	The need is for specific friendships.
Belief is placed in group decisions.	Belief is placed in individual decisions.
Value standards differ for in-groups and out-groups (particularism).	Value standards should apply to all.

2.1.3.1 Relating individualism factors to accident rates.

As noted above, Hofstede [16] pointed out that higher individualism might increase safety. The collectivist attitude, which is similar to Geller's "active caring" [8], may lead to improved safety. As long as groupthink [23] does not become a serious factor, safety could also be a factor that is enhanced by group decisions.

2.1.4 Masculinity – Femininity (MAI)

Masculinity . . . encompasses its opposite pole, Femininity. Measurements in terms of this dimension express the extent to which the dominant in values in society are "masculine" that is, assertiveness, the acquisition of money and things, and not caring for others, the quality of life, or people. These values were labeled "masculine" because, within nearly all societies, men scored higher in terms of the values' positive sense than of their negative sense (in terms of assertiveness, for example, rather than its lack) even though the society as a whole might veer toward the "feminine" pole. Interestingly, the more of an entire society scores to the masculine side, the wider the gap between its "men's" and "women's" values [18] (p. 196).

Hofstede [18] gives the following examples in Table 4.

2.1.4.1 Relating masculinity factors to accident rates.

As noted in Table 4, the masculine ranked cultures had more aggressive drivers [16], which led to more fatal highway accidents. Similar to individualistic societies, masculine societies are performance and production oriented. This again leads to a greater potential for accidents. Ambition and drive could also lead one to break the rules. Sutherland & Cooper [37, 38] found that Type A personalities were more likely to be involved in accidents.

Table 4. The Masculinity Dimension [18]

Feminine	Masculine
Men needn't be assertive, but can also assume nurturing roles.	Men should be assertive. Women should be nurturing.
Sex roles in society are more fluid.	Sex roles in society are clearly differentiated.
There should be equality between the sexes.	Men should dominate in society.
Quality of life is important.	Performance is what counts.
You work in order to live.	You live in order to work.
People and environment are important.	Money and things are important.
Interdependence is the ideal.	Independence is the ideal.
Service provides the motivation.	Ambition provides the drive.
One sympathizes with the unfortunate.	One admires the successful achiever.
Small and slow are beautiful.	Big and fast are beautiful.
Unisex and androgyny are ideal.	Ostentatious manliness ("machismo") is appreciated.

2.1.5 Confucian Dynamism: or Long Term Outlook (LTO)

In later studies, a fifth dimension was identified [17, 19]; it was called Confucian dynamism and was eventually renamed to Long Term Outlook (LTO). This dimension discriminates between long-term outlook and short-term outlook.

Hofstede [17] gives the following examples in Table 5.

Table 5 Key differences between short term and long-term orientation societies [17]

Short-term orientation	Long-term orientation
Respect for traditions	Adaptation of tradition to a modern context
Respect for social status obligations regardless of cost	Respect for social status obligations within limits
Social pressure to 'keep up with the Jones' even if it means overspending	Thrift, being sparing with resources
Small savings . . . little money for investment	Large savings . . . funds available for investment.
Quick results expected	Perseverance towards slow results
Concern with 'face'	Willingness to subordinate oneself for a purpose
Concern with possessing the Truth	Concern with respecting the demands of virtue.

2.1.5.1 Relating long-term outlook to accident rates.

There have been no studies that relate LTO to accidents. There may be some connection with the perseverance aspect of LTO. This may act to reduce the general level of production pressure.

3. Variables that May Moderate the Effect the Cultural Dimensions have on Safety

There are several factors that could override or act in concert with the cultural propensity toward accidents. These would include stress factors, safety training, level of education, level of financial attainment, personality type, immigration status, marriage status, age, gender, mental health, emotional disposition, emotional state, government rules regulation and enforcement, and degrees of hazard, communications failure due to non-native language communications. In general, with stress being a significantly related to accident rates [37, 38] anything that affects the stress levels might be seen as a modifier in the rate of accidents befalling an individual or an organization.

Lee and Ashforth [27] in their meta-analysis study of burnout (which is related to high stress levels) reviewed 61 studies on burnout. In their study, 33 burnout correlates were looked into. The correlates were associated with the dimensions of burnout, the associated stressors, and the moderators of the burnout levels. Further, the degrees of burnout were then related to the outcome variables of job satisfaction, commitment, and involvement. Many of these correlates were significantly related to the three job burnout stages. The job stressors were: role ambiguity, role clarity, role conflict, role stress, stressful events, workload, work pressure, and physical comfort. The modifiers of the job stress were: social support, supervisor support, coworker support, work friends, community bond, family resources, peer cohesion, and team cohesion.

Individual levels of stress certainly seem to be a large factor in a person's involvement in an accident [4, 6]. Safety training and safety programs have an effect on the level of accidents in an organization [8, 35].

Cooper [6] (p.60) in citing of the U.K Office of Population Census and Surveys (the 1970 to 1972 listing of a standardized tabulation of deaths by occupation) points out that the accidental death rate generally progresses (increases) from professionals, to semi professionals, skilled non-manual, to skilled manual, to partly skilled, to unskilled labor. In addition, on an overall basis the incidences of ill health increased in the same general direction. He used the illness incidents as a correlate with the manifestations of stress.

Meng and Smith [31], in their examination of several hundred Canadian workers conclude that low wage earners, immigrants, males, and married people have a higher probability of collecting workers compensation benefits (being significantly injured) than other workers. This is not an indictment for stress but there are some leanings in that direction.

In their studies of injured individuals Hirschfeld, and Behan [13, 14] surmise that the mental and emotional health of the individual play a large part in their propensity to be injured.

Cooper [6], citing a 1975 University of Michigan study by Caplan, Cobb, French, Van-Harrison, and Pinneau, points out that accident levels differ from industry to industry and from white collar to blue-collar positions. Zohar [41] states that below stress, the degree of danger is dictated by the job itself: “A second predictor is level of risk in individual jobs” (p. 590).

In the brief description of findings for the National Council on Compensation Insurance Inc. study “ Gender in Workers Compensation Claims” [39], it is pointed out that age, gender, and degree of hazard have a modifying affect on accident rates. It is also generally well known that young males are more prone to automobile accidents than young females.

3.1 Communications failure due to non-native language communications. (Contributing Author - Personal correspondence 12/3/13 Jiun-Yin Jian, PhD, CPE)

Another component impacting safety culture is language. With the multi-ethnic multi-lingual workforce in process industry where in this example English is the primary language, the organizational culture becomes more vibrant and yet it is also more challenging to manage task-specific safety protocols and policies with non-English or English-as-a-second-language (ESL) speakers. The structure (e.g., grammar) and semantic meaning of various languages may contribute to non-English speakers' misunderstanding and result in unsafe human behavior.

Acquisition of first and second languages involves in processes of transfer, code switching, storage, and chunking [30]. In addition, based on the information processing model in traditional cognitive psychology, cue mapping and conceptualization using a new language lead to message formulation and comprehension. In MacWhinney's example [30], in English, the agent in the sentence, "the eraser hits the cat", is 'the eraser'; however, in Spanish, the agent would be 'the cat'. Consequently, the variance in cue marking in different languages may cause confusion.

Although an interpreter may be helpful to monolingual speakers, he or she may have difficulties in translating, especially verbal materials, in real time as the number of chunking that can be stored in short-term memory is believed to be four - while he or she continues to comprehend incoming materials [30] and to build a proper mental model.

Antonovsky et al. [1] showed that communication contributes 66% of maintenance failure in petroleum industry. For example, Centers for Disease Control and Prevention (CDC) reported a 2011 chlorine release at a plant was caused by a Spanish-speaking worker who did not understand the labeling written in English. A study in maritime industry illustrated that out of 14 vessels, the common working language is English, which is the second language for everyone onboard [11].

2007 US Census data released in 2010 shows that about 20% of individuals speak a language other than English at home. For individuals between age 18 and 65, only approximately 11% of the population is fluent in more than one language. According to US Bureau of Labor Statistics (BLS), for foreign born Hispanic/Latino workers, the fatality rate in 2012 was 64%. OSHA has estimated 25% of job related accidents are due to language barriers (http://www.strategy-business.com/advertorial/Rosetta-Stone-Language-Means-Business).

One possible approach to mitigate the language barrier is to structure communication or human-system interface using "task languages" [34]. The representation of tasks reflects users' categorization and recollection of semantic comprehension. Other potential interventions to overcome "lost in translation" are: train the trainer; toolbox safety meetings; safety awareness posters in different languages; demonstrating all steps to accomplish a task; having trainees repeat the task; and briefings and debriefings. Oregon OSHA also recommends that it may also be helpful to include a technical leader, a language leader, and a social leader in any safety meetings or training programs.

4. Organizational Relationships to Hofstede's Cultural Variables

Hofstede states [17] (p. 181) that the derived national culture dimensions are based on societal values and these societal values are clearly different from what is normally looked at as organizational culture. The description of an organizational culture includes symbols, heroes, rituals, and values. The national dimensions relate more to the values aspect of organizational culture. He states that symbols, heroes, and rituals (he calls these practices) are relatively superficial to the underlying values described by his dimensions. In his earlier work [16], Hofstede points out that "factors which lead to correlations across individuals are psychological (characteristics of persons); the factors which lead to correlations across countries are sociological and cultural (characteristics of social systems)" (p. 125).

> Hofstede [17] describes the socialization of the values and practices.
>
> Values are acquired in one's early youth, mainly in the family, in the neighborhood, and later in school . . . . Organizational practices, on the other hand, are learned through socialization at the workplace, which most people enter as adults, that is, with their values firmly in place. (p. 182)

The values differed more according to nationality, age, and education while the practices varied as a function of the organization that one belonged. The link to organizational practice culture comes from the hiring practices of the company [16, 17]. The hiring process may, for many practical reasons, bias the value culture of the organization by hiring people of a certain nationality, age, education, and sex. Neither the legality nor the legitimacy of these practices will be discussed. As a simple example, the type of work might require a particular level of education.

These national culture dimensions do not apply to individuals but to groups of individuals. "[C]ultures are formed by the interaction of different personalities, both conflicting and complementary, forming a whole which is more than the sum of its parts" [17] (p. 255). The national culture dimensions were developed to discriminate between national, regional and ethnic cultures, they were not suitable for discriminating among individuals. "The national culture questionnaire is no personality test and should not be used as such" (p.254).

4.1 Organizational level Dimensions

Hofstede [17] as the leader of a 1985-1987 research project, under the auspices of the Institute for Research on Intercultural Cooperation (IRIC), studied 20 organizations, using a factor analysis of 61 survey questions. The questions from the original national culture survey were used plus other questions which were developed from pilot interviews and surveys. The survey included questions were used to identify the organizational practice culture. From this project six deeper organizational culture dimensions "cross organizational values" (p. 193) were developed. The dimensions were:

1. "Process oriented vs. results oriented
2. Employee oriented vs. job oriented
3. Parochial vs. professional
4. Open system vs. closed system
5. Loose control vs. tight control
6. Normative vs. pragmatic" (p. 188).

It should be noted that the numbering of the dimensions has no theoretical meaning.

> Hofstede [17] described the dimensions in the following ways:
>
> Dimension 1 opposes a concern with means (*process oriented*) to a concern with goals (*results oriented*) . . . . the process-oriented cultures people perceive themselves as avoiding risk and making only a limited effort in their jobs . . . the result–oriented cultures people perceive themselves as comfortable in unfamiliar situations. (p. 189)
>
> Dimension 2 opposes a concern for people (employee oriented) to a concern for completing the job (job oriented) . . . . the employee-oriented cultures people feel that their personal problems are taken into account, that the organization takes responsibility for employee welfare . . . . In job-oriented units people experience strong pressure to complete the job, they perceive the organization as only interested in the work employees do. (p. 190)
>
> "Dimension 3 opposes units whose employees who derive their identity largely from the organization (parochial) to units in which people identify with their type of job (professionals)" (p. 190).
>
> Dimension 4 opposes open systems to closed systems . . . . the open system units members consider both the organization and its people open to newcomers and outsiders . . . . In the closed system units, the organization and its people are felt to be closed and secretive, even among insiders; only very special people fit into the organization. (p. 191)

> "Dimension 5 refers to the amount of internal structuring in the organization . . . . People in loose control units feel that no one thinks of cost . . . . People in tight control units describe their work environment as cost conscious" (p. 191).
>
> Dimension 6 . . . deals with the popular notion of 'customer orientation'. Pragmatic units are market driven; normative units perceive their task towards the outside world as the implementation of inviolable rules . . . . the major emphasis is on correctly following organizational procedures, which are more important than results. (p. 191)

As with many factor analyses there may be correlations between the dimensions. From, the above descriptions it appears that they are interrelated. What is important to note is that these six key dimensions differed from group to group within an organization and that they were related to the original national value dimensions in various ways. Dimension 1 correlated to the PDI dimension. Dimensions 4 correlated to the UAI dimension. Dimension 3 correlated with IDV and MAI [17].

For dimension 1, it is noted that larger power distance scores are associated with process orientation while low power distance scores were associated with results orientation. For dimension 4, the open communications correlated with weak uncertainty avoidance. Dimension 3 was found to be correlated to individuality and masculinity. No links were found between the national value dimensions and dimensions 2, 5, or 6 [17].

5. Cultures Verses Fatality Rates

Infortunio [20] reconfirmed Hofstede's [16] analysis that showed that that national level fatality rates were associated with UAI, which is the stress related national culture variable. His position it that, as a stress surrogate, the UAI measure should be able to identify organizations that are at risk for realizing higher accident and fatal injury rates than organizations that rank lower on this scale. Reasonably, the, IDV, PDI and MAI [8, 16] rankings appear that they may also be related to societal or organizational injury rates. Looking at the Hofstede's [17] cross-organizational values, it appears that some of these are stress related and they might be useful as a predictor of organizational accident levels.

With stress as a major correlate to injury [13, 14, 16, 17, 28, 37, 38, 40] it may be reasonable to propose the following hypotheses:

1. A results oriented organization would have higher injury rates than a process oriented organization (dimension 1).
2. A job oriented organization would have higher injury rates than an employee oriented organization (dimension 2).
3. A parochial oriented organization would have higher injury rates than a professional oriented organization (dimension 3).

4. A closed system oriented organization would have higher injury rates than an open system oriented organization (dimension 4).
5. A tight control organization would have higher injury rates than an open system oriented organization (dimension 5).
6. A normative culture organization would have higher injury rates than pragmatic culture organizations (dimension 6).

5.1. Modifiers to the Hypotheses.

The six dimensions listed above can actually modify each other or otherwise interact creating modifiers that increase or decrease injury rates. Beyond these, several other modifiers that need to be accommodated in determining the potency of the dimensions as they relate to injury rates.

5.1.1 Industry group.

The first modifier to organizational accident rates would be the industry that the company or division belongs to. It is obvious that some industries are more dangerous than others [6, 38, 40, 41].

5.1.2 Functional group.

The sub groups within a particular industry may be particularly prone to injuries (maintenance, operations, warehousing, vehicle delivery, etc.) [6, 38, 40, 41].

5.1.3 Organizational or group stress level.

The next modifier that shroud be tested for within an industry grouping is the organizational and group level stress. Stress might be the overriding factor that is modified by the above dimensions or the dimensions may be modified by the organizational stress level [4, 6, 38, 40, 41].

5.1.4 Peer and supervisory social support within the organization.

The social support provided by the peers or the supervisors might have moderating effects on the accident rates [36].

5.1.5 General organizational and group education level.

The education level attained by the management or the general group employees may have moderating effects on accident rates [6, 37, 38].

5.1.6 Organization, division, and group size.

The organization size, division size, or group size may act as a modifier on accident rates [4, 41].

5.1.7 Organizational or division profitability.

The organizational or divisional profitability levels may modify the accident rates [36].

5.1.8 Quality of the safety program.

The quality of the safety program may interact with the other variables to modify the accident rates [8].

5.1.9 Gender and age.

Average age of the group and the gender makeup of the group may act as modifiers at the group level [16, 17, 36].

6. A Suggested Study

To determine which of the above factors are correlated to accident rates, the 61 questions used by the Hofstede IRIC team along with questions that would identify the level of strength of the modifiers. A good study would chose at least 10 and possibly 20 companies with a minimum of 200 people per company larger the company the more statistically significant the results will be [12]. Extraordinarily large companies would be very costly to study, so it is suggested that the company sizes be limited to between 200 and 1000 employees. Large units within diverse companies (large enough to have an "SIC" industry code) would satisfy as a different company [17].

The best thing to do [17] would be to study a single industry (single SIC code) within a single country. If this is impractical, the author of this paper suggests that it would be reasonable to normalize the organizational injury rates to the average injury rates for the SIC code. From this, z scores can be obtained [12]. These z scores can compared using various statistical tests (t, factorial ANOVA, multivariate regression) the Z scores (accident rates) can be tested in a multivariate stepwise regression analysis to determine which main effect variables are important. Once the main effect variable(s) are identified then a factorial ANOVAs can be run to determine the strength of any modifiers.

6.1 The Organizational Development Process: The Intervention.

Once the variables of interest are identified, the companies that are in need of assistance can be introduced to an Organizational Development (OD) process that will help them to improve their accident rates. Hodge, Anthony, & Gales [15] state that OD

> is the area of organizational change that aims at improving social functioning of organizations. OD is a subgroup of organizational change strategies that uses knowledge of the behavioral sciences for planned interventions into organizational processes with the goal of increasing organizational health and effectiveness . . . . OD change efforts . . . typically focus on improving and facilitating interactions among people in organizations. The goals of OD have

> been to improve communications among workers, to develop norms of trust and acceptance and to move towards collaboration and consensus. (p. 370)

In order to make a permanent long-term change in a culture that possibly extends to the values level of the workers, the OD process will have to be applied in its entirety and with full knowledge and acceptance by those involved [15].

The change process for an organizational improvement program must be implemented at eight separate organizational levels in order to be made a part of the permanent organization culture (J.D. Theodore, personal communication, January 5-9, 1998 & July 31 – August 4, 2000).

6.1.1 Principles of Organization

The initial level of any organizational change must include the review or inclusion of the principles of organization, which are planning, organizing, staffing, directing and controlling [26]. Unless these principles are addressed properly, the organizational change will not have been implemented properly, and it will not be long lasting [15].

6.1.2 Planning

The project should be planned in detail. Here where the management or the consultant begin to plan the implementation of the change in culture which would lead to the desired outcomes Prototype or pilot arrangements can be set up at different plants or in different sections of the organization [15].

6.1.3 Organizing.

Organizing includes setting up the proper reporting relationships between the appropriate departments, the line management and the upper management [15]. Changes in the organization may be required to properly incorporate long-term changes to the organization. In all likelihood, it will be the safety officers that will have the central responsibility for organizing the necessary changes. Organizing includes setting up the proper reporting relationships between the safety department, the line management and the upper management. The safety leadership should preferably be at or very near the VP home office functional or staff level [8].

6.1.4 Directing.

The plant safety officers should have reporting structure to the corporate safety officer, and to the local plant manager (matrix arrangement). Other alternatives exist. The objective would be to have a group that has a coordinated voice in addressing top management, and a coordinated center of knowledge dissemination. It should also have a voice in potentially overriding local line management. Those who will be acting should be given the resources and resources to do so [15].

6.1.5 Staffing.

Staffing will include the hiring or placing of qualified personnel at the various plants who have safety knowledge and leadership backgrounds. Directing the effort should be made the responsibility of the corporate safety leadership. The staffing principle (in particular) is critical to the success of the OD process. The inclusion of the proper people with the proper skills in critical positions will help insure the desired long-term outcomes [15].

6.1.6 Controlling.

Controlling the effort should done through appropriate feedback measures and benchmarks. Once the appropriate success measures are in place, the appropriate reward systems can be put n place [8, 15].

6.1.7 Policies and Procedures

The second organizational level that would be addressed is the policies and procedures level. The policies and procedures for the implementation of the appropriate value change or stress reduction programs or interventions should be put in place; they should be spelled out in organizational procedure manuals [15].

6.1.8 Performance Appraisals

The performancc appraisals for the plant upper and supervisory management, the corporate executives, the various departments, and staff employees should have rewards or incentives associated with the implementation of the intervention program [15].

6.1.9 Management Development

The plant management, supervisory and upper management and the corporate leadership should all be given training in the value and implementation of the chosen change programs They should be shown the bottom line and employee involvement and commitment potentials [8, 15].

6.1.10 Employee Development

Employee involvement and training may be needed. The levels of employee involvement and commitment to the decided upon policies and programs will increase if the training, techniques, programs and processes are instituted properly [8, 15].

6.1.11 Intradepartmental Communications

This is a necessary part of a value change program [15]. The communications between the supervisor and the employees themselves is necessary to create a social support relationship between the employees [15, 36] This communication is essential to the operation and success of the process.

6.1.12 Interdepartmental Communications

Inter-department communications between, upper management, the various departments (plant and corporate), the local upper plant management and the personnel departments are necessary to allow each to understand the position of the other department with arrangements being necessary to accommodate each other's needs [15].

6.1.13 Individual Training of Upper Management

Without the commitment and understanding of upper management the installation of any organizational change, program or process will not have a good chance for success [15].

7. Conclusion

If all goes well, the organizational study will help target the appropriate organizational culture value or dimensions and the appropriate modifiers will allow for the reduction of losses in organizations.

The Organizational Development Intervention takes into account that a systems approach is needed and should be implemented.

8. References

[1] Antonovsky, A., Pollock, C., Straker, L. "Identification of the Human Factors contributing to maintenance failures in a petroleum operation," *Human Factors*, 1-16, 2013

[2] Bartlett, C. A., Ghoshal. S. *Transnational management: texts, cases, and readings in cross-border management* (2nd Ed.). Chicago: Irwin. 1995

[3] Barron, R. A., Greenberg, J. *Behavior in organizations: understanding and managing the human side of work* (3rd Ed). Boston: Allyn and Bacon. 1989

[4] Cartwright, S., Cooper, G. L., Barron, A. "The company car driver: Occupational stress as a predictor of motor vehicle accident involvement." [48 paragraphs]. *Human Relations* [On-line serial.]. Available: HTTP://ehostvgw21.epnet.com. 1996

[5] Cattell, R. B. *Handbook for the 16 P.F.* Illinois: I.P.A.T. 1964

[6] Cooper, C. L. *The Stress Check*. Englewood Cliffs: NJ. Prentice-Hall. 1981

[7] Etzioni, A. *A comparative analysis of complex organizations: On power, involvement, and their correlations*. New York: Free Press. 1975

[8] Geller, E. S. *The Psychology of safety: How to improve behaviors and attitudes on the job*. Radnor, PA: Chilton. 1996

[9] Golembiewski, R. T. "Burnout also has gone global, it seems: Mounting worldwide evidence based on the phase model." Unpublished manuscript, The University of Georgia, at Athens. 1998

[10] Golembiewski, R. T., & Boss, R. W. "Phases of burnout in diagnosis and intervention: individual level of analysis in organization development and change." *Research in Organizational Change and Development*, Volume 6, 115-152, 1992

[11] Hetherington, C., Flink, R., Mearns, K. "Safety in shipping: the human element," *Journal of Safety Research*, Volume 37, 401-411, 2006
[12] Hair, J. E. Jr., Anderson, R. E., Tatham, R. L., & Black, W.C. *Multivariate data analysis* (5th ed.). Upper Saddle River, NJ: Prentice Hall. 1998
[13] Hirschfeld, A. H. and Behan, R. C. "The Accident Process: I. Etioloical Considerations of Industrial Injuries," *Journal of the American Medical Association*, Volume 186, Issue 3. 193-199, 1963
[14] Hirschfeld, A. H. and Behan, R. C. "The Accident Process: III. Disability: Acceptable and Unacceptable," *Journal of the American Medical Association*, Volume 197, Issue 2, 85-89, 1996
[15] Hodge, B. J., Anthony, W. P. & Gales, L. M. *Organization Theory: A Strategic Approach* (5th ed.). Upper Saddle River, NJ: Prentice Hall. 1996
[16] Hofstede, G. *Culture's consequences: International differences work-related values* (Abridged edition), Beverly Hills: Sage. 1984
[17] Hofstede, G. *Cultures and organizations: Software of the mind* (Rev. ed.), New York: McGraw-Hill. 1991
[18] Hofstede, G. "Motivation, leadership and organization: Do American organizational theories apply abroad?" In C. A. Bartlett, & S. Ghoshal (Eds.), *Transnational management: text, cases, and readings in cross-border management* (2nd ed., pp.193-211). Chicago: Irwin. (Reprinted from Organizational Dynamics, Summer, 1980), 1995
[19] Hofstede, G, & Bond, H. M. "The Confucius connection: from cultural roots to economic growth," *Organizational Dynamics*, Volume 16, Issue 4, 4-21, 1988
[20] Infortunio, F. A. *An exploration of the correlations between fatal accident rates across nations and the cultural dimensions of power distance, uncertainty avoidance, individuality, and masculinity*. D.B.A. dissertation, Argosy University/Sarasota, United States -- Florida. Retrieved July 29, 2010, from Dissertations & Theses: Full Text.(Publication No. AAT 3220609), 2003
[21] International Labor Organization *Yearbook of labor statistics* (39th Ed.). Switzerland. 1979
[22] International Labor Organization *Yearbook of labor statistics* (53rd Ed.). France. 1994
[23] Janis, I. L., "Groupthink," In. W. E. Natemeyer & J. S. Gilberg (Eds.), *Classics of organizational behavior* (2nd ed.) (pp. 179-187). Danville, IL: Interstate Printers & Publishers. (Reprinted from Psychology Today, November 1971, pp. 43-46, 74-76), 1989
[24] Kellog, W. N. "The effects of emotional excitement upon muscular steadiness," *Journal of Experimental Psychology*, Volume 15, 142-166, 1932
[25] Kojhler, S. & Kamp, J. *American workers under pressure*. Technical Report: St. Paul: MN. St. Paul Fire and Marine Insurance Company. 1992
[26] Kreitner, R. *Management* (4th ed.). Boston: Houghton Mifflin. 1989
[27] Lee, R. T., & Asforth, B. E. "A meta-analytic examination of the correlates of the three dimensions of job burnout," *Journal of Applied Psychology*, Volume 81, Issue 2, 123-133, 1996
[28] Lynn, R. *Personality and national character*. Oxford: Sage. 1971

[29] Lynn, R. & Hampston, S. L. "National differences in extraversion and neuroticism," *British Journal of Social and Clinical Psychology*, Volume 16, 131-137, 1975
[30] MacWhinney, B. *A Unified Model of Language Acquisition*. Department of Psychology. Paper 195. http://repository.cmu.edu/psychology/195. 2004
[31] Meng, R. & Smith, D. "Union Impacts on the Receipt of Workers' Compensation Benefits," *Industrial Relations (Canadian)*, Volume 48, Issue 3, 503-517, 1993
[32] Millendorfer, J. *Mechanism of social-psychological development*. Vienna: STUDIA (Studiengruppe fur Internationale Analysen.) 1976
[33] Neuman, G. A., Edwards, J. E., & Raju. N. "Organizational development interventions: A meta-analysis of their effects on satisfaction and other attitudes," *Personnel Psychology*, 461-483, 1989
[34] Payne, S. J. & Green, T. R. G. "Task-action grammars: a model of the mental representation of task languages," *Human-Computer Interaction*, Volume 2, 93-133, 1986
[35] Perrow, C. *Normal accidents: Living with high-risk technologies*. Princeton: Princeton University Press. 1999
[36] Quick, J. C., Quick, J. D., Nelson, D. L., & Hurrell, J. J., Jr. *Preventive stress management in organizations*. Washington, DC: American Psychological Association. 1997
[37] Sutherland, V. J. & Cooper, C. L. *Man and accidents offshore: An examination of the costs of stress among workers on oil and gas rigs*. Colchester: Lloyds of London Press, Ltd. 1986
[38] Sutherland, V. J. & Cooper, C. L. *Stress and accidents in the offshore oil and gas industry*. Houston, TX: Gulf Publishing Company. 1991
[39] "Study: Men are more prone to work injury than women," *Best's Review*, Volume 101, Issue 2, 90, 2000
[40] Warshaw, L. J. *Managing stress*. Reading: MA. Addison-Wesley. 1979
[41] Zohar, D. "A group-level model of safety climate: Testing the effect of group climate on micro accidents in manufacturing jobs," *Journal of Applied Psychology*, Volume 85, Issue 4, 587-596, 2000

Benchmarking safety culture in major hazards industries in the Rotterdam area (The Netherlands)

Prof. Gerard I.J.M. Zwetsloot PhD
Netherlands Organisation for Applied Scientific Research TNO
P.O.Box 718, 2130 AS Hoofddorp The Netherlands
gerard.zwetsloot@tno.nl

Robert A. Bezemer M.Sc. MTD
Netherlands Organisation for Applied Scientific Research TNO
P.O.Box 718, 2130 AS Hoofddorp The Netherlands
robert.bezemer@tno.nl

Prepared for Presentation at
American Institute of Chemical Engineers
2014 Spring Meeting
10th Global Congress on Process Safety
New Orleans, LA
March 30 – April 2, 2014

UNPUBLISHED

Benchmarking safety culture in major hazards industries in the Rotterdam area (The Netherlands)

Prof. Gerard I.J.M. Zwetsloot PhD
Netherlands Organisation for Applied Scientific Research TNO
P.O.Box 718, 2130 AS Hoofddorp The Netherlands

Robert A. Bezemer M.Sc. MTD
Netherlands Organisation for Applied Scientific Research TNO
P.O.Box 718, 2130 AS Hoofddorp The Netherlands

Keywords: process safety culture, benchmarking, major hazards

Abstract

Introduction

Several incidents in the Netherlands in companies dealing with large quantities of chemicals got a lot of attention. It was broadly recognized that the safety cultures in these cases were relatively poorly developed. Against this background, the regional environmental inspectorate DCMR asked TNO in 2012 to assess the safety culture in fourteen major hazard companies in four related sectors. The two main objectives were (1) to gain insight into the process safety culture of 14 companies in four industrial sectors: refineries, (petro)chemical industry, bulk storage (tank parks), and chemical warehousing and logistics and (2) to allow benchmarking of process safety culture between those sectors and companies. It was the intention that this research also generated useful insights for the participating companies into the strengths and weaknesses of their process safety culture.

Methods

The TNO Quick Scan Safety Culture was developed and conducted in 14 major hazard companies. The selection included four (petro) chemical plants, two refineries, four bulk storage and four chemical warehousing and logistics companies. The Quick Scan measures 14 dimensions of process safety culture.

The applied method consisted of a limited review of relevant documentation, a two day site visit and a close out meeting. During the site visit two researchers made a walk through and conducted a series of interviews. In the close out meeting they presented the results to management, followed by discussion.

Main findings

The safety culture scores of the (petro)chemical companies and the refineries were all good or acceptable. In contrast, in the bulk storage and chemical warehousing/logistics sectors several companies scored the acceptable minimum or even below.

Strengths and weaknesses of process safety culture were assessed for each company individually, but are presented at sector level. This implies that there are opportunities for improvement both at company and sector level.

Conclusion and discussion
The chemical industries and refineries had well-developed process safety cultures, although there were certainly opportunities for further improvement. The sectors at the end of the chain, the bulk storage and chemical warehousing and logistics do not have such well-developed process safety cultures; in a few cases their cultures were below what was regarded by the researchers as the minimum for major hazard industries.
There are clearly limitations of the quick scan approach. Nevertheless, the methodology proved to be useful for benchmarking purposes between sectors and between companies. It led to useful feedback on the strengths and weaknesses of the companies' safety culture.

1. Introduction

Ever since the Chernobyl disaster [1] and the Piper Alpha disaster in 1998 in the North Sea [2], safety culture is on the agenda. In the analyses of the disasters a poor safety culture was regarded as an important causal factor that referred to intangible and often ambiguous issues like information difficulties, violations, failure to recognize emerging danger, role ambiguity, management complacency, poor communication, low prioritization of safety, etc. [3]. Such factors are regarded as 'latent conditions' for incident causation [4].
Nowadays, safety culture is widely recognized as important for accident prevention. There are many definitions of safety culture. Guldenmund [5] discusses 18 different definitions of safety culture and the related concept of safety climate. The concept of safety culture is usually used as a generic term, which does not differentiate between process safety and personal safety. However for an assessment of process safety culture, such a distinction is very relevant, and is a consequence of the lessons learned from the BP Texas explosion [6].
In this research we focused on 'process safety culture' which we defined as the attitudes, values, explicit or implicit assumptions, perceptions and habits of the members of an organization relevant for dealing with process safety risks (elaborated in [7]). Safety culture is an aspect of the organizational culture, and can also be understood as the 'unwritten rules' [8] in the organization, which are reproduced and enforced through socialization processes. It is important to note that values, implicit assumptions and habits do not only refer to conscious behavior but also to unconscious behavior of the members of the organization. Safety culture is a multi-layered concept. There is often a focus on the expressions of safety culture. In practical terms that implies a focus on activities, behavior, policies, and procedures that are relevant for safety.

Major hazard industries have to comply with rather detailed legislation with respect to the safety organization and technology. In the European Union this concerns the so called 'Post Seveso (III)' legislation. There are, however, no legal requirements with respect to safety culture, except for nuclear power plants. Safety culture is and remains therefore primarily the responsibility of the companies involved.

In the Netherlands several incidents in companies dealing with large quantities of chemicals got a lot of media and political attention. It was broadly recognized that in these cases, the organizational safety cultures were poorly developed. Though most of these incidents took place in the sectors of chemical logistics and trade and bulk storage, the chemical industry as a whole was confronted with criticism in the mass media, while the authorities were criticized for being too flexible and industry friendly. This urged the authorities to be stricter in enforcing the legislation for major hazard industries in general, and to take new initiatives to guarantee chemical safety.
Against this background, the regional environmental inspectorate DCMR asked TNO in 2012 to assess the safety culture in 14 major hazard companies in four related sectors. The focus was on those characteristics of safety culture that are relevant for environmental safety and the management of major hazards. The two main research objectives were: (1) to gain insight into the (process) safety culture of 14 companies in four industrial sectors: refineries, (petro) chemical industry, bulk storage (tank parks), and chemical warehousing and logistics, and (2) to allow benchmarking of process safety culture between those sectors and between companies. A secondary aim was to provide the participating companies with useful insights into the strengths and weaknesses of their (process) safety culture.

2. Materials and Methods

2.1 The selection of companies

The research was conducted in 14 major hazard companies, selected by the environmental inspectorate out of the more than one hundred companies or plants that fall under the regime of the Post Seveso Directive in the Rotterdam area. The selection included four (petro)chemical plants, two refineries, four bulk storage and four chemical warehousing and logistics companies. The inspectorate selected organizations with a good as well as a not so good safety reputation. The researchers were not informed about their reputations. All companies approached by the environmental inspectorate agreed to co-operate. For two organizations with several major hazard plants or units, it was agreed between the organizations and the inspectorate to select the chemical plant or storage unit which was, according to the company, their unit with the most advanced safety culture. This implies that the selection was somewhat to the positive compared to the total population of major hazard industries in the area. This was accepted, because for the companies involved the benefit was that they would obtain an independent check by objective outsiders, whether these units rightly had an exemplary role within their organizations.

2.2 The dimensions of process safety culture assessed

The TNO Quick Scan Process Safety Culture measures 14 dimensions of process safety culture; 9 dimensions are taken from the Hearts & Minds methodology (which comprises

18 dimensions [9], see Table 1) as that methodology, originally developed for Shell, is broadly used by the Dutch chemical industry. However, the Hearts & Minds methodology was developed before the BP Texas disaster [6] and is more focused on safety culture in general than on process safety culture specifically. Five complementary dimensions were included because the scientific literature and industrial experience demonstrate that they are very relevant for process safety (see Table 1).

Table 1 Fourteen dimensions of process safety culture measured in the research

Nine dimensions from the Hearts & Minds methodology [9] used in this research	*Five complementary dimensions used in this research*
• Leadership and commitment • Opinion of management about the causes of incidents • Profit versus safety • Safety communication • Participation and commitment of employees • Contractor management • Procedures and rule management • Incident reporting and analysis • Execution and follow-up of audits	• Personal versus process safety • Functioning and roles of supervisors • Maintenance management • Learning from incidents • Dealing with complexity

About the five complementary dimensions:

- The process safety versus personal safety dimension was included because since the BP Texas disaster [6] it is broadly recognized that different scenarios are relevant, with consequences for safety management and culture.
- The functioning and roles of supervisors was included because the supervisor plays a key role in safety communication, and in promoting safe behavior on the shop floor. It is also well-known that supervisors are often overloaded with information and tasks, which may affect their safety role.
- Maintenance management is crucial for the technical condition of installations, and a key determinant of process safety. Delay of preventive maintenance can be financially attractive, but easily undermines safety margins [10]. Especially the balance between planned preventive maintenance and trouble shooting is an important indicator of the importance attributed to process safety, and therefore a key issue for process safety culture.
- The learning from incidents dimension was included because it is known to be a challenge in industrial practice. Research shows that there are many bottlenecks in the learning process [11] and it is certainly related to the organizational culture.
- Complexity due to variation in a process or in process handling may lead to unexpected events, which can be critical for safety [12]. Safety culture and especially 'organizational mindfulness [12] is important to recognize the meaning of unexpected events, in order to be able to deal with them adequately. Alertness

for exceptional situations and recognizing their meaning for safety is therefore a characteristic of a good (process) safety culture.

2.3 The assessment procedure

The actual assessment of process safety culture comprised three stages for each company. In the preparatory stage historic data were received from the local environmental safety authorities about safety incidents, including non-compliances and fines of the last years (maximum five years). From the companies we received their current safety programs, their last management review of the process safety management system, and any additional information which was relevant according to the company. It was avoided to become overloaded with safety documentation about the existing process safety management system, as the assessment aimed to focus on culture and behavior in daily practice. In this way, the research was fairly complementary to regular audits of the process safety management system.

The on-site part of the assessment was carried out in two days, involving two researchers. In total, for the fourteen assessments, a team of seven researchers carried out the assessments. They were experienced safety and psychosocial experts having a broad experience with safety culture. They were trained in advance to be prepared to deal adequately with socially desirable responses of the interviewees, and to make sure they would get insight into the real safety practices. The data about historic incidents were very valuable in this respect, as were observations from the walk-through, and the information from the management review and safety plans. These sources of information helped to discuss relevant culture issues. Indeed, checks and balances between different sources were essential for the reliability of the methodology. This involved four types of triangulation [13], i.e. between observations, documents, interviews and also between the members of the research team.

The assessment started with a walk-through (observations) followed by a series of interviews of about one hour each. The interviewees included the plant manager, the HSE manager, one or two middle managers and supervisors, operators from day and continuous shifts, personnel of contractors, the maintenance manager or maintenance personnel, and a member of the Works Council. In each interview all dimensions of safety culture relevant for the interviewee were addressed; this implies that with e.g. the plant manager and the HSE manager all 14 dimensions were addressed, while in other interviews only a selection of the dimensions was addressed. The researchers individually documented the findings per safety culture dimension addressed, directly after each interview, also documenting other remarkable findings.

At the end of the second day the results of observations and interviews were compiled and a close-out session was organized with management and other company representatives (often the interviewees). An inspector from the regional environmental safety inspectorate DCMR was also present. The research team presented the findings using a standardized power point format, see table 2.

Table 2 The findings of the assessment presented at the close-out sessions (per company)

A score on the safety culture ladder (similar to that in the Hearts & Minds methodology), a scale from 1 (pathologic safety culture) till 5 (generative safety culture)
The relatively or absolutely strong and weak dimensions of process safety culture
The consistency in the process safety culture: the variation in quality of safety culture among the 14 dimensions and the degree of consistency between the various interviews
Often: useful observations about the functioning of the Hazard and Risk Control System in practice

This, of course, triggered a dialogue between company management and representatives and the research team, wherein the research team was challenged to present evidence, especially for results which were not as good as company management had expected. In these cases management and representatives were challenged to take the findings of the research team seriously, and commit themselves to improve the absolute or relative weaknesses in the company's process safety culture. Of course, the way the dialogue with management and company representatives takes place during the close-out, is to a certain degree also reflecting the organization's safety culture. The dialogue therefore was an opportunity for the researchers to get a deeper understanding of the company's safety leadership and culture, while for the company it was an opportunity to deepen their understanding of the observations and findings of the research team. The dialogues did, however, not lead to significant adaptations of the outcomes of the assessments.

2.4 Documentation and reporting

The companies nor the environmental safety inspectorate received a written report about the findings per company, other than a pdf version of the power point presentation presented at the close out. This was to avoid that confidential information would become publicly available after the research. Information from the governmental authorities, including projects contracted out, can in the Netherlands become publicly available in case somebody makes a dedicated request to get that information. In this case it was assumed that some newspapers would make such requests.
The results of the fourteen process safety culture assessments where used to arrive at conclusions per sector and for benchmarking purposes. The results thereof were reported to the environmental safety inspectorate, and also used as a basis for presentations of the research to the regional industry association and its members.

3. Benchmark findings for process safety culture

3.1 Quantitatively benchmarking

For each organization, the process safety culture was rated, using a range from 1 – 5, similarly to the Hearts & Minds safety culture ladder [9]. A score of 3.0 (comparable to a calculative safety culture in the Hearts & Minds methodology) was regarded as the minimum acceptable score for a major hazard company. The results are presented in table 3.

Table 3 Scores on the process safety culture ladder per sector

Sector	Average culture score per company	Range
Refineries (N=2)	3.7	3.4 – 4.0
(Petro)Chemical (N=4)	3.9	3.5 – 4.0
Bulk storage (tank parks) (N = 4)	3.3	3.0 - 4.0
Chemical warehousing and logistics (N=4)	2.8	2.5 – 3.0

As table 3 shows, the (petro)chemical companies and the refineries all had a good or acceptable safety culture. In the bulk storage sector the inter-company variation was substantial, but all cultures were assessed at acceptable, while in one case it was good[1]. In the sector of chemical warehousing and logistics the average score was below standard, with two companies below and the other two at the minimally acceptable level.

In five of the 14 companies the score on each of the 14 dimensions as well as all interviews indicated a good safety culture. The average culture score of these five companies was 3.8. These companies can be regarded as frontrunners in the development of safety culture. Three companies achieved an average 3.8 score, while in these organizations there were remarkable inconsistencies between the various interviews, i.e. there seemed to be subcultures or ambiguities in the safety culture. Unfortunately, within the limited timeframe available for the Quick Scan, it was not possible to further investigate these inconsistencies. Therefore, the good scores of these three companies remain more uncertain than those of the other 5 companies with well-developed safety cultures. One company got a very consistent score of 3.0.

Three other companies also got an average score of 3.0, but in these cases the consistency among the dimensions and interviews was less. Finally, two companies had a score of no more than 2.5. In these companies there was a range of weakly developed dimensions of safety culture; there were complementary observations of some specific unsafe practices as well. Table 4 gives an overview of these findings per sector.

Table 4 Consistency in process safety scores

	Good or acceptable score on all 14 dimensions	Average score acceptable but with some weak dimensions	Many weaknesses, average is below standard
Refineries	2		
(Petro)chemical	4		
Bulk storage	3	1	

[1] This bulk storage unit served as the internal benchmark in the company as a whole, having a better-developed process safety culture than the other units of the same company.

Warehousing and logistics		2	2

In a few companies there were observations of some specific unsafe practices. Again, we were not able to further investigate this, which means that for these cases the safety culture scores are associated with substantial uncertainties.

3.2 Qualitatively benchmarking

The benchmark was also carried out qualitatively, i.e. assessing the strong and weak dimensions per sector. Obviously, in the sectors with a good process safety culture more dimensions were found to be strong, while in the organizations with weaker process safety culture, more dimensions were weak.

In the four (petro)chemical companies and the two refineries no weak dimensions were found at all: all dimensions where either good or excellent. However, for the bulk storage and the chemical warehousing several weak dimensions were identified, while excellently-developed dimensions in these sectors, especially in the chemical warehousing and logistics, were scarce.

Figure 1 Qualitative variations of process safety culture dimensions per sector

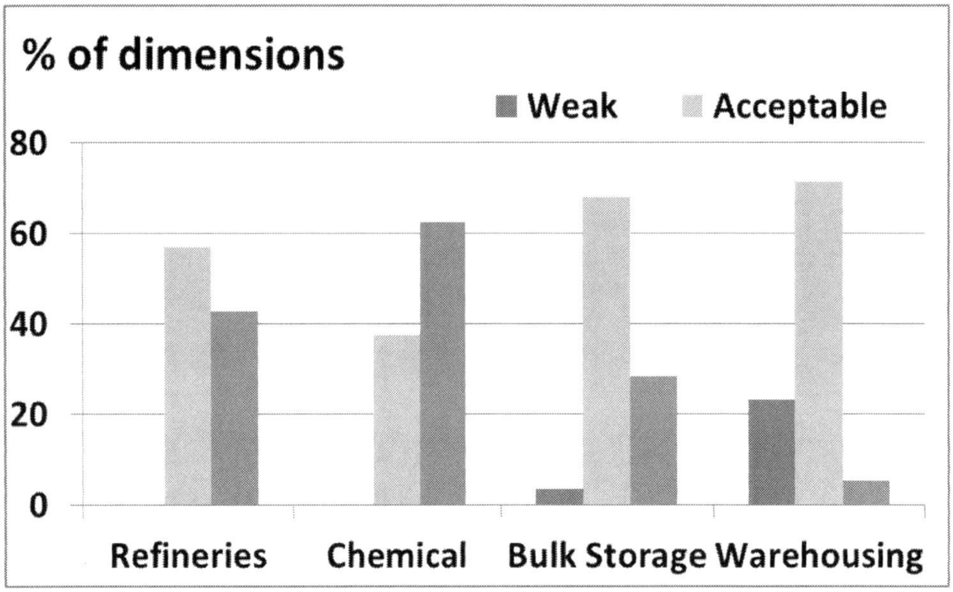

In each company and each sector opportunities were identified for improving the safety culture. As shown in figure 1, the refineries and the chemical companies could improve their relative weaknesses, while the bulk storage and chemical warehousing and logistics sectors several absolute weaknesses needed improvement.

Three dimensions could be improved in all four sectors: execution and follow-up of audits, learning from incidents, and dealing with complexity.
Additionally, in the tank parks and the chemical warehousing the management vision on accident causation as well the reporting and analyses of incidents were relatively or absolutely weak, and can certainly be improved.
In the chemical warehousing and logistics, three other dimensions also require dedicated efforts: safety leadership and commitment, safety communication, and participation and commitment of employees. These three dimensions were often excellently developed in the chemical industry and the refineries, implying great opportunities to learn from each other.
There were more opportunities to learn from each other on specific dimensions of process safety culture, especially across sectors. We identified some companies with inspiring safety leadership while in a few others safety leadership seemed to be absent. We noted examples of excellent safety communication, but also cases of poor safety communication. We saw major hazard industries with a well-developed focus on process safety, but also a few examples where safety was centered on personal or transport safety issues. In fact in a few warehousing and logistics companies the control of major hazards was almost limited to formal procedures and technical facilities. In the population of 14 organizations there were also significant differences regarding the dimensions 'Procedures and rule management', 'Functioning and roles of supervisors', 'Maintenance management', and 'Learning from incidents'.
Finally in some cases relevant observations were made outside the fourteen cultural dimensions.
We noticed examples of process safety management systems that were excellently functioning with strong involvement of managers and workforce, but also a few examples where this was mostly paper work. There were a few organizations where scenario-thinking was not practiced (only in the paper work), while awareness about such scenarios is crucial for the prevention of major accidents. Finally we saw good examples of safety knowledge management, where relevant information was broadly shared and discussed, and a few examples where almost all knowledge was concentrated into the head of one person and information files on his personal computer hard disk.

4 Conclusions

The first conclusion is that the process safety culture in major hazard industries should be stratified per sector. The chemical industries and refineries had well-developed process safety cultures, although there were certainly opportunities for further improvement. The sectors at the end of the chain, the bulk storage and chemical warehousing and logistics do not have such well-developed process safety cultures; in a few cases their cultures were below what was regarded by the researchers as the minimum for major hazard industries.
The process safety culture of 9 of the 14 companies was rather consistent, i.e. the variation in quality between the fourteen dimensions taken into account was minimal. All companies with a consistent process safety culture had a good or an acceptable safety

culture level. In three companies the average score was acceptable, but with significant variations between the fourteen dimensions. This means that (1) the assessment of their process safety culture is associated with more uncertainties. It also implies that some of the dimensions of process safety culture in these companies were below what was generally regarded as acceptable. There were two companies with a process safety culture that was regarded as substandard; in these companies the majority of dimensions were below standard, but some dimensions were sufficient or good. Consistency of process safety culture among the fourteen dimensions apparently is a characteristic of a well-developed safety culture, while inconsistency is an indication of a less mature or even substandard safety culture.

We found significant differences in strong and weak aspects of process safety culture among the organizations. It was possible, however, to identify shared strong and weak dimensions per sector.

The methodology used in this research, the TNO Quick Scan Process Safety Culture, turned out to be useful for an assessment of process safety culture, both at company/plant level and at sector level.

5. Discussion

5.1 Practical relevance

Though since the BP Texas report [6] it is broadly acknowledged that process safety and personal safety should be clearly distinguished, such distinction is usually not yet made in the area of safety culture. In this project we focused explicitly on process safety culture. However, we are still far from a generally accepted definition of process safety culture.

The main strengths of the methodology used, the TNO Quick Scan Process Safety Culture, are that the method leads to an assessment of process safety culture which can be given in short notice, at relatively low costs. The method turned out to be useful for assessing the general state of the process safety culture, as well as for identifying its strong and, perhaps relatively, weak dimensions. The latter imply distinct areas for improvement. Indeed, from several companies we got the feedback that the benchmark provided valuable insights into specific issues in their safety culture, which was used by management in their plans to (further) improve their safety culture.

5.2. Limitations to the assessment of process safety culture

The process safety culture was measured through the 14 dimensions given in table 1. The available budget allowed only 'quick scans' as assessments of the process safety culture. In-depth analyses were not possible, though were desirable in the cases were the process safety culture showed inconsistencies among the various dimensions. Therefore, the quantitative scores are more reliable for the companies with a well-balanced process safety culture; for the other companies the scores imply greater uncertainties. Also, it

could be argued that the quantitative 'score' which represents the average of fourteen dimensions, is not sufficient to give a good picture of the safety culture in case of companies with a less–balanced culture, as one number does not adequately reflect the strengths and weaknesses thereof.
Finally, one can discuss whether a quantitative score of the process safety culture should be based on the average of all dimensions, as was practiced in this research. During the presentation of results for the industry, some safety consultants criticized the scores for being too positive; they advocated a culture score based on the lowest score, as they regarded the weakest dimension as the critical dimension for process safety. As we felt that limitations in our 'quick scan' approach did not allow us to come up with a reliable score for each individual dimension, that was not an option here.

The methodology has its limitations. For the organization the 'burden' of the research is limited to the interviews during two days. In this period strengths and weaknesses of the process safety culture can be assessed. The method also implies clear limitations: though it is possible to assess whether each of the 14 dimensions are (relatively or absolutely) strong or weak, it is not possible to give a representative score on each of the safety culture dimensions with sufficient reliability. We give therefore a quantitative score based on the average of the 14 dimensions. This is a score for the process safety culture as a totality.

5.3. Challenges for major hazard industries

For individual industries it is difficult to assess their own (process) safety culture. The members of the organization are likely to share habits and perceptions with the other members of the organization, implying potential blind spots for some relevant aspects of process safety culture. An assessment by an independent qualified outsider can therefore be very valuable for the industry. Of course, it is up to the industry then to evaluate and value the results thereof, for the further development of their safety culture and policy.
Assessing process safety culture is one, but improving that safety culture is something else. That will require safety leadership, a collective learning process, it requires persistency over time, and consistency in management efforts and messages.
The research showed significant variations in process safety culture between the sectors within the major hazard industries. There seem to be two implications. Firstly there are many opportunities for cross sectional learning. Secondly the chemical and petrochemical industries seem to have a specific interest and responsibility in this respect. An interest because major incidents in the bulk storage or chemical warehousing and logistics have a direct impact on the industry's reputation and 'license to operate'. A responsibility because the bulk storage sector and the chemical warehousing and logistics are business partners that deal with their raw materials and/or products. The chemical or petrochemical industries are usually the most powerful players in these business chains, implying a responsibility to trigger safety improvements at their business partners.
Finally, governmental inspections increasingly follow a risk-based approach. While process safety culture is likely to remain a matter of the industry, the authorities and the public at large increasingly regard it as a factor that is relevant for risk-based inspections

as part of an effective enforcement strategy. The implication seems to be that the development of a good process safety culture is important for the major hazard industry in order to develop or sustain a good relationship with the authorities.

6. References

[1] "Summary Report on the Post-Accident Review on the Chernobyl Accident", *Safety Series* No. 75-INSAG-1 (International Nuclear Safety Advisory Group), IAEA, Vienna, 1986.

[2] The Hon. Lord W. Douglas Cullen. "The public inquiry into the Piper Alpha disaster". ISBN 0101113102. 488 pages, 2 volumes. London: H.M. Stationery Office, 1990.

[3] Cox, SJ & Flin, R, "Safety culture: philosopher's stone or man of straw?" *Work and Stress*, Vol 12, no. 3, pp. 189-201, 1998.

[4] Groeneweg, J. *Controlling the Controllable, preventing business upsets.* 5th revised edition, Global Safety Group, Leiden: 2002.

[5] Guldenmund F.W. "The nature of safety culture: a review of theory and research". *Safety Science* , Vol. 34 (1-3), p. 215-257, 2000.

[6] Baker J, *The Report of the BP US Refineries Independent Safety Review Panel*, 2007

[7] Zwetsloot G.I.J.M., "Prospects and limitations of process safety performance indicators", Letter to the editor, *Safety Science* 47 (2009), 495-497, 2009

[8] Scott Morgan, P., *The unwritten rules of the game*, McGraw Hill. 1994

[9] Parker D., M. Lawrie & P. Hudson, "A framework for understanding the development of organisational safety culture", *Safety Science*, Vol. 44 (6), p. 551-562, 2006.

[10] Rasmussen J., "Risk management in a dynamic society: a modelling problem", *Safety Science* Vol. 27 (2/3), p. 183-213, 1997.

[11] Drupsteen L., J. Groeneweg, G.I.J.M. Zwetsloot, "Identifying critical steps in learning from incidents", *Journal of Occupational Safety and Ergonomics* Vol 19 (1), p. 63-77, 2013.

[12] Weick K.E. & K. Sutcliffe. *Managing the unexpected*, Jossey-Bass, San Francisco, 2007.

[13] Yin R.K., *Case study research – design and methods*, 3rd edition, Sage, Thousand Oaks, 2009

Process Safety Culture - making this a reality

ChoNai Cheung
Accident Release Prevention Supervisor
Health Services Division, Contra Costa County
4585 Pacheco Blvd.
Martinez, CA 94553
Phone: (925) 335-3212
chonai.cheung@hsd.cccounty.us

Gerald Burch, Senior Engineer and William G. Bridges, President
Process Improvement Institute, Inc. (PII)
1321 Waterside Lane
Knoxville, TN 37922
Phone: (865) 675-3458
Fax: (865) 622-6800
e-mail: gburch@piii.com
e-mail: wbridges@piii.com

Prepared for Presentation at
10th Global Congress on Process Safety
New Orleans, LA
March 31-April 2, 2014

UNPUBLISHED

10TH GLOBAL CONGRESS ON PROCESS SAFETY

Process Safety Culture - making this a reality

ChoNai Cheung
Accident Release Prevention Supervisor
Health Services Division, Contra Costa County

Gerald Burch, Senior Engineer and William G. Bridges, President
Process Improvement Institute, Inc. (PII)

Keywords: process safety culture, process safety management, near miss reporting, human factors, leadership, accountability, employee participation

Abstract

Process Safety Culture (PSC) has received considerable attention recently. Many implementers see this as an intangible attribute of a company or site. Some workers see PSC as ‘code words" for management not wanting to take responsibility for process safety management. Others see PSC as something that can be affected directly by the actions of management or by an active program targeted directly at the site culture. This paper shows what effects the true "culture" at a site. It shows what Contra Costa County (one regulator) is doing to encourage establishment and measurement of process safety culture. And it shows that tangible, real, activities within a site are what make safety culture a reality. The paper also reviews the approaches to direct and indirect measurement of process safety culture, and the value of these.

Background

On April 2, 2010, the Tesoro Anacortes (WA) refinery experienced a catastrophic rupture of a heat exchanger, causing fatally injuring seven Tesoro employees working in the immediate vicinity. The Draft Investigation Report (CSB, January, 2014) lists "Process Safety Culture" as a key issue and offers the following observations about Tesoro's culture at the time of the accident:

- "Refinery management had normalized the occurrences of hazardous conditions,"
- "The refinery process safety culture required proof of danger rather than proof of effective safety implementation."

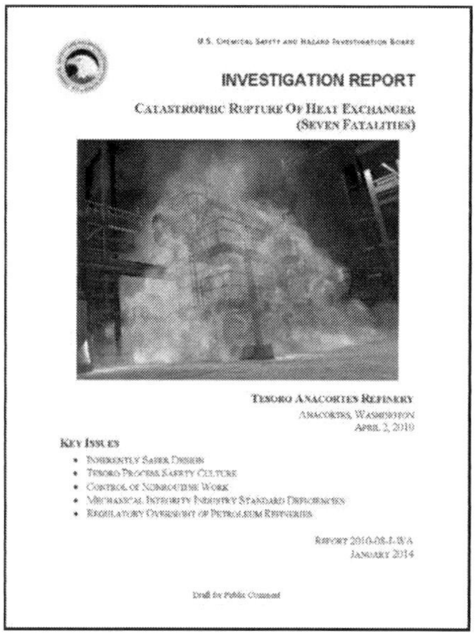

On March 23, 2005, the BP Texas City (TX) refinery experienced the most serious U.S. workplace disaster of the past two decades, resulting in 15 deaths and more than 170 injuries. The Baker Panel reported that deficiencies in "BP's corporate safety culture, corporate oversight of process safety, and process safety management systems" were contributing factors to this and other incidents which had previously occurred at BP facilities.

On January 28, 1986, the space shuttle Challenger exploded killing all 7 astronauts on board. The Rogers Commission reported that NASA's organizational culture failed to prevent this accident. Seventeen years later, on February 1, 2003, the space shuttle Columbia disintegrated upon re-entry of the Earth's atmosphere killing all 7 astronauts on board. The Columbia Accident Investigation Board (CAIB) reported that "In our view, the NASA organizational culture had as much to do with this accident as the foam." The CAIB also found "disturbing parallels remaining" from seventeen years earlier, making the determination that "NASA had not learned from the lessons of Challenger."

Our collective history is filled with tragic accidents, life altering events, fatal injuries, and incidents affecting entire communities and industries. The origin of the root causes of many of these can be traced to a flawed leadership decisions which is exhibited as poor organizational culture.

In June 2006, the Contra County Board of Supervisors recognized the importance of a strong process safety culture in minimizing accidents. An amendment was adopted to the Contra Costa County and City of Richmond Industrial Safety Ordinances requiring that all covered facilities perform an initial Safety Culture Assessment within one year, and at least once every five years thereafter.

Defining Process Safety Culture

Culture is defined by Merriam-Webster as "the set of shared attitudes, value, goals, and practices that characterizes an institution or organization." Process safety culture is defined by the Center for Chemical Process Safety (CCPS) as "How we do things around here." "What we expect here." And "How we behave when no one is watching."

Culture exists and is

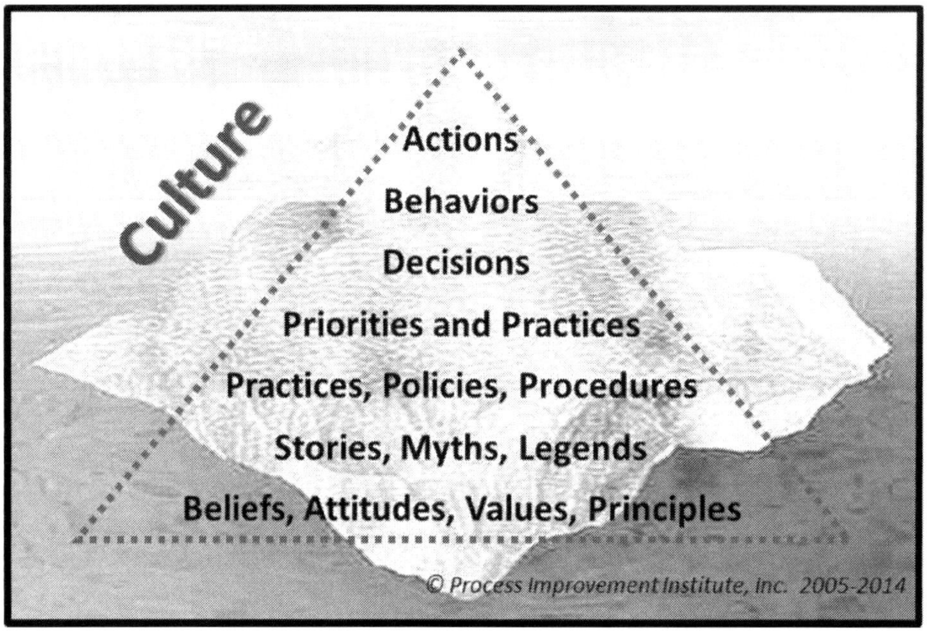

influenced on many levels. Individuals within an organization have their own set of beliefs and values based on their personal biases and experiences. Leaders within an organization have their own set of individual beliefs and values as well, but they also have a broader accountability for the organization's well-being. Leaders influence others on a daily basis through their conversations, their decisions, their behaviors, and their actions. Collectively, groups within an organization shape their own cultural identity through the perceptions, interactions, and behaviors of their group members. Individual values may be suspended in certain circumstances in order to maintain favor, status, and even membership within the group. Organizational culture may reflect the collective values of individuals and groups, the personalities of charismatic leaders, shifting priorities based on the perceived scarcity of resources or actions necessary for survival, or a long storied history of successes, behaviors, myths, and legends.

In short, culture is a very complex concept, and it can be very difficult to measure, influence, and manage. Most would agree that it is difficult to assign an absolute, quantitative measure of good or bad Culture. However, it is possible to identify, measure, analyze, and improve certain activities and characteristics that are recognized as key components of a positive process safety culture.

In studying the NASA Space Shuttle Disasters, the CCPS (March, 2007) identified six core principles necessary to maintain a positive safety culture. These are:

- Maintain a Sense of Vulnerability
- Combat Normalization of Deviance
- Establish an Imperative for Safety
- Perform Valid/Timely Hazard/Risk Assessments
- Ensure Open and Frank Communications
- Learn and Advance the Culture

The Baker Panel report organized their findings into slightly different, and perhaps more measurable, categories for evaluating process safety culture. These are:

- Process Safety Leadership
- Employee Empowerment
- Resources and Positioning of Process Safety Capabilities
- Incorporation of Process Safety into Management Decision-Making

Regardless of the model structure chosen or the descriptions assigned to different categories, rest assured that process safety culture is a very real, finite, and tangible concept which can be measured, monitored, and improved over time. Proactively working toward a more positive process safety culture will minimize process safety incidents, and this will save lives.

Examples of Process Safety Culture

Process Safety Leadership

Some organizations choose to develop their leaders through broad but brief exposure to many different business situations and challenges. While this may create successful leaders conversant in a wide variety of disciplines and business models, it may not create leaders who are connected with the people they lead and who are committed to long-term, sustainable process safety goals. It also may not give them time to build a core competency or appreciation for process safety. Their rapid fire, goal oriented decisions may yield strong quarter to quarter results, but too often these leaders have moved on to their next challenge long before the true impact of short-term thinking has been realized.

> ***Example:*** A management decision to reduce costs (staffing, maintenance, and capital budgets). While this may increase near-term profits, what are the longer range impacts of neglected equipment, deferred investments, and lost human talent? To paraphrase a renowned process reliability expert, there has never been a case where cutting critical talent has saved an organization; in fact it speeds its demise. (7)

Rapid management turnover may also create a frenetic "programs of the day," characterized by slogans, banners, key chains, and revised individual goals and objectives. As revolving managers compete for boardroom attention, scarce organizational resources are redeployed on the latest management fads. An overloaded and confused workforce is left with minimal guidance and no long term direction; given enough changes like these, workers distrust any new initiative. Both short-term thinking and frequent site management changes (every 10 months on average) were listed as safety culture deficiencies at BP Texas City in the Baker Panel report (10) and CSB investigation (9).

Effective leadership seeks to understand true root causes and permanently resolve issues with solid solutions, not temporary and ineffective "bandaids". CSB concluded that at Tesoro (11), management complacency had normalized the routine occurrence of hazardous conditions by using steam to mitigate heat exchanger leaks rather than investigating and resolving their root cause. Leaks and other mechanical issues had become so common that additional operators were routinely assigned to assist with leak detection and leak management during startup of the unit. "This past practice contributed to the presence of the six additional workers in the unit during the April 2010 incident."

> ***Example:*** *Effective management would take these lessons learned and turn them into concrete key performance indicators. In the Tesoro example this could include measuring and monitoring Temporary Leak Repairs (TLR), those repairs that are not permanent, with an overall target of zero TLRs that are older than one month. Then, one category of TLR would be heat exchanger leaks, another clamps on lines, and another clamps on flanges.*

Effective management is committed to the underlying health and well-being of the organization. Slogans and banners are replaced by high standards, shared values, and core-beliefs which sustain the organization regardless of who is in charge. Safety, quality, and efficiency become the cornerstone upon which all decisions rest. Effective management leads by asking the questions that matter about the things that matter, by recognizing, rewarding, and engaging those who uphold the right values and beliefs, and by setting the example for others to follow through their actions, deeds, and decisions.

Employee Empowerment

The term "empowerment" may be a favorite buzzword of Dilbert's pointy haired boss, but it remains a critical component of a positive safety culture in spite of the word's overuse. Front line employees are already heavily invested in process safety since they are the people with the highest potential for exposure and the greatest personal risk. Involving employees at all levels of the organization in process safety management ensures that it is widely understood and supported. This has the added benefit of recruiting additional resources (with intelligent minds and experienced hands) to support the often overwhelming multitude of tasks that must be completed, such as hazard evaluations, incident investigations, procedure validation, and training. Spread the wealth, share the knowledge, release the reins (allow workers to be in charge of key activities/programs), and reap the benefits.

In a positive process safety culture, empowerment often begins with instilling the authority in EVERYONE to question decisions which are being made about process safety.

> ***Example:*** *At one refinery, one of the most widely supported initiatives is called "Stop Work Authority" (SWA). Every employee has the right to call time-out whenever they are uncomfortable with the safety of anything that might be going on. Work is immediately stopped, a meeting of knowledgeable people is convened, the concerns are discussed, and work proceeds only when everyone agrees that appropriate measures have been taken and it is safe to move ahead. There are no repercussions to invoking SWA and in fact, favorable stories are frequently shared where the worker is cast as hero by preventing a potential incident from occurring. Unfortunately, the implementation of SWA for maintenance tasks (typically related to occupational safety impacts) has worked great while the implementation of SWA for operational issues (typically related to process safety impacts) may not be as effective. Management will sometimes start, restart, or fail to stop a process even when workers are very concerned over the lack of adequate safeguards for the current situation. In the eyes of the employees, management's unwavering support of SWA appears to wane as the financial consequences of stopping work (i.e. halting production) increase.*

Some less effective organizations embraced a "whatever works" approach where the ends justify the means. Standard operating and maintenance procedures and stop work policies were ignored in the interest of saving time or money (short-term). Errors and failures went unreported because doing so will get a worker or their friend in trouble. Supervisors looked the other way when shortcuts are taken, reacting only when required to do so because a poor result, an injury, or a loss event could no longer be hidden.

> ***Example:*** *Operations has a chattering high level alarm. Some individuals might shelve or disable the alarm for now, deferring action for as long as possible, hoping it will either go away or at least wait until the next shift arrives. Unfortunately, deferred action often increases the potential process risks (eliminating an independent protective layer) and results in even greater losses, since the real issue (a failed bottom's pump for example) continues undetected until it can no longer be ignored.*

In a positive process safety culture, individuals acknowledge the problem immediately, manage the risks appropriately, investigate the potential causes, and take corrective action to restore the system to a safe operating posture. **They know that timely and permanent fixes are less expensive and less hazardous than temporary measures**, and they don't need a supervisor to tell them to follow procedures and rules. They hold themselves and each other accountable because it is the right thing to do. They value the training they receive, they freely share their knowledge and experience with new colleagues, and they proactively take action to resolve any problems they encounter.

Some less effective organizations were characterized by a culture of making excuses and assigning blame. An unhealthy climate existed where the first question asked was "who is responsible" and the automatic response was "not me, I didn't see a thing." In this environment people lacked trust in each other and in their leaders. Real issues were hidden as long as possible, people were afraid to show weakness by asking for help, and people were hesitant to accept positions of accountability. Leaders were often the first ones blamed when things inevitably went wrong

> ***Example:*** *An incident occurs which appears to be caused by a shortcut taken from the written procedures. The individual is disciplined, the entire work group is retrained in the correct procedure, and management gets to show how tough they are on enforcing adherence to procedures and on quick correction of poor safety performance. Never mind the fact that 25% of the procedure steps are inaccurate so no one uses the written procedure anymore, and/or a missing tool has been on back order for two months, and/or the supervisor himself had taken the same shortcut last month when he thought no one was looking.*

In a positive process safety culture, the first questions following an incident are what and why, not who. Those are followed by asking how we can improve the procedures, task design, practices, equipment, and management systems to prevent this from ever happening again. All incidents are reported and investigated without blame. Root causes (management system failures) are identified and corrective measures are taken to implement sustainable and permanent solutions. Trust and collaboration are high because the ***culture*** is blame free when it comes to mistakes. People ask for help when they need it and give help without being asked.

One of the keys is for Management to commit to a blame free ***culture*** for near misses reported and for investigation findings (except for sabotage, of course). If this is set and management sticks to it, then near miss reporting will go up and empowerment to fix problems will go up, because people are no longer ducking for cover knowing someone will be blamed for every human error. Management needs to learn and teach to all workers that Errors cannot be avoided when humans are involved and that not recognizing and reporting errors and failures will simply increase the overall risk. If errors and failures are reported, then the root causes of the human errors can be found and the error rates lowered or the human errors mitigated in some way.

Resources and Positioning of Process Safety Capabilities

Almost all organizations talk about their commitment to maintain a safe workplace. Their walls are covered with banners that say “Safety First”, “Zero Incidents”, and “Think Before You Act”. One true test of organizational culture lies in the meaningful commitment of time and resources necessary to make those slogans mean something. Do they “put their money where their mouth is”?

In a strong process safety culture, accountability for process safety is assigned in a direct line that runs from the CEO through the unit managers and straight to the process operator. An unwavering commitment is made to develop and provide expertise at every level of the organization, decisions are consistently guided by their potential impact on process safety, and individual goals incorporate process safety metrics (leading indicators, not lagging indicators like number of losses and injuries) as a primary measure of performance. Promotions and rewards are unquestionably connected to demonstrated commitment to process safety.

Weaker organizations might commit resources on paper, but then fill those roles with ineffective or incompetent people. For example, trainers may show up on the organizational chart as a full time position. In reality though, trainers are often used as utility staff, handling special projects, greeting visitors, leading tours, and working on new initiatives. Training materials become obsolete, drills and training sessions get delayed, CBT's replace instructor led classes in the interest of "efficiency," and over time the knowledge and preparedness of staff at all levels erodes. Stronger cultures understand the importance of training and they invest in developing and maintaining the skills of their people.

Strong organizations will measure resources and financial investment as a leading indicator of process safety culture. Examples are inspection, testing and preventive maintenance (ITPM), maintenance budgets, capital investment, operating procedures, and emergency preparedness. A quick walk around the plant can give some measure of process safety culture. Are fences in good repair? Are tanks and lines labelled and painted? Are pipe supports clearly intact and maintained? If not, first impressions are often a good early indicator of a crumbling infrastructure and deterioration within. Of course, for a deeper measure of culture, an in-depth audit of the ITPM plans, programs, and procedures is necessary. Measure the timeliness of correcting deficiencies, adherence to ITPM schedules/procedures, and the expansion of problems detected in one area to investigate similar applications in other areas.

Incorporation of Process Safety into Management Decision-Making

Weaker organizations focus primarily on lagging indicators of safety performance, without considering how results are achieved. Loss of containment events, occupational injuries, and regulatory citations are used to measure performance and reward success. Safety prizes are handed out whenever a new milestone is achieved. Compliance with the regulations is considered to be good enough. They may be reluctant to conduct safety audits because audits find things that need to be fixed and fixing things costs money. Training is done only when required, it is done as expediently as possible and always with an eye on the costs. Programs that focus on employee behaviors are popular because they cost very little to implement and they may be effective in the short term in reducing some types of injuries. Unfortunately, these also inadvertently and wrongly blame error rates on "individual behaviors and attitudes" rather than management systems and controls designed to minimize human error rates. Each individual is responsible for their own behavior, right? Well in a general sense, this is true enough; but the statement is often misused to imply that perfect control of human error is a possibility and even a valid expectation. Zero human error rates are NOT a possibility. That is why we need multiple layers of protection against major accidents.

> ***Example:*** *The lowest measured limits for human error rates are in the airline industry. Pilots performing routine tasks multiple times per day with excellent control of human factors have demonstrated error rates as low as 1/200 (0.5%). In the process industry it*

is generally believed that error rates as low as 1/100 (1%) are achievable with excellent control of human factors, although this has not been demonstrated. The actual error rate increases with the influence of human factors such as high task complexity (up to 5 times), low experience and training (up to 10 times), or high fatigue (up to 20 times). Fortunately, both airplanes and process units have multiple layers of protection which limit the probability that a single human error will become a major loss event. These layers of protection must include effective control of human factors to allow humans to perform as close as possible to optimum levels.

In a positive process safety culture, organizations understand that the "process" (the path) is inseparable from the "product" (the result); if they do the right things right, then the results will follow. They monitor leading performance indicators through periodic audits of compliance and ongoing measurement of activities – and they tie individual promotions, recognitions, and rewards to achieving these goals. They evaluate the accuracy of procedures, the completeness of permits, and the condition of the workplace. They measure completeness, timeliness, and effectiveness of training, preventative maintenance tasks, action items and repairs, and hazard evaluations. They make sure that inherently safer, long term fixes are expedited rather than continuing to rely on temporary measures. They measure the level of employee engagement, the number of safety meetings held, the ratio of near loss to loss incidents reported, and the number of investigations completed. When an accident does occur, they don't panic and run for cover – they learn from their failures and they strengthen their defenses.

Example: Process Safety Metrics - leading indicators of a positive PSC

Mechanical Integrity

- *Timely completion of ITPMs including documentation/analysis/follow-up on "as found" data*
- *Backlog of process safety and process integrity work orders*
- *Maintenance emergency repairs and break-in work versus planned maintenance*
- *Temporary leak repairs*
- *PMI inspections performed*
- *Effective "bad actors" program*

Action Item Follow-up

- *Timely completion of recommendations and action items (from all sources including PHAs, IIs, PSSRs, MOCs, Compliance Audits)*
- *Audit of closure process for completeness and effectiveness*

Management of Change

- *Timely completion of temporary and permanent MOCs*
- *Audit/review of maintenance and project work orders for proper use of MOC system*
- *Audit/review of DCS programming changes fort proper use of MOC system*

Process Safety Competence

- *Timely completion of process safety trainings*
- *Evaluation of operator response to process safety events*
- *Field verification and validation of operating and maintenance procedures*
- *Reporting ratio of near loss incidents to loss incidents (NLI/LI)*

- *Timely completion of incident investigations*
- *Analysis (trends) and follow-up on incident investigation results*
- *Contractor compliance audits*

Human Factors Control

- *Compliance with fatigue management guidelines (overtime hours, consecutive days worked)*
- *Observation of pre-job planning activities, shift turnovers, and radio transmissions (communications)*
- *Evaluation of Human Machine Interfaces (HMI)*
- *Staffing, vacancies, absenteeism*
- *Job experience, certifications, and training levels*
- *Effective use of "Management of Organizational Change" processes*

Real Examples of GREAT Process Safety Culture

Kemya (Al-Jubail Petrochemical Company; SABIC affiliate, partnered with 50/50 with Exxon) - example of GREAT culture from day one, continuity; sustainability … stemming from initial and continued excellent leadership.

- Operating since 1985. Petrochemical process in Jubail, Saudi Arabia, makes ethylene, polyethylene
- Established strong process safety culture in first year(s) and maintained consistent management vision from the initial startup
- PSM designed around Exxon Operating Integrity Management System (OIMS)
- Invested in strong initial process safety engineering and process safety management competencies
- Invested (continual) in empowering workers to take leads on writing procedures, helping to manage changes, investigations, near miss reporting, etc.
- New workers immediately begin build competencies in process safety.
- Workers help train each other to "fix" a procedure or task instruction rather than "oh, let's not follow that because the steps are not correct in it"
- Thorough risk review of all hazards and tasks
- Leadership today sustains the SAME vision as was established originally; no-one has come in and slashed the programs that sustain the equipment and competencies and maintain the equipment (but that could happen; it has happened in other sites we have been to that went from great PSM to poor PSM
- They score high on process safety competencies and PS culture on interviews
- Same for occupational competencies (workers learn to do JSA in their first few months onsite)
- Their 29 year record in safety is best in class!

BP Cherry Point Refinery - example of GREAT culture from day one, continuity; sustainability ... stemming from initial and continued excellent leadership.

- Cherry Point, WA, USA
- 800 workers onsite; produces gasoline, diesel, and other fuel products from crude oil
- Operating since 1971
- Establish strong process safety culture early (24 years ago, under Arco) and maintained consistent management vision since then
- Refinery management did not compromise on budgets for critical competencies and activities
- PSM designed initially around WISHA (Washington State, USA) PSM standard
- Invested 12-7 years (continual) in empowering workers to take leads on writing procedures, helping to manage changes, investigations, near miss reporting, manage changes; provided skill training and coaching on all topics for the workers involved in these activities
- One operator from each refinery unit rotates every two years into a process safety group that manages process safety day-to-day (MOCs, risk reviews for MOCs, investigations, procedure updates, human factors evaluations, etc.)
- Provided training to all staff at levels on process safety, human factors, tailored to each group:
    - Process engineering
    - Management/Leadership
    - Operators
    - Maintenance
- New workers immediately begin build competencies in process safety.
- Workers help train each other to "fix" a procedure or task instruction rather than "oh, let's not follow that because the steps are not correct in it"
- Thorough risk review of all hazards and tasks
- Leadership today sustains the SAME vision as the past 15 years; no-one has come in and slashed the programs that sustain the equipment and competencies and maintain the equipment (but that could happen; it has happened in other sites we have been to that went from great PSM to poor PSM)
- They score high on process safety competencies and PS culture on surveys (highest scores on PS Culture during the Baker Panel surveys across all of BP USA)
- Their process safety performance in the past 15 years is best in class!

Process Safety Culture Qualitative Assessment

Some quantitative measures of process safety performance, such as the process safety leading indicators discussed earlier, are also indicators an organization's process safety culture. Often, these can be compared across different facilities and even across different industries to establish standards and gauge the relative effectiveness of organizations. However, many of the more qualitative elements of culture may be unique to an individual facility or even to a specific work group within the facility. It is very difficult to compare results and establish qualitative standards across different work units, but it is still possible to establish a baseline and measure progress over time within the same (or similar) work units.

When conducting a culture assessment, data is best collected in a variety of ways designed to encourage complete and accurate disclosure while minimizing any bias imposed by the collection process. Result can then be compared to eliminate (or explain) inconsistencies. Various collection methods consist of:

- Written surveys
- Focus Group Interviews
- Individual Interviews
- Field Observations

Written surveys, such as the Baker Panel Survey Instrument, are useful for comparing and measuring shifts in employee perceptions over time. Written surveys are easy to administer and analyze, and they can be useful in measuring perceptions. However, written surveys are limited by their impersonal nature, the limited scope of the questions, and the inability to follow up with additional questions to better understand why people feel the way they do.

EXAMPLE: Example Questions from the Baker Panel Written Survey

- *I have received training on hazard identification, control and reporting in the last 12 months.*
- *I believe a culture exists as this facility that encourages raising process safety concerns.*
- *Management puts a high priority on process safety through actions and not just empty slogans.*
- *There is usually sufficient staff in my work group to perform my job safely.*
- *In my work group, process safety concerns are secondary to achieving production goals.*
- *Written operating procedures (or checklists/job aids) are regularly followed.*
- *Interlocks, alarms, and other process safety-related devices are regularly maintained.*
- *Workers sometimes work around process safety concerns rather than report them.*
- *The process safety training that I have received allows me to recognize when a process should be shut down if safety critical interlocks, alarms or other process-safety devices fail or become unavailable during operation.*

Focus group interviews are a potentially valuable tool for obtaining qualitative clues and insights into how people feel or think about an issue or concept. The participants are selected because of a common relationship to the topic being analyzed. For example, people with similar organizational roles (supervisors, mechanics, or operators) or people within a common work group (a common process unit or department). The moderator's role is very important in guiding the discussion, asking open-ended questions, staying on task, encouraging full engagement, avoiding dominance, and seeking alternate perspectives. Participants often build on each other's ideas bringing issues to the table which might otherwise lay dormant, but some individuals may not be as forthcoming with conflicting or sensitive information in this setting.

Individual interviews often provide more candid and concrete examples of behaviors, actions, events, and stories that support individual perceptions and feelings. The moderator must be able to establish rapport, assure confidentiality, ask open-ended questions, and avoid judgmental or leading responses. A sufficient number of interviews are necessary to ensure that the findings are representative of the work group being studied and to help protect the identity of individuals. A sample size of 10-15% is usually necessary to establish statistically significant results.

When conducting safety culture interviews, it is useful to provide the moderator with an interview guide to help keep the conversation flowing on relevant topics. Following is an example of an interview guide developed by PII for use with a specific client:

EXAMPLE: Question Categories for Individual Safety Culture Interviews

- *Accountability (Responsibilities well defined, challenges for meeting them?)*
- *Learning (Competence, time allocated to training, too much, too little, cancellations?*
- *Corrective action program (Issues are addressed, timely appropriate?)*
- *Commitment (Management support, importance of safety, personal involvement?)*
- *Reporting and environment for raising concerns (Near misses, Willingness, practices, hesitation, and retaliation?)*
- *Change Management (reorganization, organizational changes preparedness, effectiveness)*
- *Work control, work practices (empowerment, being able to stop processes, direct instructions, procedure quality)*

Field observations are another source of data for evaluating an organization's process safety culture. These may be performed as a general walk through of a unit, observing behaviors, human factors, and the work environment – but the best data is collected with a specific purpose in mind such as an evaluation of operating procedures, human machine interfaces, shift exchange processes, or operator response to critical alarms. PII has developed and used a variety of guidelines and checklists to assist in the collection of field data.

Process safety culture can be qualitatively measured through a combination of methodologies such as surveys, interviews, and observations. However, it is important to note that the perceptions and opinions are NOT always reflective of the culture itself. Attempts to compare perception data across different organizations may be misleading. However, culture assessments conducted over time within an organization can identify trends and relative changes in perception which point to successes and improvement opportunities. Process safety culture perception measurements coupled with a strong set of process safety metrics will (especially those metrics targeted on investments and leadership), over time, establish a solid baseline for monitoring and improving organizational performance.

PSC Measurements Required by CCHMP

The **Contra Costa County Hazardous Materials Program** (CCHMP) is considered to be one of the elite regulatory agencies in the United States. Located in a densely populated and environmentally sensitive region on the Northeastern shores of the San Francisco Bay, CCHMP is charged with the mission of protecting "human health and the environment by promoting pollution prevention, increasing process safety knowledge and environmental awareness, responding to incidents, and implementing consistent regulatory compliance and enforcement programs."

CCHMP oversees compliance with both the Contra Costa County Industrial Safety Ordinance (ISO) and the Richmond Industrial Safety Ordinance (RISO). These require regulated facilities to implement programs to prevent chemical accidents from occurring that could have a detrimental impact to the surrounding communities.

Recognizing the importance of PSC in preventing process accidents, ISO was amended in June 2006 to include the requirement that facilities conduct an initial Safety Culture Assessment, and then at least once every five years thereafter. The amendment also allows CCHMP to "perform its own Safety Culture Assessment after a Major Chemical Accident or Release or the occurrence of any incident that could reasonably have led to a Major Chemical Accident or release, or based on CCHMP audit results."

The Contra County Guidance Document for Safety Culture Assessment (Section F, June 15, 2011) allows a great deal of latitude to facilities in selecting an assessment methodology, as long as the following attributes are assessed:

- Management Commitment and Leadership
- Individual Performance and Accountability
- Peer Perception and Accountability
- Safety Program Performance

The first three components (leadership, individual, and peer-to-peer) can be measured qualitatively through interviews, surveys, and observations of the organization. The fourth component (performance) can be measured directly through leading and lagging performance metrics such as timely closure of process safety recommendations and action items, completion of scheduled inspections, tests, and preventative maintenance tasks, the ratio of near loss incidents to loss incidents reported and investigated, and the number of process safety related events.

<u>ISO/RISO Covered Facilities</u>

- *Shell Martinez Refinery*
- *Tesoro Golden Eagle Refinery*
- *Phillips 66 Rodeo Refinery*
- *Air Products at Shell*
- *Air Products at Tesoro*
- *Air Liquide Large Industries at Phillips 66*
- *General Chemical Bay Point (now Chemtrade)*
- *Chevron Richmond Refinery*
- *General Chemical Richmond (now Chemtrade)*

Early Results and Future Plans at CCHMP for PSC

It is still too early to tell what impact the ISO PSC assessment requirements will ultimately have in preventing chemical process accidents.

All of the regulated facilities have completed their initial PSC assessments, but most were already conducting periodic assessments on their own. Written surveys are the norm with the Baker Panel survey serving as the model for most. Some companies have developed their own corporate instruments. None have reported the use of focus groups, individual interviews, or observations for their PSC assessments.

All companies have also developed both leading and lagging process safety metrics based primarily on data they were already collecting. They are measuring and monitoring this data internally to identify improvement opportunities and measure the success of their process safety initiatives.

CCHMP is continuing to work with our facilities to develop better process safety metrics, better PSC assessment tools, and a deeper understanding of the dynamic relationship between workers and management which helps explain why some prevention programs are more successful than others. Continued measurable success will encourage facilities to further expand their process safety improvement ideas and extend their programs beyond the minimum regulatory requirements.

Acronyms Used

AIChE – American Institute of Chemical Engineers
CCHSD -- Health Services Division of Contra Costa County, California, USA
CCHMP – Contra County Hazardous Materials Program, California, USA
CCPS – Center for Chemical Process Safety (of AIChE)
CFR – Code of Federal Regulations
EPA – Environmental Protection Agency (USA)
ITPM – Inspection, Testing, and Preventive Maintenance
LOPA – Layer of Protection Analysis
MOC – Management of Change
PHA – Process Hazard Analysis
P&ID – Piping & Instrumentation Diagram
PSI – Process Safety Information
PSM – Process Safety Management
RAGAGEP – Recognized and Generally Accepted Good Engineering Practice
RBPS – Risk-Based Process Safety
RCA – Root Cause Analysis
US EPA – United States Environmental Protection Agency
US OSHA – United States Department of Labor, Occupational Safety and Health Administation

References

1. U.S. Chemical Safety and Hazard Investigation Board, *Investigation Report: Catastrophic Rupture of Heat Exchange (Seven Fatalities),* Tesoro Anacortes Refinery, Anacortes, WA, Report No. 2010-08-I-WA, Draft for Public Comment, January, 29, 2014
2. Baker, James A., et.al, The Report of the BP U.S. Refineries Independent Safety Review Panel, January, 2007.
3. *Columbia Accident Investigation Board (CAIB) Report*, NASA, August 2003.
4. Contra County Hazardous Materials Program (CCHMP), *Contra County Industrial Safety Ordinance, Section F Safety Culture Guidance*, June 15, 2011.
5. Center for Chemical Process Safety (CCPS), *Guidelines for Risk Based Process Safety*, American Institute of Chemical Engineers (AIChE), 2007.
6. Moore, Ron, *What Tool? When? A Management Guide for Selecting the Right Manufacturing Improvement Tools,* Elsevier, 2007.
7. U.S. Chemical Safety and Hazard Investigation Board, Investigation Report: Refinery Fire and Explosion, BP, Texas City, TX, Report No. 2005-04-I-TX, March, 2007.
8. Center for Chemical Process Safety (CCPS), *Process Safety Leading and Lagging Metrics ... You Don't Improve What You Don't Measure*, American Institute of Chemical Engineers (AIChE), January 2011.
9. Bridges, W., *Gains in Getting Near Misses Reported (Updated),* 8th GCPS, AIChE, 2012.
10. Tew, R. and Bridges, W., *Human Factors Missing from PSM,* 6th GCPS, AIChE, 2010.
11. Bridges, W., and Collazo-Ramos, G., *Human Factors and their Optimization*, 8th GCPS, AIChE, 2012.

Influence of Customers on PSM Programs – Hitting a Moving Target

Hope A. Luebeck
Veolia Environmental Services
700 East Butterfield Road, Suite 201
Lombard, IL 60148
Hope.luebeck@veolia.com

Prepared for Presentation at
American Institute of Chemical Engineers
2014 Spring Meeting
10th Global Congress on Process Safety
New Orleans, LA
March 30 – April 2, 2014

UNPUBLISHED

Influence of Customers on PSM Programs – Hitting a Moving Target

Hope A. Luebeck
Veolia Environment North America
700 East Butterfield Road, Suite 201
Lombard, IL 60148
Hope.luebeck@veolia.com

Keywords: Process Safety Management (PSM), Hazardous Waste Industry (HWI), Hazardous Waste Facility (HWF), Highly Hazardous Chemicals (HHCs), Reclaim, Fuel Blending, Incineration, Operational Discipline

Abstract

Any responsible organization should strive for continual improvement in their Process Safety Management (PSM) program. However, lack of resources in the wake of multiple and conflicting priorities can serve as impediments to progress. Additionally, some organizations, such as those in the Hazardous Waste Industry (HWI), must also contend with a diverse customer base (each with their own unique demands and requirements) which can further complicate matters.

This document presents ideas and lessons learned in how PSM programs may be improved in spite of these impediments. In particular, it discusses how Hazardous Waste Facilities (HWF) respond to the challenges associated with processing a wide range of materials on a daily basis. It describes how the HWI has developed a robust PSM program by effectively managing inherent variations in the type and number of hazards associated with such operations. The concepts described herein are applicable to organizations with similar challenges that also seek to strengthen their PSM programs.

1. Why Aren't Mature PSM Programs Optimal?

Since the PSM regulation has been in existence for over 20 years, one would think that general compliance with it would be relatively good. And indeed, there are recognized stalwart leaders of PSM in many industries. But they stand in contrast to the many more who still struggle with compliance, perhaps due to PSM's necessary descriptive rather than prescriptive nature. Consequently, many aspects of such a program must be implemented as best interpreted by and applied to each unique organization.

Safety culture plays a pivotal role with any compliance program. According to the Center for Chemical Process Safety (CCPS), this is "how the organization behaves when no one is watching"[1]. Undesirable safety cultures have been cited as contributing factors to a number of major and deadly chemical process industry accidents such as the 1984 Bhopal, India toxic gas release and the March, 2005 Texas City refinery explosion. Companies' safety cultures infiltrate every aspect of compliance programs, since they set the tone for what is acceptable or not.

Ongoing major process accidents [2] indicate that the PSM goal of eliminating catastrophic incidents has a long way to go. Things are still going wrong and incidents keep repeating because we do not share our experiences or we forget the lessons learned [3]. Furthermore, a number of the 14 PSM elements can be challenging to manage, particularly Process Safety Information, Management of Change and Mechanical Integrity. Thus, companies continue to increase their understanding and administration of such areas and constantly work to improve their compliance.

Impediments to progress also include a lack of resources (e.g. people, time and money) and conflicting priorities. Events within the business world in the last decade have caused companies to reduce their workforces, spend less capital and focus on other things rather than invest fully in their PSM programs. Consequently, improvement in such programs has slowed and progress has taken longer to occur.

2. Hazardous Waste Industry Issues

The HWI must endure the added complications of having to process variable and unpredictable waste streams from an equally variable customer base which is comprised of a plethora of industries and activities. This significantly influences how the development of a solid PSM program accommodates continual improvement from the influx of myriad data. Some of these considerations were described in a September, 2013 article in Chemical Engineering Progress by this author [4].

As noted above, companies in the HWI receive an ever-changing array of materials for processing, recycling, recovery, and/or destruction. These materials may originate from customers' normal and upset conditions, may be the result of one-time events from a particular run, or be from a discontinued operation. Thus, the materials handled by the HWI are rarely the same composition, in the same quantities or the same combinations. This necessitates PSM programs which are built upon this variability.

Among the activities which are affected by this customer variability are reclaim, fuel-blending, and incineration operations. Reclaim facilities accept contaminated solvents for purification, often by separation or distillation processes. Fuel-blending operations combine certain classes of waste liquids or solvents which are not fit for reclaim, but still have value as a source of BTUs for cement kilns. Incineration facilities accept different kinds of materials in assorted phases (liquids, solids, gases, sludges, etc.) for safe destruction by thermal treatment. In each type of operation, facility personnel must remain vigilant about properly characterizing the physical and chemical properties of process feedstock and identifying incompatibilities and hazards on top of the challenges associated with safe processing.

Incoming wastes for a HWF are often comingled prior to being processed, especially in the case of materials which will be incinerated. HWI facilities must combine compatible materials in order to stage them for their targeted processing path. Proper identification of wastes and thorough compatibility testing are critical to making competent decisions about safely consolidating materials. Consideration must also be given to the potential effect of comingled solutions on the ability to separate key components of mixtures by distillation without creating unnecessary azeotropes which could interfere with distillation schemes and processing objectives.

3. Continual Improvement Tools

PSM programs don't improve by chance. They improve by concerted and targeted efforts. Companies must enhance their arsenal of tools to support continual improvement in their PSM programs. They must provide a safety culture which encourages program growth and vitality. Strengthening their communication with customers and employees, applying flexibility to their business approaches, and streamlining their operations according to their experiences are valuable assets to a thriving compliance program.

Communication

Effective communications are required for virtually any successful work situation and affect the success of all PSM elements. These have a significant impact on our operational discipline which must include performing tasks correctly and recognizing, responding to and seeking help, as needed, to unanticipated situations or conditions [5]. Responsibly employing operational discipline principles facilitates effective responses to uncommon customer or situational requirements.

Communication among affected parties

When organizations' PSM programs are significantly influenced by external variables such as customers' requirements, it is critical that there be an excellent channel of communications with them. Details derived from these interactions can affect how PSM is approached, how deadlines are met, and what resources are allocated to structured implementation plans. All of these are important to ensure success in not only meeting the goals of the customer, but also the company's PSM goals. Indeed, PSM and business goals are inextricably woven in many ways.

In the HWI, much of the due diligence related to customer needs must be done early in the process, well before any material is delivered for processing. It is critical to understand what must be processed, how it must be handled and transported, and what the expected product is. Sometimes, the focus on a specific compound to be reclaimed or destroyed is so great that it can commandeer focus from other contaminants which also pose hazards and risks. Thus, facilities must remain vigilant in obtaining thorough information (from sales agents and customer contacts) about even minor constituents in a waste stream and their associated hazards/incompatibilities so that the materials can be handled effectively with proper safeguards in place. Failure to thoroughly characterize a waste stream has led to unexpected consequences and effects prior to and during processing.

Since customers affect PSM programs in the HWI, it is imperative that resources are applied to analyze, recognize and then manage the identified challenges posed as an ongoing vigil. This is often necessary because the analytical needs of the customer or client are not the same as that of an HWF and does not always include all the properties that must be known. Personnel responsible for client accounts must be engaged with both the customer and the receiving processing facility. The communications bridge between both sides should support the free flow of all information needed to ensure success of the operation. This includes reaching out to all necessary groups such as laboratory, safety and technical personnel to minimize or mitigate the known risks. Of course, success means accomplishing the customer goals with zero safety incidents.

Strong relationships between clients and HWI representatives are the basis of the kind of frank and factual communications which are needed to ensure proper and safe handling of wastes. This success is bolstered by employees who feel engaged within the organization. They are more willing to offer input on situations and generally perform better because positive and cooperative relationships improve job performance [6] and strengthen a company's safety-based culture. Continual efforts to improve these sorts of collaboration can help a company's PSM program to thrive despite changing customer needs.

<u>Communication of Process Safety Information (PSI)</u>
PSI has three main components: chemical hazards; process technology; and equipment knowledge. These serve as a foundation for many of the other PSM elements. It is important for any company to manage PSI well since this will help them to be in better control of their PSM programs.

Preventing process accidents requires vigilance [7]. Vigilance is also needed to manage PSI and keep assorted records and documents current and accurate. This is a particular challenge when what the customer needs to be done requires continuous updating of relevant PSI for a given set of activities. Effectively applying operational discipline principles will enhance the proper management of PSI.

In the HWI, the chemical hazards posed by materials received for processing from customers drives the other two components of PSI. If a site completely understands the hazards posed by a given material, then the proper technology and controls needed to manage the hazards can be defined. Thus, the equipment and operating conditions required can also be determined. Unfortunately, the information needed by the HWF facility tasked to handle the waste is sometimes insufficient, so laboratory and compatibility studies must be conducted to provide additional details for understanding the hazards which must be mitgated.

Once the hazards and required technology are adequately outlined, then what systems are needed for success can be established. Most often, this results in repurposing existing equipment and systems and modifying handling and/or operating conditions as required. Sometimes though, teams must be assembled to brainstorm possibilities and determine the most time- and cost-effective manner to approach a given situation. This may result in specialized procedures, unique equipment or PPE, and an outline of things to avoid.

Flexibility
Flexibility is another crucial tool for improving one's safety programs. Companies must be willing to be responsive to both meet customer goals and achieve requirements of regulatory programs. They have to be nimble in their safety programs to make appropriate changes

according to circumstances and learnings from past experiences. By doing this, they will improve their compliance as they conduct daily business.

Repeatedly Reviewing Hazards

Analyzing the hazards of a process (Process Hazards Analysis or PHA) is a key element of PSM. This is commonly accomplished according to HAZOP or What-If analysis methodologies. When process hazards change due to a shift in incoming materials, additional evaluations must be performed. This typically occurs via the Management of Change (MOC) process, which is discussed in more detail below. These hazard evaluations may be as encompassing as revalidating a process if enough radical changes to the assumed risk may occur, or as simple as renewing a subset of an existing PHA.

There may also need to be flexibility in hazard review timing and content. A process may need significant modifications to suit the intent of the customer project. The perceived hazards which are introduced from these changes must be determined. New safeguards may be required to be successful in processing the new or different hazardous waste while maintaining PSM compliance.

As facilities respond to customer-driven situations, they need to be certain that their risk tolerance is never compromised under the guise of 'responding to customer needs'. When there is a lack of sense of vulnerability, 'normalization of deviation' [8] can occur. This is where changes are made over time, usually to optimize cost and productivity, and can lead to drifts in safety practices as well as assumed risk in operations.

Timely Training

Effectively dealing with customer issues often results in changes which require employees to be informed of modifications in work plans and be trained in these. Companies must be able to identify these opportunities and provide appropriate new or refresher training to employees (and perhaps even contractors) so they have the skills they need to do their jobs well given the new circumstances. Systems must exist to ensure that proper communication is made to affected employees and appropriate training sessions occur.

Employees in the Hazardous Waste Industry tend to be heavily trained in handling hazardous materials and hazardous material cleanup through HAZWOPER and other specialized methodologies. However, customer guidance is often needed to help a facility safely store and handle new hazardous substances when waste streams are particularly unusual or complex. This collaboration between client and supplier educates novices in particular hazards of their job and supports both getting the job done right for the customer as well as promoting compliance success.

Keeping Operating Procedures Current

As stated earlier, numerous discussions must occur to plan and execute activities to fulfill customer needs. But when two or more people speak, errors in understanding are possible. One way to improve comprehension is to have detailed written instructions [9], otherwise known as operating procedures. These are crucial to guarantee accuracy and consistency in how a group of employees conduct a specified task.

In the case of the HWI, these written instructions are especially critical due to the unique nature of what needs to be accomplished with a given waste stream and to control the hazards of the materials to be processed for an HWI customer. The instructions may be formal operating procedures, checklists, guidelines or other documentation which must be followed to achieve the planned goal(s). Existing procedures may need to be updated on a case-by-case basis, or entirely new ones developed according to need.

Operators and maintenance personnel also need to understand any changes in equipment design/operating parameters which may be required to mitigate hazards. Even the way the equipment is maintained may need to be modified. Appropriate written procedures for these activities provide important information for personnel to utilize while minimizing the inherent risks due to the hazardous materials which are processed.

Optimizing Change Management

Change management often comes into play via Management of Change (MOC) and/or Pre-Start Up Safety Review (PSSR) when HWI companies respond to customer requirements. MOC helps facilities document changes to covered processes and ensures that appropriate scrutiny is given to these changes. PSSR ensures that certain requirements are met before new Highly Hazardous Chemicals (HHCs) are fed to any processes.

In the HWI, planned processing of new or different customer waste streams can drive change management activities due to either temporary or permanent setups of new or modified equipment. MOCs are also required for new waste streams, which need thorough review for potential new risks to a process or site. This process helps to guarantee that either no new risks are assumed by an organization or identifies additional safeguards required to control the new risks at an acceptable level.

PSSR goes hand-in-hand with MOC. Changes in customers' business which necessitates modifications to equipment and systems can lead to special shutdown or turnaround considerations to address perceived risks or enhance processing capability. The PSSR element will ensure that particular requirements are met. This includes verification that: the changes were implemented as designed; that safety, operating, maintenance and emergency procedures are in place and are adequate; that hazard analyses were performed as required; and that training of affected personnel operating the process has occurred. Of course, all these will require evaluations to distinguish between items which are required for safe startup and those which may be completed at a later time.

Properly applying a company's change management principals will help ensure that the MOC and PSSR items are appropriately aligned and completed to facilitate new equipment startup and to ensure that safety, environmental and operational requirements are met.

Evolving Mechanical Integrity (MI) Programs

MI is probably the most confounding of all the PSM elements. Sites can have difficulty identifying which of their equipment in their covered processes should be part of the MI program (so called critical equipment). This, in turn, influences which equipment is subject to more rigorous inspection and upkeep practices and what maintenance and inspection, testing, and preventive maintenance (ITPM) written procedures must exist.

In the HWI, customers generally provide little input to MI programs but their business does affect its content. The HWI needs to be cognizant of proper materials of construction and applicable design codes required to safely process customer streams. Changes in composition (e.g. liquids versus abrasive solutions), characteristics (e.g. pH or corrosivity), or variations in other stream qualities can affect equipment limitations and maintenance requirements. These issues can also impact inspection regimens. Some vessels might require protective coatings or passivation or be subject to other specialized treatment regimens. Parts of a system may need to be replaced with more durable materials of construction. There might even be a need to upgrade equipment to ASME Section VIII pressure vessels or other more strict design codes.

The aforementioned items have implications on ITPM schedules, Quality Assurance practices and adhered-to design codes. As these evolve, changes must be made to equipment files, Piping and Instrument Diagrams (P&IDs) and other important PSI information. Depending on the degree of changes and number of PSM elements which may be affected by the changes, more personnel and facility systems may need to be involved to effectively manage the required actions.

Maintenance practices must be consistent with the desired degree of ITPM dictated by the customer materials which are processed. If there are more aggressive substances being contained in process equipment, there should be more frequent and more thorough inspections to guarantee that equipment is not degrading prematurely. These may influence repair or replacement plans and capital expenditures for a site. It is important to be proactive with inspecting, repairing, and replacing equipment. Although this may cost a facility some time and money, it is far less expensive and safer than waiting until a system fails because wear or degradation accelerated due to changes in materials being processed.

Streamlining of Operations

Even if we manage the 14 PSM elements well, improve our communications and flexibly manage our operations, our PSM programs may still need more work. We also need to learn from our experiences and apply these lessons to our operations to streamline how we manage these programs.

Improving Communications and Training

As learning opportunities arise from experiences with customers and particular waste streams, these must be communicated throughout an organization so that lessons can be applied to future operations and the PSM elements which may be touched by these developments. If something failed or worked right, these need to be identified and noted appropriately. Employees must be encouraged to utilize new practices that went particularly well. Otherwise, memories may be short and what was learned may not be incorporated into future operations. Failures will be doomed to repeat themselves and wins may be lost if not shared with affected parties.

Strengthened procedures

Original or modified procedures used for new situations must be critiqued afterward to determine if they properly described how the activity unfolded. We need to determine if appropriate variables and process limits were utilized and sufficient direction was given to operators for required tasks. We also need to verify if enough process information was collected along with adequate test data and interpretations of what happened during the run. During trials and new runs, it is possible to become so consumed in the activity that we fail to thoroughly record enough results or critical occurrences. We must update applicable procedures per the above scrutiny and modify control parameters based on results of the process run. It is impossible to improve regulatory compliance if we are not diligent in applying experiences to future operations. We need to apply lessons learned from trials so they can be permanently built into operating schemes as appropriate.

Learning from Change Activities

Many of us have closed out an MOC or PSSR and then remembered something else we should have done or done differently. When this occurs, we should document what to modify for the next change occasion. It is important to update our local MOC or PSSR procedures per experience to help both ourselves and others to better execute change in the future.

The HWI tends to execute fewer PSSRs than in other industries, so it has to capitalize on every opportunity to optimize its use of this review process. More time needs to be spent with employees who are assigned to PSSR teams to remind them of their role and the significance of properly assessing the items on the PSSR checklist. This helps develop fellow employees to more effectively participate on PSSR teams and be better prepared for the next change management situation.

Enhancing Incident Investigation Practices

George Santayana said that 'Those who cannot remember the past are condemned to repeat it [10].' We need to find ways to apply learnings from incidents and near misses to prevent recurrence, either for the same sort of situation or one that is very similar to it. We should not ignore past events. History often repeats itself, thus we must be wary to prevent incidents, accidents and injuries from complacency.

We must communicate identified lessons, especially why the event happened, to all levels of the organization. Often the most valuable lessons are those within one's own company. Thus, we must review these regularly to prevent smaller incidents from escalating to a catastrophe [11]. We should likewise update our emergency preparedness and training programs per noteworthy incident findings so that we enhance the robustness of them.

Common incidents due to handling new customer streams in the HWI include unexpected reactivity, fouling of process equipment, and unpleasant or toxic odors which could cause illness or injury to employees and neighbors of a facility. Conducting thorough due diligence before processing a new waste stream can help to prevent undesired circumstances, near-misses, and potential incidents.

Improving Emergency Preparedness and Response (EPR)
PSM facilities tend to have robust EPR plans and a wider variety of circumstances for which they prepare. When customer business leads to more variable process operations, companies must consider the possibility of a greater breadth of potential adverse events. While all of these cannot be practiced through real drills, many can be thought-through by conducting tabletop drills and brainstorming conceived situations. The HWI also must consider potential mistakes by customers, including improperly labeled or commingled wastes. Even waste that must be temperature-controlled to remain stable may be received at non-optimal conditions, leading to potential handling and/or safety issues.

Debriefings with operations personnel before and during special customer runs can help to avoid adverse situations by keeping employees aware of changes in customer receipts and associated hazards of these. Employees must be advised of best practices to be followed so that EPR plans do not have to be enacted.

Smarter Auditing
Since the inception of the PSM standard over 20 years ago, strong and effective management system reviews have been shown to drive continuous program improvement and reduce the likelihood of significant process safety incidents [12]. It is important to identify issues with our systems which need more attention or resources to resolve problem items.

Auditors in the HWI search for patterns in how sites handle unconventional customer material receipts, address change management, and investigate incidents, among other foci. Audit teams may review if appropriate controls were established to handle unusual situations. If there appears to be a series of historical audit findings which could be related to how certain customer business was handled, then auditors may attempt to uncover the challenging issues which led to repeated failures in one or more elements of an organization's PSM program. These are intended to identify 'weak links' in the PSM program so that organizations have the proper tools to be compliant and continually improve. A method of ferreting out root causes for audit findings rather than simply identifying causal factors associated with these can be very beneficial as noted by D. Heller [13] in a recent article.

4. Conclusions

Do not let changing customer needs or similar moving targets make you 'miss your mark' in achieving PSM goals. Successful organizations plan and remain ready for assorted situations and the HWI is no different. The nature of its business is variety and change. The economy, customer production rates/product mix, customer process upsets, expired products or raw materials, etc. all influence how much and what kind of feedstock is available for its facilities at a given time and thus the hazards which must be managed.

Defining a cohesive plan, capably identifying and reacting to new hazards, exhibiting flexibility in processing needs, and learning from experience are likewise paramount. Thorough due diligence with new or changed hazardous waste streams enables proper planning for safe material handling, effective processing, meeting customer goals and keeping HHCs properly managed in the HWI. But it takes an inordinate amount of focus for an organization to streamline its methodologies used in these situations to achieve and surpass business, safety, and environmental obligations. The CCPS Publication "Guidelines for Risk Based Process Safety" [14] provides many practical tips for reducing risk and is a great tool for organizations working toward continual improvement in their PSM programs.

Dennis Hendershot reminded us that we know how to improve our process safety performance. Our biggest challenge is not technical, it is cultural. We need to actually do what we already know how to do, we need to do it well, and we need to do it everywhere all of the time [15]. This requires concentration on the things we do well and emphasis on the things which need development. Anything is possible when targeted and concerted efforts are driven by a responsible organization!

5. References

[1] *Building Process Safety Culture: Tools to Enhance Process Safety Performance*, New York: Center for Chemical Process Safety of the American Institute of Chemical Engineers, 2005.

[2] U.S. Chemical Safety Board, 2012. Available at: http://www.chemsafety.gov.

[3] T.A. Kletz, What Went Wrong? Case Histories of Process Plant Disasters and How They Could Have Been Avoided, 5th Edition, Butterworth-Heinemann/IChemE, 2009.

[4] H. Luebeck, *Tackling Process Safety Challenges in the Hazardous Waste Industry*, Chemical Engineering Progress, September, 2013, pp. 41-47.

[5] Bruce K. Vaughn and Trevor A. Kletz, Continuing Our Process Safety Management Journey, Process Safety Progress (Vol. 31, No. 4), p. 337 (2012).

[6] Joseph F. Louvar, How to Communicate to Create A Safety Culture and Improve PSM Results, Process Safety Progress, (Vol. 32, No. 1), 2013.

[7] James A. Klein and Seshu Dharanavaram, Improving the Performance of Established PSM Programs, Process Safety Progress, (Vol. 31, No. 3), p. 261.

[8] D. Vaughn, The Challenger Launch Decision: Risky Technology, Culture, and Deviance at NASA, Chicago University of Chicago Press, 1996.

[9] John W. Herber, Quality Problems as Process Safety Warning Signs, Process Safety Progress, (Vol. 32, No. 2), p. 175, 2013.

[10] G. Santayana, The Life Of Reason, Vol. 1, Dover Publications, Mineola, N.Y., 1905.

[11] John W. Herber, Quality Problems as Process Safety Warning Signs, Process Safety Progress, (Vol. 32, No. 2), p. 175, 2013.

[12] Alan C. Brackey, Process Safety Management: 21 Years Plus or Minus – What I Wish I'd Known Then and What We Can't Afford to Forget Now, Process Safety Progress, (Vol. 32, No. 3).

[13] D.M. Heller et al, *Identifying Root Causes in Auditing*, Chemical Engineering Progress, 2013 (need to complete citation).

[14] Guidelines for Risk Based Process Safety, AIChE Center for Chemical Process Safety, Wiley, 2007.

[15] D.C. Hendershot, Process Safety Management – You Can't Get It Right Without A Good Safety Culture, Process Safety Progress, (Vol. 31), p. 2-5, 2012.

Process Safety: Are You Managing Your Risks or Are You Managing Your Activities?

Stephen Gill
SABIC
Wilton Centre, Redcar, TS10 4RF, United Kingdom
stephen.gill@sabic-europe.com

Prepared for Presentation at
American Institute of Chemical Engineers
2014 Spring Meeting
10th Global Congress on Process Safety
New Orleans, LA
March 30 – April 2, 2014

UNPUBLISHED

Process Safety: Are You Managing Your Risks or Are You Managing Your Activities?

Stephen Gill
SABIC
Wilton Centre, Redcar, TS10 4RF, United Kingdom
stephen.gill@sabic-europe.com

Keywords: Process Safety Management

Abstract

In process safety, effectively protecting against major accident risk scenarios often hinges on small details such as the correct setting on a trip system, being able to recognize and respond to a critical alarm in the face of all the other control room distractions, and strictly adhering to the few key process safety related steps in a complex procedure. Ensuring that these key details are clearly identified, understood and kept intact through the design, installation, and operating phases of the process safety lifecycle can be challenging.

One reason for the difficulty is that process safety management frameworks are often built around delivering specific activities such as process hazard reviews, safety integrity level assessments, incident investigations, training material development, and reviewing and updating operating procedures. These different activities are done by different people in different departments at different times. The critical details or understanding can easily be lost in the transitions between the activities or the departments responsible for them. The result is the whole being less than the sum of the parts, leading to the all too common management question: "Why do we keep having these incidents after all the time and resources we put into process safety?"

This paper outlines a framework for focusing on 'risk scenario management' rather than 'activity management'. With this approach, the key information on the risk scenarios and the required controls or protection layers becomes the central focus, and all of the supporting activities either feed into developing this information, or use this information as a basis for setting up the maintenance and operating information and systems. This provides clear links between the related activities and work processes and maintains the integrity of the information, keeping it 'evergreen'. The framework can be further strengthened by a process safety auditing approach which not only identifies gaps in the protection layers, but also the gaps in linkages between the different activities or in the activities themselves.

The results of shifting the framework emphasis can be a much clearer focus on managing the risks, with the supporting activities much better aligned. As the activities themselves are largely the same, operating with a 'risk scenario management' focus does not take any more resource than an 'activity management' focus.

1. Introduction

Process safety in an operating company is about ensuring that possible risk scenarios are identified, appropriate controls are put in place, and that those responsible for the process understand the risks and make sure the controls remain effective. It is a game of details—the key barriers that stand between normal operation and a major incident are often the small things like the correct set point in a trip system, the operator being able to identify a key alarm and respond appropriately, and the adherence to the few critical steps in a start-up procedure. In order to be effective, these crucial details need to be understood and maintained by every shift and every plant manager operating every plant on every site. Those who have a direct supporting role for the assets also need a clear understanding. .

A recurring frustration for operating companies is having incidents where the risk scenario is understood at some level in the organization and in the wider industry but the knowledge or key control was somehow lost or compromised at the operating front. This is particularly difficult when the company has put significant money and resource into process safety management.

Depending on how a company is organized, there may be a tendency to focus on the process safety related activities rather than focusing on ensuring that the results of the activities are effective at the operating face where they matter most.

Process safety management is obviously a very broad topic; some definitions and scope limits for this paper are provided here:

Major Risk Scenario: A sequence of events that may lead to the loss of containment of hazardous material (e.g. flammable, toxic) with subsequent potential for fatality or major injury impact on individuals, major asset damage, or major environmental impact. This paper is aimed at the management of specific risk scenarios related to the process that can cause loss of containment (e.g. vessel overfill, runaway reaction or explosion in a fired heater firebox) rather than generic scenarios associated with mechanical integrity (e.g. pipe or vessel corrosion or pump seal failure) or safe systems of work (e.g. control of open ends or preparation for maintenance).

Risk Controls or Layers of Protection: Mechanisms that recognize major risk scenarios and provide barriers that can prevent a scenario from starting, or stop the scenario from progressing further. Examples would be safety instrumented systems, mechanical interlocks that control valve opening/closing sequences, alarms with associated operator response, or administrative or procedural controls which ensure safety critical tasks are carried out correctly.

2. The Process Safety Lifecycle

Figure 1 is a generic view of the process safety lifecycle.

The lifecycle starts with the design. As part of the design process, risk scenarios are identified through hazard and operability studies (HAZOP) or other methodologies and sufficient risk controls are defined to meet the company risk criteria. These controls are then installed as part of the plant construction. Responsibility is handed over to the operating team and training is provided. A program is put in place to make sure that the controls are maintained (e.g. a test schedule for trip systems).

As time goes on, the need may arise for modifications to the facility. As part of a management of change process, the modifications are evaluated to see if they introduce new or impact on previously identified risk scenarios.

If incidents occur, the investigation may discover previously unidentified risk scenarios, or may conclude that the existing safeguards were inadequate and additional or more robust protection layers are required.

Figure 1.

Additional risk scenarios may be identified through other reviews or may be highlighted as part of learning from wider industry incidents. These scenarios are typically evaluated by semi-quantitative or quantitative risk analysis and may result in the revision or addition of protection layers.

At some regular interval, a wider scale process hazard review (PHR) or process hazard analysis (PHA) is conducted. These studies review the risk scenarios (whether from the original design, projects or modifications, or learning from internal or external incidents) and either confirm that the controls or protection layers are adequate, or assign an action for a more thorough analysis.

The process safety lifecycle concept is fairly widely accepted in the industry and most operating companies would recognize and are practicing the different elements.

The intent of this lifecycle or collection of activities is to ensure that the risk scenarios are clearly understood by the relevant personnel and that the appropriate controls or protection layers are in place and maintained. While this may seem obvious, there is a real danger that the focus of the organization ends up on completing the activities (the inputs) rather than on ensuring the quality delivery of the end product.

3. 'Activity Focused' Management of the Process Safety Lifecycle

The lifecycle shown above is comprised of a number of different activities. In many instances, process safety is addressed by managing each activity as essentially a separate exercise. This may make sense organizationally, as responsibility for the different activities may reside within different groups or divisions (e.g. project execution, operations, technical safety, etc.). However, managing the activities separately without an over-arching connecting framework can lead to disconnects and fragmentation resulting in gaps or holes in the understanding of the risk scenarios or with the risk controls themselves. Some of the difficulties with an 'activity focused' management approach are outlined below.

3.1 Some Challenges with the Process Safety Lifecycle

3.1.1 Different Parts of the Organization, Different Times

One challenge with managing the lifecycle by managing the individual activities is that they are done by different parts of the organization (with different repositories for information generated), and at different times. This is illustrated below with a few examples.

Handover Between Design and Operations:
A key activity during the design phase of a project is the hazard analysis and evaluation of risk controls. This might be done using a hazard and operability (HAZOP) methodology followed by layer of protection analysis (LOPA) to set the safety integrity level (SIL) of the trip systems or evaluate non-instrumented protection layers to ensure that the risk meets the corporate threshold. This is typically done as part of the front end engineering design (FEED) phase.

As the project progresses through detailed design and construction (possibly 2 or more years after the HAZOP and LOPA were completed) a different group of people write the operating procedures and the training packages for the operators. This transition is a prime place for content and understanding to be lost. The information on the risk scenarios and the key controls may be distributed among the HAZOP and LOPA reports, in the instrument specifications, and the design package process description. That may present a fragmented picture to those translating the design intent into operating procedures and training modules. Scenarios where operator action forms part of the risk control (e.g. response to a critical alarm, specific steps or checks in a start-up method) are particularly vulnerable to being lost or obscured.

Projects and Modifications
Through the life of the plant, projects and modifications will be installed which impact on risk scenarios from the original design and which add new scenarios. The information on the 'basis

for safety' is then spread between the original design book and a long series of project and minor modification documents.

Periodic reviews, learning from external incidents, incident investigations
As the plant continues to operate, various safety studies or reviews will be conducted. These may be wide scale reviews such as periodic Process Hazard Reviews, with the accompanying actions and subsequent analysis. There will also be occasional smaller reviews prompted by changing standards or to address specific legacy issues. One-off LOPA or quantitative risk assessments will be conducted occasionally as concerns arise about existing scenarios or potential new scenarios as a result of incident investigations, incident reports from wider industry, etc. Each of these reviews will have its own documentation or output which refines the understanding of the risk scenarios, but also increases the number of different places where important information is stored.

3.1.2 Optimization of individual activities without understanding the connections

The different activities in the process safety lifecycle are owned by different parts of the organization. The owner of the individual activity doesn't always have a clear understanding of how their activity interacts with other activities or processes in the lifecycle. As a result, activities or processes are often optimized for the primary user and not necessarily to facilitate related activities.

For example, a HAZOP is often seen as a design activity. The HAZOP work process may focus on ensuring that all the arising actions are closed out in a timely manner so they don't hold up the project, rather than producing a clear summary of the major risk scenarios and their controls for future use in writing operating procedures and training materials.

Similarly, LOPA documents may be owned by the instrument or control teams, who view them as design basis documents showing the probability of failure on demand (PFDave) that the system needs to meet. The fact that they provide a summary of initiating events, other protection layers, and the assumptions around conditional modifiers and enabling event and are therefore contain valuable information that should be readily available and visible to the operating teams may not be appreciated.

In many companies, the output of a PHA or PHR is largely focused on identifying potentially under-mitigated risk scenarios and assigning actions to address them. The PHA report (which should summarize the major risk scenarios and their associated controls) doesn't often find a place in the operating documentation. The closure of the actions (which may generate valuable information) is typically spread over several years and the learning is captured in a collection of individual action close-out reports.

3.2 The Results of an 'Activity Focused' Approach

As described above, the activities that make up the process safety lifecycle are often owned by different parts of the organization, occur at different times, and are designed for a specific purpose rather than being seen as one part of a more complex process.

This approach can lead to

- Information being lost in transitions between activities or over time
- Fragmented documentation of the major risk scenarios and the associated controls
- Lack of understanding of the major risk scenarios and controls at the operational level where it matters most
- Lack of visibility for the management line that the major risks are being managed adequately

In the aftermath of an incident, the information about the risk scenario and what should have been the key controls can often be found in a PHR report, in the original design book, or buried in the operating procedures. When a company has invested significant time and resource in doing the individual activities, this can be very frustrating—an expensive, damaging incident occurred that was understood somewhere in the organization, but not where (or when) it mattered most. Without clear and specific connections, the system is less than the sum of the parts. It is a bit like the child's game of 'Telephone', where each person whispers a phrase in the ear of their neighbor with the message becoming progressively scrambled as it goes down the line.

4. A 'Risk Scenario Focused' Management Framework

The previous section highlighted some of the weaknesses in an approach where the individual activities that make up the process safety lifecycle are managed independently. The following sections suggest an approach which provides much clearer connections between the activities and keeps the organization focused on controlling the risks.

The main feature of this approach is a set of living, evergreen, master documents which summarize the major risk scenarios and associated controls. This set of documents becomes the hub or focal point connecting all of the activities in the lifecycle, and ensures that the scenarios are always clearly visible to the relevant parts of the organization and especially to the operating teams.

This approach is explained in more detail by giving an overview of the concept, providing an example risk scenario summary, and then describing the role of these documents through the lifecycle.

4.1 'Evergreen' Major Risk Scenario Summaries as the Connection between Activities

Regardless of when they occur in time or which department is responsible for them, all of the activities in the process safety lifecycle are associated with either

1) Defining the risk scenarios and the controls required to manage the risks to an acceptable level based on the corporate criteria, or

2) Operating the assets safely by understanding the risk scenarios and ensuring the required controls are maintained and effective.

All of the ‘defining’ and ‘operating/maintaining’ activities can be connected together by one central ‘master set’ of summary documents that detail the major risk scenarios and the required controls or layers of protection.

Any activity that defines a new risk scenario or refines the understanding of an existing scenario updates the information in this ‘master set’ of documents. This extracts the key information from the individual activity documentation and makes it readily visible and available in a central ‘master’ location.

Similarly, if the organization is undertaking any activity that may impact on a major risk scenario or that requires an understanding of a scenario, there is one central source which provides the ‘master’ current information.

This is illustrated in Figure 2 below:

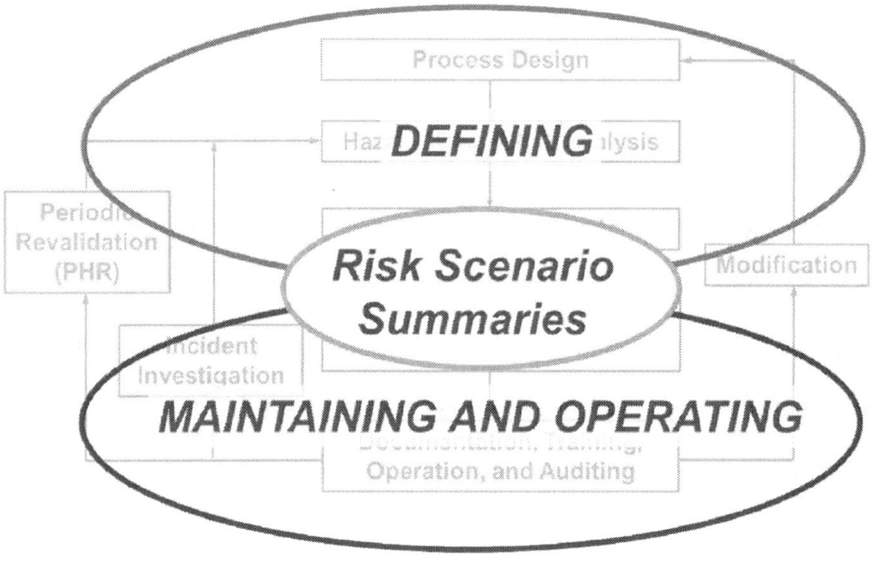

Figure 2.

These major risk scenario summary documents then become the ‘backbone’ for managing the risks from the initial design phase all the way through the operation and maintenance life of the plant.

They provide the bridge between asset design and operation by capturing the ‘safe operating envelope’. They are first generated in the design phase, and then become part of the ongoing revision controlled operating documentation. They are maintained and updated throughout the life of the asset, incorporating any additional learning, changes due to projects or modifications, or revisions to corporate criteria or standards.

Because all of the other activities either feed into these documents or use the documents as their basis, they connect the different activities in time and between departments and always provide the *current* definition of risk scenarios and required controls to those operating the assets on a day to day basis.

4.2 An Example Risk Scenario Summary—A Fired Process Heater

This section illustrates the risk scenario summary concept using the simplified example of a fired heater being used to heat a hydrocarbon feed stream. See Figure 3 below.

The example was chosen to include the three main types of risk controls:

- Safety instrumented systems
- 'Critical' alarms with associated operator response
- Administrative or procedural controls.

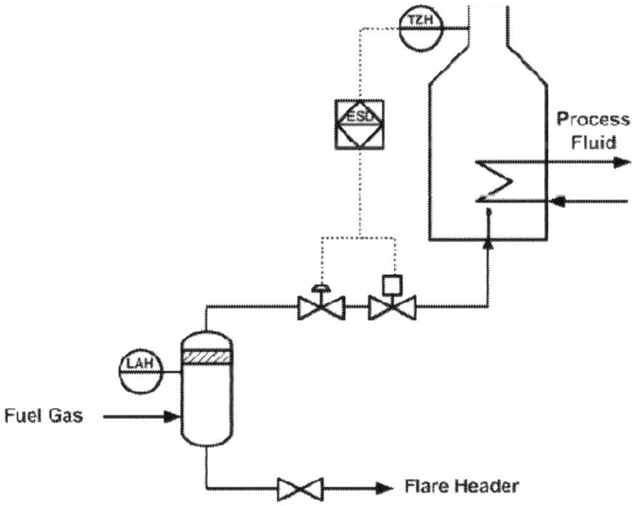

Figure 3.

Consider three major risk scenarios for this example:

1. Overheating and failure of the furnace tubes: This could be due to either loss of process flow or failure of the heat input controls, and could lead to loss of process containment with a subsequent fire and explosion. This scenario is controlled by a high flue gas temperature trip which shuts off the fuel supply to the heater.

2. Liquids in the fuel gas stream extinguishing the flame at the burner tip: This could result in furnace flame out with the subsequent re-ignition of accumulated gas on a hot surface

in the firebox, leading to an explosion. This scenario is controlled by a high level alarm on a knock-out drum in the fuel gas line. This alarm is deemed 'critical' in the annunciation system and there is a written procedure requiring the operator to drain any accumulated liquids from the knock out drum to the flare header if the alarm comes in.

3. Lighting the pilots with a gas accumulation in the firebox: This explosion risk is controlled by two administrative requirements in the startup procedure—doing a leak test to ensure that fuel gas isn't getting through the shutoff valves and into the firebox, and a flammable gas test of the firebox atmosphere before the pilots are lit.

For this example, the risk scenario summary might look something like:

Major Risk Summary for the Feed Heater

Risk Scenario	Typical Causes	Controls
Over-firing heater with subsequent tube failure and fire/explosion	• Temperature controller fails open • Low process flow	<u>Trip System</u> • High flue gas temperature trip TZA-101 set at 350 °C, tripping XCV-123 and TCV-456. SIL 1 Required
Burner flame out with continued fuel flow and subsequent re-ignition and explosion	• Liquid slug in fuel gas	<u>Critical Alarm</u> • High level in fuel gas KO pot set at 20% <u>Operator Response</u> • Drain liquid to flare header
Ignition of pilots with a flammable atmosphere in the firebox leading to an explosion	• Leaking gas valves • Insufficient purge prior to startup	<u>Administrative controls</u> Start-up procedure must include the following safety critical steps: • Test that XCV-123 and TCV-456 aren't leaking gas into the firebox when closed • Flammable gas test at the bottom of the convection bank and at the burner location prior to light off

The summary succinctly defines the major risk scenarios with some potential causes and the 'minimum set' of controls that are required to meet the corporate risk criteria.

4.3 The Risk Scenario Approach through the Process Safety Lifecycle

This section follows the life of the risk scenario summaries through the process safety lifecycle, illustrating how they connect the different activities together and maintain the continuity of focus on the risks rather than just the activities.

<u>4.3.1 The Initial Design Process</u>

During the front end engineering design phase, some form of hazard study or risk analysis will be conducted to identify the significant risk scenarios and confirm that sufficient controls or layers of protection are provided to meet the corporate risk criteria. Where the study identifies gaps, the design team takes actions to close the gap by modifying the design or by doing a more detailed analysis of the scenario.

At this point, the information on the significant risk scenarios might be distributed throughout various design documents (P&IDs, instrument data sheets, process description) and study reports HAZOP, LOPA, etc.).

In a 'risk scenario' focused management approach, a key deliverable from the design phase is the original set of risk scenario summary documents. The scenarios are then clearly visible in a central place and not scattered throughout the project documentation. The summary documents move from being design documents to being part of the operating documentation.

The major risk scenario summary documents are the key information transfer mechanism from the design team to the commissioning and operating teams.

4.3.2 Pre-Commissioning

Pre-commissioning activities include writing operating procedures and training materials and subsequently training the operating teams.

As stated above, the major risk scenario summary documents are the key information transfer mechanism. Using the summaries, the pre-commissioning team ensures that the major risk scenarios and associated controls are included in the operating procedures and training materials. This would include an overall description of the scenario and the controls for the operating procedures and training material.

For safety instrumented functions there is typically a fairly robust commissioning and testing procedure, with the design intent clearly available in the design documentation.

For critical alarms with associated operator response and for administrative/procedural controls, however, the pre-commissioning phase is where the design intent of what is required from the operator is more clearly defined. At this point, the risk summaries are used to ensure that:

- For 'critical alarms', clear guidance is provided on the required operator response (e.g. in the form of an operator response checklist). In this example, that would be the requirement to drain the fuel gas knockout pot to flare if the high level alarm comes in.

- For administrative controls, the key safety related steps in the job checklist or other document that will be used to control the operating team on the plant need to be highlighted as being part of the controls against a major risk scenario. Depending on company policy, these steps may require supervisor sign off or other confirmation that they have been adequately completed. In this example, the steps of testing that the fuel gas valves aren't leaking flammables into the firebox, and checking that the firebox is free of flammables before lighting the pilots need to be clearly identified as safety critical.

The major risk scenario summaries are the basis for transferring the key information into the operating procedures, training materials, and front-line operating documentation such as alarm response and task checklists or procedures.

4.3.3 Commissioning and Start-Up

There can be significant pressure during commissioning to make changes on the fly in order to get the plant started up. The major risk scenario summary documents need to be readily available as references for those making decisions during commissioning and startup. The summaries provide very clear visibility of the scenarios and key controls, reducing the risk of an important safeguard being bypassed or being compromised by a 'start-up workaround'.

The major risk scenario summaries provide clear risk scenario visibility during a time of significant change and organizational pressure.

4.3.4. Normal Operation

In normal operation, the risk scenario summaries essentially define the minimum process safety standards for the operating and maintenance teams. The scenarios for the facility are readily visible along with the required controls. The documents should be key 'go-to' references when questions or issues arise with scenario related instruments, valves, alarms, procedural steps, etc. If there is a need to bypass a trip system initiator, for example, it will be clear which scenarios the trip system addresses. This provides guidance on what temporary measures might be implemented during the duration of the bypass.

They are the basis for safety information provided in any updated training modules or operating procedures. The fact that they are living 'evergreen' documents ensures that the plant teams have a single source for the key safety information. While there may be additional detail available in the collection of project, PHR, LOPA, etc. documents, the information directly relevant to the operating teams resides in one place.

Audits are periodically conducted to make sure that the required controls are in place and robust and that the requirements laid out in the summary documents are being met. Note that effective auditing is discussed in more detail in a later separate section.

The risk scenario summaries provide an ongoing 'evergreen' view for the operating teams on the major risks and controls. The information is in a concise format and available in a central location rather than distributed through a number of documents and systems.

4.3.5 Management of Change

Hardware Changes or Projects:
When a hardware change is being considered, the risk scenario summaries provide a very clear view on whether the proposed change impacts on any of the controls for an existing scenario or increases the demand rate on the controls. If, for example, a change was made in the upstream

fuel system for the heater which made it more likely that liquid would carry over into the fuel system, it would be much easier to identify the issue. If additional controls are required as a result, the summary document is updated. This prevents key information being fragmented in a series of modification or project design documents.

If a new scenario is introduced as part of a project or hardware change, the scenarios and associated controls are either added to an existing summary, or a new summary is provided.

The major risk scenario summaries provide clear visibility of the risk scenarios for those evaluating the impact of hardware changes. This is key to ensuring modifications don't degrade the required controls.

If scenarios are modified or new scenarios are introduced the summaries are updated, providing a clear transfer of information from the project or modification documentation into the operating documentation. This prevents key information from being dispersed through various project files.

<u>Personnel Changes:</u>
When there are changes in the operating team or operations management, the scenario summaries and the associated training material is used to ensure that the new individuals understand the risk scenarios and controls before they have full operating responsibility for the assets. There are many other things that may be important for them to learn, but the summary documents ensure that the most important elements aren't obscured by all the rest of the information.

Risk scenario summaries make the key safety information readily available to new personnel and can easily be included in the induction or training material.

<u>4.3.6 Periodic Process Hazard Reviews</u>

The risk scenario summary documents are the 'current view' of the main risks and controls for the plant. They therefore become the starting point for the PHR. The PHR team can then review whether they consider the scenarios to be adequately described and whether sufficient controls are in place. The team may consider whether there are additional initiating events that may not be addressed by the existing controls, or whether the demand rate on a trip system or critical alarm activation rate is higher than is reasonable for the SIL level in place, etc. The PHR team will also look for additional scenarios not currently covered by the summary documents, or consider learning from other plants with a similar technology.

Any gaps that are identified are typically assigned as actions. Closure of the actions may be spread over several years. Any analysis done as part of the action closure may change the view on a scenario or the associated controls. The scenario summary documents will be updated accordingly. Updating the scenario summaries ensures that they reflect the current view, and that the key information generated during the PHR and subsequent action closure is clearly transferred to the operating teams and doesn't stay buried in the PHR documentation.

The scenario summaries are living documents which are updated based on the PHR findings, whereas the PHR report and subsequent action closure documentation are snapshots in time.

4.3.7 Incident Investigations

Incident or 'near miss' investigations may highlight gaps in the protection layers against a major risk scenario. When the investigation actions which address the gaps are completed, the risk scenario summaries are updated with the new information. This captures the key outputs of the investigation into the central living document, rather than having the information dispersed through the investigation follow up paperwork.

The same applies to learning from external incidents. If an incident reported in the industry is relevant to the facility, appropriate safeguards can be put in place and the scenario understanding and safeguards documented in the summaries.

The summaries allow for clear capture of learning from incidents or near misses that are related to major risk scenarios in a central master document. This ensures the operating personnel have clear access and visibility.

4.4 Activity Focused Management vs. Risk Scenario Focused Management

With the minor expense of generating and maintaining a master set of summary documents, the risk scenario focused approach provides much better connections between the various lifecycle activities by providing either a central destination for the information generated, or a central source of information to be used. More importantly, the approach gives the relevant personnel and particularly the operating teams a clear, centrally located, revision controlled, up to date view of the major risk scenarios and the required controls.

There is also a wider advantage to this approach. Having clear summary documents available for the major risk scenarios and controls facilitates consistency throughout the organization. For example, a review of the way the risks associated with fired heaters are being managed can be done by pulling out the fired heater summary documents for the various plants of interest and comparing them. If someone has a regional or business aligned process safety role, they have the tools available to look at the consistency of approach across a range of assets with similar technologies or risk potential, identifying best practice or where there might be deficiencies. That kind of cross-asset analysis is very difficult to do if the information is dispersed through a number of different design documents, study reports, and operating procedures.

4.5 Transitioning to a 'Risk Scenario Focused' Management Framework

How would a company move from an 'Activity Focused' approach to a 'Risk Scenario Focused' approach?

The starting point is generic to any organizational change:

Pull together a clear vision of what the 'Risk Scenario Focused' approach would look like in the organization—each company will be slightly different. Identify the activities that define the scenarios, and those that require an understanding of the scenarios and controls. Generate a sales pitch that shows the 'as-is' and the 'to-be' and why the transition is worth doing. Once the organization acceptance (or at least lack of active resistance) is in place,

1) Generate risk scenario summaries for all of the major risk scenarios on the plant(s). The information might currently be dispersed in a number of different places such as the original plant HAZOP, the latest PHR report, LOPA records, operating procedures, etc. Note that the more difficult it is to pull the information into one place, the more likely it is that the organization can benefit from making the transition!

2) Put the summaries in the same retention system as the other revision controlled plant documentation (e.g. P&IDs, operating procedures, etc.)

3) Update the project and management of change work processes to require generating new or updating existing risk scenario summaries as part of the project or modification safety review exercises.

4) Update the PHR methodology or work process to require an update (or at least a check whether an update is required) to the risk scenario summaries based on the review and the subsequent action close out.

5) Work toward making the documents an integral part of the day to day working on the plants. This is as much about a cultural shift as it is about documentation or procedures. There are a number of strands that might be important:
    a. Reviewing the training material to ensure that the scenarios are adequately covered. This gets the training department familiar with the use of the documents.
    b. Review the job method or checklists and highlighting the steps that are safety critical, referencing the relevant scenario summary. This imbeds the scenario summaries into the day to day operating routines.
    c. Provide training for the operating teams, supporting technical teams, project teams, etc. in the need for and use of the scenario summaries.
    d. Reinforce the importance of the scenario summary documents by auditing against them. The audit approach is covered in more detail in the next section.

5. Effective Auditing

Auditing is a key part of every process safety framework. It is required to check that what is required by all of the programs, procedures, etc. is actually in place. This section looks at typical auditing practices in an 'Activity Focused' framework, and then suggests an alternative approach which is greatly facilitated by a 'Risk Scenario Focused' framework.

5.1 Deficiencies in 'Activity Focused' Auditing

If an organization is set up to manage process safety by managing the different activities that make up the process safety lifecycle, the tendency is to also audit each activity separately. An audit on the PHR process, for example, might check that the last PHR had been done within the required frequency, confirm that the actions have been closed out, and check whether any high priority actions had gone beyond the target time frame. The audit might also look at the team makeup to check the right level of participation and that the person leading the study was on the approved list. When auditing the training program related to process safety, the auditor might check the training records for the operating teams to make sure they are in date and that any new personnel had received their training prior to being given full responsibility.

While auditing each activity is very important, such a piecemeal approach does not provide any information on whether the collection of activities has given the right results—***that the major risk scenarios are well understood at the plant level and that the required controls are in place and robust.*** It also doesn't provide information on how well the different activities are connected together and whether the collection of activities is likely to continue to deliver the desired results.

5.2 The 'End to End' Auditing Concept

To address the deficiencies noted in the previous section, an audit approach is proposed which is designed to do two things:

1) Provide information on how effectively the risk scenarios are actually being managed
2) Provide information on whether there are any significant gaps in the activities or the connections between the activities that make up the process safety lifecycle

The concept is called 'end-to-end' auditing as it looks at representative risk scenarios from hazard identification and risk analysis right through to how well the scenarios are understood at the plant level and whether the required risk controls are in place. The concept is shown against the process safety lifecycle in Figure 4.

For this approach, some representative risk scenarios for the plant are selected. The audit can start at the plant level and work backwards, or start with the hazard identification and risk analysis report and work forwards to the plant level, or start in the middle and work both directions. At the plant level, the audit looks at whether the scenarios are well understood and whether the controls in place and robust. For the activities or connections between the activities, the audit looks at whether the information is complete and consistent as it moves between the different areas—essentially 'are the results at the plant level consistent with what was intended to be at the plant level'?

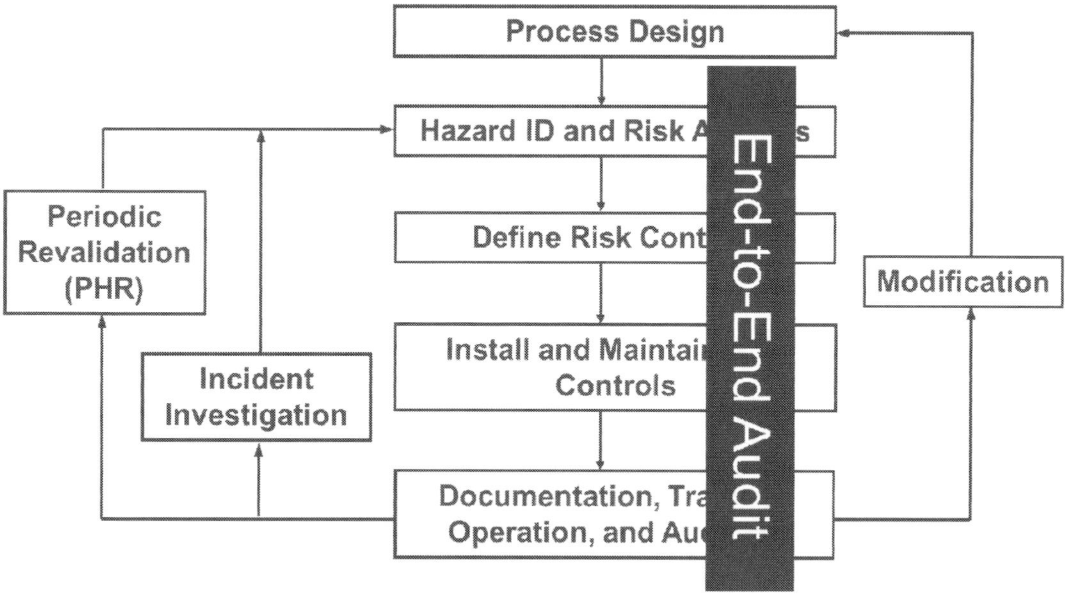

Figure 4.

5.2.1 Illustrating the End to End Auditing Concept

A fired process heater example was used in Section 4.2 to show what a major risk scenario summary might look like. This section uses that example and the associated risk summary to illustrate the end-to-end audit concept.

The example had three risk scenarios which were controlled by a trip system, a critical alarm and associated operator response, and some administrative or procedural controls.

For the overfiring scenario protected by a trip system, the audit path might be:

- What is the risk scenario this trip was designed to protect against? Where is that documented?
- What is the target SIL requirement and how was that conclusion reached (risk graph, LOPA, QRA report)?
- Can you demonstrate that the equipment as installed meets the SIL requirement (installed per design, test frequency matches what was assumed for the design, system has been tested at the appropriate frequency and the loop functioned properly)?
- Does the trip setting in the field match the design?
- Does the demand rate on the trip match with what was assumed during design (either records or interviews)?
- Does the training material for shift personnel and plant management adequately cover the risk scenario and the trip as the protection?

- Have the appropriate personnel been trained (records check)?
- Do the appropriate personnel actually understand the risk scenario and the controls (interviews)?
- Is the scenario clearly documented in the operating instructions?

For the scenario of liquid in the feed where a critical alarm and operator response provide the protection, the audit path might be:

- What is the risk scenario the alarm is designed to protect against? Where is that documented?
- How was the conclusion reached that an alarm with associated operator response was sufficient control?
- Does the alarm installation broadly meeting the company criteria for critical alarms (design, visibility, alarm hierarchy, ability to be detected in an alarm flood)?
- Has the alarm been tested at the appropriate frequency and has it functioned properly?
- Does the alarm setting in the field match the design intent?
- Does the operator have written response documentation?
- Is there evidence that when the alarm comes in the proper response is followed?
- Is the actual response time likely to be fast enough to provide sufficient protection?
- How often does the alarm come in? Is that within the assumed frequency in the hazard analysis? Is there a concern that the operators use the alarm as an operating guide (i.e. 'running to the alarm')?
- Does the training material adequately cover the risk scenario, the alarm, and the intended operator response?
- Have the appropriate personnel been trained (records check)?
- Do the appropriate personnel actually understand the risk scenario and the controls (interviews)?
- Is the scenario clearly documented in the operating instructions?

Similarly, for the furnace lighting scenario where administrative/procedural controls provide the protection, the audit path might be:

- What is the risk scenario the administrative/procedure controls are designed to protect against? Where is that documented?
- How was the conclusion reached that administrative controls were sufficient to address the risk?
- Is there a written procedure for the furnace start-up?
- Are the safety critical steps (valve leak test and flammable atmosphere checks) clearly identified and visible in the start-up procedure?
- Is there evidence that the start-up procedure is followed and that supervisor checks or signatures (if required) are in place for the critical steps?
- Have the appropriate personnel been trained (records check)?
- Do the appropriate personnel actually understand the risk scenario and the controls (interviews)?

- Are the scenario and required administrative safeguards clearly documented in the operating instructions?

5.2.2 What to do with the Audit Results

Remember that this audit approach has two objectives:

- Identify gaps in the understanding of the risk scenarios and the availability of the controls or protection layers at the plant level where it matters most
- Identify gaps in either the individual activities in the lifecycle or the way the different activities fit together

Findings from the audit are therefore useful at different levels and in different parts of the organization:

Plant or Operating Teams
Any deficiencies in the risk scenario understanding or availability/suitability of the required controls (incorrect trip setting, trip test out of date, alarm response procedure not consistently followed, etc.) should be rectified. This would normally be the responsibility of the local plant or site team. Note that this auditing approach reinforces the message that having the risk scenarios understood and the controls in place is a priority for the company and that process safety isn't just about attending reviews and generating actions.

Functional Management
There is typically a functional management layer a step removed from the plant. These are the people with responsibility for technical safety assessments, trip and alarm systems, training programs, and general EHSS management. The audit results may highlight systemic issues in the outputs or work processes of these groups. As examples, there may be issues across all the scenarios reviewed:

- With training material that doesn't adequately cover the risk scenarios or where the information is fragmented
- With the way trip and alarm testing is carried out or defects rectified
- With operating instructions or job methods where the key process safety related steps aren't clearly visible or adequately covered.

These systemic issues can then be addressed by the relevant functional managers.

Process Safety 'Architect'
At some level above the functional leadership there will be a 'process safety architect' with corporate, regional, or business level responsibility for making sure that the whole process safety lifecycle works. While the results of a single audit may not be useful, the audit findings from an 'end-to-end' approach across a site or across a business line may highlight gaps in the way the overall process safety framework is designed or implemented. Again, this type of auditing will not just identify whether each activity is being carried out appropriately; the approach will highlight whether the whole framework is consistently and effectively delivering risk scenario understanding and robust risk controls at the plant level.

6. Summary and Conclusion

The aim of process safety management is to ensure that risk scenarios are well understood and that adequate control measures for those scenarios are in place and maintained. While the different activities and processes that make that happen are important, there is an organizational danger in having the focus on the individual activities, and not on whether the whole series of activities is consistently delivering the right results at the operating level.

Management frameworks which treat each activity as a separate exercise may be susceptible to losses of understanding in transitions between activities or departments, local optimization of activities without an appreciation of how that activity fits into the overall safety lifecycle, and difficulties in maintaining a clear overall view of the major risk scenarios and controls as the plant evolves and personnel change over time.

A management framework that revolves around controlled evergreen summary documents for the major risk scenarios and their controls provides consistent, clear visibility for the operating teams. The use of these summaries also ties together the activities that are done by different parts of the organization at different times, and provides a mechanism for keeping the understanding of the risk scenarios and key controls current as the plant is modified and the understanding of the risks is refined.

Auditing is a key element in making sure that process safety systems are effective. An 'end-to-end' audit approach can check that both the understanding of the risk scenarios and controls are in place at the operating level where it counts most, *and* provide information on how well the different process safety work processes and activities are connected together to give those results.

In conclusion, those with a responsibility for the process safety lifecycle need to ask ***'are we managing the risks, or are we just managing the activities'?*** One way to answer that question is to do end-to-end audits on a couple of major risk scenarios at the plant, or at a couple of plants on a site, or at plants on a couple of different sites. See what the results tell you....

Walk the Line

Jerry J. Forest
Celanese
225 E. John Carpenter Freeway Suite 1200, Dallas, TX 75234
Jerry.Forest@Celanese.com

Prepared for Presentation at
American Institute of Chemical Engineers
2014 Spring Meeting
10th Global Congress on Process Safety
New Orleans, LA
March 30 – April 2, 2014

UNPUBLISHED

Walk the Line

Jerry J. Forest
Celanese
225 E. John Carpenter Freeway Suite 1200, Dallas, TX 75234
Jerry.Forest@Celanese.com

Keywords: Conduct of Operations, COO, Operational Discipline, OD, Engineering Discipline ED, Management Discipline, MD, walk the line

Abstract

Analysis of incident and event sharing data from sources such as the American Fuels and Petrochemical Manufactures (AFPM) process safety portal indicate that a large number of process safety incidents have control of energy causes. The most fundamental responsibility of the process plant operator is to control the energy between any two points in the process. Failure to control this energy due to improper line-up can lead to inadvertent mixing, challenges to safe operating limits, and loss of primary containment.

This paper discusses a variety of practical conduct of operations topics that help the process operator bring consistency to operation in order to minimize line-up errors and the process safety incidents that result from them. The concepts are: setting and reinforcing the expectation for energy control and establishing operational discipline and operational readiness. The operational discipline topics discussed are: improved communication through shift notes, shift instructions, and operator evaluation, returning equipment to operation from maintenance, line labeling, and commissioning and process safety start-up review activities. A brief case study is presented that discusses results from these activities - collectively called *Walk the Line*. Walk the line represents a culture change for operations where understanding of energy control each time a change is made in the process eliminates loss of primary containment incidents.

1. Introduction

The most fundamental thing that an operator does is to control the energy between any two points in a process. Energy can take many forms including the transfer of momentum, heat, and mass as well as energizing equipment through addition of electricity or initiation of chemical reactions by various means. Despite this fundamental of operation, industry continues to experience loss of primary containment (LOPC) incidents related to energizing systems where the transfer of energy results in an unintended consequence. A common cause of loss of primary containment is transfer of material to an unintended point or initiating a transfer to a system that contains an open end [1]. Adopting a belief that all process safety incidents are preventable leads us to analyze data to find and eliminate causes of the incidents. Effective root cause analysis doesn't stop at the top level event, but continues to explore the conditions and behaviors associated with the event [2]. For example, with regard to line-up causes, we should go beyond the "operator left valve open" cause, and ask why the valve was left open.

LOPC incidents due to control of energy causes commonly involve an operator starting a pump or opening a valve thereby transferring energy to a system that has an open end. The transfer was "lined-up" incorrectly. Operator line-up causes can include lack of training, poor communication, lack of discipline in monitoring the process, returning equipment from maintenance, and changes in operational state. These causes can be eliminated by setting and reinforcing an expectation of energy control, and emphasizing sound conduct of operations and equipment readiness practices. This paper discusses a strategy to change culture to ingrain operational discipline practices in the organization in order to achieve repeatable results and eliminate operator line-up causes.

2. Case Study

Analysis of LOPC data indicate that about half of the causes are related to conduct of operations and the majority of those are operator line-up issues. These incidents are 100% preventable. Therefore, a deeper analysis of the causes was completed in order to eliminate them.

Although the causes varied from site to site, there were many common elements:

- Energy control expectation not set
- Line-up not formally included in operator training
- Lack of discipline and consistent format around operating instructions, shift notes, and shift relief
- Failure to follow or lack of specific SOP's for complex line-ups

In order to address these causes, each site around the globe was asked to analyze their own data and complete pareto analysis of the line-up causes. For those sites that had no incidents related to operator line-up, best practices were shared in quarterly regional conduct of operations meetings. We then kicked off the Walk the Line initiative that incorporated each of the activities described in this paper.

The three components of the Walk the Line are:

1. Set and constantly reinforce the expectation for energy control;
2. Establish or improve operational discipline fundamentals affecting energy control; and,
3. Reinforce established operational readiness practices.

The theme of Walk the Line was purposefully chosen with a Johnny Cash motif to raise awareness to this preventable cause. Several techniques were used at the corporate level to set the expectation, train the organization on the expectation, and raise awareness. These include:

- Constant reinforcement of the expectation
- Use of regional meetings for best practice sharing
- Newsletters that describe the topics discussed in this paper
- Toolbox meeting white papers that describe an actual incident, then ask operators to define what they would do in a similar circumstance
- Short videos that reinforce the Walk the Line concept in a memorable way.

Individual sites then customized their Walk the Line improvement plan with one or more activities. A breakdown of the activities completed in 2013 is shown in Figure 1.

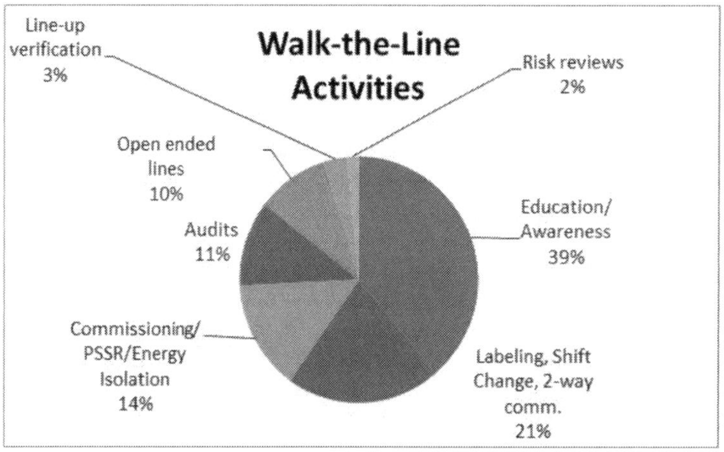

Figure 1 – Walk the Line Activities

It is too early to declare that operator line-up causes have been eliminated; the trend of the number of LOPC's with "walk the line" causes is certainly in the right direction as illustrated in Figure 2.

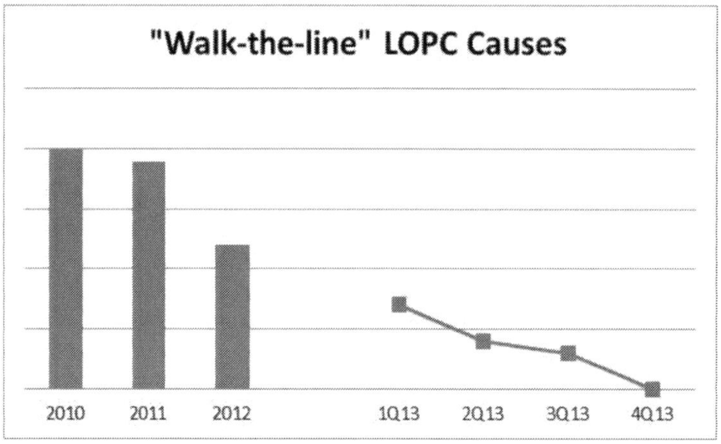

Figure 2 – Number of LOPC with Walk the Line Causes

The year on year improvement shown from 2011 to 2012 is believed to be caused by a focus on conduct of operations and the quarterly trend due to the 2013 Walk the Line initiative.

3. Walk the Line

Walk the line is a description of all of the things that operators do to make certain that they understand where energy will go when transferred. It represents the fundamental duties of operators. Walk the line means that operators have completed a field assessment of a system before changing the operational state. The assessment includes inspection of key operational components and the mechanical condition of equipment to ensure that the system integrity is robust enough to complete the change without failure. The intent of the assessment is to answer the question, "can I proceed safely with the planned change?" That goes for every change including things like changes in operational mode (startup, shutdown, etc.) pump operation, and valve position.

If an operator does not know with certainty where a material will flow when a change such as turning a valve or starting a pump occurs, then a complete assessment of the flow path should be made. However, there are several practices that help the operator ensure the operational state of the plant including setting the expectation, communications, operator evaluations, returning equipment to operation from maintenance, line labeling, and commissioning and operational readiness activities.

4. Set the Expectation

Changing the culture relating to energy control begins with a clear definition of the operator's role and responsibilities associated with plant operation. In his book *The Seven Habits of Highly Effective People*, Steven Covey states that performance is a function of knowing what to do, knowing how to do it, and wanting to do it [3]. Operators know what to do when we tell them what to do, how to do things by training and experience and they want to do it when management communicates that it is important to them. The expectation for energy control should be set during initial operator training and certification and reinforced at routine intervals with refresher training. Constant reinforcement of the expectation to understand equipment readiness engrains the walk the line concept in the organization and drives a culture that stresses energy control.

4.1 Operator Training & Refresher Training

Elements of a good operator training program include both knowledge and skill demonstration [4]. The knowledge element is achieved through required reading, classroom presentations, computer-based training, and other traditional learning tools. Initial operator training should include the many fundamentals of being an operator including basics such as choosing the proper personal protective equipment, how to open and close a valve, how to start up a pump, how to install and remove a plug, and how to line up equipment. Do not assume that operators know these things. Clearly define the operator's responsibility and train for various operational discipline tasks [5]. In addition to those previously mentioned other tasks related to energy control are discussed below and include (but are not limited to):

- Proper development of shift notes and making shift relief
- Reading and following operating instructions
- Practicing two-way communication
- Developing comprehensive line labeling
- Learning how to complete an effective equipment evaluation round
- Returning equipment from maintenance
- Bypassing equipment
- Communication of abnormal operations

These same topics should be covered in refresher or recertification training. The training and training material should be documented.

Equally important as the knowledge component of training is the skill demonstration. This can be effectively modeled by assigning new operators to more experienced persons to mentor the operator. Mentors, or on-the-job training if done by individuals who exhibit the correct behaviors, can be a powerful learning venue to teach energy control concepts with young operators. The mentor must clearly understand the basic principles that are part of the training objectives in order to understand what it is that must be taught.

4.2 Constant reinforcement

Process safety culture can be defined as a commonly held set of values, norms and beliefs [6]. One way to mold the culture to a desirable state is to constantly reinforce the values, norms and beliefs that drive the desirable behavior [7]. Each shift, front line supervisors have an opportunity to drive the desirable behaviors through shift instructions, toolbox meetings, and personal dialog with the operator. For example, shift instructions might be left to return a pump to service from maintenance on the next shift. This is an opportunity to reinforce the expectation of a field assessment to ensure that there are no open ends and that the equipment is ready for service. When the expectation is constantly reinforced, it eventually becomes "the way we do business around here."

4.3 Raising Awareness

Another effective way to set the expectation for energy control is by producing videos, newsletters, posters, and other visual aids that help remind the operator of the fundamentals to walk the line.

5. Operational Discipline Topics for Energy Control

If an operator does not know with certainty where the material will travel during a transfer then the line must be physically inspected prior to the transfer. There are however, several operational discipline topics that help the operator understand with certainty the current state of equipment and flow of materials [8].

5.1 Operator Shift Log & Shift Relief

The best practice is to have front line supervisors, specialists, and each operator position complete a shift log. Shift logs introduce discipline or repeatability in operations. This discipline is introduced in many ways. Logs should be structured to include a minimum amount of required information that the operators must address whether applicable or not. If the log is structured correctly, it becomes the basis of shift relief and that relief covers the same minimum topics each time. In order to record the most accurate information and have time to write a complete log, the operator should complete it throughout the shift and take a few minutes at the end of the shift before shift handover to make sure it is complete.

The best practice for shift logs is to provide a form or checklist where operators complete notes on pre-defined topics [9]. These pre-defined topics include:

- Abnormal events reporting
- Incidents on shift – safety, process safety, environmental, and reliability
- Significant alarms on shift
- Interlocks tripped
- Walk the line section that includes-
    - Abnormal line-ups
    - Maintenance completed on shift
    - MOCs/ PSSRs completed on shift
    - Equipment bypassed
    - Equipment taken out of service or returned to service

A good shift log will list these and other topics. Each shift, the operator addresses each pre-defined topic. If there is nothing significant to report, that is noted. If there is something to report from the pre-defined topics the operator should describe the condition, what action was taken if any, and the status of the situation at the end of the shift. The log becomes the basis for shift change communication. Each topic is discussed each shift. This provides structure and discipline to ensure that all relevant topics are discussed.

Another best practice is return to work activities. On any return to work all manufacturing personnel assigned to the unit (operators, specialist, front line supervisors, engineers, etc.) should be required to read all logs related to their position since they were last at work and provide positive acknowledgement that the log has been read with an initial or signature.

5.2 Operating Instructions

Similar to the operator logs, operating or shift instructions can be completed on a form or checklist that includes the same pre-defined sections listed above to help the operator identify those areas where line-ups may have changed.

5.3 Informal Unit Walk Through

Either during or immediately following shift relief, operators should make an informal walk through of their unit. The purpose of this walk-through is to become familiar with the unit line-up and verify the items discussed during shift relief. These items include abnormal line-ups, equipment bypassed or out of service, equipment readied for maintenance, and other normal operating observations. Abnormal observations not discussed during shift change should be recorded in the operator log. As part of the informal walk-through, the operator should make observations on open ended lines and correct those that are found. Informal unit walk-through and open-ended equipment evaluation isn't limited to shift change. A good practice is to establish a walking route and complete that circuit more than once per day.

5.4 Operator Evaluation Rounds

Often called operator rounds, gauge sheets or reading sheets, a fundamental of being an operator is to understand the condition of the unit by completing a formal unit walk-through and recording pertinent information on the condition of the unit. These walk-through's can be called evaluation rounds because the real purpose is for operators to evaluate the equipment. Therefore, in order to be effective we need to specifically identify which areas of the plant need to be frequently observed due to activities that may lead to line-up errors or open ends and specifically call this out on the reading sheet. The operator puts a √ or x for observations; checks mean everything is good and an x means there is an irregularity. If an x is recorded the operator notes both the condition and what action taken to correct.

5.5 Free Issue

In order to immediately correct open ends, the operator should have plugs easily available. Train on the correct material of construction and have a positive material identification procedure in place. Also, train operators on the correct installation and removal of plugs.

5.6 Unit Audits

Another engineering discipline topic that can be used to verify that there are no open ends is to conduct periodic performance audits. Create an audit checklist specific to each unit and complete a unit walk through with the responsible front line supervisor and operator. Correct open ends and other issues as they are found. This is another form of constant reinforcement of the expectation of no open ends and of proper unit line-up prior to changes.

6. Operational Readiness

With many companies the equipment owner defines when equipment is ready for maintenance and when it is ready to return from maintenance. The equipment owner is usually defined as operations. In order to eliminate line-up causes of process safety incidents, this role must be clearly defined and the expectations set for responsibilities. Otherwise, it may lead to the classic "not by job" error. For the purpose of this discussion, it is assumed that operations is the equipment owner and has final responsibility at all times for the energized state of the manufacturing unit.

6.1 Maintenance/ Operations Turn-Over

Loss of primary containment can occur on equipment turnover to maintenance due to improper decontamination of equipment. There is an excellent opportunity to use the communication tools listed above such as instructions, shift logs, shift relief, and one-on-one dialog to document the current state of equipment to be turned over to maintenance. Maintenance activities should not begin until formally approved by the operator on shift. Formal approval includes safe work permitting. The operator is responsible for the energized state of equipment until the permit is signed on, and signed off at the completion of work.

6.2 Verification after Maintenance

The purpose of verification after maintenance is to ensure equipment readiness and proper line-up prior to putting equipment back in service after maintenance. Equipment readiness is verified by field walk-down to ensure the proper line-up. While some companies may define the maintenance person as responsible to return certain pieces of equipment to an operational state, it is the equipment owner's responsibility to verify that line-up is correct prior to introducing energy to the process.

6.3 Independent Verification after Maintenance

Where complex equipment line-ups that increase the risk of a wrong position in the piping system exist, independent verification after maintenance may be necessary. Independent verification includes a second qualified person to perform independent verification of the line-up. Independent means that both time and distance separate the initial verification and the independent check. Items such as pressure relieve devices, complex safety instrument systems, and complex piping manifolds should be considered for independent verification.

6.4 Other Operational Readiness Topics

6.4.1 Standard Operating Procedures

For equipment that is routinely taken in and out of service, process safety critical equipment, and other complex line-up situations, standard operating procedures (SOPs) should be developed to ensure equipment readiness. A useful format in these situations is to structure the SOP as a checklist and require operator sign-off for each step. A deviation process is executed when the SOP isn't followed as written, and completion of the checklist is verified by a qualified person.

6.4.2 Formal Commissioning

A formal commissioning should be done on all equipment returned to service. The methodology used may vary depending upon the possible hazards, but commissioning should be completed nevertheless. Various commissioning activities may include:

- P&ID walk down, highlighting systems verified
- Pneumatic leak checks, soap checks, and pressure tests
- Loop checks from the DCS to the field and back
- Safety system operational checks
- Formal field audits that validate equipment readiness

6.4.3 Pre-Startup Safety Reviews (PSSR)

Formal PSSRs are completed and documented for changes in a process, but should also be done for startup of new equipment and when larger systems are returned to service. The fundamentals of PSSR are to verify that equipment is as per design, that procedures are in place and adequate, that any recommendations produced as a result of the change are complete, and that employees have received the proper training. PSSRs are usually done in addition formal commissioning activities.

6.4.4 Equipment Line-up Aids

A well designed line-labeling program can help aid operator line-up. Line labeling can include a description of the material in a line and the destination with the direction of flow. When done correctly and maintained line labeling is a valuable aid to help the operator ensure all transfers of energy reach the intended destination.

7. Conclusion

Causes of process safety incidents related to operator line-up have their origin in process safety management systems. These include culture, leadership, training, and conduct of operations. By having strong systems that constantly reinforce the expectation to understand energy control and Walk the Line - we can eliminate these causes.

8. References

[1] AFPM Safety Portal Event Sharing Database, www.afpm.org/safetyportal (accessed December 13, 2013) – Login credentials required.

[2] Kletz, Trevor. *Learning From Accidents in Industry*, Butterworth & Co. Ltd, 1990

[3] Covey, Steven R, The 7 Habits of Highly Effective People. Simon & Schuster, New York, 1990

[4] Idrees and Aslam. Prevent major failures and accidents with well-trained and empowered operators. Process Safety Prog 29 (2010), 174-178

[5] J. Haesle, C. Devlin, and J. McCavit. Improving process safety by addressing the human element, Process Safety Prog 28 (2009), 325-330.

[6] Center for Chemical Process Safety, Conduct of Operations and Operational Discipline, American Institute of Chemical Engineers, Newark, NJ. 2011.

[7] J. Forest. How to Evaluate Process Safety Culture, Process Safety Prog 31 (2012), 195-197

[8] J. Forest. Improving conduct of operations using best practice tools from operating, engineering, and management discipline. Presented at the 8th global congress on process safety, April 1-4, 2012

[9] J. Klein and E. Francisco. Focus on personal operational discipline to get work done right. Process Safety Prog ZZ (2012), XX-YY

10TH GLOBAL CONGRESS ON PROCESS SAFETY

Is There an Acceptable Level of Competence in Process Safety for a Graduate Chemical Engineer?

Brian Dickson
Senior Teaching Fellow
University of Strathclyde
Glasgow UK G1 1XW

Daniel A. Crowl
Herbert H. Dow Professor for Chemical Process Safety
Michigan Technological University
Houghton, MI 49931

Prepared for Presentation at
American Institute of Chemical Engineers
2014 Spring Meeting
10th Global Congress on Process Safety
New Orleans, LA
March 30 – April 2, 2014

UNPUBLISHED

Is There an Acceptable Level of Competence in Process Safety for a Graduate Chemical Engineer?

Brian Dickson
Senior Teaching Fellow
University of Strathclyde
Glasgow UK G1 1XW

Daniel A. Crowl
Herbert H. Dow Professor for Chemical Process Safety
Michigan Technological University
Houghton, MI 49931

Keywords: process safety, education, competency, accreditation

Abstract

In considering who and what process safety education is delivered in universities, two international bodies have the influence over what is taught. The two standards in chemical engineering dominating the process chemical industry in USA, and in UK, SE Asia, and the Middle East are: ABET (set in conjunction with AIChE) and UKSPEC (devolved to IChemE), respectively, and these can be looked upon as having a common set of chemical engineering competences at the graduating student stage since they offer equivalency of status to each other.

This paper will review the different approaches each institution takes to measure competence in process safety education. This paper will look at how process safety education is provided to students, and question its breadth and depth, challenge the competence level set by the two qualifying authorities and suggests a structure of minimum/ acceptable/ gold standard training that graduate student chemical engineers need to be able to demonstrate. It will also comment on some of the challenges such a proposal faces.

The paper will address the position of the graduating chemical engineer not the specialist safety engineer.

1. Introduction: History of Process Safety Education

1918: Lewis DeBlois, **DuPont's first corporate safety manager**, wrote "*...safety engineering, with its interests in design, equipment, organization, supervision, and education ... bears as well a very definite and important relation to all other branches of engineering. This relation is so close, and its need so urgent, that I am convinced that some instruction in the fundamentals of safety engineering should be given a place in the training of every young engineer. He should be taught to think in terms of safety as he now thinks in terms of efficiency. Conservation of life should surely not be rated below the conservation of energy. Yet, few of our technical schools and* *universities offer instruction in this subject, and the graduates go out to their profession with only vague surmises on 'what all this talk on safety is about.'"*

1932: **In the U. S., engineering accreditation was originally founded as the Engineers' Council for Professional Development (ECPD),** an engineering professional body dedicated to the education, accreditation, regulation, and professional development of the engineering professionals and students in the United States. ECPD was comprised of seven founding engineering societies, including AICHE.

1980: ECPD changes its name to Accreditation Board for Engineering and Technology (ABET).

1984: Bhopal, India tragedy occurs.

1985: In response to the Bhopal accident, the AICHE forms the Center for Chemical Process Safety (CCPS). One of its first charter items is to improve process safety instruction for undergraduate chemical engineers.

1986: CCPS forms the Undergraduate Education Committee charged with improving process safety education. The Committee decides to develop a book of homework problems and supporting instructional materials that can be used in all undergraduate chemical engineering courses. The Committee awards a contract to the University of Arkansas under the direction of Reid Welker and Charles Springer.

1986: IChemE Safety and Loss Prevention Study Group decides to make the problem of safety education a topic for one of its meetings at London University. It identified the need for a training course called latterly "Teaching the Teachers Course" and a safety curriculum which was integral to the undergraduate course.

1998 The first "Teaching the Teachers Course" at which there was about 50% representation of schools of chemical engineering took place in Exeter University lead by Trevor Kletz, and covered the main steps of hazard studies (HAZOP), quantification of effects and quantification of the frequency by event trees.

The same week the Piper Alpha disaster occurs.

1990: CCPS Undergraduate Education Committee publishes the book S*afety, Health and Loss Prevention in Chemical Processes* (AICHE/CCPS, 1990). The problem set is distributed free to all undergraduate chemical engineering students. An instructor's manual containing the problem solutions and a substantial amount of supporting material is distributed free to all chemical engineering faculty. The introduction to this book states:

Studies done by the Safety and Health Division of AIChE, by the CCPS Undergraduate Education Subcommittee, and by ABET revealed that the most effective way to introduce safety, health, and loss prevention concepts to undergraduate engineering students was through their integration into existing courses. It was found that adding a safety and health course as a core requirement to an already crowded curriculum was next to impossible. As an elective course, ABET criteria could not be satisfied. Consequently, the Subcommittee designed this material for use in many existing courses leading to chemical and other engineering degrees.

The problems teach safety, health, and loss prevention as an integral part of many engineering solutions and can provide new engineers with insights to industrial situations they are likely to encounter. This material may also be used as a reference for graduates as they begin industrial careers and by industrial in-house courses for new engineering employees. They further demonstrate that safety and health issues can be handled by basic engineering principles and logic, and are not foreign to basic engineering practices.

Finally, through use of this material, we hope to instill increased recognition and acceptance of the professional and ethical responsibilities which engineers must have to provide safe chemical plants, processes and products."

1990: Crowl and Louvar publish first edition of process safety textbook *Chemical Process Safety, Fundamentals with Applications*, published by Prentice Hall (Crowl and Louvar, 1990). One of the primary objectives in this textbook is to demonstrate the rich fundamental basis for process safety to convince faculty that process safety was more than just "hardhats and safety glasses." The textbook takes a risk based approach rather than a regulatory approach.

1992: **The CCPS Undergraduate Education Committee forms SACHE** – Safety and Chemical Engineering Education. A kick-off breakfast is held at the Fall AICHE Annual Meeting in Los Angeles. Over 40 schools sign on as members. The purpose of SACHE is to provide a flexible environment to support chemical engineering departments in adding process safety content to the undergraduate chemical engineering curriculum. The material produced includes problem sets, instructional modules, workshops and also a bi-annual newsletter.

1996: Lees Preface to the 2nd Edition (Lees, 1996) refers to Mexico City, Flixborough, Pasadena, Bhopal, Chernobyl and Piper Alpha, for instance, and concludes with the comment "*The institutions which educate the engineers who will be responsible for the design and operation of plants handling hazardous materials have a duty to make their students aware of the hazards and at least to make a start at gaining competence in handling them."*

1997: ABET adopts Engineering Criteria 2000 (EC2000). This new approach replaced the previous prescriptive method. EC2000 focused on what is learned rather than what is taught. At its core was the call "for a continuous improvement process informed by the specific mission and goals of individual institutions and programs."

2007: The Baker Report on the BP accident (Baker, et al., 2007) includes "Recommendation #3 – Process Safety Knowledge and Expertise" suggested:

"BP should develop and implement a system to ensure that its executive management, its Refining line management above the refinery level, and all U.S. refining personnel, including managers, supervisors, workers, and contractors, possess an appropriate level of process safety knowledge and expertise."

2009: CSB report on the T2 Laboratories, Inc. reactive chemical explosion (CSB, 2007) recommended that ABET *"Work with the American Institute of Chemical Engineers to add reactive hazard awareness to baccalaureate chemical engineering curricula requirements"*

2010: SACHE issues *SACHE Recommendations for ABET Safety Content in Chemical Engineering.*" This guidance document was prepared with input from over 100 industrial and academic contributors and contains eight specific content recommendations. This document is available at www.sache.org and is shown in Table 2.

2011: ABET adopts new accreditation requirements for chemical engineers (ABET, 2014) by adding the following bolded words to the chemical engineering program criteria statement:

"The curriculum must provide a thorough grounding in the basic sciences including chemistry, physics, and/or biology, with some content at an advanced level, as appropriate to the objectives of the program. The curriculum must include the engineering application of these basic sciences to the design, analysis, and control of chemical, physical, and/or biological processes, ***including the hazards associated with these processes."***

This new wording means that all accredited chemical engineering programs must now include content on process safety. The specific process hazards and the kind of analysis and control procedures are not described or constrained in the wording – but must be consistent with the educational outcomes and objectives of the program. Outcomes represent student abilities at the time of graduation while objectives are the student abilities five years after graduation. It is up to the individual department to define these outcomes and objectives.

In addition to the chemical engineering program criteria, ABET also specifies general criteria that applies to all baccalaureate programs in engineering. These eight general criteria discuss issues on student performance, continuous improvement, faculty quality, facilities and institutional support. General criterion 3 relates to student outcomes and is composed of eleven student outcomes, listed as items a through k. Item c on this list states: "An ability to design a system, component, or process to meet desired needs within realistic constraints such as economic, environmental, social, political, ethical, health and safety, manufacturability, and sustainability."

2011: IChemE sets out the "Requirements for Safety Teaching" through its accreditation procedures and in its 2011 IChemE Guidelines (IChemE, 2011b) accreditation of chemical engineering degrees. (To be discussed in the next section of this paper.)

2012: New ABET process safety requirements go into effect in the US.

2. Establishing a Standard for comparison

In addition to these historical events, many studies and reports have been completed by different organizations. Chemical engineers also operate across many industries and in a variety of roles, many progressing to senior management. An IChemE salary survey and report (IChemE, 2011a) found that 65% of chemical engineers fill the roles of senior engineer, site/plant manager or senior manager. It would seem pertinent to consider whether universities are fulfilling such an industrial need in providing graduate engineers with appropriate process safety education.

It could be argued that universities are not capable of teaching all subjects in depth to graduate engineers and industry has a duty to train its employees in the skills required to complete their work. Yet, as Lee's suggest (Lees, 1996), universities do have a role to direct their students to subjects that will expand their knowledge base and in the context of ethical and professional development understand their role in society. In addition, one of the authors (Crowl) has experience educating students and also training industrial folks. His experience shows clearly that it is much easier to educate and train students on both process and laboratory safety as early as possible in their academic program.

It is in this context that this paper considers "safety" as a paramount duty of the chemical engineering role of "design and operate chemical processes efficiently, effectively and economically in a safe manner" and explores what that standard might be. IChemE as the accrediting authority for UK, some mainland Europe, SE Asia and Australia, have begun to suggest minimum subject criteria in their latest *Guide to Accreditation 2011* (ICheme, 2011b) and suggests a standard class in process safety teaching shown in Table 1. This potential class requires that 2% of all BEng teaching must be in safety in order to achieve accreditation (1.7% at MEng).

This compares well with the AICHE Safety and Chemical Engineering Education (SACHE) recommendations shown in Table 2.

Table 1: Definition of a standard class in process safety teaching.

Program/unit	**Safety and loss prevention**	
University code, credits (ECTS):	CHE 2003	5 ECTS
Aims:	To provide introduction to hazard identification and quantification as applicable to process plant.	
Syllabus:	(Described by department)	
Learning outcomes		
Specific underpinning core chemical engineering	Understand the general tools used in designing safe processes. Able to perform simple hazard ID exercises. Able to make assessments of event frequency, magnitude and effect. Understand how risk assessment is carried out. Understand the concept of LOPA.	
Embedded learning (SHE, economic, societal, ethical)	Be aware of legislation governing safety in design and operation of chemical plant etc.	

Note on Credit Levels: 5 ECTS (European Credit Transfer and Accumulation System) is equivalent to 100 hours of Student Output. This is typically 40-60 hours of lectures, so there is equivalence to 5 US Class Credits.

Table 2: SACHE Recommendations for ABET Safety Content in Chemical Engineering, February 18, 2010

The Safety and Chemical Engineering Education (SAChE) Committee makes the following recommendations with regard to student understanding to meet ABET curriculum requirements. 1. The graduate must understand the importance of process safety and the resources and commitment required. This should include the important incidents that define process safety, and how these incidents affected the practice of chemical engineering. 2. The graduate must be able to characterize the hazards associated with chemicals and other agents. This must include toxic, flammable, and reactive hazards. 3. The graduate must understand and be able to apply concepts of inherently safer design. 4. The graduate must understand how to control and mitigate hazards to prevent accidents. This should include generally accepted management systems, plant procedures and designs to prevent accidents. 5. The graduate should be familiar with the major regulations that impact the safety of chemical plants. 6. The graduate should understand the consequences of chemical plant incidents due to acute and chronic chemical releases and exposures. 7. The graduate should be reasonably proficient with at least one hazard identification procedure. 8. The graduate should have an introduction to the process of hazard evaluation and risk assessment

Compare this approach now to the 2013-14 ABET process safety requirements, which are "outcome based" that: "The curriculum must include the engineering application of these basic sciences to the design, analysis and control of chemical, physical and/or biological processes, including the hazards associated with these processes."

The questions that the authors ask are:

"Do you think that this worldwide approach is adequate?"

"Should there be such a disparity in approaches to content?"

"Should premier chemical engineering universities have embraced Lee's call and over the following 16 years, moved safety teaching practice on at a dynamic pace to keep the subject at the same level as 'Underpinning Math & Science' or 'Chemical Engineering Laboratory Practice'"

"Is it reasonable to suggest that process safety is fundamental to all branches of chemical engineering and should have re-imminence in the chemical engineering curriculum?"

3. A Look at Current Practice

If we look at what may be happening in current teaching practice, a survey by one of the authors in 2010 (Crawley and Dickson, 2013) of 8 UK and 12 US university chemical engineering departments suggests at best 50% of universities feature safety and loss prevention as a named topic in their undergraduate chemical engineering programs. The remainder only integrate safety material into unit operations and other classes, although all incorporate the topic into design projects too. Only a few departments reported any specialist teaching and use of specialist staff, the common practice being to rotate the subject across all teaching staff.

If this is added to a review of 2011 IChemE accreditation reports shown in Table 3, which cover a wider international audience, then the adherence to high levels of process safety teaching becomes even weaker.

4. What does this tell us about Teaching Practice?

Only 7 from 29 institutions meet this minimum 5 ECTS standard since they either don't feature a discrete course or dilute safety content with sustainability, ethics or economics topics.

Examining Table 3, and based on the ratings assessed from the surveys, we can complete scores by territory as follows:

Mean Score: (out of 4)
- UK: 3
- World: 2
- US: 2

Best in Class:
- UK: 5 institutions from 10
- World: None
- US: 2 institutions from 12

Unacceptable:
- UK: 3 institutions from 10
- World: 3 institutions from 7
- US: 7 institutions from 12

We can therefore take a view based on this analysis that:

"Teaching safety is not a chemical engineering priority"

Table 3: Institution Survey. Compiled by BR Dickson 2010 and 2011.

Institution[#]	Class	ECTS Credits *Part Only	Elective	Use of HAZOP in design *Assessed	Rating against IChemE Standard* 1-4
EU1[2]	Environment, Safety & Society	3*	N	Y	2
EU2 [2]	Industrial Safety & Hygiene	4.0*	N	Y	1
EU3[2]	Process Safety & Sustainability	2*	N	Y	1
AS[2]	Safety Engineering	5	N	Y	3
AS[2]	Plant Safety & Loss Prevention	3	N	N	1
AS[2]	Economics, Safety & Ethics	5*	N	Y	2
CA1[1]	QA & HSE engineering	7.5*	N	Y	2
UK1[2]	Safety & Loss Prevention	5	Y	Y	3
UK2[2]	Environmental & Safety Engineering	7.5*	N	Y	4
UK3[2]	Safety & Loss Prevention	5	Y	Y	4
UK4[1]	Safety	5	Y	Y	4
UK5[1]	Safety & Loss Prevention	5	N	Y	4
UK6[1]	SLP & Environmental Control	5*	Y	Y	3
UK7[1]	Chem. Eng. Safety	5	Y	Y	4
UK8[1]	Integrated	0	N	Y	1
UK9[1]	Integrated	0	N	Y	1
UK10[1]	Integrated	0	N	Y	1
US1[1]	Chem. Eng. Process Safety	1	N	?	2
US2[1]	Chem. Eng. Safety	2	N	?	2
US3[1]	Process Safety Design	3	N	?	3
US4[1]	Integrated	0	N	?	1
US5[1]	Integrated	0	N	?	1
US6[1]	Integrated	0	N	?	1
US7[1]	Integrated	0	Y	?	1
US8[1]	Integrated	0	Y	?	1
US9[1]	Integrated	0	N	?	1
US10[1]	Integrated	0	Y	?	1
US11[1]	Process Safety	5	N	Y	4
US12[1]	Process Safety	5	N	Y	4

Notes:
The US survey was done prior to the implementation of the new ABET safety requirements.
[#] Institution names have been withdrawn in the published material.
[1]Based on questionnaire in 2010 [2]Survey of Accreditation reports 2011
*Rating assigned against the IChemE Core Process Safety Standard for an integrated course is a nominal score of 1 if evidence suggests less than 10 hours of lectures, and 4 if there is compliance.

5. What is Best in Class?

5.1 Examples of Best Practice (1): The Strathclyde Approach

At the outset there is no claim here that this approach is novel or unique since it uses internationally available material as a resource and is not untypical with 2/3 of other UK universities in style.

However, the approach is recognized within industry as the best grounding in safety training for graduates coming from UK universities by Exxon, BP, Wood Group, Conoco, Total, Proctor & Gamble, Unilever and Diageo amongst top line graduate recruiters and there is a consistent flow of recruits directly into these corporation's health and safety functions.

5.1.1 What do we do?

At the University of Strathclyde safety is taught twice in a five year/two semester p.a. MEng program and is also incorporated particularly in design teaching and comprises of:

Year 1: Mainly science and math based with an introduction to chemical engineering subjects that will be taught in later years (this would include **safety**)
Year 2: Basic principles of chemical engineering
Year 3: Based around unit operations and process design and **safety principles**
Year 4: Split between chemical engineering processes and a major detailed design project incorporating **HAZOP**

There is an exit point here with the award of BEng.

Year 5: Again divided equally between advanced chemical engineering topics often based on research areas and an industrial placement. **Safety management** is taught here.

The exit point here is the award of MEng which is the majority of students (classed as an integrated first degree Masters).

The rational for the delivery of safety classes is two-fold; first, by paralleling teaching process design and safety principles in Year 3, there is an opportunity for collaboration and cross referencing material especially plant layout, control/alarm and relief systems. Safety is not therefore seen as a specialist subject and is an integral part of the chemical engineering discipline.

It is also a preparatory class for the year 4 detailed design project where students face the challenge of the chemical engineering role of "design and operate chemical processes efficiently, effectively and economically in a safe manner" and continue what was applied in years 1-3 with a demonstration of HAZOP in their design submission which is independently assessed from the whole project submission

Second, in year 5, we are preparing students for their future responsibilities as professional engineers and as suggested previously safety management is a core responsibility of all engineers no matter their grade, and the class takes an approach of challenging industry practice.

5.2 Examples of Best Practice (2): The Michigan Technological University Approach

The safety experience at Michigan Tech developed initially out of the unit operations (UO) laboratory. The primary motivation at that time was to prepare the students to meet the safety needs of industry and to insure that the students understand the importance and commitment of safety prior to graduation.

5.2.2 What do we do?

The degree content on process safety contains two critical components. This includes, first, lectures on the technical content related to process safety, and second, putting the technical content into practice using pilot scale equipment in the UO lab. The management system in the UO lab is student based and faculty/staff audited, and is very important in demonstrating the commitment and resources required for process safety..

The UO lab at Michigan Tech is a 4,800 sq. foot facility with three floors and an open bay area. The laboratory is comprised of 17 experiments that are mostly pilot-scale using modern processing equipment and control systems. The largest pilot scale equipment includes a 30 gallon reactor and a 3-story distillation column with structured packing. The students are assigned open-ended objectives for each experiment and work in teams of 3 or 4.

The Michigan Tech process safety program evolved over a period of time. A history of this evolution will now be presented to show the evolution process and the effort involved. Where does your university fit into this timeline?

From 1946 when the UO laboratory was initially established up through the early 1980s, the safety program was autocratic. The lab supervision made the rules, set the example and expected students to follow. Bad safety behavior resulted in a grade penalty and threat of removal from the course. This set up a regulatory environment with respect to safety and did nothing to advance the safety program or demonstrate the commitment required. It was entirely dependent on continuous inspections / audits by laboratory supervision. Most universities operate today in this fashion.

In 1982 the first safety experiment was added to the laboratory. Typical assignments for this included: serving as safety officers, studying and reporting on an industrial accident, reviewing safety regulations, and conducting laboratory safety inspections, to name a few.

In 1983, the students wrote the first *Safety Manual for Use in the Unit Operations Laboratory* under the direction of Dr. Pintar. This manual still exists, but is reviewed and updated annually.

In the late 1980s, an elective process safety course was offered. The course was developed as a collaborative effort with Wayne State University in Detroit, and BASF Corporation in Wyandotte, MI. The effort was funded by the U.S. National Science Foundation (NSF).

Also in the late 1980s, a student hazard reporting system was developed for the UO laboratory. A student contest resulted in the name for this program as PAWS – Prevent Accidents with Safety. Coincidentally, the Michigan Tech mascot is the husky dog. PAWS cards are submitted by the students during laboratory operation and are tracked by an on-line system until resolution. At the end of each academic year, all the PAWS cards are reviewed by faculty and laboratory staff and recommendations are made for improving the safety program the following year.

In 1990 Michigan Tech added a required 3 credit quarter course on process safety to the curriculum. The course was offered in the last quarter of the junior year. The course used the recently published textbook on process safety by Crowl and Louvar (Crowl and Louvar, 1990). The course outline was as follows: 1: Introduction, 2: Toxicology, 3: Industrial Hygiene, 4: Source Models, 5: Toxic Release and Dispersion Modelling, 6: Fires and Explosions, 7: Preventing Fires and Explosions, 8: Introduction to Reliefs, 9: Relief Sizing, 10: Hazards Identification, 11: Risk Assessment.

In 1993, one of the authors (D. A. Crowl) joined Michigan Tech. In 1995, in collaboration with the faculty and staff, a new, student-run safety program was initiated in the UO laboratory. This included a number of important components. First, the course syllabus listed the first leaning objective: “Students will develop a constant awareness of safety so that all work is performed in a safe manner.” Second, the Job Safety Assessment (JSA) procedure was required for all laboratory experiments. Third, a required safety meeting was held after every UO laboratory run day. The safety meetings were organized and run by the students.

The student organizational structure in the UO laboratory is as follows. Each student team designates a team member to serve as a safety liaison. All of the safety liaisons join together to form the student safety committee. This committee has several functions, including managing the required safety meetings, pre and post laboratory inspections, and providing technical assistance for any safety issues that arose.

The safety meetings have several important components. First, all PAWS cards submitted during the run day are reviewed and discussed with, hopefully, resolution of the issue. Second, the student committee does a presentation on some safety topic, with discussion afterward.

In 2000, Michigan Tech changed from a 10-week quarter to a 14-week semester. The required safety course was changed from a 3 credit quarter course to 3-credit semester course. The combined course now included ten weeks of lecture on process safety and 4 weeks on green engineering. The process safety content was now equivalent to 2 semester credits. At that time the course was moved to the first semester of the senior year. Additional lectures on reactive chemicals were added via the new chapter 8 in the 3rd edition of Crowl and Louvar (Crowl and Louvar, 2011).

The PAWS program also became an evaluation tool under the new ABET 2000 requirements.

The Michigan Tech experience has been very successful at providing the technical knowledge required for process safety – primarily through the required course – but also provides through the student run UO laboratory safety program an excellent demonstration of the importance, commitment, and resources required for process safety.

Is there more that should/can be done?

6. The Extended Syllabus for Process Safety

The IChemE *Accreditation Guide* also provides a matrix exemplar, shown in Table 4, of how learning outcomes in SHE topics may be realised. It is extensive, well beyond the scope of a 5 ECTS credit module and closer to 30 credits of teaching.

Is this the level of process safety competence that Baker Panel (Baker, 2007) was looking for in *"BP should develop and implement a system to ensure that its executive management, its refining line management above the refinery level, and all U.S. refining personnel, including managers, supervisors, workers, and contractors, possess an appropriate level of process safety knowledge and expertise."*

It certainly would display an individual's competence to carry out the paramount duty of the chemical engineering role of "design and operate chemical processes efficiently, effectively and economically in a safe manner."

Certainly, Du Pont seems to agree with this through their PSM Competency Leadership Program, having taken on the Baker recommendation, if the following is considered:

DuPont PSM Standard; S21A (Rain, 2011):
"Site leaders shall play a key role and be accountable for establishing and sustaining site safety culture, PSM expectations, accountabilities and objectives; resourcing program elements; providing reward and recognition; and measuring ongoing performance on a routine basis. Site management shall be responsible for demonstrating ongoing management leadership, commitment, and personal involvement in site PSM programs plus providing for and encouraging employee involvement."

We do not expect universities to produce fully competent graduates in process safety, fully prepared for industry. But we should expect a much better prepared graduate than is generally produced now. Perhaps universities should focus more on the fundamentals of process safety since this is what faculty are excellent at and it is unlikely that graduates will learn this in any unified fashion in industry?

An important part of the undergraduate experience should be a demonstration of the importance and commitment required for process safety since this is much easier to do with youthful students and before they learn how to do things the wrong way.

Table 4: Safety, health and environmental guidelines from IChemE *Accreditation Guide* (IChemE, 2011b).

	SHE specific topics	SHE integral with other technical topics	Management related SHE topics
SHE basics	- Flammability - Explosivity - Toxicity - Ecosystems (atmospheric, aqueous, solid and integrated) - Bio-hazards - COSHH: Control of Substances Hazardous to Health - Thermal radiation	- Principles of dispersion of pollutants (with maths and mass, heat and momentum transfer) - Energy conservation and the global impact of the process industries. - Principles of process containment (particularly in biotechnology)	- Risk assessment and management - Safe systems of work - History of SHE for process industries - SHE element in engineers social responsibility - Introduction to statutory requirements in SHE; national and international
Bachelor level SHE topics	- Inherent safety - Environmental impact assessment - Integrated pollution control - Safety in plant operation, maintenance and modification - Identification of Hazards (HAZID) - Risk assessment and QRA including frequency estimation, criteria, inherent safety - Good manufacturing practice (GMP) for production of GMOs - Principles of area classification - Principles of plant safety including ESD, isolation, fire protection, plant layout, pressure hazards, reaction hazards, fires and explosions	- Behaviour of gaseous and vapour releases - Dispersion in water - Physics and effects of noise - Particle behaviour and dust explosions - Pressure relief, venting and flaring - Choice of plant construction materials and corrosion /erosion protection - Runaway reactions - Multiphase properties - Case studies to illustrate SHE hazards consequences and prevention	- Application of environmental impact assessment (EIA) and Hazard and operability studies (HAZOP) to design project - Prevention and control in SHE; management systems and the management of SHE - Principles of loss prevention - Life cycle analysis - Cost benefit principles in SHE control; (e.g. BATNEEC, BPEO, ALARP or equivalents) - Legislative framework, relevant codes and standards for loss prevention (e.g. Control of Major Hazards, Control of Hazardous Substances) - Waste minimisation
Advanced SHE topics (depth and breadth)	- Health and safety law and codes of practices - Environmental regulation - Human factors and human error in safety and reliability - Social responsibility - Further HAZOP - Safety and reliability including fault tree analysis - Communicating SHE objectives	- Further advanced methods for estimating effects - Advanced dispersion (air, water, ground) - Principles of eco-systems; evaluating eco-system effects	- Further management of SHE management systems, standards procedures and auditing, - Further safety related standards and codes and their application (e.g. NFPA) - Product stewardship in relation to transportation and the environment - Clean technology and the holistic approach - Regulation of genetically modified organisms (GMO's) - Economic impact of SHE - SHE incident and near-miss reporting, investigation and management.

7. What is to be Done?

7.1 Where to teach in the curriculum?

Most chemical engineering programs are already at the maximum credits allowed by their university. How to add additional content? There are a number of instructional approaches to add content – what works depends on the particular situation and resources.

Process safety in an industrial facility is an important part of every activity. Thus, the ideal situation is to integrate process safety throughout the entire chemical engineering curriculum. Much of this content can be added to existing design and unit operations classes, and the UO laboratory. Another approach is a dedicated safety course. A hybrid approach is also possible, with some content in existing courses, i.e. design or UO lab, and a required 1 or 2 credit course to provide a unifying overview with additional topics that don't fit well elsewhere.

7.2 Who should teach process safety?

Ideally the content should be taught by full-time faculty with some knowledge of industrial process safety practice and/or industrial experience. There is a large community of very committed and willing industrial folks with significant safety experience that can assist - this assistance can be in the form of help with content preparation, answering questions about specific safety issues, assisting students, and even delivering part or all of the lectures.

The challenge to achieve process safety competence needs to be a three way partnership between accrediting bodies, universities and industry

The accrediting bodies have a duty to ensure that the accreditation process team members are competent to assess the standards of the taught subject. It is necessary that the status of the safety element, the detail and the hours expended are reviewed during the accreditation visits. The question may arise of course as to whether the assessors are competent to carry this out in detail.

Academia also needs the resources to deliver, yet a disturbing picture is emerging on this. Only three UK departments from 8 UK University chemical engineering departments reported any specialist who was teaching, whereas the common practice is to rotate the subject across all teaching staff.

7.3 Why is this a problem?

To understand this it is necessary to understand academic career progression and how to become a top academic. It requires a focus on research linked to organizational and funding body themes – process safety is not part of that focus. As a result, the number of safety specialist teachers is on the decline.

Faculty right now, in general, do not have adequate background in process safety. We believe that this issue will eventually improve if all undergraduates have adequate safety instruction. Then, when these graduating students return to academia as faculty they will understand the importance of process safety, just as they understand the importance of thermodynamics, reactor design, process control, etc.

Finally, industry has a lobby role to fill. Since it is represented on the accrediting bodies by individual members, it needs to lobby for a better standard of accreditation in process safety. It can add resources to academia through case studies, visiting lecturers, site visits, and internships and projects linked to process safety.

8. Conclusions

Our study has reached the following conclusions:

1. We cannot expect universities to produce fully competent graduates in process safety, fully prepared for industry. But we should expect a much better prepared graduate than is generally produced now.
2. Faculty right now, in general, do not have adequate background in process safety. We believe that this issue will eventually improve if all undergraduates have adequate safety instruction.

9. Recommendations

What we are recommending here:

1. A prescriptive set of safety instructional requirements is necessary for all chemical engineering departments. This will insure that all chemical engineering graduates have adequate background no matter where they find employment - either industry, government or academia - or if they go to large or small employers. These requirements should be specified by accreditation agencies.
2. Accreditation reviewers and visitors must have adequate safety background to insure that the department audits are valid and consistent between departments.
3. Safety should be integrated as much as possible into all undergraduate courses but must be a required part of the design and laboratory experiences. A dedicated course or hybrid approach may be required based on faculty expertise and interest.
4. An important part of the undergraduate experience should be a demonstration of the importance and commitment required for process safety since this is much easier to do with youthful students and before they learn how to do things the wrong way.
5. Universities and industry should team together to assist chemical engineering departments in meeting safety requirements. This should include visitor exchanges going both ways.

10. Closing comments

1. We think these recommendations should be worldwide.
2. ABET criteria needs to be more specific on the fundamentals of process safety. AICHE is the only US engineering society that offers an annual symposium (Global Congress on Process Safety) dedicated to the topic (except nuclear). Should AICHE put greater leverage on ABET?
3. IChemE needs to be more rigorous in applying its *Accreditation Criteria* (IChemE, 2011b) perhaps by encouraging cross-university collaboration.

11. References

ABET, Criteria for Accrediting Engineering Programs, 2014 – 2015, available at www.abet.org, 2014.

AICHE/CCPS, *Safety, Health, and Loss Prevention in Chemical Processes, Problems for Undergraduate Engineering Curricula,* American Institute for Chemical Engineers, NY (1990).

Baker. J. A. The Report of the BP U.S. Refineries Independent Safety Review Panel (2007).

Crawley, F. K., Dickson, B. R., University of Strathclyde, Glasgow, G1 1XJ: Preparing Graduates for the real safety world – the past and the future (2013).

Crowl, Daniel A. and Louvar, J. F., *Chemical Process Safety, Fundamentals with Applications,* 1st ed., Prentice Hall, Englewood Cliffs, NJ (1990).

Crowl, Daniel A. and Louvar, J. F., *Chemical Process Safety, Fundamentals with Applications,* 3rd ed., Prentice Hall, Englewood Cliffs, NJ (2011).

CSB, *Investigation Report: T2 Laboratories, Inc., Runaway Reaction*, U.S. Chemical Safety and Hazard Investigation Board, Washington, DC (2007).

IChemE, *IChemE Salary Survey*, Institution of Chemical Engineers, UK (2011a).

IChemE, *Accreditation of Chemical Engineering Degrees: A Guide for University Departments and Assessors*, Institution of Chemical Engineers, UK (2011b).

Lees, F. P., *Loss Preevention in the Process Industries,* 2nd ed., Butterworth – Heinemann, Amsterdam (1996).

Rain, B., *Operation Discipline: Panacea for Performance,* Mary Kay O'Connor International Symposium, Texas A&M University, College Station, TX (2013).

COMPETENCE REQUIREMENTS FOR PROCESS HAZARD ANALYSIS (PHA) TEAMS

Paul Baybutt
Primatech Inc.
Columbus, Ohio, USA
paulb@primatech.com

Prepared for Presentation at
American Institute of Chemical Engineers
2014 Spring Meeting
10th Global Congress on Process Safety
New Orleans, LA
March 30 – April 2, 2014

UNPUBLISHED

COMPETENCE REQUIREMENTS FOR PROCESS HAZARD ANALYSIS (PHA) TEAMS

Paul Baybutt
Primatech Inc.
Columbus, Ohio, USA
paulb@primatech.com

Key words: Process hazard analysis; process safety management; PHA quality, PHA team qualifications, PHA certification.

Abstract

Process hazard analysis (PHA) is the foundation of process safety management programs. The quality of the PHA performed for a process directly affects the level of risk tolerated for the process. The lower the quality of a PHA, the more likely higher risk will be tolerated. There are few requirements in OSHA's process safety management regulations for PHA teams other than that the PHA shall be performed by a team with expertise in engineering and process operations, the team shall include at least one employee who has experience and knowledge specific to the process being evaluated, and one member of the team must be knowledgeable in the specific PHA method being used. Clearly, more detailed requirements for participation in a PHA are desirable.

A competence management program should be used to ensure PHA practitioners are appropriately qualified. Criteria for selecting PHA team leaders, or facilitators, and other team members are key to such a program and are described in this paper. The criteria cover both technical and personal attributes. Application of the criteria is described and team performance metrics, which can be used to correlate performance with the assessment of competence to validate the criteria and methods used, are discussed.

Owing to the importance of the role played by team leaders, certification of their competence is desirable. Criteria for certification are described and their application is discussed.

1. Introduction

Process hazard analysis (PHA) is used to identify hazard scenarios that can have adverse impacts on such receptors as people, property and the environment so that a determination can be made as to whether the level of risk posed by a process is tolerable [1, 2]. If not, recommendations can be made for appropriate risk reduction measures. PHA is a foundation element of process safety management (PSM) programs. PHA studies are conducted by teams of people. Consequently, they are subject to various possible human failures influenced by human factors [3, 4]. In particular, the qualifications, experience, and personal characteristics of participants play a critical role in the quality of PHA studies. Teams with members who are poorly qualified, possess inadequate experience, or have undesirable personal characteristics will produce poor quality studies in which scenarios may be missed or described inadequately, and scenario risks may be estimated incorrectly. Such PHA faults can lead to higher risk being accepted than should be tolerated. Consequently, individuals who possess the appropriate competencies to perform their role as a member of a PHA team should be selected by screening them against suitable criteria. Not only is the competence of individuals important but also the competence of individuals acting as a team is critical to the success of PHA. This paper suggests appropriate criteria for selecting PHA participants and teams and describes how they can be applied.

Selection of PHA team members should be governed by a competence management program. Each aspect of a competence management program is described in this paper. Of course, successful performance of PHA is the desired outcome so performance metrics are discussed to help determine the degree of success in applying selection criteria to PHA participants.

Team leaders or facilitators guide PHA teams in applying the chosen PHA method. Consequently, they play a critical role in PHA and should be subject to certification requirements. Criteria for certification are described and their application is discussed.

2. Regulatory Requirements for PHA Teams

Regulations around the world have established few requirements for PHA participants. For example, the process safety management (PSM) standard of the US Occupational Safety and Health Administration (OSHA) requires only that [5]:

- The PHA shall be performed by a team with expertise in engineering and process operations.
- The team shall include at least one employee who has experience and knowledge specific to the process being evaluated.
- One member of the team must be knowledgeable in the specific process hazard analysis methodology being used.

Clearly, these are minimal requirements and more detailed requirements are desirable owing to the critical role that people play in PHA.

In the preamble to the PSM standard, OSHA states:

- In order to conduct an effective, comprehensive PHA, it is imperative that the analysis be performed by competent persons, knowledgeable in engineering and process operations.
- Those persons be familiar with the process being evaluated.
- A team approach is the best because no one person will possess all of the knowledge and experience necessary to perform an effective PHA and when more than one person

is performing the analysis, different disciplines, opinions, and perspectives will be represented and additional knowledge and expertise will be contributed to the analysis.

- Some companies include an individual on the team who does not have any prior experience with the particular process being analyzed to help insure that a fresh view of the process is integrated into the analysis.

- Employees and other experts may be brought onto the team on a temporary basis to contribute their specialized knowledge.

In another publication [6], OSHA stated that:

- The team leader needs to be fully knowledgeable in the proper implementation of the PHA methodology to be used.

- The team leader should be impartial in the evaluation.

- The team leader needs to be able to manage the team and the PHA study.

- The other full or part-time team members need to provide the team with expertise in areas such as process technology; process design; operating procedures and practices; alarms; emergency procedures; instrumentation; maintenance procedures, both routine and nonroutine tasks, including how the tasks are authorized; procurement of parts and supplies; safety and health; and any other relevant subjects.

- The ideal team will have an intimate knowledge of the standards, codes, specifications, and regulations applicable to the process being studied.

- The selected team members need to be compatible.

- The team needs to be able to work together while benefitting from the expertise of others on the team or outside the team to resolve issues and to forge a consensus on the findings of the study and recommendations.

- Some team members may only be a part of the team for a limited time.

These criteria go beyond those specified in the PSM standard and begin to identify important selection criteria for PHA participants.

The PSM standard does not specify formal training requirements, such as attending short courses, for PHA team members. However, in a letter of clarification [7], OSHA stated that if an OSHA representative needs to determine whether a team member or the facilitator is knowledgeable in the chosen PHA method, the representative may choose to review training records (formal, non-formal, on-the-job training, etc.) and interview team members to collect information to determine if the team leader or other members have acquired a level of understanding that enables them to effectively use the chosen PHA method.

Thus, OSHA recognizes that PHA participants should be qualified and competent to participate in studies, in both their technical capabilities and personal characteristics, although the PSM standard does not specifically require the screening of participants against such criteria.

3. Competence of Personnel and Teams

Competence of personnel is defined as the ability of personnel to perform tasks according to expectations [8]. Competence implies appropriate qualifications, training, skills, physical and mental capabilities, knowledge, understanding, behavior and attitudes as well as the ability to perform tasks according to defined performance standards (see Figure 1).

Not only must PHA team members be competent individually, but also the team of which they are a part, must function competently. Of course, the ability of the team to perform PHA constructively and cooperatively depends on the interactions of the team members, which in turn is a function of the personal attributes of team members, including their personalities, backgrounds, behaviors and attitudes. Thus, team competence can be controlled by appropriate selection of team members, taking into account not only their competence as an individual but also how their personal characteristics may influence the performance of the team.

4. Competence Management Program

The competence of PHA team members and teams should be addressed using a competence management program that covers these topics:

- Responsibilities
- Competence requirements
- Team composition
- Selection criteria for teams and team members
- Development of competencies
- Assessment of competence
- Documentation of competence
- Maintaining competence
- Reassessing competence
- Monitoring competence

- Defining and using performance metrics
- Certification of practitioners
- Continuous improvement

Each topic is now described.

4.1 Responsibilities

Company management is responsible for the selection of team leaders and ensuring that they are appropriately qualified. Usually, the manager responsible for a PHA study, or the team leader assigned to a study, selects the other team members. They should advise each other and jointly approve the other team members.

4.2 Competence requirements

Competencies should be defined for the various types of PHA team members together with assessment criteria. They should be realistic and appropriate for the tasks to be performed. The levels of competence required should match the complexity of the process being studied and the PHA method used. Thus, competence requirements should be graduated into levels such as practitioners, specialists, and experts. Of course, there are minimum requirements. Various tools can be used to identify competence requirements including task analysis and skill and knowledge inventories. Competence-job matrices are sometimes used to define requirements.

4.3 Team composition

Different types of team members require different types of competencies, although there is overlap in those required. This section describes the various roles that must be played by PHA participants. They include:

- Leader / facilitator - Responsible for study preparation and organization, guiding the team in the use of the chosen PHA technique, managing the team and the study, and preparing a study report.
- Scribe / technical secretary - Records PHA sessions and may prepare the study report under the guidance of the team leader. Experienced team leaders may act as their own scribe.
- Technical team members - Brainstorm the identification of hazard scenarios. Identify initiating events, intermediate events, consequences, safeguards, and enablers. Perform risk ranking and may identify recommendations for risk reduction measures. There are two types of team members:
    - Core team members - Participate in the study on a full-time basis. Their involvement is critical to the success of the study. They help to achieve consistency across study

sessions. They may need to be present to satisfy regulatory requirements. Typical core team members are shown in Table 1.

- Specialty team members - Possess technical expertise in a particular area and attend only certain sessions where their expertise is needed. Typical specialty team members are shown in Table 1.

- Other personnel - May be representatives from vendors of self-contained processes, such as skid-mounted units, and licensed technology; contractors who perform activities such as maintenance; the design company responsible for a new process; or the engineering company that will construct a new process. Their role is to provide information on equipment design, maintenance, operation, etc. that may not be known by company personnel.

- Interpreter - PHAs may be conducted by teams who are unable to communicate in a common language or cannot do so well enough to perform the study properly. Thus, the role of an interpreter is to enable team members to communicate effectively.

- Site coordinator - Acts as a liaison between the team and the process facility. Helps to ensure that adequate facilities, such as a meeting room, are provided and that other team needs are met. Not a participating member of the PHA team.

4.4 Selection criteria for teams and team members

Different types of team members require different types of competencies, although there is some overlap in requirements. Selection criteria are needed for both individual team members and teams. Each type of team member should be selected based on technical qualifications and personal characteristics, or attributes, that are appropriate for each type of team member. It is useful to define both critical and recommended attributes as ideal team members are unlikely to be found in the real world.

4.4.1 Team selection

This section provides some suggested overall criteria that should be met by PHA teams. PHA teams necessarily must be multi-disciplinary. Team members are needed who together can provide the information needed to define the design intent completely for a process including how it is operated, controlled and maintained. Suggested technical disciplines are shown in Table 2. More than one person from the same technical discipline may be needed, e.g. operators, to reflect different levels of experience, ways of performing their jobs, attitudes, etc. Team members may cover more than one technical area if their expertise allows. Team dynamics are very important for an effective and efficient study and they are determined by the selection of team members. Suggested criteria that should be met by PHA team members as a group are:

- Collectively possess the knowledge and skill necessary to identify hazard scenarios for the process.

- Be compatible and able to work together effectively.
- Be able to reach a consensus on PHA worksheet entries.
- Comply, individually and collectively, with applicable regulatory, industry and/or company requirements.

A team should not consist entirely of people who know the process as groupthink can be a problem [4]. It is a good idea to have an independent senior engineer to challenge assumptions that otherwise would be made by the team and to contribute knowledge that may not be possessed by the team. This role can be played by an independent experienced team leader.

It is not unusual for people to be included on a team for training purposes to gain process knowledge and PHA experience. Other people may be included for other purposes, such as a union shop steward as an observer. Such people should not be relied upon as full team members and must not interfere with achieving the study objectives.

4.4.2 Team member selection

This section provides suggested criteria for each type of PHA team member. There are various positive and negative attributes for team members that influence their effectiveness as a team member. The criteria for positive attributes cover both technical qualifications and personal characteristics of participants which are divided into critical and recommended categories. The criteria also include negative attributes.

4.4.2.1 Leader / facilitator

Team leaders must have a thorough technical understanding of PHA and excellent facilitation skills. Suggested positive technical and personal attributes for team leaders are provided in Table 3. Negative attributes are provided in Table 4. The team leader does not need to be a technical expert on the process under study and, arguably, should not be, as they would undoubtedly have mindsets about the process that could inhibit their independence in facilitating the study.

4.4.2.2 Scribe / technical secretary

Scribes must be able to understand and accurately record the team discussions without interfering with the process of conducting the PHA. Suggested positive technical and personal attributes for scribes are provided in Table 5. Negative attributes are provided in Table 6. Young engineers are often selected as scribes. They should not be someone whose input to the PHA is critical.

4.4.2.3 Technical team members

Team members must possess technical knowledge in some aspect of the process, be able to contribute that knowledge to the PHA team, and be capable of participating in the identification of hazard scenarios for the process. Suggested positive technical and personal attributes for technical team members are provided in Table 7. Negative attributes are provided in Table 8. They apply to both core and specialty team members.

4.4.2.4 Other personnel

Their technical and personal attributes are similar to core and specialty team members.

4.4.2.5 Interpreter

Interpreters may provide interpretation simultaneously, or on an as-needed basis. Owing to the effort involved, more than one interpreter may be needed. Interpreters need a high level of technical ability and need to know technical terms in the languages used. They must have high stamina owing to the intellectual effort involved and the durations of PHA sessions.

4.4.2.6 Site coordinator

Requirements for site coordinators are straightforward. They need to be available for the duration of the study and should be well-connected within the company so as to be able to procure whatever may be needed by the team. They should be helpful and responsive to the team.

4.4.3 Practical considerations in selecting team members

There may be only one choice of person for a particular role on the team, e.g. the process engineer. To the extent that the person does not meet the selection criteria, usually in their personal attributes, the team leader needs to be ready to deal with any issues that may arise [4].

Often, the pool of core and specialty team members is limited, thus preventing the selection of an ideal team and compromises likely will be needed. Of course, the availability of personnel also must be taken into account. The division of competencies into critical and recommended categories assists in making compromises.

4.5 Development of Competencies

PHA participants must possess both technical and personal competencies. The ability to perform tasks competently can be developed through training and experience. Competence for performing a task is often developed through initial training followed by coaching and supervision by experienced personnel combined with periodic refresher training. Thus, PHA team leaders can gain competency in the technical requirements of the position through attendance at suitable short courses, on-the-job mentoring by more experienced PHA team leaders, and refresher training. Typically, technical team members develop technical competency through job experience. Scribes and interpreters can gain technical competence through training and practice.

Competencies in personal attributes largely are inherent to the person and may be difficult to change, particularly in team members whose participation in PHA studies is a small part of their overall job responsibilities. Team leaders who lead studies frequently may be motivated to modify their personal attributes, at least insofar as they are perceived by other team members during a study. However, some behaviors are easier to change than others, for example, a team leader may be able to teach themselves to be a better listener but they may have difficulty overcoming a tendency to be impatient.

4.6 Assessment of competence

Competence should be assessed before personnel are allowed to participate in a PHA study. Methods used should be appropriate, valid and reliable. Technical competencies and some personal competencies, such as leadership skills, can be assessed by verbal or written tests, demonstrations, and observation of task performance. Other personal competencies, such as the ability to read people, can be more difficult to assess.

In some fields of endeavor, aptitude tests and psychometric personality tests are used to assess individual competencies, particularly those relating to personal characteristics. However, they are not yet employed routinely in selecting PHA participants. Behavioral observation and personality typing can be employed and are desirable for prospective team leaders. However, such methods may discourage participation in PHA studies by prospective technical team members. Reliance may need to be placed on prior knowledge of and prior experience with their personal competencies.

The assessment of competence should be correlated with subsequent task performance to validate the method used. The assessment requires the use of metrics to gauge the performance of teams and participants. They are described in a later section. Assessors should themselves be competent in assessing competence as well as credible, consistent, and independent.

4.7 Documentation of competence

Companies should formally designate personnel as meeting required competencies through documentation. Some participants, such as team leaders, should be certified, licensed or accredited by external organizations owing to the importance of the role they play during a PHA study. Care should be exercised to ensure such certification, licensing or accreditation is meaningful.

4.8 Maintaining competence

Maintenance of competence by PHA participants should include refresher or on-going training provided at a frequency based on the anticipated deterioration in competence. The provision of PHA procedures and requirements is also an important part of helping to ensure tasks are performed consistently and correctly. Other forms of assistance, such as supervision, coaching and job aids, should be provided, as appropriate.

4.9 Reassessing competence

Competencies should be re-assessed periodically, such as through re-examination and observation of performance by a qualified assessor. Actual performance may be reviewed, or other forms of assessments, such as demonstrations, may be used. The frequency of reassessment should be based on the frequency of task performance and the anticipated decay rate for competencies. There must be suitable responses in the event of substandard performance, for example, improvement in training, personnel selection, etc. Reassessment should also be performed to qualify an individual to return to a designated role on a PHA team after the elapse of a significant time period.

4.10 Monitoring competence

Competency levels should be tracked over time in order to determine if systemic problems develop in the competence management program and to permit continuous improvement. Such tracking is facilitated by the use of performance metrics which can be used to help gauge the effectiveness of PHA performance.

4.11 Defining and using performance metrics

The purpose of ensuring the competence of PHA participants and teams is to help ensure that PHA studies are performed effectively and efficiently. In particular, they should identify hazard scenarios as thoroughly and completely as possible. Thus, metrics for team performance are more meaningful than metrics for the performance of individual team members.

Unfortunately, there is no way of knowing if all credible scenarios have been identified in a study. Some people may argue that the occurrence of an actual incident that is not identified in a PHA study is an indicator that the study was not performed as well as possible. However, catastrophic incidents, which are the type of most concern in PHA, occur infrequently and not often enough to provide sufficient data to validate effective performance, the details of real-world incidents often are far more complex than can be identified by PHA, and it is highly unlikely that any PHA study will identify all possible scenarios owing to the difficulty of doing so. Hence, the unidentified incident rate is not a good measure of the quality of a PHA study.

Possible metrics to judge the quality of a PHA, and indirectly the effectiveness of a competence management program for PHA teams, include:

- The average amount of time taken to identify a hazard scenario. Values that are significantly lower or greater than the norm for a particular company and type of process may indicate issues with the competence of the participants. Of course, such determinations must be made in the context of each study.

- The ratio of the number of hazard scenarios identified to the size and complexity of the process using a suitable measure of size and complexity such as capital cost, or an index representing the number and size of each piece of equipment present. Significant deviations from this ratio may indicate issues with the competence of team members. Of course, the measure used for size and complexity of the process must be meaningful and consistent across processes.

- Number and type of departures from PHA guidelines governing the study. They can be identified by peer review.
- Number and type of findings from periodic audits. Audits may identify omissions and deficiencies in PHA studies.
- Number and type of findings from audits of teams during the performance of a PHA study by an expert PHA practitioner, although the results are subjective and qualitative in nature and necessarily are based on a sampling of the PHA sessions actually conducted during a study.

Metrics for the performance of individual team members are more difficult to specify. Direct observation of team members in action is required by an observer who is an expert PHA practitioner. Checklists of desired attributes can be used to score performances of individuals but the approach suffers from the subjective and qualitative nature of the results as for performance audits of the team as a whole. Also, the presence of an observer may influence the behavior of people.

All of these metrics require the application of expert judgment to determine how meaningful they are in the context of a particular PHA study. Such assessments are in their infancy and need further development. When the need for corrective action is identified, it may be straightforward, for example, an instruction to a team member to be punctual, or it may be difficult, for example, the need for a team member to avoid being argumentative.

4.12 Certification of practitioners

Often, companies assume the competence of team leaders based on the adequacy of their experience or training, possession of certain qualifications, or the availability of a procedure for them to follow. These assumptions may be not only invalid but also they are insufficient to qualify a person to facilitate the performance of PHA studies where people's lives are at stake. Sometimes PHA team leaders are qualified on poor grounds, for example, they have read a book on PHA, been a member or scribe for a study, or attended a short course. Maybe they are just a team member who volunteers to lead a team in the absence of anyone else willing to do so. These may be useful attributes for a PHA team leader but they are not nearly sufficient to qualify a person to lead a study.

Owing to the importance of the role played by team leaders, formal certification of their competence against a set of defined criteria is desirable. Certification at different levels should be possible. Novice practitioners without actual experience facilitating a PHA need to be capable of certification at a low level so they can begin facilitating simple studies. Once they have actual PHA leadership experience, they can move to higher certification levels after meeting criteria appropriate to each level. Companies should restrict the type of study facilitated according to the certification level of the leader.

Certification criteria should include:

- Academic qualifications - A technical degree in a relevant discipline will demonstrate that the individual has the needed intellectual capacity and basic technical knowledge. Independent confirmation that the individual has earned the claimed qualifications is essential.

- Experience - Relevant on-the-job work experience provides a background and knowledge that is useful for a team leader. Confirmation of the claimed work experience by peers and/or supervisors should be required.

- Facilitation skills - For higher certification levels, feedback on the performance of the team leader should be sought from team members from more than one PHA study on the skills of the leader in managing the team. Team members providing feedback should owe no allegiance or obligation to the team leader to help ensure objectivity in the feedback provided.

- Work products - For higher certification levels, examples of PHA studies facilitated by the individual can be utilized to demonstrate competence. Expert PHA practitioners should review the work products against the PHA guidelines that were used to produce them to determine their validity.

- Training - Team leaders should have successfully completed appropriate training courses recently. Courses should be designed by a certified course designer and delivered by a certified instructor.

- Professional references - References should attest to the academic qualifications, experience, work products and professional ethics of the individual.

- Successful completion of a written examination that meets prevailing standards for tests of competence.

University engineering degrees alone do not provide the level of knowledge or the skills to act as PHA team leaders. Consequently, attendance at continuing education courses is necessary for all prospective team leaders.

Certification should be provided and administered by an independent third party using a defined set of publicly-available standards and certification criteria so that there is transparency in the process.

4.13 Continuous improvement

Competence management programs should be subject to continuous improvement. Performance metrics for PHA studies should be monitored and ways sought not only to achieve tolerable levels of performance but also to improve performance steadily over time.

5. Conclusions

PHA studies play a critical role in PSM programs. They must be performed by competent practitioners. PHA studies are performed by teams of people. Thus, both the competence of individual team members and the entire team are important. Participants and teams should be screened against appropriate criteria to demonstrate their competence as part of a competence management program.

There are few formal regulatory requirements for team members. Of course, PHA teams must comply with those requirements that do exist but more detailed requirements are desirable owing to the importance of PHA.

Competencies for PHA team members can be categorized as technical and personal. Technical competencies include education, experience, skills, and training. Personal competencies address behavior, attitudes, etc. Competencies can be classified as critical and recommended to assist in selecting team members in the real world where ideal team members likely will not exist. Competencies were defined using this categorization and classification scheme for the key participants in a PHA study, specifically, team leaders, scribes and technical team members.

Technical competencies can be developed through training, coaching, mentoring, supervision by experienced personnel, job experience and practice. Competencies in personal attributes largely are inherent to the person and may be difficult to change. However, team leaders who facilitate studies frequently may be motivated to try and modify their personal attributes.

Competence assessment is an essential aspect of competence management and involves such approaches as verbal or written tests, demonstrations, observation of task performance, behavioral observation, personality typing, and prior knowledge of behaviors. Assessing technical competencies usually is easier than assessing personal competencies.

Competencies should be maintained, for example, through refresher or on-going training, and re-assessed periodically, for example, through examinations. Competence management programs should be subject to continuous improvement, for example, by monitoring performance metrics for PHA studies. Metrics should be used for both individual and team performance. Other aspects of a competence management program include responsibilities for managing competence and documentation of competence.

Owing to the critical role played by team leaders, they should be certified formally using criteria that address academic qualifications, experience, facilitation skills, work products, training, professional references, and examination.

6. References

[1] "Guidelines for Hazard Evaluation Procedures", 3rd Edition, Center for Chemical Process Safety / American Institute of Chemical Engineers, 2008.

[2] Baybutt, P., Analytical Methods in Process Safety Management and System Safety Engineering – Process Hazards Analysis, in Handbook of Loss Prevention Engineering, Haight, J. M., (ed), Wiley-VCH, 2013.

[3] Baybutt, P., "On the Ability of Process Hazard Analysis to Identify Accidents", Process Safety Progress, Vol. 22, No. 3, pps. 191 - 194, September, 2003.

[4] Baybutt, P., The Role Of People And Human Factors In Performing Process Hazard Analysis And Layers Of Protection Analysis, J. of Loss Prevention in the Process Industries, Vol. 26, pps.1352-1365, 2013.

[5] Process safety management of highly hazardous chemicals, 29 CFR Part 1910.119, US Department of Labor, Occupational Safety and Health Administration, 1992.

[6] OSHA Publication 3133, Process Safety Management Guidelines for Compliance, 1994.

[7] OSHA letter to Mr. Robert Summers, The Norac Company Inc., November 19, 2001.

[8] Baybutt, P., Competence Management, Chapter 29, Human Factors Methods for Improving Performance in the Process Industries, AIChE / CCPS, New York, 2007.

Table 1. Typical Core and Specialty Team Members

Core Team Members	Specialty Team Members
Design engineer	Instrumentation / electrical engineer
Process engineer	Mechanical engineer
Operator (cover inside and outside activities). More than one may be needed	Programmer
Maintenance technician or engineer. More than one may be needed	Inspection / materials engineer
Controls engineer	Research scientist / chemist
Safety engineer	Environmental engineer / regulatory specialist
	Quality assurance / quality control
	Human factors specialist
	Industrial hygienist
	Industrial engineer
	Emergency responder

Note: There can be some overlap between core and specialty team members.

Table 2. Suggested Technical Disciplines for a PHA Team

- Design engineering
  - Knowledge of how the process is intended to operate
  - Knowledge of applicable standards, codes, specifications and regulations
- Process engineering
  - Understanding of the process science and technology
  - Ability to judge the adequacy of existing safeguards
- Process controls engineer
  - Knowledge of the process instrumentation, controls, alarms and interlocks
- Operations and maintenance
  - "Hands on" operating and maintenance experience
  - Knowledge of how the process responds to upsets
- Safety engineering
  - Knowledge of process hazards, safety systems, relevant company policies
- Other
  - Specialty technical areas
  - PHA facilitation and recording
  - Quality control
  - Etc.

Table 3. Positive Attributes for Team Leaders

Technical	Personal
Critical	
Technical education	Impartial in the evaluation
Formal training in PHA and leadership / facilitation skills	Organized
Knowledge of and expertise in the PHA method to be used	Seeks consensus
Knowledge of regulatory, industry and company requirements for PHA	High stamina
Understands processes and their operation quickly	Able to focus on multiple items simultaneously
Reads engineering drawings easily	Respected by the team
Motivational / interpersonal skills	Friendly and cooperative
Communication skills	Strong personality to control the team and drive the study but not overbearing
People and project management skills	
Recommended	
Experience as a team member	Patient
No day-to-day responsibilities for the process being studied, e.g. the team leader should not be the process engineer or the designer for the process	Able to read people
	Diplomatic
	Imaginative
	Quick thinking
	Gently authoritative
	Able to help team members see matters in a new light

Table 4. Negative Attributes for Team Leaders

- Wants to participate in brainstorming
- Poor listener
- Insensitive to team members
- Projects sense of superiority
- Rushes the team
- Rushes the judgment
- Biased
- Opposite of positive attributes

Table 5. Positive Attributes for Scribes

Technical	Personal
Critical	
Technically-oriented	Able to establish a good working relationship with the team leader
Understand the PHA process	Responsiveness to the team leader
Expertise with the means used to record PHA sessions, typically software on a computer	Attention to detail
Good typing, spelling and grammar skills	
Recommended	
Familiarity with terms and acronyms used	Capable of being a helper for the team leader, not just a recorder
Knowledge of processes	Good listener

Table 6. Negative Attributes for Scribes

- Challenges the team leader for control of the team
- Makes worksheet entries before the team has reached a consensus
- Plays with recording software
- Jumps around the worksheet unnecessarily
- Participates in the study at the expense of recording it
- Does not know or understand guidelines for worksheet entries
- Opposite of desirable attributes

Table 7. Positive Attributes for Team Members

Technical	Personal
Critical	
Knowledge of and expertise in some aspects of the process	Creative
Work with the process to be studied	Good memory
Knowledge of codes, standards, regulations and other requirements that apply to their area of expertise	Willing to participate
Able to read P&IDs and understand other process documentation	Willing to listen to others
	Able to communicate technical issues in their area(s) of expertise to other team members
	Able to express themselves clearly
	Open-minded
	Unbiased
	Willing to tolerate a detailed and thorough study
	Committed to time required
Recommended	
Knowledge and experience with the PHA method to be used	Sense of ownership and responsibility for the process to help ensure their commitment and motivation for the study
	Logical
	Alert
	Patient
	Able to concentrate
	Focused
	Not afraid to express their opinion
	Not intimidated by working in group made up of different disciplines
	Pays attention
	Comes to the point quickly

Table 8. Negative Attributes for Team Members

• Talkative • Over-enthusiastic • Argumentative • Sarcastic • Dominant • Arrogant • Skeptical • Wants to control the study	• Will not let go of an issue • Wants to go beyond the defined scope and objectives for the study • Promotes a personal agenda • Poor attendance / punctuality • Uses offensive language • Brings other work to sessions • Opposite of desirable attributes

Figure 1. Elements of Competence for Individuals

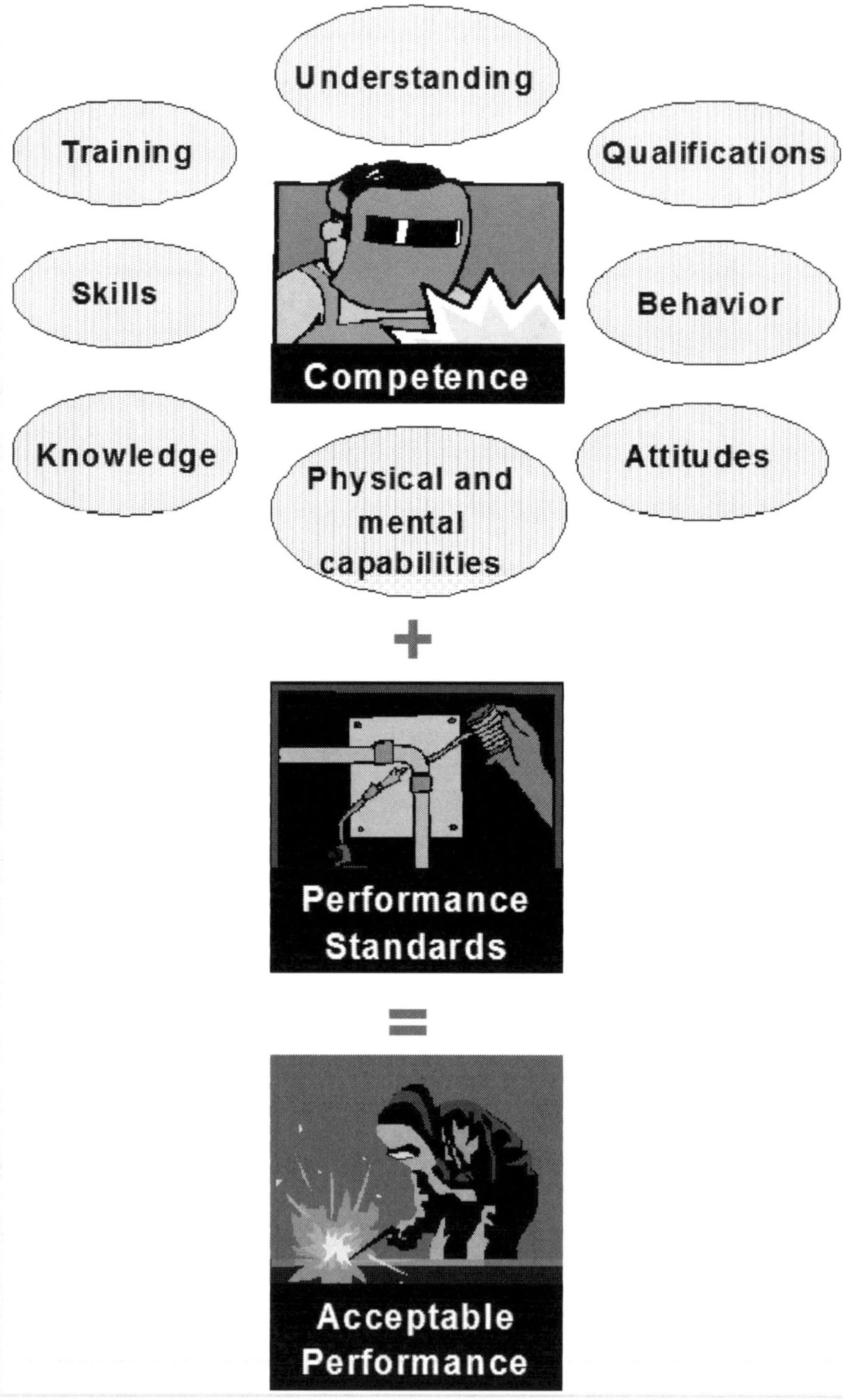

Process Safety Competency - effective approaches to creating and judging competency on process safety

Revonda Tew, Senior Engineer; Gerald Burch, Senior Engineer; and William Bridges, Senior Engineer
Process Improvement Institute, Inc. (PII)
1321 Waterside Lane
Knoxville, TN 37922
Phone: (865) 675-3458
Fax: (865) 622-6800
e-mail: rtew@piii.com
e-mail: gburch@piii.com
e-mail: wbridges@piii.com

Prepared for Presentation at
10th Global Congress on Process Safety
New Orleans, LA
March 30 - April 2, 2014

UNPUBLISHED

Process Safety Competency - effective approaches to creating and judging competency on process safety

Revonda Tew, Senior Engineer; Gerald Burch, Senior Engineer; and William Bridges, Senior Engineer
Process Improvement Institute, Inc. (PII)
Knoxville, TN 37922
e-mail: rtew@piii.com

Keywords: process safety competency, process safety management, incident investigation, root cause analysis, near miss reporting, process hazard analysis, human factors, leadership, accountability, employee participation

ABSTRACT

Process safety is a deep topic and requires the involvement of nearly ALL staff at a site. But, how do you make sure your staff are up to the task? And how do you judge the competency of subcontractors or third party experts? This paper describes the basics of building competencies in each aspect of process safety, including those tasks that require expert levels of competencies. It shows how different companies plan for the progression to levels of competencies and the typical requirements to reach a new level. One focus of the paper is that competency cannot be "specified" (at least not completely). Competency must instead be judged by those who are already competent in that skill or role.

Background

The increasingly complex and technical demands of process safety management (PSM) have placed a large demand on existing resources across the process industries. Success requires the utilization, involvement, and full support of nearly ALL staff at a site. Success also demands that a substantial portion of staff be competent and capable of contributing.

Unfortunately, the process safety competencies required for people to become fully involved are not easy to achieve quickly. New engineering graduates may have academic exposure to safety engineering principles and critical thinking skills, but they generally lack specific training on process safety fundamentals, though that is changing in man engineering curriculum. More importantly, new graduated always lack the practical experience required to make sound risk management judgments and they lack the skill required for many tasks in PSM, such as leading process hazard analyses (PHAs).

Similarly, operators and mechanics may have the practical, hands-on experience required, but they often lack the technical skills necessary to perform some of the skill-based tasks that fall to them, and they may also lack the analytical background to evaluate and manage risk effectively.

The challenge facing nearly all companies today is how to identify, develop, nurture, and maintain competent individuals across the organization with the right combination of natural abilities, experience, education, and training.

Developing, Measuring, and Maintaining Individual Competency

Competency is the skill or ability to do something well enough to meet a standard. An individual gains competency through the combination of natural abilities, general education, experience, and specific skill/task training (classroom and hands-on), as should in Figure 1.

Some of these components, such as education or experience (knowledge), can be measured directly and minimum standards established. However, evaluation of the competency level requires demonstration of the ability and judgment by others who are already competent. How can someone who NOT an expert judge the competency of someone in a new skill or activity?

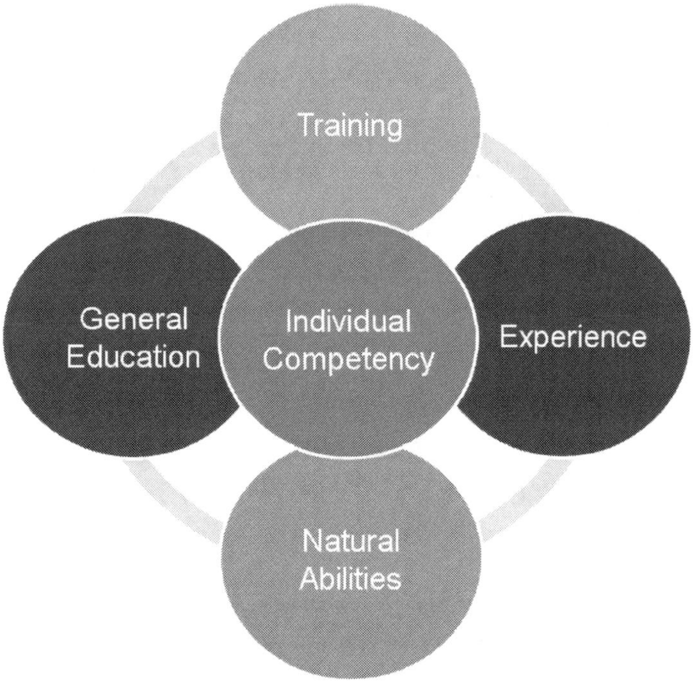

Figure 1: The Make-Up of Competency in an Individual
© Process Improvement Institute, Inc. (2013)

Progressive competency levels are achieved over time by participation in specific activities and achievement of specific milestones. Increasing levels of expertise are mastered and maintained through training and mentoring others, cross-functional and cross-departmental sharing of resources and information, stewardship of organizational guidelines and standards, and external engagement with other experts in sharing and developing new knowledge.

Figure 2 shows how this progression applies to building competency in incident investigation / root cause analysis (II/RCA). Two case studies (actual examples from two company) using the model below is provided after the figure.

Figure 2: Building Competencies for II/RCA Leadership

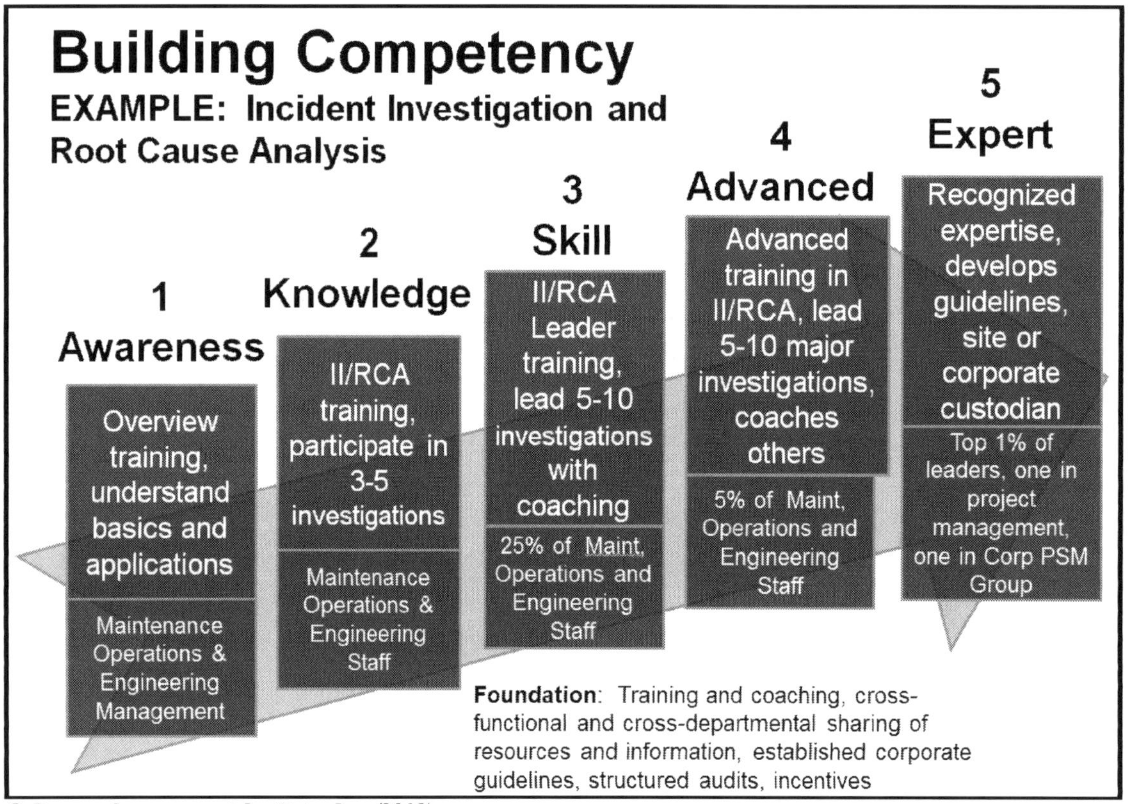

Case Study: Building Competencies in II/RCA at AMOCO Oil Offshore Business Unit

The ratio of near loss incidents to loss incidents reported is a critical leading indicator of process safety performance. Near loss incidents provide an opportunity to learn about and correct management system weaknesses before losses are incurred. The desired ratio is 50 to 100 reports of near loss incidents for every loss incident incurred.

In 1996, Amoco Oil's offshore business unit on Louisiana (about 650 workers and other staff) determined that an increase in near loss reporting was desired to improve their process safety performance. They had historically reported less than 2 near loss incidents per loss incident incurred. First they educated superintendents and managers in how to get near misses reported, including how to create and maintain a blame free environment. This change in management approach included having workers lead all investigations to reduce the fear caused by investigations led by bosses. This changed worked great and the near miss report jumped from the prior ratio to about 80 in one month! (Near miss reporting increased from about 20 a month to about 900 a month, in one month of implementation.)

The only way to manage the huge increase in incident investigations was to develop additional resources to participate in and lead investigations. With help from staff now at PII, AMOCO trained about 60 staff (~10% of the work force in that business unit) in the incident investigation and root cause analysis leadership. The training included examples from within AMOCO to add realism and pertinence. Some coaching of new leaders also occurred via phone and e-mail. Then about 9 months later, Refresher training was provided for the 60 leaders, using incidents that they had investigated; this training help make mid-course corrections to the approach/techniques used by the newly trained leaders.

These in-house staff carried on the training for many years thereafter. This case study illustrates progression of the II/RCA leaders from the Knowledge level through the Advanced Level (and perhaps Expert Level).

Case Study: Building Competencies in II/RCA at A Large Pharmaceutical Company

As mentioned earlier, the desired ratio of number of near misses reported to number of actual loss should be 50 to 100 in a company with excellent management systems for process safety. This also extends to quality, reliability, and productivity.

In 1997, the pharmaceutical company (with more than 50,000 workers and staff world-wide) determined that an increase in near loss reporting was desired to improve their process safety performance. They combined all incident databases (there had been 5 different ones) into one database covering reliability, quality, safety, process safety, productivity, and environmental protection. Then, they developed a customized training course (with help from staff now at PII) and trained 7 internal staff of the fundamentals taught in this course; these 7 had been identified as future trainers and they began applying the new methods/approach right away. About a year later, the future trainers appeared expert enough in the methods (as judged by a expert investigator now at PII) and training of 10% of the site staff began across all of the company. Originally, the site staff was trained by the expert at PII, but gradually, the 7 Lilly instructors began to gain skill and confidence in teaching this topic. After about 20 courses in the USA and UK, the training was turned over entirely to the 7 internal trainers.

These changes worked great and the near miss report jumped from the prior ratio to about 105 (the highest ratio we have ever measured).

This case study illustrates progression of the II/RCA leaders from the Knowledge level through the Advanced and Expert Levels.

Developing and Maintaining Organizational Competency

An organization gains competency through the identification and development of the requisite **Skills**, but skills alone are not sufficient. Organizational competency also requires that **Information** is developed and shared, that a learning **Culture** (supportive, nurturing, and encouraging) is maintained to ensure that the skills are developed and applied in an effective manner, and that **Performance Measures** are monitored to continuously evaluate performance and reevaluate organizational needs.

Figure 3: The Make-Up of Competency for an Organization
© Process Improvement Institute, Inc. (2013)

Step 1: Skills – What are the organization's needs?

Understanding the organization's process safety skill requirements is the first building block of developing competency across the organization. What are the organization's needs?

A gap analysis is essential to develop these skill requirements. Begin by identifying the organization's current strengths and weaknesses in process safety. What is going well and should be maintained? What improvements are necessary and how will those be achieved? Then, establish organizational goals and objectives for maintaining and improving process safety performance.

Once process safety goals and objectives have been established, it is possible identify what skills, individual competency levels, and resources are needed to successfully meet those objectives. Comparing the available inventory of resources with the organization's needs identifies those gaps which must be filled.

> *Example: An organization has determined that it is underreporting "near misses" by a factor of 10. Leaders are confident that increased reporting and investigation of near loss incidents will identify and resolve management system weaknesses (root causes) common to both near loss incidents and loss incidents. By eliminating these common causes the risk of incurring losses will be decreased. However, increasing the number of incidents reported by a factor of 10 will also increase the number of investigations which must be performed. Managing this increased workload requires that additional resources (shift operators and mechanics) be trained to lead effective and efficient investigations as quickly as possible after the incident occurs. Involving workers has the added benefit of increasing process safety awareness, employee engagement, and ownership of the proposed resolutions. To achieve this goal, the management of the company needs to become proficient in process safety management and especially become expert in reducing the barriers to near miss report (the biggest barrier being fear of blame for the mistakes that led to the near miss or accident).*
>
> *This change requires awareness training and then knowledge by management; skill and advanced training by 10-15% of the workers, and expert level for the workers assigned to be the chief investigators (gatekeepers of quality). This progression was followed at Amoco Oil and Eli Lilly mentioned earlier, but has also been implemented in dozens of other companies world-wide.*

Table 1 is one example of a "process safety skills" inventory for an organization; this one is for II/RCA only; the table which identifies resources required and training/skills needed.

Table 1: Typcial Competencies needed for II/RCA across an Organization

Training need	Group				
	Upper management	Supervisors	Engineers/Operators/ Maintenance Staff (general)	Staff leading investigation of complex events	Staff leading investigation of low complexity events
Incident/problem reporting definitions and procedure/forms	Yes	Yes	Yes	Yes	Yes
Objectives of II/RCA, Near miss reporting, Gathering data (including interviewing of peers)	**Proficiency**	**Proficiency**	**Proficiency**	**Proficiency**	**Proficiency**
Analysis techniques (CFC, FTA, Root Cause Charting)	Awareness	Awareness	Awareness	**Proficiency**	**Moderate proficiency**
Investigation follow-through requirements	Their responsibilities, including measuring performance of II	Their responsibilities	Their responsibilities	Their responsibilities	Their responsibilities
Communicating analysis results and resolutions	Yes	Yes	Yes	Yes	Yes
Trending Techniques	Awareness	Awareness	Awareness	Detailed	Awareness
Typical length of course segment	3 hours	3 hours	3 hours	3 to 5 days	1½ to 2 days
Typical number of personnel at an operating facility	All	All	All	3 per site of 100 total operations and maintenance staff	15 per site of 100 total operations and maintenance staff

Similar inventory tables can and should be developed for each aspect of process safety engineering:

- Relief valve sizing
- Safety instrumented systems
- Corrosion engineering
- Selecting the right materials of construction
- Etc.

and each activity/skill within process safety management

- Investigators
- PHA
- Auditing
- Procedure Writing
- Etc.

Just like with specific skills such as II/RCA and relief valve sizing, company leadership also needs to have a set of knowledge, and selected leadership may need to also advance to skill, advanced, and expert levels as should in Figure 4 below.

Figure 4: Building Competencies for Process Safety Leadership

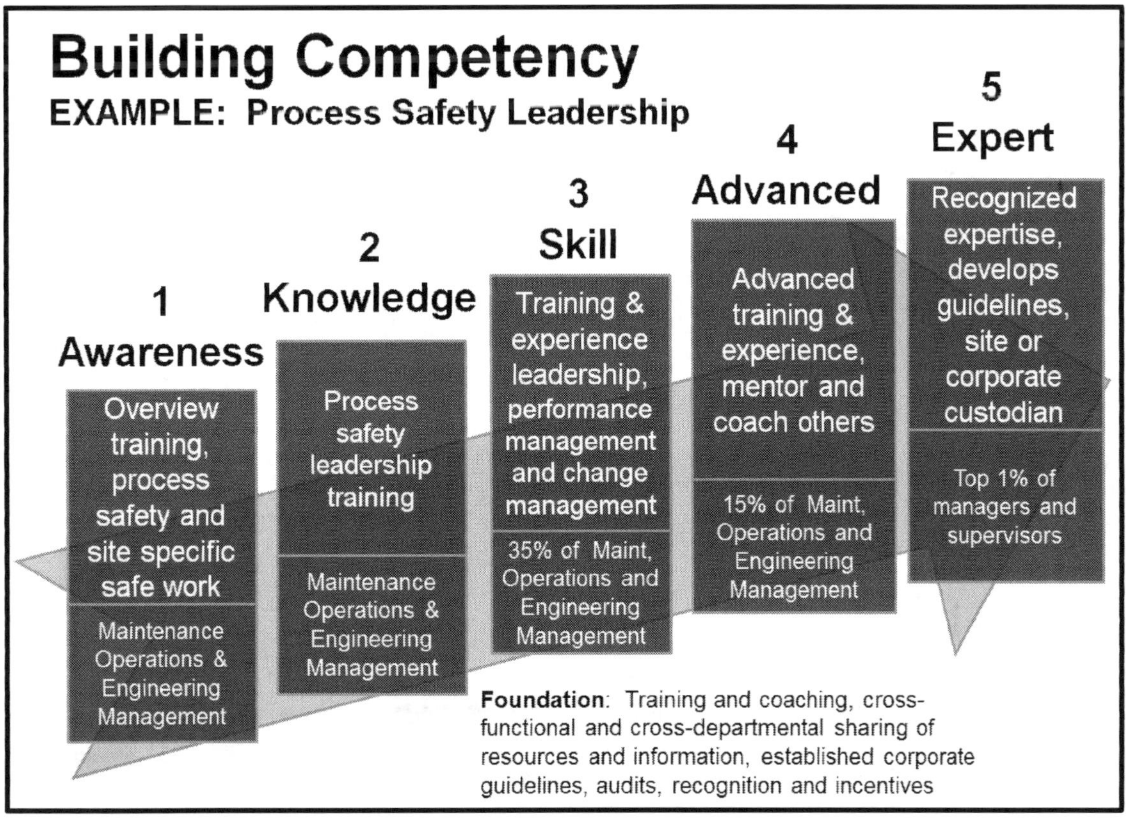

© Process Improvement Institute, Inc. (2013)

Step 2: Identify Candidates

After the organization has identified the inventory of process safety skills and resources needed, it must now determine how those needs will be met. Some organizations may decide to develop a relationship with external service providers for much of the process safety engineering expertise required and even for some of the process safety skills. Others may decide to develop this expertise internally, relying on external support only until internal resources can be developed (it is hoped that this is the approach all companies adopt). There are many factors outside the scope of this paper that must weigh into this decision for each organization. For instance, it is not practical or wise to have someone other than plant workers develop operating procedures and maintenance procedures; having outsiders develop these may produce well written documents, but the content will not be as accurate and the language and jargon will be somewhat different if outsiders download the information from the heads of workers and write the procedures. And, there is a practical necessity to have shift staff trained on how to perform root cause analyses of near misses, if the plant gets 100 near misses reported per loss event.

Regardless, it is important to maintain a healthy mix of internal participation to ensure ownership and consideration of specific process technology issues and external engagement to ensure continuous learning and best practice standards.

If internal resources will be used, then it is important to continually look for and identify good candidates early in the process for further development as advanced and expert resources. Awareness and knowledge training provide ideal opportunities to begin this process.

For formal training, ensure all sessions include interactive workshops and other small group activities designed to apply the skills being taught. Have the instructors observe and evaluate participants based on their interest level, their natural ability to grasp and apply the skill, and their interaction with others as participant and leader. Ideal candidates will quickly grasp the technical aspects of a skill, learn to apply them in an appropriate way, serve as a positive role model for others, gravitate to leadership roles within the group, and gently begin to encourage, coach, and teach others who may not grasp the concepts as quickly.

At this stage, it is important to keep an open mind about the possibilities that lie within each individual. At the risk of reinforcing stereotypes, engineers sometimes bring technical knowledge and critical thinking skills, but sometimes lack the communication skills and personal empathy needed to become a good teacher. Operators may bring the practical knowledge and credibility necessary to be a good facilitator, but sometimes lack the detail orientation and education necessary to conduct a thorough analysis. People bring a wide variety of education, experience, abilities, and biases to the table. Your competency development job is to build on their strengths and fill their gaps.

Step 3: Culture - Develop Career Paths and Opportunities

Incorporate the process safety skills inventory and competency requirements (Table 1 was an example of one of about 20 such inventories necessary) into the normal career progression for all roles. When considering individuals for advancement or promotion, give consideration to the development and demonstration of process safety competencies appropriate to the role. For example, incident investigation, management of change, and operating procedures are important within the operations ranks. Management of change, mechanical integrity, and writing of maintenance procedures are important for the progression of mechanics. Process technology and hazard evaluation are important for the progression of process engineers. In each discipline, establish the minimum competencies required to progress to the next level. Reward those who exceed the minimum requirements with additional learning opportunities, special projects, coaching assignments, individual recognition, and promotions. With higher levels of progression, should come higher compensation as well.

It is often desirable to develop a formal career path for engineers which includes advancement and demonstration of process safety competency. Table 2 is an example of a technical professional career progression path for a key competency identified by the organization. In this example, competence levels are described for the skill "Pressure Relief Device Adequacy Assurance." which is one of elements of "Process Safety Risk Assessment." A complementary skill of "relief valve design" is necessary for "process safety engineering."

> ***Example:*** *A process engineer has shown aptitude and expressed interest in advancement in process safety. Work with this engineer's manager to build a developmental plan which incorporates process safety competency developmental goals. Identify learning opportunities such as an advanced class in inherently safe process considerations, participation/leadership of a process hazard analysis, an opportunity to lead a complex incident investigation, or a mentoring assignment to coach and develop others in leading human factor audits.*

Step 4 – Build individual competency

Individuals build competency through their natural abilities, education, training, and experience. Formal training classes provide the basic awareness and knowledge necessary, but advanced skills require application and hands-on experience in real-world applications. Mentors who have already demonstrated advanced skills should be assigned to each new learner to provide guidance, feedback, encouragement and support. This assignment is important to the mentor as well since real expertise is developed by coaching and teaching others. Experts gain and enhance their competence by acting as stewards over the discipline, developing standards and guidance documents, overseeing coaches and trainers, and interacting with experts outside the organization to continuously learn, improve, and create new knowledge.

Some organizations reward Experts at the same compensation level as vice presidents of the company because they recognize the potential vulnerability and loss of investment in losing Experts.

Table 2: Example of a Generic Skill Progression Path (courtesy of PETRONAS)

		DISCIPLINE COMPETENCE MATRIX				
	Competence Level	Awareness	Knowledge	Skill	Advanced	Expert
Technology		Competence Level Descriptors				
1	Pressure Relief Device Adequacy Assurance					
			Awareness +	Knowledge +	Skill +	Advanced +
1.1	This technology is concerned with the skills required to ensure safety in the design of processes and plants through avoidance of over or under pressure.	Aware of the different types of pressure relief devices.	Able differentiate and identify the application of different type of pressure relief devices.	Able to select suitable type of pressure relief device for specific application. Able to identify type of pressure relief device at site. Understand the terms associated with ensuring reliability of pressure relief devices such as "failure to danger", pop test, pre-pop test, bench test and Trevi test.	Able to interpret the pressure relief device datasheet. Able to train other on the selection of suitable pressure relief devices for a specific application. Able to elaborate on the terms associated with ensuring reliability of pressure relief devices and conduct site verification.	Acts as an internal consultant on all aspects of pressure relief issues (selection, design and installation); stays current on relief valve literature. Able to advise the plants on ensuring reliability of pressure relief devices to prevent Loss of Containment (LoC).
1.2	This includes specification of equipment design pressure, fundamental understanding of the causes of over/under pressure, development of overall relief philosophy and detailed specification of hardware and systems to protect the process.	Aware of causes of over and under pressure in plant operation and design. Understands importance of safe disposal of vented material.	Has received formal training and able to describe common causes of over and under pressure in plant operation and design. Aware of the key codes and international standards for pressure system design requirements (e.g. API 520, API RP 521 and DIERS).	Able to identify relief case for specific relief system. Able to identify the necessary information and performing pressure relief device sizing calculation. Understand and apply of the key codes and international standards for pressure system design requirements.	Able to review pressure relief device sizing calculations and provide assurance to stake holder. Interpret the outcome of pressure relief device sizing calculation and recommend suitable corrective measures, as required.	Train and coach others to carry out calculations for pressure relief device sizing.
1.3	This technology describes the ability to calculate the provision for pressure/vacuum relief as ultimate protection to prevent loss of containment.	Aware of the general configuration and different type of flare header design.	Able to describe the function of the various components of flare system. Understand the calculation used for flare network analysis and aware of suitable tools/software for the calculation.	Detailed understanding of flare systems and its adequacy inclusive of pressure drop across the system and potential for backpressure.	Perform flare header calculations to determine adequacy for design relief case.	Train and coach others to carry out calculations to determine adequacy of flare header system.
1.4	Ability to calculate the loading on flare header during design relief case, to ensure adequate capacity and minimize back pressure.	Aware of the general configuration and different type of flare header design.	Able to describe the function of the various components of flare system. Understand the calculation used for flare network analysis and aware of suitable tools/software for the calculation.	Detailed understanding of flare systems and its adequacy inclusive of pressure drop across the system and potential for backpressure.	Perform flare header calculations to determine adequacy for design relief case.	Train and coach others to carry out calculations to determine adequacy of flare header system

The list below shows how this process works for building competence in Incident Investigation and Root Cause Analysis:

> ***Level 1 (Awareness)*** *- attend a 1-2 day classroom training covering human factors, data collection, causal factor charting, and root cause categories.*
>
> ***Level 2 (Knowledge)*** *- attend a 3-4 day classroom training which reviews the basics and then includes advanced investigation topics such as conducting interviews, collecting, organizing and analyzing more complex data, and teaches fault tree analysis in an investigation setting. Practice workshops are followed by participating in 3-5 investigations led by others.*
>
> ***Level 3 (Skill)*** *- lead 5-10 investigations of low to moderate complexity (99% of the incidents at a facility will typically fall into this category) with the active support and oversight of a coach. Additional training on human factors, procedures, mechanical integrity, and hazard evaluation may also be required.*
>
> ***Level 4 (Advanced)*** *- coach Level 3 investigators and lead 5-10 major, more complex investigations. Participate in or lead investigations at other sites to gain the experience and practice needed (typically, fewer than 1% of the incidents at a facility are major).*
>
> ***Level 5 (Expert)*** *- serve as the site or corporate steward, developing policies, procedures, and guidelines for conducting investigations and analyzing incidents. Monitor and analyze organizational performance, evaluate effectiveness, identify and implement improvements. Interact with other experts outside the facility through involvement in corporate teams, industry groups, site visitations, and other opportunities to expand and grow their skills.*

Step 5 – Maintain Proficiency and Extend Skills

Organizations maintain and extend process safety competency by providing a supportive culture, making information available, and evaluating performance.

A supportive process safety culture begins with management's understanding, commitment, and unwavering support of process safety management as a critical risk management tool. Resources are made available, priorities are clearly established, and key performance indicators are monitored and discussed to ensure that process safety related activities are completed. A supportive process safety culture also values and recognizes its expertise. Process safety stewards are visible, engaged, and available. They maintain up-to-date training materials and searchable databases, teach and attend training classes, and maintain high visibility by pushing information out to the organization on relevant achievements, results, findings, and future plans. They ensure that everyone understands the important role of process safety management in preventing serious incidents and injuries.

Step 6 - Measure and evaluate results, reevaluate needs

A strong set of leading and lagging metrics is essential to measuring and monitoring performance. Most of the metrics associated with process safety competency are activity based, such as the following examples:

- Ratio of near misses report to number of accidents that occur
- Training sessions completed
- Achievement of specific competency goals
- Timely completion of investigations or all causal factors and root causes
- Timely completion of high quality MOCs
- Timely completion of high quality PHAs
- Timely completion of compliance and best practice audits
- Timely closure of recommendations and action items
- Timely completion of and good results from human factors audits
- Procedure quality and accuracy audits

Periodically, it is also important to review and reevaluate the organization's process safety goals and objectives to determine how things are going. Are these still relevant and are they being achieved? Are the identified competency needs being met and have new needs become apparent?

Ensure that the process safety plans and strategies are completely aligned with the organization's overall goals. If not, determine why and adjust accordingly to remain a relevant and contributing part of the organization's success.

Additional Case Studies – Results Achieved

PHA Leaders at UNITED (SABIC affiliate). *This petrochemical site (ethylene, polyethylene, alpha-olefin) of about 600 employees determined that 4 PHA Leaders are needed for this size and nature of complex to handle PHAs and Revalidations and large MOC risk reviews.*

- *Experts (PII, in this case) trained 15 process engineers to the knowledge level in leading and scribing PHAs (5 days)*
- *The PII instructor then chose the 8 best candidates, and coached them for 2 weeks while they led and scribed 3 different PHAs (re-do of existing units; initial PHAs were poor)*
- *Training and coaching included all methods (HAZOP, What-If, Checklist, FMEA) with analysis of continuous mode, along with procedure modes for startup and shutdown*
- *During the PHA report-writing phase, PII coached, reviewed, and edited the draft PHA/HAZOP reports, with the UNITED leaders/scribes finishing*
- *By the end of the process, PII certified 4 leaders and 3 scribes*
- *These leader/scribes completed the PHAs of all modes of operation for all 4 plants and for utilities*

- *In the 5 years since, the leaders have been backfilled as a couple were promoted to superintendent*
- *Certified leaders/scribes have ensured that each PHA and each large MOC risk review was performed well*
- *See Figure 5 below for the overall competency progression for PHA*

Figure 5: Building Competencies for Process Hazard Analysis

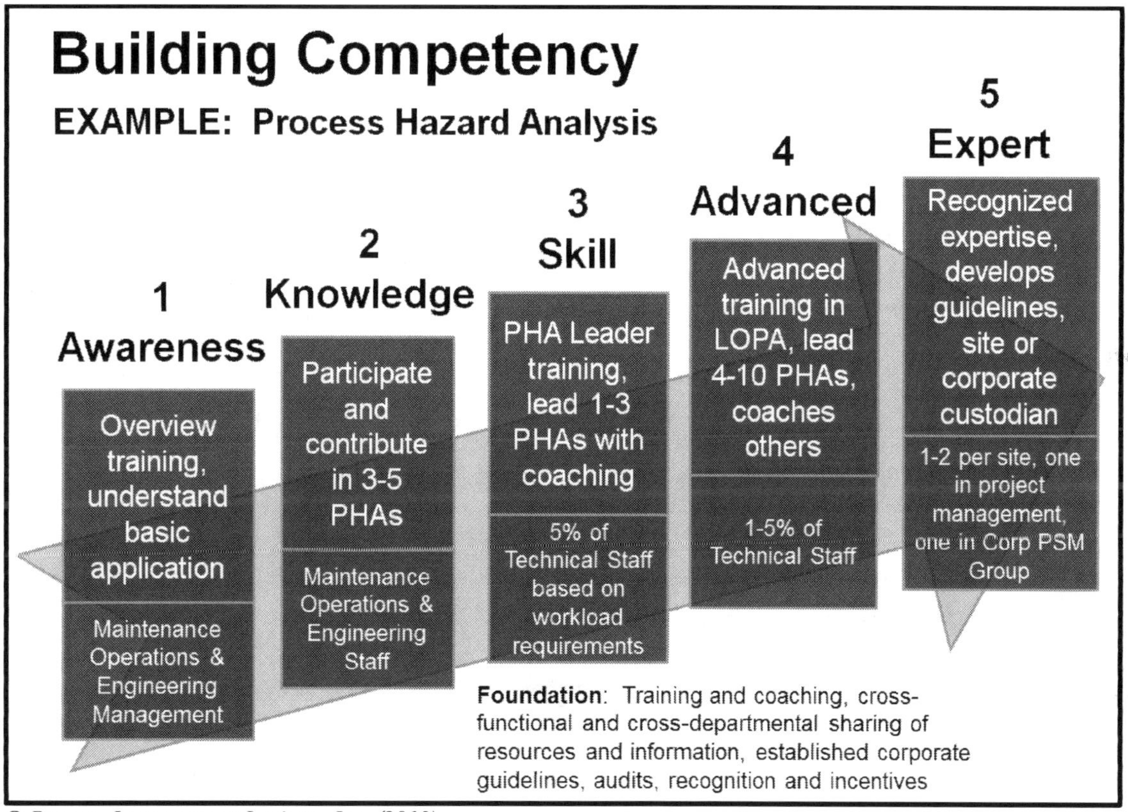

Operating Procedures Writers at a Medium Size pharmaceutical site with about 250 operatora: *The site identified deficiencies in their operating procedures as a root cause of a high percentage of incidents. They developed procedure writing guidelines and trained 3 engineers, 45 operators and 10 mechanics in the procedure writing process. Now, subject matter experts (SMEs; operators and maintenance technicians) write and review the procedures, and evaluate accuracy, completeness, and clarity. The clarity is judged versus the best practices for controlling human errors in written procedures. Technical resources (engineers, etc.) continue to provide editorial comments, but SMEs are the owners of their procedures. This took less than 50% of the effort originally estimated. PII provided the initial training and coaching:*

- *2 days of classroom training by experts in procedure writing and human factors, for 3 sessions to train the 58 folks listed above (6 training days total)*

- *3 days of hands-on coaching by experts for each group, where pairs of graduates wrote procedures from walk-downs in the field. This was followed by an additional 5 days of coaching onsite about one month later. Total of 14 days of courses and hands-on coaching at the site*
- *2 days of coaching by experts from a distance (by e-mail and telephone)*
- *3 days assistance by experts on development of management system for writing procedures and management of changes to procedures*
- *See Figure 6 below for the overall competency progression for writing procedures*

Figure 6: Building Competencies for Procedure (work instruction) Writing

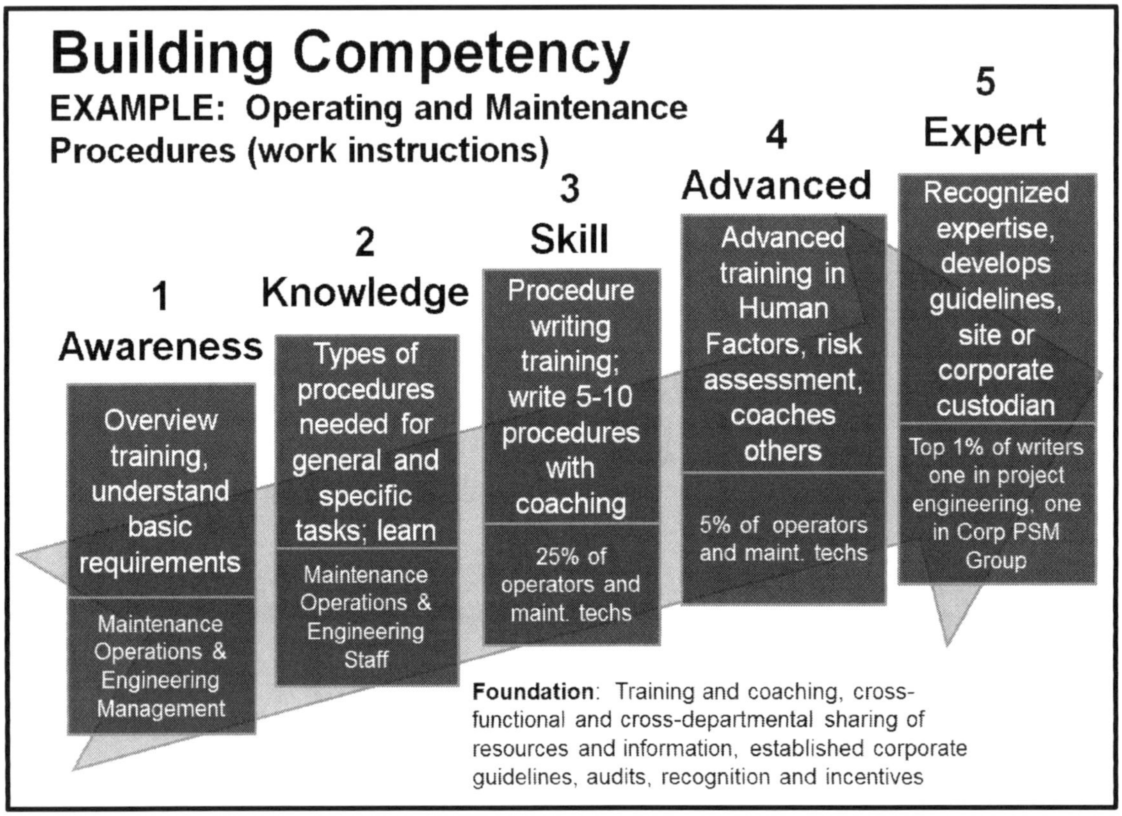

© Process Improvement Institute, Inc. (2013)

Full Set of Process Safety Competencies for Operators at a Refinery. *This refinery on the West Coast of the USA, has about 800 employees. Management established a very strong process safety culture early (20+ years ago, under Arco) and maintained consistent vision since then. There has been continual investment in empowering workers to take the lead on writing procedures, incident investigations, near miss reporting, and change management; with no compromise on budgets for critical competencies and activities.*

- *One operator per unit was identified, trained, and assigned responsibility for PSM unit leadership on a day-to-day basis including coordination of the MOC process, risk reviews for MOCs, incident investigations, procedure updates, human factors evaluations, etc. This assignment rotates every two years.*

- *PII provided skill training and coaching on all topics for these unit leaders in MOC, mini-PHA of procedure changes, procedure writing and analysis (what-if and 2 guide-words), and deep training on human factors.*
- *New workers immediately begin building competencies in process safety.*
- *Workers help train each other to "fix" a procedure or task instruction rather than "oh, let's not follow that because the steps are not correct in it"*
- *There is a thorough risk review of all hazards and tasks*
- *The organization scores high on process safety competencies and process safety culture on surveys (they had the highest scores across all of BP USA on Process Safety Culture during the Baker Panel surveys)*
- *Their process safety performance in the past 15 years is best in class*

CONCLUSIONS

Competencies can be achieved and maintained at chemical process plants and companies. Organizations need to recognize their gaps in competencies first; this first step is a major failure of many companies, even the largest and oldest, because for some process safety needs, the organization does not know how to judge competencies. An organization needs to have existing experts in order to have the coaches/mentors available to build the missing process safety competencies; for the short-term, this may require a company to contract in the experts. Building competencies requires sustaining staff in a role long enough to build competency; and the organization must get a few of these competent staff to reach Advanced and Expert levels. Eventually, competency levels can be self-sustaining; this is possible if the organization recognizes the importance of sustaining competencies and rewards those with very high level of competency accordingly.

ACRONYMS USED

AIChE – American Institute of Chemical Engineers
CCPS – Center for Chemical Process Safety (of AIChE)
CFR – Code of Federal Regulations
GCPS -- Global Congress of Process safety
HAZOP – Hazard and Operability; as in HAZOP Analysis or HAZOP Study
II/RCA – Incident Investigation / Root Cause Analysis
LOPA – Layer of Protection Analysis
MOC – Management of Change
PHA – Process Hazard Analysis
PSM – Process Safety Management
RAGAGEP – Recognized and Generally Accepted Good Engineering Practice
RCA – Root Cause Analysis
US EPA – United States Environmental Protection Agency
US OSHA – United States Department of Labor, Occupational Safety and Health Administation

REFERENCES

1. Bridges, W., *Gains in Getting Near Misses Reported (Updated),* 8th GCPS, AIChE, 2012.
2. Williams, T. and Bridges, W., *Effective Procedure Writing – A Two-Part Series (Part 1: Reducing Risk of Worker Errors; Part 2: Addressing Process Safety Management Requirements for Operating Procedures),* Chemical Engineering Progress, 1997.
3. Excerpts on Learning Objectives and Competency Requirements from various PII Training Manuals.

354011 Executive Order Panel

Abstract:

On August 1, 2013, President Obama signed Executive Order 13650, entitled Improving Chemical Facility Safety and Security. The Executive Order (EO) directs the Federal Government to improve operational coordination with state and local partners; improve Federal agency coordination and information sharing; modernize policies, regulations, and standards; and work with stakeholders to identify best practices. This panel will be hosted by representative from the three agencies that chair the EO Working Group – U.S. Environmental Protection Agency, U.S. Department of Homeland Security, and U.S. Department of Labor –OSHA. The panelist will give an overview of the EO and the working group's progress and welcome input from the audience on key issues.

What Have We REALLY Learned?
(25 years after Piper Alpha)

Michael P. Broadribb
Baker Engineering & Risk Consultants, Inc.
3330 Oakwell Court, Suite 100, San Antonio, Texas 78218-3024
mbroadribb@bakerrisk.com

Prepared for Presentation at
American Institute of Chemical Engineers
2014 Spring Meeting
10th Global Congress on Process Safety
New Orleans, LA
March 30 – April 2, 2014

UNPUBLISHED

What Have We REALLY Learned?
(25 years after Piper Alpha)

Michael P. Broadribb
Baker Engineering & Risk Consultants, Inc.
3330 Oakwell Court, Suite 100, San Antonio, Texas 78218-3024
mbroadribb@bakerrisk.com

Keywords: process safety management, offshore, incidents, permit to work, safety culture, human factors, emergency response, regulatory regime

Abstract

Twenty five years ago, a major disaster in the North Sea took the lives of 165 persons on board the Piper Alpha oil production platform as well as 2 people from the rescue crew. To this day, the incident represents the worst offshore oil industry disaster ever and has become an industry-changing watershed event.

The subsequent Public Inquiry report made a number of recommendations that were related to inherently safer design (ISD), including the mandatory use of a systematic fire and explosion analysis, analysis of smoke and gas ingress to living quarters, and the requirement for a temporary (safe) refuge area capable of surviving the initial fire/explosion. Other recommendations addressed analysis of the vulnerability of safety critical equipment / elements, and evacuation, escape, and rescue in the event of major incidents.

Implementation of the above recommendations aimed to reduce the residual risk of the design through the use of an ISD approach. The residual risk then had to be managed by a safety management system (SMS), which is another inquiry recommendation that comprises many of the elements of process safety. Finally, the inquiry recommended developing a Safety Case to describe and justify the design, the inherent hazards and residual risk, and the SMS. This Safety Case is now the basis for periodic audits.

Over the past twenty five years this author has had the opportunity to visit offshore production platforms and drilling rigs and to review new offshore projects in many regions of the world. This paper addresses both the strengths and weaknesses observed, and a number of common themes involving management systems, human factors, process safety and integrity management at such facilities. Comparisons will be drawn to the evidence aimed at preventing future major incidents that the author and others presented at the Piper Alpha Inquiry, and to the Inquiry recommendations.

1. Introduction

The Piper Alpha platform (Fig. 1) was located in 474 feet of water in the UK sector of the North Sea about 120 miles north east of Aberdeen, Scotland, and was commissioned by the Operator, Occidental Petroleum (Caledonia) Ltd., in 1976.

Production peaked at 300,000 barrels of crude oil per day (bbls/day) from 24 wells, but had declined to 125,000 bbls / day by the time of the incident in 1988.

Figure 1: The Piper Alpha Platform

Piper Alpha (Piper) was connected to the Flotta oil terminal in the Orkney Islands with a 30-inch diameter crude oil subsea pipeline it shared with the Claymore platform (Fig. 2).

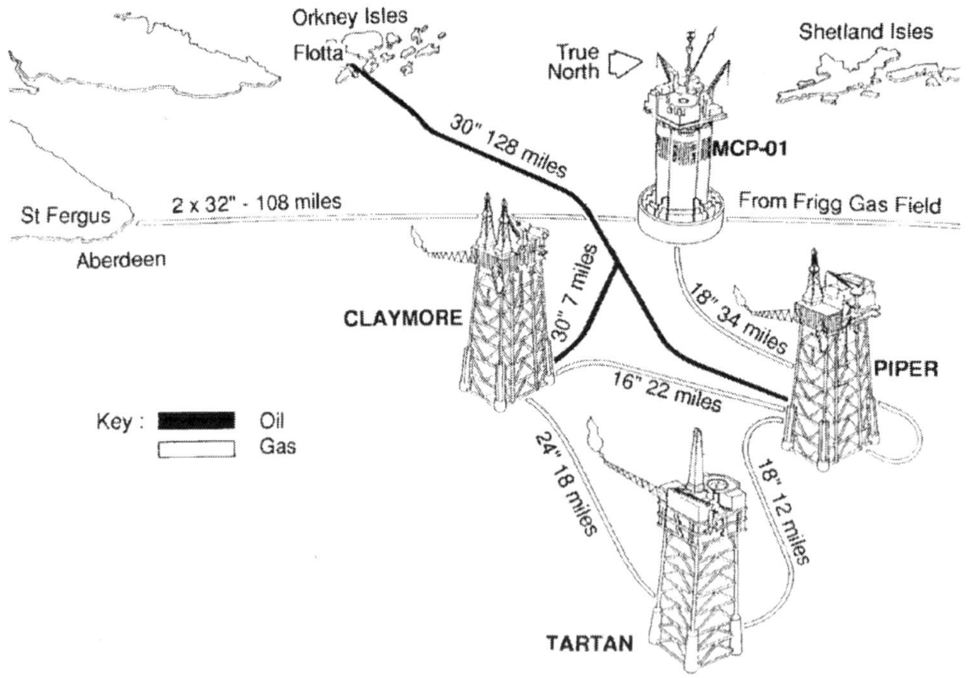

Figure 2: Pipeline Connections to Piper Alpha [1]

In 1980 Piper had been modified to produce gas as well as oil. After those modifications Piper's normal operating mode was known as Phase 2, i.e. operating with a gas conservation module. Separate 16 and 18-inch diameter subsea gas pipelines connected the Piper to the Texaco Tartan and Occidental Claymore platforms, as well as the Total MCP-01 gas processing platform, thus giving the Piper four main transport risers: the oil export riser, the Claymore gas riser, the Tartan gas riser, and the MCP-01 gas riser.

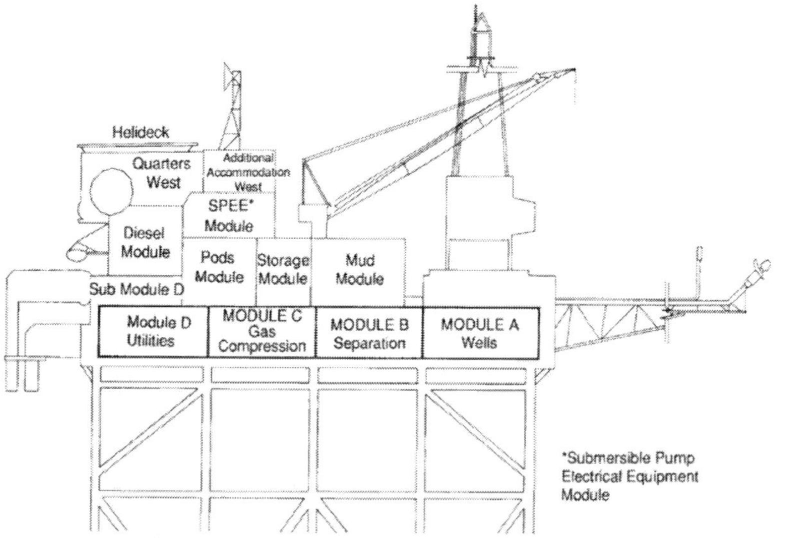

Figure 3: Piper Alpha (west elevation) [1]

The Piper platform was comprised of four main modules:

A. Wells,
B. Oil/gas separation,
C. Gas compression,
D. Utilities.

Firewalls separated those four modules from the living quarters and the power generation located above the utilities module (Figs. 3 and 4).

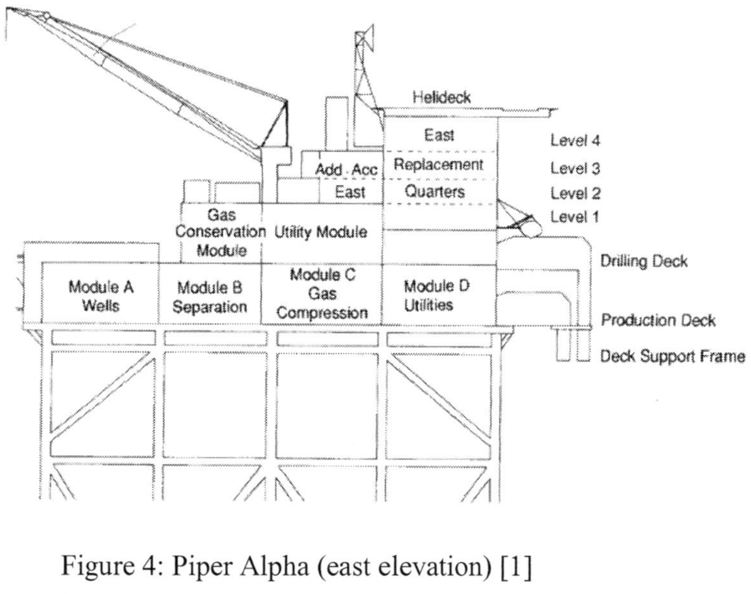

Figure 4: Piper Alpha (east elevation) [1]

Two high pressure condensate pumps were located in the gas compression module, with one operating and one on standby. As well as exporting gas to MCP-01, the processed gas was also used to fuel the main power generators that supplied the electrical power to the platform.

On July 6, 1988 a major disaster took the lives of 165 persons on board the Piper Alpha platform and 2 of the rescue crew from the Sandhaven, a standby vessel for a mobile offshore drilling unit (MODU) nearby. To this day, the incident represents the worst offshore oil industry disaster ever, and has become an industry changing watershed event.

2. Sequence of Events

A major turnaround and upgrade modifications commenced in July 1988, including the replacement of the gas conservation module. Occidental decided not to shut down the platform during the turnaround as originally planned and the Piper Alpha reverted to its initial Phase 1 mode of operation, i.e. operating without a gas conservation module.

Since the detailed sequence of events has been extensively reported, most notably in the report of the subsequent public inquiry (known as the Cullen Report) [1], the following is a brief summary of the major contributory events:

On the morning of July 6, 1988, pressure safety valve (PSV #504) on one of two Condensate Pumps (pump A) was removed for routine maintenance. The open condensate pipe was temporarily sealed with a blind flange that was either not rated for the normal pressure or was not tightened properly. Because the routine maintenance work was not completed by the end of day shift (6:00 pm), the engineer noted on the work permit that the job was incomplete and removed the permit from the job site to the control room, but neglected to inform the night shift.

The diesel and electric fire pumps were on manual control during the evening of July 6 as divers were working on the platform jacket structure. At 9:45 pm, Condensate Pump B stopped due to a suspected hydrates blockage and could not be restarted. Power supply for construction work depended on Pump B, and there were only a few minutes to get the pump back on line. Therefore, a decision was made by the night shift crew to start Pump A, as no one knew the actual status of the permit and the PSV location was not visible from the pump.

At 9:55 pm gas leaked from the blind flange at high pressure and activated numerous gas alarms. Before anyone could act, the gas cloud exploded destroying the firewalls of the gas compression module. Missiles and debris from the explosion ruptured condensate and oil pipework, creating secondary fires that escalated generating thick black smoke and toxic fumes. An emergency shutdown (ESD) was activated to immediately stop production and processing from Piper's wells before the damaged control room was abandoned. Unfortunately, the public address system likely failed in the initial explosion, and several managers and supervisors had already been killed during the initial blast. No orders were given to evacuate the platform.

In an emergency the normal procedure was to muster at the lifeboat stations, but since they were exposed to the large fires, the crew had no option but to muster in the fireproof living quarters. Although the living quarters were situated under the helideck, smoke from the fires prevented helicopters from landing. The ingress of combustion products (smoke, carbon monoxide, carbon dioxide) into the living quarters through doors and penetrations caused the situation to deteriorate. With the chain of command having failed, no instructions were given to the crew.

However, on their own initiative, two men tried to reach the fire pumps that were on local manual start, but were apparently killed doing so.

The Claymore and Tartan platforms continued to pump oil despite being aware of an emergency, as they had no voice communication with Piper Alpha and neither platform manager had permission to shut down from his respective leadership. The back pressure from this pumping continued to feed the fires through ruptured pipework, although the fires would have burnt out had Claymore and Tartan shut down when they first became aware of Piper's emergency.

At 10:20 pm the Tartan gas riser failed catastrophically, releasing in excess of 15 tonnes of gas per second at 120 bar pressure. While most of the Piper employees were still sheltering in the smoke-filled living quarters, some desperately tried to jump into the sea from various levels of the platform including the helideck.

The fire from the Tartan gas line was dwarfed at 10:50 pm when the gas riser for MCP-01 platform also exploded catastrophically. With flames in excess of 300 ft high, the extreme heat started to melt the surrounding structure. Two crewmen on a fast rescue boat and some of the jumpers they had rescued were killed in the explosion. At this point the Claymore platform stopped pumping oil.

At 11:20 pm the Claymore gas riser also failed, and 30 minutes later the living quarters and the power generation/ utilities module collapsed into the sea and sank. Within another hour the remaining platform structure collapsed with the exception of the wells module.

Later that year, the living quarters were recovered from the sea bed, and the bodies of 87 men were found inside. Thirty bodies were never found. There were only 61 survivors, mainly those who took matters into their own hands and jumped into the sea.

3. The Aftermath

A public inquiry was convened by the UK government under the chairmanship of The Hon. Lord W. Douglas Cullen, a senior member of the Scottish judiciary. He was ably assisted by three assessors, most notably an experienced process safety expert, Professor Frank Lees (of *"Loss Prevention in the Process Industries"* [2] fame). The inquiry was conducted in two parts to determine (i) the circumstances and causes of the disaster, and (ii) examine means to prevent future major accidents.

The oil and gas production companies' trade association, the UK Offshore Operators Association (UKOOA, later renamed Oil & Gas UK), provided 34 expert witnesses, including this author, to the second part of the inquiry. The author's evidence focused on process safety management. More specifically, with a regulator proposing mandatory installation of subsea isolation valves (SSIVs) on all platform/pipeline interfaces, the author presented a quantitative risk assessment (QRA) approach to demonstrate that SSIVs did not universally provide the most effective risk reduction and that other alternative safety measures should be considered.

The subsequent public inquiry report (known as the Cullen Report) [1] determined that the initial cause was due to a leak of gas condensate as a result of pressurizing the pipework where the PSV had been removed. Lord Cullen was highly critical of the Operator's inadequate maintenance and safety practices. He also determined that there were a number of contributory factors, including a breakdown in Piper's chain of command and communication to the crew, failure of firewalls, and absence of blast walls when Piper was upgraded to handle gas, and the continued operation of other platforms that were not shut down when Piper raised the initial emergency.

Even before these findings were published, most prudent offshore Operators conducted internal assessments of their own platforms and management systems as these causes began to emerge from the evidence heard by the public inquiry. Some of those assessments included evaluations of work permit systems; pipeline ESD systems; ingress of smoke into living quarters; evacuation, escape and rescue systems; and developments in Formal Safety Assessment (FSA) methodologies for offshore platforms [3, 4].

The Cullen Report made 106 far-reaching recommendations, which were all accepted by the North Sea industry, and set new standards for the management of safety on offshore oil and gas platforms that are still in force today. Nearly half (48) were directed at operating companies, while, with the exception of one for the Standby Ship Owners Association, the remainder (57) fell to the regulator.

In fact, Lord Cullen recommended that a specialist division of the UK Health and Safety Executive (HSE) should oversee health and safety offshore, in place of the Department of Energy (DEn), as having both production and safety overseen by the same agency provided a conflict of interest. The inquiry directed that the HSE should be well-resourced (in both competencies and numbers) to fulfill its duties, unlike its predecessor that was under-resourced. Furthermore, the existing prescriptive regulatory framework was replaced with a goal-setting framework driving toward continuous improvement. Central to this framework was a Safety Case that Operators/owners of every fixed and mobile installation operating in UK waters are required to submit and receive acceptance from the regulator before proceeding with activities.

Subsequently new regulations were promulgated in the UK sector of the North Sea mandating the Cullen Report recommendations. Similar regulations have been adopted in other offshore areas such as Norway and Australia. Even before some regulations had been finalized, the UK industry acted immediately to carry out their recommendations, which expedited learning of many of the lessons below.

Finally and often overlooked in the aftermath of the world's worst offshore disaster, Lord Cullen reassured stakeholders (industry, government, regulator, public) that offshore oil and gas operations could be conducted safely if a logical, goal-oriented approach was implemented, together with effective application of technology and rigorous inspection practices.

4. Lessons Learned

Significant improvements have been made in the UK offshore industry since Piper Alpha, including improvements in both hardware and safety culture. The following details some of the lessons learned:

4.1 Safety Case

The Safety Case regulations [5] came into force in 1992, and by late 1993 Safety Cases for every platform and rig had been submitted to the HSE, who accepted the last of them by the end of 1995. The Safety Case must describe and justify the platform/rig design, the inherent hazards and residual risk (ALARP[1]), and the means of managing the residual risk to ensure health and safety. These means are known as the Safety Management System (SMS) which embraces the elements of both process safety and occupational safety. The Safety Case must be maintained up to date over the lifecycle of the platform/rig, and is the basis for periodic audits. The regulations were revised in 2005 to reflect the experience gained from the early Safety Cases.

The Safety Case is backed up by a regulator with adequate and competent resources. The former regulator's inspections had been superficial and had not revealed latent deficiencies in safety management when Piper's work permit system was audited in 1988 before the incident. The HSE's inspections by contrast are now thorough and conducted in depth by knowledgeable and experienced inspectors.

The Safety Case has forced a more robust and systematic assessment of major hazards and has driven risk reduction much more so than OSHA PSM [6], EPA RMP [7], MMS SEMP [8], or what BSEE SEMS [9] is likely to achieve. This has probably reduced the likelihood of a similar incident occurring in UK waters by several orders of magnitude. It is a best in class regulatory regime second to none, a lesson for other regimes around the world, and a fitting legacy in memory of those who died on Piper Alpha.

While the safety case ***regime*** is best in class, the practice of developing the Safety Case is sometimes not without fault. Some operating companies have allowed the Safety Case to become a paper exercise produced by consultants. It has still driven risk reduction, but much of the potential learning and understanding has been lost as a result of the operating companies employing consultants to write the document instead of doing it themselves. Some of the consultants in turn have developed voluminous documents that, at the very least, would benefit from a concise but comprehensive summary manual. Such a manual could be read in a day by a new manager or employee, who would then have a good understanding of the platform design, its residual risk and how the SMS manages those risks. Thus the Safety Case could become an excellent training tool.

4.2 Forthwith Studies

A number of Cullen's recommendations are related to inherently safer design (ISD), including the mandatory use of:

[1] As low as reasonably practicable.

- Systematic analysis of fire and explosion hazards,
- Analysis of smoke and gas ingress into living quarters, and the requirement for a temporary (safe) refuge capable of surviving the initial fire/explosion and any escalation for a reasonable duration to permit evacuation and escape,
- Analysis of the vulnerability of safety critical equipment or elements (SCE), such as emergency shutdown valves (ESDV),
- Analysis of evacuation, escape and rescue in the event of major incidents.

The sum of these recommendations has reduced residual risk of the design through use of an ISD approach. The residual risk then has to be managed by the SMS. These four analyses, known as the *Forthwith Studies*, were subsequently promulgated in detailed goal-setting regulations [10], which set out the objectives that must be achieved. This approach allows flexibility in the choice of methodology or equipment that may be used by Operators to meet the requirements.

The Forthwith Studies have been responsible for a fundamental change in platform design, and to a lesser extent in drilling rig design. Early platform design, even in the North Sea, owed allegiance to designs used in the less hostile and shallower environment of the Gulf of Mexico (GoM). As such the platform plot plan was essentially square, and, although the modules were normally organized so that the most hazardous operations were separated from the living areas, the Piper conversion to gas processing ruined this safety concept. The end result brought together sensitive areas with gas compression, probably the highest risk, next to the control room. As a result of the Forthwith Studies, future production platform designs invariably have a more exaggerated rectangular plot plan/layout than the Piper Alpha's essentially square layout. This increases the separation of accommodation from hazardous modules.

The layout of Piper Bravo, the platform designed for Elf Enterprise Caledonia Ltd. to replace Piper Alpha when they took over the license block from Occidental, was rectangular (fig. 5). Piper Bravo set the standard for a new generation of offshore platforms with safety as an inherent feature [11].

An alternative to the rectangular layout is the provision of a separate bridge-linked accommodation platform, which creates an even greater separation to hazardous modules. However it is sometimes difficult to justify the extra cost of a second platform for living quarters, especially in lower production oil/gas fields and/or greater water depths.

Other design lessons learned were:

- Blast walls (as well as firewalls) can aid separation of sensitive modules, unlike Piper where firewalls were not upgraded to blast walls, and were destroyed in the first gas explosion.
- The design of the temporary refuge, invariably the living quarters, should be protected from fire and explosion, pressurized, and specified with an HVAC system equipped with automated fire dampers to prevent the ingress of smoke and gas.
- Unlike Piper's control center and radio room that were devastated in the first explosion, control rooms (process and emergency) should be remote from hazardous modules.

- It is important to avoid common mode failure of communication systems (radio, public address). Piper's systems were both knocked out by loss of the control room, radio room, and uninterruptible power supply (UPS).
- Emergency Shutdown Valves (ESDVs) are necessary to protect against large hazardous inventories in pipelines, and should be protected from fire and explosion by enclosures.
- Exposure to huge inventories in hydrocarbon pipelines can be limited by using 'not normally manned' platforms as collector hubs. The Forties Unity riser platform is a good example of this concept.
- Diesel should not be stored above the process plant. This was responsible for some of the fire escalation on Piper.
- Fire pumps must have spatial separation and remote start, unlike Piper where the fire pumps were in the same location and on local start.

Figure 5 clearly shows the rectangular design of the Piper Bravo supporting jacket and each level of the platform. Accommodation is located as far as possible from the hazardous modules (wells/drilling, gas compression, oil/gas separation). The non-hazardous utilities module separates the accommodation and production modules.

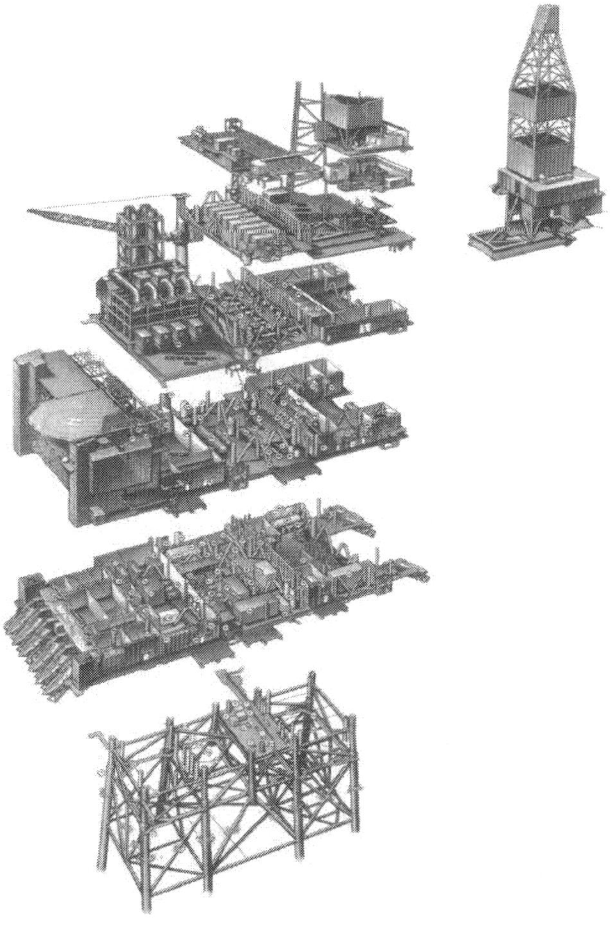

Figure 5: Piper Bravo Layout [12]

The accommodation is designed as the temporary refuge with at least 2 hours of fire/explosion protection, and has shielded escape routes (an industry first) leading to free-fall lifeboats.

Walls and decks separating each module were analyzed and designed to contain any blast within the immediate area. In an emergency, the process equipment has a rapid blowdown capability to remove hydrocarbons. Diesel firewater pumps supplying the deluge systems are located in the wellhead and utilities areas.

Piper Bravo has 6 risers under the wells module (remote from accommodation) connecting it to subsea oil and gas pipelines. Each pipeline has duplicated ESDVs (topsides and subsea), closing automatically on loss of hydraulic pressure.

4.3 Safety Management System (SMS)

The ISD approach recommended by the inquiry indirectly places less reliance on the human than was the case before the Piper Alpha disaster. However the human element is still important, especially in delivering an effective SMS performance. Indeed the Cullen report recognized the need for better leadership competencies for Offshore Installation Managers (OIMs), especially during emergencies. It is especially important that OIMs, supervisors and operators are empowered to shut down operations in the event of significant safety issues, unlike the crews of the Tartan and Claymore platforms who did not believe they had authority to stop production, even though they could see that Piper was burning.

Other important human factors offshore include correctly and diligently implementing operating and maintenance procedures, permits to work, management of change, and emergency preparedness, i.e., many of the traditional elements of process safety. However, the regulator also expects the SMS to address some of the additional process safety elements in Risk-Based Process Safety [13].

Workforce involvement is an important facet of the safety management system, and was mandated in 1989 through regulation [14]. This has been a major driver of improvements in offshore health and safety, and safety representatives have made a valuable contribution and are actively engaged with the industry's "Step Change in Safety" [15] campaign to share best practices. This initiative aims to continue the post-Piper Alpha improvements, avert complacency, and drive even greater improvement through employee participation.

Some of the specific operations and maintenance lessons learned include:

- There must be written operating procedures, unlike Piper's Phase 1 operation that had only been implemented once since 1979.
- Temporary promotion is undesirable during critical operations, such as a non-routine operating mode. The Operations Superintendent, deputy Operations Superintendent, and Lead Production Operator on Piper were all temporary promotions during a non-routine operation.
- Training of contractors is essential. It had been the first day on the platform for the Production Operator (a contractor) on the gas plant, who was on his own without any operating procedures.
- Work permit systems must be foolproof. If two or more jobs are being performed on the same equipment or in the same area, the chance for confusion as to which permit applies exists. Multiple work permits must be properly controlled. The permit to remove the PSV was suspended after PSV removal and had been removed from the job site without any communication to others. Work on the second permit had not commenced.
- Shift logs and handover must be comprehensive. The night shift didn't know the PSV had been removed or that the condensate pump had been electrically isolated.
- Blanks must be rated for the full service pressure and fully bolted. It is unlikely that this was done properly.
- Fire pumps, especially diesel emergency pumps, must always be available and accessible. Both fire pumps on Piper were on local manual control due to diving operations near their suctions. An earlier audit recommendation had not been implemented to keep the

fire pumps in automatic mode if divers were not working directly in the vicinity of their intakes.

- Safety equipment needs maintenance, too. Only 2 life rafts were launched - neither inflated.

4.4 Safety Culture

Culture is the result of everything that happened or failed to happen. The quality of leadership and commitment to safety can drive or limit the safety culture of an organization. The essential features of a sound safety culture include enforcing high standards of operation while maintaining a sense of vulnerability, open and effective communication, and timely response to known issues and workforce concerns. Managers must develop and sustain a culture that embraces both process safety and occupational safety.

There are a number of safety culture lessons from the Piper Alpha:

- Safety yields greater production in the long term. The operator tried to keep the gas plant running after a high level alarm, and didn't know there was no trip in the Phase 1 mode.
- Safety must not be compromised by production or cost priorities. The Claymore and Tartan platforms continued to pump hydrocarbons, as the OIMs did not have permission to shut down due to the huge cost, since it would take several days to restart production with substantial financial consequence. If they had shut down after the initial blast, the rupture of the gas risers responsible for Piper's destruction would likely have been prevented enabling the crew's evacuation.
- Procedures are of no use if they are not followed. Adequate safety policies and procedures were in place, but their implementation was deficient. The work permit system was often knowingly and flagrantly disregarded, e.g. missing signatures and gas tests, and there was failure to inspect jobsites or return permits to the responsible person at end of shift. These problems were not reported, and management did not verify operation of the system.
- Periodic safety audits must be performed with rigor to identify management system weaknesses before a major incident occurs. Piper's audits were superficial, few if any issues were found, and not all were reported.
- Known problems with safety systems must be resolved in a timely manner. Problems with the firewater deluge system had been known since 1984, and 50% of firewater sprinklers in the gas compression module were corroded and blocked. Even if firewater had been available, its effectiveness would have been limited. Management also knew that the Piper structure had no fireproofing and that structural integrity could be lost in minutes in the event of a large hydrocarbon fire.

It is clear from these examples that Piper management adopted a superficial attitude to risk management and did not personally probe problems or verify compliance to company practices. Other Piper cultural issues are described in a CCPS publication [16].

4.5 Emergency Response

It is unclear if there was sufficient time for a more effective emergency evacuation. Most of the personnel with authority to order evacuation were killed in the first blast which destroyed the control room. The second explosion occurred twenty two minutes later. By the time rescue helicopters reached the scene, flames over 300 feet in height prevented safe approach.

The Piper area of the North Sea had a dedicated firefighting and rescue semi-submersible vessel, the Tharos, which included an extending gangway bridge. By chance, the Tharos was moored in the vicinity of Piper, but its arrival at Piper was delayed by 30 minutes due to its thrusters cutting out. Then Tharos's fire monitors did not discharge water until 14 minutes after the fire pump was started, due to too many monitors being opened. Tharos attempted to extend its gangway to Piper but the slow moving mechanism failed, probably due to the high temperatures.

The massive fire, high temperatures and thick smoke prevented escape by Piper's own lifesaving equipment (lifeboats, life rafts). Eventually, due in part to spray from Tharos's monitors, some men climbed down the platform and jumped into the sea. Most survived, but many suffered horrific burns from burning oil on the sea. Other men trapped higher up jumped from the helideck at the 174 feet level, but few survived the fall. Those who did survive jumping into the sea were picked up by fast rescue craft and Piper's dedicated standby safety boat, Silver Pit, a converted trawler. The Silver Pit, whose searchlight didn't work, suffered significant damage from the heat and missiles, but picked up 37 of the 59 survivors from the burning sea.

The above description reveals a number of specific emergency response lessons learned including:

- Management and platform crews need to be prepared for a major emergency. Piper's emergency training was cursory, and needed to be much more thorough. Management failed to ensure that emergency training was being provided as intended.
- Emergency plans need to be exercised and understood by all concerned. A critique of drills and exercises should be used as a basis for continuous improvement. Emergency drills need to include major incident scenarios in which the chain of command may be impaired. The Tharos could have responded sooner if the operation of its fire monitors had been better understood.
- Emergency equipment needs to be designed for the emergency scenarios under which it may be required to function and meet its design intent. The Tharos's gangway was too slow and not designed to operate under intense heat. Like many standby safety boats at the time, the Silver Pit was essentially unsuitable for the purpose of rescuing survivors.
- More than one route to helicopters and lifeboats must be present at any given time to ensure evacuation of the platform in a crisis situation. The lifeboats and other escape routes were inaccessible due to the fires.
- Emergency equipment needs to be maintained and periodically tested to assure it is reliable and fit for purpose. Two life rafts were launched but neither inflated.

As a result of these lessons, many platforms today have a greater number of self-inflating life-rafts, ladders, knotted ropes, lifebuoys and other escape devices in diverse locations on all deck levels. However these systems are only backup to the main escape method, totally enclosed lifeboats, with at least two protected routes to the lifeboats from anywhere on the platform.

Some Operators in the North Sea have opted for free-fall lifeboats rather than conventional davit launched boats. These free-fall boats are located in a heat shield, designed to withstand temperatures as high as 1000°C (1832°F), and can be launched from inside the boat when everyone is strapped in. The lifeboat hits the sea at an angle such that its momentum takes it underwater away from the platform until it emerges about 150 feet clear of the platform structure (so that it cannot strike the structure due to wave action) or any fires at sea level. Piper Bravo was the first platform in the UK sector equipped with free-fall lifeboats.

5. Conclusions

Twenty five years ago a truly horrific catastrophic disaster took many lives in the space of less than three hours, and to this day most of the relatively few survivors still bear mental scars, not to mention horrendous burns. Today all that is visible of what in 1988 was the North Sea's largest and oldest platform is a wreck buoy. The Piper Alpha disaster has become a watershed event that provided the impetus for change in the management of offshore safety, and justly changed the industry in the North Sea and further afield.

The Piper Alpha tragedy is rich in lessons involving many aspects of engineering design, risk management, management systems, human factors, process safety and integrity management. Many of these lessons are as equally applicable to the onshore refining and chemical industries as they are to offshore oil and gas drilling and production. The North Sea industry embraced many of the lessons and Public Inquiry recommendations in the immediate aftermath of the disaster, and significant improvements have been achieved. There were so many lessons to learn that the author may have missed some, while the industry, particularly outside the North Sea, has forgotten others that are being re-learned the hard way. Figure 6 lists some of the major incidents that have occurred around the world since Piper, and that share some similar lessons.

The tremendous foresight of Lord Cullen and his Public Inquiry assessors paved the way for a new rigorous goal-setting Safety Case regime that is second to none, and has driven risk reduction in the North Sea by several orders of magnitude. It is a lesson for other regulatory regimes around the world, but alas too few have heeded these lessons and are still committed to outdated prescriptive approaches.

Nevertheless, even in the North Sea and elsewhere around the world, incidents on offshore production platforms and onshore refineries and chemical plants continue to occur that contain a similar cocktail of leadership and organizational failures, poor behaviors and operating discipline, deficient asset integrity, and an absence of coherent safety management. Those incidents were all foreseeable and preventable. We can never become complacent.

Similar weaknesses in leadership, human factors, culture, and safety management were apparent from the author's recent visits to drilling rigs [17, 18] around the world in the aftermath of the Deepwater Horizon incident [19]. Unless the drilling industry changes its focus on monitoring individual worker injuries, while ignoring warning signs of process safety problems that have the potential to lead to major incidents, then the industry is likely destined for more disasters. While nothing in the drilling industry may ever match the enormity of Piper, every life lost is unacceptable.

YEAR	MAJOR INCIDENT
1989	Exxon Valdez, Prince William Sound, Alaska
1989	Phillips Polymer Plant, Pasadena TX
1992	Fuji Sekiyu Refinery, Sodegaura, Japan
1993	Oil Platform, Lake Maracaibo, Venezuela
1994	Terra Fertilizer Plant, Port Neal IO
1998	Esso Gas Plant, Longford, Australia
2000	KNPC Refinery, Mina Al-Ahmadi, Kuwait
2001	Total AZF Fertilizer Plant, Toulouse, France
2001	Petrobras P-36 Semi-Submersible, Brazil
2003	PetroChina Gas Field, Chongqing, China
2004	Sonatrach LNG Plant, Skikda, Algeria
2004	West Pharmaceutical Services Plant, Kinston NC
2005	BP Refinery, Texas City TX
2005	ONGC Mumbai High North Platform, India
2005	Total/Texaco Oil Storage Depot, Buncefield, UK
2007	Formosa Plastics PVC Plant, Illiopolis IL
2008	Imperial Sugar Plant, Port Wentworth GA
2010	Transocean Deepwater Horizon Drilling Rig, GoM
2012	Paraguana Refinery, Falcon State, Venezuela
2013	West Fertilizer Storage Facility, West TX

Figure 6: Selected Major Events from Piper to Present

While safety on production platforms has undoubtedly improved in the last 25 years, we owe it to the memory of those who lost their lives in the Piper incident to continue to make safety absolutely the first priority. A disaster like Piper must never be allowed to happen again. That would be a fitting legacy in memory of those who died.

6. References

[1] The Hon. Lord Cullen, *"The Public Inquiry into the Piper Alpha Disaster,"* HMSO, London, November 1990.

[2] Lees, Frank P., *"Loss Prevention in the Process Industries, Hazard Identification, Assessment and Control,"* Elsevier, London, 1979 (and subsequent editions)

[3] Broadribb, Michael P., *"The Application of Formal Safety Assessment to a New Installation,"* paper presented at Risk Analysis in the Offshore Industry Workshop, Aberdeen, UK, October 1990.

[4] Broadribb, Michael P., *"The Application of Formal Safety Assessment to an Existing Offshore Installation,"* International Conference proceedings - Management and Engineering of Fire Safety and Loss Prevention, Aberdeen, UK, February 1991.

[5] HM Government, *"Offshore Installations (Safety Case) Regulations 1992,"* Statutory Instruments, 1992 No.2885, Health and Safety, UK, 1992.

[6] US Department of Labor, Occupational Safety and Health Administration, *"Process Safety Management of Highly Hazardous Chemicals,"* 29 CFR Part 1910.119, 1992.

[7] US Environmental Protection Agency, *"Accidental Release Prevention Requirements: Risk Management Programs Under Clean Air Act Section 112(r)(7),"* 40 CFR Part 68, 1996.

[8] US Department of the Interior, Minerals Management Service, "*Safety & Environmental Management Program",* voluntary program based upon American Petroleum Institute *"Recommended Practice for Development of a Safety and Environmental Management Program (SEMP) for Offshore Operations and Facilities,"* API RP 75.

[9] US Department of the Interior, Bureau of Safety and Environmental Enforcement, *"Safety and Environmental Management Systems,"* 30 CFR Part 250 Subpart S, 2010.

[10] HM Government, *"The Offshore Installations (Prevention of Fire and Explosion, and Emergency Response) Regulations 1995,"* Statutory Instruments, 1995 No.743, Health and Safety, UK, 1995.

[11] *"New Piper B Reflects Piper A Safety Lessons,"* Oil & Gas Journal, Vol.91 No.8, Feb. 23, 1993.

[12] Talisman Energy (UK) Ltd., *"Piper 'B', Infrastructure Code of Practice"*, Aberdeen, UK, December 2004

[13] CCPS, *"Guidelines for Risk Based Process Safety,"* New York, 2007

[14] HM Government, *"The Offshore Installations (Safety Representatives and Safety Committees) Regulations 1989,"* Statutory Instruments, 1989 No.971, Health and Safety, UK, 1989.

[15] Step Change in Safety, *"Strategic Plan 2010-2015,"* Aberdeen, UK, 2010. www.stepchangeinsafety.net (accessed Dec. 21, 2013)

[16] CCPS, *"Building Process Safety Culture: Tools to Enhance Process Safety Performance; Piper Alpha Case History,"* New York, 2005.

[17] Broadribb, Michael P., *"What Have We REALLY Learned Bit by Bit?"* presentation at the Oil & Gas UK Piper 25 Conference, Aberdeen, UK, June 2013.

[18] Broadribb, Michael P., *"Well, Well! Are We Drilling in the Right Direction?,"* paper presented at the 5th CCPS Latin America Conference on Process Safety, Cartagena de Indias, Colombia, August 2013.

[19] National Commission on the BP Deepwater Horizon Oil Spill and Offshore Drilling, *"Deep Water; The Gulf Oil Disaster and Future of Offshore Drilling,"* January 2011.

Lac-Mégantic Accident: What we learned.

Jean-Paul Lacoursière, Eng.
Associate Professor
University of Sherbrooke
2500 boul. de l'Université, Sherbrooke, QC J1K 2R1, Canada
jpla@sympatico.ca

Prepared for Presentation at
American Institute of Chemical Engineers
2014 Spring Meeting
10th Global Congress on Process Safety
New Orleans, LA
March 30 – April 2, 2014

UNPUBLISHED

Lac-Mégantic Accident: What we learned

Jean-Paul Lacoursière. Eng.
Associate Professor
Université de Sherbrooke
2500, boul. de l'Université, Sherbrooke, QC J1K 2R1, Canada
jpla@sympatico.ca

Keywords: Lac-Megantic accident; Train derailment; Bakken Crude; DOT-111; Hazardous good classification

Abstract

A tragic train derailment in Lac-Megantic, a small Quebec community caused 47 fatalities, the destruction of part of the town and huge cleaning costs.

This paper reviews information available at the time of publication of this paper and extracts lessons that could be used by the chemical process industry.

1. Introduction

On July 6, 2013 at 1:14 A.M., a catastrophic railroad accident occurred in Lac-Mégantic, Quebec, Canada when an unattended freight train stationed on the main line and containing hazardous materials (Bakken Crudes from North Dakota) rolled down a descending grade and subsequently derailed. The derailment resulted in multiple explosions and fires (Figure 1), which caused the confirmed death of forty-two people and presumed death of five more, extensive damage to the town center, clean-up costs, and the evacuation of 2000 people from the surrounding area.

Lac-Mégantic is located 250 km from Montreal and at about 48 km from the State of Maine border.

The Lac-Mégantic derailment pushed the issue of rail safety into the spotlight. This accident triggered a debate in Canada and the United States on the safety of hazardous material transport.

The multiple layers to the July 6 catastrophe attracted interest with its gripping stories of survival and loss, the vast environmental consequences and the stunning images of the devastation. It also launched new debates about the continent's boom in oil-by-rail traffic.

The Lac-Megantic disaster was one which caused everyone who lives in a small Canadian city, or town that has freight trains rumbling through it, to worry. People say "That could happen here someday".

Credit: David Charron

Figure 1 Downtown Fire Viewed from Church Porch

In Lac-Megantic, the mayor believes the crash that struck her community of 6,000 has potential repercussions for people across Canada and beyond. "I think this tragedy allowed the whole world to reflect on rail safety," Colette Roy-Laroche told The Canadian Press. Roy-Laroche hopes all the attention generated by the disaster will inspire governments and companies around the globe to push for improvements to train security.

2. Facts

2.1 The accident

The Transportation Safety Board of Canada on July 19, 2013 issued a Rail Safety Advisory Letter that summarizes the incident as follows.

At about 22:45 Eastern Daylight Time (EDT) on 05 July 2013, Montreal Maine & Atlantic (MMA) freight train MMA 2 (the train) was proceeding eastward on the MMA Sherbrooke Subdivision, enroute from Montreal (QC) towards Saint John (NB). The train was 4701 feet long and weighed 10,287 tons. It was comprised of 5 head-end locomotives, a VB car used to house the locomotive control equipment necessary for MMA's single engineer train operation, 1 loaded box car used as a buffer followed by 72 non-pressure dangerous goods tank cars (DOT-111) loaded with petroleum crude oil (Class 3, UN 1267). At approximately 23:00, the train stopped at the designated MMA crew change point at Mile 7.40 near Nantes, Quebec. The single operator secured the train and departed for the evening leaving the lead locomotive unlocked, running and the train unattended on mainline track with a descending grade of 1.2%.

At about 23:50, a local resident reported a fire on the lead locomotive (MMA 5017) to the 911 emergency call centre. (Figure 2) Subsequently, the local fire department responded along with another MMA employee. At about midnight, similar with established operating practice, emergency shutdown procedures were initiated on the lead locomotive and the fire was extinguished. After extinguishing the fire, the second MMA employee and the fire department departed the site again leaving the train unattended.

Credit: Nancy Cameron

Figure 2 Locomotive Fire, in Nantes, Quebec

At approximately 1:00 a.m. the next day (the early morning of July 6th) it appears that the train began rolling and picking up speed down the descending grade toward the town of Lac-Mégantic, Québec. Near the center of town, the box car and 63 of the loaded tank cars derailed. The locomotives, which separated from the train, traveled an additional 1/2 mile before coming to a stop. A number of derailed tank cars released product resulting in multiple BLEVEs, (Figure 3) explosions and subsequent fires. (Figure 4)

Credit: David Charron

Figure 3 BLEVE a BAKEN Crude Tank Car

The BLEVEs instantly incinerated the patrons of a local pub (Musi-Café) and nearby residents. There were 42 confirmed fatalities and 5 persons are still missing. There was also extensive damage to the town, and approximately 2,000 people were evacuated from the surrounding area. The fire lasted for 40 hours.

Credit PC/ M. Charron

Figure 4 Tank Cars in Fire

Canadian Prime Minister Stephen Harper described the site as a war zone. (Figure 5)

Credit PC/ M. Charron

Figure 5 Building Destruction

While the investigation is ongoing and the Transportation Safety Board of Canada has not reached any final conclusions, it has made a determination that the braking force applied to the train was insufficient to hold it on the 1.2-percent descending slope between Nantes and Lac-Mégantic.

2.2 Detailed Sequence of Events

The following provides a detailed sequence of the events that lead to the accident and the resulting response of emergency responders and public authorities.

FRIDAY, July 5

- **11:25 p.m.:** An engineer from the Montreal, Maine & Atlantic Railway (MMA) parks a train carrying 72 tankers, each carrying thousands of litres of crude oil (roughly 100,000 litres), and five locomotives in Nantes, approximately 11 kilometres outside of Lac-Mégantic. According to MMA, he had stopped for a crew change and then retired to a nearby motel for the night.
- **11:30 p.m.:** A resident in Nantes calls 911 after seeing a parked locomotive on fire between Nantes and Lac-Mégantic.
- **11:42 p.m.:** 12 firefighters arrive on the scene. Nobody from MMA was there.

SATURDAY, July 6

- **12:12 a.m.:** Fire in locomotive is extinguished.
- **12:13 to 12:15 a.m.:** Two MMA employees arrive on the scene. Firefighters leave soon after establishing that the situation is under control.
- **1:14 a.m.:** The first explosion in Lac-Mégantic is reported, followed by at least two others. Initial reports suggest 30 buildings are destroyed. Much of downtown is flooded with crude oil and fire. Patrons of a crowded bar flee. Many remain unaccounted for.
- **1:15 a.m. to 4 a.m.:** Explosions continue to rock the town as tanker cars containing oil ignite. Firefighters and volunteers are alerted and begin assessing the scene. They receive backup from Montreal and surrounding areas. Firefighters from Nantes are alerted, and firefighters come from as far as Farmington, ME., to help manage the crisis. About 1,000 people are removed from the area because of toxic fumes and dangerous particles.
- **9 a.m. (approximately):** The Canadian Red Cross begins setting up a temporary shelter for evacuees at the Polyvalente Montignac — a local high school. Later in the day, a spokesperson confirms that 250 beds are available but the waiting list exceeds 500. A Facebook group is set up to track people who are missing and those who have been accounted for. As of early Sunday morning, there are more than 15,000 members active on the group, a reported 199 people missing and 158 found.
- **1 p.m. (approximately):** A second evacuation order is issued for another 1,000 people from a community downwind from the blast site. An analysis by the Sécurité civile de l'Estrie detects dangerous levels of toxic fumes.
- **2 p.m.:** Firefighters contain the blaze, preventing it from spreading further. Denis Lauzon, spokesperson for the volunteer fire department, said later: "We've stopped the fire, the (rest of) downtown is safe."
- **3:30 p.m.:** The Sûreté du Québec (SQ) holds its first news conference and confirms that one person is confirmed dead. Police say they expect more casualties, but won't speculate about how many people are still missing. Reports suggested upwards of 60 people are still unaccounted for.
- **4 p.m.:** Quebec Premier Pauline Marois surveys the scene and holds a news conference. She says she is profoundly saddened by the devastation in Lac-Mégantic.
- **10 p.m.:** A boil water advisory is issued to residents of Lac-Mégantic and Mayor Colette Roy-Laroche warns municipalities downstream from the blast site to closely monitor their water quality.

SUNDAY, July 7

- **9 a.m.:** Authorities announce that two more people are confirmed dead and say the death toll is expected to rise.
- **Noon:** Authorities announce that an additional two bodies have been recovered, raising the official death toll to five. They also give the first official count of missing: 40 people are unaccounted for. The perimeter around the disaster site is reduced, allowing some people to return home.
- **3 p.m.:** The Transportation Safety Board of Canada (TSB) posts on Twitter saying it has recovered the train's event recorder.
- **4:30 p.m.:** Prime Minister Stephen Harper holds a news conference after touring the devastation. "It's like a war zone," he says.
- **5 p.m.:** Montreal, Maine & Atlantic Railway issues statement saying company officials haven't been able to conduct their own investigation and pledge support to the community.
- **5 p.m.:** Fire extinguished with the help of foam from Ultramar Refinery (Valero) 40 hours after derailment.
- **7 p.m.:** Authorities hold a news conference and explain that investigators still haven't been able to tour the full blast zone because many areas are still too hot. The Transportation Safety Board has brought in nine investigators and the coroner's office has a special forensic identification team on site to help identify remains.

MONDAY, July 8

- **7 a.m.:** The bodies of two of the five people confirmed dead have been taken to Montreal but identification may prove a lengthy process, says Geneviève Guilbault, a spokesperson for the Quebec coroner's office. The number of people missing remains at about 40, says Sgt. Benoit Richard of the Sûreté du Québec.
- **8:30 a.m.:** Government officials ask area residents to "be patient" as Environment Quebec workers try to determine whether it is safe for people evacuated from the Fatima sector to return home. "We're doing everything we can to get them home as quickly as possible," says the Lac-Mégantic mayor. A boil-water advisory remains in effect, as a precaution.
- **5 p.m.:** At a press conference, the SQ announces that the official death toll has risen to 13, and that the number of missing now officially stands at 50. Coroner's office spokesperson Geneviève Guilbeault urges families of the missing to present themselves at their makeshift office and bring personal items that could contain the missing person's DNA, such as baseball caps, toothbrushes, razors and combs.

TUESDAY, July 9

- **8:30 a.m.:** The mayor announces that about 1200 residents who had been evacuated from their homes are allowed to return, beginning immediately. The homes of about 600 people, near the epicentre, remain off-limits. The boil-water advisory remains in effect.
- **10:15 a.m.:** Just prior to the catastrophic derailment, the train had been rolling — unattended — down a 1.2-degree grade and picking up speed, head investigator Donald

Ross of the TSB says. When the runaway train reached a curve in the heart of Lac-Mégantic, he adds, it was going faster than recommended for that turn. That's where the tanker cars tumbled off the track. "We're examining all the brake systems," Ross says. "But until we had a chance to do our work, were not going to comment on what was or wasn't working."

- **5 p.m.:** At a press conference, the SQ announces that two more bodies have been recovered, bringing the total number of confirmed deaths to 15. Fifty people in total are still unaccounted for.

WEDNESDAY, July 10

- **8:30 a.m.:** Many residents of the Lac-Mégantic area are back at work. "All industries will be operational as of this morning," Mayor Colette Roy-Laroche announces at a media briefing. "Tourism is an important part in our rebuilding," she adds. "I also want to thank all the people for their messages of support from all over the world. These messages give us the strength and courage to keep going." The boil-water advisory remains in effect.
- **9:30 a.m.:** The estimated number of missing is now hovering around 60, including the 15 confirmed dead, Insp. Michel Forget of SQ says. Investigators won't make names public, noting that some persons earlier considered missing have since been located. Forget warned of the emotional toll a wrong name on the list has on those working and waiting in the town.
- **11:30 a.m.**: The Quebec government promises an immediate $60 million to help the people of Lac-Mégantic, its municipal government and its businesses to deal with the emergency and begin the rebuilding process. "All of Quebec is in mourning," Premier Pauline Marois says. Flags at the National Assembly and government buildings throughout the province will fly at half-mast for one week, beginning Thursday.
- **p.m.**: At a press conference in Lac-Mégantic, MMA chairman Edward Burkhardt says the company has not had time to do a complete investigation of the red zone, and revealed that the MMA employee working early Saturday has been suspended without pay. Burkhardt no longer believes that the employee appropriately applied the handbrakes on the cars.
- **6 p.m.**: The SQ announces that 20 people are confirmed dead. The coroner's office announces that it has been able to identify a first victim, but out of respect for the person's family, would not reveal any more details about the person's identity. The total number of missing people is now 50, down from the 60 announced on earlier today.

THURSDAY, July 11

- **1 p.m.**: In her second visit to the town, Premier Pauline Marois speaks to media in Lac-Mégantic and says social services, civil security services "are being well done" and carried out on a coordinated basis. Marois also called MMA chairman Burkhardt's attitude "deplorable" and "unacceptable."
- **3:30 p.m.**: Lac-Mégantic Mayor Colette Roy-Laroche announces that 600 more residents in the Cousineau sector will be allowed back home on Friday. About 200 residents remain barred from their homes.

- **5:30 p.m.**: At a press conference, the SQ increases the confirmed death toll to 24. The coroner's office confirms that the first identified victim was 93-year-old Éliane Parenteau.

FRIDAY, July 12

- **3:30 p.m**.: The TSB holds a press conference to update the status of the investigation into what caused the explosion. They say they are looking into several factors that could have contributed to the train running off the rails. "This will be a complex investigation, and I want to be clear — it will take months or more," said TSB chair Wendy Tadros.
- **4:30 p.m.**: The SQ's and coroner's office holds a press conference to let the public know more bodies had been recovered from the disaster site, bringing the death toll to 28. Seven more of the dead had been identified, but were not made public by the authorities.
- Some areas cannot be investigated due to toxic fumes coming from the crude oil the train was carrying through Lac-Mégantic.
- "It was putting our lives in danger, even though our workers have masks," said SQ spokesperson Michel Forget.
- **9:00 p.m.**: Residents hold a candlelight vigil to mourn those lost in the explosion. More than 50 cities across the province held vigils as well.

SATURDAY, July 13

- **4:30 p.m.:** The remains of five more people have been recovered from the rubble, bringing the confirmed death toll to 33, the SQ says.

SUNDAY, July 14

- **Noon:** SQ officers on the site observe a minute of silence in memory of the dead.
- **4 p.m.:** Two more bodies have been pulled from the carnage, the SQ says, bringing the total number of victims to 35.
- The area's boil-water alert is lifted.
- Two commercial buildings are knocked down using heavy equipment because they were unsafe to enter. Blazing heat and dangerous materials in the so-called red zone have made for slow going during the search. In some areas, crews wearing protective gear and breathing apparatus are working in 15-minute shifts to avoid heat exhaustion.

MONDAY, July 15

- **10:15 a.m.:** A class-action lawsuit announced against MMA and railway chairman Edward Burkhardt, freshly filed in the Sherbrooke courthouse, could prove one of the largest in Canadian history if Quebec Superior Court gives it the green light. Proposed class representatives are Guy Ouellet, whose wife died, and Yannick Gagné, owner of Musi-Café. Three of his employees were killed when his popular bar was completely engulfed in flames. Jeff Orenstein, of Consumer Law Group Inc. in Montreal, said there is no dollar figure on the suit yet, but he guessed it would be among the largest ever filed in Canada.

- **4 p.m.:** The SQ announces two more bodies have been found, bringing the confirmed death toll to 37. Eleven of the victims have been identified by the coroner's office, eight of whom were named publicly. But Geneviève Guilbault, a spokesperson for the coroner's office, adds that none of the remains recovered have been released to the families, since tests are under way to confirm the cause of death.

TUESDAY, July 16

- **4 p.m.:** The SQ announced it has found the remains of one more person in the rubble of the Lac-Mégantic train derailment, to bring the total of confirmed dead to 38. A spokesperson for the coroner's office announced that investigators have identified one more victim, bringing the total of identified victims to 12. The total number of presumed victims remains at 50.

WEDNESDAY, July 17

- **4 p.m.:** A spokesperson for the Quebec coroner's office said Wednesday afternoon that the identities of five more victims of the Lac-Megantic derailment have been determined, bringing the number of identified victims to 17. Names will likely be made public Friday morning on Quebec coroner's website, said Genevieve Guilbault. No additional bodies were recovered from the rubble Wednesday, with the total number of confirmed victims remaining at 38, among the 50 presumed dead.

THURSDAY, July 18

- **3:30 p.m.:** Mayor Roy-Laroche said that U.S. President Barack Obama expressed his condolences to the people of Lac-Mégantic through the office of Prime Minister Stephen Harper.
- **4 p.m.:** The SQ said that four more bodies have been found, bringing the official death toll to 42. The coroner's office announced that 19 victims have been identified but their identities will be made public once families have been notified.

FRIDAY, July 19

- **10:30 a.m.:** TSB investigators offer a briefing updating their investigation. Among the major revelations are that the quality of the handbrakes engaged on the tanker cars are under investigation as well as the number applied by the engineer, said Ed Belkaloul, a TSB team member. Tests are also being conducted on the light crude oil that was spilled to look for the presence of any other substances. The TSB also flagged two immediate safety concerns to Transport Canada, which is in charge of regulating the rail industry. First, the rule for the securement of unattended locomotives and second, the securement and surveillance of unattended trains carrying dangerous goods.
- **4 p.m.:** The number of confirmed victims in the Lac-Mégantic disaster has jumped to 47, the SQ announced Friday afternoon. A spokesperson for the Quebec coroner's office said

three more victims have been identified, bringing that total to 22. Police now believe 47 people have been killed in the July 6 train derailment and explosion.

2.3 The Operator

Montreal Maine and Atlantic (MMA) was the owner of the railway line and operator. MMA was a Class II[1] freight railroad that operated in the U.S. states of Maine and Vermont and the Canadian provinces of New Brunswick and Quebec between 2002-2014. It was headquartered in Hermon, Maine.

Its Canadian subsidiary was named the Montreal, Maine and Atlantic Canada Company with offices in Farnham, Quebec.

MMA and its Canadian subsidiary entered Chapter11/CCAA bankruptcy protection in August 2013 as a direct result of the Lac-Mégantic derailment, a runaway train incident in July 2013 which resulted in an estimated $200 million in damage and the deaths of 47 people.

MMA's assets were sold at auction to Railroad Acquisition Holdings, LLC, a subsidiary of Fortress Investment Group, LLC on January 21, 2014. Fifteen locomotives worth $1.6 million were excluded from the deal and will be sold separately. The sale was approved by bankruptcy judges on January 23, with the transfer of assets expected to occur on or before March 31, 2014. Railroad Acquisition Holdings, LLC has established a new railroad named Central Maine and Quebec Railway (CMR) to operate the former MMA rail lines.

MMA was formed in October 2002. The railway was 72.8% owned by Rail World ($54.7 million) and 12.8% by Caisse de dépot et placement du Québec ($7 million) with the remaining minority stakes held by fifteen other private investors.

MMA purchased the assets arising out of the 2002 bankruptcy of the holding company Iron Road Railways.

Closing of the sale by the trustee for Iron Road Railways took place in January 2003 and MMA began operations over 745 mi (1,199 km) of track. An average of 25 trains was operating each day and the company employed roughly 350 people at that time.

Almost immediately after MMA began operations, two of the railway's largest customers, a pair of Great Northern Paper mills in Millinocket and East Millinocket, declared bankruptcy and shut down.

MMA responded by reducing its workforce to 275 employees and cut salaries by 40 percent. The decline of the pulp and paper industry as well as the lumber industry, which originally represented 60% of the railway's traffic base, had rendered a company worth $60 million in 2003 effectively worthless. Caisse de dépôt et placement du Québec's accountants wrote that

[1] A Class II (Class 2) railroad is a freight-hauling railroad that is mid-sized in terms of operating revenue. As of 2006, a railroad with revenues greater than $20.5 million but less than $277.7 million for at least three consecutive years is considered a Class II railroad. Switching and terminal railroads are excluded from Class II status.

company's investment in MMA down to a nominal $1000 from 2008 onward. MMA laid off additional workers in 2006 and in 2008 as the company struggled to become profitable.

In the spring of 2010, MMA cut train crews from two workers to one, installing remote control and cutting the total workforce to 175 people (down from twice that number in 2006). These steps were estimated to save $4.5 million annually in wages. Later that year, it sold the northern 233 miles (375 km) line to the state of Maine; this track was in very poor condition but its sale brought $20 million, $7 million of which was used to repay a 2003 loan from la Caisse de dépôt et placement du Québec, an MMA stockholder. MMA was the only Maine railway to use a single-person crew and a remote-control device instead of the more common two-person crews.

From 2003-2013 MMA recorded a higher accident rate than other North American railroads, according to data from the Federal Railroad Administration. In 2012 the railway had 36.1 accidents per million miles travelled, in comparison to a national average of 14.6 accidents.

Freight volumes continued to drop as manufacturers, needing to meet customer demands for smaller quantity just in time delivery in order to reduce inventories, began to prefer trucking firms as faster and more reliable than MMA.

The MMA had been constantly struggling financially since it was formed in 2002 due to economic problems facing its primary customers in the forestry industry. The forestry industry in Maine and Quebec fared poorly in the Great Recession of 2008-2010 and in 2010, MMA cut its workforce dramatically and sold hundreds of miles of track in northern Maine to control operating costs and manage its debt. The company's fortunes began to change in 2012 when MMA embraced oil-by-rail as a path to profitability. Transporting oil allowed it to re-hire laid-off workers and purchase additional locomotives. The oil being transported over MMA was destined for Saint John, New Brunswick's Irving Oil Refinery. By early 2013, environmental groups and Maine state officials were expressing concern about the increasing amounts of crude oil on both the MMA and its rival Pan Am Railways in the wake of a string of minor derailments on both railroads. According to Maine Department of Environmental Protection director Scott Whittier, "The transportation of crude oil across rail lines is a concern because many times, rail lines are very close to sensitive water bodies, so it does present a potential threat that we need to prepare for."

In a March 2013 Associated Press interview, MMA chief executive Robert Grindrod dismissed these concerns, stating "The statistics tell you how much has been transported, but to the best of my knowledge, there hasn't been any spilled or released,"

On June 11, 2013 an MMA locomotive spilled 13,000 litres (3,400 US gal) of diesel oil from its fuel tank at Frontenac (approximately 5 kilometres (3.1 mi) east of Lac-Mégantic, Quebec) after the tank was perforated while using a switch.

2013 Lac-Mégantic derailment

On July 6, 2013, an MMA train with 72 DOT-111 cars carrying crude petroleum products derailed in the centre of Lac-Mégantic, Quebec. The runaway train, without anyone at the controls after a stop at Nantes for a crew change, derailed at 1:14 AM, and four of its cars exploded. The city's downtown core suffered catastrophic damage, with many businesses and residential properties completely destroyed. Many of the residents lost their homes or jobs; 42 people are confirmed dead and 5 more people are missing and presumed dead.

Rail World president and CEO Edward Burkhardt first visited Lac-Mégantic after the accident on 10 July 2013, and was heckled by residents. After the accident the railway's safety record was called into question. The train had been left unattended and improperly secured overnight on the mainline, with one locomotive running, on a 1.2% grade leading 18 kilometres (11 mi) from Nantes, Quebec downhill to a 10 miles per hour (16 km/h) curve where the line entered downtown Lac-Mégantic. The railway air brakes appear to have released after a local fire brigade shut down the locomotive to extinguish a fire in the engine, allowing the oil-laden cars to roll downhill into the town as an insufficient number of hand brakes had been applied to hold the train.

Burkhardt refused to publicly disclose the amount of liability insurance but acknowledged in mid-July "Whether we can survive is a complex question. We're trying to analyze that right now." Questions were also raised about the condition of the line; on July 11 a Magog newspaper reported one in ten railroad ties to be rotten with many spikes loose enough to be removed by hand.

Following the accident, the MMA ceased operations on its lines between Lac-Mégantic and Jackman, Maine, effectively severing rail transport on its lines between Maine and Quebec, though rail traffic continued outside the affected area. In Quebec, MMA had continued operation from Farnham with a skeleton staff after the derailment, having laid off 24 Canadian employees and 64 workers in Maine in July 2013 as the line remains blocked by wreckage at Lac-Mégantic. MMA aggregate gross revenue dropped by two thirds, leaving $1 million/month in revenue on the broken line. According to CEO Burkhardt, "This may cost us our company. This may cost us our investment, cost the employees their jobs, the customers in Quebec, in Maine their rail service."

On August 6, 2013, Burkhardt stated that MMA has no further plans to carry oil by rail.

Following the tragedy, parent company Rail World faces potentially hundreds of millions of dollars in lawsuits and environmental cleanup costs due to the destruction of downtown Lac-Mégantic. It is unclear how much of the disaster's cost will be paid by liability insurance; laid-off MMA workers have reported problems obtaining the severage pay owed them and cheques to suppliers have been rejected due to non-sufficient funds.

On August 7, 2013, the company filed for bankruptcy protection in both the Quebec Superior Court in Montreal (under the Companies Creditors Arrangement Act) and the United States Bankruptcy Court in Bangor, Maine (Chapter 11). According to Burkhardt, the financial obligations of both U.S. and Canadian subsidiaries "now exceed the value of their assets, including prospective insurance recoveries, as a direct result of the tragic derailment at Lac-Megantic. A process under Chapter 11 and the CCAA is the best way to ensure fairness of treatment to all in these tragic circumstances."

MMA had $50–100 million in U.S. assets and less than $18 million in Canadian assets, including the track itself. It owed $27.5 million on a 2005 Federal Railroad Administration loan, $6 million on a 2009 line of credit issued by the Wheeling and Lake Erie Railway and $3.5 million to various other creditors and tax authorities.

The total cost of the Lac-Mégantic derailment is likely to exceed $200 million; MMA had $25 million in liability insurance and $274,000 in its bank accounts at the time of bankruptcy filing.

Eighty-five MM&A employees remained on duty after the bankruptcy filing to provide service to all stations on the line except Lac-Mégantic, where all tracks remained blocked since the July 6 derailment and fire. Fifteen of those workers were laid off on August 20.

On August 13, 2013, the Canadian Transportation Agency announced it planned to revoke the MMA's certificate of fitness effective August 20, 2013, citing inadequate liability insurance; this effectively would have banned the MMA from operating in Canada. The August 20 date has been extended until February 1 and subsequently to April 1. As the insurance policy's franchise provides no coverage for liability claims under a quarter-million dollars, the Québec Superior Court has set aside this sum in MMA's bankruptcy as available to victims of a future disaster involving MMA.

On September 12, Transport Canada shut down part of the MMA line after an inspection of six track segments found substandard rail conditions which included a concentration of defective ties on a section near a propane storage facility. A failed October 11 inspection led Transport Canada to embargo the entire line between Lennoxville and Lac-Mégantic.

On December 18, 2013, the rail line from Sherbrooke was reopened through Lac-Mégantic with numerous restrictions, such as a prohibition on transport of dangerous cargo; a train's manifest being released no less than four hours ahead; no parking on tracks within 4 km (2 mi) of the town centre; a conductor and engineer must both be on board; and a train's speed must not exceed 16 km/h (10 mph). The first test train carrying particle board from the local Tafisa factory to Sherbrooke rolled through the town centre. There are plans to reroute the tracks outside the town by changing the track's route between Nantes and Frontenac; no time table has been set.

MMA's assets were sold at auction to Railroad Acquisitions Holdings, LLC, a subsidiary of the Fortress Investment Group, LLC, in January 2014 in an attempt to keep the line open as a going concern . That group's initial bid for US$14.25 million, placed in December 2013, was the only to offer to buy the entire MMA system. The terms of the sale were announced in mid-February 2014 when the new railroad named Central Maine and Quebec was registered in the United States and Canada. The sale is to be finalized in March 2014.

A few hours after court approval of the sale, Great Northern Paper announced a 16-week closure of its mill in East Millinocket. On January 30, trustees filed a US lawsuit on behalf of the bankrupt railway, alleging that World Fuel Services was negligent in failing to properly label the train's dangerous cargo and that, had its true volatility and inflammability been honestly disclosed on shipping documents, MMA procedures would have prevented it being left unattended on a main rail line.

2.3 The Rolling Stock

The deadly accident has given new urgency to a long-standing debate about the need for safer tanker cars, a reform that Mr. Hunter, CP CEO said has been stonewalled for decades by petroleum and chemical producers and other commodity shippers who own the vast majority of North America's tankers.

2.3.1 Vulnerable Cars

The Quebec accident also revived a debate over the type of cars used to haul oil. For years, regulators and watchdogs have sought improvements to a common car design shown to be susceptible to rupture when derailed. The NTSB estimates that 69 percent of today's rail tank-car fleet has "a high incidence of tank failure during accidents," Chairman Deborah Hersman wrote in letter last year. The agency recommended thicker shells and other modifications to strengthen the cars.

Senator Charles Schumer, a New York Democrat, called on U.S. regulators to phase out the older cars, known as DOT-111s, saying they've contributed to spills of hazardous materials.

"The DOT-111 tank car has proven particularly prone to spills, tears and fires in the event of a derailment, and it's simply unacceptable for New York's communities along the rail lines to face that risk when we know thicker, tougher cars could keep us safer," Schumer said in an e-mailed statement.

The rail industry is fighting a proposal to retrofit existing cars, saying it could cost as much as $1 billion.

DOT Rule Making Petition

- That the structure around the top fittings should be of such strength as to afford protection for the top fittings in case of a collision
- That the tank head and shield or combination of the tank head, shield and external metal shield, should be of such strength as to afford puncture protection of the tank car in case of a collision
- That the strength of the tank car carrying crude oil and ethanol should be adequate to withstand the internal pressure as will allow the safety valve to pass the contents of the tank car in vapor form before bursting pressure of the tank car is reached
- That the flow capacity of the safety valve will relieve the tank from the internal pressure below the bursting pressure of the tank in case of an accident involving fire.

2.3 The Bakken Crudes

One month before the disaster in Lac-Megantic, a New Brunswick refinery raised concerns about a lack of testing on crude oil that is shipped by rail. In a presentation to industry an Irving Oil employee said there is an overreliance on testing at the refinery, once the crude has already been delivered, to determine the content. That testing comes "too late in the process to address any safety issues," says the presentation, which was prepared for a Crude Oil Quality Association conference in Seattle. [1]

Worries about the composition of oil drawn from the Bakken formation have grown after a train carrying the Bakken crude derailed in Lac-Megantic, setting off a series of fierce explosion not normally associated with crude oil.

The slide show is dated June 6, one month before the Lac-Megantic tragedy. Irving's refinery is one of several in North America that process oil from the Bakken fields, which straddle North Dakota and parts of Manitoba and Saskatchewan. The oil is now being investigated by Canada's Transportation Safety Board to determine why it caused such massive explosions, which investigators said were uncharacteristic of typical crude oil.

The presentation – titled "Crude by Rail Quality Issues and prepared by Gary Weimer, quality assurance and control manager at Irving's refinery in Saint John – said the company was encountering tank cars that contained "sludge" and "contaminants," while some cars had "unknown substances." [1] According to the presentation, Irving also had concerns with oil from various sources being mixed in tank cars, making it difficult to pinpoint the makeup of the crude being shipped. "Samples are not from a homogenous mixture," the documents say. "Rail cars after 5-7 days delivery time have begun layering; water and sediment on bottom; light products on top." Concerns have been raised since Lac-Mégantic about the light crude oil being shipped from the Bakken oilfields. Some observers believe the product is so light that it vaporizes easily, making it more susceptible to combustion. More testing at oil loading sites would help "identify issues related to safety of personnel or specification while the rail car is in transit," the Irving document says. "This will give time to plan action from the data." Other industry members have raised concerns about the oil.

If true, the Irving Oil contentions would mean that North American railroads have knowingly been given erroneous hazmat shipping documents to accompany crude oil shipments. As common carriers, railroads are legally bound to accept crude oil certified by the shipper to be correctly classified and legally contained. The railroads have no control over loading practices or the interior condition of tank cars, which are owned and maintained by the shippers and receivers.

Misrepresenting to railroads and customs officials the true nature of hazardous material shipments is an offense in both Canada and the U.S. Nonetheless, Irving's Weimer told his industry colleagues that oilfield sampling of outbound crude "is almost non-existent."

Cameron Conecnik [2] states that a new standard is required for vapor pressure testing: D323 is not suitable for crude with significant light ends components. There is little to no documentation to support the validity of the monographs published in API Bulletin 2517.

- Correlation dates back from the 50's or 60's with little documented references as to how data was analyzed and correlation made
- Based on "weathered crude" and therefore there has questionable applicability for today's stabilized crudes vs. those in which the correlation was derived
- Uncertain as to whether certain stabilized crudes would better fit the refined product nomograph or the Crude Oil nomograph.

In July, the U.S. Federal Railroad Administration notified the petroleum industry about high levels of corrosion in rail cars, believed to be caused by hydrogen sulfide in oil from the Bakken area. The corrosion is believed to make the tank cars more susceptible to rupture or explosions.

3. Regulatory Aftermath

The Transportation Safety board of Canada (TSB) is conducting an in-depth investigation of the causes of Lac-Mégantic accident and has formulated emergency recommendations

The catastrophic consequences of the Lac-Mégantic accident and the known increase over the last several years in rail transportation of Class 3 hazardous materials has made it clear the need to review the existing regulations and industry practices to such transportation . The U.S, Pipeline and Hazardous Material Safety Administration (PHMSA) and the Federal Railroad Administration (FRA) are working closely to prevent similar incidents from occurring.

This regulatory process has to be coordinated between Canada, Mexico and the United States since it will eventually affect the three countries.

The regulatory development in Canada and the United States are described in the following paragraphs.

3.1 Canada

In response to this accident, Transport Canada (the Canadian government department responsible for regulating transportation safety in Canada) issued a series of directives.

3.1.1 Rail Safety Advisory Letter Securement of Train Left Unattended

An emergency railroad safety advisory letter pursuant to Section 33 of the Canadian Railway Safety Act was issued by the National Transportation Safety Board [6]:

Over the years, the TSB has investigated a number of similar runaway accidents.

- On 12 August 1996, all 3 three occupants in the operating cab of the lead locomotive of Canadian National (CN) westward freight train No. 117 were fatally injured when their train, which was travelling at about 54 mph, collided head-on with a cut of 20 runaway cars moving eastward at about 30 mph, some six miles east of Edson, Alberta. The runaway cars had been left on a track in Edson Yard by a crew who had applied hand brakes to 2 hopper cars. The performance of the hand brakes on that type of car was found to be highly variable, and the variability was not commonly known amongst operating employees. Although the crew thought the cars had been secured, the resultant brake shoe force on the two cars was insufficient to prevent movement.

- On 11 February 2009, the Southern Ontario Railway 0900 Hagersville Switcher, consisting of 4 locomotives and 43 cars, ran uncontrolled from Mile 0.10 to Mile 1.9 of the Hydro Spur track. The train reached a speed of 20.7 mph before travelling over a split switch derail and derailing 9 loaded dangerous goods tank cars. Three tank cars loaded with gasoline (UN 1203) were breached and released approximately 31 000 litres of gasoline. Two nearby homes were evacuated; there were no injuries. The investigation determined that the accident occurred when the crew left the train unattended on a one per cent descending grade, without the train being properly secured.
- On 11 December 2011, as freight train LIM-55 was descending a long steep grade, the locomotive engineer, unable to control the train speed using the dynamic and automatic brakes, applied the emergency brakes at Mile 68.00 of the Wacouna Subdivision to stop his movement. One hour later, the train ran away, descending the grade for a distance of almost 15 miles and reaching a maximum speed of 63 mph. The train finally came to a stop at Mile 52.80. No one was injured and there was no derailment. The investigation determined that 1 hour after the emergency brakes were applied and the train came to a stop, the air brakes released and, because the braking force applied by the hand brakes was insufficient, the train ran away.
- On 18 January 2012, 13 loaded coal cars, which were running uncontrolled northward from the Hanlon siding, Mile 41.7 on the Grande Cache Subdivision, collided with stationary train A45951-16 at Mile 44.5. Nine of the 13 cars and the 3 leading locomotives from the train derailed. Two crew members sustained minor injuries and were treated on site. The 3rd crew member was seriously injured and was air lifted to hospital in Hinton, Alberta. Approximately 2800 litres of fuel and 740 tons of coal were spilled. About 250 feet of track was damaged. The investigation determined that the conductor's belief that the cars were on level ground likely led to the conclusion that 1 hand brake was sufficient to secure the cut of cars. When the braking force from the train brakes was sufficiently reduced due to bleeding off of the air in the cars' brake cylinders, the retarding force of the 1 applied hand brake could not resist the gravitational force of the 13 loaded coal cars. In addition, because the brake effectiveness test was conducted by attempting to push the cars upgrade, it did not adequately verify if the braking force of the hand brake application was sufficient to hold the cars.

Each of these investigations brings into question the effectiveness of the Transport Canada (TC) approved Canadian Rail Operating Rules (CROR) Rule 112 entitled "Securing Equipment" and the standard railway operating procedures and practices utilized for performing that task

3.1.2 Rail Safety Advisory Letter Securement of Train Left Unattended

An emergency railroad safety advisory letter pursuant to Section 33 of the Canadian Railway Safety Act was issued by the National Transportation Safety Board. The directive ordered railroad companies in Canada to ensure that [6]:

- Within five days of the issuance of the directive, all unattended controlling locomotives on a main track and sidings are protected from unauthorized entry into the cab;
- The directional controls, commonly known as reversers, are removed from any unattended locomotives, preventing them from moving forward or backward, on a main track or sidings;
- Their company's special instructions on hand brakes are applied to any locomotive attached to one or more cars that are left unattended for more than one hour on a main track or sidings;
- In addition to complying with their company's special instructions on hand brakes referred to in the item immediately above, the automatic brake is set in full service position and the independent brake is fully applied for any locomotive attached to one or more cars that are left unattended for one hour or less on a main track or sidings;
- No locomotive attached to one or more loaded tank cars transporting dangerous goods is left unattended on a main track; and
- No locomotive attached to one or more loaded tank cars transporting dangerous goods is operated on a main track or siding with fewer than two persons qualified under their company's requirements for operating employees.

3.2 United States

3.2.1 Emergency Order Establishing Additional Requirements for Attendance and Securement of Certain Freight Trains and Vehicles on Mainline or Mainline Siding Outside of a Yard or Terminal – FRA Emergency Order 28, Notice No 1 – Safety Advisory 2013-06

The Federal Railroad Administration (FRA) of the United States Department of Transportation (DOT) issued on August 7, 2013 an Emergency Order (EO), which requires railroads operating on the general system to implement additional processes and procedures to ensure that certain unattended trains and vehicles on mainline track or mainline siding outside of a yard or terminal are properly secured against unintended movement. [3]

FRA has found there is significant non-compliance among the railroads with respect to FRA's securement regulations. With limited resources, FRA can inspect only a small percentage of trains and vehicles for regulatory compliance. However, even with its limited resources, FRA has recorded nearly 4,950 securement defects in the course of its inspections since January 2010, an average of approximately 1,483 defects per year. With increased shipments of hazardous materials such as crude oil and ethanol, securement non-compliance, particularly on mainline track and mainline sidings outside of a yard or terminal, has become a serious, immediate safety concern. Therefore, additional measures are necessary to protect the health and safety of railroad employees, the general public, and the environment.

As a result, FRA is ordering that each railroad take the following actions on mainline track or mainline siding outside of a yard or terminal to ensure the safe transportation by rail of hazardous material:

1. No train or vehicle transporting hazardous materials shall be left unattended on a mainline track or mainline siding outside of a yard or terminal until the railroad develops, adopts, complies with and makes available to FRA upon request, a plan that identifies specific locations and circumstances when such trains or vehicles may be left unattended. The plan shall contain a sufficient safety justification for any determination allowing such trains or vehicles to be unattended.
2. Develop processes for the securement of unattended trains or vehicles transporting hazardous materials on mainline track or mainline siding outside of a yard or terminal if permitted by the railroad's plan required by this order that contains the following requirements:
   a. The controlling locomotive cab must be locked or the reverser on the controlling locomotive must be removed and secured, and
   b. Employees who are responsible for securing trains and vehicles transporting hazardous materials must communicate to the train dispatcher the number of hand brakes applied, the tonnage and length of the train or vehicle, the grade and terrain features of the track, any relevant weather conditions, and the type of equipment being secured; train dispatchers must record the information provided; and train dispatchers or other qualified railroad employees must verify and confirm with the train crew that the securement meets the railroad's requirements.

<u>3.2.2 Hazardous Materials: Rail Petitions and Recommendations To Improve the Safety of Railroad Tank Car Transportation (RRR) - Advance Notice of Proposed Rule Making</u>

PHMSA is considering revisions to the Hazardous Materials Regulations (HMR) to improve the regulations applicable to the transportation of hazardous materials by rail. The revisions are based on eight petitions received from the regulated community and four National Transportation Safety Board (NTSB) recommendations which are referenced by a petition. [4]

Under the HMR, hazardous materials are categorized by analysis and experience into hazard classes and packing groups based upon the risks that they present during transportation. The HMR specify appropriate packaging and handling requirements for hazardous materials based on this classification, and require a shipper to communicate the material's hazards through the use of shipping papers, package marking and labeling, and vehicle placarding. The HMR also require shippers to provide emergency response information applicable to the specific hazard or hazards of the material being transported. Finally, the HMR mandate training requirements for persons who prepare hazardous materials for shipment or who transport hazardous materials in commerce.

In this safety advisory, PHMSA and FRA are reinforcing the importance of proper characterization, classification, and selection of a packing group for Class 3 materials, and the corresponding requirements in the Federal hazardous materials regulations for safety and security planning. In addition, PHMSA and FRA are reinforcing that they expect offerors by rail and rail carriers to revise their safety and security plans required by the Federal Hazardous Materials Regulations, including the required risk assessments, to address the safety and security issues identified in FRA's Emergency Order No. 28 and the August 7, 2013, joint Safety Advisory.

Specifically, these amendments propose to:

(1) Relax regulatory requirements to afford the Federal Railroad Administration (FRA) greater discretion to authorize the movement of non-conforming tank cars;
(2) Impose additional requirements that would correct an unsafe condition associated with pressure relief valves (PRV) on rail cars transporting carbon dioxide, refrigerated liquid;
(3) Relax regulatory requirements applicable to the repair and maintenance of DOT Specification 110, DOT Specification 106, and ICC 27 tank car tanks (ton tanks);
(4) Relax regulatory requirement for the removal of rupture discs for inspection if the removal process would damage, change, or alter the intended operation of the device;
(5) Impose additional requirements that would enhance the standards for DOT Specification 111 tank cars used to transport Packing Group (PG) I and II hazardous materials. The NTSB recommendations directly relate to the enhancement of DOT Specification 111 tank cars.

Table 1 provides a brief summary of the petitions and NTSB Recommendations addressed in the ANPRM.

Table 1. Summary of petitions

Party submitting petition	Summary
Eastman Chemical Co.	Revise the wording of § 174.50 to afford FRA greater discretion in authorizing car movement.
The Compressed Gas Association (CGA).	Revise § 173.314 Note 5 to clearly indicate that the liquid portion of the gas must not completely fill the tank prior to reaching the pressure setting of the regulating valves or the safety relief valve, whichever is lower.
Carroll Welding Supply	Revise the ton tank repair, maintenance, and marking regulations for consistency with existing regulations for DOT 3-series cylinders since ton tanks share more in common with these cylinders than tank cars.
American Chemistry Council (ACC)	Proposes a change to the wording in § 173.31(d)(1)(vi) intended to prevent damage or loss of effectiveness of rupture discs removed from their initial placement in the relief device by adding language that would except them from removal if the inspection itself would damage, change, or alter the intended operation of the device.
Association of American Railroads (AAR)	Proposes a new standard for newly-constructed DOT Specification 111 tank cars used to transport PG I and II materials.
Village of Barrington, Illinois and The Regional Answer to Canadian Nation	Stresses the importance of adopting P–1577 for newly-constructed and existing DOT Specification 111 tank cars in accordance with NTSB Recommendations R–12–5 and R–12–6. In addition, the petition urges PHMSA to adopt NTSB Recommendation R–07–4.
ACC, American	Proposes that PHMSA apply requirements related to top fittings

Petroleum Institute (API) and The Chlorine Institute, Inc. (CI)	protection, reclosing pressure relief devices, and head and shell thickness requirements as suggested in P–1577 and P–1587 for DOT Specification 111 tank cars used to transport ethanol and crude oil in PG I and II.
API, ACC, CI, and The Renewable Fuels Association (RFA)	The Petitioners request that PHMSA separate new tank car regulatory requirements from any potential retrofits for the timely adoption of revised regulatory requirements for the construction of new DOT Specification 111 tank cars used for the transportation of ethanol and crude oil.
NTSB	With the assistance of the FRA, require that railroads immediately provide to emergency responders accurate, real-time information regarding the identity and location of all hazardous materials on a train.
NTSB	Require that all newly-manufactured and existing general service tank cars authorized for transportation of denatured fuel ethanol and crude oil in PGs I and II have enhanced tank head and shell puncture resistance systems and top fittings protection that exceed existing design requirements for DOT Specification 111 tank cars.
NTSB	Require that all bottom outlet valves used on newly-manufactured and existing non-pressure tank cars are designed to remain closed during accidents in which the valve and operating handle are subjected to impact forces.
NTSB	Require that all newly-manufactured and existing tank cars authorized for transportation of hazardous materials have center sill or draft sill attachment designs that conform to the revised Association of American Railroads' design requirements adopted as a result of Safety Recommendation R–12–9.

The petitioners stress the importance of promulgating enhanced standards for existing and new tank cars used to transport PG I and II materials in accordance with the NTSB Railroad Accident Report — *Derailment of CN Freight Train U70691– 18 With Subsequent Hazardous Materials Release and Fire,* Cherry Valley, Illinois, June 19, 2009 (RAR–12– 01). [7]. The following paragraphs list some of the findings and recommendations from this NTSB Railroad Accident Report:

Findings

16 If enhanced tank head and shell puncture-resistance systems such as head shields, tank jackets, and increased shell thicknesses had been features of the DOT-111 tank cars involved in this accident, the release of hazardous materials likely would have been significantly reduced, mitigating the severity of the accident.

17 The safety benefits of new specification tank cars will not be realized while the current fleet of DOT-111 tank cars remains in hazardous materials unit train service, unless the existing cars are retrofitted with appropriate tank head and shell puncture-resistance systems.

18 Requirements for protection of the top fittings of the DOT-111 tank cars involved in this accident are inadequate because the protective housings were not able to withstand the forces of the derailment.
19 The existing standards and regulations for the protection of bottom outlet valves on tank cars do not address the valves operating mechanisms and therefore are insufficient to ensure that the valves remain closed during accidents.
20 Tank car design standards for the attachments of draft sills to sill pads and of sill pads to the tanks are insufficient to protect the integrity of the tanks in accidents in which the draft sills are subjected to significant downward deformation.
21 Had an effective safety management system been implemented at the CN, the inadequacies and risks that led to the accident would have been identified and corrected and, as a result, the accident may have been prevented.

Recommendations

To the Federal Railroad Administration:

Require that safety management systems and the associated key principles (including top-down ownership and policies, analysis of operational incidents and accidents, hazard identification and risk management, prevention and mitigation programs, and continuous evaluation and improvement programs) be incorporated into railroads risk reduction programs required by Public Law 110-432, Rail Safety Improvement Act of 2008, enacted October 16, 2008. (R-12-3)

To the Pipeline and Hazardous Materials Safety Administration:

Require that all newly manufactured and existing general service tank cars authorized for transportation of denatured fuel ethanol and crude oil in Packing Groups I and II have enhanced tank head and shell puncture-resistance systems and top fittings protection that exceeds existing design requirements for DOT-111 tank cars. (R-12-5)

Require that all bottom outlet valves used on newly manufactured and existing non-pressure tank cars are designed to remain closed during accidents in which the valve and operating handle are subjected to impact forces. (R-12-6)

Require that all newly manufactured and existing tank cars authorized for transportation of hazardous materials have center sill or draft sill attachment designs that conform to the revised Association of American Railroads design requirements adopted as a result of Safety Recommendation R-12-9. (R-12-7)

To the Federal Railroad Administration:

Assist the Pipeline and Hazardous Materials Safety Administration in developing regulations to require that railroads immediately provide to emergency responders accurate, real-time information regarding the identity and location of all hazardous materials on a train. (R-07-2) (Reiterated)

To the Pipeline and Hazardous Materials Safety Administration:

With the assistance of the Federal Railroad Administration, require that railroads immediately provide to emergency responders accurate, real-time information regarding the identity and location of all hazardous materials on a train. (R-07-4) (Reiterated)

Safety and Security Plans as They Pertain to Class 3 Materials

Each person who offers for transportation in commerce or transports in commerce certain hazardous materials, including Class 3, packing group (PG) I or II materials that are offered for transportation or transported in a large bulk quantity, must develop and adhere to a transportation safety and security plan that conforms to the requirements of the HMR.

A safety and security plan must include components addressing personnel security, unauthorized access, and en route security.

The HMR require offerors of hazardous materials to properly classify and describe the hazardous material being offered for transportation. As part of this process, proper characterization of a hazardous material (considering the material's underlying chemical properties, corrosivity, and other characteristics) is fundamental to ensuring the selection of proper packaging and that the hazards of the materials are properly described in the required shipping documentation. Proper characterization will identify properties that may not affect classification, but will affect the integrity of the packaging or present additional hazards, such as corrosivity, sulfur content and dissolved gas content. Ensuring the proper classification, characterization, and PG assignment of a hazardous material is a key building block of the HMR, and is especially important for assessing risks and developing a safety and security plan.

3.2.4 Safety Alert – January 2, 2014 – Preliminary Guidance from Operation Classification

PHMSA issued a safety alert to notify the general public, emergency responders and shippers and carriers that recent derailments and resulting fires indicate that the type of crude oil being transported from the Bakken region may be more flammable than traditional heavy crude oil. [9]

Based upon preliminary inspections conducted after recent rail derailments in North Dakota, Alabama and Lac-Mégantic, Quebec involving Bakken crude oil, PHMSA is reinforcing the requirement to properly test, characterize, classify, and where appropriate sufficiently degasify hazardous materials prior to and during transportation.

PHMSA stress that "Proper characterization will identify properties that could affect the integrity of the packaging or present additional hazards, such as corrosivity, sulfur content, and dissolved gas content. These characteristics may also affect classification. PHMSA stresses to offerors the importance of appropriate classification and packing group (PG) assignment of crude oil shipments, whether the shipment is in a cargo tank, rail tank car or other mode of transportation. Emergency responders should remember that light sweet crude oil, such as that coming from the Bakken region, is typically assigned a packing group I or II. The PGs mean that the material's flashpoint is below 73 degrees Fahrenheit and, for packing group I materials, the boiling point is below 95 degrees Fahrenheit. This means the materials pose significant fire risk if released from the package in an accident."

4 Lessons for the chemical process industry

The Center for Chemical Process Safety published *Guidelines for Chemical Transportation Risk Analysis* [10] [11]. The primary goal of transportation risk analysis (TRA) is to provide tools for reducing the risks when transporting hazardous materials. The publication also may be used to acquaint an engineer or manager with the terminology and procedures that will be followed by an outside group of consultants. It may also be used by management to define the skills required to develop an "in-house" risk analysis capability. In addition, the book collects and presents, in one spot, a wide array of calculation procedures often used by risk analysts.

The objective of *Guidelines for Chemical Transportation Risk Analysis* is to enable the practicing engineer to be able to:

- Respond to a request to conduct a transportation risk assessment
- Convert that request into a definable problem statement
- Develop a scope of work
- Understand the types of data and sources of information required
- Analyze the results for reasonableness
- Present the results in a useful format.

There are many problems for which a transportation risk analysis study may be warranted. Three of the more popular studies include:

- Route selection (What is the safest way to ship?)
- Mode selection (Should the shipment be by truck, rail, barge or pipeline?)
- Facility location (Where should a new plant or warehouse be built?

When a routing type of TRA study is completed various paths between the origin and destination are studied. Each path is composed of a number of segments characterized by such parameters as length, population exposure and condition of link (roadway, railbed, waterway, pipeline, etc.). These parameters influence the accident frequency along the route. In addition, the conditional likelihood of release given that an accident has occurred is often needed. Such a conditional probability is strongly influenced by equipment design, wall thickness of package, and speed of shipment. If a toxic chemical is being shipped, the downwind consequences of a release may need to be estimated using release rate and air dispersion models. The result of a routing TRA study will be a preferred route along with any special precautions that need to be taken during shipment. Such special requirements might include:

- Restrictions on the type of equipment to be used
- Time of day control for the movement through highly congested areas
- Special requirements on driver training
- Special requirements for route emergency responder
- Radio, mobile telephone or satellite communication requirements
- Restrictions on the movement during bad weather.

When a mode selection TRA study is completed, the goal is to define the method of shipment. The modes discussed in *Guidelines for Chemical Transportation Risk Analysis* include:

- Pipeline
- Tank truck
- Rail tank car
- Barge
- Ocean Going Vessel

The equipment used in each of these shipment modes is unique. *Guidelines for Chemical Transportation Risk Analysis* presents an overview of commonly used transportation systems for each of these modes. In addition to the selection of mode, the need for special design considerations may become apparent. These special needs can be incorporated into the final set of recommendations to reduce the risk of movement of the chemical.

The use of transportation risk analysis has been quite limited. Historically, this is due in part to the manner in which transportation activities are managed in many organizations. Purchasing and distribution are generally housed together and kept quite separate from the manufacturing or production operations, and transportation safety is restricted to DOT compliance issues. In addition, the ownership of and liability for many transportation movements is shared among many parties.

Given that transportation activities occur in areas beyond a facility's control and that other parties may be involved (such as railroads and other carriers), attention to transportation risk analysis is critical.

5 Conclusions

The absence of pipeline to transport petroleum crudes to refineries has forced the upstream industry to use rail facilities. This resulted in a huge traffic of rail cars. DOT 111 tank cars make close to 70% of the fleet of tank cars. These tank cars are fragile and they are known to be fragile. Light crudes from the Bakken formation often improperly classified were loaded in DOT 111 tank cars. Trains including 70 to over 100 tank cars were assembled. In Lac-Mégantic, Montreal Maine and Atlantic (MMA), a company facing financial difficulties with an infrastructure in a deteriorated state, was the train operator. The company had obtained from Transport Canada a waiver to operate with one train driver. The train was allowed to park on the main line on top of a slope and left without supervision during a shift change. A locomotive fire caused the failure of the break system. The train rolled down the slope, derailed with punctures/ruptures of tank cars, fire, explosions, loss of 47 lives, destruction of city downtown and a massive pollution of the area, lake and river. MMA insurance was $25 million, far less then what is required to cover the damages. As a result of this limited insurance coverage allowed by regulation, the Canadian and Quebec Governments are left with the clean up and rebuilding costs. This is a real catastrophic event.

Pressure to move crude oil to the market has forced the upstream petroleum industry to take short cuts.

It does not appear that the transportation risk analysis commonly used in the Chemical Process Risk Analysis was used or even considered.

Tougher regulations by the Canadian and U.S. Transport Agencies are required to discipline an industry that has shown poor performance since it started shipping crude oil by trains in large volume.

5. References

[1] Weiner, Gary "Crude Quality Issues", Proceedings Crude Oil Quality Association Meeting, Seattle, WA, 6 June 2013 http://www.coqa-inc.org/20130606_Weimer.pdf (accessed 2013-12-30)

[2] Conecnik, Cameron "Proposal for an Improved Method of Crude Oil Vapor Pressure Determination," Proceedings Crude Oil Quality Association Meeting, Seattle, WA, 6 June 2013, http://www.coqa-inc.org/20130606_Segato_Vapor%20Pressure_[Compatibility_Mode].pdf (accessed 2013-12-30)

[3] Federal Register. "Emergency Order Establishing Additional Requirements for Attendance and Securement of Certain Freight Trains and Vehicles on Mainline Track or Mainline Siding Outside of a Yard Terminal" Vol 78, No 152 / Wednesday August 7, 1913 / Notices pp. 48218 – 48224, https://www.federalregister.gov/articles/2013/08/07/2013-19215/emergency-order-establishing-additional-requirements-for-attendance-and-securement-of-certain (accessed 2014-01-02)

[4] Federal Register. "Hazardous Materials: Rail Petitions and Recommendations to Improve the Safety of Railroad Tank Car Transportaqtion (RRR) " Vol 78, No 152 / Friday Septembert 6, 1913 / Proposed Rules pp. 54849 – 54861, http://www.gpo.gov/fdsys/pkg/FR-2013-09-06/pdf/2013-21621.pdf (accessed 2013-09-06)

[5] Federal Register. "Safety and Security Plan for Class 3 Hazardous Material Transported by Rail" Vol 78, No 152 / Wednesday November 20, 2013 / Notices pp. 69745 – 69746, http://www.gpo.gov/fdsys/pkg/FR-2013-11-20/pdf/2013-27785.pdf (accessed 2014-01-02)

[6] Transportation Safety Board of Canada "Rail Safety Advisory Letter 09/13 regarding the securement of equipment and trains left unattended Ottawa, Canada, 2013/09/13 http://www.bst-tsb.gc.ca/eng/medias-media/sur-safe/letter/rail/2013/r13d0054/r13d0054-617-09-13.asp (accessed 2014-01-02)

[7] National Transportation Safety Board "NTSB Railroad Accident Report — Derailment of CN Freight Train U70691–18 With Subsequent Hazardous Materials Release and Fire Cherry Valley, Illinois June 19, 2009", Washington, D.C, 2012/02/14

http://www.ntsb.gov/doclib/reports/2012/RAR1201.pdf
(accessed 2014-01-05)

[8] Pipeline and Hazardous Material Safety Administration "Safety Alert – January 2, 2014 – Preliminary Guidance for Oeration Classification" Washington, D.C., Canada, 2014/01/02
http://www.phmsa.dot.gov/staticfiles/PHMSA/DownloadableFiles/1_2_14%20Rail_Safety_Alert.pdf
(accessed 2014-01-05)

[9] CBC New "Runaway trains almost triple reported rate, CBC finds
Uncontrolled cars careen down mountains, come across the U.S. border", 2013/12/03
http://www.cbc.ca/news/canada/runaway-trains-almost-triple-reported-rate-cbc-finds-1.2448494
(accessed 2013-12-03)

[10] AIChE/CCPS *Guidelines for Chemical Transportation Risk Analysis* (1995)

[11] Bendixen, Lisa, Raymond A. Freeman, Dennis Hendershot, "Using the CCPS Guidelines for Chemical Transportation Risk Analysis, Process Safety Progress, Vol. 16, No 1, pp 32-36, American Institute of Chemical Engineers

10TH GLOBAL CONGRESS ON PROCESS SAFETY

Case Study of the Domino Effect in a Catastrophic Solid Oxidizer Fire

Russell A. Ogle, Juan C. Ramirez, and Todd M. Hetrick
Exponent, Inc.
4580 Weaver Parkway, Suite 100
Warrenville, Illinois 60555 USA
rogle@exponent.com, jramirez@exponent.com, thetrick@exponent.com

Prepared for Presentation at
American Institute of Chemical Engineers
2014 Spring Meeting
10th Global Congress on Process Safety
New Orleans, LA
March 30 – April 2, 2014

UNPUBLISHED

Case Study of the Domino Effect in a Catastrophic Solid Oxidizer Fire

Russell A. Ogle, Juan C. Ramirez, and Todd M. Hetrick
Exponent, Inc.
4580 Weaver Parkway, Suite 100
Warrenville, Illinois 60555 USA
rogle@exponent.com, jramirez@exponent.com, thetrick@exponent.com

Keywords: domino effect, solid oxidizer, sodium chlorate, NFPA 430, case study

Abstract

This case study examines a succession of three separate fires which occurred in a period of four days at an agricultural chemical manufacturing facility. The facility was located in a small chemical park. The succession of fires illustrates the concept of the domino effect: the first fire caused the second fire, and the second fire caused the third. The final fire was responsible for the total destruction of two businesses and the interruption of a third business. Each fire involved the solid oxidizer sodium chlorate.

Despite having manufactured agricultural chemicals for over 25 years, the owner and management of the facility had a poor understanding of the hazards of sodium chlorate. The fundamental root cause of these fires was the facility owner's inadequate knowledge of the hazards of sodium chlorate. Workers at the facility had little comprehension of the ability of sodium chlorate to cause organic materials to spontaneously ignite. Housekeeping at the facility was poor. And finally, too large of a quantity of sodium chlorate was stored inside the facility, and this large inventory of solid oxidizer was placed adjacent to combustible materials. Simple procedural safeguards would have been sufficient to prevent the ultimate property damage.

1. Introduction

ABC Company was a small, privately owned business that manufactured a variety of herbicide products. The herbicides were solid granular mixtures with sodium chlorate, a powerful oxidizer as their primary active ingredient. The company was located in a small business park with other manufacturers. ABC Company experienced a series of three distinct fires that ultimately destroyed two businesses, caused significant business interruption, and resulted in a serious environmental release.

Two key issues were identified in this series of incidents. First, it was clear that the owner of ABC Company did not understand or appreciate the hazards of sodium chlorate ($NaClO_3$). A simple comparison of the manufacturing practices at this facility with the safety data sheet (SDS) for sodium chlorate revealed serious unsafe practices. Specifically, this was less of a problem of not understanding the hazards; it was more the result of becoming too comfortable with near misses and warning signs of poor process safety management.

Second, the three distinct fires were causally related in a manner similar to the domino effect. In the realm of process safety the domino effect refers typically to a sequence of failures which cascade from a single initiating event. A common example is a sequence of storage tank or pressure vessel failures caused by explosions or fires. The sequence of failures is typically idealized as a prompt series of events. In this case study the domino effect was caused by a combination of poor housekeeping practices with sodium chlorate, improper storage of sodium chlorate, and facility siting. The consequences were delayed and not immediately obvious to emergency response personnel.

For more detail on the theoretical framework used to describe domino effect accidents, as well as investigation and analysis techniques, the interested reader is referred to references [1-5] as well as references therein.

2. Background

The ABC Company facility was a steel clad building set upon a concrete slab foundation. The building included an office area, a laboratory, a warehouse, and several distinct manufacturing areas. Each manufacturing area corresponded to a particular herbicide product. ABC Company received bulk solid and liquid chemicals and blended the ingredients according to recipes. The manufacturing process for any one of their products consisted largely of weighing solid ingredients, blending, coating, agglomeration, drying, screening, and packaging.

Solid sodium chlorate was the primary active ingredient in many of these herbicides. Sodium chlorate is a white, granular, water soluble solid. It is a strong solid oxidizer (NFPA 430 Class 3) "that will cause a severe increase in the burning rate of combustible materials with which it comes into contact" [6]. If exposed to contamination or heat, sodium chlorate has the potential to undergo a violent, self-sustained, thermal decomposition generating heat and oxygen [8]. The thermal decomposition can be so rapid, that such an event can be classified as an explosion, and the TNT equivalent of sodium chlorate has been estimated to be about 0.035 kg TNT per kg of sodium chlorate [11]. Finally, when mixed with organic materials it can cause spontaneous ignition or it can cause the mixture to become shock or friction sensitive [12]. A typical safety data sheet contains all these warnings [13], so facilities that handle and store sodium chlorate should be aware of the hazards.

3. Incident Summary

The First Fire: *Super Sack Fire Suppressed by Automatic Fire Suppression System*

ABC ordered the sodium chlorate because they had scheduled a production campaign of "Monobor Chlorate," one of their herbicide products. Figure 1 is a sketch showing the layout of the two rooms of interest in this investigation, the Pellets Room and the Granules Room.

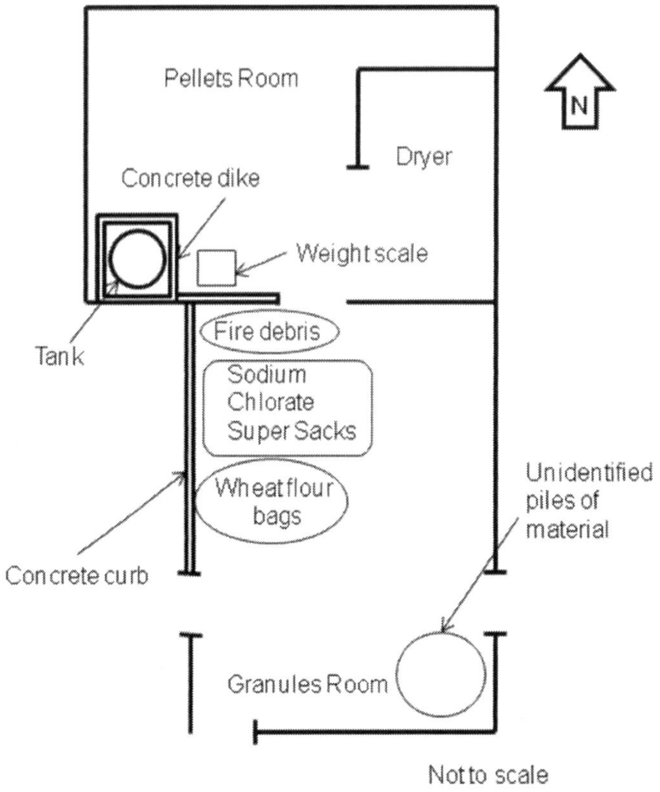

Figure 1. Approximate layout of the Pellets and Granules Rooms (post-fire)

On Day 1, the Pellet Room Operator was instructed to bring a single "super-sack" of sodium chlorate from the Granules Room into the Pellets Room. The operator later stated that he set the plastic pallet with the super sack down at a location roughly four to six feet from the scale in the Pellets Room. He dismounted his fork lift and began to walk away. As he walked away from the fork lift, he turned and reported seeing sparks and flames below the super-sack at the bottom of the pallet. His assistant also saw the flames underneath the pallet.

As the fire grew in intensity, as many as 30 sprinkler heads activated and extinguished the fire before the local fire department arrived. Production activities were suspended and workers spent the next two days cleaning up the fire debris and water.

Examination of the fire scene revealed fire damage to the ceiling in this approximate location. Figure 2 shows the roof damage from an aerial view, and Figure 3 shows the view of the damage from ground level. Figure 4 is a view of the roof damage from inside the Pellets Room, and Figure 5 is a view of the weight scale in the Pellets Room. Aside from the roof damage, the extent of building damage was limited.

Figure 2. Aerial photograph showing roof damage from first fire

Figure 3. Roof damage from the first fire as observed from the Granules Room looking north

Figure 4. Roof damage from the first fire as observed from the Pellets Room looking southeast

Figure 5. Location of weight scale in Pellets Room looking southwest

The Second Fire: ***Extensive Property Damage, Environmental Release Due to Surface Runoff***

On Day 3, ABC employees continued the cleanup activities. Shortly before the end of the work day, several ABC employees reported hearing two or more loud sounds that they identified as an explosion coming from the Granules Room. All workers in the building safely evacuated the building.

The local fire department arrived within minutes of the explosion and extinguished the fire five hours later. Figure 6 shows damage following the Day 3 explosion and fire.

Figure 6. Aerial photograph showing aftermath of Day 3 fire and explosion

Large volumes of fire water drained in an uncontrolled manner to the municipal sewer system and into a nearby creek.

The Third Fire: ***Extensive Property Damage***

In the background of Figure 6 the smoke from a third fire can be seen. This fire started on Day 4 at a facility adjacent to the ABC property. This was the XYZ Company, a manufacturer of wood preservatives and insecticides. Although it was not possible to establish the exact cause of this fire, there were indications that the explosion from the second fire expelled flaming debris to the surrounding area including the XYZ facility.

The XYZ facility stored large quantities of flammable liquids and toxic chemicals. Because the fire department was not able to get information on the hazardous materials stored at the XYZ facility, the incident commander decided to adopt a defensive strategy and allowed the fire to self-extinguish.

4. Causal Factors for the First Fire

The first fire was caused by a forklift operator setting down a plastic pallet of sodium chlorate onto waste sodium chlorate pellets that had fallen to the floor. As he walked away from the fork lift, the operator reported having turned and seeing sparks and flames below the super sack at the bottom of the pallet, followed by flames rising up the super sack of sodium chlorate. There was no hot work being performed in the vicinity of the location of the super sack; no one was reported to be using smoking materials; and there were no electrical equipment or heat sources in this location. The only potential ignition source remaining is the impact- or friction-induced ignition of the plastic pallet facilitated by the presence sodium chlorate.

This fire and its extinguishment (via activation of as many as thirty sprinkler heads) generated a considerable quantity of waste material that was left inside the building. The slope of the floor allowed the wet waste material to drain towards the remaining inventory of solid oxidizer.

5. Causal Factors for the Second Fire

After the first fire, production activities at ABC were suspended and workers were dedicated to cleaning up the fire debris and residual fire suppression water. During these cleanup activities, the plant fire sprinkler system was not returned to an operable condition and was left disabled. Note that, as for the first fire, there was no hot work being performed; no one was reported to be using smoking materials; and there were no electrical equipment or heat sources in the location of the second fire.

The natural slope created a preferred pathway for runoff contaminated with sodium chlorate to flow into the Granules Room and contact organic matter, fire debris, packaging materials, and other combustible items. Organic matter infused with sodium chlorate is susceptible to self-heating and spontaneous ignition. A second fire originated in the Granules Room and was readily able to spread because the sprinkler system was left inoperable following the first fire.

NFPA 430 requires oxidizers to be protected with an automatic sprinkler system [6]. Following the first fire, the sprinkler system was not returned to an operable condition. NFPA 25, *Standard for the Inspection, Testing and Maintenance of Water-Based Fire Protection Systems*, includes specific guidance and requirements regarding impairments to fire protection systems. The purpose of a fire protection system impairment program is to take measures during the impairment to ensure (1) that increased risks are minimized and (2) to limit the duration of the impairment. An emergency impairment is defined as when the system is out of service for more than 4 hours in a 24-hour period. During an emergency impairment, as resulted from activation of the system during the first fire event, NFPA 25 states that "emergency actions shall be taken to minimize potential injury and damage" [7]. The lack of an automatic fire protection system for an extended period of time in the Granules Room did not meet the requirements of NFPA 430 or NFPA 25, and allowed for the uninhibited incipient growth of a second fire event.

The spontaneous ignition of a fire in the Granules Room spread to nearby combustibles and eventually attacked the eight super sacks of sodium chlorate causing its violent thermal decomposition.

The excessive quantity of sodium chlorate (8,000 kg) stored inappropriately inside the building contributed to the magnitude of the damage. One of the two approved methods for storing sodium chlorate is in an open outside area containing no incompatible materials and located away from all other structures [6]. Research has shown that even a small degree of confinement, such as that offered by a three-sided roofed enclosure, is sufficient to cause an initially small fire to produce an explosion [11].

Segregated storage of the eight super sacks of sodium chlorate inside the Granules Room was contrary to the requirements of NFPA 430. NFPA 430 specifically requires that the storage of sodium chlorate shall only be cutoff or detached; segregated storage is not permitted. NFPA 430 also requires that oxidizers shall be stored to avoid contact with incompatible materials such as ordinary combustibles [6]. ABC Company failed to comply with these requirements.

Based on the conditions present after the first fire and the hazards associated with Class 3 oxidizers, removing the eight super sacks of sodium chlorate from the area where the fire protection system was impaired would have been the most effective means of preventing the additional damage caused by the second fire event.

6. Causal Factors for the Third Fire

Within hours of extinguishing the second fire, a third fire started on the adjacent business property, a manufacturer of wood treatment chemicals.

This third fire was determined to be caused by contact with hot sodium chlorate and burning debris scattered by the explosion during the second fire. This projection of fire via burning debris is an example of a fire cascading to yet another stage, in a domino-effect fashion.

7. Discussion

There has been an increasing level of interest in recognizing, evaluating, and controlling chemical reactivity hazards [9]. The Center for Chemical Process Safety (CCPS), a technical directorate of the American Institute of Chemical Engineers (AIChE), published a guideline intended for general audiences in the manufacturing community [10]. This text specifically identifies solid oxidizers as a chemical reactivity hazard, and offers the following caution on page 52:

> *"Inadvertent contact of oxidizers with reducing agents, including combustible materials, is the most important issue regarding what can go wrong when handling oxidizing substances. This contact will increase the burning rate of the combustible*

materials; it may also cause a fire to ignite without any additional ignition source. Some oxidizers can also undergo self-sustained decomposition, vigorously or explosively, when contaminated or exposed to heat or shock."

The management practices presented in the CCPS guidance would have been sufficient to prevent this series of fires.

The origin of the first fire was determined to be near the weight scale in the Pellets Room. The first fire was caused by the ignition of a plastic palled induced by friction or impact in the presence of sodium chlorate.

The origin of the second fire was in the Granules Room, and it was due to self-heating and spontaneous ignition of organic matter infused with sodium chlorate. The thermal decomposition of the super-sacks resulted in an explosion.

ABC was solely responsible for compliance with the requirements of NFPA 430, or with complying with generally accepted good engineering practices regarding storage and handling of sodium chlorate. The damages from the second fire could have been limited by removing the super-sacks from the Granules Room after the first fire. Likewise, the damages from the second fire could have been limited by returning the automatic sprinkler system back to service. Finally, ABC could have prevented the second fire with better housekeeping/waste disposal.

8. Conclusion

This case study examined a succession of three separate fires which occurred in a period of four days at an agricultural chemical manufacturing facility. The facility was located in a small chemical park. The succession of fires illustrates the concept of the domino effect: the first fire caused the second fire, and the second fire caused the third. The final fire was responsible for the total destruction of two businesses and the interruption of a third business. Each fire involved the solid oxidizer sodium chlorate.

Awareness on behalf of ABC of the hazards and simple requirements for the use and storage of sodium chlorate as outlined in readily available resources [6, 8, 11, 12, 13] would have dramatically reduced the magnitude of the damage associated with the fires, if not prevented them altogether.

9. References

1. Darbra, R. M., et al., "Domino effect in chemical accidents: Main features and accident sequences," *J. Hazard Mater* 183 (2010) 565-573

2. Zhang, X and Chen G., "Modeling and algorithm of domino effect in chemical industrial parks using discrete isolated island method," *Safety Sci* 49 (2011) 463-467

3. Abdolhamidzadeh, B. et al., "Anatomy of a domino accident: Roots, triggers and lessons learnt," *Process Saf Environ* 90 (2012) 424-429

4. Bernechea, E., et al., "A model for estimating the impact of the domino effect on accident frequencies in quantitative risk assessments of storage facilities," *Process Saf Environ* 91 (2013) 423-437

5. Kadri, F. et al., "Method for quantitative assessment of the domino effect in industrial sites," *Process Saf Environ* 91 (2013) 452-462

6. NFPA 430, *Code for the Storage of Liquid and Solid Oxidizers*, National Fire Protection Association, Quincy, 2000.

7. NFPA 25, *Standard for the Inspection, Testing and Maintenance of Water-Based Fire Protection Systems*, National Fire Protection Association, Quincy, 2000.

8. HSE, *Storage and Use of Sodium Chlorate and Other Similar Strong Oxidizers*, Guidance Note CX3, Fourth Edition, 1998.

9. CSB, *Hazard Investigation: Improving Reactive Hazard Management*, U.S. Chemical Safety and Hazard Investigation Board, Washington, DC, October 2002.

10. CCPS, *Essential Practices for Managing Chemical Reactivity Hazards*, American Institute of Chemical Engineers—Center for Chemical Process Safety, New York, 2003.

11. Hazardous Installations Directorate, *Safety Report Assessment Guide: Chemical Warehouses – Hazards*. October 31, 2002.

12. Urben, P. G. (Editor), "*Bretherick's Handbook of Reactive Chemical Hazards,*" 6th edition, 1999.

13. J.T. Baker®, *Sodium Chlorate Material Safety Data Sheet*, MSDS Number S3314, October 8, 2004.

Complex Explosion Development in Mines: Case Study – 2010 Upper Big Branch Mine Explosion

Scott G. Davis
GexCon US
4833 Rugby Avenue, Suite 100
Bethesda, MD 20814
sgdavis@gexcon.com

Derek Engel and Kees van Wingerden
GexCon US
4833 Rugby Avenue, Suite 100
Bethesda, MD 20814

Prepared for Presentation at
American Institute of Chemical Engineers
2014 Spring Meeting
10th Global Congress on Process Safety
New Orleans, LA
March 30 – April 2, 2014

UNPUBLISHED

Complex Explosion Development in Mines:

Case Study – 2010 Upper Big Branch Mine Explosion

Scott G. Davis, Derek Engel & Kees van Wingerden
GexCon US
4833 Rugby Avenue, Suite 100
Bethesda, MD 20814
sgdavis@gexcon.com

Keywords: *explosion dynamics, mine explosions, accident investigations*

Abstract

On April 5th, 2010 a methane explosion occurred within the Upper Big Branch mine south of Charleston, WV. Twenty-nine men lost their lives as a result of a flammable concentration of methane that built up in the enclosed space and ignited, resulting in a methane explosion that transitioned into a coal dust explosion. This study used the FLACS CFD solver to conduct a detailed explosion analysis to evaluate the complex overpressure development throughout the mine as a result of the flammable cloud ignition. As a result of the accident investigation, unique explosion patterns were found in the mine where certain "blast indicators" within the mine shafts were deformed in such a manner that was inconsistent with the likely flow of the expanding blast wave. The FLACS analysis will analyze the explosion dynamics and shed light on the damage observations made after the blast.

1 Introduction

A powerful explosion tore through the Upper Big Branch (UBB) mine, killing 29 miners and injuring 2, at approximately 3:02 p.m. on April 5, 2010 in Montcoal, West Virginia. The UBB mine was owned by Massey Energy and operated by its subsidiary Performance Coal Company (PCC). This explosion was the deadliest coal mine disaster in the United States in 40 years.

At the time of the accident, the mine had four sets of drift[1] openings and a fan shaft. Miners generally entered the mine from the Ellis portal or the North or South (known as UBB) portals. Four active sections were producing coal, including the 1 North Longwall Panel, Headgate 22 (HG 22), Tailgate 22 (TG 22), and one advancing room-and-pillar section (known as the "Barrier Section"). A detailed mine map is shown in Figure 1.

[1] Drift - a horizontal opening into a coal mine or "tunnel"; miners and supplies enter at the drift; also can provides a source for ventilation

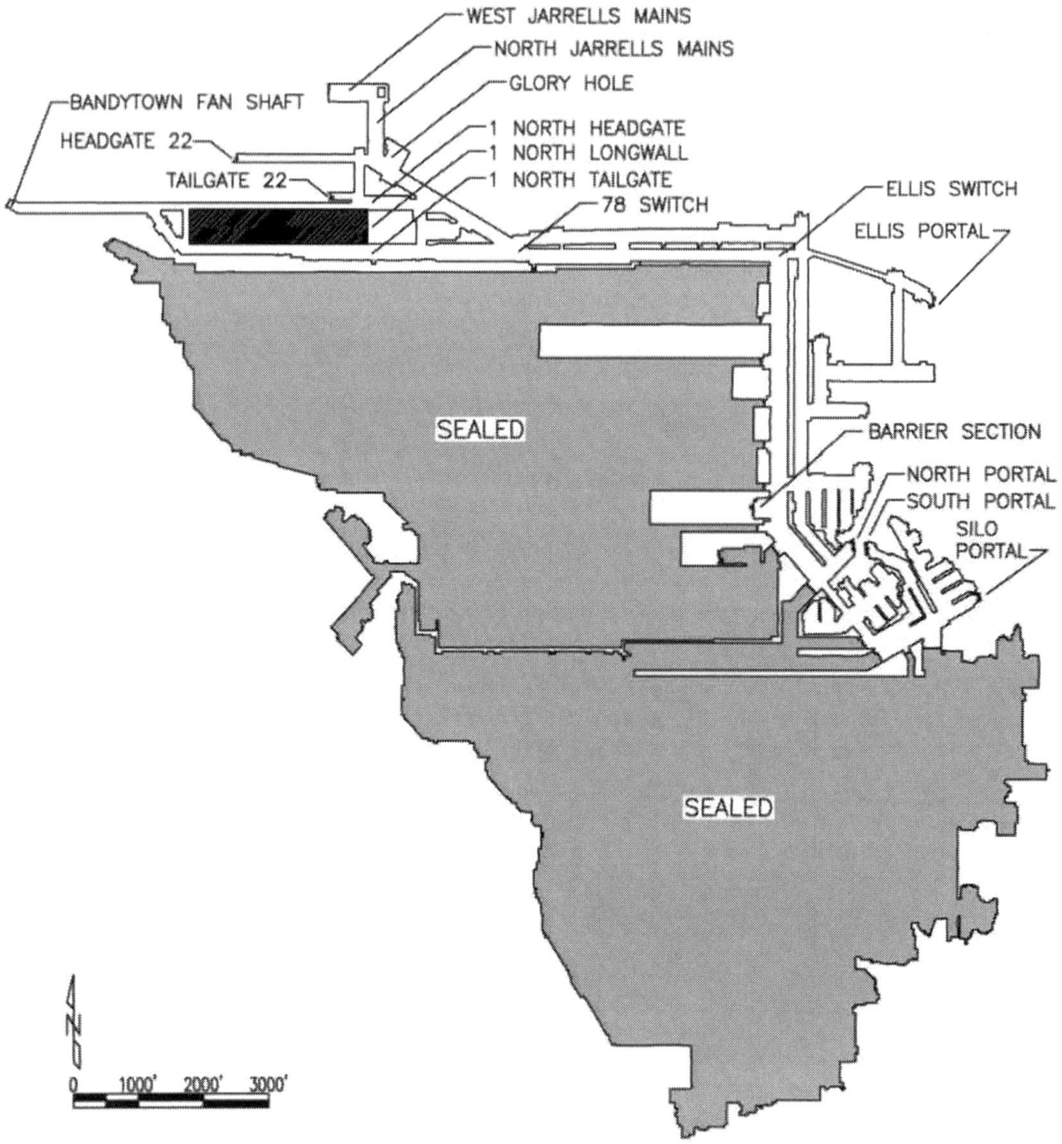

Figure 1: UBB Mine Map

At UBB, coal was mined from the Eagle coal seam, which had an average thickness of 54 inches. A longwall was utilized at UBB, which is a method of mining in which a cutting machine known as a "shearer" cuts coal in a long, single slice (see Figure 2). A number of hydraulic jacks or "shields" (typically 1.75 meters wide) are placed side-by-side to support the coalface (see Figure 2). As the coal is extracted, the shields, which are tapered downward towards the rear, advance and the roof collapses behind them. This collapsed material behind the shields is commonly referred to as the "gob".

Figure 2: Example longwall with shearer and shields

The approximately 1,000-foot wide active longwall was supported by 176 shields, which extended across the full longwall face. The headgate is the start of the longwall face (shield 1). At the time of the incident approximately 5,450 feet of the longwall panel had already been mined, with about 1,240 feet remaining in the panel. Figure 3 shows a diagram of the 1 North Longwall Panel, Headgate #22 and Tailgate #22. The grey area in the figure is the region that had been previously mined (gob), the red lines are the conveyor belts at the time of the incident, the longwall face at the time of the incident is the region just to the right of the grey area, the white region (between the longwall face and stop) is the region to be mined, and the drifts or "tunnels" are outlined by grid like layout in the figure. The longwall was being mined from left to right in the figure, whereas the Headgate #1 is at the top and Tailgate #1 is at the bottom of the region to be mined.

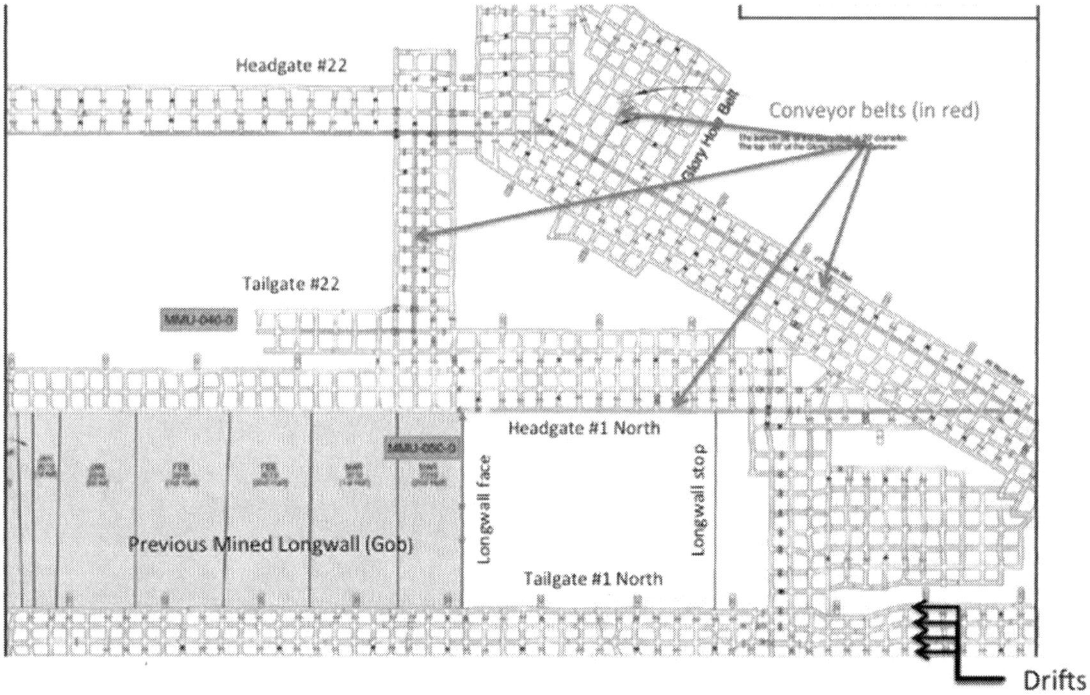

Figure 3: UBB Longwall, Headgate #22 and Tailgate #22 areas

The goal of this paper was to evaluate and better understand the unique explosion patterns that can develop within a mine after an explosion. To accomplish this, we used FLACS, a 3D CFD code that can be used to study the development of explosions in complex geometries. While GexCon was part of an investigation team that analyzed the cause of the accident, the results from that work are proprietary and confidential. Therefore, the results from the two main publically available investigation reports were used in the present analysis, namely that conducted by the US Department of Labor Mine Safety and Health Administration – Coal Mine Safety and Health (MSHA) [1] and by Governor Tomblin's Independent Investigation Panel (GIIP) [2]. As the goal of this report is to evaluate the complex explosion development in mines, GexCon makes no comment as to the validity of the details provided in investigation reports and simply adopts the findings in these reports. For more details regarding the investigations, the reader is referred to the publically available reports.

The two public agency reports [1][2] generally concluded that methane that had accumulated in the gob behind the shields and subsequently migrated to the longwall face. As the shearer cut into the sandstone mine roof, the resulting sparks ignited a pocket of methane, creating a fireball that ignited the methane that had accumulated in the gob behind the shields. The flame traveled into the tailgate area, where it ignited an explosive mixture of methane that had also accumulated in the tailgate. The methane explosion lifted accumulated coal dust, and the explosion transitioned into a coal dust explosion, which propagated through more than two miles of the mine.

The next section will describe the main findings of the MSHA and GIIP reports regarding the cause of the blast and the development of the blast. Using this data, we applied a FLACS analysis of the explosion dynamics to shed light on the observations after the blast. Due to time constraints and the computational demands of the mine geometry, a ventilation and methane release study were not conducted to evaluate the methane accumulation in the present paper. The specifics of the explosion study will be discussed later.

2 Description of the Incident Findings

2.1 Chain of Events

The reports [1][2] concluded that methane accumulated on the tailgate end of the longwall. MSHA concluded that methane was likely liberated from floor fractures at the longwall shields near the tailgate on April 5, the day of the explosion. In addition, the lack of roof supports in the tailgate area caused the roof to fall in this region. The fallen roof restricted the airflow leaving the longwall face, allowing methane to accumulate instead of being diluted or ventilated from the mine. MSHA identified the longwall shearer as the likely ignition of the methane accumulated in the tailgate region.

Their investigation indicated that the longwall shearer was being used to mine the area near the tailgate at the time of the explosion. Their findings state that methane likely migrated from behind the longwall shields to the longwall shearer that was cutting coal and sandstone. MSHA found that the ignition was likely caused by two worn drill bits on the tail drum, which lost their carbide tips, and created "hot streaks" (smears of metal found on rock when metal is heated to near its melting point from friction caused by the shearer's bits hitting into layers of rock above or below the coal seam) while cutting the sandstone. In addition, seven water-spray nozzles were missing on the drum, hence failing to move the methane away from the shearer bits and cool the hot streaks created during the mining process. As a result of this heating methane ignited. The GIIP report concluded that mechanical sparks generated by the shear caused the ignition, and the inadequately maintained water-spray nozzles could not extinguish the flame.

Their investigation indicated that at approximately 3:02 p.m., the flame from the initial methane ignition ultimately encountered a larger methane accumulation in the tailgate area, triggering a localized explosion. This methane explosion near the tailgate immediately encountered fuel in the form of dangerous accumulations of coal dust, which propagated a dust explosion originating in the tailgate entry. The resulting coal dust explosion killed the 29 miners. Loose coal, coal dust and float coal dust[2] were abundant in all areas of the mine, including the area affected by the explosion.

MSHA further concluded that inadequate levels of inert rock dust[3] were applied. U.S. Bureau of Mines conducted tests that demonstrated that the explosive nature of coal dust can be rendered

[2] Float coal dust - fine coal particles (typically less than 0.75 microns) that can remain airborne for longer periods of time than coarse dust, and which can be transported by the ventilating air relatively long distances before settling.

[3] Rock dust is pulverized limestone, dolomite, gypsum, anhydrite, shale, adobe, or other inert material, preferably light colored, 100 percent of which will pass through a sieve having 20 meshes per linear inch and 70 percent or more of which will pass through a sieve having 200 meshes per linear inch.

inert by the application of rock dust. Federal Standards state that in areas where rock dust is required, the incombustible content of the combined coal dust, rock dust, and other dust shall be not less than 65% at intake airways and no less than 80% at the return airways. Analyses of 1,803 dust samples taken from UBB after the explosion showed that more than 78% were out of compliance with the federal standard.

2.2 Flame and forces

Along with determining percent of incombustible materials, MSHA also had the dust samples tested for coke using the "Alcohol Coke Tests", which identify trace, small, large or extra-large quantities of coke in each sample. MSHA used the coke results, along with other physical evidence, as an indicator of flame travel. They stated that, "Research on alcohol coke testing indicates that coke is found whenever coal particles are dispersed into a flame, and therefore the presence of coke is a good indication of flame travel." Figure 4 shows the extent of flame travel as outlined in red in the figure.

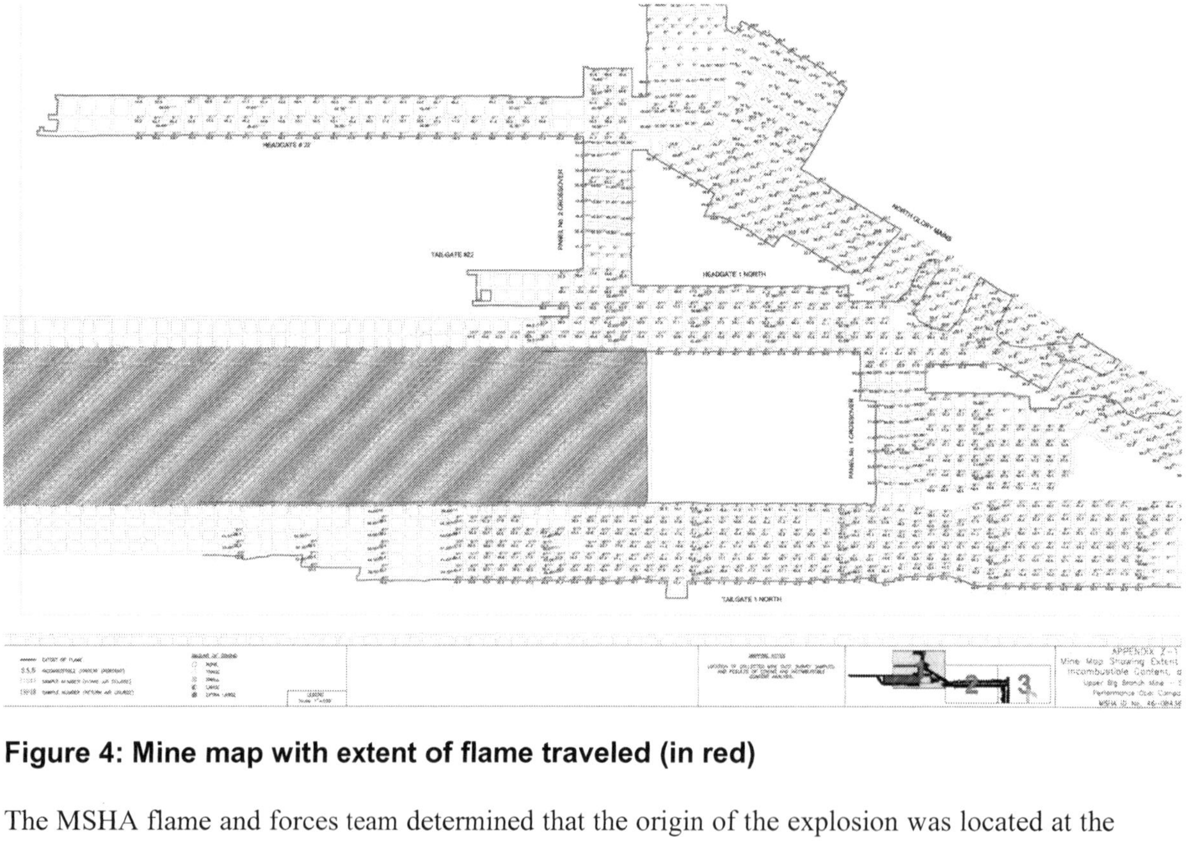

Figure 4: Mine map with extent of flame traveled (in red)

The MSHA flame and forces team determined that the origin of the explosion was located at the intersection of the active longwall face and the No. 7 entry of Tailgate 1 North. This location was near crosscut 48 and marked with a yellow arrow in Figure 5. MSHA concluded that the localized methane explosion suspended and ignited coal dust, which did not have sufficient incombustible content to prevent a dust explosion. The next paragraphs provide a brief description of the flame travel and pressures determined by MSHA, which are also summarized in Figure 5 utilizing the large red arrows.

MSHA also determined that the subsequent coal dust explosion began approximately at the intersection of crosscut 48 and entry No. 6 in Tailgate 1 North. Explosion forces were generated by the flame in all directions, including back across the longwall face where they determined that an approximately 14 psi (pounds per square inch) pressure wave traveled back to the longwall tailgate. MSHA determined that the explosion damage indicated that the coal dust explosion initially propagated in all directions, where the flame eventually traveled through all entries and crosscuts in Tailgate 1 North from as far as crosscut 77 (west of the longwall). They estimated that the flame traveled in this direction at about 1,000 feet per second (fps) while generating a pressure exceeding 18 psi.

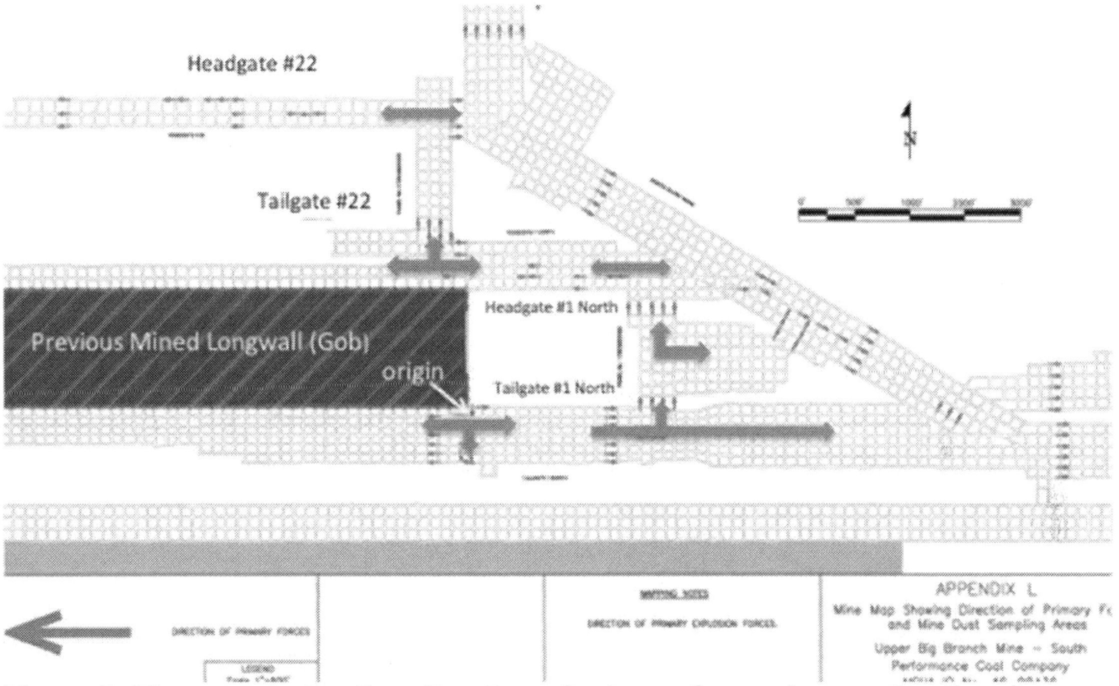

Figure 5: Mine map indicating direction of primary forces (arrow directions)

The explosion also traveled into the tailgate entries east of the longwall. The flame initially traveled in the tailgate entries at about 600 fps, generating a pressure of nearly 6 psi, whereas MSHA estimated that additional coal dust caused the flames to accelerate to over 1,000 fps in all tailgate entries several hundred feet before reaching the crossover entries. Eventually the flame propagated to the locations where the tailgate intersects the North Glory Mains (the horizontal 45° angle drifts), where the flame extinguished in this direction. MSHA also determined that the flame turned 90° to the left and entered the crossover between Tailgate 1 North and Headgate 1 North. As flame entered Headgate 1 North, MSHA determined that flame propagated with at a speed of about 1,200 fps, generating over 20 psi (estimated from damage to several monorail sections). Although flame did not enter the longwall face, pressures ranging from 7 to 14 psi did travel along the longwall from the headgate.

MSHA also noted that the flame of the coal dust explosion also traveled toward Tailgate 22, where explosion pressures increased to near 20 psi. Just before entering TG 22, the flame also turned 90° right and accelerated through the crossover entries between TG 22 and HG 22. The flame entered HG 22 and turned 90° left and right. The portion of the flame that turned left traveled into HG 22 and propagated to the faces at speeds approaching 1,500 fps generating a pressure of approximately 25 psi. MSHA calculated that explosion pressures were on the order of 52 to 65 psi, due to pressure piling that occurred as the flame and forces continued to push against the dead end of HG 22. This resulted in a reflected shock wave traveling back though HG 22 that could have reached a maximum pressure of 105 psi, whereas the flame extinguished after reaching the faces as it consumed the available oxygen in HG 22.

MSHA concluded that the flame extent and the pressures generated by the explosion are consistent with a coal dust explosion not a massive methane explosion. They stated that for a methane explosion to cover the area where flame passed at UBB, it would have generated pressures far in excess of what was observed and calculated for this explosion. MSHA estimated that the explosive accumulation of methane that was eventually ignited contained approximately 300 cubic feet of methane. When diluted with air to 10 percent, this volume of methane would form an explosive volume of 3,000 cubic feet. The methane explosion propagated away from the longwall face. With a flame speed of approximately 300 feet per second, the methane explosion would have extinguished in about 1/2-second while generating a maximum pressure of about 4 psi.

The flame zone that actually occurred at UBB had a volume of about 31 million cubic feet. MSHA concluded that his flame zone would easily be achieved in a coal dust explosion that generates limited pressure. They also concluded that the ignition of such a volume of methane at stoichiometric or ideal conditions in an underground mine could have resulted in a detonation with possible explosion pressures exceeding 600 psi, many times greater than the explosion forces that actually occurred underground.

2.3 Blast and drag forces

We used the GIIP report, along with the blast survey data in the MSHA report, to evaluate blast and drag (dynamic pressure) forces were observed in the mine. The combined blast effects were comprised of both the pressure loads and drag forces. Overpressure damage is mainly associated with large objects, where a differential in pressure loading can cause structural damage, while drag forces are associated with the speed at which the gases/air are flowing past an object, sometimes referred to as the explosion or blast "wind". Figure 6 shows the remains of a cinder block wall that was destroyed by overpressure generated by the explosion or shock wave within the mine. Figure 7 and Figure 8 show damage that is more typically associated with the blast wind or drag forces (dynamic pressure), which in these photos deformed the steel "pie pans" in the direction that the wind was flowing past the objects. Pie pans are round plates bolted against the rock surface – the rock bolts are used to reinforce the overhead rock.

Figure 6: Remains of a cinder block wall [2]

Figure 7: Damage to the rock bolted metal plates (pie pans) on the ceiling [2]

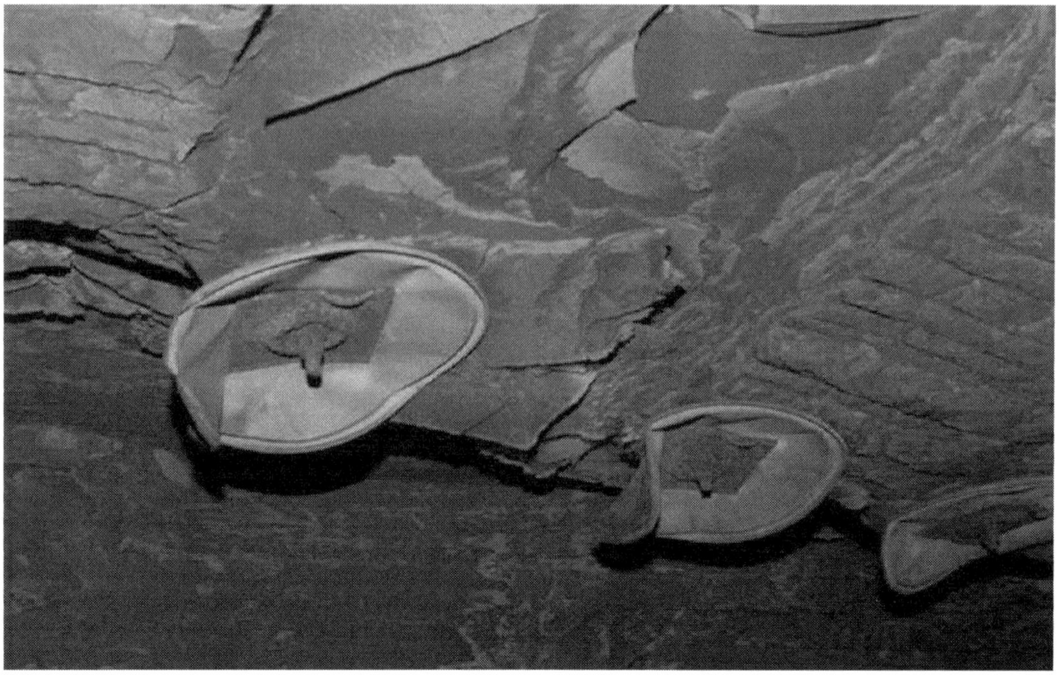

Figure 8: Damage to rock bolted plates or "pie pans" [2]

Depending on how the pressure is generated and propagated within the mine, unique explosion patterns can be found in the mine where certain "blast indicators" within the mine shafts were deformed in such a manner that was inconsistent with the likely flow of the expanding blast wave. One example that was found in the mine was that of the scoop, or a rubber-tired haulage vehicle, found in Headgate 22 after the explosion (Figure 9). Explosion forces were observed which originated through a cross cut and blowing the mandoor[4] towards the scoop, while the pie pan is deformed in another manner not consistent with the blast originating from the cross cut. The detailed layout scoop within Headgate 22 can be seen in the MSHA survey shown in Figure 10.

[4] A stopping is a ventilation control that is used to separate fresh air from return air. The mandoor is a doorway through the stopping.

Figure 9: Scoop located in Headgate 22 after the explosion

Other evidence of inconsistent blast forces were observed in Headgate 22. The mantrip[5] found Headgate 22 had blast damage that confirmed the direction of the expanding flame front and blast wave, and at the same time had damage that apparently contradicted the blast direction. More specifically, the force of the explosion was great enough to blow two metal canopies off of the section mantrip and hurl them for hundreds of feet in opposite directions. Figure 11 shows an example of a mantrip from the UBB mine with the metal canopies still present. Figure 12 shows the general location of the mantrip in HG 22, while Figure 13 shows a more close up view of the mantrip and the remains of the steel canopies blown over 200 feet east and west of the mantrip.

The physical damage observed on the mantrip does not intuitively follow the direction of the blast, and in fact presents apparently contradictive blast indicators. This is an ideal case that can be evaluated with more advanced methods, such as CFD.

[5] A vehicle on a track used to take miners to and from the working places. Due to the low ceilings in mines, mantrips tend to be of reduced height.

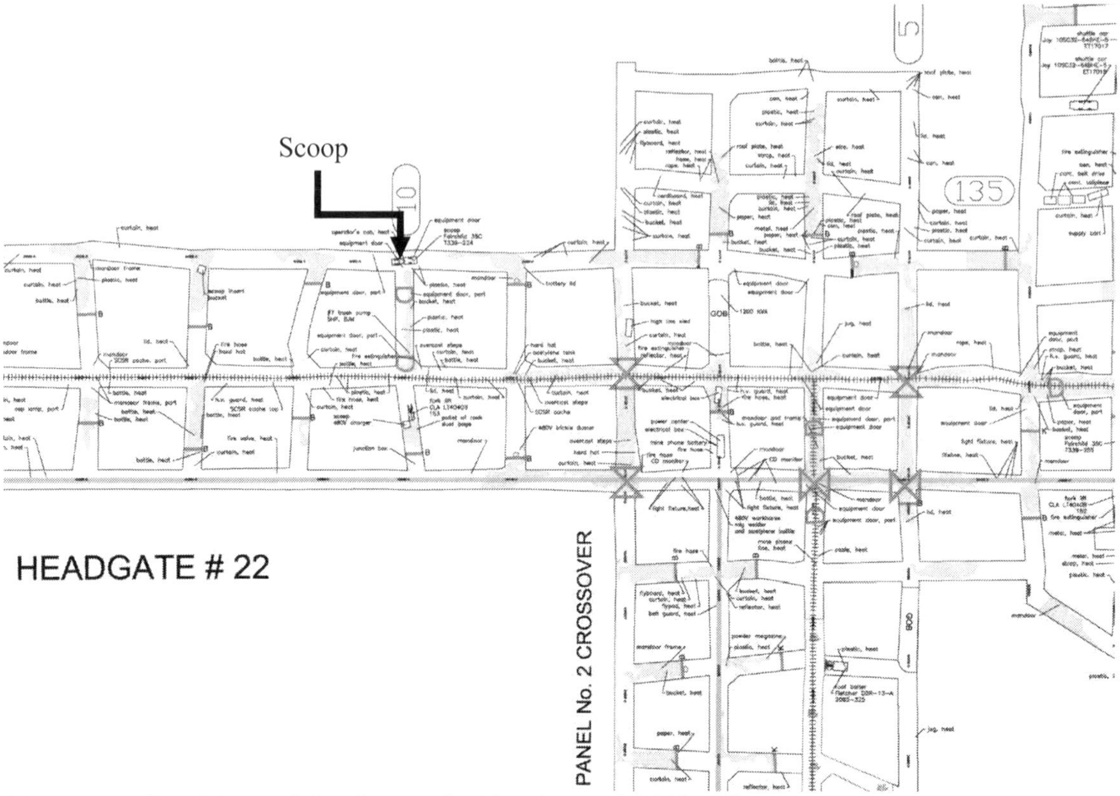

Figure 10: Position of the Scoop in Headgate 22 [1]. The green areas are the debris fields found in the mine, while the red lines damaged concrete block walls, the red "D" are damaged equipment doors.

Figure 11: Example of a mantrip from the UBB mine [2]

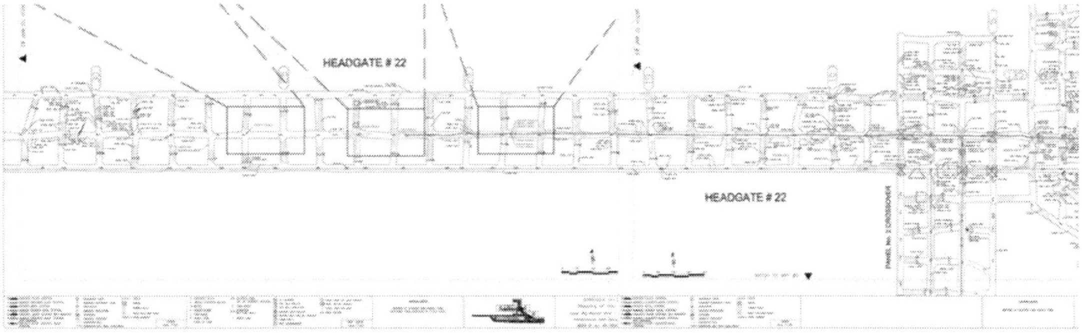

Figure 12: Mine survey of HG 22 showing the location of the mantrip [1]

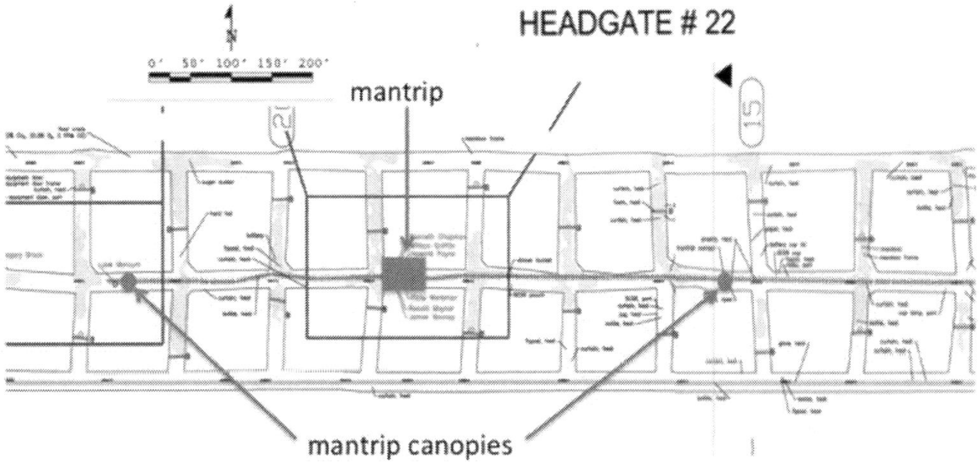

Figure 13: Mine survey of the HG 22 mantrip and the canopy remains blown over 200 feet both east and west of the mantrip [1]

3 CFD modeling Analysis

GexCon undertook CFD simulations to determine scenarios that may explain the complex explosion dynamics that were observed in the mine. For the present paper, the goals of the simulations were not to identically reproduce the exact failure scenario, but to identify and understand the governing explosion dynamics that can explain the observed explosion damage.

To understand which mechanisms could lead to the accident and observed damage, a number of explosion simulations were performed using the computational fluid dynamics (CFD) model FLACS. FLACS is a leading CFD tool for the simulation of gas dispersion and vapor cloud explosion scenarios, and is extensively used for safety studies in the petrochemical industry. FLACS is capable of modeling gas and aerosol releases, dispersion of vapors, ventilation in structures, and ignition of flammable fuel-air mixtures to evaluate the flame front progression and the overpressures due to explosions. Sensor points can be placed within the computational domain to track the time history of relevant variables such as static or dynamic pressure, velocities, fuel concentration, etc.

3.1 FLACS description

FLACS solves the compressible Reynolds-Averaged Navier–Stokes (RANS) equations (conservation equations for mass, momentum, enthalpy and species) on a 3D Cartesian grid using the finite volume method [3]. The RANS equations are closed using the k-ε turbulence model with the standard set of constants taken from Launder and Spalding [4]. The SIMPLE pressure correction method [4] is applied, and extended for compressible flows with source terms for the compression work in the enthalpy equation.

FLACS contains a flamelet-based combustion model with one-step reaction kinetics, where the laminar burning velocity is one important measure of the reactivity of a given mixture. A model for flame growth that describes how the local reactivity changes with parameters like concentration, temperature, pressure, turbulence, etc. is implemented to model combustion and explosion dynamics. The real flame area is then properly described and corrected for curvature at smaller scales than the grid, using sub-grid models to account for the interaction of the flame with objects smaller than the grid size [5]. Flame acceleration is included, due to flame instability, flame-folding by obstacles and turbulent mixing [6]. Also, it is worth mentioning that FLACS assumes a reduced set of combustion products (CO_2, H_2O, CO, H_2, OH) based on the thermodynamic equilibrium of the beginning state of the reactants (e.g., air, fuel, pressure, temperature).

Since even small objects can have a significant impact on gas dispersion and – even more so – on flame acceleration, the accurate representation of the 3D geometry has always been a focal point in the development of FLACS. A so-called "distributed porosity concept" [7] was developed to allow the efficient handling of complex geometries using a Cartesian grid: large objects and walls are represented on-grid, while smaller objects are represented as sub-grid elements. Sub-grid elements contribute to flow resistance, turbulence generation and flame folding in the simulation, which allows the effect of small obstacles to be accounted for accurately, while maintaining reasonable simulation times.

3.2 NIOSH LLEM Experiments – FLACS validation

While FLACS has been extensively validated against explosions in petrochemical facilities, it has also been validated against mine experiments. One such set of tests was performed by NIOSH, they conducted a series inside the tunnels of the Lake Lynn Experimental Mines [8] to evaluate blast wave propagation in confined mines and evaluate pressure loading on seals. The purpose of including these experiments is to illustrate how pressure wave decay will become significantly different when confined inside a tunnel system. Two different experimental configurations were simulated, and consisted of a single drift (single tunnel) and a three-drift (three tunnels). The model of the Lake Lynn Experimental Mines tunnels are shown in Figure 14, the one-drift tunnel is referred to as the D-drift, whereas the three-drift tunnel is referred to as the B-drift tests. GexCon participated in a blind test study, were we simulated a total of six experiments in these two test configurations as part of a commissioned study for NIOSH in 2006 without prior knowledge of the outcome. The following is an overview of FLACS' performance:

Table 1: FLACS blind prediction performance reported by (Zipf *et al.*, 2007) for two different sensors in each experiment

Test	Cloud Size	Geometry	Ignition	Test (kPa)	FLACS (kPa)	Deviation
468	3.66x6x 2m	D-drift	Inner end	#1 21.5	18.7	-13%
				#10 18.6	17.6	-5%
469	8.23x6x2m	D-drift	Inner end	#1 58.9	57.5	-3%
				#10 48.1	51.9	+8%
470	12.2x6x2m	D-drift	Inner end	#1 75.0	76.3	+2%
				#10 74.8	71.0	-5%
484	12.2x6x2m	B-drift	Inner end	#10 83.9	71.2	-15%
				#526 29.4	28.9	-3%
485	18.3x6x2m	B-drift	Inner end	#10 101.3	97.6	-4%
				#526 34.1	42.3	+24%
486	18.3x6x2m	B-drift	Center	#10 48.1	57.1	+19%
				#526 24.9	23.1	-7%

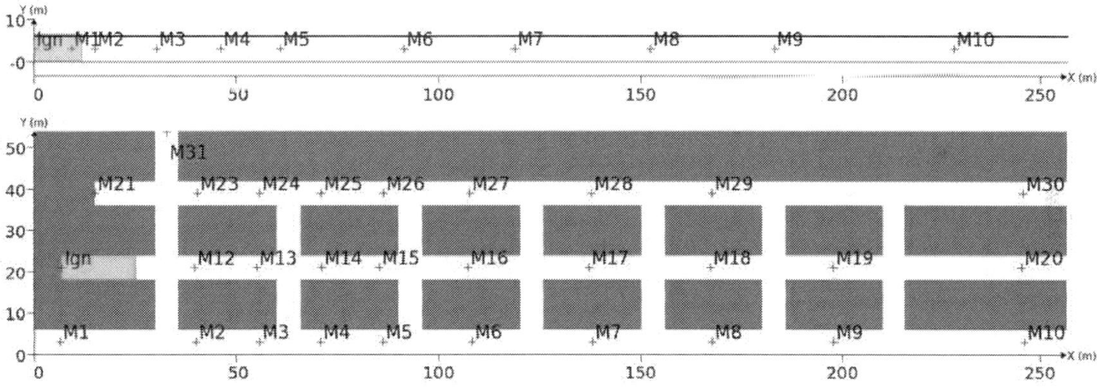

Figure 14: Illustration of LLEM geometry model with pressure sensors, the D-drift tunnel is shown in the upper plot and the B-drift tunnel in the lower plot. The numbering of sensors may deviate from that used in the experiments. The green sections are the locations of the flammable clouds.

To illustrate the pressure decay in tunnels, and FLACS' ability to model this phenomenon, we have selected two of the experiments listed above. These are the strongest explosion tests in each scenario, test 470 in the D-drift and test 485 in the B-drift. In Figure 15, the pressure development in the FLACS simulation as the pressures are propagating down the tunnels is compared to the observations. It can be seen that the simulated pressure decay follows the experimental observations closely. In Figure 16 three pressure curves are selected for each of the two tests and compared to simulated pressures. The first point is chosen in the inner end of each drift, where the highest explosion pressures were seen, the third point around 230 m from the ignition, and the point in the middle was around 130-150 m from ignition. By studying these curves in detail, it is evident that the physics reproduced in the FLACS-CFD software closely resembles the observed physics from the experiments. One interesting observation is that the losses for the pressure wave propagation in the single tunnel are much less than those observed in the more complex three-drift tunnel.

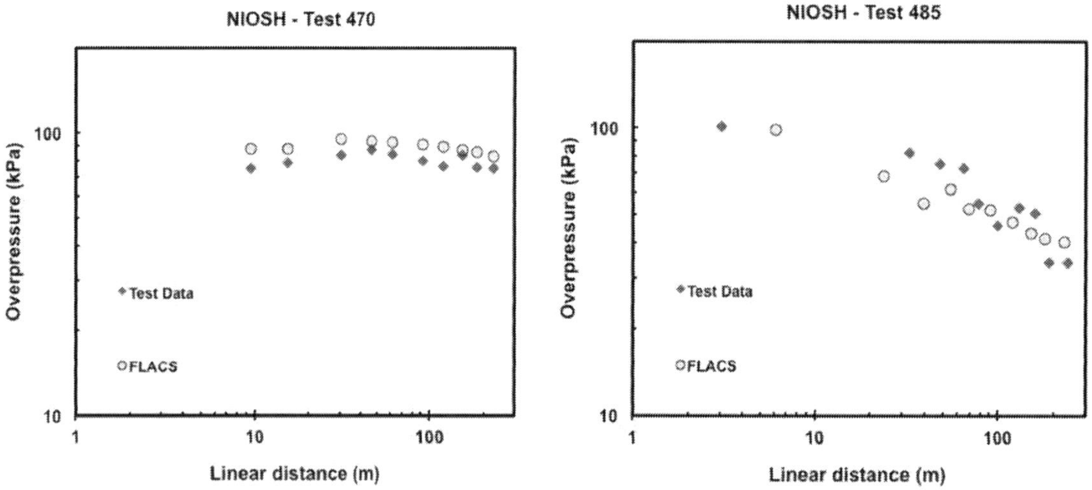

Figure 15: Example of pressure decay with distance, test 470 (left, one drift tunnel with 150 m^3 gas cloud) and test 485 (right, 3-drift tunnel with 220 m^3 gas cloud)

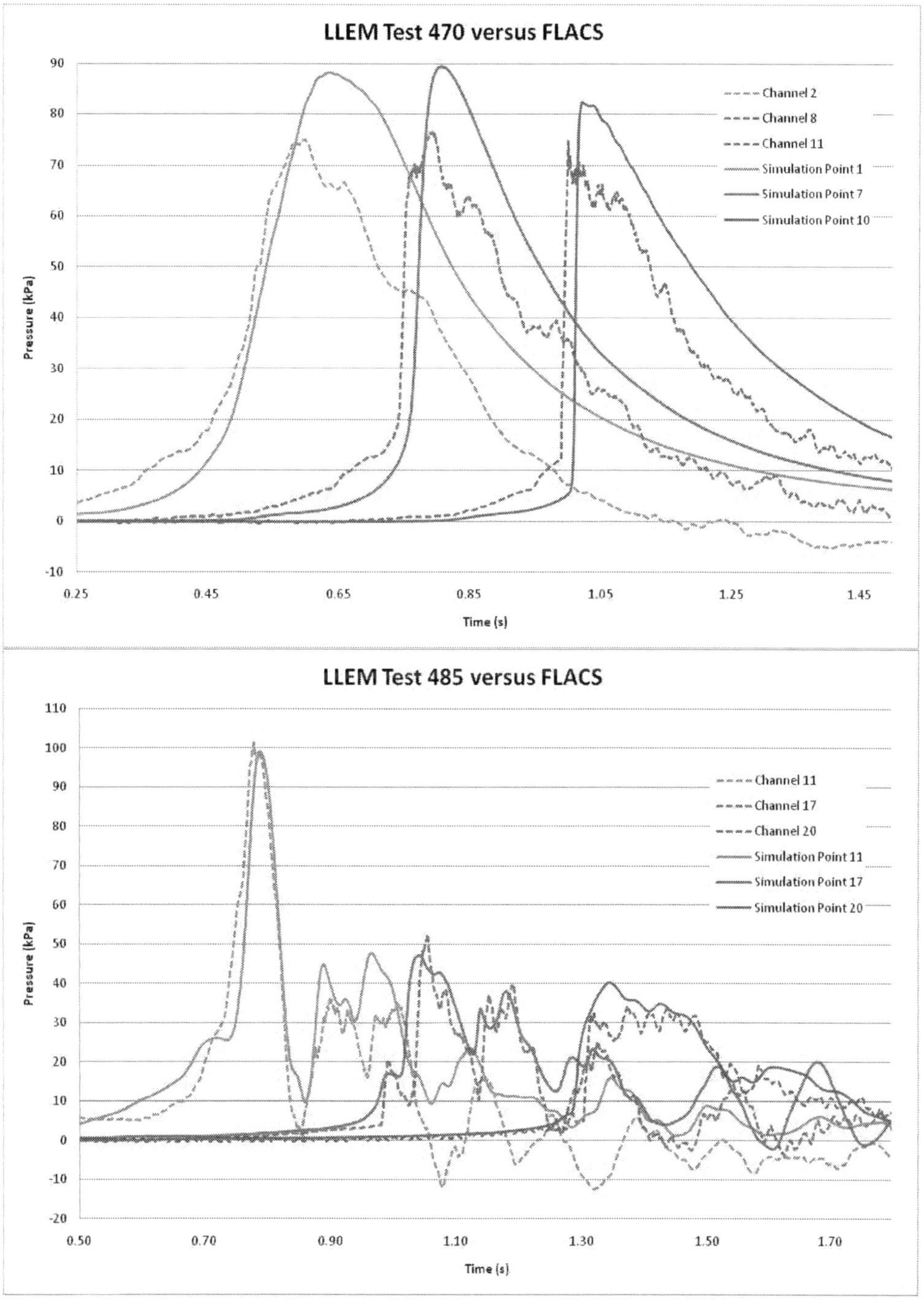

Figure 16: Pressure decay with distance NIOSH LLEM experiments 470 (upper) and 485 (lower) versus FLACS

3.3 Explosion physics in mines

The development of explosions within mines is a very complex process and can develop pressures much larger than those observed in small closed volume experiments. For example, the theoretical largest explosion pressure of a burning methane-air deflagration in a closed volume system is given by:

$$\frac{P_{max}}{P_{in}} = \left(\frac{n_{exp}}{n_{in}}\right)\frac{T_{max}}{T_{in}}$$

where P_{max} is the explosion pressure, P_{in} is the starting atmospheric pressure, n_{exp}/n_{in} is the molar ratio of products to reactants (typically around 1), and T_{max}/T_{in} is the temperature ratio of burned to unburned gases. For mixtures near the flammability limits, the maximum explosion pressure is around 1 barg (14.7 psig) and for mixtures near stoichiometric the theoretical value increases to approximately 8 barg (117 psig).

In mines however, the flame starts as a slow deflagration and the burned gases expand and "push" the unburned gases ahead of the flame. The velocity of the unburned gas ahead of the flame increases, which increases turbulence ahead of the flame. While the degree of turbulence is dependent on fuel, concentration and geometry, the increased turbulence in the unburned gases increases the surface area of the flame and hence increases the combustion rate. This will cause the burned gases to expand more rapidly and results in compression/pressure waves ahead of the flame. The pressure waves will "pre-compress" the unburned gases or P_{in}, hence resulting in pressures that can be 7 times this initial value due to the expansion ratio. These transient pressures can be much higher than the equilibrium pressures and can be the result of very fast deflagrations. Under certain conditions, these deflagrations can even cause transition to detonation, or a shock wave induced combustion front.

In a dispersed phase deflagration (e.g. dust), the flame front begins propagating through the suspended fuel, which in turn results in rapid heating and expansion of the burned gases. This expansion creates compression waves, and if they gather enough strength, a shock wave will be created. The expanding flow and shock wave can further suspend and ignite dust, which continues the dust explosion. The magnitude of the shock wave is typically determined by the speed of the propagating flame. The faster the flame travels, the higher the pressures from the shock wave become, where flame speeds as high as 5,000 feet per second have been measured in experimental explosions.

As mentioned earlier, destructive forces in mines can be summarized by static pressure (overpressure), dynamic pressure (blast wind associated with drag forces) and reflected pressures. The forces can also manifest themselves very differently depending on the type of explosion. For example, for more reactive mixtures where the flame speed transitions to fast deflagrations, the pre-pressurization can result in flow reversals that result in notable drag forces opposite to the flame and blast wave direction. In addition, as the blast wave hits a dead end, it can reflect and result in pressures 2-8 times higher than the incident wave. FLACS can capture many of these effects and can be used to evaluate destructive forces in a complex mine.

The combustion of coal dust behaves very similarly to methane-air mixtures, except with the following notable differences: (1) the blast wave must perturb and suspend the dust in order to burn coal-air mixtures; (2) the explosion pressures are slightly lower than methane-air, and (3) coal dust can burn well into the fuel-rich region. Therefore, if we assume that the blast is strong enough to continually suspend dust into the mine, the governing explosion physics (overpressure and drag forces) can be evaluated using a methane/air deflagration as a surrogate for a dust/air deflagration. The reactivity of the methane/air mixture can be varied, similar to varying the concentration of dust in order to assess the deflagration of varying dust/air mixtures.

3.4 Analysis of complex blast loads

In order to more easily illustrate the complex phenomena observed in the UBB mine, several simplified geometries were created. In total four geometries were used, one single drift tunnel, a three drift tunnel (similar to the LLEM geometry), a simplified "T" model where three drifts come to a "T" intersection and an actual model of the UBB mine.

As discussed earlier, there were very complex blast loads that can be encountered when a gas or dust explosion propagates within a mine. One of these phenomena is a strong flow reversal that is observed when there is evidence of a strong deflagration or detonation. These phenomena have been observed in both unconfined and confined explosions and are due to the pre-pressurization of the unburned gases ahead of the flame front, followed by the expansion of those gases as the flame passes through the mixture. This creates a very high-pressure transient zone, which can be much higher than that calculated in theoretic closed volume. While this high pressure region "pushes" the unburned gases ahead of the flame as expected, it also is much higher than the equilibrium burned gas pressure behind the flame.

LLEM geometries

This flow reversal is illustrated in Figure 17 where the methane-air mixture was ignited on the far left, and the flame is consuming methane from left to right. From the below figure, we can see that as fuel (methane) is consumed via combustion along the length of the drift, there are areas at the flame front where the horizontal velocity is very high (>500 m/s) in the positive or right direction (red color) and areas behind the flame where the velocity is very high (>500 m/s) yet negative or left direction (blue). These high velocities result in very large dynamic pressures or drag forces ($\frac{1}{2}\rho v^2 \geq 2$ bar or 30 psi). These velocities and drag forces that oppose the travel of the flame front and blast wave often reach a maximum as the flame passes a cross cut containing fuel.

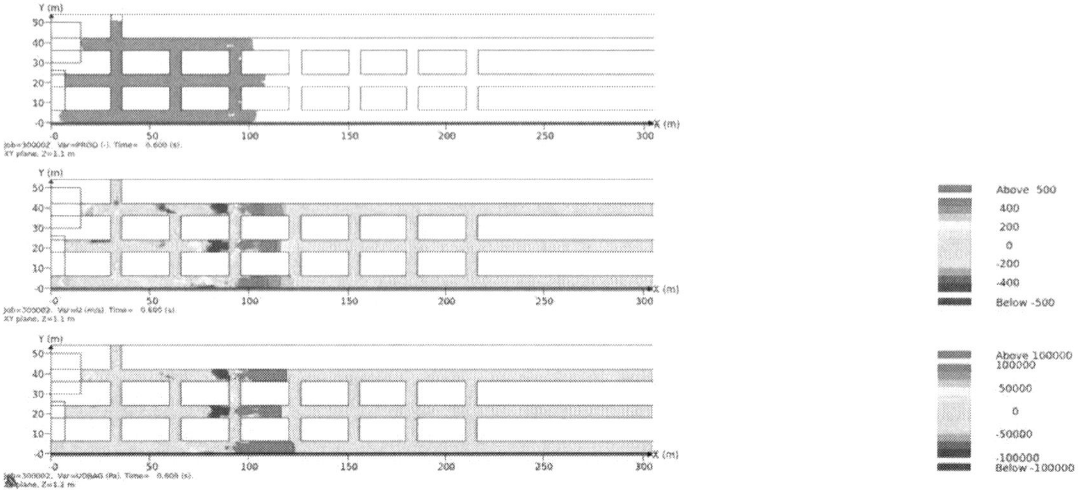

Figure 17: Full stoichiometric methane. From top to bottom: flame front, X direction velocity in m/s and X direction drag in Pa.

While this same pre-pressurization is also observed in an example of a single drift mine, which causes a both velocities in the positive and negative direction, the magnitude of the negative velocities is much lower. Figure 18 shows a comparable one-drift mine configuration. While there is negative velocity present, the corresponding drag is significantly lower and the flow reversal is generally an order of magnitude lower than the flow reversal in the three-drift mine configuration.

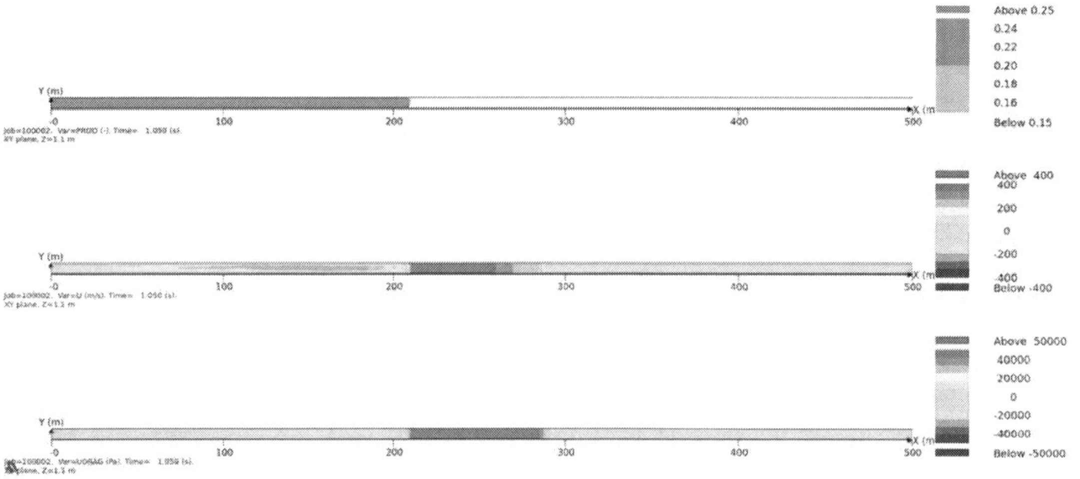

Figure 18: Full stoichiometric methane. Flame front (top), X direction velocity in m/s (middle) and X direction drag in Pa (bottom) at 1050 msec after ignition.

Further examination of the velocity vectors within the center most cross cut in the three drift example above is shown in Figure 19. While the velocity vectors show strong vectors going in both the positive (right) and negative (left) horizontal direction, there are also vectors of high magnitude originating and flowing from within the cross cuts.

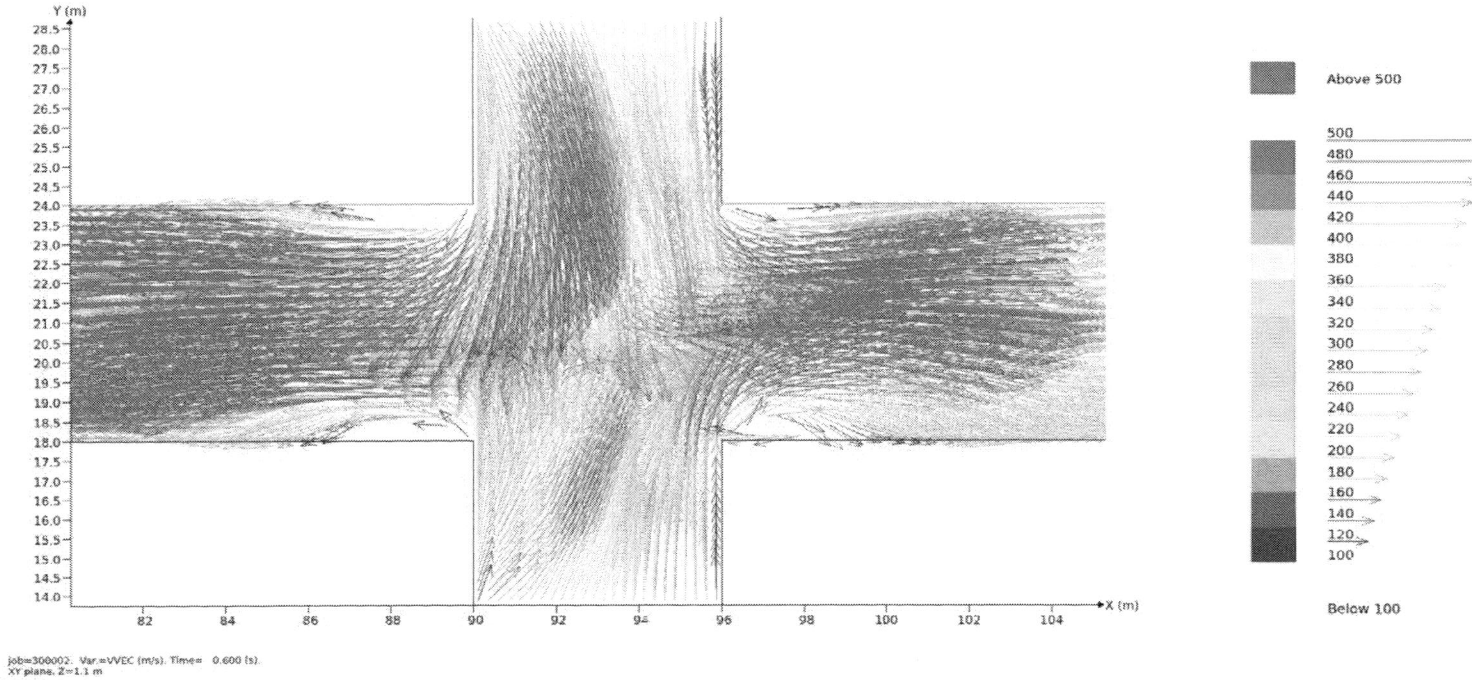

Figure 19: Velocity vectors in m/s at region of interest.

Further examination of the maximum pressures in the three drifts reveals that very high pressures (>15 barg) are found within the crosscuts within the drifts and are much higher than general pressures within the mine. Figure 20 shows the flame front progressing down the mineshaft along with the maximum recorded pressure at each location 750 msec after ignition. From this image, it should be observed that the maximum pressures recorded in the mine were in the crosscuts.

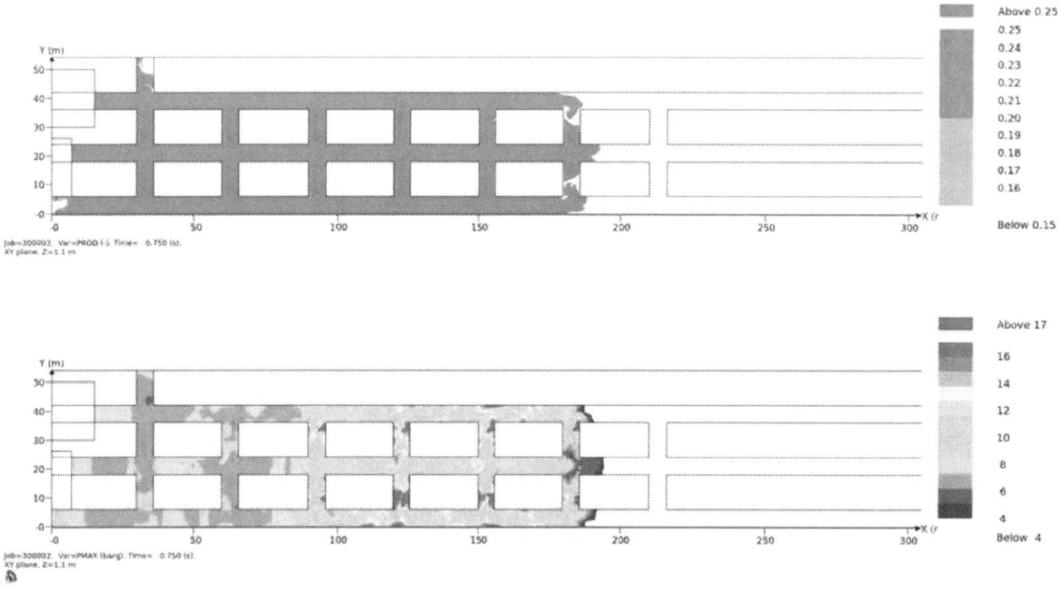

Figure 20: Full stoichiometric methane. Flame front (top) and maximum pressure in bar (bottom) at 750 msec after ignition.

Further examination of the crosscut region reveals that this significantly higher pressure within the cross cut is due to the combination of the region within the crosscut becoming pressurized from the approaching from the upper and lower drifts, which ultimately burns resulting in the thermal expansion of this pre-pressurized region. This can be visualized in Figure 21.

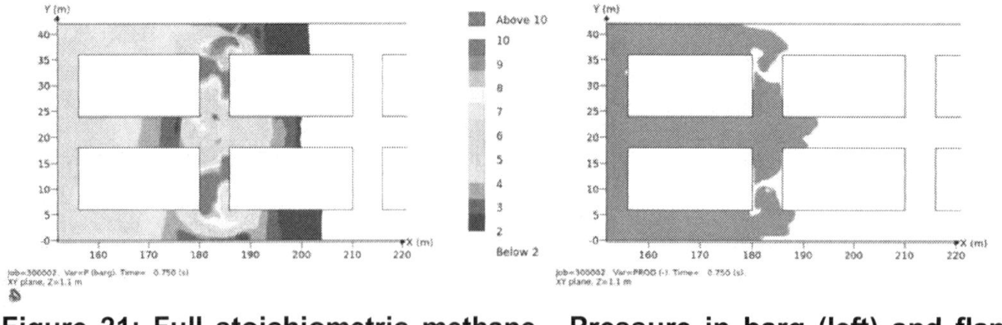

Figure 21: Full stoichiometric methane. Pressure in barg (left) and flame front (right).

As the pressure behind the flame equilibrates, the high pressure region at the flame and from the crosscut causes a significant flow reversal, against the direction of the flame propagation. In contrast, the single drift does not have the contribution of the high pressure cross cut region, resulting in much lower negative direction velocity.

This magnitude of the flow reversal is also correlated to the reactivity of the fuel (e.g. varying coal dust concentration). As a trial, the equivalence ratio was changed from stoichiometric in the above examples to nearer 1.4 for the following plots, simulating a less reactive dust concentration. An equivalence ratio of 1.4 is near the UFL of methane and the flame speeds are generally about one-third lower than the stoichiometric values. As a result of this reduced flame speed and expansion ratio, the transient pressurization is lower and the flow reversal effect was less significant. Figure 22, as compared to Figure 17, shows noticeably lower areas of negative velocity and especially for drag in the negative direction. Therefore, if the reactivity of the fuel-air mixture is low enough, the complex flow reversals and associated damage opposite to the propagating blast wave on items like the pie pans in the UBB mine will eventually cease.

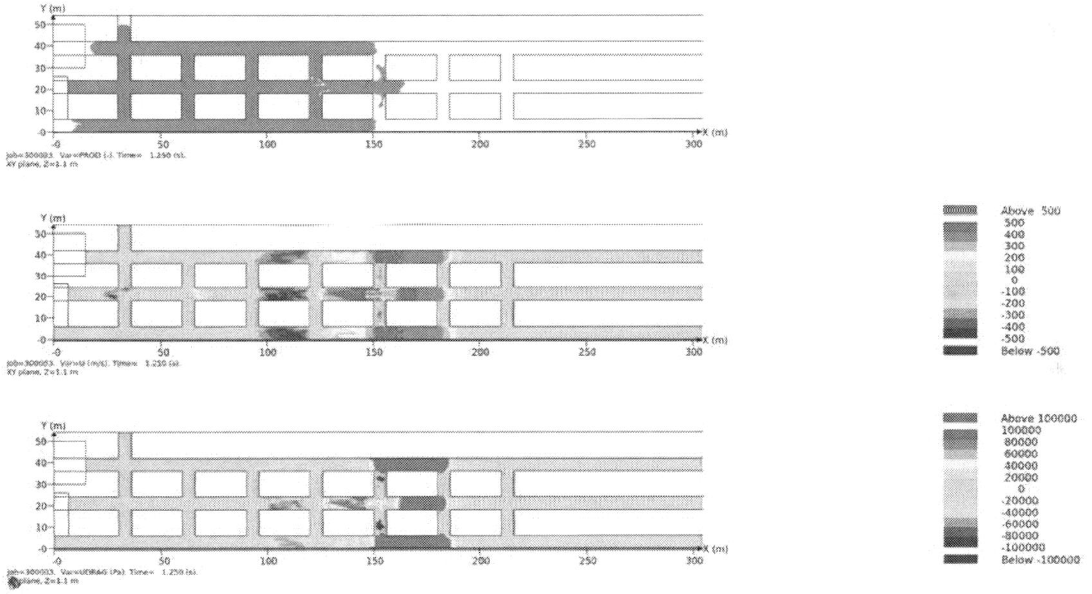

Figure 22: Equivalence ratio 1.4 methane. From top to bottom: flame front, X direction velocity in m/s and X direction drag in Pa.

Not only are the drag forces lower, but the resulting overpressures within the mine are also lower, as expected. More specifically, if there was a very reactive coal dust concentration or flammable gas concentration, the pressures within the mine can become greater than 10 barg (150 psig) in less than 100 meters (see Figure 20), while for a less reactive mixture the pressures within the mine never exceeded approximately 5 barg (75 psig) as shown in Figure 23. Therefore, the resulting pressure damage can also provide a good indicator of the reactivity of the fuel/air mixture as well as the volume of the combustible mixture. For example, the more reactive the fuel/air mixture, the higher the resulting overpressures from the explosion for equal size volumes. Similarly, larger explosion pressures are generally observed (up to a certain limit) for large combustible volumes.

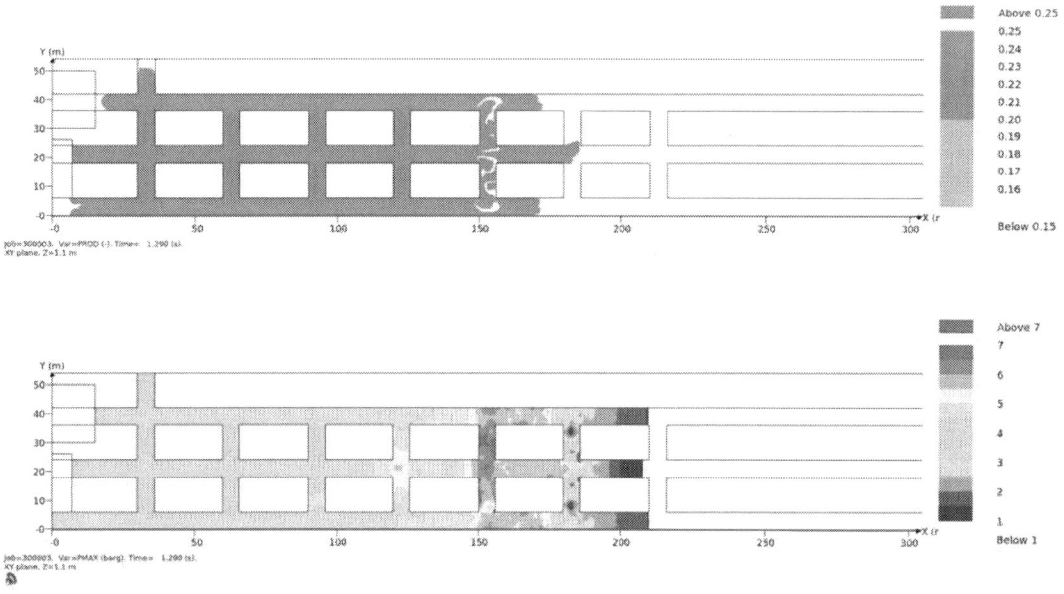

Figure 23: Equivalence ratio 1.4 methane. Flame front (top) and maximum pressure (bottom) at 1290 msec after ignition.

As the geometry becomes more complex, additional dynamics develop, which can include pressure reflections and flame arrival from different locations. In order to evaluate the additional explosion dynamics, a perpendicular set of mine drifts was introduced to the three-drift geometry. While the flow reversal remains present with the same governing physics, we begin to see complications with pressure reflections at the end walls and its associated flow reversals, as well as local high-pressure regions due to flame arrival from various directions. This "tee" geometry can be found in Figure 24. The geometry is closed on all sides except the bottom end of the figure in this orientation. In this geometry configuration, the "upward" direction is –Y to take advantage of the parallel solver in the most recent version of FLACS.

Modified T-geometry

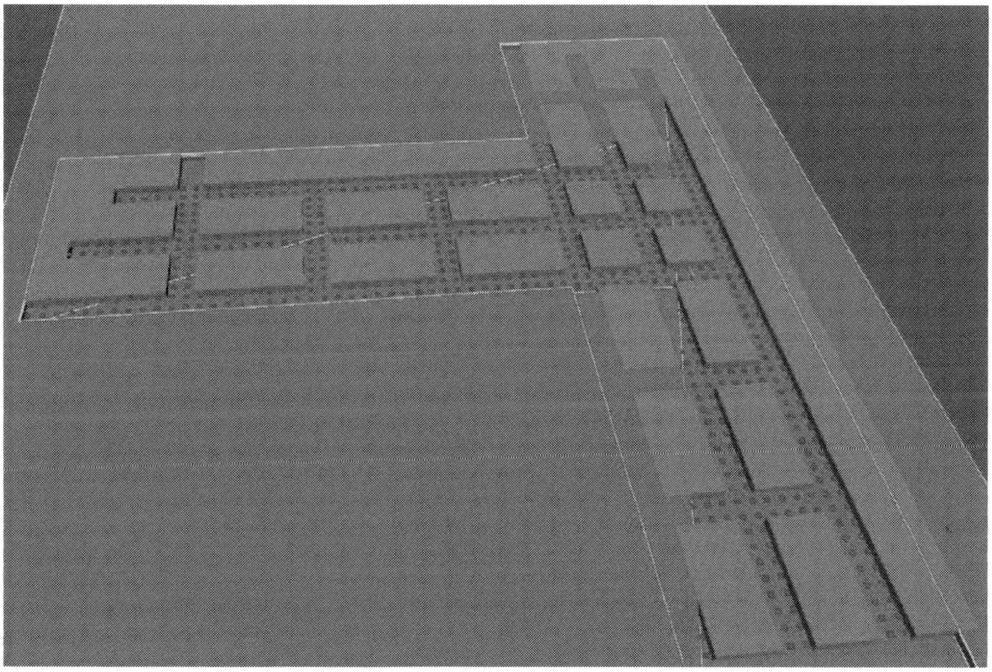

Figure 24: "Tee" mine geometry configuration.

In the following Figure 25 through Figure 29, a reactive combustible cloud (stoichiometric methane) fills the entire mine geometry and is ignited. The figure sequence is different points in time as the explosion develops. In each of the figures from left to right and top to bottom the images are: flame front, pressure in bar, drag in the X direction in Pa and drag in the Z direction in Pa.

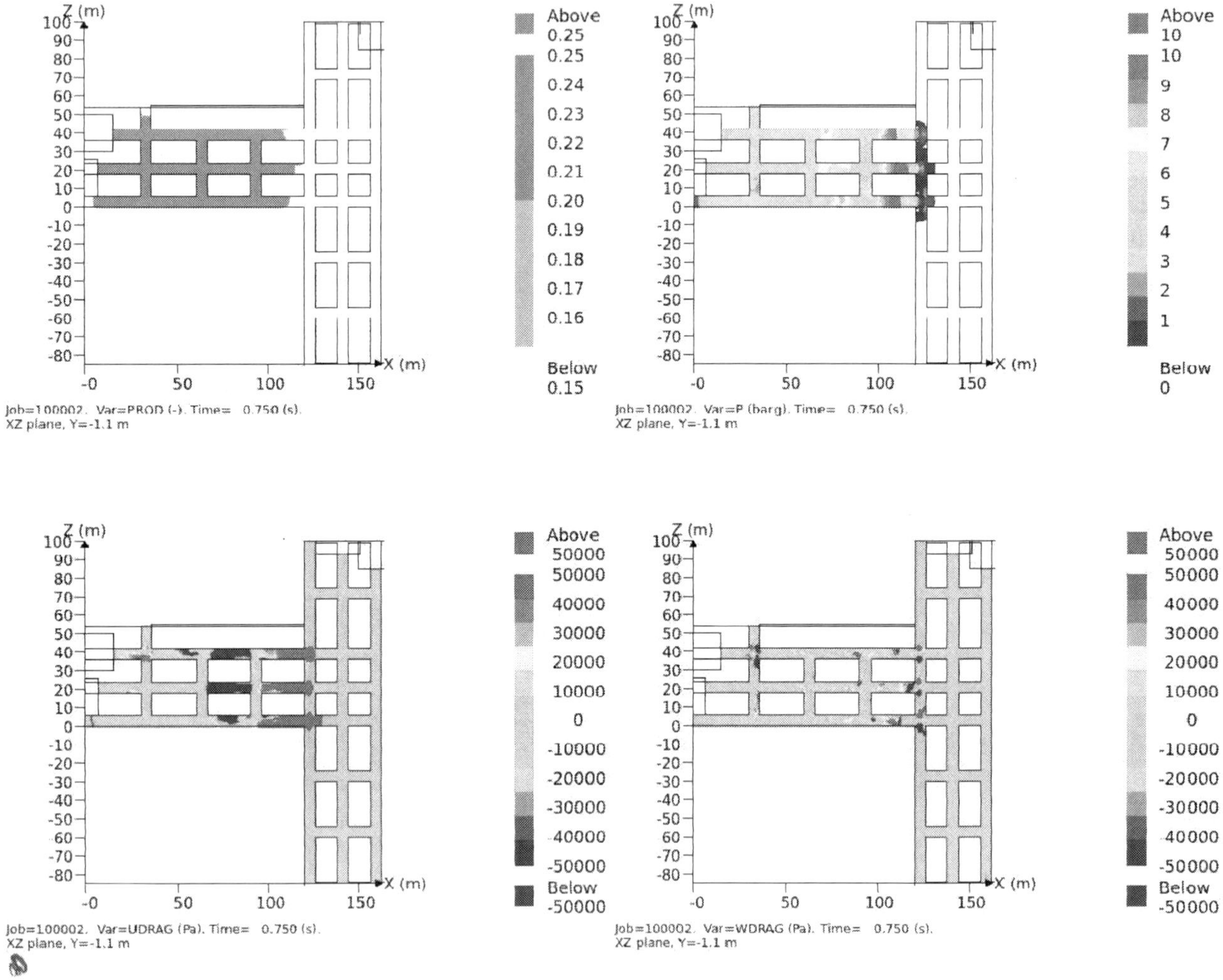

Figure 25: "Tee" mine, full stoichiometric methane at 750ms after ignition.

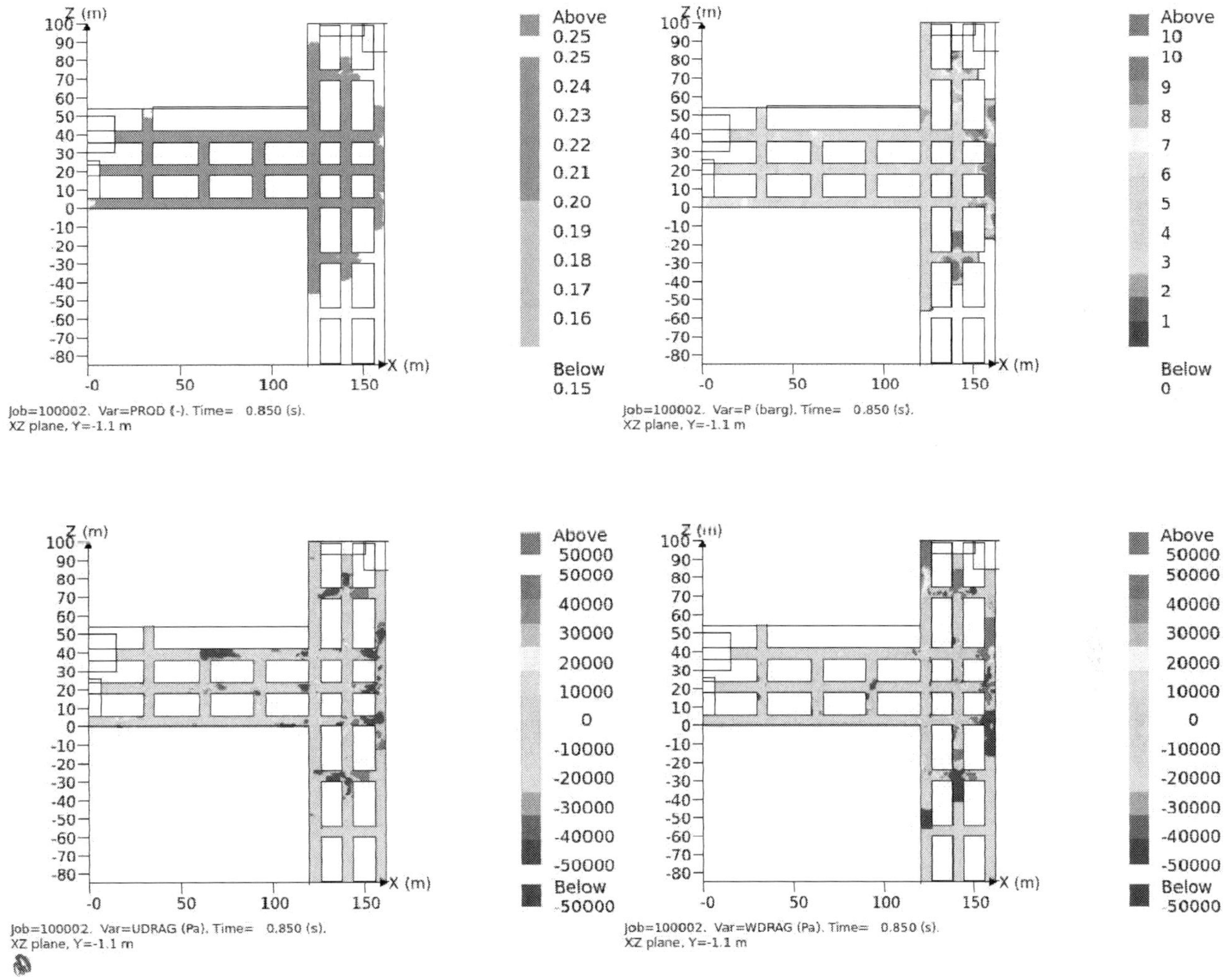

Figure 26: "Tee" mine, full stoichiometric methane at 850ms after ignition.

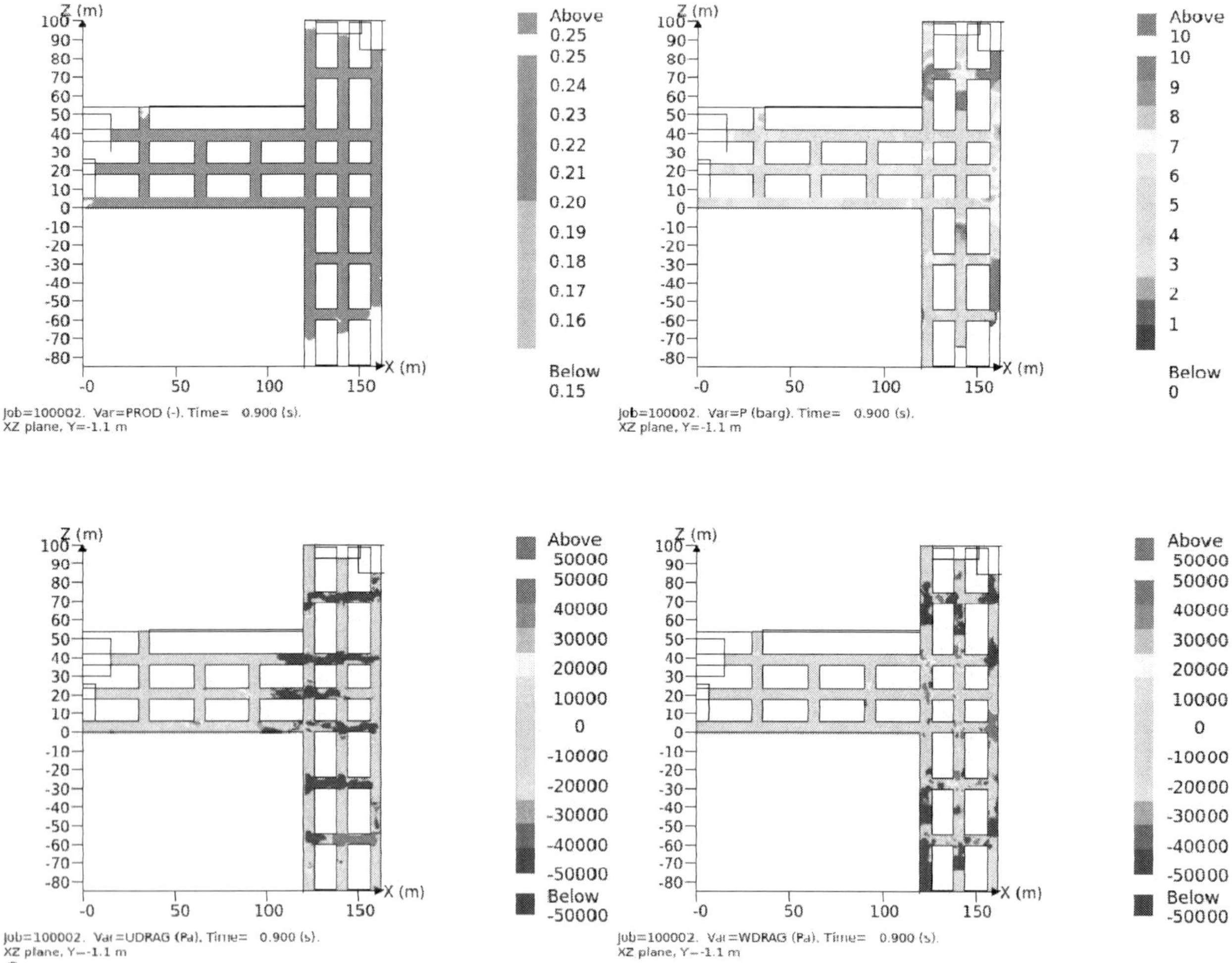

Figure 27: "Tee" mine, full stoichiometric methane at 900ms after ignition.

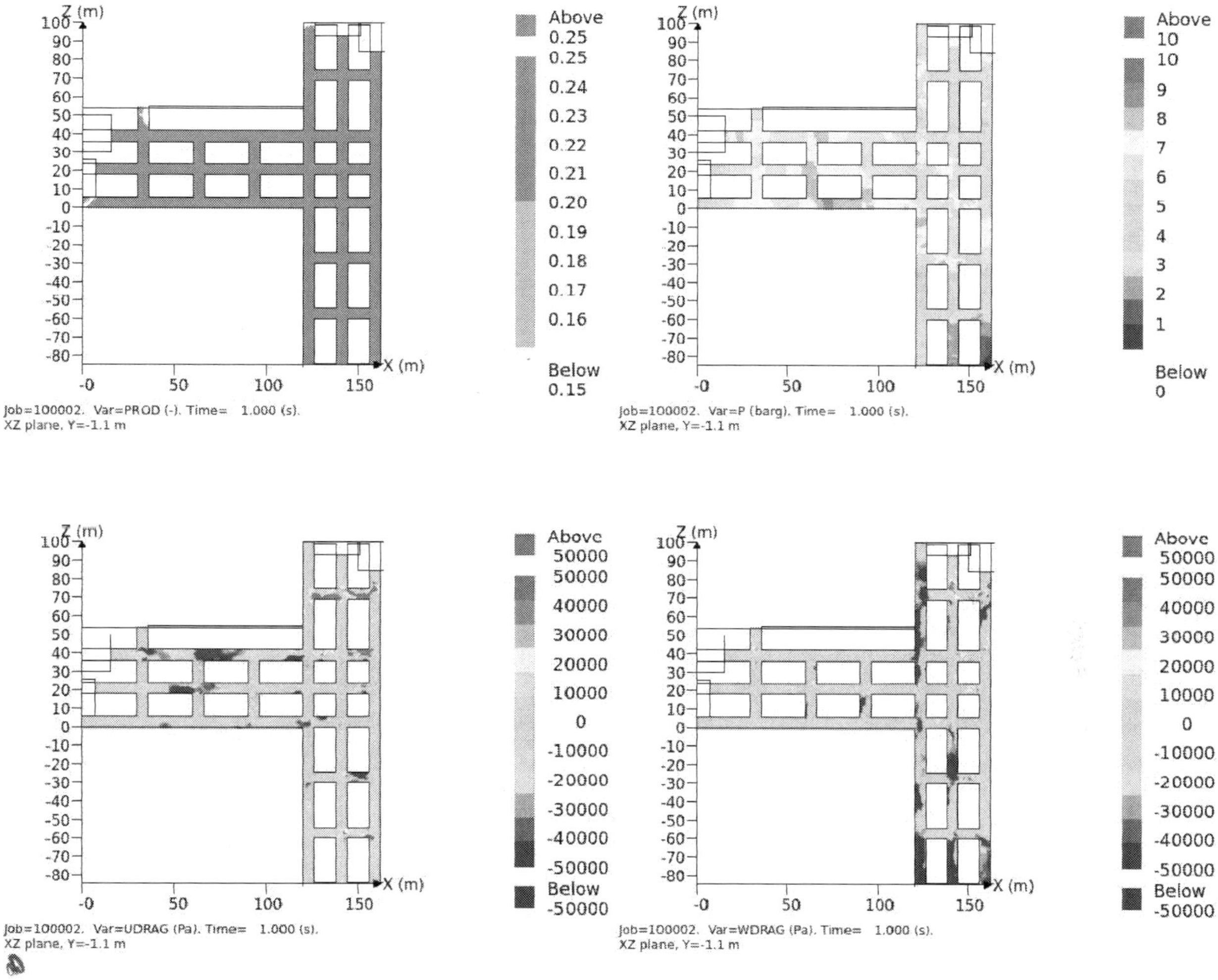

Figure 28: "Tee" mine, full stoichiometric methane at 1000ms after ignition.

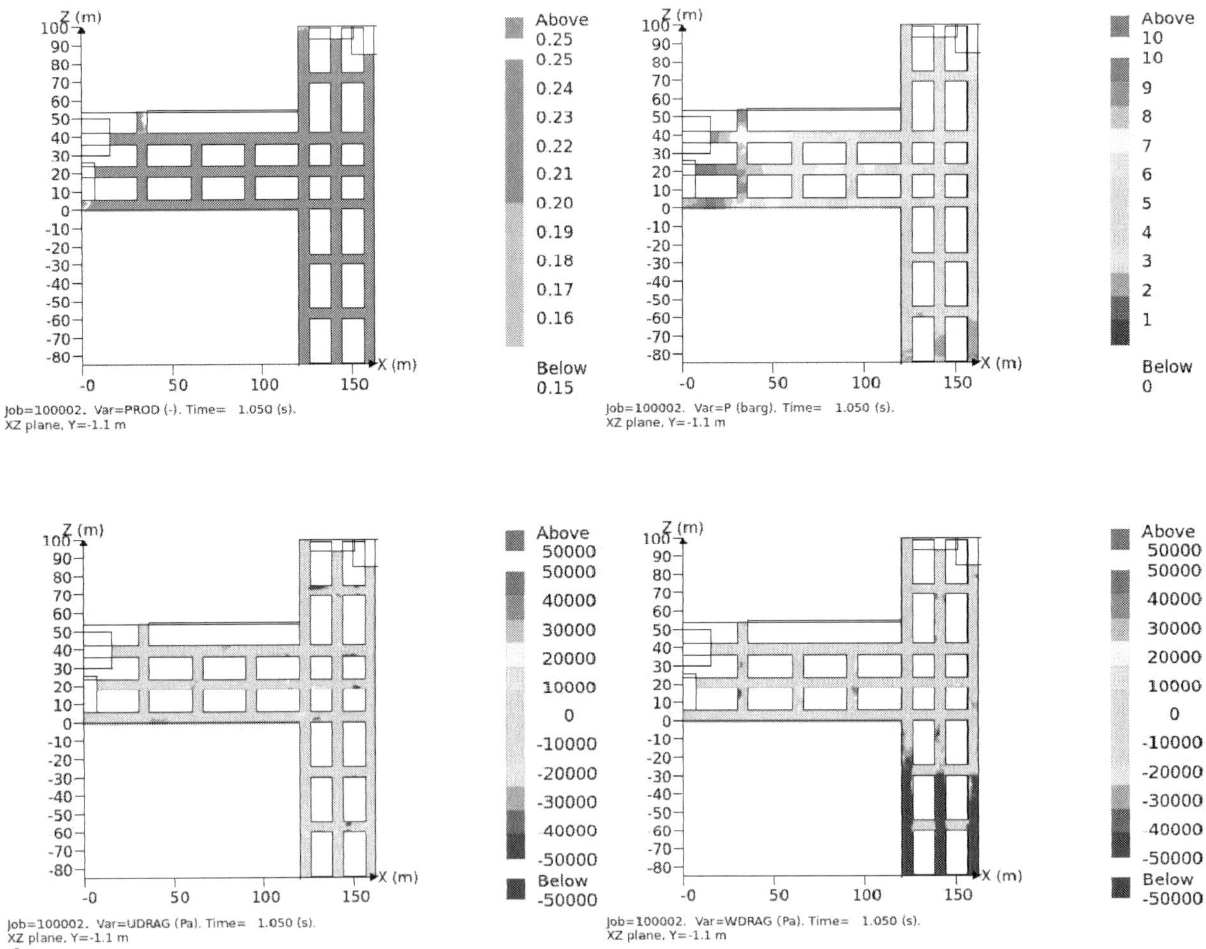

Figure 29: "Tee" mine, full stoichiometric methane at 1050ms after ignition.

Figure 25 shows that there is still flow reversal present in the early explosion development, which is unrelated to the pressure reflection off a closed wall in front of the flame. As the explosion progresses to Figure 26, there is a region of interest forming at the right most end of the center X direction drift. Not only is there a significant pressure rise due to the flame arrival at each of the cross cuts, but there is a region where the center drift reaches the right wall where the flow is in opposite directions at the intersection (see Figure 26). This figure also shows that the pressure front (top right image) on the –X most side of the Z direction drifts has propagated the fastest. As the explosion develops into Figure 27, the pressure reflects off the +Z closed surface and begins to travel down to the open –Z boundary. As the explosion finishes in Figure 28 and Figure 29, the tunnel that contained the ignition sources experiences a pressure reflection from the perpendicular drift with however no associated flow (e.g., just a shock wave).

The tee mine configuration allowed the flame front in the left-most vertical drift to progress further down the mine than the right-most drift. This then allows for the flame front to progress through the cross-cuts to meet the right-most drift flame at various locations in the drifts. More specifically, Figure 27 shows that there is a localized high pressure region near the open end of the mine (in excess of 20 bar), which is caused by the flame arriving from both middle Z drift via the crosscut and from above in the right-most Z drift, resulting in pre-pressurization prior to flame arrival and thermal expansion.

With the tee mine we also see various areas of localized pressure and pressure reflections. With the full cloud at stoichiometric concentration mine configuration, in the upper closed section of the mine we see extremely high pressures as a result of pre-pressurization and pressure reflections compounding with one another. The following sequence of images in Figure 30 show flame front (left) and pressure right) at select times during the simulation.

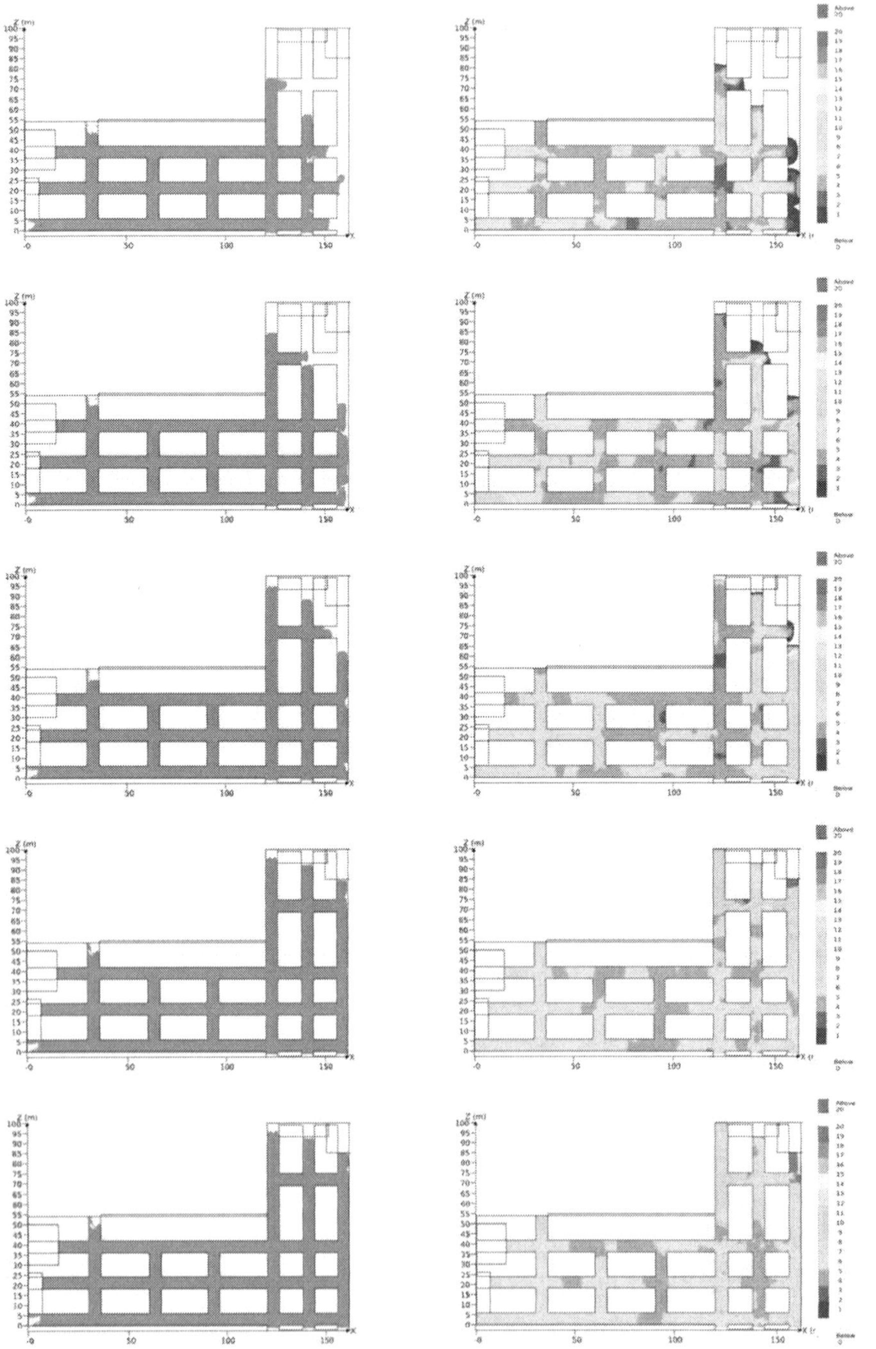

Figure 30: "Tee" mine, ER=1.0 methane at 840ms, 860ms, 890ms, 900ms after ignition.

As the flame front reaches the upper right corner in these graphics we can see that the pressures eventually exceed 20 bar with the peak pressures on the walls and ceilings reaching over 50 bar. We can see that two flame fronts approach the upper right most area at approximately the same time.

As was the case with the simplified three-drift geometry, the effects are lessened when the reactivity of the fuel is reduced due to the reduced flame speed and thermal expansion ratios. The results for this were not presented for the sake of clarity.

UBB geometry

Lastly to evaluate how these governing physics interact in a full mine, a replica of the UBB mine was constructed to be used in FLACS. This can be seen in Figure 31.

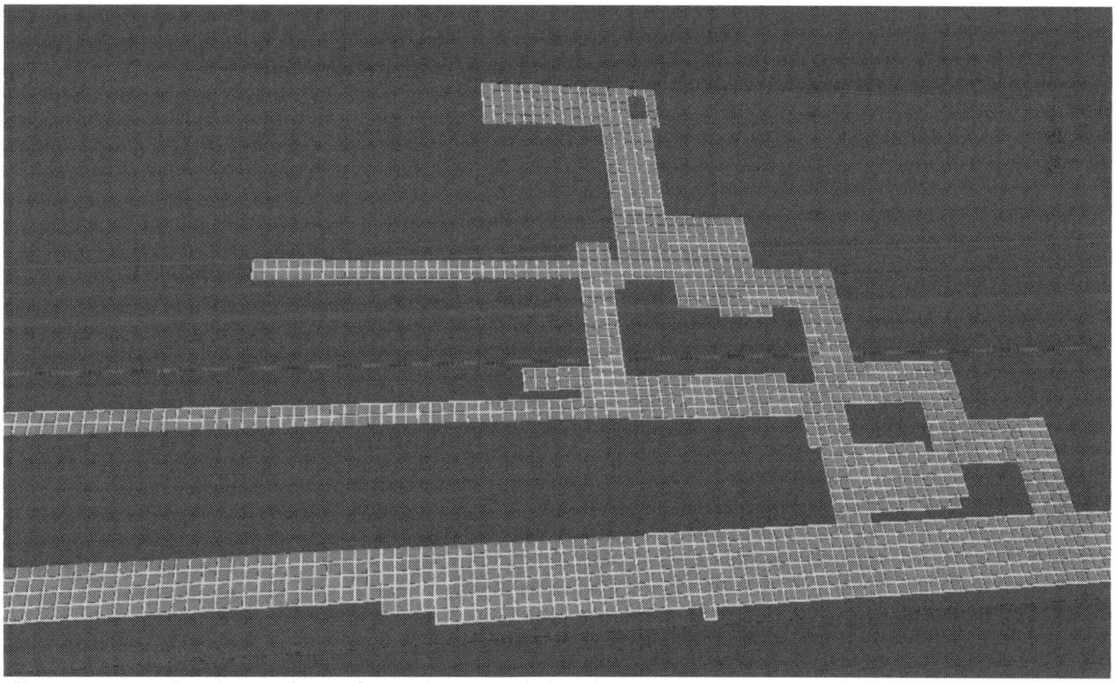

Figure 31: Vertical drag impulse and horizontal drag impulse at 1476ms after ignition in UBB mine geometry.

Due to the computationally intensive requirements of the full mine, only preliminary simulations have been performed at the time the paper was written. These have included a very reactive fuel-air volume throughout the mine (stoichiometric methane/air) as well as a less reactive fuel-air mixture (methane/air equivalence ratio of 1.4). Additional sensitivities will need to be conducted to evaluate varying reactivities in different sections of the mine.

Similar to the simplified geometries, the same flow reversals and reflections were observed. For the more reactive mixtures, these resulted in extremely high flow reversals, where the dynamic pressure exceeded (2 bar or 30 psi) as shown in Figure 32. The MSHA report indicated that the flame travel very quickly in Headgate #22, therefore this was a reasonable preliminary assumption. The magnitude of the drag forces determined from the simulations can easily account for the damage observed on the mantrip, where one section of the roof canopy was blown 200ft to the west and in the direction of the advancing blast wave, while another section was blown 200 to the east against the advancing blast wave. In addition, according to a recent report on the Sago Mine incident, they determined that the minimum dynamic pressures necessary to deform the pie plates is around 2 bar (30 psi). Therefore, the CFD results indicate that for reactive fuel-air volumes the dynamic pressures in the direction opposite to the propagating blast wave can be in excess of those needed to deform pie plates in a mine. This suggests that they may be locations in the mine where the blast indicators appear to point in directions opposite to the propagation of the blast wave. GexCon is currently investigating this in further detail.

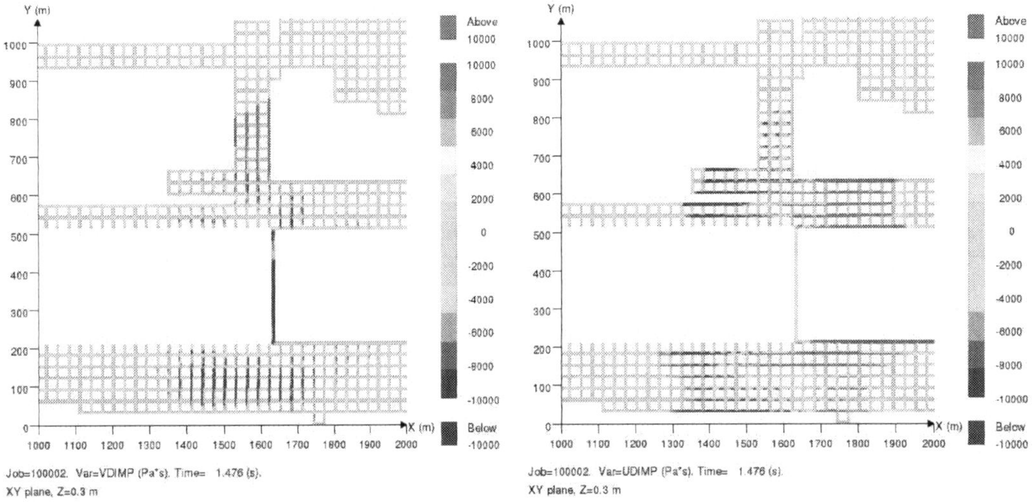

Figure 32: Vertical drag impulse in Pa*s (left) and horizontal drag impulse in Pa*s (right) at 1476ms after ignition in UBB mine geometry.

4 Conclusions

This study used a state of the art CFD solver to conduct a detailed explosion analysis to evaluate the complex overpressure development in a mine explosion. Using the incident at Upper Big Branch mine as a basis, it was discovered that mine explosions may contain unique damage patterns where certain "blast indicators" within the mine shafts were deformed in such a manner that was inconsistent with the likely flow of the expanding blast wave.

Using a simplified geometry with three drifts over 200 meters long and crosscuts every 20 meters; we demonstrated that transient regions of very high pressure could occur within the crosscuts between the drifts. This was due to the flame front arriving from both below and above, which resulted in a significant pre-pressurization this region prior to the flame arrival. The increased initial pressure followed by the flame expansion, resulted in pressure higher than the equilibrium values. This caused a significant flow and drag forces (dynamic pressure) both in the direction of advancing blast wave and against it. The magnitude of such dynamic pressure exceeded 2 bar (30 psi) for very reactive fuel-air mixtures. This strong flow reversal can account for the apparent contradictory blast indicators observed in the UBB mine. In addition, changes in the direction of the mine drifts allow for the flame front in each drift to separate and cause additional explosion dynamics that may be counterintuitive. It was also shown that the magnitude of these unique dynamics is dependent upon the reactivity of the fuel, which, with more analysis, could help to identify regions of high or reactive concentration throughout the UBB mine incident.

The results further suggest that as the complexity of the mine increases, the explosion dynamics and resulting damage may become increasingly difficult to interpret intuitively. This is the focus of our ongoing work.

5 References

[1] United States Department Of Labor Mine Safety And Health Administration, Coal Mine Safety And Health, Report Of Investigation, Fatal Underground Mine Explosion, April 5, 2010 Upper Big Branch Mine-South, Performance Coal Company. Montcoal, Raleigh County, West Virginia, ID No. 46-08436, Norman Page et al.

[2] Upper Big Branch: The April 5, 2010, explosion: a failure of basic coal mine safety practices. Report to the Governor. Governor's Independent Investigation Panel. McAteer and associates. May 2011.

[3] Hjertager, B.H. (1986). Three-dimensional modelling of flow, heat transfer, and combustion. Handbook of Heat and Mass Transfer, Gulf Publishing Company, Houston: 1303-1350.

[4] Launder, B.E. & D.B. Spalding (1974), *"The numerical computation of turbulent flows", Computer Methods in Applied Mechanics and Engineering"* **3** (2): 269-289

[5] Middha, P. & Hansen, O.R. (2008). Predicting deflagration to detonation transition in hydrogen explosions. Process Safety Progress, 27 (3): 192-204.

[6] Abdel-Gayed, R.G., Bradley, D., and Lawes, M. 1987, Proc. of the Royal Society of London, 414, 407

[7] Arntzen, B.J. (1998). Modelling of turbulence and combustion for simulation of gas explosions in complex geometries. Dr. Ing. Thesis, NTNU, Trondheim.

[8] Zipf, RK, Sapko, MJ and Brune, JF (2007), Explosion Pressure Design Criteria for New Seals in U.S. Coal Mines, NIOSH IC 9500, July 2007.

AUTHOR INDEX

AUTHOR INDEX